Methods in Enzymology

Volume 186
OXYGEN RADICALS IN BIOLOGICAL SYSTEMS
Part B
Oxygen Radicals and Antioxidants

METHODS IN ENZYMOLOGY

EDITORS-IN-CHIEF

John N. Abelson   Melvin I. Simon

DIVISION OF BIOLOGY
CALIFORNIA INSTITUTE OF TECHNOLOGY
PASADENA, CALIFORNIA

FOUNDING EDITORS

Sidney P. Colowick and Nathan O. Kaplan

Methods in Enzymology

Volume 186

Oxygen Radicals in Biological Systems

Part B
Oxygen Radicals and Antioxidants

EDITED BY

Lester Packer
Alexander N. Glazer

DEPARTMENT OF MOLECULAR AND CELL BIOLOGY
UNIVERSITY OF CALIFORNIA, BERKELEY
BERKELEY, CALIFORNIA

Editorial Advisory Board

ACADEMIC PRESS, INC.
Harcourt Brace Jovanovich, Publishers

San Diego   New York   Boston
London   Sydney   Tokyo   Toronto

ACADEMIC PRESS, INC.
San Diego, California 92101

United Kingdom Edition published by
ACADEMIC PRESS LIMITED
24-28 Oval Road, London NW1 7DX

LIBRARY OF CONGRESS CATALOG CARD NUMBER: 54-9110

ISBN 0-12-182087-4 (alk. paper)

PRINTED IN THE UNITED STATES OF AMERICA
90 91 92 93   9 8 7 6 5 4 3 2 1

Table of Contents

Section I. Production, Detection, and Characterization of Oxygen Radicals and Other Species

v

Section II. Assay of Enzymes or Substances Involved in Formation or Removal of Oxygen Radicals and Derived Products

A. Proteins

B. Small Molecules

Section III. Assay and Repair of Biological Damage Due to Oxygen and Oxygen-Derived Species

A. Lipid Peroxidation

B. Protein Oxidation

C. Modification of Nucleic Acids and Chromosomes

Section IV. Leukocytes and Macrophages

Section V. Organ, Tissue, and Cell Damage and Medical Applications

Section VI. Oxygen Free Radicals in Ischemia and Reperfusion-Associated Tissue Damage

Contributors to Volume 186

Article numbers are in parentheses following the names of contributors.
Affiliations listed are current.

BONG-WHAN AHN (49), *Department of Biochemistry, Chonnam National University Medical College, Kwangju, Republic of Korea*

KAZUAKI AKASAKA (13), *Department of Food Chemistry, Faculty of Agriculture, Tohoku University, Sendai, 981 Japan*

BRUCE N. AMES (31, 38, 54), *Division of Biochemistry and Molecular Biology, University of California, Berkeley, Berkeley, California 94720*

ADOLFO AMICI (49), *Istituto di Biochimica, Facolta di Medicina e Chirurgia, Universita de Ancona, 60100 Ancona, Italy*

G. A. S. ANSARI (45), *Department of Human Biological Chemistry and Genetics, University of Texas Medical Branch, Galveston, Texas 77550*

F. S. ARCHIBALD (22), *Biological Chemistry Section, Pulp and Paper Research Institute of Canada, Pointe Claire, Quebec, H9R 3J9 Canada*

KLAUS-DIETER ASMUS (15), *Hahn-Meitner Institut Berlin, Bereich S, Abteilung Strahlenchemie, D-1000 Berlin 39, Federal Republic of Germany*

STEVEN D. AUST (48), *Biotechnology Center, Utah State University, Logan, Utah 84322*

CHARLES F. BABBS (11), *The William A. Hillenbrand Biomedical Engineering Center, Purdue University, West Lafayette, Indiana 47907*

RUMYANA A. BAKALOVA (37), *Institute of Physiology, Bulgarian Academy of Sciences, Sofia 113, Bulgaria*

GIORGIO BELLOMO (66), *Dipartimento di Medicina Interne e Terapia Medica, Clinica Medica I, University of Pavia, 27100 Pavia, Italy*

JOSEPH N. BENOIT (80), *Department of Physiology and Biophysics, Louisiana State University Medical Center, Shreveport, Louisiana 71130*

J. BEREITER-HAHN (79), *Institut für Kinematische Zellforschung, Fachbereich Biologie, Johann Wolfgang Goethe Universität, 6000 Frankfurt am Main 70, Federal Republic of Germany*

M. BERGER (52), *Laboratoires de Chimie, Département de Recherche Fondamentale, Centre d'Etudes Nucléaires, 85X, F-38041 Grenoble Cedex, France*

WAYNE F. BEYER, JR. (23), *Department of Biochemistry, Duke University Medical Center, Durham, North Carolina 27710*

BENON H. J. BIELSKI (4, 5, 6), *Department of Chemistry, Brookhaven National Laboratory, Upton, New York 11973*

H. CHAIM BIRNBOIM (57), *Ottawa Regional Cancer Center and Departments of Biochemistry and Microbiology/Immunology, University of Ottawa, Ottawa, Ontario, K1H 8L6 Canada*

ROBERT C. BLAKE II (27), *Department of Biochemistry, Meharry Medical College, Nashville, Tennessee 37208*

WOLF BORS (36), *Institut für Strahlenbiologie, GSF Research Center, D-8042 Neuherberg, Federal Republic of Germany*

JEFF A. BOYD (29), *Laboratory of Molecular Carcinogenesis, National Institute of Environmental Health Sciences, Research Triangle Park, North Carolina 27709*

GARRY R. BUETTNER (8, 9), *ESR Center, College of Medicine, University of Iowa, Iowa City, Iowa 52242*

GREGORY B. BULKLEY (78), *Department of Surgery, The Johns Hopkins Hospital, Baltimore, Maryland 21205*

PAULA BURCH (67), *Department of Cell Biology, Baylor School of Medicine, Houston, Texas 77251*

RAYMOND F. BURK (84), *Division of Gastroenterology, Department of Medicine and Center in Molecular Toxicology, Vanderbilt University School of Medicine, Nashville, Tennessee 37232*

THEODORE J. BURKEY (72), *Department of Chemistry, Memphis State University, Memphis, Tennessee 38152*

DIANE E. CABELLI (6), *Department of Chemistry, Brookhaven National Laboratory, Upton, New York 11973*

ENRIQUE CADENAS (16, 30), *Institute for Toxicology, University of Southern California, Los Angeles, California 90033*

J. CADET (52), *Laboratoires de Chimie, Département de Recherche Fondamentale, Centre d'Etudes Nucléaires, 85X, F-38041 Grenoble Cedex, France*

KEVIN H. CHEESEMAN (42), *Department Biochemistry, Brunel University, Uxbridge, Middlesex, UB8-3PH, England*

DANIEL F. CHURCH (72), *Biodynamics Institute, Louisiana State University, Baton Rouge, Louisiana 70803*

ISABEL CLIMENT (49), *Laboratory of Biochemistry, National Heart, Lung, and Blood Institute, National Institutes of Health, Bethesda, Maryland 20892*

MINOR J. COON (27, 28), *Department of Biological Chemistry, University of Michigan Medical School, Ann Arbor, Michigan 48109*

KENNETH C. CUNDY (54), *Sterling Research Group, Great Valley Corporate Center, Great Valley, Pennsylvania 19355*

JOHN T. CURNUTTE (59), *Department of Molecular and Experimental Medicine, Research Institute of Scripps Clinic, La Jolla, California 92037*

KELVIN J. A. DAVIES (51), *Department of Biochemistry and Institute for Toxicology, University of Southern California, Los Angeles, California 90033*

C. DECARROZ (52), *Rhone-Mérieux Laboratoires IFFA, F-69007, Lyon, France*

HERBERT DE GROOT (46), *Klinische Forschergruppe Leberschädigung, Institut für Physiologische Chemie und Abteilung für Gastroenterologie, Heinrich-Heine-Universität Düsseldorf, D-4000 Düsseldorf 1, Federal Republic of Germany*

MIRAL DIZDAROGLU (55), *Center for Chemical Technology, National Institute of Standards and Technology, Gaithersburg, Maryland 20899*

H. H. DRAPER (43), *Department of Nutritional Sciences, University of Guelph, Guelph, Ontario, N1G 2W1 Canada*

EDWARD A. DRATZ (40, 41), *Department of Chemistry, Montana State University, Bozeman, Montana 59717*

DAVID A. EASTMOND (61), *Environmental Toxicology Graduate Program, University of California, Riverside, Riverside, California 92521*

INGRID EMERIT (58), *Laboratoire de Genetique, Institut Biomedical des Cordeliers, Universite Paris VI and CNRS, 75006 Paris, France*

LARS ERNSTER (16, 30), *Department of Biochemistry, University of Stockholm, S-106 91, Stockholm, Sweden*

HERMANN ESTERBAUER (42), *Institute of Biochemistry, University of Graz, A-8010 Graz, Austria*

SPENCER B. FARR (68), *Laboratory of Toxicology, Harvard University, Boston, Massachusetts 02115*

PETER W. F. FISCHER (21), *Nutrition Research Division, Health and Welfare Canada, Ottawa, Ontario, K1A 0L2 Canada*

E. N. FRANKEL (39), *Department of Food Science and Technology, University of California, Davis, Davis, California 95616*

BALZ FREI (38), *Division of Biochemistry and Molecular Biology, University of California, Berkeley, Berkeley, California 94720*

IRWIN FRIDOVICH (23), *Department of Biochemistry, Duke University Medical Center, Durham, North Carolina 27710*

WILLIAM H. FRIST (81), *Department of Cardiac and Thoracic Surgery, Vanderbilt University, Nashville, Tennessee 37232*

JÜRGEN FUCHS (17, 73, 79), *Zentrum der Dermatologie und Venerologie, Abteilung II, Klinikum der Johann Wolfgang Goethe Universität, 6000 Frankfurt am Main 70, Federal Republic of Germany*

EWA GAJEWSKI (55), *Center for Chemical Technology, National Institute of Standards and Technology, Gaithersburg, Maryland 20899*

DONITA GARLAND (49), *Laboratory of Mechanisms of Ocular Diseases, National Eye Institute, National Institutes of Health, Bethesda, Maryland 20892*

CARLOS J. GIMENO (54), *Division of Biochemistry and Molecular Biology, University of California, Berkeley, Berkeley, California 94720*

ALEXANDER N. GLAZER (14, 31), *Division of Biochemistry and Molecular Biology, Department of Molecular and Cell Biology, University of California, Berkeley, Berkeley, California 94720*

D. NIEL GRANGER (80), *Department of Physiology and Biophysics, Louisiana State University Medical Center, Shreveport, Louisiana 71130*

CARLO GREGOLIN (47), *Department of Biological Chemistry, University of Padova, 35121 Padova, Italy*

MATTHEW B. GRISHAM (80), *Department of Physiology and Biophysics, Louisiana State University Medical Center, Shreveport, Louisiana 71130*

JOHN M. C. GUTTERIDGE (1), *Division of Chemistry, National Institute for Biological Standards and Control, Potters Bar, Hertfordshire EN6 3QG, England*

M. HADLEY (43), *Department of Food and Nutrition, College of Home Economics, North Dakota State University, Fargo, North Dakota 58105*

ROBERT D. HALL (10), *Development Division, SYVA Company, Palo Alto, California 94303*

BARRY HALLIWELL (1), *Department of Medical Biochemistry, King's College, University of London, London WC2R 2LS, England*

HOWARD J. HALPERN (64), *Department of Radiation Oncology, University of Chicago School of Medicine, Chicago, Illinois 60637*

PHILIP E. HARTMAN (32), *Department of Biology, The Johns Hopkins University, Baltimore, Maryland 21218*

JEFFREY J. HAYES (56), *Department of Chemistry, The Johns Hopkins University, Baltimore, Maryland 21218*

WERNER HELLER (36), *Institut für Biochemische Pflanzenpathologie, GSF Research Center, D-8042 Neuherberg, Federal Republic of Germany*

JAMES S. HENDERSON (76), *Department of Pathology, Faculty of Medicine, University of Manitoba, Winnipeg, Manitoba, R3E 0W3 Canada*

KRISTINA E. HILL (84), *Division of Gastroenterology, Department of Medicine and Center in Molecular Toxicology, Vanderbilt University, Nashville, Tennessee 37232*

PAUL HOCHSTEIN (30), *Institute of Toxicology, University of Southern California, Los Angeles, California 90033*

HELEN HUGHES (75), *Center for Experimental Therapeutics and Section on Hypertension/Clinical Pharmacology, Baylor College of Medicine, Houston, Texas 77030*

ROLF D. ISSELS (77), *Institut für Klinische Haematologie and Medizinische Klinik III, Klinikum Grosshadern, Ludwig-Maximilians-Univeristät, D-8000 München 70, Federal Republic of Germany*

HARTMUT JAESCHKE (75, 83), *Center for Experimental Therapeutics and Section on Hypertension/Clinical Pharmacology, Department of Medicine and Pharmacology, Baylor College of Medicine, Houston, Texas 77030*

VALERIAN E. KAGAN (37), *Institute of Physiology, Bulgarian Academy of Sciences, Sofia 113, Bulgaria*

B. KALYANARAMAN (35), *Department of Radiology, National Biomedical ESR Center, Medical College of Wisconsin, Milwaukee, Wisconsin 53226*

LAURANCE KAM (56), *University of Pennsylvania, School of Medicine, Philadelphia, Pennsylvania 19104*

JAMES P. KEHRER (65), *Department of Pharmacology and Toxicology, College of Pharmacy, The University of Texas at Austin, Austin, Texas 78712*

IL HAN KIM (50), *Department of Chemistry, Baijae University, Taejon, Republic of Korea*

KANGHWA KIM (50), *Department of Food and Nutrition, College of Natural Science, Chonnam National University, Gwangju, 500-05, Republic of Korea*

W. H. KOPPENOL (12), *Biodynamics Institute, and Departments of Chemistry and Biochemistry, Louisiana State University, Baton Rouge, Louisiana 70803*

RICHARD J. KULMACZ (44), *Department of Biochemistry, University of Illinois at Chicago, Chicago, Illinois 60612*

MARY R. L'ABBÉ (21), *Nutrition Research Division, Health and Welfare Canada, Ottawa, Ontario, K1A 0L2 Canada*

WILLIAM E. M. LANDS (44), *Department of Biochemistry, University of Illinois at Chicago, Chicago, Illinois 60612*

JONATHAN A. LEFF (69), *Webb-Waring Lung Institute and the University of Colorado Health Sciences Center, Denver, Colorado 80262*

ANKE-G. LENZ (49), *Inhalation Project/Biochemistry Group, Gesellschaft für Strahlen-und Umweltforschung, D-8042 Neuherberg, Federal Republic of Germany*

RODNEY L. LEVINE (49), *Laboratory of Biochemistry, National Heart, Lung, and Blood Institute, National Institutes of Health, Bethesda, Maryland 20892*

CHRISTINA LIND (30), *Department of Biochemistry, University of Stockholm, S-106 91 Stockholm, Sweden*

MATILDE MAIORINO (47), *Department of Biological Chemistry, University of Padova, 35121 Padova, Italy*

STEFAN L. MARKLUND (25), *Department of Clinical Chemistry, Umeå University Hospital, S-90185 Umeå, Sweden*

JOSEPH P. MARTIN, JR. (19, 67), *Bioprocess Research and Development, The Upjohn Company, Kalamazoo, Michigan 49001*

Z. MASKOS (12), *Biodynamics Institute, Louisiana State University, Baton Rouge, Louisiana 70803*

RONALD P. MASON (9, 33), *Laboratory of Molecular Biophysics, National Institute of Environmental Health Sciences, Research Triangle Park, North Carolina 27709*

LAURA A. MAYO (59), *Department of Internal Medicine, University of Michigan Medical Center, Ann Arbor, Michigan 48109*

ANTONY F. MCDONAGH (31), *Department of Medicine, and the Liver Center, University of California, San Francisco, San Francisco, California 94143*

HIROSHI MEGURO (13), *Department of Food Chemistry, Faculty of Agriculture, Tohoku University, Sendai, 981 Japan*

ROLF J. MEHLHORN (17, 73), *Applied Science Division, Lawrence Berkeley Laboratory, Berkeley, California 94720*

WALTER H. MERRILL (81), *Department of Cardiac and Thoracic Surgery, Vanderbilt University, Nashville, Tennessee 37240*

CHRISTA MICHEL (36), *Institute für Strahlenbiologie, GSF Research Center, D-8042 Neuherberg, Federal Republic of Germany*

DENNIS M. MILLER (48), *Department of Environmental Health, University of Washington, Seattle, Washington 98195*

JAMES F. MILLER, JR. (44), *Department of Biochemistry, University of Illinois at Chicago, Chicago, Illinois 60612*

JERRY R. MITCHELL (75, 83), *The Upjohn Company, Kalamazoo, Michigan, and Center for Experimental Therapeutics* and Section on Hypertension/Clinical Pharmacology, Department of Medicine, Baylor College of Medicine and Physiology, Houston, Texas 77030

ALESSANDRA MOCALI (18), Istituto di Patologia Generale, Università degli Studi di Firenze, 50134 Florence, Italy

KENNETH D. MUNKRES (24), Laboratory of Molecular Biology and Department of Genetics, The University of Wisconsin, Madison, Wisconsin 53706

MICHAEL E. MURPHY (63), Institut für Physiologische Chemie 1, Heinrich-Heine Universität Düsseldorf, 4000 Düsseldorf 1, Federal Republic of Germany

ARNO NAGELE (77), Medizinische Klinik III, Klinikum Grosshadern, Ludwig-Maximilians-Universität, D-8000 München 70, Federal Republic of Germany

MINORU NAKANO (20, 34, 62), College of Medical Care and Technology, Gunma University, Maebashi-shi, Gunma 371, Japan

W. E. NEFF (39), Northern Regional Research Center, Agricultural Research Service, U.S. Department of Agriculture, Peoria, Illinois 61604

ETSUO NIKI (3, 34), Faculty of Engineering, Department of Reaction Chemistry, University of Tokyo, Hongo, Tokyo 113, Japan

GERALD D. NORDBLOM (27), Pharmaceutical Research Division, Warner Lambert-Parke Davis, Ann Arbor, Michigan 48105

HIROSHI OHRUI (13), Department of Food Chemistry, Faculty of Agriculture, Tohoku University, Sendai, 981 Japan

CYNTHIA N. OLIVER (49, 60), Laboratory of Biochemistry, National Heart, Lung, and Blood Institute, National Institutes of Health, Bethesda, Maryland 20892

STEN ORRENIUS (66), Department of Toxicology, Karolinska Institutet, S104-01 Stockholm, Sweden

ROBERT E. PACIFICI (51), Department of Biochemistry and Institute for Technology, University of Southern California, Los Angeles, California 90033

LESTER PACKER (17, 73, 79), Department of Molecular and Cell Biology, University of California, Berkeley, Berkeley, California 94720

FRANCESCO PAOLETTI (18), Istituto di Patologia Generale, Università degli Studi di Firenze, 50134 Florence, Italy

JEEN-WOO PARK (54), Division of Biochemistry and Molecular Biology, University of California, Berkeley, Berkeley, California 94720

ROBERT B. PENDLETON (44), Department of Biochemistry, University of Illinois at Chicago, Chicago, Illinois 60612

WILLIAM A. PRYOR (72), Biodynamics Institute, Louisiana State University, Baton Rouge, Louisiana 70803

D. N. RAMAKRISHNA RAO (33), Laboratory of Molecular Biophysics, National Institute of Environmental Health Sciences, Research Triangle Park, North Carolina 27709

JOHN E. REPINE (69), Webb-Waring Lung Institute and the University of Colorado Health Sciences Center, Denver, Colorado 80262

SUE GOO RHEE (50), Laboratory of Biochemistry, National Heart, Lung, and Blood Institute, National Institutes of Health, Bethesda, Maryland 20892

GERALD M. ROSEN (64), Department of Pharmacology and Toxicology, University of Maryland School of Pharmacy, Baltimore, Maryland 21201

ALBERTA B. ROSS (5), Radiation Chemistry Data Center, Radiation Laboratory, University of Notre Dame, Notre Dame, Indiana 46556

J. D. RUSH (12), Biodynamics Institute, Louisiana State University, Baton Rouge, Louisiana 70803

VICTOR M. SAMOKYSZYN (48), School of Pharmacy, Department of Pharmaceutical Chemistry, University of California, San Francisco, San Francisco, California 94143

MANFRED SARAN (36), Institut für Strahlenbiologie, GSF Research Center, D-8042

Neuherberg, Federal Republic of Germany

K. M. SCHAICH (7), Department of Food Science, Cook College, Rutgers University, New Brunswick, New Jersey 08903

HEINZ-PETER SCHUCHMANN (53), Max-Planck-Institut für Strahlenchemie, D-4330 Mülheim an der Ruhr 1, Federal Republic of Germany

ELENA E. SERBINOVA (37), Institute of Physiology, Bulgarian Academy of Sciences, Sofia 113, Bulgaria

SHMUEL SHALTIEL (49), Department of Chemical Immunology, The Weizmann Institute of Science, Rehovot, 76100 Israel

MARK SHIGENAGA (54), Division of Biochemistry and Molecular Biology, University of California, Berkeley, Berkeley, California 94720

HELMUT SIES (63), Institut für Physiologische Chemie I, Heinrich-Heine Universität Düsseldorf, 4000 Düsseldorf 1, Federal Republic of Germany

MICHAEL G. SIMIC (2), Center for Radiation Research, National Institute for Standards and Technology, Gaithersburg, Maryland 20899

AJIT SINGH (76), Radiation Applications Research Branch, Atomic Energy of Canada Ltd, Whiteshell Nuclear Research Establishment, Pinawa, Manitoba, R0E 1L0 Canada

HARWANT SINGH (76), Radiation Applications Research Branch, Atomic Energy of Canada Ltd, Whiteshell Nuclear Research Establishment, Pinawa, Manitoba, R0E 1L0 Canada

LELAND L. SMITH (45), Department of Human Biological Chemistry and Genetics, University of Texas Medical School, Galveston, Texas 77550

MARTYN T. SMITH (61), School of Public Health, University of California, Berkeley, Berkeley, California 94720

EARL R. STADTMAN (49, 50), Laboratory of Biochemistry, National Heart, Lung, and Blood Institute, National Institutes of Health, Bethesda, Maryland 20892

MELISSA GALE STEINER (11), The William A. Hillenbrand Biomedical Engineering Center, Purdue University, West Lafayette, Indiana 47907

ROBERT J. STEPHENS (40, 41), Cell Biology Program, Life Sciences Division, SRI International, Menlo Park, California 94025

JAMES R. STEWART (81), Department of Cardiac and Thoracic Surgery, Vanderbilt University, Nashville, Tennessee 37232

ROLAND STOCKER (31), The Heart Research Institute, Camperdown, New South Wales 2050, Australia

TSANKO S. STOYTCHEV (37), Institute of Physiology, Bulgarian Academy of Sciences, Sofia 113, Bulgaria

SATOSHI SUMIDA (17), Department of Sports and Sciences, Osaka Gakuin University, Suita, Osaka 564, Japan

MARC S. SUSSMAN (78), Department of Surgery, The Johns Hopkins Hospital, Baltimore, Maryland 21205

LANCE S. TERADA (69), Webb-Waring Lung Institute and the University of Colorado Health Sciences Center, Denver, Colorado 80262

DAVID W. THOMAS (40, 41), Cell Biology Program, Life Sciences Division, SRI International, Menlo Park, California 94025

HJÖRDIS THOR (66), Department of Toxicology, Karolinska Institutet, S104-01 Stockholm, Sweden

T. THÜRICH (79), Zentrum der Biologischen Chemie, Zentrum der Johann Wolfgang Goethe Universität, 6000 Frankfurt am Main 70, Federal Republic of Germany

DANIELE TOUATI (68), Genetique et Membranes, Institut Jacques Monod, C.N.R.S., Paris VII, 75251 Paris Cedex 05, France

THOMAS D. TULLIUS (56), Department of Chemistry, The Johns Hopkins University, Baltimore, Maryland 21218

FULVIO URSINI (47), *Department of Biological Chemistry, University of Padova, 35121 Padova, Italy*

FREDERICK J. G. M. VAN KUIJK (40, 41), *Department of Chemistry, Montana State University, Bozeman, Montana 59717*

J. E. VAN LIER (52), *MRC Group in Radiation Sciences, University of Sherbrooke, Sherbrooke, Quebec, J1H 5N4 Canada*

ALFIN D. N. VAS (28), *Department of Biological Chemistry, University of Michigan Medical School, Ann Arbor, Michigan 48109*

CLEMENS VON SONNTAG (53), *Max-Planck-Institut für Strahlenchemie, D-4330 Mülheim an der Ruhr 1, Federal Republic of Germany*

J. R. WAGNER (52), *Department of Biochemistry, University of California, Berkeley, Berkeley, California 94720*

D. WEISLEDER (39), *Northern Regional Research Center, Agricultural Research Service, U.S. Department of Agriculture, Peoria, Illinois 61604*

ALBRECHT WENDEL (74), *Faculty of Biology, Department of Biochemical Pharmacology, University of Konstanz, D-7750 Konstanz, Federal Republic of Germany*

RONALD E. WHITE (27), *Department of Drug Metabolism, Squibb Institute for Medical Research, Princeton, New Jersey 08543*

G. MELVILLE WILLIAMS (82), *Department of Surgery, The Johns Hopkins University, School of Medicine, Baltimore, Maryland 21205*

CHRISTINE C. WINTERBOURN (26), *Department of Pathology, School of Medicine, Christchurch Hospital, Christchurch, New Zealand*

YORIHIRO YAMAMOTO (38), *Department of Reaction Chemistry, Faculty of Engineering, University of Tokyo, Hongo, Bunkyo-ku, Tokyo 113, Japan*

TOSHIKAZU YOSHIKAWA (70, 71), *First Department of Medicine, Kyoto Prefectural University of Medicine, Kamigyo-ku, Kyoto 602, Japan*

D. M. ZIEGLER (65), *Department of Chemistry, Clayton Foundation Biochemical Institute, The University of Texas at Austin, Austin, Texas 78712*

G. ZIMMER (79), *Zentrum der Biologischen Chemie, Klinikum der Johann Wolfgang Goethe Universität, 6000 Frankfurt am Main 70, Federal Republic of Germany*

Preface

Interest in oxygen radicals continues to grow with the accumulation of evidence that oxygen-derived free radicals play a role in a wide variety of pathological conditions such as atherosclerosis, emphysema, ischemia, radiation injury, and cancer and may be important in the aging process. On the positive side, reactive oxygen species produced by leukocytes contribute to the killing of microbes.

The contributions to this volume describe methods for the generation and determination of various radical species and for the study of the products of their attack on proteins, lipids, and nucleic acids. A number of the articles deals with the assessment of free radical-mediated ischemia and reperfusion-associated tissue damage. Fundamental studies of free radical damage on tissues, isolated organs, and whole organisms are essential for the design of protective and therapeutic medical intervention.

The methods used for the assessment of free radical damage in complex systems need to be applied with caution and appreciation for potential pitfalls. Where roles in pathogenesis are ascribed to free radicals, it is important to ascertain whether such roles are primary or secondary. We are indebted to Drs. B. Halliwell and J. Gutteridge for the opening contribution to this volume which provides a critical assessment of these basic questions.

We extend grateful appreciation to the advisory board members, Bruce Ames, Anthony Diplock, Lars Ernster, Irwin Fridovich, Rolf Mehlhorn, William Pryor, Helmut Sies, and Trevor Slater, who provided much valuable advice and contributions.

LESTER PACKER
ALEXANDER N. GLAZER

METHODS IN ENZYMOLOGY

[1] Role of Free Radicals and Catalytic Metal Ions in Human Disease: An Overview

By BARRY HALLIWELL and JOHN M. C. GUTTERIDGE

1. Reactive Oxygen Species: An Introduction

Several definitions of the term free radical exist, but the authors adopt a broad approach and define a free radical as any species that has one or more unpaired electrons. This definition embraces the atom of hydrogen (one unpaired electron), most transition metal ions, and the oxygen molecule. The definition of a free radical has sometimes been qualified by specifying that the unpaired electron should be in an orbital in the outermost electron shell, a definition which excludes several transition metal ions. However, we prefer the simpler definition since the biological roles played by transition metal ions frequently relate to their ability to participate in one-electron transfer reactions (reviewed by Hill[1] and by the authors[2,3]).

It must first be appreciated that the ground state diatomic oxygen molecule (O_2) is itself a radical, with two unpaired electrons each located in a π^* antibonding orbital (for a detailed explanation, see Ref. 3). The two unpaired electrons have the same spin quantum number (parallel spin), and so if O_2 attempts to oxidize another atom or molecule by accepting a pair of electrons from it, both new electrons must be of parallel spin to fit into the vacant spaces in the π^* orbitals. Most biomolecules are covalently bonded nonradicals, and the two electrons forming a covalent bond have opposite spins and occupy the same molecular orbital. Hence the reaction of oxygen with biomolecules is spin restricted.[2]

The human body with its large "pool" of carbon and hydrogen in biomolecules is thermodynamically unstable in air, yet we do not spontaneously combust. What factors contribute to protection against ongoing oxidation? First, the spin restriction itself is of benefit for life in an oxygen-rich environment since it slows down the reaction of molecular oxygen with biological molecules. There is also an important orbital restriction. Survival, however, ultimately depends on the constant repair of oxidative damage. This is supported by specific (enzymatic) antioxidant

[1] H. A. O. Hill, *Philos. Trans. R. Soc. London Ser. B* **294,** 119 (1981).
[2] B. Halliwell, *FASEB J.* **1,** 358 (1987).
[3] B. Halliwell and J. M. C. Gutteridge, "Free Radicals in Biology and Medicine." 2nd Ed., Oxford Univ. Press (Clarendon), Oxford, 1989.

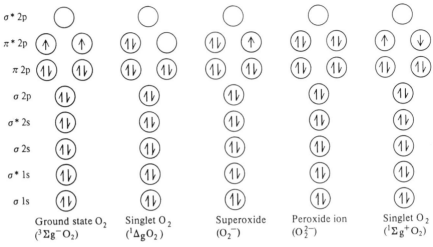

FIG. 1. Singlet states of O_2.

protection and less specific scavenger molecules which protect key sites and limit overall damage. Transition metals are found at the active sites of most oxidases and oxygenases because their ability to accept and donate single electrons can overcome the spin restriction of oxygen.[1-3]

1.1. Singlet Oxygen

Another way of increasing the reactivity of O_2 is to move one of the unpaired electrons in a way that alleviates the spin restriction. This requires an input of energy and generates the singlet states[4] of O_2 (Fig. 1). Singlet O_2 $^1\Delta_g$, the most important in biological systems, has no unpaired electrons and thus does not qualify as a radical. Singlet O_2 $^1\Sigma_g^+$ (Fig. 1) usually decays to the $^1\Delta_g$ state before it has time to react with anything. Excitation of O_2 to the singlet states can be achieved when several pigments are illuminated in the presence of O_2. The pigment absorbs light, enters a higher electronic excitation state, and transfers energy onto the O_2 molecule to make singlet O_2. Singlet O_2 formation is thus likely to occur in many pigmented systems exposed to light; the lens of the eye[5] and the illuminated chloroplast[6] are examples.

A few diseases can lead to excessive singlet O_2 formation. For exam-

[4] H. Wefers, *Bioelectrochem. Bioenerg.* **18,** 91 (1987).

[5] J. S. Zigler and J. D. Goosey, *Photochem. Photobiol.* **33,** 869 (1981).

[6] B. Halliwell, "Chloroplast Metabolism: The Structure and Function of Chloroplasts in Green Leaf Cells." Oxford Univ. Press (Clarendon), Oxford, 1984.

ple, the porphyrias are defects, usually inborn, in porphyrin metabolism. Porphyrins are often excreted in the urine and accumulate in the skin, exposure of which to light can lead to unpleasant eruptions, scarring, and thickening.[7] The severity of the damage depends on the exact structure of the porphyrins accumulated and thus will differ in different forms of porphyria. It has also been observed that certain porphyrins are taken up by cancerous tumors. After injection of a porphyrin preparation known as HPD (hematoporphyrin derivative), fluorescent products are strongly retained by tumor tissues, and this can be used to detect the presence of the tumor by observing the fluorescence.[8] Irradiation with light of a wavelength absorbed by HPD can damage the tumor, and such reactions are of potential use in cancer chemotherapy, especially for skin and lung cancer. Both hydroxyl radicals and singlet oxygen have been suggested to cause the damage to the tumor cells (e.g., Ref. 8).

Another application of photosensitization reactions in medicine is the use of psoralens in the treatment of skin diseases such as psoriasis.[9] The treatment consists of the combined application of ultraviolet light in the wavelength range 320–400 nm (UVA) and a psoralen, and is often referred to as PUVA therapy (psoralen–ultraviolet). Psoralens are a class of compounds produced by plants, and they are powerful photosensitizers of singlet O_2 production. Some drugs (e.g., tetracyclines and the nonsteroidal antiinflammatory drug benoxaprofen) and constituents of cosmetics may also damage skin by photosensitization reactions.[10–12]

It is often stated that singlet O_2 is formed by the dismutation of O_2^- radicals (see below) and during the respiratory burst of neutrophils. At best, the evidence for both these statements may be described as inconclusive (e.g., Ref. 13). Foote[14] has emphasized the careful procedures necessary to establish singlet O_2 formation in a biological system, procedures which are often ignored. Frequently, the only evidence presented for formation of singlet O_2 in a system is that the system produces light (*luminescence*) or that allegedly "specific" singlet O_2 scavengers such as 1,4-diazabicyclo[2.2.2]octane (DABCO), diphenylisobenzofuran, histidine, or azide inhibit the observed process. First, there are no specific

[7] B. Franck, *Angew. Chem. Int. Ed. Engl.* **21**, 343 (1982).

[8] J. F. Evensen, S. Sommer, J. Moan, and T. Christensen, *Cancer Res.* **44**, 482 (1984).

[9] M. A. Pathak and P. C. Joshi, *Biochim. Biophys. Acta* **798**, 115 (1984).

[10] R. H. Sik, C. S. Paschall, and C. F. Chignell, *Photochem. Photobiol.* **38**, 411 (1983).

[11] A. G. Motten, C. F. Chignell, and R. P. Mason, *Photochem. Photobiol.* **38**, 671 (1983).

[12] T. Hasan and A. U. Khan, *Proc. Natl. Acad. Sci. U.S.A.* **83**, 4604 (1986).

[13] R. L. Arudi, M. W. Sutherland, and B. H. J. Bielski, *J. Lipid Res.* **24**, 485 (1983).

[14] C. S. Foote, *in* "Biochemical and Clinical Aspects of Oxygen" (W. S. Caughey, ed.), p. 603. Academic Press, New York, 1979.

scavengers of singlet O_2. All react with hydroxyl radical, often with a greater rate constant than for their reaction with singlet O_2. A number also react with at least one organic peroxyl radical[15] and with hypochlorous acid formed by the action of myeloperoxidase in activated neutrophils.[16] If, when the above scavengers are added to a system, none of them inhibits the reaction under study, then one may conclude that singlet O_2 is not required for it to proceed. The reverse assumption cannot be made. However, the products of the reaction of cholesterol[17] and tryptophan[18] with singlet O_2 are different from those obtained on reaction of these molecules with hydroxyl radical, so isolation and characterization of such products could provide better evidence for singlet O_2 formation. The design of spin traps suitable for the specific detection of singlet O_2 in biological systems is awaited with interest.

As they decay to the ground state, molecules of singlet oxygen emit light. Individual molecules emit in the infrared (1270 nm), but light is also emitted at 634 and 703 nm by a "dimol emission" process that involves cooperation of two molecules of singlet O_2.[4,14,19] Production of light by activated phagocytic cells in the presence of "enhancers" such as luminol or lucigenin is a very convenient method of following the respiratory burst,[20,21] and it has sometimes been speculated that the light arises from singlet O_2. In fact, lucigenin-dependent light emission seems to involve O_2^-,[22] whereas luminol-dependent light emission involves the myeloperoxidase system generating HOCl both inside and outside the phagocyte.[20,23] Mixtures of myeloperoxidase, H_2O_2, and Cl^- ion can be shown to produce singlet O_2, but only under conditions unlikely to occur *in vivo*[24] (although good evidence for $^1O_2$ production during the oxidation of bromide, but not chloride, ions by peroxidase isolated from human eosinophils[25] has been presented).

[15] J. E. Packer, J. S. Mahood, V. O. Mora-Arellano, T. F. Slater, R. L. Willson, and B. S. Wolfenden, *Biochem. Biophys. Res. Commun.* **98**, 901 (1981).
[16] J. E. Harrison, B. D. Watson, and J. Schultz, *FEBS Lett.* **92**, 327 (1978).
[17] L. L. Smith, *Chem. Phys. Lipids* **44**, 87 (1987).
[18] A. Singh, H. Singh, W. Kremers, and G. W. Koroll, *Bull. Eur. Physiopathol. Respir.* **17**, 31 (1981).
[19] H. Sies, *Arch. Toxicol.* **60**, 138 (1987).
[20] P. Roscher, W. Graninger, and H. Klima, *Biochem. Biophys. Res. Commun.* **123**, 1047 (1984).
[21] G. Briheim, O. Stendahl, and C. Dahlgren, *Infect. Immun.* **45**, 1 (1984).
[22] P. Stevens and D. Hong, *Microchem. J.* **30**, 135 (1984).
[23] C. Dahlgren, *Agents Actions* **21**, 103 (1987).
[24] J. R. Kanofsky, J. Wright, G. E. Miles-Richardson, and A. I. Tauber, *J. Clin. Invest.* **74**, 1489 (1984).
[25] J. R. Kanofsky, H. Hoogland, R. Wever, and S. J. Weiss, *J. Biol. Chem.* **263**, 9692 (1988).

Hence, luminol- or lucigenin-dependent light production by phago-cytes is not due to singlet O_2 formation. In any case, the reactivity of hypochlorous acid produced in the myeloperoxidase system is so high that there is no need to invoke singlet O_2 to explain the cytotoxicity of this system.[26-29] Myeloperoxidase plays only a secondary role in bacterial killing by human neutrophils, since an inborn defect of this enzyme rarely produces clinical problems (e.g., Ref. 30). However, HOCl generated by myeloperoxidase might inactivate α_1-antiproteinase at sites of inflamma-tion and contribute to proteolytic damage (reviewed in Ref. 31).

Low-level (background) chemiluminescence is a useful assay of "ox-idative stress" in isolated organelles, whole cells, and perfused organs (e.g., Refs. 4, 19, 32, and 33). Some of the light emitted might come from singlet O_2[4,19] [perhaps generated from lipid peroxides (Section 3.2)], but some probably comes from other sources. These include excited state carbonyls[4,19] and Fenton reactions.[34]

1.2. Superoxide Radical

One-electron reduction of oxygen produces superoxide radical, O_2^-. This is frequently written as $O_2^{\cdot-}$, where the dot denotes a radical species, that is, an unpaired electron. This nomenclature is slightly illogical; one would then write oxygen as $O_2^{\cdot\cdot}$, since it has two unpaired electrons instead of superoxide's one (Fig. 1). The authors avoid the inconsistency by writing O_2^-.

Superoxide is formed in almost all aerobic cells (for reviews, see Refs. 3, and 35–39), a major source being "leakage" of electrons onto O_2 from various components of the cellular electron transport chains, such as

[26] J. M. Albrich, C. A. McCarthy, and J. K. Hurst, *Proc. Natl. Acad. Sci. U.S.A.* **78,** 210 (1981).

[27] S. T. Test, M. B. Lampert, P. J. Ossana, J. G. Thoene, and S. J. Weiss, *J. Clin. Invest.* **74,** 1341 (1984).

[28] E. P. Brestel, *Biochem. Biophys. Res. Commun.* **126,** 482 (1985).

[29] M. Wasil, B. Halliwell, M. Grootveld, C. P. Moorhouse, D. C. S. Hutchison, and H. Baum, *Biochem. J.* **243,** 867 (1987).

[30] W. M. Nauseef, R. K. Root, and H. L. Malech, *J. Clin. Invest.* **71,** 1297 (1983).

[31] S. J. Weiss, *Acta Physiol. Scand. Suppl.* **548,** 9 (1986).

[32] E. Cadenas, H. Wefers, and H. Sies, *Eur. J. Biochem.* **119,** 531 (1981).

[33] E. Cadenas, A. Muller, R. Brigelius, H. Esterbauer, and H. Sies, *Biochem. J.* **214,** 479 (1983).

[34] B. R. Andersen and L. A. Harvath, *Biochim. Biophys. Acta* **584,** 164 (1979).

[35] I. Fridovich, *Adv. Enzymol.* **41,** 35 (1974).

[36] I. Fridovich, *Annu. Rev. Biochem.* **44,** 147 (1975).

[37] I. Fridovich, *Science* **201,** 875 (1978).

[38] I. Fridovich, *Annu. Rev. Pharmacol. Toxicol.* **23,** 239 (1983).

[39] J. Diguiseppi and I. Fridovich, *CRC Crit. Rev. Toxicol.* **12,** 315 (1984).

those of mitochondria, chloroplasts, and the endoplasmic reticulum. The amount of leakage, and hence the rate of O_2^- production, increases as the O_2 concentration is raised.[40,41] It is also clearly established that O_2^- is produced during the respiratory burst of phagocytic cells (neutrophils, monocytes, macrophages, and eosinophils). The K_m for O_2 of the NADPH oxidase complex that produces the O_2^- in neutrophils may be within the range of O_2 concentrations in body fluids, so that an elevated O_2 concentration might increase O_2^- production by activated phagocytic cells.[42] Superoxide production by phagocytes plays a key role in the killing of several bacterial strains; if it is not produced, as in the inborn defect known as chronic granulomatous disease (CGD), then many bacterial strains are not killed properly, resulting in a syndrome of persistent and multiple infections.[43] Many other strains are killed perfectly normally by CGD phagocytes, however, and so other antibacterial mechanisms must be important as well.

In organic solvents, O_2^- is a strong base and nucleophile; for example, it can displace Cl^- from such unreactive chlorinated hydrocarbons as CCl_4.[44] In aqueous solution, O_2^- is extensively hydrated and much less reactive, acting as a reducing agent (e.g., it will reduce cytochrome c or nitro blue tetrazolium) and as a weak oxidizing agent to such molecules as adrenalin and ascorbic acid. It also undergoes the dismutation reaction, which can be written overall as

$$2 \, O_2^- + 2 \, H^+ \rightarrow H_2O_2 + O_2 \tag{1}$$

although at physiological pH it is largely the sum of the following two stages

$$O_2^- + H^+ \rightarrow HO_2 \cdot \tag{2}$$
$$HO_2 \cdot + O_2^- + H^+ \rightarrow H_2O_2 + O_2 \tag{3}$$

The overall rate of dismutation at pH 7 is about $5 \times 10^5 \, M^{-1} \, sec^{-1}$, and any reaction undergone by O_2^- in aqueous solution will be in competition with this dismutation reaction.[45] There is no good evidence that dismutation in aqueous solution produces singlet O_2 (see above).

Despite the moderate chemical reactivity of O_2^- in aqueous solution, aqueous O_2^--generating systems (chemical, enzymatic, or phagocytic) have been observed to do a considerable degree of biological damage (for

[40] B. A. Freeman and J. D. Crapo, *J. Biol. Chem.* **256**, 10986 (1981).

[41] H. Nohl, D. Hegner, and K. H. Summer, *Biochem. Pharmacol.* **30**, 1753 (1981).

[42] S. W. Edwards, M. B. Hallett, D. Lloyd, and A. K. Campbell, *FEBS Lett.* **161**, 60 (1983).

[43] A. I. Tauber, N. Borregaard, E. Simons, and J. Wright, *Medicine* **62**, 286 (1983).

[44] J. L. Roberts, T. S. Calderwood, and D. T. Sawyer, *J. Am. Chem. Soc.* **105**, 7691 (1983).

[45] B. H. J. Bielski and A. O. Allen, *J. Phys. Chem.* **81**, 1048 (1977).

reviews, see Refs. 3, 35–39, 46, and 47). Further evidence that O_2^- is a species worth removing *in vivo* comes from the following: (i) superoxide dismutase (SOD) enzymes are catalysts that have evolved a surface charge arrangement to facilitate the specific use of O_2^- as a substrate,[48,49] and (ii) superoxide dismutases are important antioxidants, required for the growth of aerobes without excessive DNA damage in the presence of O_2.[35–39,50–53] The SOD enzymes accelerate reaction (1) by about four orders of magnitude. Hence O_2^- must be worth removing even at the expense of forming H_2O_2,[54] although SOD enzymes in human cells work in conjunction with H_2O_2-removing enzymes such as catalases and glutathione peroxidases (reviewed in Ref. 3).

Because SOD is specific for O_2^- as a catalytic substrate, it is often assumed that inhibition of a reaction on addition of SOD means that O_2^- is required for that reaction to proceed. Before reaching such a conclusion, controls with heat-denatured protein or the apoenzyme must be performed (e.g., see Ref. 55). The SOD protein is a good scavenger of singlet O_2 and hydroxyl radical by chemical reaction, but the apoenzyme is just as good as the holoenzyme.[56] Inhibitions by SOD must also be interpreted with caution in systems containing quinones. Many semiquinones react reversibly with O_2:

$$\text{Semiquinone} + O_2 \rightleftharpoons O_2^- + \text{quinone} \qquad (4)$$

Addition of SOD, by removing O_2^-, will accelerate the loss of semiquinone, and a reaction that is actually caused by the semiquinone might be mistakenly attributed to O_2^- as a result of observed inhibition on addition of SOD.[57]

Misinterpretation may also arise when O_2^- is generated by the oxidation of a molecule such as adrenaline, dihydroxyfumarate, or hydroxydo-

[46] B. Halliwell and J. M. C. Gutteridge, *Biochem. J.* **219,** 1 (1984).

[47] B. Halliwell and J. M. C. Gutteridge, *Mol. Aspects Med.* **8,** 89 (1985).

[48] E. D. Getzoff, J. A. Tainer, P. K. Weiner, P. A. Kollman, J. S. Richardson, and J. C. Richardson, *Nature (London)* **306,** 287 (1983).

[49] J. Benovic, T. Tillman, A. Cudd, and I. Fridovich, *Arch. Biochem. Biophys.* **221,** 329 (1983).

[50] P. P. G. M. Van Loon, B. Pesold-Hurt, and G. Schatz, *Proc. Natl. Acad. Sci. U.S.A.* **83,** 3820 (1986).

[51] S. B. Farr, R. D'Ari, and D. Touati, *Proc. Natl. Acad. Sci. U.S.A.* **83,** 8268 (1986).

[52] D. O. Natvig, K. Imlay, D. Touati, and R. A. Hallewell, *J. Biol. Chem.* 14697 (1987).

[53] D. Touati, *Free Radical Res. Commun.* **8,** 1 (1989).

[54] H. E. Schellhorn and H. M. Hassan, *Can. J. Microbiol.* **34,** 1171 (1988).

[55] B. Halliwell and C. H. Foyer, *Biochem. J.* **155,** 697 (1976).

[56] I. B. C. Matheson, R. D. Etheridge, N. R. Kratowich, and J. Lee, *Photochem. Photobiol.* **21,** 165 (1975).

[57] C. C. Winterbourn, *FEBS Lett.* **136,** 89 (1981).

pamine, in which the oxidation produces O_2^- that then participates in further oxidation of the molecule. For example, oxidizing dihydroxyfumaric acid is cytotoxic, and addition of SOD protects the cells to some extent. This cannot be taken to mean that the toxicity is caused by O_2^-. Superoxide plays a role in the mechanism of dihydroxyfumarate oxidation, and addition of SOD slows that oxidation.[58] It could equally well be some other product of dihydroxyfumarate oxidation (e.g., a dihydroxyfumarate-derived radical) that is the real cell-damaging agent.

1.3. Hydrogen Peroxide

A system generating O_2^- would be expected to produce H_2O_2 by nonenzymatic or SOD-catalyzed dismutation [Eq. (1)]. Several oxidase enzymes produce H_2O_2 directly, examples being 2-hydroxyacid and urate oxidases (reviewed in Ref. 3). If we accept that O_2^- is formed *in vivo* in human cells and is scavenged by SOD, then we must accept that H_2O_2 is also produced. H_2O_2 production, probably mainly via O_2^-, has been observed from whole bacteria of several species, from phagocytic cells, from spermatozoa,[59] and from mitochondria, microsomes, and chloroplasts (reviewed in Ref. 3). The lens of the human eye contains micromolar concentrations of H_2O_2,[60] and H_2O_2 vapor has been detected in exhaled human breath.[61] It probably arises from pulmonary macrophages, although a contribution from oral bacteria[62] cannot be ruled out. H_2O_2 at concentrations up to micromolar[63] is present in most natural water supplies. However, reports that human blood plasma contains micromolar concentrations of H_2O_2 have not been substantiated.[64]

H_2O_2 has no unpaired electrons and is not a radical (Fig. 1). Pure H_2O_2 has limited reactivity, but it can cross biological membranes, which the charged O_2^- species can do only very slowly[65] unless there is an anion channel through which it can move.[66] The only example of such a channel known to date is in the erythrocyte membrane,[66] although it is possible that such a channel also exists in the membranes of vascular endothelial cells (Section 4.5.3).

Reports of the toxicity of H_2O_2 to cells and organisms are variable;

[58] B. Halliwell, *Biochem. J.* **163**, 441 (1977).
[59] M. K. Holland, J. G. Alvarez, and B. T. Storey, *Biol. Reprod.* **27**, 1109 (1982).
[60] K. C. Bhuyan and D. K. Bhuyan, *Biochim. Biophys. Acta* **497**, 641 (1977).
[61] M. D. Williams, J. S. Leigh, and B. Chance, *Ann. N.Y. Acad. Sci.* **45**, 478 (1983).
[62] E. L. Thomas and K. A. Pera, *J. Bacteriol.* **154**, 1236 (1983).
[63] O. C. Zafiriou, *Nature (London)* **325**, 481 (1987).
[64] B. Frei, Y. Yamamoto, D. Niclas, and B. N. Ames, *Anal. Biochem.* **175**, 120 (1988).
[65] M. A. Takahashi and K. Asada, *Arch. Biochem. Biophys.* **226**, 558 (1983).
[66] R. E. Lynch and I. Fridovich, *J. Biol. Chem.* **253**, 4697 (1978).

some bacteria and animal cells are injured by H_2O_2 at micromolar concentrations, whereas other bacteria and photosynthetic algae generate and release large amounts of it (reviewed in Ref. 3). The variability can be accounted for both by the activity of H_2O_2-removing enzymes and by the rate of conversion of H_2O_2 into more highly reactive radicals (see below).

If cells are exposed to O_2^--generating systems and protection against damage is seen by added catalase but not by SOD, this does not necessarily mean that damage is being done directly by H_2O_2. Superoxide generated outside the cells cannot easily enter them, whereas H_2O_2 can do so and might give rise to more reactive radical species inside the cells[67] (see below). Similarly, H_2O_2 generated within the cell may be removed by the addition of catalase outside the cell, the enzyme being able to disturb the diffusion equilibrium and cause H_2O_2 to leave the cell through the plasma membrane. SOD added externally cannot remove internally generated O_2^-, however. In any experiments with catalase, controls with denatured enzyme must also be carried out. Catalase for use in experiments with whole animals must also be carefully checked for contamination with endotoxin.[68]

Microsomal fractions prepared from many animal tissues have been shown to produce O_2^- and H_2O_2 at high rates when incubated in the presence of NADPH, and "microsomes" are frequently used for studies of the redox cycling of drugs and of lipid peroxidation. Despite their attractive name, however, microsomes are not intracellular organelles but are formed by the process of cell disruption and fractionation. When cells are disrupted in a homogenizer, the plasma membrane and endoplasmic reticulum are torn up, and the microsomal fraction, obtained by high-speed centrifugation, is a heterogeneous collection of vesicles from both these membrane systems.[68a]

The O_2^- and H_2O_2 generated by microsomes largely arise from the NADPH–cytochrome-P-450 reductase/cytochrome P-450 system.[69,69a] Increasing the amount of cytochrome P-450 and its reductase by pretreating animals with phenobarbital increases the rates of H_2O_2 production by liver microsomes prepared later from the animals.[69] In perfused rat liver, however, the basal rate of oxidized glutathione (GSSG) release, as an index of H_2O_2 production, is smaller than expected from the rates of H_2O_2 production by microsomes *in vitro* (e.g., Ref. 70), and GSSG release was not increased if the animals were pretreated with phenobarbital. Hence

[67] B. Halliwell and J. M. C. Gutteridge, *Arch. Biochem. Biophys.* **246**, 501 (1986).
[68] T. Gordon, *J. Free Radicals Biol. Med.* **2**, 373 (1986).
[68a] C. de Duve, *J. Cell Biol.* **50**, 200 (1971).
[69] A. G. Hildebrandt and I. Roots, *Arch. Biochem. Biophys.* **171**, 385 (1975).
[69a] Y. Terelius and M. Ingelman-Sundberg, *Biochem. Pharmacol.* **37**, 1383 (1988).
[70] N. Oshino, D. Jamieson, and B. Chance, *Biochem. J.* **146**, 53 (1975).

H_2O_2 formation and, by implication, O_2^- formation, by the endoplasmic reticulum *in vivo* does not occur as rapidly as would be expected from experiments on microsomes. Perhaps, during the fragmentation and membrane vesicle formation that occur on cell disruption to produce microsomes, the arrangement of the components of the cytochrome *P*-450 system within the membrane is altered so that electrons escape more easily to oxygen.[3] The oxygen concentration adjacent to the endoplasmic reticulum *in vivo* must also be much lower than that seen by microsomes incubated *in vitro*. The fact that microsomes are an artifact of subcellular fractionation, apparently producing O_2^- and H_2O_2 at abnormally high rates, should always be borne in mind in studies on them.

1.4. Hydroxyl Radical and $HO_2 \cdot$ Radical

The moderate reactivity of O_2^- and H_2O_2 in aqueous solution makes it unlikely that the damage done by O_2^--generating systems can often be attributed to direct actions of O_2^- or H_2O_2, although some targets of these species have been identified. Thus, O_2^- has been reported to inactivate *Escherichia coli* dihydroxy-acid dehydratase[71] and cardiac creatine kinase,[72] and it reacts slowly with desferrioxamine[73] (but see ii below). H_2O_2 inactivates fructose-bisphosphatase from spinach chloroplasts[74] and can have some direct effect on enzymes of cellular ATP synthesis.[75] In general, however, the damage done to cells by O_2^- and H_2O_2 is probably due to their conversion into more highly reactive species. Suggestions have included the following:

(i) *Singlet oxygen.* There is no good evidence for formation of the singlet states of oxygen in physiological O_2^--generating systems or by neutrophils (see above), although it can be formed in peroxidizing lipid systems[4,19,32,33] (Section 3.2).

(ii) $HO_2 \cdot$ *radical.* Protonation of O_2^- yields the hydroperoxyl radical ($HO_2 \cdot$) [Eq. (2)]. The pK_a of $HO_2 \cdot$ is 4.7–4.8,[45] and so, if physiological pH is assumed to be around 7.4, only about 0.25% of any O_2^- generated will exist as $HO_2 \cdot$. In close proximity to membranes, however, the pH might be considerably lower than this, and more $HO_2 \cdot$ will form. The pH be-

[71] C. F. Kuo, T. Mashino, and I. Fridovich, *J. Biol. Chem.* **262**, 4724 (1987).

[72] J. M. McCord and W. J. Russell, *in* "Oxy-Radicals in Molecular Biology and Pathology" (P. A. Cerutti, I. Fridovich, and J. M. McCord, eds.), p. 27. Alan R. Liss, New York, 1988.

[73] B. Halliwell, *Biochem. Pharmacol.* **34**, 229 (1985).

[74] S. A. Charles and B. Halliwell, *Planta* **151**, 242 (1981).

[75] I. U. Schraufstatter, P. A. Hyslop, J. Jackson, and C. C. Cochrane, *Int. J. Tissue React.* **9**, 317 (1987).

neath activated macrophages adhering to surfaces has been reported to be 5 or less,[76] and so a considerable amount of any O_2^- that they generate may exist as $HO_2 \cdot$.

There is no clear evidence as yet that $HO_2 \cdot$ plays a cytotoxic role in any biological system, but its potential importance arises from two factors. First, it is less polar than O_2^- and ought to be able to cross biological membranes about as effectively as can H_2O_2.[67] Second, $HO_2 \cdot$ is somewhat more reactive than is O_2^-. Unlike O_2^-, $HO_2 \cdot$ can attack fatty acids directly, and evidence for conversion of linolenic, linoleic, and arachidonic acids to peroxides by $HO_2 \cdot$ has been presented.[77] Jessup *et al.*[77a] have suggested that $HO_2 \cdot$ can initiate peroxidation of the lipid component of low density lipoproteins. It should also be noted that reports of the reaction of enzymes or small molecules (e.g., desferrioxamine) with "superoxide" (see above) do not usually distinguish between a true reaction with O_2^- and a much faster reaction with the small amount of $HO_2 \cdot$ in equilibrium with O_2^-.

(iii) *Hydroxyl radical.* Hydroxyl radical, $\cdot OH$, is produced when water is exposed to high-energy ionizing radiation, and its properties have been well documented by radiation chemists. It is highly reactive (reviewed in Refs. 3, and 78–80), and so any hydroxyl radical produced *in vivo* would react at or close to its site of formation. Thus, type of damage would depend on its site of formation; for instance, production of $\cdot OH$ close to DNA could lead to modification of purines or pyrimidines or to strand breakage,[81] whereas production of $\cdot OH$ close to an enzyme molecule present in excess in the cell, such as lactate dehydrogenase, might have no biological consequences.[82] Further, reaction of $\cdot OH$ with a biomolecule will produce another radical, usually (because of the extremely high reactivity of $\cdot OH$) of lower reactivity. Such less reactive radicals can cause their own problems, since they can sometimes diffuse away from the site of formation and attack specific biomolecules. For example, uric acid reacts with $\cdot OH$ radical; it protects the enzyme lactate dehydrogenase against inactivation by $\cdot OH$ but accelerates inactivation of yeast alcohol dehydrogenase, an enzyme known to be sensitive to oxidant at-

[76] D. J. Etherington, G. Pugh, and I. A. Silver, *Acta Biol. Med. Ger.* **40**, 1625 (1981).

[77] B. H. J. Bielski, R. L. Arudi, and M. W. Sutherland, *J. Biol. Chem.* **258**, 4759 (1983).

[77a] S. Bedwell, R. T. Dean, and W. Jessup, *Biochem. J.* **262**, 707 (1989).

[78] M. Anbar and P. Neta, *Int. J. Appl. Radiat. Isot.* **18**, 495 (1967).

[79] G. Scholes, *Br. J. Radiol.* **56**, 221 (1983).

[80] R. L. Willson, *in* "Biochemical Mechanisms of Liver Injury" (T. F. Slater, ed.), p. 123. Academic Press, London, 1978.

[81] A. C. Mello Filho, M. E. Hoffmann, and R. Meneghini, *Biochem. J.* **218**, 273 (1984).

[82] B. Halliwell, J. M. C. Gutteridge, and D. Blake, *Philos. Trans. R. Soc. London Ser. B* **311**, 659 (1985).

tack.[83] The radicals produced when $\cdot$OH attacks uric acid are less reactive than is $\cdot$OH, so more of them survive to reach those sensitive sites on alcohol dehydrogenase with which they can react.[83] Perhaps the best example of the importance of such "secondary" radicals *in vivo* is the ability of $\cdot$OH to initiate lipid peroxidation by abstracting hydrogen atoms to form carbon-centered and peroxyl radicals (Section 3).

Most of the $\cdot$OH generated *in vivo*, except during excessive exposure to ionizing radiation, comes from the metal-dependent breakdown of H_2O_2, according to the general equation

$$M^{n+} + H_2O_2 \rightarrow M^{(n+1)+} + \cdot OH + OH^- \tag{5}$$

in which M^{n+} is a metal ion. M^{n+} can be titanium(III) or iron(II) (reviewed in Ref. 84). Cobalt(II) reacts with H_2O_2 to form a species that appears similar to $\cdot$OH,[85,86] as do some chromium, vanadium, and nickel complexes.[84,87–91] Copper(I) may also react with H_2O_2 to form $\cdot$OH; a copper–oxygen complex in which the copper has an oxidation number of 3 has been suggested to be formed as well as or instead of $\cdot$OH.[90,92–99a]

Generation of reactive species such as $\cdot$OH from complexes of vanadium, cobalt, chromium, and nickel may be important when considering the toxicology of metal poisoning,[89] but probably only the iron(II)-dependent formation of $\cdot$OH actually happens *in vivo* under normal condi-

[83] K. J. Kittridge and R. L. Willson, *FEBS Lett.* **170**, 162 (1984).
[84] C. Walling, *in* "Proceedings of the 3rd International Symposium on Oxidases Related Redox Systems" (T. E. King, H. S. Mason, and M. Morrison, eds.), p. 85. Pergamon, Oxford, 1982.
[85] J. M. C. Gutteridge, *FEBS Lett.* **157**, 37 (1983).
[86] C. P. Moorhouse, B. Halliwell, M. Grootveld, and J. M. C. Gutteridge, *Biochim. Biophys. Acta* **843**, 261 (1985).
[87] R. J. Keller, R. P. Sharma, T. A. Grover, and L. H. Piette, *Arch. Biochem. Biophys.* **265**, 524 (1988).
[88] S. Liochev and I. Fridovich, *Biochim. Biophys. Acta* **924**, 319 (1987).
[89] F. W. Sunderman, Jr., *Toxicol. Environ. Chem.* **15**, 59 (1987).
[90] M. Masarwa, H. Cohen, D. Meyerstein, D. L. Hickman, A. Bakac, and J. H. Espenson, *J. Am. Chem. Soc.* **110**, 4293 (1988).
[91] S. Kawanishi, S. Inoue, and S. Sano, *J. Biol. Chem.* **261**, 5952 (1986).
[92] E. Shinar, T. Navok, and M. Chevion, *J. Biol. Chem.* **258**, 14778 (1983).
[93] D. A. Rowley and B. Halliwell, *Arch. Biochem. Biophys.* **225**, 279 (1983).
[94] D. A. Rowley and B. Halliwell, *J. Inorg. Biochem.* **23**, 103 (1985).
[95] S. H. Chiou, *J. Biochem.* (*Tokyo*) **94**, 1259 (1983).
[96] Y. N. Kozlov and V. N. Bardnikov, *Russ. J. Phys. Chem.* (*Engl. Transl.*) **47**, 338 (1973).
[97] G. R. A. Johnson, N. B. Nazhat, and R. A. Saalalla-Nazhad, *Chem. Commun.*, 407 (1985).
[98] R. Stoewe and W. A. Prutz, *Free Radical Biol. Med.* **3**, 97 (1987).
[99] H. C. Sutton and C. C. Winterbourn, *Free Radical Biol. Med.* **6**, 53 (1989).
[99a] M. K. Eberhardt, G. Ramirez, and E. Ayala, *J. Org. Chem.* **54**, 5922 (1989).

tions.[2,3,47,67,100] Debate continues as to the physiological significance of copper-dependent radical production *in vivo*.[90–99,101,102]

The Fe^{2+}-dependent decomposition of H_2O_2 (the so-called Fenton reaction; reviewed in Ref. 84) is usually written as

$$Fe^{2+} + H_2O_2 \rightarrow Fe^{3+} + \cdot OH + OH^- \tag{6}$$

Some Fe^{3+} complexes can react further[103–104a] with H_2O_2, although these reactions probably proceed by several stages,[103–106] one possibility being

$$Fe^{3+} + H_2O_2 \rightarrow ferryl \xrightarrow{H_2O_2} perferryl \xrightarrow{H_2O_2} \cdot OH \tag{7}$$

Additional reactions can occur, for example,

$$\cdot OH + H_2O_2 \rightarrow H_2O + H^+ + O_2^- \tag{8}$$
$$O_2^- + Fe^{3+} \rightarrow Fe^{2+} + O_2 \tag{9}$$
$$\cdot OH + Fe^{2+} \rightarrow Fe^{3+} + OH^- \tag{10}$$

Although it is often stated that $\cdot OH$ radicals only arise from a mixture of Fe^{2+} and H_2O_2, it can be seen from the above equations that $Fe(III)–H_2O_2$ mixtures can also produce them, especially if the $Fe(III)$ is attached to nitrilotriacetic acid.[104a] SOD, by scavenging O_2^- [Eq. (9)], usually inhibits $\cdot OH$ generation by $Fe(III)–H_2O_2$ systems.[103,104,106] It is also interesting to note that high Fe^{2+} concentrations can diminish the overall yield of $\cdot OH$, by scavenging it [Eq. (10)].

Four major criticisms (e.g., Refs. 57 and 107–110) have been leveled against claims (e.g., Refs. 3, 46, 47, 67, and 100) that Fenton reactions occur *in vivo*. These criticisms are (i) the rate constant for the Fenton reaction is too low to have biological significance, (ii) the species produced in the Fenton system is not $\cdot OH$ radical, (iii) there are no metal ion catalysts available *in vivo*, and (iv) hydroxyl radicals can be produced by direct reaction between organic radicals and H_2O_2, so that metal ion catalysts are not required. Objections (ii) and (iii) are discussed in subse-

[100] B. Halliwell and J. M. C. Gutteridge, *ISI Atlas Sci. Biochem.* **1**, 48 (1988).

[101] J. M. C. Gutteridge, P. G. Winyard, D. R. Blake, J. Lunec, S. Brailsford, and B. Halliwell, *Biochem. J.* **230**, 517 (1985).

[102] B. Halliwell, *Agents Actions Suppl.* **26**, 223 (1989).

[103] J. M. C. Gutteridge, *FEBS Lett.* **185**, 19 (1985).

[104] J. M. C. Gutteridge and J. V. Bannister, *Biochem. J.* **234**, 225 (1986).

[104a] O. I. Aruoma, B. Halliwell, and M. Dizdaroglu, *J. Biol. Chem.* **264**, 20509 (1989).

[105] P. M. Wood, *Biochem. J.* **253**, 287 (1988).

[106] M. J. Laughton, B. Halliwell, P. Evans, and J. R. S. Hoult, *Biochem. Pharmacol.* **38**, 2859 (1989).

[107] C. C. Winterbourn, *Biochem. J.* **198**, 125 (1981).

[108] C. C. Winterbourn, *FEBS Lett.* **128**, 339 (1981).

[109] C. C. Winterbourn, *Biochem. J.* **182**, 625 (1979).

[110] J. A. Fee, *in* "Oxidases and Related Redox Systems" (T. E. King, H. S. Mason, and M. Morrison, eds.), p. 101. Pergamon, Oxford, 1982.

$$PQ^{\ddagger} \rightarrow PQ + O_2^- \tag{11}$$
$$2 O_2^- + 2 H^+ \rightarrow H_2O_2 + O_2 \tag{12}$$
$$H_2O_2 + Fe^{2+} \rightarrow \cdot OH + OH^- + Fe^{3+} \tag{13}$$
$$Fe^{3+} + O_2^- \rightarrow Fe^{2+} + O_2 \tag{14}$$
$$Fe^{3+} + PQ^{\ddagger} \rightarrow Fe^{2+} + PQ \tag{15}$$

SCHEME I. Mechanisms for production of oxygen radicals from paraquat radical ion ($PQ^{\ddagger}$) [Eqs. (11)–(15)]. Hydroxyl radical production is iron dependent. Fe^{3+} can be reduced to Fe^{2+} by O_2^- (inhibitable by superoxide dismutase) or by $PQ^{\ddagger}$ (not inhibitable by superoxide dismutase). Reactions (11), (13), and (14) appear to be the major route of $\cdot OH$ generation under ambient O_2; reaction (15) becomes more significant at low O_2 concentrations. Similar schemes can be written for Adriamycin semiquinone and for several other drug-derived radicals (see Refs. 111–116).

quent sections. Objection (iv) initially arose from reports that paraquat radical[108] and Adriamycin semiquinone[57] can react with H_2O_2 in the absence of metal ions to form $\cdot OH$. Both claims were mistaken and have now been withdrawn[111,112]; the $\cdot OH$ formation *does* require the presence of metal catalysts. It seems that paraquat radicals, Adriamycin semiquinone, and several other drug-derived radicals reduce iron(III) chelates to yield iron(II), which then reacts with H_2O_2 in a Fenton reaction [Scheme I, Eqs. (11)–(15)].[111-116] The rate constants for reaction of iron(III) chelates with semiquinones derived from antitumor antibiotics have been published.[114]

Objection (i) is true to the extent that the second-order rate constant for reaction (6) is low,[84] below $10^2 M^{-1} sec^{-1}$. The concentrations of H_2O_2 and iron(II) available *in vivo* are unlikely to be much above micromolar in most circumstances, although higher H_2O_2 concentrations can sometimes be achieved (Section 1.3). If 1 μM each of Fe^{2+} and H_2O_2 are mixed, the rate (R) of $\cdot OH$ production is

$$R = k[Fe^{2+}][H_2O_2] = k[10^{-6}][10^{-6}] = 7.6 \times 10^{-11} \tag{16}$$

taking a value[84] for the second-order rate constant k of 76 $M^{-1} sec^{-1}$. The average volume of a liver parenchymal cell is somewhat greater than 10^{-12} liters,[117] and so the number (N) of $\cdot OH$ radicals generated per second is

$$N = 7.6 \times 10^{-11} \times 10^{-12} \times 6.023 \times 10^{23} \cong 46 \tag{17}$$

[111] C. C. Winterbourn and H. C. Sutton, *Arch. Biochem. Biophys.* **235,** 116 (1984).
[112] C. C. Winterbourn, J. M. C. Gutteridge, and B. Halliwell, *J. Free Radicals Biol. Med.* **1,** 43 (1985).
[113] J. M. C. Gutteridge and D. Toeg, *FEBS Lett.* **149,** 228 (1982).
[114] J. Butler, B. M. Hoey, and A. J. Swallow, *FEBS Lett.* **182,** 95 (1980).
[115] J. M. C. Gutteridge, G. Quinlan, and S. Wilkins, *FEBS Lett.* **167,** 37 (1984).
[116] J. M. C. Gutteridge and G. J. Quinlan, *Biochem. Pharmacol.* **34,** 4099 (1985).
[117] B. D. Uhal and K. L. Roehrig, *Biosci. Rep.* **2,** 1003 (1982).

Hence, there could be $46 \cdot OH$ radicals generated per cell every second, even with the conservative estimates we have made. In view of the high reactivity of $\cdot OH$, this could have enormous biological consequences, depending on its site of generation. Of course, as the reaction proceeds, both H_2O_2 and Fe^{2+} will be used up, and the rate of $\cdot OH$ production will fall unless they are continuously replenished. It must be further appreciated that the rate constant[96] for reaction of Cu(I) with H_2O_2,

$$Cu^+ + H_2O_2 \rightarrow Cu^{2+} + \cdot OH + OH^- \tag{18}$$

at $4.7 \times 10^3 \, M^{-1} \, sec^{-1}$, is much greater than that for Fe(II), although we have already pointed out that reaction of Cu^+ with H_2O_2 may produce a reactive Cu(III) species in addition to (or, possibly, instead of) $\cdot OH$.

1.5. Does the Fenton Reaction Actually Produce Hydroxyl Radical? The Case of Iron–EDTA

The Fenton reaction is widely thought to produce $\cdot OH$ as the reactive radical species, but there have been other suggestions, such as ferryl formation. Ferryl is an iron–oxygen complex in which the iron has an oxidation number of 4:

$$Fe^{2+} + H_2O_2 \rightarrow FeOH^{3+} \text{ (or } FeO^{2+}) + OH^- \tag{19}$$

Most scientists seem to agree that when iron ions are added as iron–EDTA complexes to H_2O_2 or to systems generating both O_2^- and H_2O_2, $\cdot OH$ is formed.[99,118–120] It has been identified by spin-trapping methods (e.g., Refs. 121 and 122), by aromatic hydroxylation studies,[120,123,124] and by scavenging it with added molecules at rates predictable from their rate constants for reaction with "real" $\cdot OH$, produced by radiolysis of water.[125,126,134] Recently, however (e.g., Ref. 127), Rush and Koppenol have challenged this view on the basis of detailed studies of cytochrome c degradation by mixtures of Fe^{2+}–EDTA and H_2O_2; they claim that ferryl, and not $\cdot OH$, is the reactive species formed.

[118] B. Halliwell, *FEBS Lett.* **92**, 321 (1978).

[119] J. M. McCord and E. D. Day, *FEBS Lett.* **86**, 139 (1978).

[120] M. Grootveld and B. Halliwell, *Free Radical Res. Commun.* **1**, 243 (1986).

[121] M. Fitchett, B. C. Gilbert, and M. Jeff, *Philos. Trans. R. Soc. London Ser. B* **311**, 517 (1985).

[122] B. E. Britigan, M. S. Cohen, and G. M. Rosen, *J. Leukocyte Biol.* **41**, 349 (1987).

[123] M. Grootveld and B. Halliwell, *Biochem. J.* **237**, 499 (1986).

[124] B. Halliwell, M. Grootveld, H. Kaur, and I. Fagerheim, *in* "Free Radicals, Methodology and Concepts" (C. Rice-Evans and B. Halliwell, eds.), p. 33. Richelieu Press, London, 1988.

[125] B. Halliwell, J. M. C. Gutteridge, and O. I. Aruoma, *Anal. Biochem.* **165**, 215 (1987).

[126] B. M. Hoey, J. Butler, and B. Halliwell, *Free Radical Res. Commun.* **4**, 259 (1988).

[127] J. D. Rush and W. H. Koppenol, *J. Biol. Chem.* **261**, 6730 (1986).

However, if $\cdot OH$ is not formed by reaction of iron–EDTA chelates with H_2O_2, it becomes very difficult to explain the results of aromatic hydroxylation, spin-trapping, and scavenger experiments.[118–126] Thus, the reactive species formed must react with scavengers with the correct $\cdot OH$ rate constant (as determined by pulse radiolysis),[125,126] attack spin traps to give the "$\cdot OH$" spin adduct,[121,122] and hydroxylate aromatic compounds to give the correct product distribution.[124,128] Studies on Fenton chemistry are very complicated, in part because iron ions can oxidize or reduce secondary radicals produced by the attack of $\cdot OH$ on added molecules, and all previous challenges to the $\cdot OH$ formation theory have been explicable by some such mechanism.[84,121] Also, Fe^{2+}–EDTA reduces cytochrome c directly,[129] a reaction that could interfere with kinetic studies. The authors feel that the balance of current evidence does support $\cdot OH$ as a reactive species produced in the Fenton reaction using Fe–EDTA chelates.

How else could the results of Rush and Koppenol[127] be explained? It may well be that the first product of reaction of H_2O_2 with Fe^{2+}–EDTA may be a ferryl species, or something similar, that then decomposes to release $\cdot OH$.[100] If ferryl were scavenged, for example, by cytochrome c, then $\cdot OH$ would not be detected. Thus, the equation[100]

$$Fe^{2+}\text{–EDTA} + H_2O_2 \rightarrow \underset{\text{(ferryl?)}}{\text{intermediate species}} \rightarrow Fe^{3+}\text{–EDTA} + \cdot OH + OH^- \quad (20)$$

would account for the experimental observations.

Although EDTA itself scavenges some $\cdot OH$ radicals ($k_2 = 2.8 \times 10^9$ M^{-1} sec^{-1}),[99,130] the "open" structure of its iron chelates allows many of them to escape. In general, EDTA accelerates iron-dependent formation of $\cdot OH$ by[120,131] (i) maintaining iron ions in solution, that is, preventing them from precipitating at physiological pH, and (ii) changing the redox potential of iron ions to favor reaction with O_2^- and H_2O_2. In phosphate-buffered systems, mechanism (i) seems to predominate.[120] It should be noted that if Fe^{2+} and EDTA are mixed together and incubated before adding them to oxidant-containing systems, the resulting $\cdot OH$ generation is decreased, simply because Fe^{2+}–EDTA complexes oxidize rapidly under aerobic conditions to form Fe(III)–EDTA complexes.[120,132,148]

[128] B. Halliwell and M. Grootveld, *FEBS Lett.* **213**, 9 (1987).

[129] C. Bull, J. A. Fee, P. O'Neill, and E. M. Fieldon, *Arch. Biochem. Biophys.* **215**, 551 (1982).

[130] C. Walling, *Acc. Chem. Res.* **8**, 125 (1975).

[131] W. Flitter, D. A. Rowley, and B. Halliwell, *FEBS Lett.* **158**, 310 (1983).

[132] D. C. Harris and P. Aisen, *Biochim. Biophys. Acta* **329**, 156 (1973).

1.6. Hydroxyl Radical Production in Superoxide-Generating and in Other Biological Systems

Formation of $\cdot$ OH radicals in a wide range of O_2^--generating systems, including activated phagocytes, "autoxidizing" compounds, and mixtures of hypoxanthine and xanthine oxidase, has been demonstrated by a number of techniques including the ability of $\cdot$ OH to hydroxylate salicylate,[123,118,133] benzoate,[134] phenylalanine,[124,135] or phenol,[133,136] give a characteristic electron spin resonance (ESR) signal with spin traps,[122,137,138] oxidize methional or related substances to ethene (ethylene) gas,[139,140] attack tryptophan,[18,119] convert dimethyl sulfoxide to methanal (formaldehyde), methanesulfinic acid, or methane,[141–143] decarboxylate [*carboxyl-*^{14}C] benzoic acid to $^{14}CO_2$,[144] attack benzoate to produce thiobarbituric acid-reactive material,[145] convert benzoate to fluorescent products,[146,134] and degrade deoxyribose to give products that form a color on heating with thiobarbituric acid.[147,148] In chemical terms, the most satisfactory methods for demonstrating $\cdot$ OH radical formation are probably ESR spin trapping and aromatic hydroxylation; both are "trapping" methods in which $\cdot$ OH reacts with a foreign "detector" molecule to form specific products. An alternative approach is "fingerprinting": identifying the damage done by $\cdot$ OH by looking at the formation of specific products from naturally occurring biomolecules.[128,149] Thus, attack of $\cdot$ OH on the purine and pyrimidine bases of DNA produces a multitude of products, which can be derivatized, separated by GLC, and identified by mass

[133] R. Richmond, B. Halliwell, J. Chauhan, and A. Darbre, *Anal. Biochem.* **118**, 328 (1981).

[134] J. M. C. Gutteridge, *Biochem. J.* **243**, 709 (1987).

[135] H. Kaur, I. Fagerheim, M. Grootveld, A. Puppo, and B. Halliwell, *Anal. Biochem.* **172**, 360 (1988).

[136] R. Richmond and B. Halliwell, *J. Inorg. Biochem.* **17**, 95 (1982).

[137] E. Finkelstein, G. M. Rosen, E. J. Rauckman, and J. Paxton, *Mol. Pharmacol.* **16**, 676 (1979).

[138] E. Finkelstein, G. M. Rosen, and J. Rauckman, *Mol. Pharmacol.* **21**, 262 (1981).

[139] C. Beauchamp and I. Fridovich, *J. Biol. Chem.* **245**, 4641 (1970).

[140] J. Diguiseppi and I. Fridovich, *Arch. Biochem. Biophys.* **205**, 323 (1980).

[141] J. E. Repine, J. W. Eaton, M. W. Anders, J. R. Hoidal, and R. B. Fox, *J. Clin. Invest.* **64**, 1642 (1979).

[142] S. M. Klein, G. Cohen, and A. I. Cederbaum, *Biochemistry* **20**, 6006 (1981).

[143] C. F. Babbs and M. J. Gale, *in* "Free Radicals, Methodology and Concepts" (C. Rice-Evans and B. Halliwell, eds.), p. 91. Richelieu Press, London, 1988.

[144] A. L. Sagone, C. Democko, L. Clark, and M. Kartha, *J. Lab. Clin. Med.* **101**, 196 (1983).

[145] J. M. C. Gutteridge, *Biochem. J.* **224**, 761 (1984).

[146] W. H. Melhuish and H. C. Sutton, *J. Chem. Soc. Chem. Commun.*, 970 (1978).

[147] J. M. C. Gutteridge, *FEBS Lett.* **128**, 343 (1981).

[148] B. Halliwell and J. M. C. Gutteridge, *FEBS Lett.* **128**, 347 (1981).

[149] M. Dizdaroglu, *BioTechniques* **4**, 536 (1986).

spectrometry.[149] This complex pattern of base modification may be a fingerprint for damage to DNA by ·OH radical.[149,149a] Spin trapping as a technique for detecting ·OH is prone to artifacts, but careful control experiments can rule them out.[122,137,138,150] After aromatic hydroxylation experiments, the hydroxylated products produced by attack of ·OH on benzene rings can be separated by HPLC or GLC, quantitated, and distinguished from the products formed by enzymatic hydroxylating systems (reviewed in Ref. 124).

The other methods for detecting ·OH radicals listed above are less specific, especially formation of ethene gas.[151] Conversion of methional to ethene can be achieved by alkoxyl and peroxyl radicals,[151] the peroxidase action of heme proteins,[152,153] ligninolytic enzymes from *Phanerochaete chrysosporium*,[154,155] and myeloperoxidase.[156] When using ethene formation, or other methods that are not specific for ·OH, it is usual to attribute the reaction observed to ·OH if it is inhibited by so-called hydroxyl radical scavengers such as mannitol, formate, thiourea, dimethylthiourea, ethanol, 1-butanol, glucose, Tris buffer,[157] or sorbitol. Experiments with scavengers can be valid if the following points are borne in mind:

(i) An inhibition by a single scavenger proves nothing, especially if that scavenger is thiourea, which reacts with H_2O_2, HOCl, and alkoxyl radicals,[29,158,159] weakly inhibits xanthine oxidase,[118] and may chelate metal ions necessary for ·OH production.[86] Dimethylthiourea also reacts with HOCl and with H_2O_2.[29,160] Ethanol reacts with alkoxyl radicals,[158] although mannitol and formate do not. In order to implicate formation of free ·OH radicals as being responsible for observable damage to a target, a range of scavengers should be used, and the degree of inhibition they produce should be correlated with the published rate constants for reaction of the scavengers with ·OH.[118] In addition, the scavenger and the molecule being used to detect ·OH (the target) should show competition

[149a] O. I. Aruoma, B. Halliwell, and M. Dizdaroglu, *J. Biol. Chem.* **264**, 13024 (1989).

[150] R. P. Mason and K. M. Morehouse, *in* "Free Radicals, Methodology and Concepts" (C. Rice-Evans and B. Halliwell, eds.), p. 157. Richelieu Press, London, 1988.

[151] W. A. Pryor and R. H. Tang, *Biochem. Biophys. Res. Commun.* **81**, 498 (1978).

[152] U. Benatti, A. Morelli, L. Guida, and A. De Flora, *Biochem. Biophys. Res. Commun.* **111**, 980 (1983).

[153] S. Harel and J. Kanner, *Free Radical Res. Commun.* **5**, 21 (1988).

[154] M. Kuwahara, J. K. Glenn, M. A. Morgan, and M. H. Gold, *FEBS Lett.* **169**, 247 (1984).

[155] T. K. Kirk, M. D. Mozuch, and M. Tien, *Biochem. J.* **226**, 455 (1985).

[156] S. J. Klebanoff and H. Rosen, *J. Exp. Med.* **148**, 490 (1978).

[157] W. Paschen and U. Weser, *Hoppe-Seyler's Z. Physiol. Chem.* **356**, 727 (1975).

[158] G. W. Winston, W. Harvey, L. Berl, and A. I. Cederbaum, *Biochem. J.* **216**, 415 (1983).

[159] A. I. Cederbaum, E. Dicker, E. Rubin, and G. Cohen, *Biochemistry* **18**, 1187 (1979).

[160] W. E. Curtis, M. E. Muldrow, N. B. Parker, R. Barkley, S. L. Linas, and J. E. Repine, *Proc. Natl. Acad. Sci. U.S.A.* **85**, 3422 (1988).

kinetics, that is, they should be competing for the same species.[119] Indeed, the degradation of deoxyribose by $\cdot OH$ generated from H_2O_2 and Fe^{2+}–EDTA has been proposed as a simple "test-tube" method[125,134] for determining rate constants for reactions of $\cdot OH$ with added scavengers, which compete with the deoxyribose for $\cdot OH$ generated in free solution by reaction of Fe^{2+}–EDTA with H_2O_2.

(ii) Reaction of $\cdot OH$ radicals with a scavenger produces scavenger-derived radicals that might themselves do damage in certain systems. For example, attack of $\cdot OH$ on thiol compounds yields thiyl radicals that can combine with O_2 to give reactive oxysulfur radicals (see Table I), formate and ethanol radicals can attack serum albumin,[161] and azide radical (formed by reaction of $\cdot OH$ with azide anion, N_3^-) is a reactive species that can attack tryptophan and tyrosine.[162,163] The production and reactivity of secondary radicals may sometimes account for failure to "protect" by one scavenger,[164] as in the action of uric acid radicals on yeast alcohol dehydrogenase, discussed previously.[83] If buffers such as Tris or "Goods" buffers are present in radical-generating systems, they can be attacked by $\cdot OH$ to give buffer-derived radicals.[157,165] The "Good" buffer HEPES has been claimed to stimulate copper-dependent formation of $\cdot OH$ in certain reaction mixtures.[166] Some "Good" buffers can also stimulate flavin-dependent photochemical reactions.[166a]

(iii) $\cdot OH$ radicals are often formed in biological systems by reaction of H_2O_2 with metal ions bound at specific sites. Some or all of the $\cdot OH$ then reacts with the binding molecule and is not accessible to added scavengers of $\cdot OH$. For example, H_2O_2-dependent damage to the DNA of mammalian cells may be due to reaction of H_2O_2 with metal ions bound to the DNA.[2,81,167–169] This concept of site specificity is further discussed in Section 1.6.2.

1.6.1. Role of Metal Ions. In all the nonradiolytic biochemical sys-

[161] H. Schuessler and K. Freundl, *Int. J. Radiat. Biol. Relat. Stud. Phys. Chem. Med.* **44,** 17 (1983).

[162] A. Singh, G. W. Koroll, and R. B. Cundall, *Radiat. Phys. Chem.* **19,** 137 (1982).

[163] B. Kalyanaraman, E. G. Janzen, and R. P. Mason, *J. Biol. Chem.* **260,** 4003 (1985).

[164] G. G. Miller and J. A. Raleigh, *Int. J. Radiat. Biol. Relat. Stud. Phys. Chem. Med.* **43,** 411 (1983).

[165] J. K. Grady, N. D. Chasteen, and D. C. Harris, *Anal. Biochem.* **173,** 111 (1988).

[166] J. A. Simpson, K. H. Cheesman, S. E. Smith, and R. T. Dean, *Biochem. J.* **254,** 519 (1988).

[166a] B. Halliwell and V. S. Butt, *Biochem. J.* **129,** 1157 (1972).

[167] A. C. Mello Filho and R. Meneghini, *Biochim. Biophys. Acta* **781,** 56 (1984).

[168] I. U. Schaufstatter, D. B. Hinshaw, P. A. Hyslop, R. G. Spragg, and C. G. Cochrane, *J. Clin. Invest.* **77,** 1312 (1986).

[169] J. A. Imlay and S. Linn, *Science* **240,** 1302 (1988).

TABLE I

ROLE OF IRON AND OF OTHER METAL IONS IN CONVERTING LESS REACTIVE
TO MORE REACTIVE SPECIES

O_2^- plus $H_2O_2 \xrightarrow{Fe^a} \cdot OH$

Lipid peroxides (ROOH) $\xrightarrow{Cu/Fe^b}$ RO $\cdot$ (alkoxyl), $RO_2 \cdot$ (peroxyl), cytotoxic aldehydes

Thiols (RSH) $\xrightarrow{Fe/Cu \ plus \ O_2^c} O_2^-, H_2O_2$, HO $\cdot$, thiyl (RS $\cdot$) $\xrightarrow{O_2} RSO_2 \cdot$, RSO $\cdot^d$

NAD(P)H $\xrightarrow{Fe/Cu \ plus \ O_2^c}$ NAD(P) $\cdot$, O_2^-, H_2O_2, HO $\cdot$

Ascorbic acid $\xrightarrow{Cu/Fe^e}$ semidehydroascorbate radical, HO $\cdot$, H_2O_2, degradation products of ascorbate

Catecholamines, related autoxidizable molecules $\xrightarrow{Fe/Cu/Mn \ plus \ O_2^c} O_2^-, H_2O_2$, HO $\cdot$, semiquinones (or other radicals derived from the oxidizing compounds)

[a] The iron-catalyzed Haber–Weiss reaction.
[b] Lipid peroxide decomposition is metal ion dependent and eventually produces highly cytotoxic products, such as 4-hydroxy-2,3-*trans*-nonenal, and less toxic ones, such as malondialdehyde (Section 3).
[c] Most, if not all, so-called autoxidations are stimulated by traces of transition metal ions and proceed by free radical mechanisms.
[d] Thiyl radicals can react with O_2 to form reactive oxysulfur radicals such as sulfenyl (RSO $\cdot$) or thiyl peroxyl (RSO$_2 \cdot$).[170,171]
[e] Copper ions are especially effective in decomposing ascorbic acid, and ascorbate–copper or ascorbate–iron mixtures are cytotoxic. Hydroxyl radicals may be involved in this toxicity.

tems used to date in which $\cdot$ OH generation has been demonstrated *in vitro,* some form of metal ion catalyst is required.[46,67,112] Iron is the most likely candidate for stimulating $\cdot$ OH generation and other radical reactions *in vivo*[47]: Table I shows that the general effect of iron catalysts is to convert poorly reactive species into highly reactive ones.

The formation of $\cdot$ OH radicals observed in O_2^--generating systems is usually inhibited by addition of catalase or of superoxide dismutase (SOD). Inhibition by catalase is not surprising, since the $\cdot$ OH arises by metal-dependent splitting of H_2O_2 [Eq. (5)]. The inhibitory action of SOD, if accompanied by the proper controls discussed previously, shows clearly that O_2^- is also involved in the observed $\cdot$ OH production. The simplest explanation of its role would be that O_2^- reduces oxidized metal

[170] D. Becker, S. Swarts, M. Champagne, and M. D. Sevilla, *Int. J. Radiat. Biol.* **53,** 767 (1988).
[171] L. G. Forni and R. L. Willson, *Biochem. J.* **240,** 905 (1986).

ions and so promotes reaction (5). Hence, taking iron as an example, a cycle of reactions can be written[118,119,172] as follows

$$Fe^{3+} + O_2^- \rightarrow Fe^{2+} + O_2 \qquad (21)$$
$$Fe^{2+} + H_2O_2 \rightarrow Fe^{3+} + \cdot OH + OH^- \qquad (22)$$
$$Net \quad O_2^- + H_2O_2 \rightarrow O_2 + \cdot OH + OH^- \qquad (23)$$

Reaction (23) is often called a metal-catalyzed Haber–Weiss reaction or a superoxide-driven Fenton reaction. This interpretation of the role of O_2^- merely as a reducing agent is probably naive, but it does explain a number of experimental observations.[46,47,67,118,119,148] It also, however, raises the obvious question that, since other reducing agents are present *in vivo*, why should O_2^- be so important?[107,109,110]

We have already seen that paraquat radicals and the semiquinone forms of quinone antitumor antibiotics can reduce iron(III) complexes, although direct reaction with O_2^- is favored except at very low O_2 concentrations (Scheme I). Rowley and Halliwell[173,174] investigated the effect of adding other reducing agents to iron-dependent systems generating $\cdot OH$ radicals from O_2^- and H_2O_2. They concluded that NADH, NADPH, or thiol compounds such as reduced glutathione (GSH) or cysteine would be unlikely to prevent O_2^--dependent formation of $\cdot OH$ radicals *in vivo*. Indeed, both NAD(P)H[94,169,174] and thiol compounds[173,175,176] can interact with metal ions and H_2O_2 to increase O_2^--dependent $\cdot OH$ radical formation under certain circumstances. Tien *et al.*[177] criticized these conclusions on the basis of experiments showing that SOD does not inhibit thiol-stimulated peroxidation of liposomes. However, peroxidation in the systems used by them does not require $\cdot OH$ radicals (Section 3.1), so that the two sets of results are not incompatible.

Another important biological reducing agent is ascorbic acid, and this can certainly replace O_2^-, apparently as a reducing agent, in several of the systems that have been used *in vitro* to demonstrate O_2^--dependent formation of $\cdot OH$ radicals in the presence of iron or copper salts.[93,107,109,178] When ascorbate is the reducing agent, SOD does not prevent the $\cdot OH$

[172] K. L. Fong, P. B. McCay, J. L. Poyer, H. P. Misra, and B. B. Keele, *Chem.-Biol. Interact.* **15,** 17 (1976).

[173] D. A. Rowley and B. Halliwell, *FEBS Lett.* **138,** 33 (1982).

[174] D. A. Rowley and B. Halliwell, *FEBS Lett.* **142,** 39 (1982).

[175] G. Saez, P. J. Thornalley, H. A. O. Hill, R. Hems, and J. V. Bannister, *Biochim. Biophys. Acta* **719,** 24 (1982).

[176] H. P. Misra, *J. Biol. Chem.* **249,** 2151 (1974).

[177] M. Tien, J. R. Bucher, and S. D. Aust, *Biochem. Biophys. Res. Commun.* **107,** 279 (1982).

[178] D. A. Rowley and B. Halliwell, *Clin. Sci.* **64,** 649 (1983).

production, but it is still inhibited by catalase.[107,109,178] If both O_2^- and ascorbate are available, the relative contributions of each to the $\cdot OH$ production depend on their concentrations. Ascorbate at the concentrations normally present in human extracellular fluids can at best partially[178] replace O_2^- in reducing Fe(III), but ascorbate is rapidly oxidized by direct reaction with O_2^- and with $\cdot OH$, so that $\cdot OH$ production eventually becomes completely O_2^- dependent.[178] On the other hand, at the high (millimolar) ascorbate concentrations present in some mammalian tissues such as the eye,[179] nervous tissue,[180] or pneumocytes,[181] it is difficult to imagine O_2^- competing as a reducing agent unless the ascorbate and O_2^- are present in different subcellular compartments or unless large amounts of O_2^- are produced at a localized site. Indeed, at such high concentrations the ability of ascorbate to scavenge $\cdot OH$ radicals[78] may become significant.

An example of differential location is that ascorbate seems to be excluded from the phagocytic vacuole of human neutrophils, even though these cells are rich in ascorbate.[182] Hence, any $\cdot OH$ radical production that occurs within the phagocytic vacuole should depend on H_2O_2 and O_2^-. Localized production of large amounts of O_2^- and H_2O_2 can also occur. For example, in some forms of the adult respiratory distress syndrome large numbers of neutrophils accumulate in the lung, aggregate, and become activated to produce O_2^- and H_2O_2.[183] Depletion of ascorbic acid by excessive O_2^- generation might decrease ascorbate concentration to the point at which its overall effect is to increase rather than decrease $\cdot OH$ radical production. Ascorbate is known to be rapidly depleted at sites of inflammation,[184] possibly by reaction with O_2^- and with the myeloperoxidase-derived oxidant HOCl.[185]

Another property of ascorbic acid is its ability to "repair" damage caused by attack of $\cdot OH$ on some biological molecules.[186–188] Thus $\cdot OH$ generation by an ascorbate–iron–H_2O_2 system might sometimes be less

[179] S. D. Varma, T. K. Ets, and R. D. Richards, *Ophthalmic Res.* **9**, 421 (1977).
[180] R. Spector and J. Eells, *Fed. Proc., Fed. Am. Soc. Exp. Biol.* **43**, 196 (1984).
[181] V. Castranova, J. R. Wright, H. D. Colby, and P. R. Miles, *J. Appl. Physiol.* **54**, 208 (1983).
[182] C. C. Winterbourn and M. C. M. Vissers, *Biochim. Biophys. Acta* **763**, 175 (1983).
[183] R. M. Tate and J. E. Repine, *in* "Free Radicals in Biology" (W. A. Pryor, ed.), p. 199. Academic Press, New York, 1984.
[184] J. Lunec and D. R. Blake, *Free Radical Res. Commun.* **1**, 31 (1985).
[185] B. Halliwell, M. Wasil, and M. Grootveld, *FEBS Lett.* **213**, 15 (1987).
[186] P. O'Neill, *Radiat. Res.* **96**, 198 (1983).
[187] J. L. Redpath and R. L. Willson, *Int. J. Radiat. Biol. Relat. Stud. Phys. Chem. Med.* **23**, 51 (1973).
[188] K. R. Maples and R. P. Mason, *J. Biol. Chem.* **263**, 1079 (1988).

damaging than $\cdot OH$ generation by a O_2^-–iron–H_2O_2 system. This has been demonstrated experimentally in studies of uric acid degradation, to form reactive uric acid radicals, by $\cdot OH$ radicals.[188,189]

1.6.2. Site-Specific Nature of Fenton Chemistry. High-energy irradiation of an aqueous solution produces a substantial yield of hydroxyl radicals, which can react with added "scavengers" at rates that depend only on their concentrations and their second-order rate constants for reaction with $\cdot OH$. Indeed, competition kinetics are employed in this way to determine the rate constants for reaction of scavengers, using the technique of pulse radiolysis.[78] Similarly, when $\cdot OH$ radicals are generated by reaction of Fe^{2+}–EDTA with H_2O_2, those $\cdot OH$ radicals that escape direct scavenging by EDTA appear to enter "free solution" and are accessible to any added scavenger. Thus, this system can also be analyzed by competition kinetics and rate constants can be determined.[125,126,134] The iron–EDTA–H_2O_2 system has also been used to produce random fragmentation of DNA for sequencing studies.[190]

In Section 1.5, the authors reviewed the evidence which leads them to believe that $\cdot OH$ is formed when iron–EDTA chelates react with H_2O_2. Doubts have also arisen about the reactive species formed when H_2O_2 is mixed with simple Fe^{2+} salts, Fe^{3+}–ascorbate, Fe^{3+}–superoxide mixtures, or with Fe^{2+}–ADP chelates. Hydroxyl radicals can still be detected in the mixtures by chemical methods such as spin trapping[121,191] or aromatic hydroxylation.[131] However, the damage done to biological molecules by such systems often is not prevented by moderate concentrations of some scavengers of $\cdot OH$ (such as formate), whereas other $\cdot OH$ scavengers (such as thiourea) are still protective.[134,145,192,193]

There are two general explanations of such data. The first is that the damage is done to the added molecule not by $\cdot OH$ but by another reactive species, such as ferryl.[99,192,194] A second explanation, which the authors prefer,[67,100] although it has been dismissed by some,[99] relates to the site specificity of much $\cdot OH$ generation *in vivo*. Hydroxyl radicals are so highly reactive that they combine at or near the site of formation, as determined by the location of catalytic metal ions. Thus, an added scavenger of $\cdot OH$ cannot be expected to be present at the site of $\cdot OH$ genera-

[189] O. I. Aruoma and B. Halliwell, *FEBS Lett.* **244**, 76 (1989).

[190] T. D. Tullius, *Nature (London)* **332**, 663 (1988).

[191] R. A. Floyd, *Arch. Biochem. Biophys.* **225**, 263 (1983).

[192] H. C. Sutton, G. F. Vile, and C. C. Winterbourn, *Arch. Biochem. Biophys.* **256**, 462 (1987).

[193] G. Czapski and S. Goldstein, *Free Radical Res. Commun.* **1**, 157 (1986).

[194] C. C. Winterbourn, *J. Free Radicals Biol. Med.* **3**, 33 (1987).

tion at sufficient concentrations to be able to protect the "target" against attack by the $\cdot$OH generated on it.[193,195,196]

Why, then, should some scavengers still protect? A logical explanation was offered by Gutteridge[145]: they do not protect by scavenging $\cdot$OH, but by binding metal ions. Several $\cdot$OH scavengers (including mannitol and thiourea), have a weak capacity to bind metals and could withdraw catalytic iron ions from sensitive targets and direct damage to themselves.[145]

It should be noted that the two explanations (i.e., $\cdot$OH scavengers sometimes do not protect because the damaging species is not $\cdot$OH or because they cannot interfere with site-specific generation of $\cdot$OH) are not mutually exclusive. Thus, reaction of Fe^{2+} with H_2O_2 may initially form a ferryl species [Eq. (20)] that then decomposes to release $\cdot$OH (Section 1.5), and either or both of these could mediate damage in a given system. Studies on the reaction of Fe^{2+} (aq) with H_2O_2 ignore the fact that iron ions *in vivo* should always be bound to a ligand, whether a macromolecule or a low molecular mass chelator.

Ferryl radical is probably the reactive species at the active sites of horseradish peroxidase (in compounds I and II) and of cytochrome *P*-450. Ferryl must therefore have considerable reactivity, although it is less oxidizing than $\cdot$OH.[197] However, good evidence that ferryl species are formed in Fenton-type systems in the absence of heme rings has not yet been obtained, nor is there any evidence that ferryl species can be detected by spin trapping. Ferryl might be able to hydroxylate aromatic compounds; horseradish peroxidase–H_2O_2 mixtures will not hydroxylate aromatic compounds,[198,199] whereas the cytochrome *P*-450 system does, of course, hydroxylate, although the pattern of products is different from that owing to attack by $\cdot$OH.[124,128] Hence, if ferryl exists, it ought to be possible to distinguish it from $\cdot$OH by the technique of aromatic hydroxylation.[124,128]

The biological implications of "site specificity" of $\cdot$OH formation are profound. A major determinant of the actual toxicity of O_2^- and H_2O_2 to cells may well be the availability and location of metal ion catalysts of $\cdot$OH radical formation.[67] If, for example, catalytic iron salts are bound to DNA or to membrane lipids, introduction of H_2O_2 and O_2^- will in the first case fragment the DNA and in the other could initiate lipid peroxida-

[195] G. Czapski, *Isr. J. Chem.* **24**, 29 (1984).

[196] D. C. Borg and K. M. Schaich, *Isr. J. Chem.* **24**, 38 (1984).

[197] W. H. Koppenol and J. F. Liebman, *J. Phys. Chem.* **88**, 99 (1984).

[198] B. Halliwell and S. Ahluwalia, *Biochem. J.* **153**, 513 (1976).

[199] P. I. Smith and G. A. Swann, *Biochem. J.* **153**, 403 (1976).

tion.[2,67] The site of attack of the $\cdot$OH radicals will be largely determined by the site of the bound metal ion, and the nature of the damage will not be the same as that done by $\cdot$OH radicals generated in free solution and attacking the target randomly.[67,200–202a] For example, exposure of glycoprotein preparations to metal-dependent systems generating $\cdot$OH produces specific patterns of fragmentation rather than just random attack,[200,201] apparently because the metal catalysts bind readily to specific sites on the glycoproteins.[202]

1.6.3. Oxidant-Mediated DNA Damage in Vivo. When cells are subjected to oxidant stress, for example, by increased intracellular generation of O_2^- and H_2O_2, by treatment with high external concentrations of H_2O_2, or by exposure to cigarette smoke, DNA strand breakage usually occurs.[81,167–169,203–206] This could occur because the oxidant stress leads to activation of some specific DNA-cleaving mechanism,[206] such as a Ca^{2+}-dependent endonuclease.[207] An alternative explanation is that, if H_2O_2 survives in sufficient concentrations to reach the nucleus, it can react with intracellular metal ions to give $\cdot$OH, which fragments the DNA by site-specific $\cdot$OH attack.[2,81,167,169,208] These metal ions (presumably iron ions, since pretreatment of cells with permeant iron chelators can offer protection)[81,167,168,208] might be already present on the DNA *in vivo*.[2] Alternatively, the oxidant stress might liberate iron ions from their sites of sequestration within the cell, so that they can bind to DNA.[209]

The DNA damage produced when mammalian cells are exposed to oxidant stress can often be protected against by adding catalase to the suspension medium (e.g., Ref. 205). SOD or scavengers of $\cdot$OH have little effect: SOD cannot enter the cells easily, and $\cdot$OH scavengers cannot protect against site-specific DNA damage. However, SOD could offer

[200] J. M. Creeth, B. Cooper, A. S. A. Donald, and J. R. Clamp, *IRCS Med. Sci.* **10,** 548 (1982).

[201] C. E. Cross, B. Halliwell, and A. Allen, *Lancet* **1,** 1328 (1984).

[202] J. M. Creeth, B. Cooper, A. S. R. Donald, and J. R. Clamp, *Biochem. J.* **211,** 323 (1983).

[202a] J. L. Sagripanti and K. H. Kraemer, *J. Biol. Chem.* **264,** 1729 (1989).

[203] T. Nakayama, M. Kaneko, M. Kodama, and C. Nagata, *Nature (London)* **314,** 462 (1985).

[204] E. T. Borish, J. P. Cosgrove, D. F. Church, W. A. Deutsch, and W. A. Pryor, *Biochem. Biophys. Res. Commun.* **133,** 780 (1985).

[205] E. M. Link and P. A. Riley, *Biochem. J.* **249,** 391 (1988).

[206] H. C. Birnboim, *Biochem. Cell Biol.* **66,** 374 (1988).

[207] D. J. McConkey, P. Hartzell, P. Nicotera, A. H. Wyllie, and S. Orrenius, *Toxicol. Lett.* **42,** 123 (1988).

[208] M. Larramendy, A. C. Mello-Filho, E. A. L. Martins, and R. Meneghini, *Mutat. Res.* **178,** 57 (1987).

[209] P. E. Starke, J. D. Gilbertson, and J. L. Farber, *Biochem. Biophys. Res. Commun.* **133,** 371 (1985).

protection to hepatocytes under conditions where it entered the cells,[210] consistent with damage being mediated by O_2^--dependent Fenton chemistry. Also consistent with this mechanism of DNA damage are studies on *E. coli* K12. Exposure of growing cells to H_2O_2 leads to two kinetically distinguishable modes of cell killing: mode 1 killing is pronounced at 1–3 mM H_2O_2 and seems to involve DNA damage. Mode 2 killing occurs at higher (>20 mM) H_2O_2 concentrations, and the sites of injury are unclear as yet. Mode 1 killing is blocked by iron-chelating agents, and a role for $\cdot OH$ radicals has been proposed.[169] *E. coli* mutants defective in SOD activity are especially sensitive to mode 1 killing,[211] and they show enhanced mutation rates even in the absence of added H_2O_2.[51] Thus, reduction of DNA-bound metal ions seems to involve O_2^- rather than other cellular reducing agents.[2,51,169,210,211]

1.6.4. Deoxyribose Assay for "Free" Hydroxyl Radicals and for Site-Specific Hydroxyl Radical Production. The difference between damage to molecules by $\cdot OH$ generated "in free solution" and by $\cdot OH$ generated by metal ions bound to a target can be illustrated by studies using deoxyribose, introduced in 1981 as a simple "detector" molecule for $\cdot OH$ generated in biological systems.[147,148] The pentose sugar 2-deoxy-D-ribose is attacked by $\cdot OH$ radicals to yield a mixture of products (reviewed in Ref. 212). On heating with thiobarbituric acid at low pH, some or all of these products react to form a pink chromogen that can be measured by its absorbance at 532 nm; this chromogen is indistinguishable from a thiobarbituric acid–malondialdehyde (TBA–MDA) adduct.[147,148,213] Generation of the pink TBA–MDA adduct can be used as a simple assay for $\cdot OH$ generation in biological systems, provided that suitable control experiments are performed (listed in Ref. 213a).

If deoxyribose is incubated with H_2O_2 and an Fe^{2+}–EDTA complex (or an Fe^{3+}–EDTA complex in the presence of a reducing agent such as ascorbate or O_2^-), subsequent deoxyribose degradation is inhibited by any added scavenger of $\cdot OH$ to an extent that depends only on the concentration of scavenger relative to deoxyribose and on the scavenger's second-order rate constant for reaction with $\cdot OH$.[125,134,145] It seems that, when $\cdot OH$ is generated by reaction of Fe^{2+}–EDTA with H_2O_2, any $\cdot OH$ that escapes scavenging by the EDTA enters "free solution" and is

[210] M. E. Kyle, D. Nakae, I. Sakaida, S. Miccadei, and J. L. Farber, *J. Biol. Chem.* **263**, 3784 (1988).
[211] J. A. Imlay and S. Linn, *J. Bacteriol.* **169**, 2967 (1987).
[212] B. Halliwell, M. Grootveld, and J. M. C. Gutteridge, *Methods Biochem. Anal.* **33**, 59 (1988).
[213] K. H. Cheeseman, A. Beavis, and H. Esterbauer, *Biochem. J.* **252**, 649 (1988).
[213a] B. Halliwell, *Free Radical Res. Commun.* **9**, 1 (1990).

equally accessible to deoxyribose and to any added scavenger. Indeed, the deoxyribose assay in the presence of Fe^{3+}–EDTA, H_2O_2, and a reducing agent provides a simple method for determining rate constants for the reaction of substrates with $\cdot OH$.[125,134,145] This method has been used in several studies.[126,214,215]

If deoxyribose is incubated with H_2O_2 and Fe^{2+} (or Fe^{3+} plus a reductant) in the absence of EDTA, it is still degraded into products that can react to form a TBA–MDA chromogen. However, some $\cdot OH$ scavengers (such as ethanol, formate, dimethyl sulfoxide, and HEPES) no longer inhibit the deoxyribose degradation whereas others (such as mannitol, thiourea, and hydroxychloroquine), still do.[134,145,214] Two possible explanations of this observation have been advanced. One is that the deoxyribose-degrading species is not $\cdot OH$ but some other oxidant, such as ferryl.[192] It is known that oxidants other than $\cdot OH$ can degrade deoxyribose to a TBA-reactive material; for example, such a deoxyribose-degrading oxidant is produced by reaction of human oxyhemoglobin with equimolar H_2O_2.[216] An alternative explanation, which the authors prefer,[217] is that unchelated iron ions added to deoxyribose-containing reaction mixtures can become weakly associated with deoxyribose. When the bound iron ions react with H_2O_2, some or all of the $\cdot OH$ formed would be expected to attack the deoxyribose immediately, and scavengers could not easily prevent this "site-specific" attack.[145]

Consistent with such an explanation, evidence for a weak binding of both Fe^{3+} and Fe^{2+} to deoxyribose at physiological pH has been obtained.[218,219] Further, the metal-binding ability of a compound has been shown to be a major determinant of its ability to inhibit deoxyribose degradation in the presence of H_2O_2 and Fe^{2+} (or Fe^{3+} and a reducing agent).[145,214] For example, citrate is a very poor scavenger of $\cdot OH$ (rate constant about $10^7 \ M^{-1} \ s^{-1}$) but is a good inhibitor of deoxyribose degradation in the presence of H_2O_2, Fe^{3+}, and ascorbate.[214] Indeed, the deoxyribose assay has been used to test the ability of substances to interfere with site-specific iron-dependent generation of $\cdot OH$ radicals.[214,215]

1.6.5. Superoxide-Dependent Formation of Higher Oxidation States of Metals in Reactions Not Involving H_2O_2. There have been several proposals that O_2^- might be able to form reactive oxidants by combining

[214] O. I. Aruoma and B. Halliwell, *Xenobiotica* **18**, 459 (1988).

[215] O. I. Aruoma, B. Halliwell, B. M. Hoey, and J. Butler, *Biochem. J.* **256**, 251 (1988).

[216] A. Puppo and B. Halliwell, *Biochem. J.* **249**, 185 (1988).

[217] J. M. C. Gutteridge and B. Halliwell, *Biochem. J.* **253**, 932 (1988).

[218] O. I. Aruoma, M. Grootveld, and B. Halliwell, *J. Inorg. Biochem.* **29**, 289 (1987).

[219] O. I. Aruoma, S. S. Chaudhary, M. Grootveld, and B. Halliwell, *J. Inorg. Biochem.* **35**, 149 (1989).

with metal ions (reviewed in Ref. 220). For example, reaction of Fe^{3+} with O_2^- is not a simple reduction to Fe^{2+}: it produces perferryl as an intermediate,

$$Fe^{3+} + O_2^- \rightleftharpoons [Fe^{3+}-O_2^- \leftrightarrow Fe^{2+}-O_2] \rightleftharpoons Fe^{2+} + O_2 \qquad (24)$$

Perferryl appears to exist at the active site of peroxidase compound III, but this is a poorly reactive form of peroxidase,[221,222] making perferryl an unlikely candidate as a highly cytotoxic species.

Oxidation of manganese(II) ions by O_2^- can produce Mn(III) species, or manganese–oxygen complexes, that are more oxidizing than is O_2^- (reviewed in Ref. 223). For example, O_2^- reacts only slowly with NADH, but addition of Mn^{2+} speeds up the reaction. It seems that O_2^- oxidizes Mn^{2+} to Mn(III), which then oxidizes the NADH,[221] that is,

$$O_2^- + 2H^+ + Mn^{2+} \rightarrow Mn^{3+} + H_2O_2 \qquad (25)$$
$$Mn^{3+} + NADH \rightarrow NAD\cdot + Mn^{2+} + H^+ \qquad (26)$$
$$NAD\cdot + O_2 \rightarrow NAD^+ + O_2^- \qquad (27)$$

However, it has been argued that accumulation of manganese ions protects[223] lactobacilli against the toxicity of O_2^-, hence it seems unlikely that products of reaction of O_2^- with manganese ions could generally exacerbate O_2^- toxicity *in vivo*. Vanadate ions (VO_3^-) also stimulate oxidation of NADH by O_2^-, perhaps by formation of an "oxidizing complex"[88] of vanadium ions with O_2^-, but the concentrations of vanadate available in animals *in vivo* are probably minute. Another possibility[220] is that O_2^- can not only reduce Cu^{2+} to Cu^+:

$$Cu^{2+} + O_2^- \rightarrow Cu^+ + O_2 \qquad (28)$$

but can also oxidize it to the powerful oxidant Cu(III):

$$Cu^{2+} + O_2^- + 2 H^+ \rightarrow Cu(III) + H_2O_2 \qquad (29)$$

Whether any of these oxidizing metal complexes contributes to O_2^- toxicity *in vivo* has yet to be established. If their formation requires only O_2^- and metal ion, it should not be inhibited by catalase. However, catalase usually does inhibit metal ion-dependent damage by O_2^--generating systems.[3,46,47]

[220] G. Czapski, S. Goldstein, and D. Meyerstein, *Free Radical Res. Commun.* **4**, 231 (1987).
[221] B. Halliwell, *Planta* **140**, 81 (1978).
[222] B. Halliwell, R. F. Pasternack, and J. De Rycker, *in* "Oxidases and Related Redox Systems" (T. E. King, H. S. Mason, and M. Morrison, eds.), p. 733. Pergamon, Oxford, 1982.
[223] F. Archibald, *CRC Crit. Rev. Microbiol.* **13**, 63 (1985).

2. Metabolism of Transition Metals

The importance of transition metal ions in mediating oxidant damage (Table I) naturally leads to the question as to what forms of such ions might be available to catalyze radical reactions *in vivo*. Let us therefore review what is known about the metabolism of transition metals.

2.1. Iron

Although the presence of iron in blood was not discovered until the eighteenth century, treatment of human disease with iron is said to date back to 2735 B.C. in China and 1500 B.C. in Europe (see citations in Refs. 224 and 225). Records show that in 1681 a liquor made from iron filings, sugar, and wine was used successfully by the English physician Syden-ham to treat a malady later shown to be iron-deficiency anemia. By 1832, a lowered iron content in the blood of anemic patients had been observed, and iron salts were introduced for oral treatment of the disease. These are, of course, still widely available throughout the world for medically prescribed or self-prescribed treatment of anemia. Indeed, many foods and beverages are fortified with iron salts or with elemental iron from which, it has been estimated, some 10–15% of total iron intake in Western countries is derived.[226,227] There have been several suggestions that too much iron fortification is undertaken in developed countries.[228–230] This is particularly relevant in view of the high occurrence of the gene for the iron overload disease hereditary hemochromatosis (HH) (frequency of homozygosity 0.003–0.008),[230,231] in which intestinal iron uptake is increased. Indeed, the high frequency of the gene for HH has been thought to represent the result of natural selective forces. Since toxicity related to iron overload often does not occur until postreproductive age, it has been suggested that hemochromatosis might have an adaptive advantage for persons living on iron-limited diets. Only in the industrialized countries of

[224] G. Spiro and P. Saltman, *Struct. Bonding (Berlin)* **6**, 116 (1969).
[225] A. Bezkorovainy, "Biochemistry of Non-Haem Iron." Plenum, New York, 1980.
[226] D. P. Richardson, *Chem. Ind. (London)* **4**, 498 (1983).
[227] J. D. Cook and M. Reusser, *Am. J. Clin. Nutr.* **38**, 648 (1983).
[228] D. R. Blake, N. D. Hall, P. A. Bacon, P. A. Dieppe, B. Halliwell, and J. M. C. Gutteridge, *Lancet* **2**, 1142 (1981).
[229] J. L. Sullivan, *Lancet* **1**, 1293 (1981).
[230] J. D. Cook, B. S. Skikne, S. R. Lynch, and M. E. Reusser, *Blood* **68**, 726 (1986).
[231] C. Q. Edwards, L. M. Griffen, D. Goldgar, C. Drummond, M. H. Skolnick, and J. P. Kushner, *N. Engl. J. Med.* **318**, 1355 (1988).

the twentieth century, with widespread iron supplementation and longer lifespan, might the deleterious effects of the gene become apparent.[232,233]

An average adult human male contains about 4.5 g of iron, absorbs about 1 mg of iron per day from the diet, and excretes approximately the same amount when in iron balance. Since the total plasma iron turnover is some 35 mg/day, extremely efficient mechanisms for iron preservation must exist in the body. Slight disturbances of iron metabolism will readily lead either to iron deficiency or to iron overload. There appear to be no specific physiological mechanisms for iron excretion; loss occurs via the turnover of intestinal epithelial cells, in sweat, feces, and urine, and by menstrual bleeding in females.

About two-thirds of body iron is found in hemoglobin, with smaller amounts in myoglobin, various enzymes, and the transport protein transferrin. Iron not required for these is largely stored in ferritin. Ferritin consists of a protein shell surrounding an iron core that holds up to 4,500 ions of iron per molecule of protein.[234,235] Iron enters ferritin as Fe^{2+}, which becomes oxidized to Fe(III) and deposited in the interior. Similarly, iron can be removed from ferritin as Fe^{2+} by the action of a number of biological reducing agents, including cysteine, reduced flavins, and ascorbate.[235] Whether the reductants have to enter the iron core in order to achieve mobilization is uncertain.[236] Ferritin can be converted in lysosomes to an insoluble product known as hemosiderin (reviewed in Ref. 237). It is currently thought that this conversion is achieved by proteolytic attack, although this remains to be proved. O'Connell et al.[238,239] have made the interesting suggestion that free radical reactions may be involved in the ferritin-to-hemosiderin conversion.

Myoglobin and hemoglobin from animal tissues represent the chief food source of heme iron for nonvegetarians. The greatest percentage of dietary iron is nonheme iron, however, present in the Fe(III) state.[240] Heme is absorbed as such and the iron removed from it in the intestinal

[232] R. B. Johnson, *N. Engl. J. Med.* **319**, 1155 (1988).

[233] J. P. Kushner, M. H. Skolnick, C. Q. Edwards, D. Goldgar, L. M. Griffen, and C. Drummond, *N. Engl. J. Med.* **319**, 1156 (1988).

[234] P. M. Harrison and R. J. Hoare, "Metals in Biochemistry." Chapman & Hall, London, 1980.

[235] R. R. Crichton and M. Charloteaux-Waters, *Eur. J. Biochem.* **164**, 485 (1987).

[236] G. D. Watt, D. Jacobs, and R. B. Frankel, *Proc. Natl. Acad. Sci. U.S.A.* **85**, 7457 (1988).

[237] M. P. Weir, J. F. Gibson, and T. J. Peters, *Cell Biochem. Funct.* **2**, 186 (1984).

[238] M. J. O'Connell, B. Halliwell, C. P. Moorhouse, O. I. Aruoma, H. Baum, and T. J. Peters, *Biochem. J.* **234**, 727 (1986).

[239] M. J. O'Connell, H. Baum, and T. J. Peters, *Biochem. J.* **240**, 297 (1986).

[240] A. Jacobs and M. Worwood, *in* "Metals and the Liver" (L. W. Powell, ed.), p. 3. Dekker, New York, 1981.

mucosal cells,[241] but other forms of iron require solubilization and reduction to the Fe^{2+} state to facilitate absorption. The hydrochloric acid in the stomach achieves solubilization, and dietary vitamin C (ascorbic acid, a reducing agent) aids absorption. The most active sites of iron absorption are the duodenum and upper jejunum. Only a low percentage of the dietary iron is taken up into the intestinal cells (see below), and not all the iron taken up is transferred into the circulation. Some is deposited in ferritin within the mucosal cells and is eventually lost again when these cells exfoliate from the mucosal surface during normal turnover.

The iron taken up by the gut that is transferred to the circulation enters the plasma bound to the protein *transferrin,* which functions as a carrier molecule.[235,242] It has recently been suggested that the defect in hereditary hemochromatosis is an inability to attach iron to transferrin in the gut, so that the iron enters the plasma in a low molecular mass form.[243]

Transferrin is a glycoprotein, and each molecule has two separate binding sites to which Fe(III) attaches extremely tightly at physiological pH. Tight binding requires the presence at each site of an anion, usually HCO_3^-. Under normal conditions, the transferrin present in the human bloodstream is only about 30% loaded on average, so that the amount of "free" iron salts available in the plasma would be expected to be zero, a result confirmed by experiment.[244-246] A protein similar to transferrin, known as lactoferrin, is found in several body fluids and in milk and is produced by phagocytic cells (reviewed in Ref. 247). Lactoferrin also binds 2 mol of Fe(III) per mole of protein.

Iron from transferrin must enter the various cells of the body for use in synthesizing iron-containing proteins. The current view,[235] although some scientists have challenged it (e.g., see Ref. 248), is that transferrin is taken into iron-requiring cells by receptor-mediated endocytosis, so that it enters the cytoplasm in a vacuole. The contents of the vacuole are then acidified. Low pH assists release of iron from transferrin. The iron released might chelate to various cellular constituents such as citrate, ATP, GTP, or other phosphate esters, or the cell might contain some form of

[241] E. R. Morris, *Fed. Proc., Fed. Am. Soc. Exp. Biol.* **42,** 1716 (1983).

[242] S. P. Young and P. Aisen, *in* "The Liver: Biology and Pathobiology" (I. Arias, H. Popper, D. Schacter, and D. A. Shafritz, eds.), p. 393. Raven, New York, 1982.

[243] O. I. Aruoma, A. Bomford, R. J. Polson, and B. Halliwell, *Blood* **72,** 1416 (1988).

[244] C. Hershko, G. Graham, G. W. Bates, and E. A. Rachmilewitz, *Br. J. Haematol.* **40,** 255 (1978).

[245] J. M. C. Gutteridge, D. A. Rowley, and B. Halliwell, *Biochem. J.* **199,** 263 (1981).

[246] J. M. C. Gutteridge, D. A. Rowley, and B. Halliwell, *Biochem. J.* **206,** 605 (1982).

[247] H. S. Birgens, *Scand. J. Haematol.* **33,** 225 (1984).

[248] K. Thorstensen, *J. Biol. Chem.* **263,** 16837 (1988).

"iron donor protein" that weakly binds iron ions and donates them for iron-requiring processes.[249] The iron-free transferrin (apotransferrin) is then ejected from the cell while the small pool of intracellular iron (often called nonprotein-bound iron) can be used in the synthesis of iron proteins.[235,250] For example, mitochondria have been reported to take up iron salts rapidly,[251] for incorporation into heme and nonheme iron proteins. Mitochondria might contain small "pools" of chelated iron salts in the matrix.[252] Overall, however, it is fair to say that, in both cells and organelles, neither the size nor the chemical nature of the pool of "available" iron are at all clear.[235] We thus have several potential catalysts of $\cdot$OH radical formation available *in vivo*.

(i) *Lactoferrin and transferrin.* An early report[253] that iron-loaded lactoferrin [2 mol of Fe(III) per mole of protein] is an efficient catalyst of $\cdot$OH radical formation from O_2^- and H_2O_2 has now been disproved, as have similar claims for transferrin.[254-258] It would be expected that little or none of any $\cdot$OH formed by iron ions attached to a protein could escape from the protein and be measurable outside it: the $\cdot$OH would attack the protein instead,[67,259] in view of the very high reactivity of this radical species. In the studies claiming that lactoferrin and transferrin are catalysts of $\cdot$OH radical production, it is likely that iron ions became detached from the protein during the assay, and these released iron ions were the real catalyst of the observed $\cdot$OH radical production.[255]

Further evidence that lactoferrin does not participate in $\cdot$OH generation is provided by the observation that activated human neutrophils do not produce externally detectable $\cdot$OH unless iron ions are added to the reaction mixture, although the neutrophils secrete lactoferrin into the surrounding medium.[122,124,135,260] Indeed, the authors have argued that one function of the secretion of lactoferrin by activated neutrophils is to minimize $\cdot$OH production in the surrounding environment, by chelating "cat-

[249] T. Ramasarma, *Free Radical Res. Commun.* **2,** 153 (1986).

[250] A. Jacobs, *Blood* **50,** 433 (1970).

[251] T. Flatmark and I. Romslo, *J. Biol. Chem.* **250,** 6433 (1975).

[252] A. Tangeras, T. Flatmark, D. Backstrom, and A. Ehrenberg, *Biochim. Biophys. Acta* **589,** 162 (1980).

[253] D. R. Ambruso and R. B. Johnston, *J. Clin. Invest.* **67,** 352 (1981).

[254] C. C. Winterbourn, *Biochem. J.* **210,** 15 (1983).

[255] O. I. Aruoma and B. Halliwell, *Biochem. J.* **241,** 273 (1987).

[256] J. J. Maguire, E. W. Kellogg, and L. Packer, *Toxicol. Lett.* **14,** 27 (1982).

[257] G. R. Buettner, *Bioelectrochem. Bioeng.* **18,** 29 (1987).

[258] D. A. Baldwin, E. R. Jenny, and P. Aisen, *J. Biol. Chem.* **259,** 13391 (1984).

[259] H. Jenzer, H. Kohler, and C. Broger, *Arch. Biochem. Biophys.* **258,** 381 (1987).

[260] B. E. Britigan, G. M. Rosen, B. Y. Thompson, Y. Chai, and M. S. Cohen, *J. Biol. Chem.* **261,** 17026 (1986).

alytic" iron ions.[82,261] The metal ion-binding ability of lactoferrin may play an important part in its antibacterial action.[262]

However, if iron ions could be released from lactoferrin or transferrin by some mechanism, perhaps on the surface of a target cell, localized · OH production mediated by the iron ions would be possible. Thus, Jacob and co-workers[263] have suggested that lactoferrin can bind to certain target cells and release iron on their membranes, so that exposure of the cell to O_2^- and H_2O_2 from neutrophils causes site-specific damage.

Lactoferrin and transferrin are similar in many respects, but a major difference in their properties is that iron can be released from transferrin at pH values of 5.6 and below, whereas lactoferrin holds on to its iron down to pH values of 4.0.[264,265] The authors[82] have argued that the low pH in the microenvironment of activated adherent phagocytic cells[76] could lead to release of iron from transferrin, and that one biological role of lactoferrin is to bind some of this iron at low pH values as a protective mechanism. The ability of lactoferrin to resist damage by oxidants generated at sites of inflammation[266] is consistent with such a role.

(ii) *Ferritin and hemosiderin.* Ferritin is often regarded as a safe storage form of iron, yet ferritin has been shown to stimulate the formation of · OH radicals from O_2^- and H_2O_2.[267,268] This is not due to a reaction with the intact protein: superoxide mobilizes iron ions from ferritin, leading to iron-catalyzed production of · OH radicals.[269] A rate constant of 2×10^6 M^{-1} sec^{-1} has been calculated for reaction of O_2^- with horse spleen ferritin.[270] Hence, generation of O_2^- and H_2O_2 adjacent to ferritin deposits could cause extensive tissue damage.[268,269] O'Connell et al.[238] have shown that hemosiderin iron can also participate in · OH generation, although hemosiderin is usually less active than ferritin because it is more difficult to mobilize iron from it.[271] Perhaps conversion of stored ferritin to hemo-

[261] J. M. C. Gutteridge, S. K. Paterson, A. W. Segal, and B. Halliwell, *Biochem. J.* 199, 259 (1981).

[262] R. T. Ellison III, T. J. Giehl, and F. M. La Force, *Infect. Immun.* 56, 2774 (1988).

[263] G. M. Vercellotti, B. Sweder van Asbeck, and H. S. Jacob, *J. Clin. Invest.* 76, 956 (1985).

[264] M. L. Groves, *J. Am. Chem. Soc.* 82, 3345 (1960).

[265] B. C. Johansson, *Acta Chem. Scand.* 14, 510 (1960).

[266] B. Halliwell, O. I. Aruoma, M. Wasil, and J. M. C. Gutteridge, *Biochem. J.* 256, 311 (1988).

[267] J. V. Bannister, W. H. Bannister, and P. J. Thornalley, *Life Chem. Rep.* 2, 64 (1984).

[268] G. Carlin and R. Djursater, *FEBS Lett.* 177, 27 (1984).

[269] P. Biemond, H. G. Van Eijk, A. J. G. Swaak, and J. F. Koster, *J. Clin. Invest.* 73, 1576 (1984).

[270] G. R. Buettner, M. Saran, and W. Bors, *Free Radical Res. Commun.* 2, 369 (1987).

[271] J. M. C. Gutteridge and Y. Hou, *Free Radical Res. Commun.* 2, 143 (1986).

siderin (as happens during iron-overload disease) is biologically advantageous.[67,238,239,272]

(iii) *Hemoglobin and myoglobin*. There have been several claims that hemoglobin or myoglobin are catalysts of the Fenton reaction. However, such claims have been based on nonspecific assays for $\cdot OH$ formation[152] or an assumption that thiourea is a specific scavenger of $\cdot OH$.[273] It is well known that mixtures of H_2O_2 with oxyhemoglobin or methemoglobin can oxidize many molecules, including KTBA[153] (a popular detector for $\cdot OH$), phenylhydrazine,[274] and dimethyl sulfoxide.[275] Indeed, hemoglobin has been described as a "frustrated oxidase."[276] The oxidizing action of methemoglobin–H_2O_2 or metmyoglobin–H_2O_2 mixtures might be attributed to a heme-associated ferryl species resembling that present in horseradish peroxidase compounds I and II.[153,277]

Reaction of oxyhemoglobin with equimolar concentrations of H_2O_2 produces a reactive species, not identical to $\cdot OH$, that can degrade deoxyribose[216] and stimulate lipid peroxidation.[278,279] Whether this reactive species is heme-associated ferryl is uncertain.[216] Reaction of myoglobin with H_2O_2 produces a ferryl-type species[279a] that can accelerate lipid peroxidation,[277,280] but it does not result in deoxyribose degradation.[281] Equimolar mixtures of leghemoglobin (an O_2-binding protein isolated from soybean root nodules) and H_2O_2 also form an oxidizing species, but they do not promote deoxyribose degradation.[282]

Although reaction of *intact* hemoglobin and myoglobin molecules with H_2O_2 does produce some reactive higher oxidation states of the heme iron, as discussed above, there is no clear evidence that these proteins react with H_2O_2 to form $\cdot OH$ that can be detected outside the protein. However, incubation of heme proteins with a molar excess of H_2O_2 can

[272] D. J. Macey, M. H. Cake, and I. C. Potter, *Biochem. J.* **252,** 167 (1988).

[273] S. M. H. Sadrzadeh, E. Graf, S. S. Panter, P. E. Hallaway, and J. W. Eaton, *J. Biol. Chem.* **259,** 14354 (1985).

[274] B. Goldberg and A. Stern, *J. Biol. Chem.* **250,** 2401 (1975).

[275] A. Puppo and B. Halliwell, *Free Radical Res. Commun.* **5,** 277 (1989).

[276] R. W. Carrell, C. C. Winterbourn, and J. K. French, *Hemoglobin* **1,** 815 (1977).

[277] J. Kanner, J. B. German, and J. E. Kinsella, *CRC Crit. Rev. Food Sci. Nutr.* **25,** 317 (1987).

[278] M. R. Clemens, H. Einsele, H. Remmer, and H. D. Waller, *Biochem. Pharmacol.* **34,** 1339 (1985).

[279] H. Itabe, T. Kobayashi, and K. Inoue, *Biochim. Biophys. Acta* **961,** 13 (1988).

[279a] K. D. Whitburn, *Arch. Biochem. Biophys.* **253,** 419 (1987).

[280] D. Galaris, D. Mira, A. Sevanian, E. Cadenas, and P. Hochstein, *Arch. Biochem. Biophys.* **262,** 221 (1988).

[281] A. Puppo and B. Halliwell, *Free Radical Res. Commun.* **4,** 415 (1988).

[282] A. Puppo and B. Halliwell, *Planta* **173,** 405 (1988).

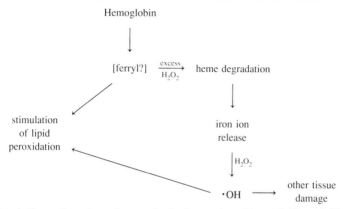

FIG. 2. Formation of reactive species by interaction of hemoglobin with H_2O_2.

cause heme degradation and release of iron ions.[283–286,286a] Recent studies have measured this iron release[283,284] and have shown that previous reports of $\cdot OH$ generation on exposing H_2O_2 to hemoglobin, myoglobin, or leghemoglobin are due either to reaction of the molecules used to detect $\cdot OH$ with other oxidizing species or to release of iron ions from the protein, followed by reaction of the released iron ions with H_2O_2 in a Fenton reaction.[216,281–284]

Figure 2 summarizes the ways in which interaction of hemoglobin with H_2O_2 might be able to do tissue damage. It is interesting to note that hemoglobin is carefully compartmentalized in cells (the erythrocytes) that have high activities of antioxidant enzymes: this may be because of the ease with which free hemoglobin can interact with H_2O_2 to produce damaging species (Fig. 2).

(iv) *Simple iron chelates.* As explained earlier, a small "transit pool" of iron chelates, probably in a low molecular mass form, is present within mammalian cells.[235,250] The exact chemical nature of this pool is not clear, but it may represent iron ions attached to phosphate esters (such as ATP, ADP, or GTP), to organic acids (such as citrate) and perhaps to the polar head groups of membrane lipids, or to DNA. All these iron complexes are capable[99,131,191,287] of decomposing H_2O_2 to form $\cdot OH$. We have already

[283] J. M. C. Gutteridge, *FEBS Lett.* **20**, 291 (1986).
[284] J. M. C. Gutteridge, *Biochim. Biophys. Acta* **834**, 144 (1985).
[285] J. T. Peters, *South. Med. J.* **40**, 924 (1947).
[286] W. N. M. Ramsay, *Biochem. J.* **49**, 494 (1951).
[286a] K. D. Whitburn, *Arch. Biochem. Biophys.* **267**, 614 (1988).
[287] R. A. Floyd, *Biochem. Biophys. Res. Commun.* **99**, 1209 (1981).

seen that the DNA damage produced by exposing mammalian cells to oxidative stress may involve $\cdot$OH generation by metal ions bound at or close to DNA (Section 1.6.3).

In view of the importance of iron in mediating oxidative damage by O_2^- and H_2O_2, it is interesting to note reports that the SOD activities of liver[267] and of the nephric folds of the larval lamprey *Geotria australis*[272] rise in approximate proportion to the content of nonheme iron.

2.1.1. Bleomycin Assay. Gutteridge *et al.*[245,246] developed the bleomycin assay as a first attempt to measure the availability of iron complexes catalytic for radical reactions in human body fluids. The assay is based on the observation that the glycopeptide antitumor antibiotic bleomycin degrades DNA in the presence of an iron salt, O_2, and a suitable reducing agent. During this reaction, base–propenals are released[288] and rapidly break down to yield malondialdehyde.[289] Binding of the bleomycin–iron–O_2 complex to DNA makes the reaction site specific, and antioxidants rarely interfere. The bleomycin assay can therefore be applied directly to biological fluids for the detection of iron available to the bleomycin molecule. The technique requires stringent removal of adventitious iron associated with laboratory water, reagents, and the bleomycin molecule itself. This can be done by using Chelex resin or, more efficiently, by employing the iron-binding properties of transferrins.[290] Iron detected in the assay is then a measure of iron available from the biological sample to bleomycin, and iron can be assayed at concentrations as low as 0.5 μM. Application of the bleomycin assay shows that blood serum or plasma prepared from healthy human subjects does not contain iron available to bleomycin. This agrees with observations that the bleomycin assay as used[245,246] does not measure iron bound to the transferrin, lactoferrin, or ferritin proteins in plasma or serum.

When the bleomycin assay is applied to plasma or serum from patients with iron overload consequent on hereditary hemochromatosis, bleomycin-detectable iron is present,[243,291] and its concentration is highly correlated to that of serum ferritin.[243] As the iron overload is brought under control by phlebotomy therapy, concentrations of bleomycin-detectable iron decline.[243] Studies by HPLC and proton Hahn spin-echo NMR suggest that most of the bleomycin-detectable iron in the plasma of these

[288] L. Giloni, M. Takeshita, F. Johnson, C. Iden, and A. P. Grollman, *J. Biol. Chem.* **256**, 8608 (1981).

[289] J. M. C. Gutteridge, *FEBS Lett.* **105**, 278 (1979).

[290] J. M. C. Gutteridge, *FEBS Lett.* **214**, 362 (1987).

[291] J. M. C. Gutteridge, D. A. Rowley, E. Griffiths, and B. Halliwell, *Clin. Sci.* **68**, 463 (1985).

patients is in a low molecular mass form, existing as complexes with citrate, and possibly as ternary iron–citrate–acetate complexes.[292]

Bleomycin-detectable iron has also been measured in human sweat,[293] in cerebrospinal fluid (CSF),[294] in synovial fluid from human knee joints,[246,295,296] and in extracts of several bacteria.[297] About 40% of synovial fluid samples withdrawn from the knee joints of arthritis patients contain bleomycin-detectable iron; the fluids in which iron is measurable have lower total antioxidant activity than the synovial fluids which register negative in the bleomycin assay.[295] Further work is needed to determine the precise molecular nature of bleomycin-detectable iron in human sweat, CSF, and synovial fluid.

2.1.2. Source of Fenton-Reactive Iron in Vivo. When the ideas that the toxicity of O_2^- and H_2O_2 involves $\cdot OH$ were first being formulated, there was much debate about the availability of "catalytic" metal ions *in vivo*,[107–110,298,299] with several investigators being dubious about their existence.[107–110] The picture emerging now seems much clearer. Organisms take great care in the handling of iron, using transport proteins such as transferrin and storage proteins such as ferritin and minimizing the size of the intracellular iron pool. Indeed, this iron sequestration may be regarded as a contribution to antioxidant defenses.[67] However, oxidant stress can itself provide the iron necessary for Fenton chemistry, for example, by mobilizing iron from ferritin (O_2^-) or by degrading heme proteins to release iron (H_2O_2).

The availability of iron to stimulate $\cdot OH$ generation *in vivo* is very limited (e.g., concentrations of bleomycin-detectable iron in human samples are rarely greater than 3 μM, except in iron-overloaded patients).[300] The authors have thus argued[47,67,301,302] that formation of $\cdot OH$ and of other damaging radicals *in vivo* may be limited by the supply of iron ions, and

[292] M. Grootveld, J. D. Bell, B. Halliwell, O. I. Aruoma, A. Bomford, and P. J. Sadler, *J. Biol. Chem.* **264,** 4417 (1989).

[293] J. M. C. Gutteridge, D. A. Rowley, B. Halliwell, D. F. Cooper, and D. M. Heeley, *Clin. Chim. Acta* **145,** 267 (1985).

[294] J. M. C. Gutteridge, D. A. Rowley, B. Halliwell, and T. Westermarck, *Lancet* **2,** 459 (1982).

[295] J. M. C. Gutteridge, *Biochem. J.* **245,** 415 (1987).

[296] D. A. Rowley, J. M. C. Gutteridge, D. Blake, M. Farr, and B. Halliwell, *Clin. Sci.* **66,** 691 (1984).

[297] J. M. C. Gutteridge and S. Wilkins, *Biochem. Int.* **8,** 89 (1984).

[298] B. Halliwell, *Biochem. J.* **205,** 461 (1982).

[299] B. Halliwell, *Trends Biochem. Sci.* **7,** 270 (1982).

[300] J. M. C. Gutteridge and B. Halliwell, *Life Chem. Rep.* **4,** 113 (1987).

[301] J. M. C. Gutteridge and A. Smith, *Biochem. J.* **256,** 861 (1988).

[302] B. Halliwell and J. M. C. Gutteridge, *Lancet* **1,** 1396 (1984).

that tissue injury, by any mechanism, can exacerbate radical reactions if it liberates "catalytic" metal ions from broken cells into the surrounding environment. This is especially true in brain, since CSF has no significant iron-binding capacity,[303] and it has been known for many years that mechanical disruption of the brain releases iron ions that can stimulate radical reactions such as lipid peroxidation.[294,304,305] Indeed, iron-chelating agents have been reported to diminish posttraumatic degeneration of the brain and spinal cord in animals,[306–308] possibly by binding iron released as a result of the trauma and thus inhibiting iron-dependent radical reactions.[306–308]

Another example of the release of iron as a result of cell injury may be provided by the observation that administration of cytotoxic drugs, during chemotherapy, to patients with acute myeloid leukemia causes sharp rises in the percentage saturation of plasma transferrin and the appearance of bleomycin-detectable iron in the circulation.[309] Since the chemotherapy of leukemias usually includes redox-cycling drugs such as anthracyclines, the released iron could interact with O_2^- and H_2O_2 produced by such drugs to do severe damage to tissues.[309]

2.2. Copper

An average adult contains about 80 mg copper. Copper is absorbed from the diet in the stomach or upper small intestine, possibly as complexes with amino acids (such as histidine) or small peptides.[234,310] Transport to the liver is thought to be facilitated by binding of Cu^{2+} to amino acids and to albumin, which has one high-affinity binding site for copper ions (except in dogs) and several binding sites of lower affinity. In the liver, copper is incorporated into the protein ceruloplasmin, which is then secreted and forms the major copper-containing protein of human plasma. Ceruloplasmin has a relative molecular mass of around 134,000 with six or seven copper ions per molecule. Six of these copper ions are tightly bound and are released only at low pH in the presence of a reducing agent (reviewed in Ref. 311). Ceruloplasmin may, however, be able to supply

[303] B. J. Bleijenberg, H. G. van Eijk, and B. Leijnse, *Clin. Chim. Acta* **31**, 277 (1971).
[304] J. Stocks, J. M. C. Gutteridge, R. J. Sharp, and T. L. Dormandy, *Clin. Sci.* **47**, 223 (1974).
[305] B. Halliwell and J. M. C. Gutteridge, *Trends Neurosci.* **8**, 22 (1985).
[306] E. D. Hall, *J. Neurosurg.* **68**, 462 (1988).
[307] E. D. Hall, P. A. Yonkers, J. M. McCall, and J. M. Braughler, *J. Neurosurg.* **68**, 456 (1988).
[308] E. D. Hall and M. A. Travis, *Brain Res.* **451**, 350 (1988).
[309] B. Halliwell, O. I. Aruoma, G. Mufti, and A. Bomford, *FEBS Lett.* **241**, 202 (1988).
[310] S. J. Lau and B. Sarkar, *Biochem. J.* **199**, 649 (1981).
[311] J. M. C. Gutteridge and J. Stocks, *CRC Crit. Rev. Clin. Lab. Sci.* **14**, 257 (1981).

copper within cells for incorporation into other copper proteins, such as superoxide dismutase and cytochrome oxidase.[312,313] This copper donor role of ceruloplasmin is often referred to as a copper transport function; however, ceruloplasmin does not specifically bind and transport copper in the way that transferrin binds and transports iron.[311,314] Ceruloplasmin may have to be irreversibly degraded to release copper from it, whereas apotransferrin is released from cells to bind more iron.[311]

According to the literature,[234,310,311] about 5% of human plasma copper is bound to albumin or to amino acids such as histidine, and the rest is bound to ceruloplasmin. However, ceruloplasmin undergoes degradation (apparently proteolytic) when serum or plasma is stored at 4°, and it has been suggested that some or all of this nonceruloplasmin serum copper pool in humans is an artifact produced by the breakdown of ceruloplasmin during the handling of the samples.[101,315] Thus, the original paper[316] describing the nonceruloplasmin copper pool referred to "serum kept at 4°C for up to 3 days and then dialysed for 48 h." These would be ideal conditions for release of low molecular mass copper ion complexes from ceruloplasmin.[315,316a]

In vitro, ceruloplasmin catalyzes the oxidation of a wide variety of polyamine and polyphenol substrates, but, with the possible exception of certain bioamines, these oxidations have no known biological significance.[311] The biological role of ceruloplasmin has been suggested to be that of a "ferroxidase"[317] which catalyzes oxidation of Fe^{2+} to $Fe(III)$ and so facilitates the binding of iron to transferrin. Unlike the nonenzymatic oxidation of Fe^{2+} in the presence of O_2,[318,319] the ferroxidase activity of ceruloplasmin (sometimes called "ferroxidase I") does not produce O_2^- or $\cdot OH$.[311] Partly because of its ability to inhibit Fe^{2+}-dependent radical reactions, ceruloplasmin is an important extracellular antioxidant.[67,85,320–323]

[312] N. Marceau and N. Aspin, *Biochim. Biophys. Acta* **293,** 338 (1973).

[313] N. Marceau and N. Aspin, *Biochim. Biophys. Acta* **328,** 351 (1973).

[314] C. T. Dameron and E. D. Harris, *Biochem. J.* **248,** 663 (1987).

[315] J. M. C. Gutteridge, *Biochem. J.* **218,** 983 (1984).

[316] P. Z. Neumann and A. Sass-Kortsak, *J. Clin. Invest.* **46,** 646 (1967).

[316a] P. J. Evans, A. Bomford, and Halliwell, *Free Radical Res. Commun.* **7,** 55 (1989).

[317] S. Osaki, D. A. Johnson, and E. Frieden, *J. Biol. Chem.* **241,** 2746 (1966).

[318] B. Halliwell, *FEBS Lett.* **96,** 238 (1978).

[319] S. F. Wong, B. Halliwell, R. Richmond, and W. R. Skowroneck, *J. Inorg. Biochem.* **14,** 127 (1981).

[320] J. M. C. Gutteridge, *Biochem. Biophys. Res. Commun.* **77,** 379 (1977).

[321] D. J. Al-Timini and T. L. Dormandy, *Biochem. J.* **168,** 283 (1977).

[322] R. A. Lovstad, *Int. J. Biochem.* **15,** 1067 (1983).

[323] J. M. C. Gutteridge, R. Richmond, and B. Halliwell, *FEBS Lett.* **112,** 269 (1980).

By contrast, copper ions attached to albumin or to amino acids can still interact with O_2^- [324] and with H_2O_2 to form $\cdot OH$ and/or some other reactive species such as Cu(III). The reactive species appears to attack the ligand to which the copper is bound and is not released into "free" solution. Binding of copper ions to albumin,[325-327] DNA,[95,202a,328-331] viruses,[332] carbohydrates,[200,202] or enzymes[92,333] can cause site-specific oxygen radical damage, a further illustration of the principle that the toxicity of O_2^- and H_2O_2 in vivo depends on the location of metal ion catalysts of $\cdot OH$ radical formation.

The ability of albumin to bind copper ions might be a biologically important protective mechanism.[67,93,334] The presence of high concentrations (40–60 mg/ml) of albumin in human plasma may chelate any copper ions that become available, such as those released from damaged cells, prevent these ions from binding to more important sites, for instance, on cell membranes.[334] Hence, if O_2^- and H_2O_2 are produced in plasma, for example, by activated phagocytic cells, reactive radicals may be formed at the albumin binding sites for Cu^{2+}, but the damage produced may be biologically insignificant because so much albumin is present.[334] Albumin is also an efficient scavenger of the myeloperoxidase-derived oxidant HOCl.[334,335]

2.2.1. Phenanthroline Assay. The chelating agent 1,10-phenanthroline degrades DNA in the presence of copper ions, oxygen, and a suitable reducing agent. Degradation results in the release of thiobarbituric acid-reactive material from DNA.[300,315] This reaction has been made the basis of a technique[315] to detect and measure "available" copper (i.e., copper chelatable by phenanthroline) in biological fluids. All reagents are carefully treated with Chelex resin to remove contaminating copper ions, and azide is added to inactivate catalase which might arise from cell lysis, since catalase inhibits the DNA degradation. Any copper that is com-

[324] R. Brigelius, R. Spottl, W. Bors, E. Lengfelder, M. Saran, and U. Weser, *FEBS Lett.* **47**, 72 (1974).

[325] J. M. C. Gutteridge and S. Wilkins, *Biochim. Biophys. Acta* **759**, 38 (1983).

[326] S. P. Wolff and R. T. Dean, *Biochem. J.* **234**, 399 (1986).

[327] G. Marx and M. Chevion, *Biochem. J.* **236**, 397 (1986).

[328] K. Shinohara, M. So, M. Nonaka, K. Nishiyama, H. Murakami, and H. Omura, *J. Nutr. Sci. Vitaminol.* **29**, 489 (1983).

[329] J. M. C. Gutteridge and B. Halliwell, *Biochem. Pharmacol.* **31**, 2801 (1982).

[330] G. J. Quinlan and J. M. C. Gutteridge, *Biochem. Pharmacol.* **36**, 3629 (1987).

[331] G. J. Quinlan and J. M. C. Gutteridge, *Free Radical Biol. Med.* **5**, 341 (1988).

[332] A. Samuni, M. Chevion, and G. Czapski, *Radiat. Res.* **99**, 562 (1984).

[333] A. Samuni, M. Chevion, and G. Czapski, *J. Biol. Chem.* **256**, 12632 (1981).

[334] B. Halliwell, *Biochem. Pharmacol.* **37**, 569 (1988).

[335] M. Wasil, B. Halliwell, D. C. S. Hutchison, and H. Baum, *Biochem. J.* **243**, 219 (1987).

plexable by the added phenanthroline is reduced by mercaptoethanol, and the resulting damage to DNA is measured and is proportional to the amount of available copper present.[315] The "phenanthroline" assay detects copper bound to the high-affinity site of albumin, to histidine, and to small peptides, but not that bound to ceruloplasmin.[315]

The precise significance of so-called phenanthroline-detectable copper in catalyzing radical reactions *in vivo* is still under investigation. One interesting observation is that phenanthroline-detectable copper is not present in freshly prepared plasma from normal humans or from small mammals, suggesting that the nonceruloplasmin copper pool is much smaller than previously supposed.[101,315,316a] Phenanthroline-detectable copper is not present even in patients with the copper-overload condition Wilson's disease, but it can be measured in plasma from some patients with fulminant hepatic failure.[316a] Phenanthroline-detectable copper has also been measured in human sweat[293] and in CSF.[336] Another suggestive study has shown an increased content of phenanthroline-detectable copper in the CSF of some patients suffering from Parkinson's disease.[337] All these studies were performed with fresh samples of plasma or CSF, to avoid any release of copper ions from ceruloplasmin on storage.[101,315]

2.3. Chelation Therapy: An Approach to Site-Specific Antioxidant Protection

If $\cdot OH$ radicals are formed from H_2O_2 by low molecular mass metal complexes in free solution and then have to diffuse a short distance to attack a target, protection by added $\cdot OH$ radical scavengers should be seen and has been reported in some *in vivo* systems.[338] On the other hand, site-specific $\cdot OH$ radical formation is almost impossible to protect against by scavenging (Section 1.6.2). Protection here can be achieved by (i) superoxide dismutase and H_2O_2-removing enzymes, to stop O_2^- and H_2O_2 from ever reaching the bound metal complexes, or (ii) chelating agents that pull the metal ions away from sensitive sites and render them inactive.

What chelating agents should be used? EDTA renders copper ions less reactive in $\cdot OH$ radical production.[334,339,340] However, iron–EDTA che-

[336] J. M. C. Gutteridge, *Med. Biol.* **62,** 101 (1984).

[337] H. S. Pall, A. C. Williams, D. R. Blake, J. Lunec, J. M. C. Gutteridge, M. Hall, and A. Taylor, *Lancet* **2,** 238 (1987).

[338] G. Cohen, *Photochem. Photobiol.* **28,** 669 (1978).

[339] B. Halliwell, *FEBS Lett.* **56,** 34 (1975).

[340] W. F. Beyer, Jr., and I. Fridovich, *Anal. Biochem.* **173,** 160 (1988).

lates are effective in catalyzing $\cdot OH$ formation from H_2O_2 and O_2^- or ascorbate *in vitro*, because of favorable alterations in the redox potential of the metal and because the chelator allows more iron to remain in solution at physiological pH (Section 1.5). Complexes of iron salts with phytate,[341] diethylenetriaminepentaacetic acid (DTPA),[118] and bathophenanthroline[318] show diminished reactivity in iron-dependent $\cdot OH$ radical production from O_2^- and H_2O_2. It has been claimed that the iron–DTPA complex is unable to catalyze $\cdot OH$ formation because it lacks an open coordination site.[341] In fact, it is well established that an Fe^{2+}–DTPA complex can react[342] with H_2O_2 to form $\cdot OH$; this chelating agent interferes with superoxide-driven Fenton chemistry because its Fe^{3+} complex is not efficiently reduced[343] by O_2^-. Reductants more powerful than O_2^- (such as paraquat radical or Adriamycin semiquinone) can reduce Fe^{3+}–DTPA and so lead to $\cdot OH$ generation in the presence of H_2O_2. Thus, as has been stressed previously,[67] DTPA is not a general inhibitor of iron-dependent radical reactions, only a partial inhibitor of superoxide-driven Fenton chemistry.

Several other chelating agents can suppress iron-dependent generation of $\cdot OH$ from H_2O_2.[344] The best-known is desferrioxamine,[73,345,346] which is also a powerful inhibitor of iron-dependent lipid peroxidation.[345,347] Desferrioxamine, available from CIBA-Geigy Pharmaceuticals as its methane sulfonate (Desferal), is widely used to prevent or treat iron overload in thalassemia and other conditions requiring repeated blood transfusion, and it has prolonged the lives of many patients. It is also very useful for treatment of acute iron poisoning. Desferal cannot be given by mouth; doses of 50–60 mg/kg body weight administered subcutaneously appear safe in the treatment of iron overload. The fact that Desferal decreases tissue injury in several animal models of human disease suggests that iron-dependent radical reactions are not without relevance *in vivo*.[346] Although Desferal has several actions in addition to its ability to inhibit radical reactions, the latter ability seems to account for the protective effects observed in most of the animal systems that have been studied (reviewed in Ref. 346).

[341] E. Graf, J. R. Mahoney, R. G. Bryant, and J. W. Eaton, *J. Biol. Chem.* **259**, 3620 (1984).
[342] G. Cohen and P. M. Sinet, *FEBS Lett.* **138**, 258 (1982).
[343] J. Butler and B. Halliwell, *Arch. Biochem. Biophys.* **218**, 174 (1982).
[344] G. J. Kontoghiorghes, M. J. Jackson, and J. Lunec, *Free Radical Res. Commun.* **2**, 115 (1986).
[345] J. M. C. Gutteridge, R. Richmond, and B. Halliwell, *Biochem. J.* **184**, 469 (1979).
[346] B. Halliwell, *Free Radical Biol. Med.* **7**, 645 (1989).
[347] E. D. Wills, *Biochem. J.* **113**, 325 (1969).

3. Lipid Peroxidation: A Radical Chain Reaction

The detection and measurement of lipid peroxidation is the evidence most frequently cited to support the involvement of free radical reactions in toxicology and in human disease. A wide range of techniques is available to measure the rate of this process, but none is applicable to all circumstances. The two most popular are the measurement of diene conjugation and the TBA (thiobarbituric acid) test, but they are both subject to pitfalls, especially when applied to human samples (reviewed in Refs. 47, 348, and 349). Let us begin by clarifying the essential principles of the peroxidation process.

3.1. Initiation and Propagation

When discussing lipid peroxidation, it is essential to use clear terminology for the sequence of events involved; imprecise use of terms such as initiation has caused considerable confusion in the literature (discussed in Ref. 349). In a completely peroxide-free lipid system, first chain initiation of a peroxidation sequence in a membrane or polyunsaturated fatty acid refers to the attack of any species that has sufficient reactivity to abstract a hydrogen atom from a methylene ($-CH_2-$) group. Hydroxyl radical can certainly do this:

$$-CH_2- + \cdot OH \rightarrow -\dot{C}H- + H_2O \qquad (30)$$

Hence, radiolysis of aqueous solutions, which produces $\cdot OH$, is known to stimulate peroxidation of any lipids present; this has been shown not only for biological membranes and fatty acids but also for food lipids (e.g., Refs. 350 and 351). The peroxidation is usually inhibited by scavengers of $\cdot OH$, such as mannitol and formate. Indeed, stimulation of lipid peroxidation is a problem in the use of ionizing radiation to sterilize foods. The rate constant for reaction of $\cdot OH$ with artificial lecithin bilayers[352] is about $5 \times 10^8 \ M^{-1} \ sec^{-1}$.

By contrast, O_2^- is insufficiently reactive to abstract H from lipids; in any case, it would not be expected to enter the hydrophobic interior of

[348] J. M. C. Gutteridge, *Free Radical Res. Commun.* **1**, 173 (1986).

[349] J. M. C. Gutteridge in "Oxygen Radicals and Tissue Injury," Proceedings of an Upjohn Symposium (B. Halliwell, ed.), p. 9. Allen Press, Lawrence, Kansas, 1988.

[350] M. J. O'Connell and A. Garner, *Int. J. Radiat. Biol. Relat. Stud. Phys. Chem. Med.* **44**, 615 (1983).

[351] H. Wolters, C. A. M. van Tilburg, and A. W. T. Konings, *Int. J. Radiat. Biol. Relat. Stud. Phys. Chem. Med.* **51**, 619 (1987).

[352] D. J. W. Barber and J. K. Thomas, *Radiat. Res.* **74**, 51 (1978).

membranes because of its charged nature. In agreement with this, O_2^- does not readily cross biological membranes, the only exception to this rule known to date being the erythrocyte membrane. Here O_2^- can travel via the anion channel, through which Cl^- and HCO_3^- ions normally pass (Section 1.3). The protonated form of O_2^-, $HO_2 \cdot$, is more reactive and appears to be capable of abstracting H from some fatty acids,[77]

$$-CH_2- + HO_2 \cdot \rightarrow -\dot{C}H- + H_2O_2 \tag{31}$$

$HO_2 \cdot$, being uncharged, should enter membranes fairly easily (in the same way that H_2O_2 crosses membranes readily). However, $HO_2 \cdot$ has not yet been proved to initiate peroxidation in cell membranes. Various iron-oxygen complexes may also be capable of abstracting H and initiating peroxidation (Section 3.2).

Since a hydrogen atom has only one electron, abstraction of H from a $-CH_2-$ group leaves behind an unpaired electron on the carbon ($-\dot{C}H-$). The presence of a double bond in the fatty acid weakens the C—H bonds on the carbon atom adjacent to the double bond and so makes H removal easier. The carbon radical tends to be stabilized by a molecular rearrangement to form a *conjugated diene;* these can undergo various reactions. For example, if two of them came into contact within a membrane they could cross-link the fatty acid molecules[352a]:

$$R-\dot{C}H + R-\dot{C}H \rightarrow R-CH-CH-R \tag{32}$$

However, by far the most likely fate of conjugated dienes under aerobic conditions is to combine with O_2, especially as O_2 is a hydrophobic molecule that concentrates into the interior of membranes.[3] Reaction with oxygen gives a peroxyl radical, $ROO \cdot$ (or $RO_2 \cdot$) (the name is often shortened to peroxy radical):

$$\diagdown CH \cdot + O_2 \rightarrow \diagdown CHO_2 \cdot \tag{33}$$

Of course, very low O_2 concentrations might favor self-reaction of carbon-centered radicals, or perhaps their reaction with other membrane components such as proteins (it must not be forgotten that most biological membranes have a high content of protein). Hence the O_2 concentration in a biological system might to some extent alter the pathway of peroxidation.[353,354]

[352a] H. Frank, D. Thiel, and J. Macleod, *Biochem. J.* **260,** 873 (1989).
[353] H. De Groot and T. Noll, *Chem. Phys. Lipids* **44,** 209 (1987).
[354] J. Kostrucha and H. Kappus, *Biochim. Biophys. Acta* **879,** 120 (1986).

The formation of peroxyl radicals is important because they are capable of abstracting H from another lipid molecule, that is, from an adjacent fatty acid side chain:

$$\text{\textbackslash CHO}_2\cdot + \text{\textbackslash CH}_2 \rightarrow \text{\textbackslash CHO}_2\text{H} + \text{\textbackslash CH}\cdot \tag{34}$$

They might also attack membrane proteins (e.g., Refs. 355, 356, and 356a).

Reaction (34) is the propagation stage of lipid peroxidation. The carbon radical formed can react with O_2 to form another peroxyl radical, and so the chain reaction of lipid peroxidation can continue. The peroxyl radical combines with the hydrogen atom that it abstracts to give a lipid hydroperoxide, sometimes shortened (not quite correctly) to *lipid peroxide*. A probable alternative fate of peroxyl radicals is to form cyclic peroxides. Of course, the initial H abstraction from a polyunsaturated fatty acid can occur at different points on the carbon chain. Thus, peroxidation of arachidonic acid gives at least six lipid hydroperoxides as well as cyclic peroxides and other products.[357]

3.2. Importance of Iron in Lipid Peroxidation

The exact role played by iron ions in accelerating lipid peroxidation is at present an area of great confusion, most of which is unnecessary. Iron ions are themselves free radicals, and Fe^{2+} can take part in electron transfer reactions with molecular oxygen[318,319]:

$$Fe^{2+} + O_2 \rightleftharpoons [Fe^{2+}\text{--}O_2 \leftrightarrow Fe^{3+}\text{--}O_2^-] \rightleftharpoons Fe^{3+} + O_2^- \tag{35}$$
$$\text{perferryl}$$

Superoxide can dismutate to form H_2O_2, giving all the essential ingredients for Fenton chemistry and the formation of $\cdot OH$ radicals:

$$2\,O_2^- + 2\,H^+ \rightarrow H_2O_2 + O_2 \tag{36}$$
$$Fe^{2+} + H_2O_2 \rightarrow Fe^{3+} + OH^- + \cdot OH \tag{37}$$

Thus, addition of an iron(II) salt to a completely peroxide-free unsaturated lipid should bring about first chain initiation of lipid peroxidation (H abstraction by $\cdot OH$). The resulting peroxidation should be inhibitable by H_2O_2-removing enzymes (catalase or selenium-containing glutathione

[355] J. V. Hunt, J. A. Simpson, and R. T. Dean, *Biochem. J.* **250**, 87 (1988).
[356] P. Hochstein and S. K. Jain, *Fed. Proc., Fed. Am. Soc. Exp. Biol.* **40**, 183 (1981).
[356a] K. S. Chio and A. L. Tappel, *Biochemistry* **8**, 2827 (1969).
[357] E. N. Frankel, *Prog. Lipid Res.* **23**, 197 (1985).

peroxidase), scavengers of $\cdot OH$ (e.g., mannitol or formate), and chelating agents that bind iron and prevent its participation in free radical reactions.

Iron-dependent, hydroxyl radical-initiated peroxidation has sometimes been demonstrated in fatty acid systems solublilized by detergents.[358] However, most scientists,[177,358–363] including the authors,[362,363] find that when catalase or scavengers of $\cdot OH$ are added to isolated cellular membrane fractions (e.g., plasma membranes, microsomes) or to liposomes undergoing peroxidation in the presence of Fe^{2+} salts or of $Fe(III)$ salts and a reducing agent (such as ascorbate), peroxidation is not inhibited. However, $\cdot OH$ radicals can usually be detected in the reaction mixtures by such techniques as aromatic hydroxylation, spin trapping, and the deoxyribose method.[358–363] Formation of the $\cdot OH$ radicals is inhibited by H_2O_2-scavenging enzymes. It follows that $\cdot OH$ is being generated in the reaction mixtures but is not required for peroxidation to proceed. How can this be explained?

It is unlikely that the lack of action of $\cdot OH$ scavengers means that the required $\cdot OH$ formation is site specific, involving iron ions bound to the membrane, so that any $\cdot OH$ formed reacts immediately with the membrane components and is not available for scavenging. Membrane-bound iron certainly does participate in lipid peroxidation.[47,364–366] However, a source of H_2O_2 would still be required, and so catalase should still inhibit. The fact that it does not has led several scientists to suggest that first chain initiation of lipid peroxidation in membrane systems incubated with iron salts in the presence of O_2 is achieved by reactive species other than $\cdot OH$. Ferryl[367] (Section 1.6) is one possibility; it has been suggested to account for the ability of myoglobin–H_2O_2 or hemoglobin–H_2O_2 mixtures to stimulate peroxidation.[277] However, ferryl formation by reaction of nonheme iron salts with H_2O_2 would still require a source of H_2O_2, and inhibition by peroxide-removing enzymes would be expected. Perferryl could conceivably be involved.[367] However, what is known of the chemis-

[358] S. D. Aust, L. A. Morehouse, and C. E. Thomas, *J. Free Radicals Biol. Med.* **1**, 3 (1985).

[359] O. Beloqui and A. I. Cederbaum, *Biochem. Pharmacol.* **35**, 2663 (1986).

[360] L. A. Morehouse, M. Tien, J. R. Bucher, and S. D. Aust, *Biochem. Pharmacol.* **32**, 123 (1983).

[361] A. Bast and M. H. M. Steeghs, *Experientia* **42**, 555 (1986).

[362] J. M. C. Gutteridge, *FEBS Lett.* **150**, 454 (1982).

[363] J. M. C. Gutteridge, *Biochem. J.* **224**, 697 (1984).

[364] B. Halliwell, *in* "Age Pigments" (R. S. Sohal, ed), p. 1. Elsevier, Amsterdam, 1981.

[365] J. M. Braughler, R. L. Chase, and J. F. Pregenzer, *Biochem. Biophys. Res. Commun.* **153**, 933 (1988).

[366] G. F. Vile and C. C. Winterbourn, *FEBS Lett.* **215**, 151 (1987).

[367] T. C. Pederson and S. D. Aust, *Biochim. Biophys. Acta* **385**, 232 (1975).

try of perferryl complexes (Section 1.6) suggests that they would be insufficiently reactive to abstract H or to insert oxygen directly into fatty acid side chains.

Studies on the kinetics of microsomal or liposomal lipid peroxidation in the presence of Fe^{2+} and/or Fe(III) salts have led Minotti and Aust[368,369] to propose that first chain initiation requires an iron(II)–iron(III)–oxygen complex, or at least a specific critical 1:1 ratio of Fe^{2+} to Fe(III). This proposal was based on observations that Fe^{2+}-dependent peroxidation in membrane systems proceeds most rapidly in the presence of Fe(III). Indeed, comparable experimental results have been obtained by several other groups.[370–373] Attempts to isolate the Fe(II)–Fe(III) complex have not been successful, however, and it has been observed that Pb^{2+} and Al^{3+} can replace Fe(III) in stimulating Fe^{2+}-dependent peroxidation in liposomes and microsomes.[373–375] If other metal ions can replace Fe(III), then a specific Fe(II)–Fe(III) complex cannot be required for the initiation of peroxidation.

It has been known for a long time[376,377] that iron plays a second important role in lipid peroxidation. Pure lipid hydroperoxides are fairly stable at physiological temperatures, but, in the presence of transition metal complexes, especially iron salts, their decomposition is greatly accelerated.[376–378] Thus, a reduced iron complex can react with lipid peroxide in a way similar to its reaction with H_2O_2: it causes fission of O—O bonds to form alkoxyl radicals (often shortened to alkoxy radicals), by the overall reaction

$$\underset{\substack{\text{lipid} \\ \text{hydroperoxide}}}{\text{R—OOH}} + \text{Fe}^{2+}\text{–complex} \rightarrow \text{Fe}^{3+}\text{–complex} + \text{OH}^- + \underset{\substack{\text{alkoxyl} \\ \text{radical}}}{\text{R—O}\cdot} \qquad (38)$$

[368] G. Minotti and S. D. Aust, *J. Biol. Chem.* **262**, 1098 (1987).

[369] G. Minotti and S. D. Aust, *Chem. Phys. Lipids* **44**, 191 (1987).

[370] J. M. Braughler, L. A. Duncan, and R. L. Chase, *J. Biol. Chem.* **261**, 10282 (1986).

[371] J. G. Goddard and G. D. Sweeney, *Arch. Biochem. Biophys.* **259**, 372 (1987).

[372] A. H. Horton, C. Rice-Evans, and B. Fuller, *Free Radical Res. Commun.* **5**, 267 (1989).

[373] O. I. Aruoma, B. Halliwell, M. J. Laughton, G. J. Quinlan, and J. M. C. Gutteridge, *Biochem. J.* **258**, 617 (1989).

[374] J. M. C. Gutteridge, G. J. Quinlan, I. A. Clark, and B. Halliwell, *Biochim. Biophys. Acta* **835**, 441 (1985).

[375] G. J. Quinlan, B. Halliwell, C. P. Moorhouse, and J. M. C. Gutteridge, *Biochim. Biophys. Acta* **962**, 196 (1988).

[376] W. A. Waters, *J. Am. Oil Chem. Soc.* **48**, 427 (1976).

[377] P. J. O'Brien, *Can. J. Biochem.* **47**, 485 (1969).

[378] M. J. Davies and T. F. Slater, *Biochem. J.* **245**, 167 (1987).

An iron(III) complex can form both peroxyl and alkoxyl radicals[378]; the overall equation written below probably represents the sum of several stages:

$$R—OOH + Fe^{3+}-complex \rightarrow RO_2\cdot + H^+ + Fe^{2+}-complex \qquad (39)$$

$$\underset{\text{radical}}{\text{peroxyl}} \qquad \underset{\substack{\text{to give alkoxyl} \\ \text{radical}}}{\text{(further reaction}}$$

The reactions of Fe^{2+} with lipid hydroperoxides are an order of magnitude faster than their reactions with H_2O_2 (rate constant for $Fe^{2+} + H_2O_2$ is about 76 M^{-1} sec^{-1}; for R—OOH + Fe^{2+}, around 10^3 M^{-1} sec^{-1})[379]; reactions of Fe^{3+} with hydroperoxides seem much slower than those of Fe^{2+}, but a precise rate constant has not yet been determined. The variable effects[345] of certain chelating agents (e.g., EDTA, DTPA) on lipid peroxidation can, at least in part, be explained by their ability to influence these different reactions. For example, EDTA can often stimulate the reaction of iron ions with H_2O_2, while slowing their reaction with lipid peroxides.[363]

Commercially available lipids are all contaminated with lipid peroxides (e.g., Ref. 380) so that liposomes or micelles made from them will already contain traces of such peroxides (the purchase of high-purity 99.9% fatty acids is no safeguard, since purity refers to contamination with other fatty acids, not to peroxide content). When cells are injured, lipid peroxidation is favored (Section 3.4), and traces of lipid peroxides are formed enzymatically in tissues by cyclooxygenase and lipoxygenase enzymes. Thus, membrane fractions isolated from disrupted cells should also contain some lipid peroxide.

When iron salts are added to isolated membrane fractions, the lipid peroxides present will be decomposed by reactions (38) and (39) to generate peroxyl and alkoxyl radicals. Both radicals can abstract H · and propagate lipid peroxidation. Thus, there is no need for ·OH formation and first chain initiation; the added metal ions are doing no more than stimulating further lipid peroxidation by breaking down lipid peroxides. It is not impossible that the putative abilities of ferryl, perferryl, and Fe(II)–Fe(III)–O_2 complexes to initiate lipid peroxidation are explicable by the abilities of these complexes to degrade traces of lipid peroxides in the membrane systems that were being studied.[349] Experiments on peroxide-free lipid systems[13] (which are very difficult to obtain; B. Bielski, personal communication) are urgently required to establish firmly whether various

[379] A. Garnier-Suillerot, L. Tosi, and E. Paniago, *Biochim. Biophys. Acta* **794**, 307 (1984).
[380] J. M. C. Gutteridge and P. J. Kerry, *Br. J. Pharmacol.* **76**, 459 (1982).

iron–oxygen complexes are really capable of abstracting hydrogen. The authors feel that more such chemical studies, and fewer repetitive biochemical experiments on isolated membranes, will lead to more progress in this field.

We saw in Section 1.6 that simple iron complexes can react with H_2O_2 to form $\cdot OH$ but that reaction of most iron proteins with H_2O_2 and/or O_2^- does not result in formation of $\cdot OH$ detectable outside the protein, unless iron is released from the protein under the reaction conditions being used. The range of iron complexes that can stimulate lipid peroxidation is wider (Table II). Thus, not only Fe^{2+} salts and simple complexes (e.g., Fe^{2+}–ADP) are effective, but also (under certain circumstances) free heme, met- and oxyhemoglobin, metmyoglobin, cytochromes (including cytochromes c and P-450), horseradish peroxidase, and lactoperoxidase (discussed in Refs. 47, 277, 349, 358, and 381–383). Ferritin stimulates lipid peroxidation to an extent proportional to the amount of iron it contains,[384] whereas hemosiderin stimulates much less strongly (on a unit iron basis).[385]

Sometimes the stimulation of lipid peroxidation by iron proteins is due to a release of iron from the protein under the conditions of the experiment. For example, stimulation of lipid peroxidation in liposomes by ferritin or hemosiderin is almost completely inhibitable by desferrioxamine, suggesting that it is mediated by released iron ions.[384,385] Stimulation of peroxidation by myoglobin and hemoglobin can involve both iron release from the protein by peroxides[283,284] and reactions brought about by the intact protein itself, such as initiation by ferryl species and/or ferryl-mediated decomposition of lipid peroxides into alkoxyl/peroxyl radicals.[279a,280] Even catalase is weakly effective in stimulating peroxide decomposition under certain circumstances, an action that has sometimes caused problems in attempts to use catalase as a probe for the role of H_2O_2 in peroxidizing lipid systems.[386]

By contrast, iron correctly bound to the two specific iron binding sites of transferrin or lactoferrin does not seem to promote peroxide decomposition.[261] Copper ions may also facilitate peroxide decomposition under

[381] P. Hochstein and L. Ernster, *Biochem. Biophys. Res. Commun.* **12,** 388 (1963).

[382] G. W. Winston, D. E. Feierman, and A. I. Cederbaum, *Arch. Biochem. Biophys.* **232,** 378 (1984).

[383] M. J. Davies, *Biochem. J.* **257,** 603 (1989).

[384] J. M. C. Gutteridge, B. Halliwell, A. Treffry, P. M. Harrison, and D. Blake, *Biochem. J.* **209,** 557 (1983).

[385] M. J. O'Connell, R. J. Ward, H. Baum, and T. J. Peters, *Biochem. J.* **229,** 135 (1985).

[386] J. M. C. Gutteridge, A. P. C. Beard, and G. J. Quinlan, *Biochem. Biophys. Res. Commun.* **117,** 901 (1983).

TABLE II
BIOLOGICAL IRON COMPLEXES AND THEIR POSSIBLE PARTICIPATION
IN OXYGEN RADICAL REACTIONS

Type of iron complex	Decomposition of lipid peroxides to form alkoxyl and peroxyl radicals	Hydroxyl radical formation by Fenton chemistry
Loosely bound iron		
Iron ions attached to		
Phosphate esters (e.g., ATP)	Yes	Yes
Carbohydrates and organic acids (e.g., citrate, picolinic acid, deoxyribose)	Yes	Yes
DNA	Probably yes	Yes
Membrane lipids	Yes	Yes
Mineral ores, e.g., asbestos, silicates[a]	Yes	Yes
Iron tightly bound to proteins		
Nonheme iron		
Ferritin (4500 mol Fe/mol protein)	Yes	Yes (when iron is released)
Hemosiderin	Weakly	Weakly (when iron is released)
Lactoferrin (iron saturated, 2 mol Fe^{3+}/mol protein)	No	No (only if iron released)
Transferrin (iron saturated, 2 mol Fe^{2+}/mol protein)	No	No (only if iron released)
Heme iron		
Hemoglobin	Yes	Yes (when iron is released)
Leghemoglobin	Yes	Yes (when iron is released)
Myoglobin	Yes	Yes (when iron is released)
Cytochrome c	Yes	Yes (when iron is released)
Catalase	Weakly	Not observed (theoretically possible if the enzyme were inactivated and iron were released).

[a] See S. A. Weitzman and P. Graceffa, *Arch. Biochem. Biophys.* **228,** 373 (1984); T. P. Kennedy, R. Dodson, N. Y. Rao, H. Ky, C. Hopkins, M. Baser, E. Tolley, and J. Hoidal, *Arch. Biochem. Biophys.* **269,** 359 (1989).

certain circumstances,[320,387–389] although more detailed studies are required as to the mechanism of copper-stimulated lipid peroxidation.[390]

[387] P. Hochstein, K. Sree Kumar, and S. J. Forman, *Ann. N.Y. Acad. Sci.* **355,** 240 (1980).
[388] J. M. C. Gutteridge, *Biochem. Biophys. Res. Commun.* **74,** 529 (1976).
[389] G. Haase and W. L. Dunkley, *J. Lipid Res.* **10,** 555 (1969).
[390] J. K. Beckman, S. M. Borowitz, H. L. Greene, and I. M. Burr, *Lipids* **23,** 559 (1988).

Singlet O_2 reacts directly with membrane lipids to give peroxides,[391] and illumination of unsaturated fatty acids in the presence of sensitizers of singlet O_2 formation, such as chlorophyll or porphyrins, promotes rapid peroxidation. Such reactions may be important in chloroplasts,[6,392] in patients suffering from porphyrias,[7] and in damage to plant tissues induced by the photosensitizing fungal toxin cercosporin.[393,394] There is some evidence (reviewed in Refs. 4 and 19) that singlet O_2 is formed during the complex degradation reactions of lipid peroxidation, by such processes as the self-reaction of peroxyl radicals:

$$LO_2\cdot + LO_2\cdot \rightarrow LOH + LO + {}^1O_2 \tag{40}$$

or the formation of excited-state carbonyls:

$$LO_2\cdot + LO_2\cdot \rightarrow LOH + LO^* + O_2 \tag{41}$$

which then transfer excitation energy onto O_2, yielding ${}^1O_2$, as they return to the ground state. This singlet O_2 could conceivably react directly with polyunsaturated fatty acid side chains to generate more peroxides.[395,396]

However, the authors are not sure that reaction of lipids with singlet O_2 to give peroxides should be described as initiation. In the absence of catalytic metal complexes to decompose peroxides, exposure to singlet O_2 will not cause the propagation of a chain reaction through the membrane. We have thus defined initiation in terms of hydrogen atom abstraction.

3.3. Enzymatic and Nonenzymatic Lipid Peroxidation

Scientists often loosely refer to iron salts and other iron complexes as initiating peroxidation, but in most cases what is really happening is that they are causing the decomposition of preformed lipid peroxides, as explained above, to generate alkoxyl or peroxyl radicals. Iron(II) and its complexes stimulate membrane peroxidation more than does iron(III). This may be explained by the greater solubility of Fe^{2+} salts in solution, the faster rate at which lipid peroxides are decomposed by Fe^{2+}, and the higher reactivity of the alkoxyl radicals so produced. Hence, the rate of peroxidation of purified membrane lipids or microsomal fractions in the presence of Fe(III) complexes can often be increased by the addition of

[391] H. R. Rawls and P. J. Van Santen, *Ann. N.Y. Acad. Sci.* **171**, 135 (1970).
[392] B. Halliwell, *Chem. Phys. Lipids* **44**, 327 (1987).
[393] M. E. Daub, *Plant. Physiol.* **69**, 1361 (1982).
[394] R. J. Youngman, P. Schieberle, H. Schnabl, W. Grosch, and E. F. Elstner, *Photobiochem. Photobiophys.* **6**, 109 (1983).
[395] E. K. Lai, K. L. Fong, and P. B. McCay, *Biochim. Biophys. Acta* **528**, 497 (1978).
[396] K. Sugioka and M. Nakano, *Biochim. Biophys. Acta* **423**, 203 (1976).

reducing agents such as ascorbic acid, certain thiols, O_2^--generating systems, or, in the case of microsomes, NADPH to provide electrons for NADPH–cytochrome-P-450 reductase.[358,387]

Peroxidation of cell or organelle membranes stimulated by Fe^{2+}, or Fe(III) plus ascorbate, *in vitro* does not require the activity of any enzymes. Another way of stimulating lipid peroxidation *in vitro* is to add lipid hydroperoxides or artificial organic hydroperoxides, such as *tert*-butyl hydroperoxide and cumene hydroperoxide. The decomposition of these to alkoxyl or peroxyl radicals accelerates the chain reaction of lipid peroxidation. Decomposition is again facilitated by metal ions and their complexes, for instance, by methemoglobin or cytochrome P-450 (see Ref. 383). Yet another way of generating peroxyl radicals to stimulate lipid peroxidation is the use of so-called azo initiators (reviewed in Ref. 397). These compounds decompose at a temperature-controlled rate to give a known flux of radicals. For example, 2,2'-azobis(2-amidinopropane) dihydrochloride (AAPH) is a water-soluble radical generator. On heating, it forms carbon-centered radicals that can react swiftly with O_2 to yield peroxyl radicals capable of abstracting H from membrane lipids

$$Cl^- H_2 \overset{+}{N} = C - C - N = N - C - C = \overset{+}{N} H_2 Cl^-$$

AAPH

$$A-N=N-A \xrightarrow{\text{heat}} N_2 + A \cdot + A \cdot \tag{42}$$
$$\text{carbon-centered}$$
$$\text{radicals}$$

$$A \cdot + O_2 \rightarrow AO_2 \cdot \tag{43}$$
$$\text{peroxyl radicals}$$
$$\text{(abstract H)}$$

Lipid-soluble azo initiators, containing more highly hydrophobic groups, also exist,[397] for example, 2,2'-azobis(2,4-dimethylvaleronitrile),

$$H_3C - CH - CH_2 - C - N = N - C - CH_2CHCH_3$$

azo group

Azo initiators have been much used in studies of antioxidants.[398–400]

[397] E. Niki, *Chem. Phys. Lipids* **44**, 227 (1987).
[398] D. D. M. Wayner, G. W. Burton, K. U. Ingold, L. R. C. Barclay, and S. J. Locke, *Biochim. Biophys. Acta* **924**, 408 (1987).
[399] G. W. Burton, A. Joyce, and K. U. Ingold, *Arch. Biochem. Biophys.* **221**, 281 (1983).
[400] T. Doba, G. W. Burton, and K. U. Ingold, *Biochim. Biophys. Acta* **835**, 298 (1985).

Peroxidation stimulated by adding Fe^{2+}, Fe(III)–ascorbate, azo initiators, or synthetic hydroperoxides to lipids is often called nonenzymatic lipid peroxidation. In nonenzymatic peroxidation the intramembrane peroxyl radicals survive long enough to be able to move to new fatty acid molecules since they can be readily intercepted and scavenged by a variety of chemically different antioxidants.[397-400] Lipid peroxide end products of the peroxidation are a complex racemic mixture of structural isomers of a wide range of fatty acyl peroxides.[357]

The term enzymatic lipid peroxidation is also found in the literature. The enzymes cyclooxygenase and lipoxygenase catalyze the controlled peroxidation of their fatty acid substrates to give hydroperoxides and endoperoxides that are stereospecific and have important biological functions (reviewed in Ref. 3). In the view of the authors, the term enzymatic lipid peroxidation should be reserved for these two types of enzyme. Unfortunately, the term has become more widely used in a way that has often led to confusion. For example, if some Fe(III) complexes are added to membrane lipids, the rate of peroxidation can be increased by generating O_2^- in the reaction mixture (e.g., by xanthine plus xanthine oxidase).[358] However, the enzyme-generated O_2^- is probably serving only to reduce Fe(III) to Fe^{2+}, which stimulates peroxidation. The function of the enzyme thus seems no different from that of, say, ascorbic acid in nonenzymatic peroxidation and the peroxidation products derived from each fatty acid side chain are complex and nonstereospecific.

The term enzymatic is also often used to describe microsomal lipid peroxidation. Microsomal fractions from several animal tissues (e.g., liver, kidney, and skin) undergo lipid peroxidation in the presence of NADPH and Fe(III) salts (often added as complexes with ADP or pyrophosphate, reagents which help to keep the iron in solution and render it redox active).[358,381,382] An antibody raised against the microsomal flavoprotein NADPH–cytochrome-P-450 reductase inhibits this peroxidation by more than 90%. During peroxidation, cytochromes b_5 and P-450 are attacked, the heme groups being degraded. Apparently the cytochrome-P-450 reductase enzyme, as well as reducing cytochrome P-450, can pass electrons to some Fe(III) complexes and so generate Fe^{2+}, which stimulates peroxidation.[358,367,381] Again, although an enzyme is involved, microsomal lipid peroxidation in the presence of iron complexes is in principle no different from nonenzymatic peroxidation, in that the enzyme is serving only to reduce the iron complexes and no stereospecific products are formed.

3.4. Consequences of Lipid Peroxidation

Extensive lipid peroxidation in biological membranes causes alterations in fluidity, falls in membrane potential, increased permeability to

TABLE III

FORMATION OF CARBONYL COMPOUNDS FROM FATTY ACIDS AND
MICROSOMAL FRACTIONS DURING LIPID PEROXIDATION IN THE
PRESENCE AND ABSENCE OF ADDED IRON SALTS[a]

System	$FeSO_4$ present	Malondialdehyde formed (nmol mg^{-1})	Other carbonyl compounds formed (nmol mg^{-1})
Oleic acid	No	—	—
	Yes	0	62
Linoleic acid	No	0	98
	Yes	12	317
Linolenic acid	No	—	—
	Yes	13	562
Arachidonic acid	No	19	588
	Yes	45	1728
Microsomes	No	5	10
	Yes	60	104

[a] Data were abstracted from McBrien and Slater.[403] Peroxidation was carried out in Tris-HCl buffer (pH 7.4) at 37° for 18 hr in shaken flasks in the presence or absence of added Fe^{2+} (20 μM $FeSO_4$). ADP–Fe^{2+}-stimulated peroxidation of microsomes was carried out for 30 min at 37°.

H^+ and other ions, and eventual rupture leading to release of cell and organelle contents,[3] such as lysosomal hydrolytic enzymes.[401] Some end products of metal ion-dependent lipid peroxide fragmentation are also cytotoxic.[402–409] Most attention is paid to malondialdehyde (sometimes called malonaldehyde), but this is only one of a great number of carbonyl compounds formed in peroxidizing systems and often is only a small percentage of the total products formed (e.g., Table III). Other toxic

[401] K. L. Fong, P. B. McCay, J. L. Poyer, B. B. Keele, and H. P. Misra, *J. Biol. Chem.* **248**, 7792 (1973).

[402] E. Schauenstein, H. Esterbauer, and H. Zollner, "Aldehydes in Biological Systems." Pion Press, London, 1977.

[403] D. C. H. McBrien and T. F. Slater, "Free Radicals, Lipid Peroxidation and Cancer." Academic Press, London, 1982.

[404] A. Benedetti, M. Comporti, R. Fulcieri, and H. Esterbauer, *Biochim. Biophys. Acta* **792**, 172 (1984).

[405] P. Winkler, W. Lindner, H. Esterbauer, E. Schauenstein, R. J. Schaur, and G. A. Khoschsorur, *Biochim. Biophys. Acta* **796**, 232 (1984).

[406] J. M. C. Gutteridge, P. Lamport, and T. L. Dormandy, *J. Med. Microbiol.* **9**, 105 (1976).

[407] S. R. Turner, J. A. Campbell, and W. S. Lynn, *J. Exp. Med.* **141**, 1437 (1975).

[408] T. W. Barrowcliffe, E. Gray, P. J. Kerry, and J. M. C. Gutteridge, *Thromb. Haemostasis* **52**, 7 (1984).

[409] H. Esterbauer, H. Zollner, and R. J. Schaur, *ISI Atlas Sci. Biochem.* **1**, 311 (1988).

aldehydes include 4,5-dihydroxydecenal[404] and 4-hydroxynonenal.[33,405,409] Lipid peroxides and/or cytotoxic aldehydes derived from them can block macrophage action, inhibit protein synthesis, kill bacteria, inactivate enzymes, cross-link proteins, generate thrombin, and act as chemotaxins for phagocytes.[402–410] Indeed, a number of "cytotoxic factors" found in sera from patients with connective tissue diseases may be, in part or in whole, lipid peroxidation products formed on storage or handling of the sera.[411,412] It has been argued that the "ferroxidase II" activity of human serum is an artifact related to peroxidation on storage and mishandling of serum samples.[101]

Not all lipid oxidation processes are harmful, and peroxidation products may play useful roles in the arachidonic acid cascade (reviewed in Ref. 3) and in the "wound response" of plant tissues (reviewed in Ref. 6). The production of lipid peroxides, and their fragmentation to toxic carbonyl compounds, in injured plant tissues may help the plant by killing bacteria or fungal spores entering the damaged site.[413]

3.5. Biomedical Significance of Lipid Peroxidation

Several assay techniques have been used to show that lipid peroxidation in animal tissues increases in a number of disease states and in poisoning by several toxins, although no one method is ideal for measuring lipid peroxidation (reviewed in Ref. 3). Behind many of these reports is the unspoken assumption that the disease or toxin causes increased lipid peroxidation, which is then responsible for the toxicity. In some cases, this assumption is probably valid (e.g., poisoning by some halogenated hydrocarbons).[414,415]

However, it was established many years ago by Barber[416] and by Stocks et al.[304] that disrupted tissues often undergo lipid peroxidation more quickly than healthy ones, especially after mechanical disruption. For example, lipid peroxides accumulate in a homogenate of brain much more quickly than they do in an isolated intact brain.[304] Reasons for this increased peroxidizability of damaged tissues include inactivation or dilution out of some antioxidants as well as the release of metal ions (especially iron) from intracellular storage sites and from metalloproteins

[410] G. Curzio, Free Radical Res. Commun. 5, 55 (1988).
[411] V. W. M. Van Hinsbergh, Atherosclerosis 53, 113 (1984).
[412] D. R. Blake, P. Winyard, D. G. I. Scott, S. Brailsford, A. Blann, and J. Lunec, Ann. Rheum. Dis. 44, 176 (1985).
[413] T. Galliard, in "Biochemistry of Wounded Plant Tissues" (G. Kahl, ed.), p. 156. de Gruyter, Berlin, 1978.
[414] M. Comporti, Lab. Invest. 53, 599 (1985).
[415] G. Poli, E. Albano, and M. U. Dianzani, Chem. Phys. Lipids 45, 117 (1987).
[416] A. A. Barber, Radiat. Res. 3, 33 (1963).

hydrolyzed by proteolytic enzymes released from damaged lysosomes.[302,304,309,416] Hence, the sequence of events:

Disease or toxin → cell damage or death → increased lipid peroxidation

can explain many of the reports of elevated lipid peroxidation in disease or toxicology.[302,417] Often, the lipid peroxidation can be inhibited without preventing the cellular injury,[417–426] as in the case of the lung damage produced by the herbicide paraquat (reviewed in Ref. 3). Thus, the increased lipid peroxidation reported as occurring in the muscles of patients with muscular dystrophy[427] or multiple sclerosis[428] may simply be a consequence of the tissue degeneration; there is no evidence that increased lipid peroxidation is in any way causative for these diseases in humans.[302,428a] The detailed experiments needed to prove that lipid peroxidation is a major cause of tissue injury, and not merely an accompaniment to it, are best illustrated by studies on the toxicity of halogenated hydrocarbons.[414,415] It must be shown that peroxidation precedes or accompanies the cell damage and that prevention of the peroxidation by antioxidants prevents the cell damage. Similarly, it must not be assumed that oxidative stress on an organism is necessarily reflected in lipid peroxidation: DNA damage (Section 1.6.3) and rises in the concentration of intracellular free calcium ions[207,424,429,429a] often seem to be earlier events mediating cell damage resulting from oxidant stress than is lipid peroxidation. The effects of oxidant stress on communication between cells is an area that deserves further investigation.[430]

[417] M. T. Smith, H. Thor, and S. Orrenius, *Biochem. Pharmacol.* **32,** 763 (1983).

[418] N. H. Stacey and C. D. Klaassen, *Toxicol. Appl. Pharmacol.* **58,** 8 (1981).

[419] P. Dogteran, J. F. Nagelkerke, J. van Steveninck, and G. J. Mulder, *Chem.–Biol. Interact.* **66,** 251 (1988).

[420] M. S. Sandy, D. Di Monte, and M. T. Smith, *Toxicol. Appl. Pharmacol.* **93,** 288 (1988).

[421] H. Muliawan, M. E. Scheulen, and H. Kappus, *Res. Commun. Chem. Pathol. Pharmacol.* **30,** 509 (1980).

[422] B. Risberg, L. Smith, M. H. Schoenberg, and M. Younes, *Eur. Surg. Res.* **19,** 164 (1987).

[423] R. S. Goldstein, D. A. Pasino, W. R. Hewitt, and J. B. Hook, *Toxicol. Appl. Pharmacol.* **83,** 261 (1986).

[424] L. Eklow-Lastbom, L. Rossi, H. Thor, and S. Orrenius, *Free Radical Res. Commun.* **2,** 57 (1986).

[425] L. Moller, *Toxicology* **40,** 285 (1986).

[426] H. Kappus. *Chem. Phys. Lipids* **45,** 105 (1987).

[427] M. I. S. Hunter and J. B. Mohamed, *Clin. Chim. Acta* **155,** 123 (1986).

[428] M. I. S. Hunter, B. C. Niemadim, and D. L. W. Davidson, *Neurochem. Res.* **10,** 1645 (1985).

[428a] A. Davison, G. Tibbits, Z. Shi, and J. Moon, *Mol. Cell. Biochem.* **84,** 199 (1988).

[429] G. Bellomo, H. Thor, and S. Orrenius, *J. Biol. Chem.* **262,** 1530 (1987).

[429a] S. Orrenius, D. J. McConkey, D. P. Jones, and P. Nicotera, *ISI Atlas Sci. Pharmacol.* **1,** 319 (1988).

[430] R. J. Ruch and J. E. Klaunig, *Toxicol. Appl. Pharmacol.* **94,** 427 (1988).

Of course, even if peroxidation is not the prime cause of damage, its occurrence as a consequence of damage may still be biologically important in worsening tissue injury, in view of the cytotoxicity of some end products of the peroxidation process, such as the unsaturated aldehyde hydroxynonenal.[402–410] Therapy with chain-breaking antioxidants or with chelating agents that can bind metal ions and prevent them from participating in free radical reactions (e.g., desferrioxamine) may be helpful in some diseases, but only to the extent that (i) further damage is actually done by the end products of the lipid peroxidation secondary to the tissue injury, and (ii) the antioxidants can reach the correct site of action. Antioxidants would certainly not be expected to cure most diseases!

However, there are two human conditions in which lipid peroxidation does appear to play a major role in the pathology. One of these is atherosclerosis, discussed in detail in Section 4.2 below. The other is the deterioration of the brain and spinal cord that occurs *after* traumatic or ischemic injury. Ischemic or traumatic injuries to the brain or spinal cord often result in more extensive tissue damage than do equivalent insults in other tissues, and free radical reactions have often been implicated in such damage.[305–308] The brain and nervous system may be especially prone to oxidant damage for a number of reasons. First, the membrane lipids are rich in polyunsaturated fatty acid side chains. Second, the brain is poor in catalase activity and has only moderate amounts of superoxide dismutase and glutathione peroxidase (reviewed in Ref. 305), perhaps because the intracellular O_2 concentration in the functioning brain is low. Third, some areas of the brain (e.g., globus pallidus, substantia nigra, circumventricular organs) are rich in iron. Brain iron has been poorly characterized, but it is likely that most of it is contained within ferritins or similar proteins.[431,432] Iron seems to play an important role in brain function since moderately severe anemia in children produces behavioral modifications and diminished learning ability.[431] Iron(II) salts are involved in the binding of serotonin to its receptors in rat brain, and there is evidence that serotonin- and dopamine-mediated behavioral responses induced by amphetamine or apomorphine are decreased in rats fed on an iron-deficient diet (reviewed in Ref. 431).

When brain cells are injured, some of the iron they contain can be easily released, but CSF has no significant iron-binding capacity (Section 2.1.2). There is a high concentration of ascorbic acid in the gray and white matter of the brain. (The choroid plexus has a specific active transport system that raises ascorbate concentrations in the CSF to about 10 times

[431] M. B. H. Youdim, "Brain Iron, Neurochemical and Behavioral Aspects." Taylor & Francis, London, 1988.

[432] M. M. Zaleska and R. A. Floyd, *Neurochem. Res.* **10**, 397 (1985).

the plasma level, and neural tissue cells have a second transport system that concentrates intracellular ascorbate even more.[180]) If the free iron content of brain or CSF were raised, the resulting ascorbate–iron salt mixtures might be expected to promote lipid peroxidation and ·OH formation. Indeed, injecting aqueous solutions of iron salts or of hemoglobin into the cortex of rats has been observed to cause transient focal epileptiform discharges, lipid peroxidation, and persistent behavioral and electrical abnormalities (reviewed in Ref. 433). Such rises in the free iron content may occur *in vivo* as a result of iron liberation from injured tissues and also as a result of bleeding in the damaged tissue; we have seen that hemoglobin can accelerate lipid peroxidation (Fig. 2). Homogenates of brain peroxidize rapidly *in vitro;* indeed, the peroxidation of ox brain homogenates has been used for many years as an assay to measure "antioxidant activity," and the peroxidation is largely prevented by iron-chelating agents such as desferrioxamine.[304] Cerebral ischemia, followed by reperfusion, could also stimulate free radical reactions, again probably because of metal release from injured cells. Since lipid peroxidation is known to generate cytotoxic products, such as hydroxynonenal, any lipid peroxidation after traumatic injury or ischemia in the brain would be expected to worsen the tissue damage.

The likelihood that iron-dependent radical reactions contribute significantly to further tissue injury in the brain and spinal cord after ischemia or traumatic injury led Hall *et al.*[306–308] to develop a series of antioxidants (the lazaroids) in which the capacity to bind iron ions in forms incapable of catalyzing free radical reactions is present within a hydrophobic steroid-based ring structure. These compounds have given promising results in several animal model systems.[306–308] For example, injury to the spinal cord often results in a drop in blood flow in the injured segment, which may lead to spinal cord degeneration. Intravenous injection of the lazaroid U74006F in cats before contusive injury to the spinal cord was found to diminish this drop in blood flow. U74006F has also been claimed to accelerate the neurological recovery of mice after concussive head injury and to diminish the effects of ischemia/reperfusion injury on the brains of cata or gerbils. These promising results seem to warrant further investigation.

It must be realized, however, that iron-dependent free radical damage need not necessarily be mediated by lipid peroxidation, and the protective effects of the lazaroids need not necessarily imply that lipid peroxidation is the tissue-damaging mechanism, since these chelators will presumably inhibit other iron-dependent free radical reactions.[433] Thus, several mech-

[433] B. Halliwell, *Acta Neurol. Scand.* **126**, 23 (1989).

anisms by which oxidant stress can damage neurons have been described (reviewed in Ref. 433); they include rises in intracellular free Ca^{2+} and increased production of excitatory amino acids. These are discussed further in Section 4.5.4.

4. Free Radicals and Human Disease

Free radicals and other oxidants are of great importance in the mechanism of action of many toxins (reviewed in Refs. 3, 414, 415, 426, and 434). Their involvement in aging[435] and in several disease states[436] has been considered in detail in recent articles: these have included iron-overload disease,[47] arthritis,[82,437] infection with malaria or other parasites,[438–440] neurological damage,[433] diabetic cataract,[441,442] dysbaric osteonecrosis,[443] silicosis,[444] asbestosis,[445] Down's syndrome,[446] and immune injury to kidney, liver, and lung.[183,436] There has also been much current interest in the role of oxidants in atherosclerosis and in ischemia/reoxygenation injury, and the authors feel it necessary to make some comments about the last two areas. Before going on to do this, it is worthwhile establishing a few basic principles. Table IV lists some of the conditions in which an involvement of radicals has been suspected (although not necessarily on the basis of good experimental evidence).

4.1. Does Increased Oxidant Formation Cause Any Human Disease?

It is tempting to attribute the cardiomyopathy seen in Keshan disease (chronic selenium deficiency) to lack of active glutathione peroxidase causing a failure to remove H_2O_2 and lipid peroxides at a sufficient rate *in*

[434] H. Sies, "Oxidative Stress." Academic Press, London, 1985.
[435] R. S. Sohal and R. G. Allen, *Adv. Free Radical Biol. Med.* **2,** 117 (1986).
[436] B. Halliwell (ed.), "Oxygen Radicals and Tissue Injury." Chapters by P. A. Ward (pp. 107, 115), I. A. Clark (p. 122), G. Cohen (p. 130), B. D. Watson (p. 81), E. D. Hall (p. 92), J. M. Braughler (p. 99), and H. S. Jacob (p. 57). Allen Press, Lawrence, Kansas, 1988.
[437] B. Halliwell, J. R. S. Hoult, and D. R. Blake, *FASEB J.* **2,** 2867 (1988).
[438] J. L. Vennerstrom and J. W. Eaton, *J. Med. Chem.* **31,** 1269 (1988).
[439] I. A. Clark, N. H. Hunt, and W. B. Cowden, *Adv. Parasitol.* **25,** 1 (1986).
[440] R. Docampo and S. J. Moreno, *Fed. Proc., Fed. Am. Soc. Exp. Biol.* **45,** 2471 (1986).
[441] S. P. Wolff, in "Diabetic Complications" (M. J. C. Crabbe, ed.), p. 167. Churchill Livingstone, Edinburgh, 1987.
[442] J. V. Hunt, R. T. Dean, and S. P. Wolff, *Biochem. J.* **256,** 205 (1988).
[443] G. R. N. Jones, *Free Radical Res. Commun.* **4,** 139 (1987).
[444] V. Vallyathan, X. Shi, N. S. Dalal, W. Irr, and V. Castranova, *Am. Rev. Respir. Dis.* **138,** 1213 (1988).
[445] S. A. Weitzman, J. F. Chester, and P. Graceffa, *Carcinogenesis* **9,** 1643 (1988).
[446] J. Kedziora and G. Bartosz, *Free Radical Biol. Med.* **4,** 317 (1988).

TABLE IV
Clinical Conditions in Which Involvement of Oxygen Radicals Has Been Suggested

Inflammatory–immune injury
  Glomerulonephritis (idiopathic, membranous)
  Vasculitis (hepatitis B virus, drugs)
  Autoimmune diseases
  Rheumatoid arthritis
Ischemia: reflow states
  Stroke/myocardial infarction/arrhythmias
  Organ transplantation
  Inflamed rheumatoid joint?
  Frostbite
  Dupuytren's contracture?
  Dysbaric osteonecrosis
Drug and toxin-induced reactions
  Iron overload
    Idiopathic haemochromatosis
    Dietary iron overload (Bantu)
    Thalassemia and other chronic anemias
      treated with multiple blood transfusions
    Nutritional deficiencies (kwashiorkor)
  Alcoholism
    Alcohol-induced iron overload
    Alcoholic myopathy
  Radiation injury
    Nuclear explosions
    Accidental exposure
    Radiotherapy
    Hypoxic cell sensitizers
  Aging
    Disorders of premature aging
  Red blood cells
    Phenylhydrazine
    Primaquine, related drugs
    Lead poisoning
    Protoporphyrin photoxidation
    Malaria
    Sickle cell anemia
    Favism
    Fanconi's anemia
    Hemolytic anemia of prematurity
  Lung
    Cigarette smoke effects
    Emphysema
    Bronchopulmonary dysplasia
    Oxidant pollutants (O_3, NO_2)
    ARDS[a] (some forms)
    Mineral dust pneumoconiosis
    Asbestos carcinogenicity
    Bleomycin toxicity
    SO_2 toxicity

    Paraquat toxicity
Heart and cardiovascular system
  Alcohol cardiomyopathy
  Keshan disease (selenium deficiency)
  Atherosclerosis
  Adriamycin cardiotoxicity
Kidney
  Autoimmune nephrotic syndromes
  Aminoglycoside nephrotoxicity
  Heavy metal nephrotoxicity (Pb, Cd, Hg)
Gastrointestinal tract
  Endotoxic liver injury
  Halogenated hydrocarbon liver injury
    (e.g., bromobenzene, carbon tetra-
    chloride, halothane)
  Diabetogenic action of alloxan
  Pancreatitis
  NSAID-induced[b] gastrointestinal tract
    lesions
  Oral iron poisoning
Brain/nervous system/neuromuscular dis-
  orders
  Hyperbaric oxygen
  Vitamin E deficiency
  Neurotoxins, including lead
  Parkinson's disease
  Hypertensive cerebrovascular injury
  Neuronal ceroid lipofuscinoses
  Allergic encephalomyelitis and other
    demyelinating diseases
  Aluminum overload (Alzheimer's dis-
    ease?)
  Potentiation of traumatic injury
  Muscular dystrophy
  Multiple sclerosis
Eye
  Cataractogenesis
  Ocular hemorrhage
  Degenerative retinal damage
  Retinopathy of prematurity (retrolental
    fibroplasia)
  Photic retinopathy
Skin
  Solar radiation
  Thermal injury
  Porphyria
  Hypericin, other photosensitizers
  Contact dermatitis

[a] Adult respiratory distress syndrome.
[b] Nonsteroidal antiinflammatory drug-induced.

vivo. Although it is not strictly proved that this is the mechanism of tissue injury, patients suffering from Keshan disease do show very low selenium-dependent glutathione peroxidase activity in blood and organs.[447,448] It is also possible that some cancers may originate as a result of faulty repair following DNA damage produced by free radicals,[3,149,449,450] and that the vessel weakness in retrolental fibroplasia is due to lipid peroxidation, since it can be ameliorated by vitamin E (reviewed in Ref. 3). The hemolysis formerly seen in some premature babies is now rare, probably because of the fortification of infant formulas with vitamin E, but the condition may be related to lipid peroxidation.

In most diseases, however, increased oxidant formation is a consequence of the disease activity. For example, infiltration of a large number of neutrophils into a localized site, followed by activation of these cells to generate O_2^- and H_2O_2, can produce an intense localized oxidative stress. This appears to happen in rheumatoid arthritis[436] and in some forms of the adult respiratory distress syndrome.[183] The question then arises as to what caused the phagocyte infiltration and activation in the first place. Oxidants generated as a result of phagocyte activation might sometimes lead to more phagocyte activation. Thus, exposure of IgG to $\cdot$OH or to HOCl in the inflamed rheumatoid joint has been claimed to generate an aggregated IgG that can cause further activation of neutrophils.[451,452] Oxidants might also lead to more activation of complement.[453] In addition, many forms of tissue injury, including mechanical disruption, can cause increased free radical reactions by inactivating cellular antioxidants and by liberating transition metal ions from their sites of sequestration within cells (Section 3.5).

4.2. Does Increased Oxidant Formation Matter?

If increased oxidant formation is usually a consequence of disease, does it then make a significant contribution to the disease pathology, or is it just an epiphenomenon? The answer probably differs in different diseases. For example, increased iron-dependent free radical generation after impact injury to the brain may well contribute to postinjury degen-

[447] A. T. Diplock, *Am. J. Clin. Nutr.* **45,** 1313 (1987).
[448] O. A. Levander, *Annu. Rev. Nutr.* **7,** 227 (1987).
[449] P. A. Cerutti, *Science* **227,** 375 (1985).
[450] G. J. Chellman, J. S. Bus, and P. K. Working, *Proc. Natl. Acad. Sci. U.S.A.* **83,** 8087 (1986).
[451] J. Lunec, D. R. Blake, S. J. McCleary, S. Brailsford, and P. A. Bacon, *J. Clin. Invest.* **76,** 2084 (1985).
[452] H. E. Jasin, *J. Clin. Invest.* **81,** 6 (1988).
[453] K. T. Oldham, K. S. Guice, G. O. Till, and P. A. Ward, *Surgery* **104,** 272 (1988).

eration of the tissue (Section 3.5). By contrast, the increased lipid peroxidation demonstrated in the wasted muscles of patients with muscular dystrophy may simply be a consequence of the tissue damage and make no contribution to it; there is no evidence that antioxidants are beneficial in this disease (Section 3.5). Bearing these points in mind, let us now examine some diseases in which the evidence for an involvement of oxidants in the disease pathology is stronger than average.

4.3. Atherosclerosis

Cardiovascular disease is the chief cause of death in the United States and Europe. Many important forms of human cardiac disease, together with many cases of localized cerebral ischemia (stroke), are secondary to the condition of atherosclerosis, a disease of arteries that is characterized by a local thickening of the intima, or innermost part of the vessel (reviewed in Refs. 454–457). In general, three types of thickening are recognized. Fatty streaks are slightly raised, yellow, narrow, longitudinally lying areas. They are characterized by the presence of foam cells, lipid-laden distorted cells that can arise both from endogenous smooth muscle cells and from macrophages (formed from monocytes that entered the arterial wall from the blood). Fatty streaks probably serve as precursors of fibrous plaques. These are approximately rounded raised lesions, usually off-white in color and perhaps a centimeter in diameter. A typical fibrous plaque consists of a fibrous cap (composed mostly of smooth muscle cells and dense connective tissue containing collagen, elastin, proteoglycans, and basement membranes) covering an area rich in macrophages, smooth muscle cells, and T lymphocytes and a deeper necrotic core that contains cellular debris, extracellular lipid deposits, and cholesterol crystals. In general, fibrous plaques have already begun to obstruct the arterial lumen. Complicated plaques are probably fibrous plaques that have been altered by necrosis, calcium deposition, bleeding, and thrombosis. Plaques cause disease by limiting blood flow to a region of an organ such as the heart or brain. A stroke or heart attack (myocardial infarction) occurs when the lumen of an essential artery becomes completely occluded, usually by a thrombus forming at the site of a plaque.[454–457]

Fatty streaks are regularly present in the arteries of children on Western diets. In the disease familial hypercholesterolemia, the accumulation of plaques is much increased, and myocardial infarcts can be observed as

[454] H. C. McGill, Jr., Clin. Chem. 33, B33 (1988).
[455] J. M. Munro and R. S. Cotran, Lab. Invest. 58, 249 (1988).
[456] R. Ross, N. Engl. J. Med. 314, 488 (1986).
[457] J. Babiak and L. L. Rudel, Baillieres Clin. Endocrinol. Metab. 1, 515 (1987).

early as 2 years of age. Very low density lipoproteins (VLDL) enter the circulation from the liver. Low density lipoproteins (LDL) are largely produced from VLDL in the circulation by several processes, including removal of some triglyceride from VLDL as they circulate through the tissues. LDL are rich in cholesterol esters and supply cholesterol to body tissues; LDL bind to receptors on the surface of cholesterol-requiring cells and are internalized, releasing cholesterol within the cells. In familial hypercholesterolemia, the receptors are defective, so that blood cholesterol (and LDL) levels become very high. The increased accumulation of atherosclerotic plaques naturally attracted attention to the possible role of cholesterol in leading to atherosclerosis, especially as it had also been observed that healthy people with high blood LDL cholesterol develop atherosclerosis at an accelerated rate.[453-457]

The origin of atherosclerosis is uncertain, but a popular current theory is that it begins with damage, by some mechanism (possibly hemodynamic), to the vascular endothelium.[456,457] This could be followed by attachment of monocytes from the circulation,[458] which develop into macrophages within the vessel wall. Activated monocytes and macrophages could injure neighboring cells by secreting O_2^-, H_2O_2, and hydrolytic enzymes, and factors released by macrophages[459] can stimulate the proliferation of smooth muscle cells. Macrophages also release platelet-stimulating factors, and adherence of platelets to injured endothelium might cause release of other agents that encourage proliferation of smooth muscle cells, although the precise role of platelets in atherosclerosis is an area of controversy (discussed in Ref. 460).

What roles could be played by oxidants in atherogenesis? First, as mentioned above, activation of macrophages or their monocyte precursors, for example, in a fatty streak, could injure neighboring cells and lead to more endothelial damage and damage to smooth muscle cells.[459] Second, normal macrophages possess some LDL receptors, but if LDL is peroxidized it is recognized by separate receptors known as the acetyl-LDL receptors or the scavenger receptors.[461] LDL bound to these receptors is taken up with enhanced efficiency, so that cholesterol rapidly accumulates within the macrophage and may convert it to a foam cell.[459,461] Arterial endothelial cells, smooth muscle cells, and macrophages are themselves known to be capable of oxidizing LDL so that

[458] M. T. Quinn, S. Parthasarathy, and D. Steinberg, *Proc. Natl. Acad. Sci. U.S.A.* **85,** 2805 (1988).
[459] M. J. Mitchinson and R. Y. Ball, *Lancet* **2,** 146 (1987).
[460] L. K. Curtiss, A. S. Black, Y. Takagi, and E. F. Plow, *J. Clin. Invest.* **80,** 367 (1987).
[461] M. E. Haberland and A. M. Fogelman, *Am. Heart J.* **113,** 573 (1987).

macrophages will internalize it faster. The modification process may involve the derivatization of lysine residues of the protein moiety of LDL by lipid peroxidation products, such as cytotoxic aldehydes.[458,462–466] LDL modification by these cells *in vitro* appears to require addition of traces of iron or copper ions, whose origin has not yet been identified *in vivo*, although they could conceivably form by breakdown of metalloproteins (e.g., hemoglobin or ceruloplasmin) within the plaque (reviewed in Ref. 102). Incubation of LDL with aqueous extracts of cigarette smoke has also been shown to cause oxidative modification, leading to increased LDL uptake by isolated macrophages.[467] Cigarette smoke is known to be rich in oxidants and metal ions.[468]

Third, any lipid peroxides present in LDL (e.g., from the diet) could conceivably contribute to the initial endothelial cell damage that is thought to start off the whole process, for example, by worsening hemodynamic damage. As an example, studies *in vitro* have shown that linoleic acid hydroperoxide increases the permeability of endothelial cell monolayers to macromolecules.[469] Fourth, it has been suggested that products formed in peroxidized LDL, such as lysophosphatidylcholine,[458] might act as chemotactic factors for blood monocytes, encouraging their recruitment into an atherosclerotic lesion. Fifth, low (submicromolar) concentrations of peroxides might accelerate cyclooxygenase- and lipoxygenase-catalyzed reactions in endothelium and in any platelets present, leading to enhanced formation of eicosanoids.[470] Oxidized LDL may also stimulate the production of eicosanoids by macrophages.[471] There have been several speculations that oxidation products of cholesterol might also be involved in atherogenesis (e.g., Ref. 472); cholesterol is oxidized to a

[462] G. Jurgens, H. F. Hoff, G. M. Chisolm, and H. Esterbauer, *Chem. Phys. Lipids* **45**, 315 (1987).
[463] H. Esterbauer, G. Jurgens, O. Quehenberger, and E. Koller, *J. Lipid Res.* **28**, 495 (1987).
[464] M. T. Quinn, S. Parthasarathy, L. G. Fong, and D. Steinberg, *Proc. Natl. Acad. Sci. U.S.A.* **84**, 2995 (1987).
[465] U. P. Steinbrecher, *J. Biol. Chem.* **262**, 3603 (1987).
[466] C. P. Sparrow, S. Parthasarathy, and D. Steinberg, *J. Lipid Res.* **29**, 745 (1988).
[467] M. Yokode, T. Kita, H. Arai, C. Kawai, S. Narumiya, and M. Fujiwara, *Proc. Natl. Acad. Sci. U.S.A.* **85**, 2344 (1988).
[468] D. F. Church and W. A. Pryor, *Environ. Health Perspect.* **64**, 111 (1985).
[469] B. Hennig, C. Enoch, and C. K. Chow, *Arch. Biochem. Biophys.* **248**, 353 (1986).
[470] M. A. Warso and W. E. M. Lands, *Br. Med. Bull.* **39**, 277 (1983).
[471] M. Yokoda, T. Kita, Y. Kikawa, T. Ogorochi, S. Narumiya, and C. Kawai, *J. Clin. Invest.* **81**, 720 (1988).
[472] A. W. Bernheimer, W. G. Robinson, R. Linder, D. Mullins, Y. K. Yip, N. S. Cooper, I. Seidman, and T. Uwajima, *Biochem. Biophys. Res. Commun.* **148**, 260 (1987).

wide variety of products in peroxidizing lipid systems,[17,473] and oxidized cholesterol has been reported to be toxic to arterial smooth muscle cells.[472,473]

It might be supposed, therefore, that elevated blood lipid concentrations could lead to elevated blood lipid peroxide concentrations, contributing to endothelial injury and accelerating the whole process of atherogenesis. Ca^{2+} accumulation in plaques may also be significant,[474] since rises in intracellular Ca^{2+} within cells are important damaging events during oxidant stress.[207,424,429a] Thus, *is* there more lipid peroxide in patients with atherosclerosis? Studies with antibodies,[474a] and the deposition of ceroid, in atherosclerotic lesions are strongly suggestive that peroxidation is occurring within the plaque,[475,476] but studies on plasma have not given clear-cut answers. One problem that has bedeviled studies on this point is that storage or mishandling of plasma samples leads to peroxidation of lipoproteins[101,477,478] because of release of low molecular mass copper complexes from ceruloplasmin.[102] Several studies identifying so-called cytotoxic factors in the plasma of patients with various diseases have been lead astray by this artifact. It must be remembered that plasma contains powerful preventative (metal ion-binding) and chain-breaking antioxidants[67,479] which limit lipid peroxidation; perhaps depletion or failure of these protective mechanisms is also involved in the pathogenesis of atherosclerosis.[102,480] A second problem is that most of the assays that have been used to examine plasma samples are flawed: this is discussed in detail in Section 4.4 below.

If oxidants do initiate atherosclerosis, or contribute to its pathology, then an increased intake of antioxidants (especially lipid-soluble chain-breaking antioxidants that accumulate in lipoproteins) might be expected to have a beneficial effect.[480] Probucol, a drug used clinically to lower

[473] A. Sevanian and A. R. Peterson, *Food Chem. Toxicol.* **24**, 1103 (1986).

[474] A. Fleckenstein, M. Frey, J. Zorn, and G. Fleckenstein-Grun, *Trends Pharmacol. Sci.* **8**, 496 (1987).

[474a] D. Steinberg, S. Parthasarathy, T. E. Carew, J. C. Khoo, and J. L. Witztum, *N. Engl. J. Med.* **320**, 915 (1989).

[475] R. Y. Ball, K. H. L. Carpenter, and M. J. M. Mitchinson, *Arch. Pathol. Lab. Med.* **111**, 1134 (1987).

[476] R. Y. Ball, K. L. H. Carpenter, J. H. Enright, S. L. Hartley, and M. J. Mitchinson, *Br. J. Exp. Pathol.* **68**, 427 (1987).

[477] D. M. Lee, *Biochem. Biophys. Res. Commun.* **95**, 1663 (1980).

[478] J. Schuh, G. F. Fairclough, Jr., and R. H. Haschemeyer, *Proc. Natl. Acad. Sci. U.S.A.* **75**, 3173 (1978).

[479] J. M. C. Gutteridge and B. Halliwell, *in* "Cellular Antioxidant Mechanisms" (C. K. Chow, ed.), Vol. 2, p. 1. CRC Press, Florida, 1988.

[480] K. F. Gey, G. B. Brubacher, and H. B. Stahelin, *Am. J. Clin. Nutr.* **45**, 1368 (1987).

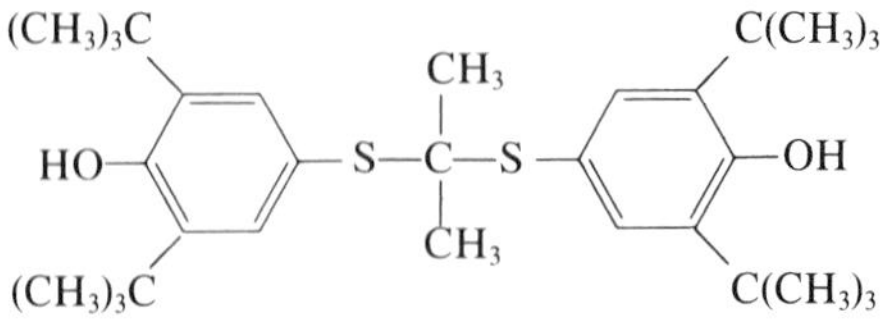

FIG. 3. Structure of probucol or 4,4'-[(1-methylethylidene)bis(thio)]bis[2,6-bis(1,1-di-methylethyl)phenol]. The phenolic OH groups confer chain-breaking antioxidant activity on the molecule.

blood cholesterol levels, is a powerful chain-breaking antioxidant inhibitor of lipid peroxidation *in vitro,* as would be expected because its structure (Fig. 3) is similar to that of several phenolic antioxidants. It has been found that the antiatherogenic effect of probucol in rabbits is far greater than expected from its cholesterol-lowering ability, suggesting that its antioxidant activity might also be biologically relevant, by inhibiting the local oxidative modification of LDL in the arterial wall.[481] LDL isolated from humans after probucol administration resist peroxidation *in vitro,* consistent with the above proposals.

4.4. What Assays Should Be Used to Measure Lipid Peroxides in Human Body Fluids?

The two most popular assays used to date to measure lipid peroxidation end products in human body fluids, diene conjugation and the TBA test, are seriously flawed.

4.4.1. Diene Conjugation. Oxidation of unsaturated fatty acid side chains is accompanied by the formation of conjugated diene structures that absorb ultraviolet light in the wavelength range 230–235 nm (reviewed in Refs. 3, 348, and 349). Measurement of this UV absorbance is an extremely useful index of peroxidation in studies on pure lipids, but it often cannot be used directly on biological materials because many of the other substances present, such as heme proteins, absorb strongly in the ultraviolet and create a high background. This can be overcome by extracting diene conjugates into certain organic solvents, although many such solvents have significant UV absorbance, thus reducing the accuracy of spectrophotometric measurements. Corongiu *et al.*[482,483] have improved the sensitivity of the diene conjugation method by using second-

[481] T. E. Carew, D. C. Schwenke, and D. Steinberg, *Proc. Natl. Acad. Sci. U.S.A.* **84,** 7725 (1987).

[482] F. P. Corongiu and A. Milia, *Chem.–Biol. Interact.* **44,** 289 (1983).

[483] F. P. Corongiu, G. Poli, M. U. Dianzani, K. H. Cheeseman, and T. F. Slater, *Chem.–Biol. Interact.* **59,** 147 (1986).

derivative spectroscopy. In the second derivative $(d^2A/d^2\lambda)$ spectrum, the shoulder that appears in the ordinary (A/λ) spectrum translates into a sharp minimum peak that is easily measurable and is a good index of the conjugated dienes present. The increased resolution of second-derivative spectroscopy may allow discrimination between different conjugated diene structures present.[483,483a]

Although conjugated diene methods have often been successfully used to study peroxidation in animal body fluids and tissue extracts, their application to human body fluids has produced serious problems. Dormandy and Wickens[484] used HPLC techniques to separate UV-absorbing "diene conjugates" from human body fluids and reported that the UV-absorbing material consists almost entirely of a non-oxygen-containing isomer of linoleic acid, 9(*cis*),11(*trans*)-octadecadienoic acid. They proposed that this compound is produced by reaction of the carbon-centered radicals, obtained when H is abstracted from linoleic acid, with protein.[484] Such a reaction is certainly possible, but examination of reaction rate constants strongly suggests that reaction with oxygen is the preferred fate of carbon-centered radicals, except at very low O_2 concentrations. In any case, peroxidation of biological membranes produces carbon-centered radicals from several fatty acids, not only linoleic acid, and would not be expected to give only a 9(cis),11(trans) isomer. Further, this UV-absorbing product is not found in the plasma of animals subjected to oxidant stress (e.g., rats given the hepatotoxin bromotrichloromethane, a potent inducer of lipid peroxidation). Thus, Thompson and Smith[485] and the authors[47,349] have argued that 9,11-octadecadienoic acid is most unlikely to arise by lipid peroxidation. It may be ingested in food and produced by the metabolism of gut bacteria.[349,485,486]

The fact that the major UV-absorbing material in human body fluids does not appear to arise by lipid peroxidation means that application of simple diene conjugation methods to human body fluids, or to extracts of them, is probably not measuring lipid peroxidation, although diene conjugation methods still seem applicable to animal body fluids.[484] Thus, the authors cannot recommend use of diene conjugation techniques on human material.

4.4.2. Thiobarbituric Acid (TBA) Test. The TBA test is one of the oldest and most frequently used tests for measuring the peroxidation of fatty acids, membranes, and food products. It is the easiest method to use

[483a] M. Grootveld and R. Jain, *Free Radical Res. Commun.* **6,** 271 (1989).
[484] T. L. Dormandy and D. G. Wickens, *Chem. Phys. Lipids* **45,** 353 (1987).
[485] S. Thompson and M. T. Smith, *Chem.–Biol. Interact.* **55,** 357 (1985).
[486] A. J. E. Green, B. J. Starkey, S. P. Halloran, G. McKee, C. J. G. Sutton, B. T. B. Manners, and A. W. Walker, *Lancet* **2,** 309 (1988).

(the material under test is merely heated with thiobarbituric acid under acidic conditions, and the formation of a pink color measured at or close to 532 nm), and it can be applied to crude biological systems. Unfortunately, the simplicity of the test has led many scientists into using it as an index of peroxidation without understanding exactly what it measures, and it has been widely applied to measure "lipid peroxides" in plasma or tissue samples from patients with various diseases (reviewed in Ref. 487).

Small amounts of free malondialdehyde (MDA) are formed during the peroxidation of most membrane systems; greater amounts have been detected in peroxidizing rat liver microsomes[488] (Table III). MDA can react in the TBA test to generate a colored product, an adduct of 2 mol of TBA with 1 mol of MDA. In acid solution, the product absorbs light (λ_{max} 532 nm) and fluoresces (λ_{max} 553 nm), and it is readily extractable into organic solvents such as 1-butanol.[349]

Because the TBA test is calibrated with MDA, the results are often expressed in terms of the amount of MDA produced in a given time. This has sometimes given the impression that the TBA test detects only free MDA, and so measures the amount of free MDA in the peroxidizing lipid system. Thus, the molar extinction coefficient of the MDA–TBA adduct (1.54×10^5 at 532 nm) has been used to calculate the amount of "MDA formed." However, the amount of free MDA produced in most peroxidizing lipid systems (with the possible exception of rat liver microsomes)[488] is low and would be insufficient to give a significant color yield in the TBA test. Indeed, it was shown many years ago that the bulk of the "MDA" detected in the TBA test was nonvolatile and therefore could not be present in the sample being assayed, but instead was formed by decomposition of lipid peroxides and further peroxidation during the acid-heating stage of the test itself (reviewed in Refs. 489–492). Peroxide decomposition requires the presence of iron ions contaminating the TBA reagents: this can lead to artifacts in studies of the action of metal-chelating agents on lipid peroxidation, since they can also affect color development in the TBA test itself.[489,493] Some scientists add antioxidants, such as butylated hydroxytoluene, to the TBA test to minimize peroxidation during the test (reviewed in Ref. 494).

[487] K. Yagi, *Chem. Phys. Lipids* **45**, 337 (1987).
[488] H. Esterbauer and T. F. Slater, *IRCS Med. Sci. Biochem.* **9**, 749 (1981).
[489] J. M. C. Gutteridge and G. J. Quinlan, *J. Appl. Biochem.* **5**, 293 (1983).
[490] J. M. C. Gutteridge and T. R. Tickner, *Anal. Biochem.* **91**, 250 (1978).
[491] J. M. C. Gutteridge and T. R. Tickner, *Biochem. Med.* **19**, 127 (1978).
[492] E. N. Frankel and W. E. Neff, *Biochim. Biophys. Acta* **754**, 264 (1983).
[493] E. D. Wills, *Biochim. Biophys. Acta* **84**, 475 (1964).
[494] J. A. Buege and S. D. Aust, this series, Vol. 30, p. 302.

In addition to this problem, several compounds other than MDA give products that absorb at, or close to, 532 nm on heating with TBA. For example, bile pigments in plasma react with TBA to produce colored products,[491] urine contains several TBA-reactive substances,[490] and various unsaturated aldehydes also react in the TBA test.[495] Thus, simple measurement at 532 nm after a TBA test could include contributions from these substances, although fluorescence measurements can usually distinguish the products they form from the real TBA–MDA adduct. A similar distinction can be achieved by separating the TBA–MDA adduct from the reaction mixture before measurement. This is best done by modern HPLC or gas-chromatographic methods.[496–500] Even so, it must be noted that exposure of several carbohydrates and amino acids to hydroxyl radicals, produced by ionizing radiation or metal ion–H_2O_2 systems, yields products that do give a genuine TBA–MDA adduct on heating with TBA.[134,489,491,501] The exact amount of color developed with these alternative compounds depends on the type and strength of acid used in the TBA test, and on the time of heating.[134,489,501] Application of the TBA assay to human body fluids will also measure endoperoxides produced enzymatically by the prostaglandin synthesis pathway.[502]

The lack of specificity of the TBA assay when applied to human body fluids, such as plasma, is shown by the work of several groups. Lands and co-workers,[503] using a specific enzyme method based on the ability of lipid peroxides to activate the enzyme cyclooxygenase, measured a mean peroxide content in human plasma of around 0.5 μM, whereas expression of the results from a TBA test in terms of peroxide equivalents gave a mean value of 38 μM. Largilliere and Melancon[496] found considerable TBA reactivity, but no free MDA, in human plasma. In animals, plasma TBA reactivity varies with species, strain, age, sex, food intake, and stress.

[495] H. Kosugi, T. Kato, and K. Kikugawa, *Anal. Biochem.* **165,** 456 (1987).
[496] C. Largilliere and S. B. Melancon, *Anal. Biochem.* **170,** 123 (1988).
[497] L. W. Wu, L. Latriano, S. Duncan, R. A. Hartwick, and G. Witz, *Anal. Biochem.* **156,** 326 (1986).
[498] D. R. Janero and B. Burghardt, *Lipids* **23,** 452 (1988).
[499] H. Hughes, C. V. Smith, J. O. Tsokos-Kuhn, and J. R. Mitchell, *Anal. Biochem.* **152,** 107 (1986).
[500] R. P. Bird, S. S. O. Hung, M. Hadley, and H. H. Draper, *Anal. Biochem.* **128,** 240 (1983).
[500a] K. L. H. Carpenter, R. Y. Ball, K. M. Ardeshna, J. P. Bridman, J. H. Enright, S. L. Hartley, S. Nicolson, and M. J. Mitchinson, *in* "Lipofuscin: State of the Art" (I. Z. Nagy, ed.), p. 245. Elsevier, Amsterdam, 1987.
[501] J. M. C. Gutteridge, *Int. J. Biochem.* **14,** 649 (1982).
[502] T. Shimizu, K. Kondo, and O. Yayaishi, *Arch. Biochem. Biophys.* **206,** 271 (1981).
[503] P. J. Marshall, M. A. Warso, and W. E. M. Lands, *Anal. Biochem.* **145,** 192 (1985).

Thus, the authors do not recommend the TBA test for measuring "lipid peroxides" or "aldehydes" in human biological material.

It should be noted that more specific assays of the peroxide content of human body fluids have not given consistent results to date. Thus, the cyclooxygenase-based assay found 0.5 μM "total lipid peroxide" in human plasma,[503] whereas an HPLC-based separation followed by a microperoxidase-dependent determination found no peroxides at all in human plasma, even in patients with adult respiratory distress syndrome[64] (upper limit 0.03 μM). The results of the application of more specific assays to samples from patients with atherosclerosis are awaited with interest.

4.5. Ischemia/Reoxygenation Injury

4.5.1. Intestinal Injury. Damage to the heart or brain by completely depriving a portion of the tissue of O_2 (ischemia) is a major cause of death in Western society. Atherosclerosis, leading to blockage of an essential artery, often by thrombus formation, is usually the culprit (Section 4.3). Severe restriction of blood flow, leading to O_2 concentrations lower than normal (hypoxia) but not complete O_2 deprivation, can also result if the blocked artery is the major, but not the only, source of blood to the tissue in question.

Tissues made hypoxic or ischemic survive for a variable time, depending on the tissue in question. Thus, skeletal muscle is fairly resistant to hypoxic injury, whereas brain is very sensitive and does not usually survive ischemia for more than a few minutes. However, any tissue made ischemic for a sufficient period will be irreversibly injured. Tissues respond to ischemia in a number of ways. Early responses may include increased glycogen degradation and anaerobic glycolysis, leading to lactate production and acidosis. ATP levels begin to fall, and AMP is degraded to cause an accumulation of hypoxanthine (Fig. 4). Eventually, glycolysis slows, and membrane damage becomes visible under the microscope.

If the period of ischemia or hypoxia is insufficiently long to injure the tissue irreversibly, much tissue can be salvaged by reperfusing the tissue with blood and reintroducing O_2 and nutrients. In this situation, reperfusion is a beneficial process overall. However, McCord[504] and Parks and Granger[505] showed in the early 1980s that reintroduction of O_2 to an ischemic or hypoxic tissue can cause an additional insult to the tissue (reoxygenation injury) that is, in part, mediated by oxygen radicals. The relative importance of reoxygenation (often called reperfusion) injury depends on

[504] J. M. McCord, *Fed. Proc., Fed. Am. Soc. Exp. Biol.* **46**, 2402 (1987).
[505] D. A. Parks and D. N. Granger, *Am. J. Physiol.* **250**, G749 (1986).

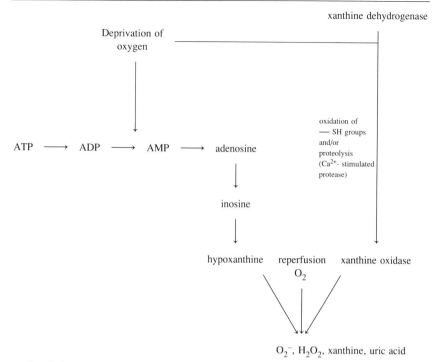

FIG. 4. A suggested mechanism for tissue injury on reoxygenation of ischemic or hypoxic tissues. (Modified from Ref. 504.)

the duration of ischemia/hypoxia. If this is sufficiently long, the tissue is irreversibly injured and will die, so reoxygenation injury is not important (one cannot further injure a dead tissue). However, for a relatively brief period of ischemia/hypoxia, the reoxygenation injury component may become more important, and the amount of tissue remaining undamaged can be significantly increased by including oxidant scavengers in the reoxygenation fluid. The meaning of "relatively brief" in this context depends on the tissue in question, whether one is dealing with ischemia or hypoxia and, if the latter, what degree of hypoxia is actually achieved.

Although the enzyme xanthine oxidase is frequently used as a source of superoxide (O_2^-) in experiments *in vitro,* almost all the xanthine-oxidizing activity present in normal tissues is a dehydrogenase enzyme that transfers electrons not to O_2, but to NAD^+, as it oxidizes xanthine or hypoxanthine into uric acid (reviewed in Ref. 504). When tissues are disrupted, the dehydrogenase becomes converted to the oxidase enzyme by oxidation of essential thiol groups and/or by limited proteolysis. Hence, on purification, the enzyme appears as an oxidase, producing O_2^-

and H_2O_2 when xanthine or hypoxanthine is oxidized. McCord and others (see Refs. 504 and 505) suggested that the conversion of dehydrogenase to oxidase occurs in ischemic/hypoxic tissue, by proteolytic attack (although conversion by oxidation of enzyme thiol groups is also possible). The depletion of ATP in hypoxic tissue causes hypoxanthine accumulation. This hypoxanthine can be oxidized by the xanthine oxidase when the tissue is oxygenated, causing rapid generation of O_2^- and H_2O_2, which might lead to severe tissue damage. The hypothesis of McCord *et al.* is summarized in Fig. 4.

The first evidence supporting the hypothesis came from studies on intestine. Partial arterial occlusion in a segment of cat small intestine (hypoxia), followed by reperfusion, causes gross histologically observable damage to the tissue and increases intestinal vascular permeability. Intravenous administration of SOD, or oral administration of allopurinol (an inhibitor of xanthine oxidase) to the animals before removal of the arterial occlusion, offered protection against damage. Infusion of a mixture of hypoxanthine and xanthine oxidase into the arterial supply of a segment of normal cat intestine greatly increased vascular permeability, an effect that was decreased by the presence of SOD or dimethyl sulfoxide (DMSO) in the infusion (reviewed in Ref. 504).

The effect of DMSO, a powerful scavenger of $\cdot OH$, perhaps suggests that this radical was involved in the damage, although inhibition by a single scavenger is insufficient evidence to prove $\cdot OH$ involvement (Section 1.6), and DMSO has many pharmacological properties unrelated to radical scavenging.[275] Desferrioxamine, which suppresses iron-dependent $\cdot OH$ generation, also protects against the injury, suggesting that iron ions are required.[506] Iron ions able to catalyze $\cdot OH$ production could arise as a result of cellular injury and metal ion release from intracellular storage sites, proteolytic digestion of metalloproteins, and/or breakdown by H_2O_2 of hemoglobin liberated as a result of bleeding on tissue reperfusion (Section 2.1.2).

It was also found that regional intestinal ischemia in cats results in an accumulation of hypoxanthine in the tissue; the hypoxanthine disappears quickly on reperfusion, and both lipid peroxidation and formation of oxidized glutathione could be measured in the reperfused intestine.[507] It has been reported that SOD diminishes the lethal effect of bowel ischemia in rats.[508] Finally, it is known that intestine contains xanthine dehydro-

[506] L. A. Hernandez, M. B. Grisham, and D. N. Granger, *Am. J. Physiol.* **253**, G49 (1987).
[507] M. H. Schoenberg, B. B. Fredholm, U. Haglund, H. Jung, D. Sellin, M. Younes, and F. W. Schildberg, *Acta Physiol. Scand.* **124**, 581 (1985).
[508] M. C. Dalsing, J. L. Grosfeld, M. A. Schiffler, D. W. Vane, M. Hull, R. L. Bachner, and T. R. Weber, *J. Surg. Res.* **34**, 589 (1983).

genase activity in most mammals, including man,[509,510] and that conversion of this enzyme to oxidase on ischemia/hypoxia does occur,[504] although there is some discussion as to how rapidly.[511]

As far as intestinal ischemia/reoxygenation is concerned, the essential features of the proposal of McCord *et al.* (Fig. 4) seem to be supported by experimental evidence. Although allopurinol is not a specific inhibitor of xanthine oxidase (see Section 4.5.2 below), other evidence for the involvement of this enzyme has been obtained. Pterinaldehyde (2-amino-4-hydroxypteridine-6-carboxaldehyde), a powerful inhibitor of xanthine oxidase that often contaminates commercial preparations of the vitamin folic acid, also offers protection against reoxygenation injury in cats.[512] Feeding animals a diet rich in tungsten decreases tissue concentrations of xanthine dehydrogenase because the tungsten is incorporated into the enzyme in place of molybdenum at the active site; tungsten-containing enzyme is inactive (e.g., see Refs. 513 and 513a). Segments of intestine from rats pretreated with tungsten showed much less reoxygenation injury after hypoxia than segments from normal rats.[513a]

The hypoxanthine–xanthine oxidase system may not be the only source of radicals to which reoxygenated intestine (and other tissues) are subjected *in vivo*. Generation of oxygen-derived species by neutrophils in reoxygenated intestine is another potential source of damage. Entering neutrophils (as well as neutrophils trapped within the hypoxic tissue) can adhere to endothelium and release products (O_2^-, H_2O_2, eicosanoids, proteolytic enzymes) that might worsen injury. Depleting animals of neutrophils, or pretreatment of the animals with an antibody that prevents adherence of neutrophils to endothelium, has been reported to diminish reoxygenation injury to intestine in studies with whole animals.[514] Mitochondria damaged by ischemia might also "leak" more electrons than usual from their electron transport chain, thus forming more O_2^-.

A role of oxidants in reoxygenation injury has been proposed for some other parts of the gastrointestinal tract. Severe bleeding can cause a rapid fall in blood pressure, leading to tissue hypoxia, so that when blood volume is restored a "shock" syndrome (hemorrhagic shock) can result,

[509] T. A. Krenitsky, J. V. Tuttle, E. L. Cattau, and P. Wang, *Comp. Biochem. Physiol. B: Comp. Biochem.* **49,** 687 (1974).

[510] R. W. E. Watts, J. E. M. Watts, and J. E. Seegmiller, *J. Lab. Clin. Med.* **66,** 688 (1965).

[511] D. A. Parks, T. K. Williams, and J. S. Beckman, *Am. J. Physiol.* **254,** G768 (1988).

[512] D. N. Granger, J. M. McCord, D. A. Parks, and M. E. Hollwarth, *Gastroenterology* **90,** 80 (1986).

[513] J. M. Brown, M. A. Grosso, G. J. Whitman, L. S. Terada, J. E. Repine, and A. H. Harken, *Surgery* **104,** 266 (1988).

[513a] D. N. Granger, *Am. J. Physiol.* **255,** H1269 (1988).

[514] M. B. Grisham, L. A. Hernandez, and D. N. Granger, *Am. J. Physiol.* **251,** G567 (1986).

which injures many tissues. Pretreatment of rats or cats with allopurinol or SOD before hemorrhagic shock has been reported to minimize the gastric lesions observed and to improve survival.[515-517] Both allopurinol and desferrioxamine have been claimed to increase survival in dogs subjected to severe hemorrhagic shock.[518,519] Oxidants and eicosanoids may also play some role in shock induced by bacterial endotoxin, for example, in sepsis caused by gram-negative bacteria.[520,521] The effects of endotoxin may be largely mediated by increased formation of tumor necrosis factor (TNF, previously called cachectin), a polypeptide hormone released from monocytes and macrophages.[522] TNF has many biological effects, including an ability to sensitize neutrophils and monocytes so that they respond to stimuli by increased oxidant formation.[523] Nathan[524] has reported that TNF can itself stimulate the production of oxidants by neutrophils that are adhering to surfaces, as can happen in ischemic/reperfused tissues. TNF has also been shown to increase transcription of the mRNA for manganese-SOD in a human cell line.[525]

4.5.2. Cardiac Injury. The pioneering studies of Parks, Granger, McCord, and co-workers[504,505,511-515,513a] naturally attracted the attention of cardiologists, since ischemic/hypoxic injury to heart muscle is a major cause of death. Again, after relatively short periods of ischemia or hypoxia, reoxygenation gives an additional insult to the tissue that can often be diminished by SOD (manganese or copper,zinc enzymes), catalase, scavengers of $\cdot$ OH such as mannitol, desferrioxamine, or allopurinol (reviewed in Refs. 526–529). For example, reperfusion of heart after periods

[515] M. A. Perry, S. Wadhwa, D. A. Parks, W. Pickard, and D. N. Granger, *Gastroenterology* **90**, 362 (1986).

[516] H. Bitterman, N. Aoki, and A. M. Lefer, *Proc. Soc. Exp. Biol. Med.* **188**, 265 (1988).

[517] M. Itoh and P. H. Guth, *Gastroenterology* **88**, 1162 (1985).

[518] G. Allan, D. Cambridge, L. Lee-Tsang-Tan, C. W. Van Way, and M. V. Whiting, *Br. J. Pharmacol.* **89**, 149 (1986).

[519] S. Sanan, G. Sharma, R. Malhotra, D. P. Sanan, P. Jain, and P. Vadhera, *Free Radical Res. Commun.* **6**, 29 (1989).

[520] N. H. Manson and M. L. Hess, *Circ. Shock* **10**, 205 (1983).

[521] J. C. Saez, P. H. Ward, B. Gunther, and E. Vivaldi, *Circ. Shock* **12**, 229 (1984).

[522] K. J. Tracey, Y. Fong, D. G. Hesse, K. R. Manogue, A. T. Lee, G. C. Kuo, S. F. Lowry, and A. Cerami, *Nature (London)* **330**, 662 (1987).

[523] I. A. Clark, C. M. Thumwood, G. Chaudhri, W. B. Cowden, and N. H. Hunt, *in* "Oxygen Radicals and Tissue Injury" (B. Halliwell, ed.), p. 122. Allen Press, Lawrence, Kansas, 1988.

[524] C. F. Nathan, *J. Clin. Invest.* **80**, 1550 (1987).

[525] G. H. W. Wong and D. V. Goeddel, *Science* **242**, 941 (1988).

[526] E. Braunwald and J. A. Kloner, *J. Clin. Invest.* **76**, 1713 (1985).

[527] P. J. Simpson and B. R. Lucchesi, *J. Lab. Clin. Med.* **110**, 13 (1987).

[528] R. Bolli, *J. Am. Coll. Cardiol.* **12**, 239 (1988).

[529] J. M. McCord, *N. Engl. J. Med.* **312**, 159 (1985).

of ischemia too short to produce any necrosis results in prolonged depression of contractile function (stunning)[528] and sometimes generation of arrhythmias,[530] and radical production has been detected by spin trapping.[531,531a,531b] Many studies showing protective effects of the above antioxidants have been done with isolated perfused hearts from rats, rabbits, dogs, and pigs.[527–530,532,533] Some studies have also been performed with organs *in situ*. For example, in open-chest dogs (chest opened, coronary artery branch partially or completely ligated, clamp removed after various periods) it has been claimed that pretreatment of animals with SOD, catalase, or allopurinol can decrease the size of the infarction produced by occlusion of a coronary artery branch, followed by reoxygenation (reviewed in Refs. 527 and 528).

The extent of the protection by antioxidants against cardiac reperfusion injury has been very variable in different experiments. At least three reasons can account for this. First, the importance of reoxygenation injury as a fraction of total tissue damage declines as the period of ischemia increases, so that in experiments with long periods of ischemia/hypoxia so much tissue injury has been done in the ischemia phase of the experiment that radical scavengers are unlikely to offer much protection (a tissue irreversibly injured by O_2 deprivation will die however it is reperfused). In dog heart, for example, coronary occlusion for less than about 20 min produces stunning on reperfusion, but essentially no infarct.[527,528] Some antioxidants (e.g., desferrioxamine) are very effective in protecting against stunning.[528] Occlusion for more than 4 hr followed by reperfusion gives an infarct about the same size as that resulting from permanent occlusion (i.e., damage is maximum and cannot be further increased), and antioxidants have no effect.

Second, in many animal studies (especially with open-chest dogs), the extent of O_2 deprivation during arterial occlusion can vary considerably because of collateral vessels. Thus, "ischemia" is often hypoxia, to variable extents. If flow does not cease completely, not only does some O_2 enter the "ischemic" tissue but some metabolic products, such as H^+, are removed and acidosis is less severe. Third, it has been shown that many

[530] S. Manning, M. Bernier, R. Crome, S. Little, and D. Hearse, *J. Mol. Cell. Cardiol.* **20,** 35 (1988).

[531] R. Bolli, B. S. Patel, M. O. Jeroudi, E. K. Lai, and P. B. McCay, *J. Clin. Invest.* **82,** 476 (1988).

[531a] P. B. Garlick, M. J. Davies, D. J. Hearse, and T. F. Slater, *Circ. Res.* **61,** 757 (1987).

[531b] I. E. Blasig, B. Ebert, G. Wallukat, and H. Loewe, *Free Radical Res. Commun.* **6,** 303 (1989).

[532] R. T. Pallandi, M. A. Perry, and T. J. Campbell, *Circ. Res.* **61,** 50 (1987).

[533] U. Naslund, S. Haggmark, G. Johansson, S. L. Marklund, S. Reiz, and A. Oberg, *J. Mol. Cell. Cardiol.* **18,** 1077 (1986).

of the "antioxidants" used in these studies show bell-shaped dose–response curves. For example, in studies on production of arrhythmias in rat hearts after a brief period of ischemia, there is an optimal concentration of SOD for protection, and higher concentrations give smaller protective effects.[534] The reason for these effects remains to be established. For example, a deleterious effect of impurities in some of the antioxidants studied might become significant at very high antioxidant concentrations.

The large number of successful reports using isolated animal hearts, and *in situ* studies on animals, naturally led to proposals that the model shown in Fig. 4 could account for reoxygenation damage in heart.[529] Experiments *in vitro* found that exposure of myocytes or whole hearts to oxidant-generating systems produces severe injury, including inactivation of the ATP-dependent Ca^{2+}-sequestering system of cardiac sarcoplasmic reticulum (e.g., Ref. 535). Ca^{2+} has also been suggested[529] to activate a protease that can convert xanthine dehydrogenase to the oxidase form (Fig. 4). Of course, sources of radicals other than xanthine oxidase, such as mitochondria and infiltrating neutrophils (Section 4.5.1), may also be important *in vivo*.[528] Neutrophils have been claimed to play little role in the stunning seen after brief ischemia followed by reperfusion, but they may be important after longer ischemic periods.[527,528] Thus, depleting animals of neutrophils, or injecting them with an antibody that prevents neutrophil adherence to endothelium, before performing ischemia/reperfusion studies *in situ* (using an open chest system) can decrease the size of the infarct zone produced.[527]

However, some fundamental problems have arisen in applying the model shown in Fig. 4 to cardiac systems. Rat and dog heart contain xanthine dehydrogenase activity, but the rate of proteolytic conversion of dehydrogenase to oxidase during ischemia is very slow[536,537] (e.g., in rat heart it has been reported that about 20% of the enzyme is present as oxidase to start with, and it takes about 3 hr of ischemia for this to increase to 30%).[537] The activity of xanthine dehydrogenase in rat heart has been reported to increase with the age of the animals.[538] Even worse, rabbit, pig, and human hearts have been reported not to contain any xanthine oxidase activity at all (reviewed in Ref. 539). If these reports are correct, how, then, can one explain the protective effects of allopurinol

[534] S. Riva, A. S. Manning, and D. J. Hearse, *Cardiovasc. Drugs Ther.* **1**, 133 (1987).
[535] E. Okabe, Y. Kato, H. Sasaki, G. Saito, M. L. Hess, and H. Ito, *Arch. Biochem. Biophys.* **255**, 464 (1987).
[536] J. P. Kehrer, H. M. Piper, and H. Sies, *Free Radical Res. Commun.* **3**, 69 (1987).
[537] T. D. Engerson, T. G. McKelvey, D. B. Rhyne, E. B. Boggio, S. J. Snyder, and H. P. Jones, *J. Clin. Invest.* **79**, 1564 (1987).
[538] B. Schoutsen and J. W. de Jong, *Circ. Res.* **61**, 604 (1987).
[539] B. Halliwell, *Free Radical Res. Commun.* **5**, 315 (1989).

reported by several groups, in, for example, rabbit heart systems? It may be that allopurinol inhibits another enzyme involved in oxidant generation. It might also protect by preventing depletion of purine nucleotides, which can be used as substrates for resynthesis of ATP on reoxygenation.[540] Pretreatment of animals with allopurinol causes formation of oxypurinol. For example, administration of allopurinol to humans can produce plasma oxypurinol concentrations of 40–300 μmol/liter.[541,542] *In vitro*, oxypurinol is a good scavenger of $\cdot$ OH and of the myeloperoxidase-derived oxidant HOCl[543,544]; the concentrations of oxypurinol achieved *in vivo* would probably be insufficient to scavenge $\cdot$ OH effectively, but might be able to remove some HOCl.[543,544]

It has also been questioned whether the established protective action of mannitol in several cardiac systems is really due to $\cdot$ OH scavenging, rather than to, say, osmotic effects. Both mannitol and glucose react with $\cdot$ OH with about the same rate constant,[3] but mannitol is far more protective against ischemia/reperfusion-induced arrhythmias in isolated rat heart than is equimolar glucose.[545]

Finally, if $\cdot$ OH is really formed in the reperfused heart by superoxide-driven Fenton chemistry, then a source of iron ions must be identified. This iron might be released as a result of cell injury (Section 2.1) or from bleeding caused by reperfusion of myocardium that has undergone vascular damage during ischemia. Another possible source is myoglobin.[546] Incubation of hemoglobin (Fig. 2) or of cardiac oxymyoglobin with excess H_2O_2 can lead to breakdown of the heme and liberation of iron ions in a form capable of catalyzing $\cdot$ OH production.[216,281,283,286a] Reactive ferryl species can also be formed by reaction of myoglobin with H_2O_2 (e.g., Refs. 277 and 280). It is interesting to note that hearts from iron-overloaded rats shown an increased susceptibility to reoxygenation injury.[547]

Despite these questions, the effectiveness of SOD, desferrioxamine, and other antioxidants in many of the studies performed seems well established.[527–529,533] In classic human myocardial infarction, it has been sug-

[540] R. D. Lasley, S. W. Ely, R. M. Berne, and R. M. Mentzer, Jr., *J. Clin. Invest.* **81,** 16 (1988).

[541] J. S. Cameron and H. A. Simmonds, *Br. Med. J.* **294,** 1504 (1987).

[542] B. T. Emmerson, R. B. Gordon, M. Cross, and D. B. Thomson, *Br. J. Rheumatol.* **26,** 445 (1987).

[543] M. Grootveld, B. Halliwell, and C. P. Moorhouse, *Free Radical Res. Commun.* **4,** 69 (1987).

[544] C. P. Moorhouse, M. Grootveld, B. Halliwell, G. J. Quinlan, and J. M. C. Gutteridge, *FEBS Lett.* **213,** 23 (1987).

[545] M. Bernier and D. J. Hearse, *Am. J. Physiol.* **254,** H862 (1988).

[546] K. I. Nomoto, N. Mori, T. Shoji, and K. Nakamura, *Exp. Mol. Pathol.* **47,** 390 (1987).

[547] A. M. M. van der Kraaij, L. J. Mostert, H. G. van Eijk, and J. F. Koster, *Circulation* **78,** 442 (1988).

gested that occlusion of a branch of the coronary artery is maintained for such a long period that ischemic/hypoxic injury is the sole cause of tissue death, reperfusion injury being insignificant. However, the early use of thrombolytic agents (such as streptokinase or tissue plasminogen activator) to produce clot dissolution is now becoming commonplace in many countries. Combined use of an antioxidant (such as SOD or desferrioxamine) and a thrombolytic agent might be expected to give enhanced benefit. Controlled trials using human recombinant Cu,Zn-SOD are underway to test this possibility. Reperfusion injury might also play a role in the depressed myocardial function sometimes seen after open-heart surgery and transplantation[548,549] and in exercise-induced angina pectoris, and again antioxidants might have therapeutic value.

4.5.3. Role of Endothelium. As stated above, human, pig, and rabbit hearts have been claimed not to contain xanthine oxidase.[539] However, debate continues as to whether the enzyme might be concentrated only in the endothelial cells of the heart, so that assays of tissue homogenates might be insufficiently sensitive to detect the small amount present (e.g., Refs. 550 and 551). Several other lines of evidence support a key role played by the endothelium in reoxygenation injury. Thus SOD, catalase, and desferrioxamine penetrate into intact cells only slowly, yet they have been observed to protect efficiently against reperfusion injury in several experiments (Section 4.5.3). A recent study[552] has shown that the radicals important in producing myocardial stunning in dog heart are generated within one minute of reoxygenation and must be scavenged in that time frame for protection to occur. These data strongly suggest that radical generation cannot occur deep within the tissue, but occurs at its surface.

The question then arises as to why the endothelium should have such apparently dangerous potential for oxidant production. Studies of isolated endothelial cells suggest that they may constantly produce small amounts of O_2^-.[550,553,554] Apart from its barrier function, the vascular endothelium

[548] F. Gharagozloo, F. J. Melendez, R. A. Hein, R. J. Shemin, U. J. Diseasa, and L. H. Cohn, *J. Thorac. Cardiovasc. Surg.* **95,** 1008 (1988).

[549] P. J. Del Nido, D. A. G. Mickle, G. J. Wilson, L. N. Benson, J. G. Coles, G. A. Trusler, and W. G. Williams, *Circulation* **76** (Suppl. 5), V174 (1987).

[550] J. L. Zweier, P. Kuppusamy, and G. A. Lutty, *Proc. Natl. Acad. Sci. U.S.A.* **85,** 4046 (1988).

[551] E. D. Jarasch, G. Bruder, and H. W. Heid, *Acta Physiol. Scand. Suppl.* **548,** 39 (1986).

[552] R. Bolli, M. O. Jeroudi, B. S. Patel, O. I. Aruoma, B. Halliwell, E. K. Lai, and P. B. McCay, *Circ. Res.* **65,** 607 (1989).

[553] S. L. Marklund, *J. Mol. Cell. Cardiol.* **20** (Suppl. 2), 23 (1988).

[554] R. E. Ratych, R. S. Chuknyiska, and G. B. Bulkley, *Surgery* **102,** 122 (1987).

synthesizes a wide range of products, including prostacyclin, tissue plasminogen activator, platelet-activating factor, and interleukin-1.[555] Endothelium has also been shown to produce a factor that relaxes smooth muscle, the so-called endothelium-derived relaxing factor (EDRF).[555] EDRF seems identical to nitric oxide, NO (reviewed in Refs. 555–557). EDRF can be inactivated by reaction with O_2^-, presumably by forming nitrate ion,

$$NO + O_2^- \rightarrow intermediates \rightarrow NO_3^- \qquad (44)$$

These observations led Halliwell[539] to hypothesize that the vascular endothelium has a system to produce O_2^- because the O_2^- interacts with EDRF and inactivates it in a physiological control mechanism. Any O_2^- that does not react with EDRF would presumably undergo nonenzymatic dismutation to form H_2O_2. We have seen (Section 1) that generation of O_2^- and H_2O_2 in small amounts is not intrinsically damaging, provided that no metal ions are present to form highly reactive species such as $\cdot OH$. Such "catalytic" metal ions are not present in normal human or animal plasma.[47,67,246]

It therefore appears that endothelium may have an inbuilt physiological mechanism for generating O_2^-. This may be xanthine oxidase, or it may be another system as yet undefined. What goes wrong on ischemia/reperfusion? First, ischemia may lead to increased O_2^- generation on subsequent reperfusion, for example, via hypoxanthine accumulation (Fig. 4). This could occur even if xanthine oxidase activity did not increase on ischemia, provided that there was some xanthine oxidase there to start with, as appears to be the case in rat heart.[537] Ischemia-dependent disruption of cellular electron transport chains, for instance, in mitochondria or plasma membrane, could also contribute to increased O_2^- generation when O_2 is reintroduced.[553] The ischemia may also do something more significant, however. Cell injury is known to cause release of iron (Section 2.1). Thus, the ischemic period may cause release into the system of iron that, for a brief period on reperfusion, (presumably until the ions are washed away by the blood or bound by plasma transferrin) is available to convert O_2^- and H_2O_2 to species such as $\cdot OH$ that should greatly exacerbate the endothelial injury. Evidence consistent with this proposed key role of iron is provided by the observation that, in ischemia/reperfusion systems in which protective effects of SOD and catalase have

[555] J. R. Vane, R. J. Gryglewski, and R. M. Botting, *Trends Pharmacol. Sci.* **8**, 491 (1987).
[556] S. Moncada, M. W. Radomski, and R. M. J. Palmer, *Biochem. Pharmacol.* **37**, 2495 (1988).
[557] J. M. Sneddon and J. R. Vane, *Proc. Natl. Acad. Sci. U.S.A.* **85**, 2800 (1988).

been demonstrated, desferrioxamine is also usually protective (Section 4.5.2). The proposal of Halliwell[539] explains why it appears to be essential to scavenge radicals or bind iron during the first few minutes of reperfusion in order to achieve protection.[552]

4.5.4. Cerebral Injury. We have already seen (Section 3.5) that traumatic injury to the brain or spinal cord can lead to tissue degeneration that may well involve iron-dependent free radical reactions. Cerebral ischemia or hypoxia, followed by reperfusion, should also stimulate free radical reactions, again probably because of metal ion release from injured cells or from bleeding in the reperfused area.[558–561] The acidosis caused by lactate accumulation may accelerate lipid peroxidation and $\cdot$ OH formation.[562] Acidosis might also facilitate O_2 unloading from blood hemoglobin on reperfusion, producing a transient abnormally high O_2 tension. It is not yet clear if ischemia in brain causes the xanthine dehydrogenase to oxidase conversion (Fig. 4). Rat and gerbil brain have been reported to contain low activities of xanthine dehydrogenase,[558–561,563] but the activity of this enzyme in human brain does not appear to have been clearly established. In any case, several groups have reported that SOD, allopurinol, desferrioxamine, or other iron chelators (e.g., the lazaroids; Section 3.5) have some beneficial effects in minimizing brain or spinal cord injury after ischemia/reperfusion studies in animals.[558–561,563] Few of the experiments with enzymes have been accompanied by controls with inactivated enzymes, however.

Another mechanism that may contribute to cerebral injury after hypoxia/ischemia is the generation within the tissue of excitatory amino acid neurotransmitters such as glutamate or aspartate (reviewed in Refs. 559 and 564). These excitatory neurotransmitters cause neurons to fire off continuously until they are damaged. Synthetic excitatory neurotoxins of this type include kainic acid and quinolinic acid. Dykens *et al.*[565] reported

[558] H. S. Mickel, Y. Vaishav, O. Kempski, D. von Lubitz, J. F. Weiss, and G. Feuerstein, *Stroke* **18**, 426 (1987).
[559] G. S. Krause, B. C. White, S. D. Aust, N. R. Nayini, and K. Kumar, *Crit. Care Med.* **16**, 714 (1988).
[560] K. Kogure, H. Arai, K. Abe, and M. Nakano, *Prog. Brain Res.* **63**, 237 (1985).
[561] B. D. Watson and M. D. Ginsberg, *in* "Oxygen Radicals and Tissue Injury" (B. Halliwell, ed.), p. 81. Allen Press, Lawrence, Kansas, 1988.
[562] B. K. Siesjo, G. Bendek, T. Koide, E. Westerberg, and T. Wieloch, *J. Cereb. Blood Flow Metab.* **5**, 253 (1985).
[563] A. Patt, A. H. Harken, L. K. Burton, T. C. Rodell, D. Piermattei, W. J. Schorr, N. B. Parker, E. M. Berger, I. R. Horesh, L. S. Terada, S. L. Linas, J. C. Cheronis, and J. E. Repine, *J. Clin. Invest.* **81**, 1556 (1988).
[564] B. Meldrum, *Trends Neurosci.* **8**, 47 (1985).
[565] J. A. Dykens, A. Stern, and E. Trenkner, *J. Neurochem.* **49**, 1222 (1987).

that the death of cultured mouse cerebellar neurons caused by treatment with kainic acid could be prevented by pretreating cells with allopurinol, or by adding SOD plus catalase to the culture medium. They therefore suggested that excitotoxin-induced neural damage may be mediated by O_2^- generated by xanthine oxidase (presumably oxidizing hypoxanthine accumulated as a result of excessive ATP degradation during repeated neuronal firing). These observations provide an interesting link between two apparently different mechanisms of brain injury during ischemia/ reperfusion: free radical reactions and generation of excitatory amino acids. A further link may be exemplified by the recent claim that free radicals can induce release of excitatory amino acids from rat hippocampal slices.[566]

4.5.5. Preservation of Organs for Transplantation: Kidney, Liver, and Skin. The sensitivity of kidney to oxidant damage is well established.[567,568] For example, kidney damage by the antibiotic gentamicin may involve increased oxidant production.[569] Deposition of antigen–antibody complexes in the glomeruli can lead to complement activation and infiltration of neutrophils. Resident phagocytic (mesangial) cells in the glomeruli can also respond to complement activation by secreting eicosanoids, O_2^-, and H_2O_2.[568] Ward and co-workers[567] found that infusion, into rats, of an antibody against glomerular basement membranes causes severe glomerular inflammation, involving both neutrophils entering the tissue and complement activation. Infusion of catalase offered protection, but infusion of SOD had only a small protective effect. Both H_2O_2 and HOCl (from the action of myeloperoxidase) might be involved in the damage, as might the action of proteases such as elastase and collagenase.[567,570,571] Several groups have reported that injection into rats of "nephrotoxic serum" containing antibody to rat glomerular basement membrane produces renal injury that can be ameliorated by treating the animals with SOD (reviewed in Ref. 567). The acute phase of the glomerular injury produced by injecting antibody against glomerular basement membrane into rabbits was reported to be suppressed by desferriox-

[566] D. E. Pellegrini-Giampietro, G. Cherichi, M. Alesiani, V. Carla, and F. Moroni, *J. Neurochem.* **51**, 1961 (1988).

[567] K. J. Johnson, A. Rehan, and P. A. Ward, in "Oxygen Radicals and Tissue Injury" (B. Halliwell, ed.), p. 115. Allen Press, Lawrence, Kansas, 1988.

[568] L. Baud and R. Ardaillou, *Am. J. Physiol.* **251**, F765 (1986).

[569] P. D. Walker and S. V. Shah, *J. Clin. Invest.* **81**, 334 (1988).

[570] M. C. M. Vissers, C. C. Winterbourn, and J. S. Hunt, *Biochim. Biophys. Acta* **804**, 154 (1984).

[571] R. T. Johnson, W. G. Couser, E. Y. Chi, S. Adler, and S. J. Klebanoff, *J. Clin. Invest.* **79**, 1379 (1987).

amine.[572] These variable results indicate the complexity of this model system of immune injury to kidney.

Reoxygenation of animal kidneys after a brief period of ischemia or hypoxia produces a reoxygenation injury that can again contribute to tissue damage. (After prolonged periods of ischemia, the ischemic damage itself may be so extensive that reperfusion injury is insignificant; Section 4.5.2.) Several groups have reported that pretreatment of whole animals or isolated organs with SOD, catalase, desferrioxamine, or allopurinol diminishes reperfusion injury.[573–578] Iron might be released from injured cells[579] by breakdown of hemoglobin from bleeding (see Fig. 2 and Refs. 579 and 580).

There are two clinical situations in which the kidney is subjected to temporary hypoxia: shock arising from severe blood loss (when all body organs are affected to variable degrees) and renal transplantation. Some degree of hypoxia in organs stored for transplantation is difficult to avoid; transplanted kidneys would thus be subjected to a limited degree of reperfusion injury after removal of the renal artery clamp.[574] Rat kidney contains xanthine dehydrogenase, although the rate of its irreversible (proteolytic) conversion to xanthine oxidase in ischemic tissues is slow (half-time ~6 hr).[537] There appears to be much less xanthine oxidase in dog or human kidney.[581] Hence, it is uncertain whether the model in Fig. 4 accounts for reperfusion injury in kidney.

Transplanted kidneys are frequently infiltrated by host phagocytic cells, and it is possible that oxidants produced by these might sometimes contribute to transplant rejection. Thus, antioxidants might have a beneficial role not only in allowing organ preservation for longer periods but also in diminishing graft rejection. These comments could equally well be applied to other transplanted organs, such as liver, bone[582] (see below),

[572] N. W. Boyce and S. R. Holdsworth, *Kidney Int.* **30**, 813 (1986).

[573] R. E. Ratych and G. B. Bulkley, *J. Free Radicals Biol. Med.* **2**, 311 (1986).

[574] C. J. Green, G. Healing, S. Simpkin, J. Lunec, and B. J. Fuller, *Comp. Biochem. Physiol. B: Comp. Biochem.* **83**, 603 (1986).

[575] B. J. Fuller, J. Lunec, G. Healing, S. Simpkin, and C. J. Green, *Transplantation* **43**, 604 (1987).

[576] G. L. Baker, R. L. Corry, and A. P. Autor, *Ann. Surg.* **5**, 628 (1985).

[577] J. F. Bennett, W. I. Bry, G. M. Collins, and N. A. Halasz, *Cryobiology* **24**, 264 (1987).

[578] R. Hansson, S. Johansson, O. Jonsson, S. Pettersson, T. Schersten, and J. Waldenstrom, *Clin. Sci.* **71**, 245 (1986).

[579] J. Gower, G. Healing, and C. Green, *Free Radical Res. Commun.* **5**, 291 (1989).

[580] M. S. Paller and B. E. Hedlund, *Kidney Int.* **34**, 474 (1988).

[581] J. H. Southard, D. C. Marsh, J. F. McAnulty, and F. O. Beizer, *Surgery* **101**, 566 (1987).

[582] A. P. C. Weiss, J. R. Moore, M. A. Randolph, and A. J. Weiland, *Plast. Reconstr. Surg.* **82**, 486 (1988).

and heart (Section 4.5.2), and to transplants of corneas or skin flaps.[583] Thus, several groups have reported that the survival time of hypoxic skin flaps is increased by antioxidants such as SOD, desferrioxamine, or allopurinol.[583–585]

Several groups have been interested in the possibility of "reperfusion injury" in transplanted livers, although no clear demonstration of the effectiveness of antioxidants in minimizing tissue damage has yet been achieved.[586–588] Rat liver contains xanthine dehydrogenase, but again this is only very slowly converted to oxidase during ischemia.[537] Dog or human liver has been reported to contain very little of either form of the enzyme.[581] The possibility of reoxygenation injury to lung during shock caused by excessive blood loss, or during heart–lung transplantation, has also been raised.[589]

4.5.6. Rheumatoid Arthritis and Reoxygenation Injury. The involvement of neutrophils in reoxygenation injury in gut (Section 4.5.1), kidney (Section 4.5.5), and heart (Section 4.5.2) implies that inflammatory responses such as neutrophil infiltration can contribute to hypoxic reoxygenation injury. It is also possible that hypoxia/reoxygenation injury might contribute to tissue damage at sites of inflammation. Blake and co-workers[590] have shown that tensing an inflamed human rheumatoid knee joint can generate intraarticular pressures in excess of capillary perfusion pressure, resulting in sharp drops in O_2 concentration within the joint. On relaxing the joint, the O_2 concentration rises, overshoots, and gradually returns to normal. Thus, a reoxygenation injury is certainly possible. Blake and co-workers[590] found some xanthine-oxidizing activity in both normal and rheumatoid synovial membranes, and the concentration of hypoxanthine has been reported to be increased in synovial fluid from patients with rheumatoid arthritis. They used these observations to claim that injury by the mechanism shown in Fig. 4 might be possible in human rheumatoid patients.

However, the xanthine oxidase activities measured were low, and it

[583] D. A. Parks, G. B. Bulkley, and D. N. Granger, *Surgery* **94**, 428 (1983).

[584] L. Huang, C. T. Privalle, D. Serafin, and B. Klitzman, *FASEB J.* **1**, 129 (1987).

[585] M. F. Angel, S. S. Ramasastry, W. M. Swartz, R. E. Basford, and J. W. Futrell, *Plast. Reconstr. Surg.* **79**, 990 (1987).

[586] C. S. McEnroe, F. J. Pearce, J. J. Ricotta, and W. R. Drucker, *J. Trauma* **26**, 892 (1986).

[587] H. Jaeschke, C. V. Smith, and J. R. Mitchell, *J. Clin. Invest.* **81**, 1240 (1988).

[588] H. de Groot and A. Littauer, *Biochem. Biophys. Res. Commun.* **155**, 278 (1988).

[589] I. Koyama, T. J. K. Toung, M. C. Rogers, G. H. Gurtner, and R. J. Traystman, *J. Appl. Physiol.* **63**, 111 (1987).

[590] J. Unsworth, J. Outhwaite, D. R. Blake, C. J. Morris, J. Freeman, and J. Lunec, *Ann. Clin. Biochem.* **25**, 85 (1988).

seems more likely to the authors that changes in O_2 concentrations might be affecting oxidant production by activated neutrophils, which are abundant in the synovial fluid of inflamed rheumatoid joints. Edwards et al.[42] pointed out that the K_m for O_2 of the O_2^--generating NADPH oxidase of rat neutrophils is within the range of physiological O_2 concentrations in body fluids. If the same is true of human neutrophils, it follows that local O_2 concentrations at sites of inflammation could modulate O_2^- and H_2O_2 production. More studies are required to investigate these possibilities.

4.5.7. *Pancreas.* Inflammation of the pancreas (pancreatitis) can be initiated by several means, including excess alcohol intake, pancreatic duct obstruction by a migrating gallstone, or by a period of hypoxia, for example, as a result of hemorrhagic shock. Bulkley and co-workers[591] found that the injury produced in isolated perfused dog pancreas by fatty acid infusion, ischemia, or partial duct obstruction could in each case be diminished by including allopurinol, or SOD plus catalase, in the perfusion medium (SOD or catalase given separately had little effect). Controls with inactive enzymes were not performed. Conflicting results have been obtained as to whether SOD and catalase are effective against pancreatitis in *in vivo* animal models,[591–594] and it is unclear why both SOD and catalase should be required to protect. A precise evaluation of the effect of antioxidants is important because current therapies for the treatment of pancreatitis in humans are seriously inadequate.

4.5.8. *Limb Ischemia.* Limb ischemia or hypoxia occurs during several cardiovascular conditions, such as atherosclerosis, embolism, arterial injury resulting in rapid blood loss, and (to a controlled extent) in the use of tourniquets to provide a "bloodless field" in orthopedic surgery. Modern microsurgical techniques for reattachment of amputated digits or limbs often result in the reattachment of tissue that has been hypoxic for a considerable time. Skeletal muscle is, compared with most tissues, fairly resistant to ischemic injury. However, after several hours of ischemia, a reoxygenation injury to muscle can be demonstrated. Experiments with rat hindlimbs have shown some protective effects of SOD combined with catalase against the injury,[595] and other experiments with isolated dog

[591] M. G. Sarr, G. B. Bulkley, and J. L. Cameron, *Surgery* **101**, 342 (1987).

[592] K. S. Guice, D. E. Miller, K. T. Oldham, K. M. Townsend, Jr., and J. C. Thompson, *Am. J. Surg.* **151**, 163 (1986).

[593] Z. J. Devenyi, J. L. Orchard, and R. E. Powers, *Biochem. Biophys. Res. Commun.* **149**, 841 (1987).

[594] P. J. Blind, S. L. Marklund, R. Stenling, and S. T. Dahlgren, *Pancreas* **3**, 563 (1988).

[595] K. R. Lee, J. L. Gronenwelt, M. Schlafer, C. Corpron, and G. B. Zelenock, *J. Surg. Res.* **42**, 24 (1987).

gracilis muscle have shown some protection by SOD, allopurinol, DMSO, or catalase.[596] Free radicals have been suggested to be involved in Dupuytrens contracture in humans.[597]

4.5.9. Concluding Comments. Controversy over whether the mechanism shown in Fig. 4 accounts for reoxygenation injury should not blind us to the very real protective effects of antioxidants in a wide range of experimental systems. The results of controlled clinical trials to evaluate the usefulness of antioxidants in human medical treatment are awaited with interest.

Acknowledgment

J. M. C. Gutteridge was a 1988 Greenberg scholar.

[596] R. J. Korthuis, D. N. Granger, M. I. Townsley, and A. E. Taylor, *Circ. Res.* **57,** 599 (1985).

[597] G. A. C. Murrell, M. J. O. Francis, and L. Bromley, *Br. Med. J.* **295,** 1373 (1987).

Section I

Production, Detection, and Characterization of Oxygen Radicals and Other Species

[2] Pulse Radiolysis in Study of Oxygen Radicals

By Michael G. Simic

Oxygen Radicals and Pulse Radiolysis

Because of the unpaired electron, oxygen radicals, like most other free radicals, are highly reactive not only with numerous and varied substrates but also with each other.[1] Consequently, their lifetimes either in model aqueous systems or especially in biosystems are very short. This particular characteristic renders studies of their physicochemical properties complex. Studies of their structural features can be simplified by generating them in a solid or a frozen matrix where their lifetimes may extend indefinitely. Their kinetics and energetics properties, however, can be determined only in liquid systems where the lifetimes are short. Hence, physicochemical properties of oxygen radicals can be studied only with detection techniques such as kinetic spectrophotometry, kinetic conductivity, time-resolved Raman and electron spin resonance (ESR) spectroscopy, and many other techniques with short time resolution.[2,3]

In addition to adequate monitoring systems, studies of free radicals require convenient free radical-generating techniques and methods. These techniques and methods must satisfy some basic prerequisites. First, the radicals should be generated in a short period of time, before they start reacting with each other and the substrates. Second, the concentrations of radicals generated should be high enough to allow accurate monitoring. Third, only a specific radical should be generated without the presence of other interfering radicals. Finally, the capability of generating diverse free radicals as required, and sometimes a combination of radicals, should exist.

Such features can be found in a technique known as pulse radiolysis.[3,4] Pulse radiolysis is accomplished by a combination of a pulsed-radiation-generating machine (electron accelerator) and a variety of time-resolved detecting systems. In many respects pulse radiolysis, in which free radi-

---

[1] W. A. Pryor (ed.), "Free Radicals in Biology," Vols. I–VI. Academic Press, New York, 1976–1985.
[2] L. M. Dorfman, *J. Chem. Educ.* **58**, 82 (1981).
[3] Farhataziz and M. A. J. Rodgers (eds.), "Radiation Chemistry." Verlag Chemie, New York, 1987.
[4] C. von Sonntag, "The Chemical Basis of Radiation Biology." Taylor & Francis, London, 1987.

METHODS IN ENZYMOLOGY, VOL. 186

cals are generated by ionizing radiations, is similar to laser and flash photolysis,[5] in which free radicals are generated by photonic processes.

In this chapter pulse radiolysis studies of hydroxyl ($\cdot$OH), peroxyl (HROO$\cdot$), superoxide ($\cdot$O$_2^-$), alkoxyl (HRO$\cdot$), and phenoxyl radicals (ArO$\cdot$), as well as their properties, are briefly discussed. Underlying principles, attributes, and limitations are emphasized; detailed accounts of pulse radiolysis have been published recently.[3,4]

Generation of Free Radicals

Ionizing radiations (high-energy electrons, X- and γ-rays, high-energy nuclei, etc.) ionize the matter through which they pass. In aqueous media, the first step is ionization of water,

$$H_2O \rightarrow H_2O^+ + e^- \qquad (1)$$

Both H_2O^+ and e^- are, in principle, free radicals because they do not have paired electrons. Within a few picoseconds (10^{-12} sec) they undergo the following reactions:

$$H_2O^+ + H_2O \rightarrow \cdot OH + H_3O^+ \qquad (2)$$
$$e^- + n\,H_2O \rightarrow e_{aq}^- \qquad (3)$$

Hence, if the radiation pulse were 1 nsec long, by the end of the pulse the reaction cell through which the radiation passed would contain relatively homogeneously distributed hydroxyl radicals and hydrated electrons, e_{aq}^-. In addition to these two extremely reactive radicals, a small amount of $\cdot$H is generated. Therefore, radiation splits water into primary water radicals (yields are indicated in parentheses):

$$H_2O \rightarrow \cdot OH\ (2.8) + e_{aq}^-\ (2.8) + \cdot H\ (0.6) \qquad (4)$$

In pure water the primary water radicals recombine at diffusion-controlled rates, for example,

$$\cdot OH + e_{aq}^- \rightarrow OH^- \qquad (5)$$
$$k = 3 \times 10^{10}\ M^{-1}\ sec^{-1}$$

Water radicals react with solutes when present to give a second generation of radicals, the solute radicals. The reactivities of diverse solutes with all three primary water radicals have been tabulated.[6]

The simultaneous occurrence of $\cdot$OH and e_{aq}^- reactions may interfere with the measurements of free radical properties. To resolve this experi-

[5] R. V. Bensasson, E. J. Land, and T. G. Truscott, "Flash Photolysis and Pulse Radiolysis." Pergamon, New York, 1983.
[6] G. V. Buxton, C. L. Greenstock, W. P. Helman, and A. B. Ross, *J. Phys. Chem. Ref. Data* **17**, 513 (1988).

mental difficulty, two methods may be used. Hydrated electrons may be converted to hydroxyl radicals in the presence of gaseous N_2O (usually 1 atm):

$$e_{aq}^- + N_2O \rightarrow \cdot OH + OH^- + N_2 \tag{6}$$
$$k = 9 \times 10^9 \ M^{-1} \ sec^{-1}$$

At 1 atm and room temperature, $[N_2O] = 25$ mM. Hence, reaction (6) is completed in about 1 nsec. This rapid generation of the radiolytic $\cdot OH$ has critical advantages over the much slower generation of $\cdot OH$ via the Haber–Weiss reaction,

$$H_2O_2 + Fe(II)EDTA \rightarrow \cdot OH + OH^- + Fe(III)EDTA \tag{7}$$
$$k = 10^4 \ M^{-1} \ sec^{-1}$$

Unfortunately, there is no convenient method to convert hydroxyl radical to hydrated electron. The experimental conditions can be simplified, however, by removing $\cdot OH$ with solutes that give redox-unreactive radicals. Examples of such solutes are *tert*-butanol and some allylic compounds:

$$\cdot OH + (CH_3)_3COH \rightarrow H_2O + \cdot CH_2(CH_3)_2COH \tag{8}$$
$$k = 6 \times 10^8 \ M^{-1} \ sec^{-1}$$

$$\cdot OH + \quad \overset{\diagdown}{\underset{\diagup}{}}C{=}C\overset{\diagdown}{\underset{\diagup}{}} \rightarrow \overset{\diagdown}{}\dot{C}{-}\overset{|}{\underset{|}{C}}OH \tag{9}$$

$$k = 2 \times 10^9 \ M^{-1} \ sec^{-1}$$

Neither radical is a good hydrogen atom or electron donor nor exhibits acid–base properties. Both radicals have weak absorption spectra below 250 nm, an important feature in kinetic spectroscopy.

When reducing radicals are required, the oxidizing $\cdot OH$ can be converted to a reducing radical, for example,

$$\cdot OH + HCO_2^- \rightarrow H_2O + \cdot CO_2^- \tag{10}$$
$$k = 3.2 \times 10^9 \ M^{-1} \ sec^{-1}$$

With an E_0 value of -1.9 V, the carboxyl radical is a powerful reductant. It is used to reduce disulfide bonds, cytochromes, quinones, nitro compounds, etc. Since superoxide has an E_7 value of -0.33 V, $\cdot CO_2^-$ readily reduces oxygen:

$$\cdot CO_2^- + O_2 \rightarrow CO_2 + \cdot O_2^- \tag{11}$$
$$k = 4 \times 10^9 \ M^{-1} \ sec^{-1}$$

In practice, solute concentrations of the order of 1 to 100 mM are generally used. The actual concentrations must be adjusted to satisfy specific experimental requirements, particularly when multiple solutes are used. At concentrations greater than 1 M, direct ionization of the solute (approximately equal to the weight percent of the solute) must be taken into account. The dose per pulse may be varied, and concentrations of free radicals from 1 nM to 0.1 mM per pulse can be generated easily. The time scale of free radical monitoring in most experiments is 0.1 μsec to 1 sec. If required, however, special equipment allows measurements on a time scale of picoseconds.

Physicochemical Properties of Radicals

Absorption Spectra

Free radical absorption spectra are usually red shifted compared with parent compounds, an important feature in monitoring. The spectra can be obtained point by point by measuring absorption at different wavelengths selected by a monochromator at a constant dose/pulse.[3] The absorbances are calculated from actual concentrations of radicals corresponding to the dose/pulse. Instead of multiple pulsing, an optical multichannel analyzer (OMA) records absorption spectra at a preselected time after a single pulse.[7] The technique is based on spectrographic resolution of analyzing light and a diode array instead of a monochromator and photomultiplier as in kinetic spectroscopy. Hydroxyl and alkoxyl radicals have poorly defined spectra; peroxyl radicals absorb at 240 to 250 nm, $\cdot O_2H$ at 225 nm, $\cdot O_2^-$ at 245 nm, and p-substituted phenoxyl radicals at about 400 nm.

Acid–Base Properties

Acid–base properties[8] are determined from spectral changes with pH. For example, α-hydroxy radicals are stronger acids than their parent alcohols:

$$\overset{\backslash}{\underset{/}{C}}\!\!\!\overset{\cdot}{-}OH \rightleftharpoons \overset{\backslash}{\underset{/}{C}}\!\!\!\overset{\cdot}{-}O^- + H^+ \tag{12}$$

$$pK_a \cong 10\text{–}12 \quad \text{(for aliphatic)}$$

The absorption spectra of the deprotonated forms are usually red shifted, as already indicated for the $\cdot O_2H$ and $\cdot O_2^-$ forms.

---

[7] E. P. L. Hunter, M. G. Simic, and B. Michael, *Rev. Sci. Instrum.* **56**, 2199 (1985).
[8] E. Hayon and M. G. Simic, *Acc. Chem. Res.* **7**, 114 (1974).

Redox Properties

Redox properties are determined from interactions of free radicals with oxidants and reductants. Redox potentials can be determined if a satisfactory equilibrium can be achieved between the radical under investigation, $\cdot X$, and a radical, $\cdot A$, with a known redox potential,

$$\cdot X + A^- \underset{k_r}{\overset{k_f}{\rightleftharpoons}} X^- + \cdot A \tag{13}$$

using the Nernst equation:

$$\Delta E = E(\cdot A/A^-) - E(\cdot X/X^-) = 0.059 \log K \tag{14}$$

The equilibrium constant K can be determined kinetically, $K = k_f/k_r$, or spectroscopically from relative absorptions of two radicals at the equilibrium, provided their spectra do not overlap. The redox potential of superoxide radical, $E_7 = -0.33$ V, was determined this way, using quinones (Q) with a known redox potential,[9]

$$\cdot O_2^- + Q \rightleftharpoons O_2 + \cdot Q^- \tag{15}$$

It is important to realize that the redox potentials of free radicals are one-electron redox potentials, and they should not be confused with two-electron redox potentials.

Miscellaneous Properties

Some other properties of free radicals are crucial for their proper identification and monitoring. Radicals may be neutral or charged, which is easily discernible by kinetic conductivity. They may be long lived, for example, the viologen radicals. They all have an unpaired electron and are consequently amenable to ESR investigations (flow or kinetic).[10] In addition to structural features, ESR provides information about the distribution of the unpaired electron within radicals.

Kinetics and Mechanisms of Oxygen Radicals

Hydroxyl Radical, $\cdot OH$

Reactions of $\cdot OH$ radicals are invariably studied in the presence of N_2O in order to eliminate hydrated electrons [reactions (6)]. In addition to

[9] D. Meisel and G. Czapski, *J. Phys. Chem.* **79**, 1503 (1975).

[10] A. D. Trifunac, *in* "Study of Fast Processes and Transient Species by Electron Pulse Radiolysis" (J. H. Baxendale and F. Busi, eds.), p. 163. Reidel, Hingham, Massachusetts, 1982.

the UV-H$_2$O$_2$ process: Quantitative EPR determination of radical concentrations. Wolfrum, E.J.[a], Ollis, D.F.[a], Lim, P.K.[a], Fox, M.A.[b] (NCS-CE[a]; TEX-C[b]) *J. Photochem. Photobiol., A* **78**(3): 259-65 (1994). (94D027)

hydrogen-atom abstraction [reaction (8)] and addition to double bonds [reaction (9)], hydroxyl radical adds readily to aromatic rings. For example, phenylalanine, Phe, reacts with $\cdot$OH to give three transients (o-, m-, and p-Phe—OH), which absorb in the 300 to 350 nm region.[11] Reaction of these transients can be followed and measured by monitoring their absorption. From both pulse radiolytic and product measurements the following scheme has been developed:

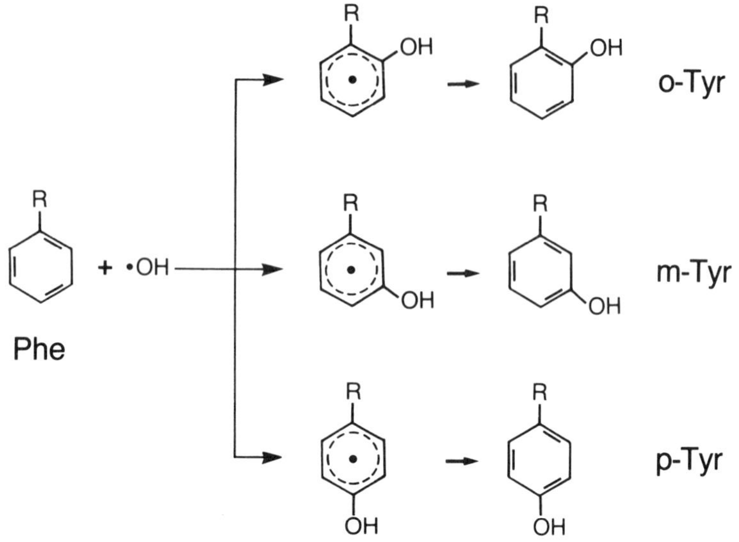

In the absence of oxygen, Phe—OH transients disappear by disproportionation and combination,[11]

$$2 \text{ Phe—OH} \longrightarrow \text{Tyr} + \text{PheH—OH} \tag{17a}$$
$$\longrightarrow \text{HO—Phe—Phe—OH} \tag{17b}$$

The hydrated benzene ring readily loses water both in the dimers and monomers, for example,

$$\text{PheH—OH} \rightarrow \text{Phe} + H_2O \tag{18}$$

In the presence of oxygen, OH adducts react with oxygen,

$$\text{Phe—OH} + O_2 \rightarrow \cdot\text{OOPhe—OH} \tag{19}$$

The reaction mechanisms of this peroxyl radical are not fully determined. Some of the products are known, however, namely, the three isomers of tyrosine. Consequently,

$$\cdot\text{OOPhe—OH} \rightarrow\rightarrow o\text{-}, m\text{-}, \text{ and } p\text{-Tyr} + \text{products} \tag{20}$$

[11] M. G. Simic, E. Gajewski, and M. Dizdaroglu, *Radiat. Phys. Chem.* **24**, 465 (1985).

Hence, both in the absence and in the presence of oxygen, $\cdot OH$ radicals lead to formation of the tyrosine isomers. Since o-Tyr has not been found in normal proteins, its presence is indicative of $\cdot OH$ radical processes.[12] Formation of o-Tyr in proteins can be used as a proof of radiation processing of foods.

Hydroxyl radical is a potent oxidant ($E_0 = 2.6$ V) capable of oxidizing numerous metal ions and their complexes. The redox component in a biosystem may be diminished owing to competition of abstraction and addition reactions.

Superoxide Radical, $\cdot O_2^-$

In addition to being formed by reactions (10) and (11), superoxide radical can be generated by the reaction of oxygen with hydrated electrons:

$$e_{aq}^- + O_2 \rightarrow \cdot O_2^- \tag{21}$$

or hydrogen atoms:

$$\cdot H + O_2 \rightarrow \cdot O_2 H \tag{22}$$

since

$$\cdot O_2 H \rightleftharpoons \cdot O_2^- + H^+ \tag{23}$$
$$pK_a = 4.8$$

as determined from spectral changes as a function of pH.[13] Superoxide radical is extremely long lived on its own because the electron transfer reaction is very slow,[13]

$$\cdot O_2^- + \cdot O_2^- \rightarrow O_2^{2-} \tag{24}$$
$$k \leq 1 \ M^{-1} \ sec^{-1}$$

The reaction between the acid and the base form is much faster,

$$\cdot O_2^- + \cdot O_2 H \rightarrow HO_2^- + O_2 \tag{25}$$
$$k = 8 \times 10^7 \ M^{-1} \ sec^{-1}$$

Because the pK_a value for $\cdot O_2^-$ is only 2 units below biological pH, superoxide would decay quite rapidly in biosystems, even in the absence of any other reactions, such as those with cytochrome c,

$$\cdot O_2^- + cyt \ c(III) \rightarrow O_2 + cyt \ c(II) \tag{26}$$
$$k \cong 10^5 \ M^{-1} \ sec^{-1}$$

[12] L. R. Karam and M. G. Simic, *Anal. Chem.* **60**, 1842 (1988).
[13] B. H. J. Bielski, *Photochem. Photobiol.* **28**, 645 (1978).

or cobalt macrocyclic complexes,[14] for example (M is 1,3,8,10-tetraene N_4),[15]

$$\cdot O_2^- + MCo(II) \rightarrow MCo(II)O_2^-$$
$$k = 1.6 \times 10^9 \ M^{-1} \ sec^{-1} \tag{27}$$

Peroxyl Radicals, ROO·

Peroxyl radicals (alkylperoxyl radicals) and their chemistry are amenable to pulse radiolytic studies. A variety of alkyl radicals can be generated in aqueous and nonaqueous solutions. In aqueous media, where most studies have been conducted, hydroxyl radical is used as an initiator. In the presence of oxygen ($N_2O : O_2 = 4 : 1$) and solute (H_2R), the following consecutive reactions occur:

$$\cdot OH + H_2R \rightarrow H_2O + HR \cdot \tag{28}$$
$$k \cong 10^9 \ M^{-1} \ sec^{-1}$$

$$HR \cdot + O_2 \rightarrow HROO \cdot \tag{29}$$
$$k \cong 10^9 \ M^{-1} \ sec^{-1}$$

Of particular interest are peroxyl radicals of amino acids, sugars, DNA bases, and fatty acids.[4]

Peroxyl radicals interact rapidly with each other, and the rate constant may vary considerably. One of the mechanisms of interaction is formation of tetroxides, which may decay according to the Russell mechanism,[15]

$$2 \ HROO \cdot \rightarrow HROOOORH \rightarrow RO + HROH + O_2 \tag{30}$$
$$k \cong 10^7 \ M^{-1} \ sec^{-1} \qquad k \cong 0.1-1 \ sec^{-1}$$

The α-halogenated peroxyl radicals, which are much more reactive than their nonhalogenated counterparts (e.g., $E_7 \simeq 0.6$ V for $CH_3OO \cdot$,[16] whereas $E_7 \geq 1.1$ V for $CCl_3OO \cdot$[17]), have been of considerable interest. The increase in the redox potential is paralleled by an increase in reactivity with electron donors (phenolic antioxidants, ascorbate). Poor electron donors, such as linoleic acid (H_2L), react fairly slowly and are not amenable to pulse radiolytic studies,[18] for example,

$$HROO \cdot + H_2L \rightarrow HROOH + HL \cdot \tag{31}$$
$$k \cong 60 \ M^{-1} \ sec^{-1}$$

[14] M. G. Simic and M. Z. Hofman, *J. Am. Chem. Soc.* **99**, 2370 (1977).
[15] G. A. Russell, *J. Am. Chem. Soc.* **79**, 3871 (1957).
[16] R. E. Huie and P. Neta, *Int. J. Chem. Kinet.* **18**, 1185 (1986).
[17] S. V. Jovanovic and M. G. Simic, *in* "Oxygen Radicals in Biology and Medicine" (M. G. Simic, K. A. Taylor, J. F. Ward, and C. von Sonntag, eds.), p. 115. Plenum, New York, 1988.
[18] J. A. Howard and K. V. Ingold, *Can. J. Chem.* **45**, 785 (1967).

An interesting reaction of some peroxyl radicals is elimination of $\cdot O_2^-$. For example, elimination of $\cdot O_2^-$ from α-hydroxyperoxyl radicals at pH values approaching the pK_a value for the hydroxyl group can be followed by kinetic conductivity because a neutral species is converted to two charged species,

$$\begin{array}{c} OO\cdot \\ \diagdown \; | \quad \diagdown \\ \diagup COH \rightarrow \diagup C{=}O + \cdot O_2^- + H^+ \\ \diagup \qquad \diagup \end{array} \qquad (32)$$

Alkoxyl Radicals, HRO·

Alkoxyl radicals (alkyloxy radicals) have not been investigated by pulse radiolysis because of their instability. The only convenient way of generating them is via[19]

$$\text{HROOH} + e_{aq}^- \rightarrow \text{HRO}\cdot + \text{OH}^- \qquad (33)$$
$$k \cong 10^{10} \; M^{-1} \; \text{sec}^{-1}$$

Alkoxyl radicals decay rapidly[19] via the so-called β-scission mechanism,

$$\begin{array}{c} \quad O\cdot \\ | \quad | \qquad\qquad | \qquad\qquad \diagup \\ -C{-}C{-} \rightarrow -C\cdot + O{=}C \\ | \quad | \qquad\qquad | \qquad\qquad \diagdown \end{array} \qquad (34)$$
$$k \cong 10^7 \; \text{sec}^{-1} \qquad \text{(in water)}$$

Alkoxyl radicals are studied more conveniently by laser photolysis, which allows faster generation.[20]

Alkoxyl and hydroxyl radicals are generic species, and their properties should be similar. For example, the following reaction,

$$\text{HRO}\cdot + H_2R \rightarrow \text{HROH} + \text{HR}\cdot \qquad (35)$$
$$k \cong 10^7 \; M^{-1} \; \text{sec}^{-1}$$

is similar to reaction (28).

Phenoxyl and Aroxyl Radicals, ArO·

Phenoxyl radicals are derivatives of the primary phenoxyl radical $(C_6H_5O\cdot)$ originating from phenol. Aromatic oxyl radicals (aroxyl radicals, ArO·) are a more general form of these important species, and no special effort will be made to distinguish them. Phenoxyl radicals originate from phenolic antioxidants [butylated hydroxyanisole (BHA), butylated hydroxytoluene (BHT), α-tocopherol]. Aroxyl radicals originate from, for example, heterocyclic physiological antioxidants (5-OH-Trp and

[19] P. Neta, M. Dizdaroglu, and M. G. Simic, *Isr. J. Chem.* **24**, 25 (1984).
[20] J. C. Scaiano, *in* "Oxygen Radicals in Biology and Medicine" (M. G. Simic, K. A. Taylor, J. F. Ward, and C. von Sonntag, eds.), p. 59. Plenum, New York, 1988.

serotonin, which are 5-hydroxyindole derivatives) and numerous other aromatic hydroxy derivatives (naphthol, 8-hydroxyguanine, uric acid).

Phenoxyl radicals can be generated from phenol using either $\cdot OH$ or oxidizing radicals (oxidants). The reactions are somewhat different[21]:

$$C_6H_5OH + \cdot OH \rightarrow \cdot C_6H_5(OH)_2 \tag{36}$$
$$k \cong 6 \times 10^9 \ M^{-1} \ \mathrm{sec}^{-1}$$

$$\cdot C_6H_5(OH)_2 \rightarrow C_6H_5O\cdot + OH^- + H^+ \tag{37}$$
$$k \cong 10^3 \ \mathrm{sec}^{-1}$$

Having a high oxidation potential $[E_7(C_6H_5O\cdot/C_6H_5OH) = 0.95 \ V]$,[17] reactions of phenol with oxidants are slow:

$$C_6H_5OH + \cdot Br_2^- \rightarrow C_6H_5O\cdot + 2\ Br^- + H^+ \tag{38}$$
$$k = 10^6 \ M^{-1} \ \mathrm{sec}^{-1}$$

The rate is increased considerably by deprotonation of the hydroxyl group or by electron-donating substituents, such as the methoxyl group (CH_3O), which reduces the redox potential to 0.73 V.[9] Hence,[17]

$$p\text{-}CH_3OC_6H_4OH + \cdot Br_2^- \rightarrow p\text{-}CH_3OC_6H_4O\cdot + 2\ Br^- + H^+ \tag{39}$$
$$k = 8.8 \times 10^7 \ M^{-1} \ \mathrm{sec}^{-1}$$

The effect of the p-methoxyl substituent is enhanced considerably by the formation of a five- or six-membered ring between oxygen of the methoxyl group and the adjacent position on the benzene ring.[22] Vitamin E (EOH) is a well-known example of a six-membered ring. The redox potential of vitamin E (E_7) is reduced to 0.48 V,[17] and its reactivity with oxidants is greatly enhanced. In cyclohexane,[23]

$$EOH + HROO\cdot \rightarrow EO\cdot + HROO \tag{40}$$
$$k = 7.9 \times 10^6 \ M^{-1} \ \mathrm{sec}^{-1}$$
$$E_a = 2.8 \ \mathrm{kcal \ mol^{-1}}$$

Rate constants for selected phenolic antioxidants and oxyl radicals are shown in Table I.

Aroxyl radicals absorb strongly ($A = 4\text{–}6 \times 10^3 \ M^{-1} \ cm^{-1}$) in the 400 nm region or, for some substituted phenols, even at longer wavelengths.[24] Hence, pulse radiolysis is an excellent experimental technique to study the properties of aroxyl radicals. One property is the decay kinetics. Most aroxyl radicals (BHA, BHT, phenol) decay rapidly,

[21] E. J. Land and M. Ebert, *Trans. Faraday. Soc.* **63**, 1181 (1967).
[22] G. W. Burton, T. Doba, E. J. Gabe, L. Hughes, F. L. Lee, L. Prasad, and K. U. Ingold, *J. Am. Chem. Soc.* **107**, 7053 (1985).
[23] M. G. Simic and E. P. L. Hunter, *in* "Radioprotectors and Anticarcinogensis" (O. F. Nygaard and M. G. Simic, eds.), p. 449. Academic Press, New York, 1983.
[24] S. Steenken and P. Neta, *J. Phys. Chem.* **86**, 3661 (1982).

TABLE I
RATE CONSTANTS FOR REACTION OF SOME RADICALS
WITH ANTIOXIDANTS[a]

Substance	$\cdot$R	ROO$\cdot$	$\cdot$CCl$_3$	CCl$_3$OO$\cdot$	$\cdot$OH
α-Tocopherol	$<10^5$	7.9×10^6	4.5×10^6	1.8×10^8	10^{10}
Ascorbate	—	2.2×10^6	—	2.2×10^8	10^{10}
BHA	—	3.4×10^6	7.8×10^6	3.9×10^7	6×10^9
BHT	$<10^2$	$\sim10^4$	$\sim10^4$	6.1×10^5	6×10^9
Phenol	—	$<10^5$	—	$<10^5$	6×10^9

[a] Rate constant (k) in M^{-1} sec^{-1}, 20°. From Ref. 23. $\cdot$R and ROO$\cdot$ in cyclohexane, $\cdot$CCl$_3$ and CCl$_3$OO$\cdot$ in CCl$_4$, $\cdot$OH in water. $\cdot$R is a carbon-centered cyclohexyl radical (c-C$_6$H$_{11}$).

$$\text{ArO} \cdot + \text{ArO} \cdot \rightarrow \text{products} \tag{41}$$
$$k \cong 10^8 \ M^{-1} \ \text{sec}^{-1}$$

An exception is the α-tocopherol radical (EO $\cdot$), which even under homogeneous conditions in cyclohexane decays very slowly,[25]

$$\text{EO} \cdot + \text{EO} \cdot \rightarrow \text{products} \tag{42}$$
$$k = 3.5 \times 10^2 \ M^{-1} \ \text{sec}^{-1}$$

In membranes the lifetime of the radical α-tocopherol should be longer yet.[26] The long life of EO $\cdot$ apparently results from multiple delocalization of the unpaired electron (resonance) and steric hindrances. An interesting consequence of its long life is the possibility of restitution by ascorbate (A),[27]

$$\text{EO} \cdot + \text{AH}^- \rightarrow \text{EOH} + \cdot \text{A}^- \tag{43}$$
$$k = 1.5 \times 10^6 \ M^{-1} \ \text{sec}^{-1}$$

Another important property of aroxyl radicals is their unreactivity toward oxygen. From pulse radiolysis experiments, and more convincingly from the steady-state oxygen uptake, it follows that most aroxyl radicals do not react with oxygen,[28]

$$\text{ArO} \cdot + \text{O}_2 \rightarrow \text{not observed} \tag{44}$$
$$k < 10^{-2} \ M^{-1} \ \text{sec}^{-1}$$

[25] M. G. Simic, in "Autoxidation in Food and Biological Systems" (M. G. Simic and M. Karel, eds.), p. 17. Plenum, New York, 1980.

[26] E. Niki, J. Tsuchiya, R. Tanimura, and Y. Kamiya, Chem. Lett., 789 (1982).

[27] J. E. Packer, T. F. Slater, and R. L. Willson, Nature (London) 278, 737 (1979).

[28] E. P. L. Hunter, M. F. Desrosiers, and M. G. Simic, Free Radical Biol. Med. 6, 581 (1988).

Conclusions

Pulse radiolysis has advanced understanding of the kinetics and reaction mechanisms of oxygen radicals. It also has provided valuable information about the redox potentials of oxygen radicals. Considerable work still lies ahead, however, before a comprehensive understanding of oxygen radical processes *in vivo* is achieved. Pulse radiolysis and laser photolysis are expected to provide crucial contributions toward those goals.

[3] Free Radical Initiators as Source of Water- or Lipid-Soluble Peroxyl Radicals

By ETSUO NIKI

Introduction

The reactions, fate, and consequence of free radicals in biological systems are in general so complicated that *in vitro* study in simplified model systems is often required. In order to study lipid peroxidation and its inhibition quantitatively, it is necessary and essential to generate free radicals at a known and constant rate and preferably at a specific site. Free radicals can be generated by a variety of methods such as irradiation, redox decomposition of hydroperoxides or hydrogen peroxide by metal ions, and thermal or photochemical decomposition of free radical initiators. To meet the above requirements, thermal decomposition of free radical initiators is preferred. Various kinds of radical initiators include peroxides, hyponitrites, and azo compounds (diazenes).

Azo compounds have been used easily and successfully as radical initiators. Azo compound **1** decomposes unimolecularly as shown in Eqs. (1) and (2) without enzymes or biotransformation to yield molecular nitrogen and two carbon radicals, $R \cdot$. The carbon radicals are formed in pairs in close proximity, some recombining to give stable products [reaction (1)], but many of them diffuse apart and react rapidly with oxygen molecules to give peroxyl radicals, $RO_2 \cdot$ [Eq. (3)],

$$R—N{=}N—R \rightarrow R \cdot + N_2 + \cdot R \rightarrow (1 - e) R—R \tag{1}$$

$$\mathbf{1} \qquad\qquad\qquad 2e\, R \cdot \tag{2}$$

$$R \cdot + O_2 \rightarrow RO_2 \cdot \tag{3}$$

where e is the efficiency of free radical production.

METHODS IN ENZYMOLOGY, VOL. 186

The structure of R determines the rate of decomposition and also the solubility in water and lipid. Thus, it is possible by choosing an appropriate R to prepare hydrophilic or lipophilic azo compounds that generate free radicals at a desirable rate and at a specific temperature. 2,2'-Azo-

$$\underset{\text{AAPH}}{HCl \cdot HN=\underset{\underset{H_2N}{|}}{\overset{\overset{CH_3}{|}}{C}}-\underset{\underset{CH_3}{|}}{\overset{\overset{CH_3}{|}}{C}}-N=N-\underset{\underset{H_3C}{|}}{\overset{\overset{CH_3}{|}}{C}}-\underset{\underset{NH_2}{|}}{\overset{}{C}}=NH \cdot HCl} \qquad \underset{\text{AMVN}}{HC-CH_2-C-N=N-C-CH_2-CH}$$

bis(2-amidinopropane) dihydrochloride (AAPH) and 2,2'-azobis(2,4-dimethylvaleronitrile) (AMVN) have often been used as hydrophilic and lipophilic radical initiators, respectively, at ambient temperatures.[1-7] Hydrophilic AAPH added to the aqueous phase generates radicals in the aqueous region, whereas lipophilic AMVN located in the lipid region of micelles or membrane generates radicals initially within the lipid region.

The rate of decomposition of AAPH is determined primarily by temperature and, to a minor extent, by solvent and pH. At 37° in neutral water, the half-life of AAPH is about 175 hr, so the rate of generation of radicals is virtually constant for the first few hours. The rate of free radical generation, R_i, from AAPH at 37° is given by Eq. (4),[6]

$$R_i = 1.36 \times 10^{-6}[AAPH] \qquad \text{mol/liter/sec} \tag{4}$$

where the concentration of AAPH is given in moles/liter. The rate of radical generation is directly proportional to AAPH concentration. The total amount of radical formed at 37° is calculated from Eq. (5),[6]

$$\text{Total amount of radical formed} = 1.36 \times 10^{-6}[AAPH] \times t \tag{5}$$

where t is time in seconds and the concentration of AAPH is given in moles/liter.

The free radicals generated from AAPH in the aqueous phase induce the chain oxidations of lipid micelles,[1,3,6,7] phospholipid liposomes,[2,4-6]

[1] Y. Yamamoto, S. Haga, E. Niki, and Y. Kamiya, *Bull. Chem. Soc. Jpn.* **57,** 1260 (1984).
[2] Y. Yamamoto, E. Niki, Y. Kamiya, and H. Shimasaki, *Biochim. Biophys. Acta* **795,** 332 (1984).
[3] L. R. C. Barclay, S. J. Locke, J. M. MacNeil, J. VanKessel, G. W. Burton, and K. U. Ingold, *J. Am. Chem. Soc.* **106,** 2479 (1984).
[4] E. Niki, A. Kawakami, Y. Yamamoto, and Y. Kamiya, *Bull. Chem. Soc. Jpn.* **58,** 1971 (1985).
[5] T. Doba, G. W. Burton, and K. U. Ingold, *Biochim. Biophys. Acta* **835,** 298 (1985).
[6] E. Niki, M. Saito, Y. Yoshikawa, Y. Yamamoto, and Y. Kamiya, *Bull. Chem. Soc. Jpn.* **59,** 471 (1986).
[7] W. A. Pryor, T. Strickland, and D. F. Church, *J. Am. Chem. Soc.* **110,** 2224 (1988).

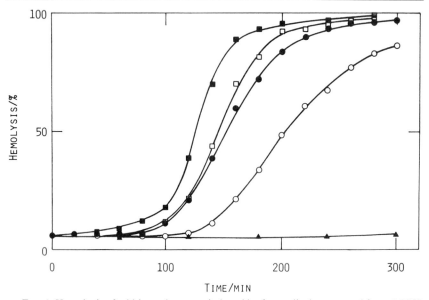

FIG. 1. Hemolysis of rabbit erythrocytes induced by free radicals generated from AAPH. The concentration of AAPH was as follows: ▲, 0; ○, 25; ●, 50; □, 75; ■, 100 mM. AAPH was added to 10% rabbit erythrocyte suspensions in physiological saline (pH 7.4) and incubated in air at 37°.

erythrocyte membranes,[8,9] and erythrocyte ghosts.[10] Figure 1 shows an example of oxidative hemolysis of erythrocytes induced by AAPH.[11] AAPH solution is prepared by dissolving an appropriate amount of AAPH in an aqueous solution containing 50 mM NaCl and 10 mM phosphate buffer (pH 7.3). Two milliliters of packed erythrocytes is suspended in 8 ml of an aqueous solution containing 125 mM NaCl and 10 mM phosphate buffer (pH 7.3). Then 10 ml of AAPH solution and 10 ml of erythrocyte suspension are mixed and incubated at 37°. An aliquot is taken periodically to measure the extent of hemolysis spectrophotometrically. AAPH induces the oxidation of lipids and proteins in the membrane and eventually causes hemolysis. Hemolysis takes place sooner if the AAPH con-

[8] Y. Yamamoto, E. Niki, Y. Kamiya, M. Miki, H. Tamai, and M. Mino, *J. Nutr. Sci. Vitaminol.* **32,** 475 (1986).

[9] M. Miki, H. Tamai, M. Mino, Y. Yamamoto, and E. Niki, *Arch. Biochem. Biophys.* **258,** 373 (1987).

[10] Y. Yamamoto, E. Niki, J. Eguchi, Y. Kamiya, and H. Shimasaki, *Biochim. Biophys. Acta* **819,** 29 (1985).

[11] E. Niki, *Chem. Phys. Lipids* **44,** 227 (1987).

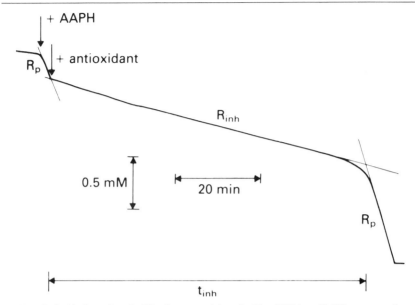

FIG. 2. Oxidation of methyl linoleate emulsions in 10 mM Triton X-100 aqueous dispersions at 37°. Addition of 10 mM AAPH induces oxidation and 10 μM uric acid suppresses
oxidation. Uric acid is consumed at a constant rate (data not shown), and when it is depleted, the induction period ends and a fast oxidation proceeds at a rate similar to that before
the addition of uric acid. From this curve, the length of the induction period (t_{inh}) and the
rate of inhibited oxidation can be measured.

centration is increased, and it was found that the extent of hemolysis was
proportional to the amount of radical formed.[9]

The inhibition of oxidation by a chain-breaking antioxidant can be also
studied quantitatively by AAPH and AMVN. Figure 2 shows the results
of oxidation of methyl linoleate emulsions in aqueous suspensions.[6]
Methyl linoleate is stable, and its spontaneous oxidation at 37° is very
slow. When AAPH is added to the aqueous phase, the radicals generated
from AAPH induce oxidation, and a constant rate of oxygen uptake is
observed. When an antioxidant such as uric acid is added, the rate of
oxygen uptake is reduced, and a clear induction (or inhibition) period is
produced. The antioxidant is consumed at a constant rate, and following
its depletion the induction period ends and a fast oxidation proceeds. It is
noteworthy that the rate of oxidation after the induction period is the
same as that before the addition of antioxidant.

The length of the induction period t_{inh} is given by Eq. (6),

$$t_{inh} = n[IH]/R_i \tag{6}$$

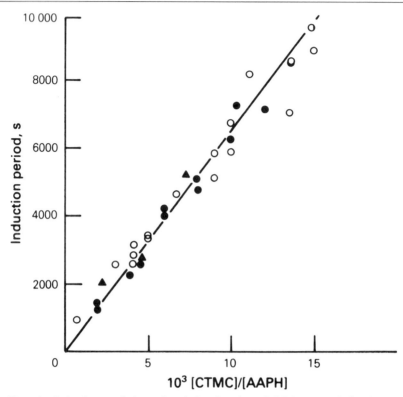

FIG. 3. Induction period produced by 2-carboxy-2,5,7,8-tetramethyl-6-chromanol (CTMC) in the oxidation of methyl linoleate emulsions (●), soybean phosphatidylcholine liposomes (○), and erythrocyte ghosts (▲) in aqueous dispersions at 37° induced by AAPH, as a function of [CTMC]/[AAPH].

where n is the stoichiometric number of radicals trapped by each antioxidant; IH, the antioxidant; and R_i, the rate of chain initiation. As shown in Fig. 3, the induction period has been observed to be proportional to [IH]/[AAPH], as predicted from Eq. (6), in the oxidation of methyl linoleate emulsions, soybean phosphatidylcholine liposomes, and erythrocyte ghost membranes in aqueous dispersions induced by AAPH and inhibited by 2-carboxy-2,5,7,8-tetramethyl-6-chromanol, a water-soluble vitamin E analog known as Trolox.

Application of AMVN to in Vitro Studies

AMVN is a lipid-soluble and water-insoluble azo compound. When AMVN is incorporated into phosphatidylcholine liposomal membranes, radicals generated from AMVN by Eqs. (1)–(3) induce chain oxidation, and a constant rate of oxygen uptake is observed. A typical experiment is as follows.[4]

Reagents

Purified soybean phosphatidylcholine
Potassium phosphate buffer (5 mM) containing 0.1 M NaCl (pH 7.4)
AMVN
α-Tocopherol

Procedure. Appropriate amounts of phosphatidylcholine, AMVN, and α-tocopherol are placed in a pear-shaped flask and dissolved in 4 ml chloroform, which is removed under reduced pressure using a rotary evaporator to obtain a thin film on the glass wall. Ten milliliters of phosphate buffer is added to the flask, and the mixture is agitated vigorously with a vortex mixer for 2 min to obtain a multilamellar liposome suspension. The final concentrations of phosphatidylcholine, AMVN, and α-tocopherol are 5.1 mM, 1.0 mM, and 3.0 μM, respectively. A portion of the solution is transferred to a vessel that is connected to an oxygen electrode, and the vessel is immersed in a water bath maintained at 37°. The rate of oxidation is measured by following the uptake of oxygen. The remaining solution is also incubated in the water bath, and an aliquot is removed periodically to analyze the formation of conjugated diene hydroperoxides and/or consumption of α-tocopherol using HPLC. The results are shown in Fig. 4.

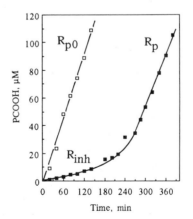

FIG. 4. AMVN-induced oxidation of soybean PC liposomes in aqueous dispersions at 37° in air, in the presence and absence of α-tocopherol. AMVN and α-tocopherol were incorporated into liposomal membranes, the final concentrations in the whole suspensions being 0.30 mM and 1.0 μM, respectively. The rates of oxidation in the absence of antioxidant (R_{p0}), and the inhibited oxidation (R_{inh}) after the induction period (R_p), can be measured from this figure.

The following points are noteworthy: (1) Soybean phosphatidylcholine is suitable as a substrate because it contains about 70% linoleic acid as an oxidizable polyunsaturated lipid whose oxidation gives conjugated diene hydroperoxides quantitatively. Therefore, the rate of oxidation can be measured quantitatively from either oxygen uptake or conjugated diene hydroperoxide formation. Quantitative measurement of the oxidation of phosphatidylcholine from other sources such as egg yolk or rat liver is more difficult.[2] (2) When AAPH is used in place of AMVN as an initiator, the liposomes should be unilamellar; these can be prepared by sonication of multilamellar liposomes. Unilamellar liposomes are required because antioxidants such as vitamin E incorporated into the inner membranes of multilamellar vesicles cannot interact with radicals generated from AAPH.[11]

One of the characteristics of a biological system is its inhomogeneity. The relative contribution of various antioxidants depends on their local concentration and the site of radical production, as well as on their inherent reactivities toward free radicals. This reactivity can be determined by using AAPH and AMVN as radical initiators in the membrane system. Figure 5 illustrates how hydrophilic antioxidants and lipophilic antioxidants function toward free radicals.[4] Hydrophilic antioxidants, such as vitamin C and uric acid, scavenge radicals in the aqueous phase and suppress the oxidation initiated with AAPH. Hydrophilic antioxidants do not suppress oxidation efficiently when induced by AMVN incorporated into the membranes, suggesting that antioxidants residing in the aqueous phase cannot scavenge radicals within the membranes efficiently.

Application of AAPH to *in Vivo* Studies

Azo compounds can be used as radical sources for the *in vivo* system, too, since they decompose thermally without biotransformation, as shown below.[12]

Procedure. A total of 70 male ICR mice weighing 20–25 g are divided into 7 groups of 10 mice each and injected intraperitoneally with 100 mg/kg AAPH dissolved in physiological saline. Ten mice injected with 0.2 ml of physiological saline into the peritoneal cavities served as controls. At intervals of 0.25, 0.5, 1, 1.5, 3, 6, and 24 hr after the administration of AAPH solution, the mice are sacrificed by cervical dislocation. Small samples of the liver, kidney, thymus, heart, and lung are removed and specimens for light and electron microscopy prepared. In order to quantify the oxidative damage, areas of fat droplets formed are measured from

[12] K. Terao and E. Niki, *J. Free Radicals Biol. Med.* **2**, 193 (1986).

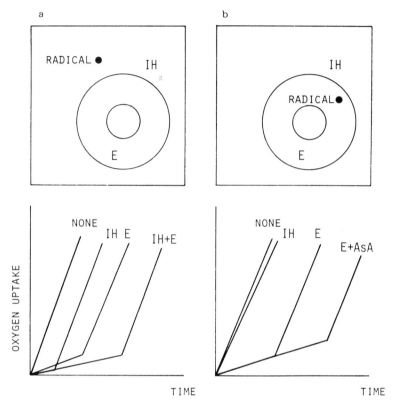

FIG. 5. Oxidation of phosphatidylcholine liposomal membranes induced by (a) AAPH and (b) AMVN and inhibition by α-tocopherol (E) incorporated into the membrane and by water-soluble antioxidants such as ascorbic acid (AsA), uric acid, and cysteine.

electron micrographs. The effects of antioxidants are examined by comparing the severity of hepatocellular injuries in mice receiving AAPH plus antioxidant with those receiving only AAPH.

AAPH caused damage to biological tissues. No specific target organ was observed. The most striking, fine structural changes were degeneration, swelling, and disruption of the endothelium lining cells of the capillaries in various organs. Death of lymphocytes in the lymphoid tissues and fatty degeneration of the liver and kidney were also observed. When water-soluble, chain-breaking antioxidants, such as 2-carboxy-2,5,7,8-tetramethyl-6-chromanol, uric acid, cysteine, and glutathione, were injected together with AAPH, the suppression of damage was dose dependent.

Conclusion

Free radicals can be generated in either the aqueous or the lipid phase as required by using water-soluble AAPH or lipid-soluble AMVN. Admittedly, such azo compounds are not present in biological systems, but they are useful tools for studying quantitatively (1) the damage induced by free radicals on biological and related molecules and membranes and (2) the inhibition in model systems. The advantages are that the radicals can be generated at a constant rate at a specific site and that the rate of radical flux can be measured and controlled. Obviously, the most important characteristic of the free radical reaction is that it proceeds by a chain mechanism, that is, the rate of the overall reaction or the extent of damage can be quite significant even if the rate of initial radical formation or the amount of attacking radical is very small. It is therefore quite important to know how long the kinetic chain lasts. The chain length can never be known without knowing the rate of chain initiation or the radical flux. In fact, in the *in vitro* experiment the kinetic chain length was as long as 100 in the oxidation of erythrocyte membranes induced by AAPH.[8] Another advantage in using azo compounds is that, unlike peroxides, they are not explosive and can be handled easily and safely.

[4] Generation of Iron(IV) and Iron(V) Complexes in Aqueous Solutions

By Benon H. J. Bielski

Introduction

In aqueous solutions simple $L_m Fe(IV)$ (ferryl) and $L_m Fe(V)$ (perferryl) complexes can be generated and studied by the pulse radiolysis technique.[1,2] Although the spectral and kinetic properties of such complexes are most conveniently studied by conventional pulse radiolysis (pr), measurement of their reactivity with substrates frequently requires the use of a modified stopped-flow (sf) spectrophotometer in which one of the flowing solutions passes through an electron beam where the desired hypervalent iron state is generated in isolation (the sf-pr method). Some studies require premixing of solutions before pulse irradiation (the pre-

[1] J. D. Rush and B. H. J. Bielski, *J. Am. Chem. Soc.* **108**, 523 (1986).
[2] B. H. J. Bielski and M. J. Thomas, *J. Am. Chem. Soc.* **109**, 7761 (1987).

mix-pr method), to minimize the reaction time between the hypervalent iron precursor, for example, Fe(III) or Fe(VI), and a substrate or radical precursor.

Ferryl Iron, Fe(IV)

Fe(IV) Hydroxide, $[(OH)_nFe{=}O]^{2-n}$

While the various forms of Fe(III) hydroxide/oxide are very insoluble (except in acid solutions), $Fe(OH)_3$ is solubilized in strong alkaline solutions with formation of the $Fe(OH)_4^-$ complex. Complex formation can be achieved at pH values as low as 10.7, but because of the rapid formation of precipitates at this pH, it is advantageous to work at pH values of at least 12. Such solutions are conveniently prepared by slow addition of a dilute $Fe(ClO_4)_3$ solution (pH 1.0) to a sodium or potassium hydroxide solution followed by removal of the gelatinous precipitate of ferric hydroxide/oxide by centrifugation. The supernatant solution containing $Fe(OH)_4^-$ is subsequently saturated with nitrous oxide and pulse irradiated:

$$H_2O \rightsquigarrow OH(2.75),\ e_{aq}^-(2.65),\ H(0.65),\ H_2O_2(0.72),\ H_2(0.45) \tag{1}$$

$$H \rightleftharpoons e_{aq}^- + H^+ \tag{2}$$

$$e_{aq}^- + N_2O + H_2O \rightarrow OH + OH^- + N_2 \tag{3}$$

$$OH \rightleftharpoons O^- + H^+ \tag{4}$$

$$O^-/OH + Fe(OH)_4^- \rightarrow FeO(OH)_n^{2-n} + (4-n)\,OH^-/(3-n)\,OH^- + H_2O \tag{5}$$

The numerical values given in parentheses in Eq. (1) are G values, that is, the number of radicals/molecules formed per 100 eV of energy dissipated in the medium.[3] The pK_a values for Eqs. (2) and (4) are 7.85 and 11.9,[3] respectively. The OH/O^- radicals are the reactive species in this system, and the lower limit for k_5 is $8.5 \times 10^7\ M^{-1}\ sec^{-1}$.[1]

The species $FeO(OH)_n^{2-n}$ decays primarily by a first-order process (k_{decay} approximately $2 \pm 1\ sec^{-1}$ in 1 M NaOH, 25°). As is apparent, the relatively, long half-life for this ferryl species allows it to be studied using a commercial stopped-flow apparatus in conjunction with pulse radiolysis. The spectrum of the $FeO(OH)_n^{2-n}$ complex is similar to that of $[(P_2O_7)_2Fe^{IV}O]^{6-}$ (see Fig. 1). Once the complex is formed under sf-pr conditions, the pH of the solution mixture can be changed by addition of appropriate buffers to the nonirradiated substrate solution. It is advantageous to add an iron-chelating agent such as phosphate, pyrophosphate, EDTA, or diethylenetriaminepentaacetic acid (DTPA) to the buffer system, as it prevents precipitation of Fe(III), that is formed during the

---

[3] H. A. Schwarz, *J. Chem. Educ.* **58**, 101 (1981).

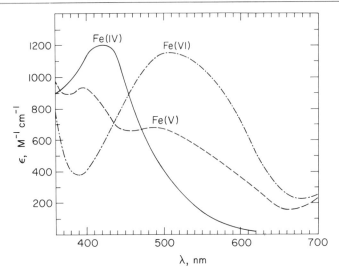

FIG. 1. Absorption spectra of K_2FeO_4 [Fe(VI)], FeO_4^- [Fe(V)], and $[(P_2O_7)_2Fe^{IV}O]^{6-}$ [Fe(IV)] in aqueous solutions at pH 10.0, 24°.

reduction of the hypervalent iron compounds [reaction (6)]. Care must be taken that the chelating agents (L_m) do not interfere with the reaction of interest.

$$Fe{=}O(OH)_n^{2-n} + substrate \xrightarrow{L_m} L_mFe(III) + oxidation\ product(s) \qquad (6)$$

Fe(IV) Pyrophosphate, $[(P_2O_7)_2Fe^{IV}O]^{6-}$

At pH 10, Fe(III) forms a pyrophosphate complex in which the bound water molecule is 71% deprotonated (pK_7 9.6)[4]:

$$[(P_2O_7)_2Fe^{III}(H_2O)]^{5-} \rightleftharpoons [(P_2O_7)_2Fe^{III}OH]^{6-} + H^+ \qquad (7)$$

Pulse experiments are routinely carried out with solutions containing 10^{-4}–10^{-3} M Fe(III), 0.1 M sodium pyrophosphate, and 26 mM N_2O. Following reactions (1)–(4), the formation and decay of Fe(IV)–pyrophosphate are described by the following set of reactions[4]:

$$[(P_2O_7)_2Fe^{III}(OH)]^{6-} + OH \rightarrow [(P_2O_7)_2Fe^{IV}O]^{6-} + H_2O \qquad (8)$$
$$2[(P_2O_7)_2Fe^{IV}O]^{6-} \rightarrow [(P_2O_7)_2FeOOFe(P_2O_7)_2]^{12-} + 2\ H_2O \qquad (9)$$
$$[(P_2O_7)_2Fe^{IV}O]^{6-} + H_2O_2 \rightarrow [(P_2O_7)_2Fe^{III}OH]^{6-} + O_2^- + H^+ \qquad (10)$$
$$[(P_2O_7)_2Fe^{IV}O]^{6-} + O_2^- + H^+ \rightarrow [(P_2O_7)_2Fe^{III}OH]^{6-} + O_2 \qquad (11)$$

where $k_8 = 8 \times 10^7\ M^{-1}\ sec^{-1}$, $k_9 \cong 10^6\ M^{-1}\ sec^{-1}$, and $k_{10} = 9 \times 10^5\ M^{-1}$ sec^{-1}. The spectrum of the $[(P_2O_7)_2Fe^{IV}O]^{6-}$ complex (see Fig. 1) has an absorption band at 420 nm with a molar absorptivity $\varepsilon_{420\ nm}$ of 1200 ± 150

4 J. D. Melton and B. H. J. Bielski, *Radiat. Phys. Chem.* (in press).

M^{-1} cm^{-1}. The 420 nm peak undergoes a blue shift below pH 10.0. Although a small amount of pyrophosphate radical is formed in this system (via OH + $P_2O_7^{4-}$), it does not interfere because of the low reactivity with the ferric–pyrophosphate complex. In addition under sf-pr conditions the pyrophosphate radicals disappear before the irradiated solution reaches the mixer (optimal total lifetime 2–7 msec; dead time ~7–12 msec). Once the ferryl–pyrophosphate complex has been formed at pH 10, it can be studied at lower pH values by the pr-sf method.

Fe(IV) Carbonate/Bicarbonate

Although there are no numerical values available describing the stability of ferric–carbonato/bicarbonato complexes, our preliminary experiments indicate that the bicarbonato complexes are more stable than the corresponding carbonato complexes. For example, significantly more Fe(III) is complexed (0.8 mM) in a 1 M NaHCO$_3$ solution at pH 8 than in a 1 M Na$_2$CO$_3$ solution at pH 12 [40 μM Fe(III)]. Although the exact composition of the Fe(III) and the corresponding Fe(IV) complexes is as yet unknown, above pH 11 the latter display the same characteristic 420 nm peak as the Fe(IV)–hydroxyl and Fe(IV)–pyrophosphate complexes. On lowering the pH, the 420 nm peak undergoes a blue shift.[5]

The Fe(IV)–carbonato complex is most conveniently generated in 0.1–1.0 M sodium carbonate/bicarbonate in the pH range 8.0–11.0. In an N$_2$O-saturated solution the OH radical is converted to the bicarbonate/carbonate radical (HCO$_3$ $\rightleftharpoons$ CO$_3$ $^{\overline{\cdot}}$ + H$^+$; pK_a 7.9),[6] which oxidizes the Fe(III)–bicarbonato/carbonato complex to the corresponding Fe(IV) state. The kinetic and spectral characteristics of the HCO$_3$·/CO$_3$ $^{\overline{\cdot}}$ radicals are well established ($\varepsilon_{600\ nm,max}$ 1860 M^{-1} cm^{-1}),[6,7] and hence the rate of formation of the ferryl complex at 420 nm is easily checked against the disappearance of CO$_3$ $^{\overline{\cdot}}$ at 600 nm. The mechanism for the formation and disappearance of the ferryl–carbonato complex (L$_m$ = HCO$_3^-$/CO$_3^{2-}$) is described by the following reactions:[5]

$$OH + CO_3^{2-} \rightarrow CO_3\ ^{\overline{\cdot}} + OH^- \qquad (12)$$
$$CO_3\ ^{\overline{\cdot}} + L_mFe(III) \rightarrow L_mFe(IV) + CO_3^{2-} \qquad (13)$$
$$2\ L_mFe(IV) \rightarrow product(s) \qquad (14)$$

The corresponding rate constants at pH 13 are 4.0×10^8, 6.0×10^7, and $1.3 \times 10^5\ M^{-1}$ sec^{-1} for Eqs. (12)–(14), respectively. Overall the ferryl–carbonato complex appears to be promising for studies with substrates as it has considerably longer lifetimes (0.5–1.0 sec) and has been used at pH values as low as 8.6.[5]

[5] B. H. J. Bielski, unpublished results, 1989.
[6] T. E. Eriksen, J. Lind, and G. Merenyi, *Radiat. Phys. Chem.* **26**, 197 (1985).
[7] G. V. Buxton and A. J. Elliot, *Radiat. Phys. Chem.* **27**, 241 (1986).

Perferryl Iron, Fe(V)

The perferryl species (FeO_4^{3-}) can be generated from ferrate (K_2FeO_4) under pulse radiolytic conditions by reduction with either the hydrated electron [reaction (15)] in the presence of *tert*-butanol (which serves as an OH radical scavenger), or by the CO_2^- radical [reaction (17)].[1,2] The use of the CO_2^- radical has the advantage that $G = 6.1$, whereas $G(e_{aq}^-) = 2.65$, thus doubling the yield of the perferryl species. Although the use of the CO_2^- radical has the disadvantage that all three hypervalent iron species [Fe(IV), Fe(V), Fe(VI)] react with $HCOO^-$, this difficulty is overcome by the use of the premix-pr technique and appropriate corrections for reactions (18) and (19),

$$
\begin{aligned}
FeO_4^{2-} + e_{aq}^- &\rightarrow FeO_4^{3-} & (15)\\
HCOO^- + OH &\rightarrow CO_2^- + H_2O & (16)\\
FeO_4^{2-} + CO_2^- &\rightarrow FeO_4^{3-} + CO_2 & (17)\\
FeO_4^{2-} + HCO_2^- &\rightarrow product(s) & (18)\\
FeO_4^{3-} + HCO_2^- &\rightarrow product(s) & (19)
\end{aligned}
$$

where $k_{15} = 2.0 \times 10^{10}\ M^{-1}\ sec^{-1}$, $k_{16} = 3.8 \times 10^9\ M^{-1}\ sec^{-1}$, $k_{17} = 3.5 \times 10^8\ M^{-1}\ sec^{-1}$, $k_{18} = 2.3 \times 10^{-2}\ M^{-1}\ sec^{-1}$, and $k_{19} = 2.5 \times 10^3\ M^{-1}\ sec^{-1}$.[2]

Overall, perferryl is a relatively stable species in alkaline solutions (pH 9–12) that has been used for rate studies with either the premix-pr or sf-pr methods.[2] Routine experiments can be carried out with solutions or solution mixtures that contain 50–200 μM K_2FeO_4, 0.5–1.0 mM HCOONa, 26 mM N_2O, and appropriate amounts of phosphate (10–100 mM) to prevent precipitation of ferric hydroxides/oxides.

The precursor of the perferryl species, K_2FeO_4, is easily synthesized from $Fe(NO_3)_3$ and NaOCl in 5–10 N NaOH and precipitated in crystalline form by addition of KOH.[8] After several recrystallizations the compound has a purity greater than 98%; the approximately 2% impurity is mostly inert K_2CO_3. The crystalline material is stable for months and, although it is a powerful oxidizing agent, is quite safe. While aqueous K_2FeO_4 solutions are relatively stable in the pH range 9–10.5 (4–6 hr), they deteriorate rapidly in more acidic media with evolution of oxygen. The decomposition process is accelerated by ferric hydroxides/oxides which accumulate as the reaction progresses. Hence, the addition of complexing agents like phosphate, EDTA, and DTPA is advantageous if they do not interfere with the reaction under study. The chemistry of Fe(VI)

[8] G. W. Thompson, L. T. Ockerman, and J. M. Schreyer, *J. Am. Chem. Soc.* **73**, 1379 (1951).

has been well reviewed,[9,10] and it has been shown that in several reactions it displays a relatively high degree of selectivity.[11,12]

Acknowledgments

The author wishes to thank Drs. D. E. Cabelli and J. D. Melton for constructive criticism of the manuscript. This research was carried out at Brookhaven National Laboratory under contract DE-ACO2-76CHOOO16 with the U.S. Department of Energy and supported by its Division of Chemical Sciences, Office of Basic Energy Sciences.

[9] R. J. Audette and J. W. Quail, *Inorg. Chem.* **11,** 1904 (1972).
[10] J. D. Carr, P. B. Kelter, A. Tabatabai, D. Spichal, J. Erickson, and C. W. McLaughlin, *in* "Proceedings of the Conference on Water Clorination and Chemical Environment Impact Health Effects" (R. L. Jolley, ed.), p. 1285. Chelsea, Bronx, New York, 1985.
[11] W. Levanson and C. A. Mcauliffe, *Coord. Chem. Rev.* **12,** 151 (1974).
[12] J. T. Groves and T. E. Nemo, *J. Am. Chem. Soc.* **105,** 6243 (1983).

[5] Sources of Compilations of Rate Constants for Oxygen Radicals in Solution

By Alberta B. Ross and Benon H. J. Bielski

Introduction

A great many rate constants for reactions of oxygen radicals with various organic and inorganic species in solution have been determined. The kinetics have been studied by time-resolved decay or buildup of intermediates produced by radiolysis, photolysis, and Fenton-type reactions, or by comparison of the reactivity of a particular species with its reactivity toward another chemical species for which absolute rate data have been established. Compilation and evaluation of rate constants for oxygen radicals reported in the literature have been carried out by several groups. These efforts have resulted in published compilations containing evaluated data for several thousands of reactions.

The most recent and/or most comprehensive publications are listed here; many earlier data compilations and reviews on the chemical kinetics of these radicals are cited in the references given below. In addition to published compilations, an online database containing some of these data has been made available by the Radiation Chemistry Data Center.

Data Compilations

Hydroxyl Radical (HO ·/O ⁻)

Rate constants for about 1,500 reactions involving the hydroxyl radical and its conjugate base in aqueous solution, complete through 1986, have been collected in a critical compilation.[1] The data are available in the RATES database, and updates will be carried out.

Perhydroxyl Radical (HO$_2$·/O$_2$⁻)

Rate constants for about 350 reactions involving the superoxide radical anion and its conjugate acid in aqueous solution have been collected in a published compilation completed in 1983.[2] Additional data for about 100 reactions have been compiled and evaluated and are available, along with data from the published compilation, in the RATES database.

Singlet Oxygen ($^1O_2$)

A comprehensive compilation of reactions of the lowest electronically excited singlet state of molecular oxygen with 690 organic and inorganic species in solution, complete through 1978, was published.[3] It is estimated that the number of reactions for which rate constants are available has at least doubled in the decade since this compilation was completed. Updating and reevaluation of the data are in progress. A new technique for observing decay of singlet oxygen, infrared luminescence, has been used to provide a new source of reliable data. A later review contains several hundred rate constants for reactions of singlet oxygen.[4]

Alkoxyl, Aroxyl, and Alkylperoxyl Radicals (RO ·, ArO ·, ROO ·)

A volume in the Landolt–Boernstein series[5] contains absolute rate constants for over 100 reactions involving alkoxyl radicals, rate constants for over 200 reactions involving aroxyl radicals, and rate constants in-

---

[1] G. V. Buxton, C. L. Greenstock, W. P. Helman, and A. B. Ross, *J. Phys. Chem. Ref. Data* **17**, 513 (1988).

[2] B. H. J. Bielski, D. E. Cabelli, R. L. Arudi, and A. B. Ross, *J. Phys. Chem. Ref. Data* **14**, 1041 (1985).

[3] F. Wilkinson and J. G. Brummer, *J. Phys. Chem. Ref. Data* **10**, 809 (1981).

[4] B. M. Monroe, *in* "Singlet Oxygen" (A. A. Frimer, ed.), Vol. 1, p. 177. CRC Press, Boca Raton, Florida, 1985.

[5] J. A. Howard and J. C. Scaiano, *in* "Landolt–Boernstein Numerical Data and Functional Relationships in Science and Technology, New Series, Group 2: Atomic and Molecular Physics" (K.-H. Hellwege and O. Madelung, eds.), Vol. 13 (H. Fischer, ed.), Part d. Springer-Verlag, Berlin, 1984.

TABLE I
RATE CONSTANTS FOR REACTIONS
OF OXY RADICALS

Radical(s)	Number of reactions	Refs.
$HO_x \cdot$	1960	1,2[a]
$RO_x \cdot$	~1300	5,6
$^1O_2{}^*$	700	3,4
O_3	250	7[a]
$CO_3{}^{\overline{\cdot}}$	180	7[a]
$RNO \cdot$	450	8
$NO_3 \cdot$	30	7[a]
$PO_4{}^{2-}$	75	7[a]
$SO_x{}^{\overline{\cdot}}$	280	7[a]
$SeO_x{}^{\overline{\cdot}}$	20	7[a]
$ClO_x \cdot$	100	7[a]
$BrO_x{}^{\cdot}$	10	7[a]

[a] In the RATES database.

volving over 1,000 reactions of alkylperoxyl radicals. Data obtained in both aqueous and nonaqueous solutions are reported. The compilation is comprehensive through 1980. Another project is underway for compilation of organic peroxyl radical rate constants.[6]

Inorganic Oxygen Radicals

Rate constants for reactions of $CO_3{}^{\overline{\cdot}}$, $NO_3 \cdot$, $PO_4{}^{2-}$, $SO_2{}^{\overline{\cdot}}$, $SO_4{}^{\overline{\cdot}}$, $SO_5{}^{\overline{\cdot}}$, $SeO_3{}^{\overline{\cdot}}$, $ClO \cdot$, $ClO_2 \cdot$, $BrO \cdot$, $BrO_2 \cdot$, and ozone in aqueous solution have been collected in a new compilation which covers the literature through mid-1986.[7]

Organic Oxynitrogen Radicals (RNO·)

Rate constants for several hundred reactions of aminoxyl and iminoxyl radicals (nitroxide radicals) in solution have been compiled.[8]

Computer-Searchable Database

The online database RATES contains aqueous solution rate constants for reactions of inorganic radicals, including oxy radicals, as summarized

[6] P. Neta, R. E. Huie, and A. B. Ross, *J. Phys. Chem. Ref. Data* **19**, in press (1990).

[7] P. Neta, R. E. Huie, and A. B. Ross, *J. Phys. Chem. Ref. Data* **17**, 1027 (1988).

[8] K. U. Ingold and B. P. Roberts, *in* "Landolt–Boernstein Numerical Data and Functional Relationships in Science and Technology, New Series, Group 2: Atomic and Molecular Physics" (K.-H. Hellwege and O. Madelung, eds.), Vol. 13 (H. Fischer, ed.), Part c, p. 166. Springer-Verlag, Berlin, 1983.

in Table I. Published versions of the data have appeared,[1,2,7] and periodic updates of the online version are carried out. Direct dial access[9,10] to the database is available. The database is integrated with a bibliographic database (RCDCbib) and a registry file (RCDCreg) which contains substance information for the chemical species.

Acknowledgments

This work was carried out under contract with the U.S. Department of Energy and supported by the Division of Chemical Sciences, Office of Basic Energy Sciences, at the Notre Dame Radiation Laboratory (A.B.R.) (Contract DE-AC02-76ER00038) and at Brookhaven National Laboratory (B.H.J.B.) (Contract DE-AC02-76CH00016). The Radiation Chemistry Data Center is also supported by the Office of Standard Reference Data of the National Institute of Standards and Technology.

[9] W. P. Helman, G. L. Hug, I. Carmichael, and A. B. Ross, *Radiat. Phys. Chem.* **32,** 89 (1988).
[10] For information on access to the RATES database, write to the author (A.B.R.) at the Radiation Chemistry Data Center, Radiation Laboratory, University of Notre Dame, Notre Dame, Indiana 46556. Some tables can be purchased from the Radiation Chemistry Data Center.

[6] Use of Polyaminocarboxylates as Metal Chelators

By Diane E. Cabelli and Benon H. J. Bielski

Introduction

The polyaminocarboxylates (PACs) discussed here all contain at least one basic unit of NCH_2COO^-. In particular, they include iminodiacetate, IDA [$HN(CH_2COO^-)_2$]; nitrilotriacetate, NTA [$N(CH_2COO^-)_3$]; ethylenediaminetetraacetate, EDTA {$[CH_2N(CH_2COO^-)_2]_2$}; diethylenetriaminepentaacetate, DTPA [$(^-OOCCH_2)_2NCH_2CH_2N(CH_2COO^-)CH_2$ $CH_2N(CH_2COO^-)_2$]; and variants of the basic EDTA structure where some of the acetate groups are substituted or the ethylene bridge is modified. The common characteristics of all of these species are that they bind metals strongly and are commonly used in *in vitro* and *in vivo* biological studies as metal (particularly iron and copper) chelators. The assumption is often made that by chelation these ligands will remove metals and thus render them inactive in relation to the mechanism under study. The purpose of this chapter is to discuss the chemistry of these metal–PAC complexes in the presence of O_2^-/HO_2, H_2O_2, and OH radicals, point out

METHODS IN ENZYMOLOGY, VOL. 186

some of the relevant reaction rates, and note where the chemistry of these metal chelates becomes important relative to the chemistry of oxy radical/peroxide systems.

The importance of using very pure metal-free chelators cannot be overemphasized. The strong chelating ability of PACs results in the tendency for metals to be introduced *into* a system along with the chelator. Even if the metal complex reacts with O_2^-/HO_2 with a rate constant of only 10^5 M^{-1} sec^{-1}, 10 μM of a metal is competitive with the nanomolar concentrations of superoxide dismutase (SOD) that are commonly used. Recrystallization of the PACs is a convenient method of purification as they all tend to be insoluble in aqueous solution as acids and are solubilized as the pH is raised. It should be noted that recrystallizations carried out without ultrapure acid, base, water are pointless. Fortunately, ultrapure acids and bases are now commercially available from a number of sources.

OH/O$^-$ Radicals and Polyaminocarboxylates

It has been convincingly demonstrated that superoxide/perhydroxyl radicals and peroxide do not react with PACs at any significant or measurable rate. OH/O$^-$ radicals, however, react very efficiently with all PACs.

$$OH \cdot \rightleftharpoons O^{\overline{\cdot}} + H^+ \qquad (pK_1 = 11.9) \tag{1}$$

The detailed mechanism of these reactions is still uncertain, but it is well established that OH radical attack on PACs ultimately leads to degradation (i.e., decarboxylation, etc.). It is relevant to note that when Fenton-type reactions are carried out in the presence of metal–PAC complexes, the PAC must be considered as a competitor for the OH radical in the same fashion as any other OH radical scavenger. The rate constants of OH with a number of PACs are given in Table I. As can be seen, PACs react very efficiently with OH/O$^-$ with rate constants that, at neutral pH, range from 2×10^8 to 5×10^9 M^{-1} sec^{-1}.

$M^{n+}/M^{(n+1)+}$ Polyaminocarboxylates and HO$_2$/O$_2^-$/H$_2$O$_2$

The reactions of the three metals of interest, manganese, copper, and iron, with superoxide/perhydroxyl and peroxide are each discussed separately.

Manganese

The overall mechanism by which Mn^{2+}/Mn^{3+}–PAC complexes react with HO_2/O_2^- [where PAC is EDTA, NTA, or 1,2-cyclohexanediamine-

TABLE I

REACTIVITY OF OH/O$^-$ RADICALS WITH POLYAMINOCARBOXYLATES

Reaction	pH	pK of PAC[a]	k, M^{-1} sec^{-1} [b]
OH + glycine	1–6	2.36, 9.57	1.5×10^7
	9.5–9.7		2×10^9
	10		5×10^9
OH + IDA	1	1.8, 2.6, 9.3	5×10^7
	7		2×10^8
O$^-$ + IDA	13		9.1×10^8
OH + NTA	0–9	0.8, 1.8, 2.5, 9.3	2.3×10^9
OH + EDTA	4.0	1.5, 2.0, 2.7, 6.1, 10.2	4×10^8
	9.0		2×10^9
O$^-$ + EDTA	Alkaline		1×10^8
OH + DTPA	5–11	1.8, 2.7, 4.3, 8.5, 10.5	5×10^9
OH + trans-CyDTA[c]	0	2.4, 3.5, 6.1, 12.4	1.2×10^{10}

[a] Data from A. E. Martell and R. M. Smith, "Critical Stability Constants, Volume I: Amino Acids." Plenum, New York, 1974.

[b] Data from G. V. Buxton, C. V. Greenstock, W. P. Helman, and A. B. Ross, J. Phys. Chem. Ref. Data 17, 513 (1988).

[c] trans-CyDTA, trans-1,2-cyclohexanediamine-N,N,N',N'-tetraacetate.

N,N,N',N'-tetraacetate (CyDTA)] occurs by the following mechanism[1,2]:

$$Mn^{2+} + O_2^- \underset{k_{-1}}{\overset{k_1}{\rightleftharpoons}} MnO_2^+ \tag{2}$$

$$MnO_2^+ \xrightarrow{2 H^+, k_2} Mn^{3+} + H_2O_2 \tag{3}$$

$$Mn^{3+} + O_2^- \xrightarrow{k_3} Mn^{2+} + O_2 \tag{4}$$

Although reaction (2) has been shown to be reversible in studies involving simple inorganic ligands,[3] no such studies have been reported using PAC chelators. The reaction between superoxide and manganic complexes is written as a simple reaction yielding Mn^{2+} and O_2; however, this does not preclude a mechanism similar to that shown by equilibrium (2) and reaction (3). The respective rates and pH values reported for these systems are given in Table II. Although both Mn_{aq}^{3+} and $MnOH^{2+}$ have been shown to react with H_2O_2 in acidic solution with rate constants of 7×10^4 and 3×10^4 M^{-1} sec^{-1}, respectively,[4] Mn^{2+}– and Mn^{3+}–EDTA complexes have

[1] J. Lati and D. Meyerstein, J. Chem. Soc. Dalton Trans., 1185 (1978).

[2] J. Stein, J. P. Fackler, Jr., G. J. McClune, J. A. Fee, and L. T. Chan, Inorg. Chem. 18, 3511 (1979).

[3] D. E. Cabelli and B. H. J. Bielski, J. Phys. Chem. 88, 3111, 6291 (1984).

[4] G. Davies, L. J. Kirschenbaum, and K. Kustin, Inorg. Chem. 7, 146 (1968).

TABLE II

REACTIVITY OF Mn^{2+}/Mn^{3+}-POLYAMINOCARBOXYLATE COMPLEXES WITH HO_2/O_2^-

PAC	k_2, M^{-1} sec^{-1} (pH)	k_3, M^{-1} sec^{-1} (pH)	k_4, M^{-1} sec^{-1} (pH)
NTA	4×10^8 (4.5); 1.2×10^8 (5.5)[a]	1.8×10^3 (4.5); 9×10^1 (5.5)[a]	1.2×10^7 (6.0)[b]
EDTA	3×10^7 (4.5); 7.5×10^6 (5.5)[a]	3×10^3 (4.5, 5.5)[a]	5×10^4 (10.0)[c]
CyDTA	—	—	7.2×10^5 (9.2)[c]

[a] Data from Ref. 1.
[b] Data from W. H. Koppenol, F. Levine, T. L. Hatmaker, J. Epp, and J. D. Rush, *Arch. Biochem. Biophys.* **251**, 594 (1986).
[c] Data from Ref. 2.

been shown to react very slowly, if at all, with peroxide[2]:

$$Mn^{3+}\text{-EDTA} + H_2O_2 \rightarrow \text{very slow} \quad \text{(pH 10.0)} \tag{5}$$
$$Mn^{2+}\text{-EDTA} + H_2O_2 \rightarrow \text{no reaction} \quad \text{(pH 10.0)} \tag{6}$$

In contrast, the complex of Mn(II) with EDDA (ethylenediaminediacetate) acts as a catalase for the disproportionation of H_2O_2 to H_2O and O_2; the k_{cat} value is 5.4 M^{-1} sec^{-1}, where $d[H_2O_2]/dt = k_{cat}[Mn(II)\text{-}EDDA_T][H_2O_2]$. The essential difference between this and other Mn(II)–PAC systems is that the manganic EDDA intermediate is binuclear and acts as a catalyst for the two-electron reduction of peroxide.[5]

Copper

The use of PAC chelators appears to render Cu^{2+} complexes unreactive toward HO_2/O_2^-. As Cu^+-PAC complexes are very reactive toward O_2, the reverse step of the cycle ($Cu^+ + HO_2/O_2^-$) can be neglected. The products of the reaction between H_2O_2 and Cu^+ complexed with phenanthroline and simple inorganic ligands is currently the subject of much research. However, no rates for the reaction of PAC complexes of Cu^+ with peroxide have been reported.

Iron

The reactivity of Fe^{2+}/Fe^{3+}-PAC complexes toward $HO_2/O_2^-/H_2O_2$ has received the most attention because of the wealth of chemistry, Fenton and otherwise, that occurs in these systems. The most extensive studies that have been reported involve EDTA as the PAC ligand and suggested the following mechanism[6]:

[5] J. D. Rush, private communication, 1989.
[6] C. Bull, G. J. McClune, and J. A. Fee, *J. Chem. Soc.* **105**, 5290 (1983).

TABLE III

REACTIVITY OF Fe^{2+}–Fe^{3+}–POLYAMINOCARBOXYLATE COMPLEXES WITH HO_2/O_2^- AND H_2O_2[a]

PAC	k_{11}, M^{-1} sec^{-1}	k_9, M^{-1} sec^{-1}	k_8, M^{-1} sec^{-1} [b]	k_7, M^{-1} sec^{-1} [b]
EDTA	2×10^4 [a], 7×10^3 [d]	$(2.5–3.5) \times 10^2$ [c]	—	1.9×10^6
EDDA	7.8×10^4 [e]	—	—	—
NTA	3×10^4 [e]	—	—	—
CyDTA	1.3×10^3 [f]	—	—	—
DTPA	1.4×10^3 [g]; 5.1×10^2 [d]	—	2×10^7 [i]	10^4
H_2O	57.8 [h]	—	—	—
HEDTA	4.2×10^4 [d]	—	—	7.6×10^5
EHPG[j]	—	—	1.4×10^7 [j]	—

[a] All reported values are bimolecular rates at the specified pH.

[b] Data from B. H. J. Bielski, D. E. Cabelli, R. L. Arudi, and A. B. Ross, *J. Phys. Chem. Ref. Data* **14**, 1041 (1985).

[c] Data from Ref. 7, pH 8–10.

[d] Data from J. D. Rush and W. H. Koppenol, *J. Inorg. Biochem.* **29**, 199 (1987), pH 7.4.

[e] Data from J. D. Rush and W. H. Koppenol, *J. Am. Chem. Soc.* **110**, 4957 (1988), pH 7.2.

[f] Data from S. Rahhal and H. W. Richter, *Radiat. Phys. Chem.* **12**, 129 (1988), pH 7.0.

[g] Data from S. Rahhal and H. W. Richter, *J. Am. Chem. Soc.* **110**, 3126 (1988), pH 7.0.

[h] Data from H. N. Po and N. Sutin, *Inorg. Chem.* **7**, 621 (1968), 1.0 M $HClO_4$.

[i] pH 7.0.

[j] EHPG, Ethylenediamine-N,N'-bis[2-(2-hydroxyphenyl)acetate]; pH 7.0; data from G. R. Buettner, *J. Biol. Chem.* **262**, 11995 (1987).

$$Fe^{3+} + O_2^- \rightarrow Fe^{2+} + O_2 \tag{7}$$
$$Fe^{2+} + O_2^- \rightarrow Fe^{3+}\text{-}O_2^{2-} \tag{8}$$
$$Fe^{3+} + H_2O_2 \rightleftharpoons Fe^{3+}\text{-}O_2^{2-} + 2\ H^+ \tag{9}$$
$$\downarrow$$
$$Fe^{2+} + O_2^- + 2\ H^+ \tag{10}$$
$$Fe^{2+} + H_2O_2 \rightarrow \text{product(s)} \tag{11}$$

where the product(s) of reaction (11) are controversial (see Table III).

Acknowledgments

This work was supported by National Institute of Health Grant R01GM23658-12 and was carried out at the Brookhaven National Laboratory under Contract DE-AC02-76CH00016 with the U.S. Department of Energy, and supported by its Division of Chemical Sciences, Office of Basic Sciences.

[7] Preparation of Metal-Free Solutions for Studies of Active Oxygen Species

By K. M. SCHAICH

Introduction

Preparation of "metal-free" solutions is, to some extent, a misnomer since what is implied is the absolute removal of all traces of metals but what is attainable with certainty is a metal content below detection levels. With graphite furnace atomic absorption spectrophotometry, these limits for the redox-active metals of greatest relevance to oxygen radical chemistry (iron, copper, and manganese) are presently less than 10^{-12} g by weight or 10^{-10} M by concentration. However, metals remain catalytically active at these levels. Thus, in using metal-free solutions the practical expectation is that the kinetic competition from metal-mediated reactions will be severely limited; however, in interpreting mechanisms it always must be remembered that metals are not completely absent or inactive.

Metal catalysis is common in oxygen radical reactions, so studies of active oxygen species require much more stringent procedures for preparing metal-free solutions than do most other areas of chemical research. Because competitive reaction kinetics are a critical feature in the detection of oxygen radicals, in their production, and in the determination of their reaction mechanisms, most conventional techniques for removal of metals from solutions (such as dithizone and bathophenanthroline extractions) cannot be used when oxygen radicals are being studied: they leave behind reagent contaminants that react with reduced oxygen species (O_2^{-}, HO·, H_2O_2, ROOH, RO·, or ROO·) or otherwise affect the course of the reaction. Furthermore, all reagents must be demetaled, not just water or buffers, so techniques must be adaptable to many kinds of chemicals and to a wide range of solution volumes, from a few milliliters to several liters. Also, it is preferable that the techniques should not alter reagent concentrations since any handling after demetaling introduces the potential for contamination.

Chelating resins, most notably Chelex 100 (Bio-Rad, Richmond, CA), are commonly used to reduce the metal content of buffer solutions for biological studies, but they have not been satisfactory for oxygen radical studies in our laboratory. Of a variety of resins tested, only -400 mesh Chelex 100 approached the required efficiency of metal removal.[1] Other

---

[1] K. M. Schaich, unpublished data, 1984.

METHODS IN ENZYMOLOGY, VOL. 186

resins and larger meshes of Chelex 100 were less effective. Chelating resins suffer several additional disadvantages: they require extensive washing and equilibration steps, all of which carry the potential for contamination; they are not effective at all pH values; cationic reagents (e.g., paraquat) as well as metals bind to them; and strong chelators such as EDTA and diethylenetriaminepentaacetic acid (DTPA) can pull metals off the column and actually *increase* solution metal concentrations.[2]

8-Hydroxyquinoline (8-HOQ) is a very efficient chelator with a high binding constant for a broad range of metals.[3–5] Solvent extraction of metals into chloroform solutions of 8-HOQ is much more efficient than is the use of chelating resins,[6] but the process is tedious and necessitates the extensive use of a very toxic solvent. Recently, however, 8-HOQ has been commercially coated onto controlled-pore glass (CPG) beads and in this form provides the best method found in our laboratory for efficient, convenient removal of metals from all volumes and classes of reagents used in oxygen radical reactions. This method is described in detail below.

Methods

General Considerations. Some very good general guidelines for handling metal-free systems have been presented previously in this series.[7,8] Briefly, when working with solutions containing submicromolar metal, ultraclean laboratory conditions must be maintained at all times to avoid inadvertent contamination. Installation of air filters or working in laminar flow hoods may be necessary for rigorous work. Containers, syringe needles, or spatulas made of metal should never be used. Quartz is optimum for preparing and storing metal-free solutions because it contains no intrinsic leachable metals that may contaminate reagents, nor does it absorb metals or other components from solutions; however, it is expensive. Pyrex glassware or plastic labware (including micropipet tips) may be used after soaking overnight in 0.1 N HCl (ultrapure or demetaled on an anion-exchange resin and redistilled), then rinsing 3 times in high-purity

[2] D. C. Borg and K. M. Schaich, *Isr. J. Chem.* **24**, 38 (1984).
[3] R. G. W. Hollingshead, "Oxine and Its Derivatives," Vols. 1 and 2. Butterworth, London, 1954.
[4] D. D. Perrin, "Organic Complexing Reagents: Structure, Behavior, and Application to Inorganic Analysis." Wiley (Interscience), New York, 1964.
[5] H. F. Walton, "Principles and Methods of Chemical Analysis," p. 81. Prentice-Hall, Englewood Cliffs, New Jersey, 1964.
[6] K. M. Schaich, manuscript in preparation.
[7] C. Veillon and B. M. Vallee, this series, Vol. 54, p. 446.
[8] H. Beinert, this series, Vol. 54, p. 435.

water (see below). Methods using more concentrated acids are common but cause greater etching of glassware and should not be used with plastics. If syringes are needed, use ones with Teflon plungers and glass or Teflon tips, along with platinum needles. Spatulas should be Teflon or ceramic, and reagents should be weighed directly into clean, acid-washed glassware.

Water and Reagents. Water for all operations (glassware washing, reagent preparation, etc.) should be ultrapure, prepared from a system that includes an ion-exchange column and that provides water with a resistivity of 18 megohms cm^{-1}. Our laboratory uses the Milli-Q (Millipore; Bedford, MA.) water purification system in an arrangement similar to that described by Veillon and Vallee.[7] Buffer, salt, acid, and alkali reagents should be of ultrapure quality with the lowest possible metal content, such as are available from Apache (Seward, IL.) and Baker (Ultrex; Phillipsburg, N.J.) Chemical Companies. EDTA, DTPA, and other organic reagents should be recrystallized 3 times[9-11] before preparing working solutions in ultrapure water. These precautions minimize the intrinsic metal contents of reagent solutions but do not render them metal free. A final rigorous demetaling of working solutions as described below is almost always required before their use in reactions involving oxygen radicals.

Monitoring of Metal Concentrations. No demetaling procedure should be taken for granted; all reagents must be analyzed for metal content before their use in oxygen radical studies. Iron is usually the metal in highest concentration in reagents, and thus is a useful general monitor, although other metals of specific interest also may be measured.

Chemical assays, such as the bleomycin test[12] for iron and the 1,10-phenanthroline test[13] for copper, have been developed, but these are time consuming and are sensitive to only about micromolar metal concentrations. The bathophenanthroline assay[8,14] is somewhat more sensitive (about 10^{-7} to 10^{-8} M) but is also time consuming. The most sensitive (10^{-10} M) and straightforward method for monitoring metal concentra-

---

[9] D. D. Perrin, W. L. F. Armarego, and D. R. Perrin, "Purification of Laboratory Chemicals." Pergamon, Oxford, 1966.

[10] L. F. Fieser and M. Fieser, "Reagents for Organic Synthesis." Wiley, New York, 1967.

[11] J. A. Riddick, W. B. Bunger, and T. K. Sakano, "Organic Solvents: Physical Properties and Methods of Purification." Wiley (Interscience), New York, 1986.

[12] J. M. C. Gutteridge and B. Halliwell, *in* "Handbook of Methods for Oxygen Radical Research" (R. A. Greenwald, ed.), p. 391. CRC Press, Boca Raton, Florida, 1985.

[13] J. M. C. Gutteridge, *in* "Handbook of Methods for Oxygen Radical Research" (R. A. Greenwald, ed.), p. 395. CRC Press, Boca Raton, Florida, 1985.

[14] H. Diehl and G. F. Smith, "The Iron Reagents: Bathophenanthroline, 2,4,6-Tripyridyl-*S*-triazine, Phenyl-2-pyridyl Ketoxime," p. 13. G. Frederick Smith Chemical Co., Columbus, Ohio, 1960.

tions uses an atomic absorption (AA) spectrophotometer with a graphite furnace and deuterium arc correction to provide maximum sensitivity and elimination of organic matrix interferences. It can be used with organic as well as aqueous solvent systems, and it is the analytical method our laboratory uses and recommends for metal analysis in oxygen radical studies.

8-Hydroxyquinoline Controlled-Pore Glass Bead Procedure for Producing Metal-Free Solutions. CPG/8-Hydroxyquinoline, available from Pierce Chemical Co. (Rockford, IL), has a capacity of 0.005 mEq Fe/g[15] (i.e., under ideal conditions 1 g can theoretically demetal 1 liter of reagent 5 μM in iron). Its extraction efficiency varies with the metal and the pH, but for iron, copper, cobalt, and manganese complete extraction occurs in the pH range 4 through alkaline.[16] Chelators such as EDTA and citrate do not interfere with the activity of unsaturated CPG/8-HOQ, either by pulling metal off or by blocking binding of metals.

CPG/8-HOQ can be used in either batch or column procedures. In the batch procedure, an amount of dry CPG/8-HOQ equivalent to about 5 times the estimated or known content of iron [or other metal(s)] is weighed into the reagent solution to be demetaled and stirred or shaken for 30 min. The chelator is removed by centrifugation or by settling, and the reagent is decanted. The metal content is analyzed by AA spectrophotometry, and the chelation procedure is repeated if necessary. This method is simple, but great care must be exercised to avoid contamination during recovery of the reagent.

Column procedures offer fewer opportunities for recontaminating the reagent, but they may be less efficient than batch extractions if the elution rate from the column is too fast. Dry CPG/8-HOQ equivalent to 5 to 10 times the metal content of the solution is packed into a column plugged with an acid-washed frit or a small amount of acid-washed glass wool. A Pasteur pipet may be used for a column when very small volumes must be demetaled. The reagent or buffer solution is applied to the column and allowed to diffuse through by gravity to allow maximum contact time with the chelator. The metal content of the eluent is analyzed by AA spectrophotometry. The column procedure described should be 100% efficient in reducing metal contents to below detectable levels, but it may be repeated if the first pass does not complete the metal extraction.

Regeneration of Controlled-Pore Glass/8-Hydroxyquinoline. After each use the CPG/8-HOQ should be washed several times with ultrapure water to remove buffer or reagent contaminants and then regenerated (to

[15] Pierce Chemical Co., personal communication, 1985.
[16] Pierce Chemical Co., technical communication, 1985.

remove the chelated metal) with 1.0 N HCl. The amount of acid required is determined by the metal loading on the column. For moderately loaded columns, 2 to 3 column volumes of acid removes more than 95% of the metal, but up to about 10 column volumes (or about 50 ml/g in batch procedures) may be necessary for complete regeneration. The beads are washed again with several column volumes of ultrapure water and dried in air. The material should remain active indefinitely.

Acknowledgments

This work was performed at the Medical Department, Brookhaven National Laboratory, Upton, New York 11973, under U.S. Department of Energy Contract DE-AC02-76CH00016.

[8] Use of Ascorbate as Test for Catalytic Metals in Simple Buffers

By GARRY R. BUETTNER

Over the years trace levels of adventitious transition metals have provided many problems for researchers studying oxidative processes, leading to the misinterpretation of many experiments.[1] For correct interpretation of these experiments, catalytic metals must be sequestered in an inactive form[2] or be removed from the solution.[1] Chelating resins are commonly employed to remove contaminating metals from buffer, salt, and protein solutions.[1] However, there always remains some doubt about the efficacy of a particular method as well as concern about subsequent recontamination of the solution. Ascorbate can be used to test for the presence of adventitious catalytic metals in simple near-neutral buffer solutions,[1] and a simple method for using this test is presented in this chapter.

Ascorbate Autoxidation

Ascorbic acid is a diacid, AH_2, with pK_a values of 4.2 and 11.6.[3] Thus, at near-neutral pH the dominant species is the monoanion, AH^-. Ascor-

---

[1] G. R. Buettner, *J. Biochem. Biophys. Methods* **16**, 27 (1988).
[2] G. R. Buettner, *Bioelectrochem. Bioenerg.* **18**, 29 (1987).
[3] S. Lewin, "Vitamin C: Its Molecular Biology and Medical Potential." Academic Press, London, 1976.

METHODS IN ENZYMOLOGY, VOL. 186

bate is an easily oxidizable reducing agent. However, the rate of air oxidation of ascorbate in aqueous solution is very pH- and catalytic metal-dependent. In the presence of catalytic metals AH_2 oxidizes very slowly, whereas AH^- oxidizes much more rapidly.[1,4] In the absence of catalytic metals, AH^- is quite stable in air-saturated buffer.[1] It appears that metals are absolutely required for this oxidation. Thus, the rate at which ascorbate solutions air oxidize can be used to monitor for the presence of contaminating catalytic metals.

Method

For the standard test a 0.100 M ascorbate stock solution is prepared using reagent grade ascorbic acid and high-purity water. This results in a colorless solution of approximately pH 2. The low pH stabilizes the ascorbate, so the solution can be kept for days or even weeks for use in the test. Also, once oxygen is consumed oxidation stops, with less than 1% of the ascorbate stock being oxidized. Thus, storage in a relatively air-tight flask helps to increase the shelf life. The appearance of a yellow color indicates significant ascorbate oxidation, and a fresh solution should be prepared.

To perform the test, 3.75 μl of the ascorbate stock is added to 3.00 ml of the solution to be tested. This results in an initial absorbance of 1.8 at 265 nm.[5] The ascorbate absorbance is followed for 15–30 min. In successfully demetaled buffer the loss of ascorbate should be 0.5% or less in 15 min. A greater loss indicates that a significant concentration of catalytic metals remains.[6,7] This test can be used in buffers with pH values ranging from approximately 4 to 8. Below pH 4 ascorbate oxidizes too slowly to provide a sensitive test; above pH 8 the concentration of the dianion becomes significant, and it oxidizes too rapidly to allow for a good test.

When a chelating resin is used to lower the catalytic metal concentration to an acceptable level, some potential problems and considerations should be noted: (1) In the column method, the flow rate should not be too high, because a slow rate of flow is essential to demetal a buffer successfully; (2) in the batch or dialysis methods,[1] the time of stirring should be long enough; (3) the chelating resin may need regeneration (in fact, the

---

[4] A. E. Martell, *in* "Ascorbic Acid, Chemistry, Metabolism, and Uses" (P. A. Seib and B. M. Tolbert, eds.), p. 124. American Chemical Society, Washington, D.C. 1982.

[5] In my experience the ε_{265} value for ascorbate in 50 mM phosphate buffer (pH 7.0) is 14500 M^{-1} cm^{-1}, although values ranging from 7500 to 20400 M^{-1} cm^{-1} have been reported.[3]

[6] If no effort is made to demetal near-neutral buffer solutions, typical ascorbate losses in the standard 15-min test can range from 1 to 30% (or more),[1,7] consistent with approximately 1 μM iron and/or 0.1 μM copper being present. After successful treatment with chelating resin these levels have been estimated to be <0.1 μM iron and ~1 nM copper.[1]

[7] G. R. Buettner, *Free Radical Res. Commun.* **1**, 349 (1986).

ascorbate test is an excellent method to determine the status of the chelating resin); (4) glassware may recontaminate the solution[8]; and (5) dust, fingers, and phantom sources of contamination may seem to appear magically.

Using this simple, inexpensive method, it is easy to verify that the procedures used to demetal buffer and salt solutions are successful. The method provides a tool for troubleshooting when problems occur, and it is an easy and repeatable way to determine and report experimental conditions.

[8] Standard laboratory glassware is used for solution storage. After this glassware has been successfully cleaned, however, it is never returned to the usual pool of laboratory glassware for general use. Rather, when fresh solutions are required, the glassware is rinsed with high-purity water and refilled with the same solution.

[9] Spin-Trapping Methods for Detecting Superoxide and Hydroxyl Free Radicals *in Vitro* and *in Vivo*

By GARRY R. BUETTNER and RONALD P. MASON

Spin trapping has become a valuable tool in the study of transient free radicals as evidenced by the many investigations in which it has been employed.[1] Oxygen-centered radicals are of particular interest because they have been implicated in many adverse reactions *in vivo*. Their short lifetimes and broad linewidths make many of these radicals difficult, if not impossible, to detect by direct electron spin resonance (ESR) in room temperature aqueous solutions. Spin trapping provides a means, in principle, to overcome these problems, but it is not without its pitfalls and limitations. We discuss some of these problems in this chapter.

Choice of Spin Trap

Two types of spin traps have been developed, nitrone and nitroso compounds. In aqueous solutions, however, oxygen-centered spin adducts of nitroso spin traps such as MNP (2-methyl-2-nitrosopropane) are, in general, quite unstable. Thus, the nitrone spin traps are by far the most popular. The most useful radical trap for the study of oxygen-centered free radicals is 5,5-dimethyl-1-pyrroline *N*-oxide (DMPO), which has

[1] G. R. Buettner, *Free Radical Biol. Med.* **3**, 259 (1987).

METHODS IN ENZYMOLOGY, VOL. 186

been used extensively to study superoxide[1,2] and hydroxyl radicals[1,3] as well as peroxyl radical formation[1,4,5] in biochemical and biological systems.

Superoxide

The spin trapping of superoxide has been of much interest because of the involvement of superoxide in many physiological processes. DMPO–·OOH (the superoxide spin adduct of DMPO) has a distinctive spectrum ($a^N = 14.2$ G, $a_\beta^H = 11.3$ G, $a_\gamma^H = 1.25$ G)[1] that is easily recognizable. However, other peroxyl adducts of DMPO will have a similar appearance. Thus, the real proof that the spectrum observed is indeed due to DMPO–·OOH is gained by using SOD (superoxide dismutase) to inhibit the signal.[6]

Although the DMPO–·OOH spectrum is distinctive, the spin trapping of superoxide is not without its problems. The actual reaction of superoxide with DMPO is very slow (k_{obs} is 60 M^{-1} sec^{-1} at pH 7 and only 30 M^{-1} sec^{-1} at pH 7.4).[7] Thus, in most superoxide-generating systems the spin trap concentration must be quite high (~0.1 M) in order to outcompete the self-decay, namely, dismutation of superoxide. In addition, the DMPO–·OOH adduct is unstable, decaying by a first-order process with a half-life of about 60 sec at pH 7.[8] Therefore, one must always be prepared to deal with a relatively weak signal, that is, [DMPO–·OOH] will, under most circumstances, be less than approximately 10 μM.

Hydroxyl Radical

The DMPO–·OH adduct is the most often reported radical adduct of DMPO ($a^N = a_\beta^H = 14.9$ G).[1] Much of the interest in the spin trapping of ·OH is due to its formation in the superoxide-dependent Fenton reaction:

$$O_2^{\bar{\cdot}} + HO_2\cdot \xrightarrow{H^+} H_2O_2 + O_2 \qquad (1)$$

$$O_2^{\bar{\cdot}} + Fe(III) \longrightarrow Fe(II) + O_2 \qquad (2)$$

$$Fe(II) + H_2O_2 \longrightarrow \cdot OH + OH^- + Fe(III) \qquad (3)$$

[2] P. J. Thornalley and J. V. Bannister, *in* "Handbook of Methods for Oxygen Radical Research" (R. A. Greenwald, ed.), p. 133. CRC Press, Boca Raton, Florida, 1985.

[3] G. R. Buettner, *in* "Handbook of Methods for Oxygen Radical Research" (R. A. Greenwald, ed.), p. 151. CRC Press, Boca Raton, Florida, 1985.

[4] M. J. Davies, *Biochim. Biophys. Acta* **964,** 28 (1988).

[5] M. J. Davies, *Chem. Phys. Lipids* **44,** 149 (1987).

[6] E. Finkelstein, G. M. Rosen, E. J. Rauckman, and J. Paxton, *Mol. Pharmacol.* **16,** 676 (1979).

[7] E. Finkelstein, G. M. Rosen, and E. J. Rauckman, *J. Am. Chem. Soc.* **102,** 4994 (1980).

[8] G. R. Buettner and L. W. Oberley, *Biochem. Biophys. Res. Commun.* **83,** 69 (1978).

Thus, SOD will inhibit DMPO–·OOH and/or DMPO–·OH formation if this reaction sequence is operative. However, catalase will inhibit the formation of ·OH in reaction (3) above. A failure of catalase to inhibit the formation of DMPO–·OH when the superoxide-driven Fenton reaction is suspected indicates that something artifactual is occurring or that another mechanism must be sought.

Two additional SOD-inhibitable routes to DMPO–·OH from DMPO–·OOH itself should be considered: (1) the reduction of DMPO–·OOH[6] (a hydroperoxide) to the alcohol DMPO–·OH, for example, by glutathione peroxidase[9]; and (2) the possible homolytic cleavage of the oxygen–oxygen bond of DMPO–·OOH to produce free ·OH, which is subsequently trapped by unreacted DMPO.[10,11] Thus, weak DMPO–·OH signals that are not catalase inhibitable should always be viewed cautiously because they quite often are artifactual rather than the result of the spin trapping of free ·OH generated by the system under study.

To establish the existence of free hydroxyl radical in spin-trapping experiments, it is necessary to perform kinetic competition experiments with hydroxyl radical scavengers.[12,13] For example, ethanol, formate, and dimethyl sulfoxide can be used in these competition experiments, because upon hydroxyl radical attack they form carbon-centered radicals that can subsequently be trapped by DMPO:

$$\cdot OH + DMPO \xrightarrow{k_1} DMPO–\cdot OH \tag{4}$$

$$\cdot OH + HCO_2^- \xrightarrow{k_2} CO_2^{\cdot-} + H_2O \tag{5}$$

$$CO_2^{\cdot-} + DMPO \xrightarrow{k_3} DMPO–CO_2^{\cdot-} \tag{6}$$

$$\cdot OH + CH_3CH_2OH \xrightarrow{k_4} CH_3\dot{C}HOH + H_2O \tag{7}$$

$$CH_3\dot{C}HOH + DMPO \xrightarrow{k_5} DMPO–CH_3\dot{C}HOH \tag{8}$$

[9] G. M. Rosen and B. A. Freeman, *Proc. Natl. Acad. Sci. U.S.A.* **81**, 7269 (1984).

[10] E. Finkelstein, G. M. Rosen, and E. J. Rauckman, *Mol. Pharmacol.* **21**, 262 (1982).

[11] Finkelstein *et al.*[10] indicate that approximately 3% of DMPO–·OOH decomposes to produce ·OH. Unfortunately, no experimental data or details are given to indicate how this estimate was made; thus, it is difficult to assess how this number should be used.

[12] A. L. Castelhano, M. J. Perkins, and D. Griller, *Can. J. Chem.* **61**, 298 (1983).

[13] Additional sources of artifactual DMPO–·OH signals are (1) hydrolysis of DMPO to produce DMPO–·OH as an impurity signal [R. A. Floyd and B. B. Wiseman, *Biochim. Biophys. Acta* **586**, 196 (1979)]; (2) the one-electron oxidation of DMPO followed by hydration of DMPO⁺ [H. Chandra and M. C. R. Symons, *J. Chem. Soc., Chem. Commun.*, 1301 (1986)]; (3) the apparently concerted hydrolysis–oxidation reaction by photochemically excited molecules [V. S. F. Chew and J. R. Bolton, *J. Phys. Chem.* **84**, 1903 (1980)]; and (4) the presence of a strong oxidant such as hypochlorous acid [E. G. Janzen, L. T. Jandrisits, and D. L. Barber, *Free Radical Res. Commun.* **4**, 115 (1987)].

Most artifacts leading to DMPO–·OH radical adduct formation will be excluded by the use of hydroxyl radical scavengers if the scavenger-derived radical adduct is detected, if a corresponding decrease in the DMPO–·OH radical adduct concentration is found, and if quantitative kinetic criteria are used.[12]

Measurement of the initial rates of formation of the DMPO–·OH and DMPO–scavenger radical adducts removes the effects of the differential radical adduct decay rates.[12] Using this approach, the relative efficiency of two hydroxyl radical scavengers is quantitatively predictable from the known rate constants for the reactions of the hydroxyl radical with these scavengers. For example, using formate (k_2) and ethanol (k_4) we can calculate k_2/k_4 from the ratio of the rates of formation of these two radical adducts:

$$\frac{k_2}{k_4} = \frac{d[DMPO-\cdot CO_2^-]/dt}{d[DMPO-CH_3\dot{C}HOH]/dt} \times \frac{[CH_3CH_2OH]}{[HCO_2^-]} \tag{9}$$

In Eq. (9), the ratio k_2/k_4 from spin trapping should agree with the ratio of rate constants for the reaction of the hydroxyl radical with these scavengers as determined from pulse radiolysis. It should be kept in mind that to arrive at this expression, it is assumed that the predominant route of scavenger radical decay is via the trapping reaction. This kinetic approach has been successfully applied to an enzyme-dependent hydroxyl radical-generating system.[14]

A similar approach has been presented by Buettner et al.[15] In this approach a ·OH scavenger is included in the spin-trapping mixture at a concentration calculated to reduce the intensity of the DMPO–·OH signal by 50%. In other words, the rate of the reaction of ·OH with scavenger (Scav) is equal to its rate of reaction with DMPO:

$$k_{scav}[Scav][\cdot OH] = k_{DMPO}[DMPO][\cdot OH] \tag{10}$$

$$[Scav] = k_{DMPO}[DMPO]/k_{Scav} \tag{11}$$

Data must be obtained under circumstances where the rate of loss of DMPO–·OH is low compared to its rate of formation. In this approach possible side reactions of the scavenger radical are not a problem, unless of course they destroy DMPO–·OH, enabling a much wider range of experimental conditions to be used.

Recently, Samuni et al.[16] have demonstrated that O_2^- reacts very

[14] K. M. Morehouse and R. P. Mason, *J. Biol. Chem.* **263**, 1204 (1988).

[15] G. R. Buettner, A. G. Motten, R. D. Hall, and C. F. Chignell, *Photochem. Photobiol.* **44**, 5 (1986).

[16] A. Samuni, C. D. V. Black, C. M. Krishna, H. L. Malech, E. F. Bernstein, and A. Russo, *J. Biol. Chem.* **263**, 13797 (1988).

efficiently with DMPO–·OH and DMPO–·CH$_3$ adducts, destroying the nitroxide and thus producing an ESR-silent species. If the flux of superoxide is high enough, the DMPO–·OH adduct may not even be observed because of its rapid removal. Thus, a high flux of superoxide would not be desirable if additional free radical reactions are expected in a superoxide spin-trapping system.

Peroxyl Radical Trapping

Peroxyl radicals have been successfully spin trapped with DMPO and PBN (α-phenyl-*N-tert*-butyl nitrone).[1,4,5] The appearance of the DMPO–·OOR spectrum is similar to that of DMPO–·OOH, albeit the splitting constants are somewhat different. As expected, DMPO–·OOR formation cannot be inhibited by superoxide dismutase.[17] The major experimental problem encountered is that DMPO–·OOR decays very quickly in aqueous solutions[4]; thus, time is a major consideration in any spin-trapping protocol.

In Vivo and In Vitro Superoxide

Many studies are pursuing the possible production of superoxide or hydroxyl radicals by cell organelles, intact cells, and organs. The detection of superoxide by spin trapping with DMPO has been achieved in all of the above. For success, however, experimental protocols must allow for the relatively short lifetime of DMPO–·OOH[8] and the possible interference by metal catalysts such as iron.[18] For example, in studying free radicals produced in myocardial ischemia/reperfusion, Arroyo *et al.*[19] immediately froze the coronary effluents in liquid nitrogen to prevent spin adduct decay. By monitoring the ESR spectra of the effluents immediately after thawing, they were successful in observing DMPO–·OOH.

DTPA, EDTA, and Desferal

The presence of transition metals (particularly iron) and various chelating agents can significantly alter the results of spin-trapping experiments.[18] Although contaminating catalytic metals can be removed from buffer and biochemical systems,[20] this would be a difficult and uncertain (perhaps impossible) process for cells and organs. Thus, chelating agents

[17] B. Kalyanaraman, C. Mottley, and R. P. Mason, *J. Biol. Chem.* **258**, 3855 (1983).

[18] G. R. Buettner, L. W. Oberley, and S. W. H. C. Leuthauser, *Photochem. Photobiol.* **28**, 693 (1978).

[19] C. M. Arroyo, J. H. Kramer, B. F. Dickens, and W. B. Weglicki, *FEBS Lett.* **221**, 101 (1987).

[20] G. R. Buettner, *J. Biochem. Biophys. Methods* **16**, 27 (1988).

are much needed tools. When studying a superoxide-generating system, EDTA will, in general, enhance the catalytic activity of iron in the reaction sequence,[1-3,18,21] thereby increasing the yield of DMPO–·OH while decreasing or eliminating the appearance of DMPO–·OOH. DTPA (diethylenetriaminepentaacetic acid) reduces or eliminates many of the problems generated by catalytic iron in superoxide-generating systems,[18] but under circumstances where a reducing agent stronger than superoxide is responsible for iron reduction, DTPA can increase DMPO–·OH formation.[14] In studying stimulated neutrophils, Britigan et al.[22,23] found DTPA (1–100 μM) to be a very useful tool; it had no effect on neutrophil superoxide production or oxygen consumption, whereas it enhanced the detection of superoxide by DMPO in their cellular experiments.

The iron chelator Desferal (deferrioxamine mesylate) renders iron essentially catalytically inactive in reactions (1)–(3) above.[21] Unfortunately the hydroxamic acid moieties of Desferal can undergo one-electron oxidation by superoxide (most likely ·OOH), hydroxyl radical, and horseradish peroxidase.[24-26] The nitroxide radical so formed is stable for a free radical, but nevertheless it reacts rapidly with cysteine, methionine, glutathione, ascorbate, and a water-soluble form of vitamin E.[25] This radical may also deactivate enzymes, as demonstrated for alcohol dehydrogenase.[25] If Desferal is present at a relatively high concentration (compared to spin trap), it can effectively compete for superoxide and hydroxyl radical.[26] Since adventitious transition metals are present at only micromolar concentrations and spin traps are used at millimolar concentrations, scavenging by Desferal is perhaps less of a problem than the interference caused by the detection of the Desferal nitroxide radical itself. In any case, the Desferal concentration should be kept as low as possible to minimize scavenging.

In Vivo and In Vitro Hydroxyl Radical

The hydroxyl radical adduct of DMPO has also been observed in cell organelles, intact cells, and organs. However, the actual determination

[21] G. R. Buettner, Bioelectrochem. Bioenerg. **18**, 29 (1987).

[22] B. E. Britigan, M. S. Cohen, and G. M. Rosen, J. Leuk. Biol. **41**, 349 (1987).

[23] Britigan et al.[22] have observed a concentration-dependent inhibition of neutrophil oxygen consumption by PBN (<10 mM). DMPO was without effect at similar concentrations. However, an apparent impurity in DMPO did produce a marked inhibition of stimulated neutrophil superoxide formation in one set of experiments, which emphasizes the necessity of always performing routine control experiments.

[24] K. M. Morehouse, W. D. Flitter, and R. P. Mason, FEBS Lett. **222**, 246 (1987).

[25] M. J. Davies, R. Donkor, C. A. Dunster, C. A. Gee, S. Jonas, and R. L. Willson, Biochem. J. **246**, 725 (1987).

[26] O. Hinojosa and T. J. Jacks, Anal. Lett. **19**, 725 (1986).

that free · OH has been trapped is somewhat problematic because the presence of classic · OH scavengers such as ethanol, dimethyl sulfoxide, or formate can have a severely perturbing influence on the system, especially at the high concentrations that are required to outcompete the reaction of any · OH formed with the numerous biochemicals present at millimolar concentrations. Thus, we believe that the unambiguous determination that free · OH has been spin trapped requires very careful experimental design and interpretation, especially when the goal is to examine free radical production in cells, organs, or whole animals.

[10] Detection of Singlet Molecular Oxygen during Chloride Peroxidase-Catalyzed Decomposition of Ethyl Hydroperoxide

By ROBERT D. HALL

Introduction

During the 1970s and 1980s, a number of researchers proposed that the production of $^1O_2$, the lowest excited state of molecular oxygen, occurs during certain biochemical reactions. However, most of the experimental evidence for these proposals has been indirect, usually relying on physical and chemical quenchers of $^1O_2$, deuterium oxide enhancement of postulated effects of $^1O_2$, or the red chemiluminescence associated with the simultaneous electronic transition of two $^1O_2$ molecules. Not surprisingly, objections have often been raised regarding the value of these indirect methods.

More recently, several groups have begun to utilize the unique luminescence of $^1O_2$ at 1268 nm to measure the production of $^1O_2$ by chemical,[1] photochemical,[2,3] or biochemical[4-7] reactions. This infrared luminescence can also be combined with the traditional methods of detection to further support or question hypotheses concerning $^1O_2$.

[1] J. R. Kanofsky, J. Org. Chem. **51**, 3386 (1986).
[2] A. U. Khan, J. Am. Chem. Soc. **103**, 6516 (1981).
[3] R. D. Hall and C. F. Chignell, Photochem. Photobiol. **45**, 459 (1987).
[4] J. R. Kanofsky, J. Biol. Chem. **258**, 5991 (1983).
[5] A. U. Khan, P. Gebauer, and L. P. Hager, Proc. Natl. Acad. Sci. U.S.A. **80**, 5195 (1983).
[6] J. R. Kanofsky and B. Axelrod, J. Biol. Chem. **261**, 1099 (1986).
[7] R. D. Hall, W. Chamulitrat, N. Takahashi, C. F. Chignell, and R. P. Mason, J. Biol. Chem. **264**, 7900 (1989).

METHODS IN ENZYMOLOGY, VOL. 186

Most biochemical research has focused on the halide-dependent reactions catalyzed by various peroxidases.[4,5] Only a small number of studies have used the emission of $^1O_2$ at 1268 nm to characterize the production of $^1O_2$ during enzyme-catalyzed decomposition of organic hydroperoxides,[6,7] a class of reactions often regarded as a source of the $^1O_2$ observed by indirect methods during lipid peroxidation.

Chloride peroxidase (chloride:hydrogen-peroxide oxidoreductase, EC 1.11.1.10) is a monomeric heme glycoprotein (MW 42,000) synthesized by *Caldariomyces fumago*. In addition to halide-dependent reactions, chloride peroxidase catalyzes the production of molecular oxygen from ethyl hydroperoxide and various perbenzoic acids.[8] We have recently published evidence that chloride peroxidase can catalyze the production of organic peroxyl free radicals and $^1O_2$.[7,9]

Instrumentation

At the present time, there is no instrument commercially available for the detection of the very weak emission of $^1O_2$ at 1268 nm. However, an instrument for steady-state detection of $^1O_2$ can be constructed with standard optical and electronic components. Figure 1 is a box diagram of an instrument currently in use in our laboratory[3] for both photochemical and chemiluminescence studies. Similar instruments are described in Refs. 2 and 4.

Our instrument utilizes optical components from Kratos Instrument Co. (Westwood, NJ). For photochemical studies, light from a mercury arc lamp (LS) passes through 10 cm of water (LF), maintained at constant temperature, and through other filters as required to isolate specific bandwidths of light for excitation of samples. During chemiluminescence experiments, the excitation light path is not used. Our sample holder (SH) accommodates a standard 10-mm quartz fluorescence cuvette.

Emission light is collected at a right angle to the excitation path and passes through a filter holder (F) mounted with a rack of interference filters and a cutoff filter that blocks all light below 950 nm. The cutoff filter can be used in conjunction with the monochromator to record the spectrum of $^1O_2$; or the monochromator can be removed and the interference filters used to observe the emission of $^1O_2$ at specific wavelengths. In the latter case, greater instrumental sensitivity can be achieved.

The light then passes through a light chopper (CH) and is collected by a large lens and focused on a germanium diode detector (GD). This is the

[8] P. F. Hallenberg and L. P. Hager, this series, Vo. 52, p. 521.
[9] W. Chamulitrat, N. Takahashi, and R. P. Mason, *J. Biol. Chem.* **264**, 7889 (1989).

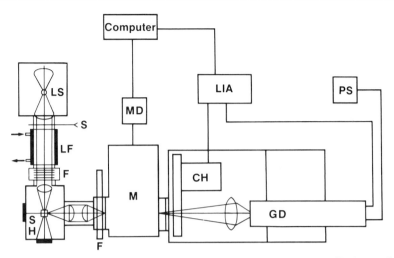

FIG. 1. Highly sensitive $^1O_2$ emission spectrometer. LS, Light source; S, shutter; LF, liquid optical filter; F, optical filters; SH, sample holder; M, monochromator; MD, monochromator driver; CH, light chopper; GD, germanium diode detection module; PS, power supply; LIA, lock-in amplifier.

heart of the instrument, and highly sensitive detection modules from Applied Detector Corp. (Fresno, CA)[2,3] and North Coast Scientific Corp. (Santa Rosa, CA),[4] complete with power supplies, have been incorporated into existing $^1O_2$ emission spectrometers.

Electronic noise reduction techniques must be used in order to maximize instrument sensitivity. Phase-sensitive detection, achieved through a combination of the light chopper and a lock-in amplifier (LIA), is used most commonly in steady-state instruments.

For chemiluminescence work, the output from the lock-in amplifier can be monitored with a standard chart recorder. However, our system is interfaced to a computer that can simultaneously drive the monochromator and acquire emission intensity data for the construction of spectra; alternatively, the computer can follow the time evolution of chemiluminescent reactions at a constant wavelength. The data can also be stored for later analysis.

Enzyme-Dependent $^1O_2$ Chemiluminescence

A 6.0-ml syringe tipped with a 19-gauge hypodermic needle, containing between 1.0 and 50.0 mM ethyl hydroperoxide (Accurate Chemical and Scientific Co., Westbury, NY) in 0.1 M sodium phosphate (p^{2}H 5.2),

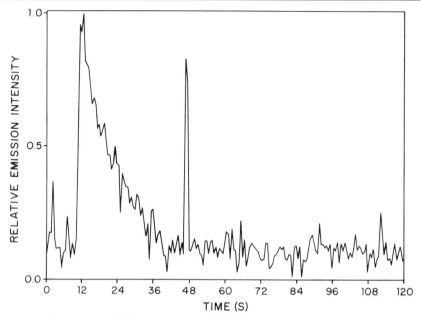

FIG. 2. Time course of $^1O_2$ luminescence following rapid mixing of a 1.0-ml aliquot of 45 mM ethyl hydroperoxide in 0.1 M sodium phosphate in deuterium oxide (p^{2}H 5.2) with a 1.0-ml aliquot of 4.0 μM chloride peroxidase in the same buffer. (From Hall *et al.*[7])

in deuterium oxide, is positioned above the sample holder of the detection system so that the solution can be injected into a standard fluorescence cuvette containing an aliquot of the same buffer with 2.0 to 10 nmol of chloride peroxidase (Sigma Chemical Co., St. Louis, MO). The volumes are chosen so that the total volume after mixing is 2.0 ml. The recording device, in our instrument an Apple IIe acquisition program, is started in order to record a baseline before the reaction. After the baseline information has been obtained, the ethyl hydroperoxide is rapidly injected into the cuvette. A typical reaction profile is shown in Fig. 2. Rapid development of luminescence occurs at the moment of injection (12 sec); then the signal slowly decays back to the baseline (40 sec). Spikes like that occurring at about 48 sec in Fig. 2 are commonly observed in $^1O_2$ luminescence data and are characteristic of the infrared detector itself; their effects can be minimized by standard statistical treatments.

The emission of $^1O_2$ at 1268 nm can be distinguished from other types of luminescence in the same spectral region by demonstrating that the emission is a relatively narrow band centered at 1268 nm. An example of the $^1O_2$ emission band obtained for the chloride peroxidase–ethyl hydroperoxide reaction can be found in Ref. 7, where we have also demonstrated that the production of $^1O_2$ requires the native enzyme.

The emission of $^1O_2$ at 1268 nm can be a reliable diagnostic for the production of $^1O_2$ in both photosensitizing and chemiluminescent systems. Its application to biochemical systems has hardly been exploited; however, as the availability of instrumentation increases and the sensitivity of the component detector improves, the detection system is likely to become a standard research instrument.

[11] Detection and Quantitation of Hydroxyl Radical Using Dimethyl Sulfoxide as Molecular Probe

By CHARLES F. BABBS and MELISSA GALE STEINER

Introduction

Numerous methods exist for trapping hydroxyl radicals (HO ·) with introduced molecular probes that yield chemically detectable products after reaction with HO ·.[1-3] Effective probes include 5,5'-dimethylpyrroline N-oxide (DMPO), benzoic acid, methional, 2-keto-4-thiomethylbutanoic acid (KTBA), p-nitrosodimethylaniline, tryptophan, and dimethyl sulfoxide (DMSO). Any of the foregoing compounds is likely to detect hydroxyl radicals in simple test-tube reaction systems that contain minimum concentrations of competing HO · scavengers. The situation becomes more difficult, however, if one wishes to measure hydroxyl radical generation in biological systems containing large amounts of proteins, nucleic acids, purines, sugars, urea, and other compounds that are readily oxidized by HO ·. Owing to their high and indiscriminate reactivity,[4] hydroxyl radicals are much more likely to react with various endogenous biological compounds in such systems than with the introduced molecular probe. This difficulty may be summarized conceptually as follows. Sup-

[1] R. A. Floyd, C. A. Lewis, and P. K. Wong, this series, Vol. 105, p. 231.

[2] B. Halliwell and J. M. C. Gutteridge, "Free Radicals in Biology and Medicine," p. 206. Oxford Univ. Press, Oxford, 1987.

[3] B. Halliwell, M. Grootveld, and J. M. C. Gutteridge, *Methods Biochem. Anal.* **33,** 59 (1987).

[4] N. N. Semenov, *in* "Some Problems of Chemical Kinetics and Reactivity," Vol. 1. Pergamon, New York, 1958.

METHODS IN ENZYMOLOGY, VOL. 186

pose that hydroxyl radicals react in a system containing both substance S and competing substance C as follows:

$$HO\cdot\ +\ S \xrightarrow{k_1} \text{product } P_1 \tag{1}$$

$$HO\cdot\ +\ C \xrightarrow{k_2} \text{product } P_2 \tag{2}$$

Based on the principle of competition kinetics,[5] the probability that HO $\cdot$ will react with S, rather than C, is given by the fraction

$$\frac{k_1[HO\cdot][S]}{k_1[HO\cdot][S]\ +\ k_2[HO\cdot][C]} = \frac{k_1[S]}{k_1[S]\ +\ k_2[C]} \tag{3}$$

If we consider S as the introduced molecular probe and C as a lumped parameter representing all the competing scavengers of HO $\cdot$ in a given system, then Eq. (3) describes how the proportion of HO $\cdot$ that is detected is influenced by the effective concentration of competing scavengers present and the rate-lumped rate constant, k_2.

Values for k_2 may range from 10^7 to 10^{10} M^{-1} sec^{-1}.[6] Hypothetically, considering [C] for the intracellular environment of a living system as at least 0.1 M, taking k_1 for the desired trapping reaction as 10^9 M^{-1} sec^{-1}, and estimating k_2 for interfering trapping reactions as 10^8 M^{-1} sec^{-1}, one finds that the fraction of HO $\cdot$ trapped by 1, 10, and 100 mM concentrations of probe, S, would be 9, 50, and 91%, respectively. Because the exact concentrations of competing HO $\cdot$ scavengers and their rate constants for reaction with HO $\cdot$ are difficult to know in most biological experiments, trapping efficiency is likely to be substantially less than 100%, to a variable and unknown extent. The only means to force efficient scavenging of HO $\cdot$ by probe S is to increase the concentration of S to about 100 mM or greater.[7,8] However, most hydroxyl radical scavengers cannot be tolerated by living systems in such high concentrations.

Dimethyl sulfoxide is the exception. It is exceedingly nontoxic and can be tolerated by living systems in up to 1 M concentrations.[9–14] The

[5] J. W. T. Spinks and R. J. Woods, "An Introduction to Radication Chemistry," 2nd Ed. Wiley (Interscience), New York, 1976.

[6] G. V. Buxton, C. L. Greenstock, W. P. Helman, and A. B. Ross, "Critical Review of Rate Constants for Reactions of Hydrated Electrons, Hydrogen Atoms and Hydroxyl Radicals in Aqueous Solution." Radiation Chemistry Data Center, Radiation Laboratory, University of Notre Dame, Notre Dame, Indiana, 1986.

[7] C. F. Babbs and D. W. Griffin, *Free Radical Biol. Med.* **6**, 493 (1989).

[8] J. E. Repine, O. W. Pfenninger, D. W. Talmage, E. M. Berger, and D. E. Pettijohn, *Proc. Natl. Acad. Sci. U.S.A.* **78**, 1001 (1981).

[9] P. E. Benville, C. E. Smith, and W. E. Shanks, *Toxicol. Appl. Pharmacol.* **12**, 156 (1968).

[10] M. M. Mason, *in* "Dimethyl Sulfoxide" (S. W. Jacob, E. E. Rosen, and D. C. Wood, eds.), p. 113. Dekker, New York, 1971.

median lethal dose (LD_{50}) of intravenous DMSO in animals ranges from about 4 g/kg body weight (51 mM) to 10 g/kg body weight (128 mM) depending on the species.[10,15] This value increases if the drug is diluted with water prior to administration in order to avoid the burning effects caused by the heat of solution of pure DMSO. The compound is rapidly absorbed by all routes, and distributes to all tissue compartments.[16] In one remarkable experiment, Benville et al.[9] studied young salmon and trout totally immersed in DMSO solutions. Fish immersed in 2% (0.26 M) DMSO for 100 days exhibited good appetite and normal weight gain.

Chemically, DMSO yields a single, stable, nonradical product, methanesulfinic acid (MSA), on reaction with hydroxyl radical,[17,18]

$$H_3C-\overset{\overset{O}{\|}}{S}-CH_3 + HO\cdot \rightarrow \cdot CH_3 + H_3C-\overset{\overset{O}{\|}}{S}-OH \qquad (4)$$
$$\text{DMSO} \qquad\qquad\qquad\qquad \text{MSA}$$

in approximately 85% yield[19] ($k = 7 \times 10^9\ M^{-1}\ sec^{-1}$).[20] Methanesulfinic acid is a primary product of the trapping reaction, and, indeed, one of the oxygen atoms in the MSA molecule is identical to the oxygen of the original trapped hydroxyl radical. Thus, measurement of MSA accumulation in DMSO pretreated biological systems provides a potential means to capture and count the hydroxyl radicals generated therein.

Sulfinic acids (RSOOH) are distinct from sulfonic acids (RSO_3H), sulfenic acids (RSOH), sulfones (R_1SOOR_2), and sulfoxides (R_1SOR_2). We have developed a simple colorimetric assay for methanesulfinic acid,[21] in the presence of high concentrations of unreacted DMSO, which is quite sensitive and easy to perform. The assay is based on the reaction of

[11] R. R. Maurer, in "The Freezing of Mammalian Embryos" Ciba Foundation Symposium 52, p. 116. Elsevier, Amsterdam, 1977.

[12] I. Wilmut and L. E. A. Rowson, Vet. Rec. **92**, 686 (1973).

[13] S. M. Willadsen, in "The Freezing of Mammalian Embryos" Ciba Foundation Symposium 52, p. 175. Elsevier, Amsterdam, 1977.

[14] M. J. Ashwood-Smith, Ann. N.Y. Acad. Sci. **243**, 246 (1975).

[15] E. R. Smith, Z. Hadidian, and M. M. Mason, Ann. N.Y. Acad. Sci. **141**, 96 (1967).

[16] C. W. Denko, R. M. Goodman, R. Miller, and T. Donovan, Ann. N.Y. Acad. Sci. **141**, 77 (1967).

[17] C. Lagercrantz and S. Forshult, Acta Chem. Scand. **23**, 811 (1969).

[18] S. M. Klein, G. Cohen, and A. I. Cederbaum, Biochemistry **20**, 6006 (1981).

[19] C. F. Babbs and M. J. Gale, in "Free Radicals: Methodology and Concepts" (C. Rice-Evans and B. Halliwell, eds.), p. 91. Richelieu Press, London, 1988.

[20] L. M. Dorfman and G. E. Adams, National Standard Reference Data Series 46 (NSRDS-NBS46), U.S. National Bureau of Standards, U.S. Government Printing Office, Washington D.C., 1973.

[21] C. F. Babbs and M. J. Gale, Anal. Biochem. **163**, 67 (1987).

sulfinic acids with diazonium salts, first described by Ritchie et al.,[22] namely,

$$CH_3SOOH + Ar—N{=}N^+ \rightarrow H^+ + Ar—NN—SO_2—CH_3 \qquad (5)$$
$$\underset{\substack{\text{sulfinic} \\ \text{acid}}}{} \qquad \underset{\substack{\text{diazonium} \\ \text{salt}}}{} \qquad \underset{\substack{\text{diazosulfone} \\ \text{(colored,} \\ \text{hydrophobic)}}}{}$$

The product is a colored diazosulfone, which can be selectively extracted into an organic solvent. Of 22 diazonium salts tested, the one generating the most intense color and having the most desirable practical attributes is Fast Blue BB salt.[21] The canary yellow reaction product (λ_{max} 425 nm) precipitates from aqueous solutions at high (>10 mM) concentrations and can be extracted into toluene–butanol (3 : 1) and measured spectrophotometrically. There are limitations to this approach, which are subsequently discussed, but it is offered as an attractive, and low-cost alternative to current methods. The sensitivity of the method is about 10 nmol per sample.

Methods

Materials

Reagent grade sulfuric acid, *n*-butanol, acetic acid, toluene, and pyridine are obtained from standard sources. Fast Blue BB salt is obtained from Aldrich Chemical Company (Milwaukee, WI). Disposable Sep-Pak chromatography columns (Cat. 51910) are obtained from Waters Associates (Milford, MA). Standard curves are prepared using authentic methanesulfinic acid, sodium salt, obtained from Fairfield Chemical Company (Blythewood, SC) or authentic benzenesulfinic acid, sodium salt (for color reaction only), obtained from Aldrich. Alternative suppliers of methanesulfinic acid can be found in Chem Sources USA (Directories Publishing Company, Orlando Beach, FL).

Sample Preparation

The practical assay of biological material has two phases. The first is the separation of sulfinic acid in the sample from potentially interfering species (sample cleanup). The second is the reaction with the diazonium salt, extraction of the colored product, and direct reading in a spectrophotometer (color reaction). For simple enzyme systems, such as the xanthine oxidase system, run in the presence of 5% DMSO, no cleanup

[22] C. D. Ritchie, J. D. Saltiel, and E. S. Lewis, *J. Am. Chem. Soc.* **83**, 4601 (1961).

procedure is required.[7] The assay of DMSO-pretreated plant or animal material requires either extraction of the sulfinic acid from interfering materials (method A, to follow) or removal of the interfering materials from the sample (method B). Plant material can be assayed after quick-freezing with liquid nitrogen, grinding with a mortar and pestle, aqueous extraction of the resulting powder, and, if necessary, concentration of the aqueous extract by lyophilization. Bacterial suspensions and animal tissues can be prepared for assay by homogenation and centrifugation. Most of the sulfinic acid remains in the supernatant. Some procedure to remove protein usually improves results. We have found that acidification with HCl and neutralization with NaOH prior to high-speed centrifugation are effective for this purpose. Trichloroacetic acid precipitation of protein cannot be used, because the trichloroacetate anion produces prodigious interference.

Acid–Butanol Extraction of Sulfinic Acid (Cleanup A)

Principle. Sulfinic acid ($pK_a \sim 2$)[23,24] is extracted from tissue homogenate into *n*-butanol at pH 0, followed by backextraction of sulfinate ions into aqueous acetate buffer at pH 4 to 5. Potentially interfering species, including proteins and amines, are precipitated or remain in the original aqueous phase. This step should be completed quickly so as to minimize the dwell time of the sulfinic acid below pH 1, since dismutation to the sulfonic acid occurs gradually under these extremely acidic conditions.[25,26]

Procedure. Tissue is homogenized in a chilled Teflon–glass homogenizer in 2 volumes of distilled water. A 2.0-ml aliquot of the aqueous sample, expected to contain 10 to 300 μM sulfinate, is placed in a test tube, and 0.2 ml of 10 N sulfuric acid is added with vortexing. The tubes are centrifuged to remove coagulated protein and the supernatants decanted into fresh tubes. Optionally, the supernatants may be extracted once or twice at this stage with 2.0 ml of toluene–butanol (3 : 1) to remove interfering detergentlike substances, and the organic layers discarded, after which methanesulfinic acid remains in the water phase. Four milliliters of butanol, previously saturated with 1 M sulfuric acid, is added and mixed thoroughly for 60 sec on a vortex mixer. The upper butanol phase

[23] C. I. M. Stirling, *J. Int. J. Sulfur Chem. Part B* **6**, 277 (1971).

[24] J. G. Baldinus, "Treatise on Analytical Chemistry, Part 2 Vol. 15: Analytical Chemistry of Inorganic and Organic Compounds" (I. M. Kolthoff and P. J. Elving, eds.), Sect. B-2, Wiley, New York, 1976.

[25] J. L. Kice and K. W. Bowers, *J. Am. Chem. Soc.* **84**, 605 (1962).

[26] E. E. Gilbert, *in* "Interscience Monographs on Chemistry" (F. A. Cotten, ed.), p. 228. Wiley, New York, 1969.

is removed with a Pasteur pipet to a second, clean test tube containing 2.0 ml of 0.5 M sodium acetate buffer, pH 5, and mixed vigorously. After centrifugation at 500 g for 3 min, the color reaction can then be run on the sulfinic acid extracted into the aqueous acetate phase. The time taken for this extraction procedure should be minimized and kept similar for all samples and standards.

Removal of Interference with Sep-Pak Columns (Cleanup B)

Principle. The ingenious Sep-Pak approach was conceived by Dr. Jean Blair Smith. Many potentially interfering species are anionic detergents, such as bile salts or free fatty acids, which form saltlike complexes with Fast Blue BB dye (λ_{max} 395 nm) that are extractable into toluene–butanol.[19] Most interfering anions are much less hydrophilic than methanesulfinic acid itself. To remove them, preliminary extraction of the sample with toluene–butanol (3 : 1) at neutral pH is helpful, but not always sufficient. Further removal of detergentlike interference can be obtained by adding Fast Blue BB dye to the sample and applying the mixture to a lipophilic (C_{18} Sep-Pak) column. When the column is subsequently eluted with water, the complexes of dye and lipophilic anions remain on the column, while sulfinic acid is readily eluted and collected. [It would appear from this and other observations that reaction (5) is quite reversible in aqueous media.] In this way detergentlike anions that are specifically capable of complexing with Fast Blue BB dye to form lipophilic salts are sequestered on the column. The color reaction with additional Fast Blue BB dye and toluene–butanol extraction is then run on the effluent fraction containing the sulfinic acid.

Procedure. Tissue samples are homogenized in 3 volumes of distilled water. To precipitate protein the pH is lowered to 1 with concentrated HCl, the sample is allowed to stand 10 min, and the pH is returned to 7.4 with NaOH. Denatured proteins are removed by centrifugation. Two-milliliter aliquots of the supernatant are extracted twice with 2 ml of toluene–butanol (3 : 1) and the organic phases discarded. Then 100 μl of 30 mM Fast Blue BB salt is added to the aqueous sample with mixing, and the tubes are allowed to stand for 10 min in the dark. A 1-ml sample is applied to a Sep-Pak (C_{18}) column that has been preeluted with 2 ml of methanol and 2 ml of water. The methane sulfinate anion is then eluted from the column with distilled water. The first 1.3 ml of effluent is discarded, and the next 1.5 ml is collected for assay by the color reaction.

Color Reaction. A consistent 1 to 2 ml volume of sufficiently clean aqueous sample is transferred to a test tube, the pH is adjusted to 2.5 by the addition of up to 0.3 ml of 0.1 or 1 N HCl, and the color reaction is begun by the addition of 0.1 ml of 30 mM Fast Blue BB salt (freshly prepared and kept in the dark). The stock solution is stable for at least 8 hr at room temperature. If necessary, stock solutions of BB dye can be extracted with chloroform to remove photooxidized products that can cause high blanks. Ten minutes is allowed for product development at room temperature in the dark. Then 1.5 ml of toluene–butanol (3 : 1) is added and mixed thoroughly with the aqueous phase for 120 sec on a vortex mixer. After low-speed centrifugation to separate the phases, the lower phase, containing unreacted diazonium salt, is removed by aspiration and discarded. The toluene–butanol phase is washed with 2 ml of butanol-saturated water for 30 sec to remove remaining unreacted diazonium salt. The tubes are centrifuged at 500 g for 3 min, and the upper phase, containing the diazosulfones, is transferred to a cuvette. One tenth milliliter of pyridine–glacial acetic acid (95 : 5) is added to stabilize the color, which otherwise fades gradually at acid pH. The bright yellow color is reasonably stable after pyridine addition, fading about 6% per day at room temperature. The absorbance as a function of wavelength from 340 to 520 nm is recorded on a strip chart recorder, using a blank prepared beginning with 2 ml of distilled water carried through the same procedure. In most experiments, the sulfinic acid content may then be calculated from the absorbance at 425 nm, with reference to a standard curve.

Derivative Spectroscopy

In some applications double-derivative analysis of absorbance spectra[27–29] may be useful to eliminate false-positive interference that does not exhibit a peak (or nadir) at 425 nm, and which can be characterized as a linear function of wavelength from 400 to 450 nm. The spectral region from 400 to 450 nm in the neighborhood of the absorption peak of authentic methanesulfinic acid is digitized, taking at least 5 points, and represented as a second-order polynomial, centered at 425 nm, calculated by a least-squares method.[30,31] The parabolic curve fit for the regression function

$$A = a_0 + a_1 \Delta\lambda + a_2(\Delta\lambda)^2 \qquad (6)$$

of absorbance (A) as a function of wavelength ($\Delta\lambda = \lambda - 425$) is accom-

[27] T. C. O'Haver and G. L. Green, *Anal. Chem.* **48**, 312 (1976).
[28] T. C. O'Haver, *Clin. Chem.* **25**, 1548 (1979).
[29] F. P. Corongiu and A. Milia, *Chem.–Biol. Interact.* **44**, 289 (1983).
[30] N. R. Draper and H. Smith, "Applied Regression Analysis," p. 17. Wiley, New York, 1966.
[31] HP-65 STAT PAC 1, Hewlett-Packard Company, 1974.

plished in our laboratory with the aid of a "C" language computer program, which calculates the least-squares regression coefficients a_0, a_1, and a_2, for $n > 1$ data points, $y = f(x)$, according to the following expressions:

$$a_2 = \frac{A - B}{[n \Sigma x_i^2 - (\Sigma x_i)^2][n \Sigma x_i^4 - (\Sigma x_i^2)^2] - [n \Sigma x_i^3 - (\Sigma x_i)(\Sigma x_i^2)]^2} \quad (7)$$

where

$$A = [n \Sigma x_i^2 - (\Sigma x_i)^2][n \Sigma x_i^2 y_i - (\Sigma x_i^2)(\Sigma y_i)] \quad (8)$$
$$B = [n \Sigma x_i^3 - (\Sigma x_i)(\Sigma x_i^2)][n \Sigma x_i y_i - (\Sigma x_i)(\Sigma y_i)] \quad (9)$$

$$a_1 = \frac{[n \Sigma x_i y_i - (\Sigma x_i)(\Sigma y_i)] - a_2[n \Sigma x_i^3 - (\Sigma x_i)(\Sigma x_i^2)]}{n \Sigma x_i^2 - (\Sigma x_i)^2} \quad (10)$$

$$a_0 = \frac{1}{n} (\Sigma y_i - a_2 \Sigma x_i^2 - a_1 \Sigma x_i) \quad (11)$$

The second derivative of the spectrum of standard methanesulfinic acid is nearly constant from 400 to 450 nm and is directly proportional to sulfinic acid concentration. From the values of the second derivative of the sample absorbance spectrum at 425 nm, obtained by differentiation of the fitted, least-squares polynomial, the methanesulfinic acid content of the sample is computed as

$$y = \frac{2a_2(\text{sample})}{2a_2(\text{standard})} \frac{V_{tot}}{V} \text{ amt}_{std} \quad (12)$$

where A is absorbance, $2a_2$ the second derivative of the parabolic curve fit at 425 nm, amt$_{std}$ the amount of standard methanesulfinate assayed (nmols), V the volume assayed, and V_{tot} the total volume of the sample. This measure of sulfinic acid content is insensitive to interference that is a constant or linear function of wavelength in the neighborhood of 425 nm.

Remarks

Diazonium Coupling Reaction. Considering the general reactivity of diazonium salts,[32] the diazonium coupling reaction for sulfinic acid is surprisingly free of interference.[21] Although many substances are known to couple with diazonium salts to produce liposoluble products,[33,34] most

[32] R. F. Muraca, in "Treatise on Analytical Chemistry" (P. J. Elving, ed.), Part 2, Sect. B-2, Vol. 15, p. 251–234. Wiley, New York, 1976.

[33] K. Venkataraman, "The Chemistry of Synthetic Dyes," Vol. 1. Academic Press, New York, 1952.

[34] K. Venkataraman, "The Chemistry of Synthetic Dyes," Vol. 4. Academic Press, New York and London, 1971.

of these coupling reactions are classically described at neutral or slightly alkaline pH.[32] The sulfinic acid coupling reaction is unusual in that it proceeds well at acid pH. The probable mechanism for the coupling reaction, by analogy with that described by Hauser and Breslow,[35] is attack of the unbonded electrons of the sulfur by the Ar—N=N$^+$ ion, followed by transfer of the oxygen-associated electron of the sulfinate anion to the sulfur atom. The availability of the unbonded electron of sulfinate anions at pH 2.5 may explain the specificity of the reaction for sulfinic acids as opposed to sulfoxides and sulfonic acids. Other well-known diazonium coupling reactions[32,36-38] with phenols, indoles, and amines (abundant in free amino acids and proteins) proceed best at slightly alkaline pH. By allowing the diazonium coupling to take place in acidic solution, the interference from these species is greatly reduced.[21] Nevertheless, false-positive results can be produced by anionic detergents and sulfite.

Causes of False-Positive Results. Anionic detergents in the sample, such as bile salts, generate the most troublesome false-positive interference with the color reaction, by forming lipophilic salts with the diazonium (Ar—N=N$^+$) cation. This type of artifactual signal typically has an absorbance peak at 395 nm (that of the native Fast Blue BB dye), clearly to the left of the 425 nm peak for authentic sulfinic acid. A good example of this phenomenon can be observed experimentally by simply mixing a detergent such as sodium lauryl sulfate with Fast Blue BB and extracting with toluene–butanol (3 : 1). A small admixture of detergent interference with true sulfinic acid signal will cause a shift of the absorbance peak to the left. Removal of detergent anions by procedures such as solvent extraction or ion-exchange chromatography may be necessary in detergent-laden systems such as rat liver. In the presence of detergents the signal-to-noise ratio is a strong function of the amount of BB dye added for the color reaction, since the final dye concentration required for half-maximal color reaction of true methanesulfinic acid is about 1 mM, whereas the dye concentration for half-maximal detergent interference is often 10 mM or greater. Exploration of the inherent trade-offs may be worthwhile in some applications.

One interesting variant of detergent interference is provided by sodium sulfite. Sulfite reacts rapidly with Fast Blue BB under acidic conditions to produce what we surmise is the sulfonic acid derivative, Ar—N=N—SO$_2$H. This species, in turn, probably forms salts with unreacted dye (Ar—N=N$^+$) to form a strongly absorbing complex, extract-

[35] C. R. Hauser and D. S. Breslow, *J. Am. Chem. Soc.* **63**, 418 (1941).
[36] R. Wistar and P. D. Bartlett, *J. Am. Chem. Soc.* **63**, 413 (1941).
[37] J. B. Conant and W. D. Peterson, *J. Am. Chem. Soc.* **52**, 1220 (1930).
[38] K. H. Beyer and J. T. Skinner, *J. Pharmacol. Exp. Ther.* **68**, 419 (1940).

able into toluene–butanol. This effect may even be exploitable to measure sulfite in some applications. It can be minimized by cleanup procedures A and B and by treatment of the final organic phase with dry, beaded ion-exchange resins.[19]

Another possible source of false-positive results that we have found is artifactual DMSO oxidation during the assay procedure itself, as opposed to biological DMSO oxidation by HO · during the experiment preceding the assay. Suitable combinations of heat, iron, oxygen, reductants such as ascorbic acid or thiols, and/or prolonged incubation or storage times can induce oxidation of DMSO by nonbiological mechanisms to give a product spectrally indistinguishable from methanesulfinic acid in the colorimetric assay.[39] Even though such autoxidation yields less than 0.1 mM methanesulfinic acid from 1 M DMSO, the effect can be of the same order as biologically induced DMSO oxidation. This experimental difficulty is easily detected, however, by a so-called late DMSO addition control experiment, in which DMSO is added to the system after biological oxygen radicals could not be produced (e.g., after depletion of substrate, after denaturation of enzymes) but prior to the assay procedure. If oxidation of DMSO during the assay is found, reasonable steps can be taken to minimize the effect, including use of fresh, reagent grade DMSO, elimination of iron from buffers and reagents,[40,41] or use of shorter, colder storage conditions for samples prior to assay.

Causes of False-Negative Results. The possibility that sulfinic acids may be degraded in biological samples before they can be measured is worthy of consideration and is easily tested by incubation of standard methane sulfinate with the biological system under study for various periods of time. Recovery of added standard from the system is an important validation step, since some sulfinate may become bound to tissue proteins and resist extraction. As previously mentioned, acid-catalyzed dismutation of methanesulfinic acid to methanesulfonic acid, which is not detected by the assay,[21] does occur slowly, so that the duration of acidification of samples during any cleanup procedure should be minimized and kept consistent for standards and samples. Certainly, samples should never be stored in an acidified state. If necessary, calibration can be done using the method of standard addition.

In some systems with high concentrations of competing scavengers of HO ·, even 1 M DMSO may trap a fraction of nascent HO · substantially less than unity.[7] This effect can be detected by studying results for a range

[39] D. W. Griffin, Masters thesis, Purdue University, Lafayette, Indiana, 1988.
[40] J. M. C. Gutteridge, *FEBS Lett.* **214**, 362 (1987).
[41] G. R. Buettner, *J. Biochem. Biophys. Methods* **16**, 27 (1988).

of DMSO concentrations added to the biological system under study. If necessary, Scatchard analysis[42] may be performed to account for the fraction of HO· radicals that react with endogenous compounds rather than DMSO.[7]

Summary

Generation of clearly harmful amounts of hydroxyl radicals in biological systems can be studied using DMSO as a molecular probe. DMSO is oxidized by HO· to form the stable, nonradical compound methanesulfinic acid, which is not normally found in living systems and which can be easily extracted from tissue and measured spectrophotometrically. The present method provides a simple, inexpensive assay for methanesulfinic acid in biological materials. As little as 10 nmol of sulfinate can be detected, and interference from diverse biological compounds is minimal. Additionally, there is no interference from a large excess of dimethyl sulfoxide, which is necessary if the assay is to be applied directly to tissues pretreated with DMSO. When straightforward cleanup procedures are utilized, there is minimal interference from glutathione or sulfate, and potentially troublesome interference from detergentlike substances can usually be minimized. Owing to its relative specificity for sulfinic acids at acid pH, the diazonium coupling reaction can thus be exploited to provide an efficient and inexpensive means of detecting methanesulfinic acid in DMSO-pretreated biological materials. The results provide a direct chemical means for measuring cumulative HO· generation.

Acknowledgments

The technical assistance of Ms. Joanne Cusumano and Ms. Meloney Cregor is gratefully acknowledged. Supported by Grants HL-36712 and HL-35996 from the National Heart, Lung, and Blood Institute, by Grant CA-38144 from the National Cancer Institute, U.S. Public Health Service, Bethesda, Maryland, and by a Focused Giving Grant from Johnson & Johnson.

---

[42] G. Scatchard, *Ann. N.Y. Acad. Sci.*, 51 (1949).

[12] Distinction between Hydroxyl Radical and Ferryl Species

By J. D. RUSH, Z. MASKOS, and W. H. KOPPENOL

Introduction

The Fenton reaction, or the reaction of ferrous ion with hydrogen peroxide at low pH, is generally considered to yield the hydroxyl radical. This radical is a strong oxidizing agent $[E^{\circ}(\cdot OH/H_2O) = 2.73$ V]$[1]$ and attacks various small molecules with rates of $10^8–10^{10}$ M^{-1} sec^{-1}, whereas its reaction with proteins is diffusion controlled.[2] The products are carbon-centered radicals that in the presence of oxygen are converted to organic hydroperoxyl radicals.[3] These radicals are rather oxidizing and can start various chain reactions.[4] The propagation reactions are well understood and are responsible for far more damage than the initiating event. It is clear that oxygen plays a dual role in this process: it makes the formation of the oxidizing species possible in the first place, and, second, it amplifies the damage through the chain reactions.

While the concept of the hydroxyl radical as an initiator has received wide support, recent evidence suggests that at neutral pH and in the presence of a chelating agent a higher oxidation state of iron might be involved. This concept is not new: as early as 1932 it was proposed that a higher oxidation state of iron, the ferryl ion (FeO^{2+}), might be involved in the decomposition of hydrogen peroxide.[5] While this appears not to be the case at low pH,[6,7] the situation is more complex at neutral pH when iron is chelated. The failure of common hydroxyl radical scavengers to inhibit, for instance, the formation of ethylene from methionine resulted in the postulation of electron-donor or "crypto-·OH" complexes.[8] Such a species should be fairly oxidizing to show more or less the same reactivity as the hydroxyl radical. For instance, a reduction potential of 1.2 V is re-

[1] W. H. Koppenol, *Bioelectrochem. Bioenerg.* **18**, 3 (1987).

[2] G. V. Buxton, C. L. Greenstock, W. P. Helman, and A. B. Ross, *J. Phys. Chem. Ref. Data* **17**, 513 (1988).

[3] K. U. Ingold, *Acc. Chem. Res.* **2**, 1 (1969).

[4] W. H. Koppenol, *in* "Oxygen and Oxy-Radicals in Chemistry and Biology" (M. A. J. Rogers and E. L. Powers, eds.), p. 617. Academic Press, New York, 1981.

[5] W. C. Bray and M. H. Gorin, *J. Am. Chem. Soc.* **54**, 2124 (1932).

[6] C. Walling, *Acc. Chem. Res.* **8**, 125 (1975).

[7] J. D. Rush and W. H. Koppenol, *J. Inorg. Biochem.* **29**, 199 (1987); J. D. Rush, Z. Maskos, and W. H. Koppenol, *FEBS Lett.* **261**, 121 (1990).

[8] R. J. Youngman and E. F. Elstner, *FEBS Lett.* **129**, 265 (1981).

quired to abstract an α-hydrogen from methanol,[9] and such a value would seem to be a lower limit. A thermodynamic derivation[10] suggests a value in excess of 0.9 V for the Fe^{IV}/Fe^{III}–EDTA couple.

The term ferryl is commonly used to describe an oxidizing iron species derived from the reaction of hydrogen peroxide by ferrous complexes. Little is known directly about the structure and reactivity of high-valent iron in aqueous solutions, however, with the exception of that of ferryl porphyrins[11,12] and some spectroscopic information and decay kinetics of ferryl and ferrate (Fe^V), FeO_4^{3-}, in alkaline solution.[13] Ferryl can be represented by Fe^{II} (H_2O_2), $Fe^{II} = O^{2+}$, $Fe^{III} O^-$, or Fe^{IV} (OH^-)$_2$, in which the formal charge is indicated by roman numerals. However, the oxidation state of iron is +4.

Methods

The methods currently used for analysis of Fenton intermediates are summarized below. In large part they rely on discrimination between the anticipated effects of the hydroxyl radical and the experimental observations. However, even in systems of much less than biological complexity the analysis is not simple, and therefore experiments involving a minimal number of reactants are best. The formation of a reactive species is rate limiting and prevents its direct observation. Rate constants for the reaction of various ferrous complexes with hydrogen peroxide are given in Table I.[14–19]

Inhibition of Peroxide Decomposition

Catalytic decomposition of hydrogen peroxide takes place if a ferrous complex, $HLFe^{II}$, is oxidized by peroxide and $HLFe^{III}$ is reduced by

[9] W. H. Koppenol and J. D. Rush, *J. Phys. Chem.* **91**, 4429 (1987).

[10] W. H. Koppenol and J. Liebman, *J. Phys. Chem.* **88**, 99 (1984).

[11] D. Ostovic and T. C. Bruice, *J. Am. Chem. Soc.* **110**, 6906 (1988).

[12] J. R. L. Smith, P. N. Balasubramanian, and T. C. Bruice, *J. Am. Chem. Soc.* **110**, 7411 (1988).

[13] J. D. Rush and B. H. J. Bielski, *J. Am. Chem. Soc.* **108**, 523 (1986).

[14] J. D. Rush and W. H. Koppenol, in "Free Radicals, Metal Ions and Biopolymers" (P. Beaumont, D. Deeble, B. Parsons, and C. Rice-Evans, eds.), p. 33. Richelieu Press, London, 1989.

[15] T. J. Hardwick, *Can. J. Chem.* **35**, 428 (1957).

[16] O. K. Borggaard, O. Farver, and V. S. Andersen, *Acta Chem Scand.* **25**, 3541 (1971).

[17] H. C. Sutton and C. C. Winterbourn, *Arch. Biochem. Biophys.* **235**, 106 (1984).

[18] B. C. Gilbert and M. Jeff, in "Free Radicals: Chemistry, Pathology and Medicine" (C. Rice-Evans and T. Dormandy, eds.), p. 25. Richelieu Press, London, 1988.

[19] S. Rahhal and H. W. Richter, *J. Am. Chem. Soc.* **110**, 3126 (1988).

TABLE I
RATE CONSTANTS FOR REACTION OF FERROUS ION AND FERROUS
AMINOPOLYCARBOXYLATE COMPLEXES WITH HYDROGEN PEROXIDE

Complex	Rate constant (10^3 M^{-1} sec^{-1}) reported by					
	H[a]	BFA[b]	SW[c]	RK[d]	GJ[e]	RR[f]
Aquo	41.5 ± 0.3[g]	60 ± 6[g]	—	—	200[g]	—
DTPA	—	—	0.51	0.8	13.5	1.37 ± 0.07
EDTA	—	9.10 ± 0.08	7	7	10	—
HEDTA	—	16.7 ± 0.1	—	42	—	—
NTA	—	18.4 ± 0.1	—	30	—	—
EDDA	—	—	—	78	—	—
Fe^{II}ATP				6.6 ± 0.3		
Fe^{II}ADP				11 ± 2		
Fe^{II}citrate				4.9 ± 0.3		

[a] Hardwick[15] (20.2°, 0.1 N HClO$_4$); this reference contains a discussion of earlier work.
[b] Borggaard et al.[16] (20°, pH 0.2 M ionic strength); values apply to unprotonated species and were determined polarographically.
[c] Sutton and Winterbourn[17] (pH 7.4, varying ionic strength).
[d] Rush and Koppenol[7] (25°, pH 7.2 ± 0.2, 38 mM ionic strength); stopped-flow study; Rush et al.[7]
[e] Gilbert and Jeff[18] (pH 7); ESR study.
[f] Rahhal and Richter[19] (pH 7.0); rapid mixing study.
[g] Data in M^{-1} sec^{-1}.

superoxide. This set of reactions [reactions (2)–(4)] is sometimes referred to as a metal-catalyzed Haber–Weiss[20] reaction[21]:

$$HLFe^{III} + H_2O_2 \rightarrow HLFe^{II} + O_2^{\cdot -} + 2\ H^+ \tag{1}$$
$$HLFe^{II} + H_2O_2 \rightarrow HLFeO^{2+} + H_2O,\ or \tag{2a}$$
$$HLFe^{III} + \cdot OH + OH^- \tag{2b}$$
$$HLFeO^{2+} + H_2O_2 \rightarrow HLFe^{III} + O_2^{\cdot -} + H_2O \tag{3a}$$
$$\cdot OH + H_2O_2 \rightarrow O_2^{\cdot -} + H_2O + H^+ \tag{3b}$$
$$O_2^{\cdot -} + HLFe^{III} \rightarrow HLFe^{II} + O_2 \tag{4}$$
$$HLFeO^{2+} + RH \rightarrow HLFe^{III} + OH^- + R\cdot \tag{5a}$$
$$\cdot OH + RH \rightarrow H_2O + R\cdot \tag{5b}$$

[20] F. Haber and J. Weiss, Proc. Roy. Soc. London A **147**, 332 (1934).
[21] The reaction O$_2^{\cdot -}$ + H$_2$O$_2$ → ·OH + O$_2$ + OH$^-$ forms part of a cycle originally proposed by F. Haber and R. Wilstätter, [Chem. Ber. **64**, 2884 (1931)] to account for the decomposition of hydrogen peroxide, but it became known as the Haber–Weiss reaction. The ability to react with hydrogen peroxide was considered an essential characteristic of the superoxide anion. It was shown later by several investigators [see W. H. Koppenol, in "Oxidases and Related Redox Systems" (T. E. King, H. S. Mason, and M. Morrison, eds.), p. 127. Pergamon, Oxford, 1982] that the rate of this reaction was too slow to be an efficient source of hydroxyl radicals.

Competition between reactions (3a) and (3b) and (5a) and (5b) reflects the relative reactivity of the hydroxyl radical and a putative hypervalent iron intermediate toward hydrogen peroxide and the organic scavenger, respectively. If the product of this reaction, R·, does not propagate the chain reaction by reducing the ferric complex, a sequence of scavengers with varying reactivity toward the hydroxyl radical can be employed to determine if the relative rates of reactions (3a) and (5a) correspond to that of the hydroxyl radical.

This system requires a substantial excess of hydrogen peroxide since its rate of reaction with the hydroxyl radical is low $[k(\cdot OH + H_2O_2) = 2 \times 10^7 \ M^{-1} \ \text{sec}^{-1}]$ compared to most organic scavengers, including the commonly used ligands of iron(II)/iron(III). Quenching aliquots of reaction mixtures in a titanium sulfate–sulfuric acid solution is a convenient method to determine hydrogen peroxide concentrations to approximately $10^{-5} \ M$ accuracy if solution components do not interfere with the absorbance of peroxotitanium at 408 nm ($\varepsilon = 740 \ M^{-1} \ \text{cm}^{-1}$).[22] Oxygen evolution may also be monitored with a Clark electrode as an index to the rate of peroxide decomposition.

The oxidation of ferrous complexes by hydrogen peroxide generates 2 mol of iron(III) per mole of peroxide. The oxidizing intermediate may be reacted with excess of iron(II), giving the full stoichiometry, or with an added scavenger. The scavenger product, which is normally a radical, may react as follows:

$$\Delta[\text{Fe(III)}]/\Delta[\text{H}_2\text{O}_2]_{\text{limiting}}$$

$R \cdot + \text{HLFe}^{II} + H^+ \rightarrow HR + \text{HLFe}^{III}$	2:1	(6)
$R \cdot + \text{HLFe}^{III} \quad \rightarrow R^+ + \text{HLFe}^{II}$	0:1	(7)
$R \cdot + R \cdot \quad\quad\quad \rightarrow \text{products}$	1:1	(8)

In their reactions toward $\text{Fe}^{II}/\text{Fe}^{III}$–aminopolycarboxylate complexes, carbon-centered radicals are usually oxidizing when the site of hydrogen abstraction is α to a carbonyl or carboxylate group as in acetone or acetate, inert if adjacent only to alkyl residues as in *tert*-butanol, and reducing if α to the hydroxyl group of alcohols. If more than one kind of radical is formed, nonintegral limiting stochiometries are found.

If $R \cdot$ reduces HLFe^{III}, the variation in stoichiometry of HLFe^{III} production with the organic scavenger concentration should follow Eq. (9a),[7] where $\alpha_{\cdot OH} = k(\cdot OH + \text{HLFe}^{II})/k(\cdot OH + HR)$:

$$\Delta\text{HLFe}^{III}/\Delta\text{H}_2\text{O}_2 = \frac{2\alpha_{\cdot OH}[\text{HLFe}^{II}]/[HR]}{1 + \alpha_{\cdot OH}[\text{Fe}^{II}L]/[HR]} \quad\quad (9a)$$

[22] M. J. Irvine and I. R. Wilson, *Aust. J. Chem.* **32**, 2283 (1979).

from which $\alpha_{\cdot OH}$ can be compared with literature values.[2] For an inert radical R·, Eq. (9a) becomes[7]

$$\Delta HLFe^{III}/\Delta H_2O_2 = 1 + \frac{\alpha_{\cdot OH}[HLFe^{II}]/[HR]}{1 + \alpha_{\cdot OH}[Fe^{II}L]/[HR]} \qquad (9b)$$

Solutions must be thoroughly deaerated and the ferrous complex maintained in excess. The stoichiometry, and hence α, can be measured with good accuracy by the stopped-flow method, using the high absorptivity of iron(III) complexes in the UV region. A current limitation is that data on the rate of hydroxyl radical reactions with iron(II) complexes is limited.

The method has been applied to iron complexes of EDTA, diethylenetriaminepentaacetic acid (DTPA), and (N-hydroxyethyl)diaminetriacetic acid (HEDTA).[7] We have used the argument that the inability of tert-butanol to scavenge the reactive intermediates formed from the interaction of a ferrous chelate with hydrogen peroxide is evidence for a higher oxidation state of iron.[23] The experimental observation is that 1.7 to 2.0 mol HLFe[II] is oxidized per mole of H_2O_2 consumed, and that tert-butanol does not decrease this ratio to 1.0 as expected for an inert tert-butanol radical [Eq. (9b)].[7] An explanation has been offered by Gilbert and Jeff,[18] who suggested that the tert-butanol radical is not inert, but oxidizes the ferrous chelate and undergoes a reductive elimination, as studied by Eberhardt,[24]

$$\cdot CH_2(CH_3)_2OH + HLFe^{II} + H^+ \rightarrow H_2C{=}C(CH_3)_2 + HLFe^{III} + H_2O \qquad (10)$$

However, the amount of 2-methylpropene formed is small, and reaction (10) appears not to account for the observed ratio of 1.7–2 ferrous complexes oxidized per hydrogen peroxide consumed (J. D. Rush and W. H. Koppenol, unpublished, 1989). Rahhal and Richter[19] also studied the oxidation of ferrous DTPA by hydrogen peroxide in the presence and absence of tert-butanol and found less than expected scavenging, irrespective of the tert-butanol concentration.

Product Analysis

In Fenton-related systems, the oxidations of organic components by either the hydroxyl radical or a higher oxidation state of iron typically give radical intermediates. Direct determination of these organic radicals by EPR (electron paramagnetic resonance) flow methods or analysis of products can give misleading results if the effect of other solution components, particularly the iron complex, is not taken into account. For in-

[23] J. D. Rush and W. H. Koppenol, *J. Biol. Chem.* **261,** 6730 (1986).
[24] M. K. Eberhardt, *J. Org. Chem.* **49,** 3720 (1984).

stance, the oxidation of alcohols in a $Fe^{II}EDTA–H_2O_2$ flow EPR experiment detected only β-carbon-centered radicals,[25] in contrast to the fact that hydroxyl radicals abstract hydrogen at the α position with about 90% efficiency. Since only the α-carbon radicals readily reduce $Fe^{III}EDTA$, however, these results were not useful in assigning the intermediate to a selective iron oxidant.[6]

A characteristic reaction of ferryl versus that of the hydroxyl radical in water has not yet been unambiguously identified. However, stopped-flow experiments showed that transient species absorbing around 300 and 410 nm are obtained when ferrous HEDTA,[7] nitrilotriacetate (NTA), and ethylenediaminediacetic acid (EDDA) react with hydrogen peroxide in the presence of formate ions.[26] Two plausible pathways for their formation have been suggested.

Pathway A[26]:

$$HLFe^{II} + H_2O_2 \quad \rightarrow HLFeO^{2+} + H_2O$$
$$HLFeO^{2+} + HCO_2^- \rightarrow \cdot LFe^{II} + CO_2^{\cdot -} + H_2O \tag{11}$$

Pathway B[27]: As in Eqs. (2b) or (2a)

$$HLFe^{II} + H_2O_2 \quad \rightarrow \cdot OH + OH^- + HLFe^{III}, \text{ or}$$
$$HLFeO^{2+} + H_2O$$
$$\cdot OH + HCO_2^- \quad \rightarrow H_2O + CO_2^{\cdot -} \tag{12a}$$
$$HLFeO^{2+} + HCO_2^- \rightarrow HLFe^{III} + CO_2^{\cdot -} + OH^- \tag{12b}$$
$$CO_2^{\cdot -} + HLFe^{II} \quad \rightarrow HLFe^{II}CO_2^- \text{ or } HLFe^{III}CO_2^{2-} \tag{13}$$

In these systems, ligand degradation of EDDA and NTA by peroxide is not inhibited by sodium formate,[26] which is consistent with one-electron oxidation of the ligand and not exclusively of formate ions.

Indicator Reactions

The use of a stable scavenger product as a chromogen to monitor the generation of oxidizing transients is a valuable method that depends on the availability of species which give selective reactions and easily detectable and stable oxidation products. Usually the progress of hydroxyl or ferryl generation is monitored spectrophotometrically. As an example, the reaction of substoichiometric amounts of radiolytically generated hydroxyl radicals with ferricytochrome c leads via a surface reaction to reduction of the heme with approximately 50% yield.[28] Since no reduction

[25] T. Shiga, *J. Phys. Chem.* **69**, 3805 (1965).
[26] J. D. Rush and W. H. Koppenol, *J. Am. Chem. Soc.* **110**, 4957 (1988).
[27] S. Goldstein, G. Czapski, H. Cohen, and D. Meyerstein, *J. Am. Chem. Soc.* **110**, 3903 (1988).
[28] J. W. van Leeuwen, A. Raap, W. H. Koppenol, and H. Nauta, *Biochim. Biophys. Acta* **503**, 1 (1978).

was observed in the presence of ferrous EDTA and hydrogen peroxide, it was concluded that formation of hydroxyl radicals in this system is unlikely.[29] Three compounds of low molecular weight commonly used are p-nitrophenol (giving a red nitrocatechol),[30] 2,2′-azinobis(3-ethyl-1,2-dihydrobenzthiazole 6-sulfonate) or ABTS[31] (giving the intensely green ABTS$^{\cdot+}$ radical cation), and methyl viologen.

ABTS$^{\cdot+}$ is formed from the parent molecule by one-electron oxidation. The reaction is effected by hydroxyl radical (but with only 58% efficiency),[32] peroxidase,[31] peroxytetrasulfenatophenylporphyrin iron(III),[33] and a variety of inorganic oxidizing radicals such as $Br_2^{\cdot-}$, which reacts with 100% efficiency.[32] Numerous other reagants, notably Fe^{3+}, can oxidize ABTS [E'(ABTS$^{\cdot+}$/ABTS) $\cong$ 0.6 V]; therefore, careful controls of separate components of a system should be performed prior to its use in Fenton-related systems.

ABTS$^{\cdot+}$ has useful absorptions at 415 nm ($\varepsilon = 3.6 \times 10^4 \ M^{-1} \ sec^{-1}$) and in the region of 600 nm that can be used to monitor the generation of oxidants in systems which contain no added reductants. We have used this system to monitor oxidizing transients during the ferric EDDA-catalyzed decomposition of peroxide. Although ABTS can be added to a concentration that scavenges efficiently in Eq. (5a) or (5b), reducing transients such as ferrous complexes and superoxide often limit the net amount of ABTS$^{\cdot+}$ formed.[26]

Two observations by this method suggest that free hydroxyl radicals are not the exclusive oxidizing transients in the iron–NTA or iron–EDDA systems.[26] (1) The rate of ABTS$^{\cdot+}$ generation increases with ABTS at concentrations well above that requisite for complete hydroxyl scavenging. (2) The addition of bromide ions in any concentration does not promote ABTS$^{\cdot+}$ formation, indicating that bromide does not scavenge the species which oxidizes ABTS as expected for the following reactions:

$$\cdot OH + 2 \ Br^- \rightarrow Br_2^{\cdot-} + OH^- \quad (14)$$
$$Br_2^{\cdot-} + ABTS \rightarrow ABTS^{\cdot+} + 2 \ Br^- \quad (15)$$

This suggests that a higher oxidation state of iron may be less reactive, for thermodynamic or kinetic reasons, than the hydroxyl radical toward bromide ions.

The ABTS/ABTS$^{\cdot+}$ system is insensitive to oxygen but very sensitive to reducing agents. The radical cation is reduced by the hydrogen perox-

[29] W. H. Koppenol, *J. Free Radicals Biol. Med.* **1**, 281 (1985).
[30] T. M. Florence, *J. Inorg. Biochem.* **23**, 131 (1985).
[31] R. E. Childs and W. G. Bardsley, *Biochem. J.* **145**, 93 (1975).
[32] B. S. Wolfenden and R. L. Willson, *J. Chem. Soc., Perkin Trans.* **2**, 805 (1982).
[33] M. F. Zipplies, W. A. Lee, and T. C. Bruice, *J. Am. Chem. Soc.* **108**, 4433 (1986).

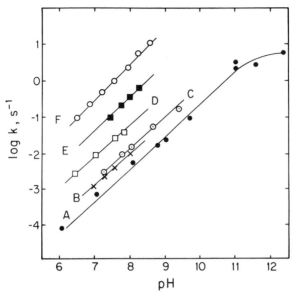

FIG. 1. First-order rate constants for reduction of the ABTS ‡ radical cation in 3.5 mM hydrogen peroxide (A) and as catalyzed by addition of 1.5 mM HLFe^{III}, where HL is DTPA (B), NTA in phosphate buffer (C), NTA in MOPS buffer (D), EDTA (E), or HEDTA (F). The concentrations of the buffers were 10 mM (MOPS, Tris, or phosphate), and rates were dependent on the buffer only for NTA.

ide anion [$k(\text{HO}_2^- + \text{ABTS}^{\ddagger}) = 1.8 \times 10^3\ M^{-1}\ \text{sec}^{-1}$], and this instability is enhanced by ferric complexes which bind with peroxide:

$$\text{HO}_2^- + \text{ABTS}^{\ddagger} \rightarrow \text{ABTS} + \text{H}^+ + \text{O}_2^- \tag{16}$$
$$\text{HLFe}^{\text{III}} + \text{H}_2\text{O}_2 \rightleftharpoons \text{HLFe}^{\text{III}}(\text{O}_2\text{H}^-) + \text{H}^+ \tag{17}$$
$$\text{HLFe}^{\text{III}}(\text{O}_2\text{H}^-) + \text{ABTS}^{\ddagger} \rightarrow \text{HLFe}^{\text{III}} + \text{O}_2^- + \text{H}^+ + \text{ABTS} \tag{18}$$
$$\text{O}_2^- + \text{ABTS}^{\ddagger} \rightarrow \text{ABTS} + \text{O}_2 \tag{19}$$

These reactions require that the iron complex used as a Fenton catalyst be fairly active so as to maintain a steady rate of ABTS ‡ formation greater than its rate of decomposition. Some experimental data are given in Fig. 1, in which the rate constants for reduction of ABTS ‡ are determined in aqueous peroxide solutions with different iron complexes. The rate of ABTS ‡ reduction exhibited by each most likely is due to the strength of peroxide binding to the ferric complex.

Solutions of approximately $2 \times 10^{-4}\ M$ ABTS ‡ can be prepared from ABTS by controlled potential electrolysis at an isolated anode. Continued oxidation leads to apparently polymeric products. These preparations should be used to check the stability of ABTS ‡ radicals with components of the experimental solutions before using the ABTS/ABTS $^+$ system as an indicator.

Measurement of Free Radical Chain Length

An approach employed by some workers involves the measurement of the chain lengths of free radical chain reactions. A known concentration of reductant is introduced into systems containing hydrogen peroxide, a metal (complex), and a scavenger such as formate[17,34] or methanol.[35,36] For instance, Sutton and Winterbourn[34] studied the following chain reactions, in which PQ indicates paraquat:

$$PQ^+ + Fe^{III}EDTA \rightarrow Fe^{II}EDTA + PQ^{2+} \tag{20}$$
$$Fe^{II}EDTA + H_2O_2 + H^+ \rightarrow Fe^{III}EDTA + \cdot OH + H_2O \tag{2b}$$
$$\cdot OH + HCO_2^- \rightarrow H_2O + CO_2^- \tag{12a}$$
$$CO_2^- + PQ^{2+} \rightarrow CO_2 + PQ^+ \tag{21}$$

Chain termination is thought to occur through the reactions:

$$Fe^{II}EDTA + H_2O_2 \rightarrow Fe^{II}EDTA(H_2O_2) \tag{2a}$$
$$Fe^{II}EDTA(H_2O_2) + PQ \cdot \rightarrow Fe^{III}EDTA + PQ^{2+} + 2 OH^- \tag{22}$$

and $Fe^{III}EDTA(H_2O_2)$ can also react with the formate anion. Reaction (2a) is thought to predominate, the ratio of hydroxyl radical versus higher oxidation state being 9:1.

Results have been presented to show that in the case of copper phenanthroline complexes that the propagation reaction is much slower than anticipated for a hydroxyl radical oxidant.[37] However, complicating factors such as cage reactions of the hydroxyl radical with the ligands might tend to obscure the interpretation. As in all the systems described, a complete knowledge of the reactions occurring among very reactive species is lacking.

Acknowledgments

This work was supported by a grant from The Council for Tobacco Research, Inc.—U.S.A.

[34] H. C. Sutton and C. C. Winterbourn, *Free Radical Biol. Med.* **6,** 54 (1989).
[35] G. R. A. Johnson, N. B. Nazhat, and A. Saadalla-Nazhat, *J. Chem. Soc. Chem. Commun.,* 407 (1985).
[36] G. R. A. Johnson, N. B. Nazhat, and R. A. Saadalla-Nazhat, *J. Chem. Soc., Faraday Trans.* **84,** 501 (1988).
[37] S. Goldstein and G. Czapski, *J. Free Radicals Biol. Med.* **1,** 373 (1985).

[13] Determination of Hydroperoxides with Fluorometric Reagent Diphenyl-1-pyrenylphosphine

By HIROSHI MEGURO, KAZUAKI AKASAKA, and HIROSHI OHRUI

Unsaturated fatty acids and their esters are easily oxidized to produce hydroperoxides at the initial stage of the peroxidation. Recently, hydroperoxides have attracted much attention as one of the factors that might be associated with some diseases and aging.[1-3] Several method have been proposed to determine lipid peroxidation products: iodometries,[4,5] thiobarbituric acid (TBA) methods,[6,7] gas chromatography,[8] high-performance liquid chromatography (HPLC),[9] and enzymatic methods.[10-12] In determining hydroperoxides in biological samples by these methods, however, problems can arise regarding sensitivity, selectivity, simplicity of procedures, and/or the effects of interfering substances.

Diphenyl-1-pyrenylphosphine (DPPP) is a new fluorescent reagent which can be used for lipid hydroperoxide determinations.[13,14] This reagent, which has almost no fluorescence, quantitatively produces a strongly fluorescent oxide (DPPP oxide) and corresponding alcohols (Fig. 1) by a reaction with lipid hydroperoxides. This reaction can be used successfully for a HPLC postcolumn detection system based on the favorable reaction rates in various organic solvents. The HPLC system can determine specific or classes of hydroperoxides. The lower detection limit of hydroperoxides using a batch method is 200 pmol and using HPLC just a few picomoles. Since the reagent in solution is unstable especially under

[1] F. Z. Meerson, V. E. Kozlov, P. Yu, L. M. Belkina, Yu, V. Arkhipenko, *Basic Res. Cardiol.* **77,** 465 (1982).

[2] T. Yamaguchi and Y. Yamaguchi, *Agric. Biol. Chem.* **43,** 2225 (1979).

[3] M. C. Barrett and A. A. Horton, *Biochem. Soc. Trans.* **3,** 124 (1975).

[4] C. H. Lea, *Proc. R. Soc. London Ser. B* **108,** 175 (1931).

[5] D. H. Wheeler, *Oil Soap* **9,** 89 (1932).

[6] K. Yagi, *Biochem. Med.* **15,** 212 (1976).

[7] M. Uchiyama and M. Mihara, *Anal. Biochem.* **86,** 271 (1978).

[8] P. K. Jarvi, G. D. Lee, D. R. Erickson, and E. A. Butkus, *J. Am. Oil Chem. Soc.* **48,** 121 (1970).

[9] H. Aoshima, *Anal. Biochem.* **87,** 49 (1978).

[10] R. L. Heath and A. L. Tappel, *Anal. Biochem.* **76,** 184 (1976).

[11] K. Kohda, K. Arisue, A. Maki, and C. Hayashi, *Jpn. J. Clin. Chem.* **11,** 306 (1982).

[12] P. J. Marshall, M. A. Warso, and W. E. Lands, *Anal. Biochem.* **145,** 192 (1985).

[13] K. Akasaka, T. Suzuki, H. Ohrui, and H. Meguro, *Anal. Lett.* **20,** 731 (1987).

[14] K. Akasaka, T. Suzuki, H. Ohrui, and H. Meguro, *Anal. Lett.* **20,** 797 (1987).

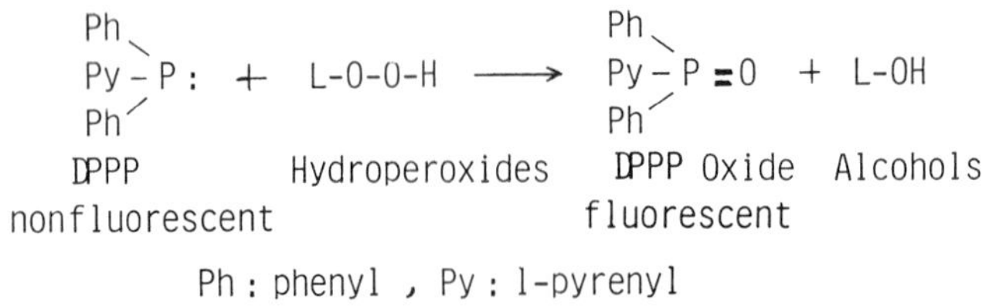

$$
\begin{array}{ccc}
\text{Ph} & & \text{Ph} \\
\text{Py} - \text{P}: + \text{L-O-O-H} & \longrightarrow & \text{Py} - \text{P} = 0 + \text{L-OH} \\
\text{Ph} & & \text{Ph}
\end{array}
$$

DPPP   Hydroperoxides   DPPP Oxide   Alcohols
nonfluorescent   fluorescent

Ph : phenyl , Py : 1-pyrenyl

FIG. 1. Reaction of DPPP with hydroperoxides.

strong light, care should be taken to use a peroxide-free solvent and not to handle the solution in strong light. Stock solutions can be kept stable for 1 week in a dark refrigerator.

Preparation of Diphenyl-1-pyrenylphosphine[13]

Diphenyl-1-pyrenylphosphine (DPPP) is prepared from triphenylphosphine and 1-bromopyrene. All the following reactions are performed under a stream of nitrogen gas. In a 500-ml, three-necked flask, 5.25 g of triphenylphosphine and 420 mg of lithium are added to 200 ml of tetrahydrofuran (THF). The mixture is stirred for 3 hr at room temperature, and then *tert*-butyl chloride (1.8 g in 20 ml of THF) is added to the mixture. After the mixture is refluxed for 10 min, it is cooled to room temperature. The residual lithium is removed by passing the mixture under a stream of nitrogen gas using a glass tube which is loosely stuffed with glass wool. The mixture is poured into another 500-ml, three-necked flask containing 1-bromopyrene (prepared by the method of D. C. Nohebel *et al.*[15]) and refluxed for 3.5 hr. The reaction mixture is poured into water, and the phosphine is extracted with chloroform (150 ml) and purified on a silica gel column (silica gel, 200 g, eluted with hexane–benzene, 7 : 3). After recrystallization from methanol–chloroform, the DPPP is obtained as a pale yellow crystal (1.05 g, 13.6%); mp 171–174°. The crystals are stable for more than 2 years in a refrigerator.

Determination of Total Lipid Hydroperoxides[14]

A lipid sample (4–200 mg) is dissolved in a mixture of methanol and chloroform (1 : 2), which contains 3 mg of butylated hydroxytoluene (BHT) per 10 ml, and the volume is adjusted to 5 ml. A 100-μl aliquot of this solution and 100 μl of DPPP solution [1 mg/10 ml in a methanol–chloroform (1 : 1)] are added to a screw-cap test tube (13 × 100 mm) and cooled on an ice bath. Then, the cap is closed tightly and the mixture

[15] D. C. Nohebel, *Proc. Chem. Soc.*, 307 (1961).

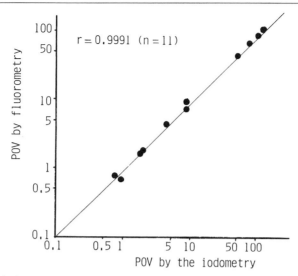

FIG. 2. Relationship between peroxide values (POV) determined by iodometry and fluorometry.

reacted in a water bath at 60° for 60 min in the dark. After cooling on an ice bath, 3 ml of methanol is added to the mixture. After shaking, the fluorescence intensity at 380 nm (excitation at 352 nm) is measured and the determination made using the calibration curve of hydroperoxides or DPPP oxide.

The relationship between peroxide values (POV) of 11 oil samples determined by iodometry and by the above method (calibration is carried out with DPPP oxide) is shown in Fig. 2. Good correlation is obtained between the two methods ($r = 0.9991$, $n = 11$). The coefficient of variation is 1.7% ($n = 6$), and the detection limit is about 200 pmol per tube. For this determination, only 0.08–3.2 mg/tube of samples (POV 0.5–100) is required by this method.

Determination of Phosphatidylcholine Hydroperoxides in Human Plasma by HPLC[16]

Sample Preparation. To 500 μl of human plasma are added 500 μl of methanol and 1.0 ml of chloroform, which contained 0.003% BHT, with mixing. The mixture is shaken vigorously and centrifuged at 1000 g for 5–10 min. The bottom layer is collected carefully and stored at 0° in the dark. One milliliter of chloroform (0.003% BHT) is added to the upper

---

[16] K. Akasaka, H. Ohrui, and H. Meguro, *Anal. Lett.* **21**, 965 (1988).

layer, and lipids are extracted 2 more times in the same manner. The extracts are combined and evaporated under reduced pressure at 20°. During the entire extraction procedure, the samples are cooled in an ice bath under nitrogen to prevent sample deterioration. To the residue, 15 μl of chloroform is added, and a 10-μl aliquot is injected into the column.

HPLC Analysis. The phosphatidylcholine hydroperoxide (PC-HPO) is determined by the method of standard addition using the fluorescent peak. The standard PC-HPO (48 pmol) prepared from egg yolk phosphatidylcholine is added to 500 μl of human plasma and was used as the standard added sample. A 10-μl aliquot of a chloroform solution is applied to a normal phase column, TSK gel silica 60 (4.6 i.d. $\times$ 250 mm), and PC-HPO is eluted with chloroform–methanol–water (9 : 21 : 1) (v/v/v) at a rate of 0.6 ml/min. The eluent is monitored by UV absorption at 235 nm before labeling, and then DPPP solution [3 mg/400 ml in acetone–methanol (1 : 3)] is mixed to it at a flow rate of 0.3 ml/min. The reaction proceeded in a reaction coil (stainless steel, 0.5 mm i.d. $\times$ 10 m) at 70°. Then it is cooled to room temperature in a second short coil in a water bath at 20°. The resultant DPPP oxide is determined by monitoring the fluorescence intensity at 380 nm (excitation at 352 nm, both slit widths 20 nm).

Amount of PC-HPO in Human Plasma. By this method, it is possible to determine PC-HPO over the range of 5–950 pmol. Figure 3 shows a typical HPLC chromatogram of a plasma extract. Peak A is that of a PC-HPO detected by the fluorometry, and peak B is a peak detected by UV at 235 nm. Peak B is used as an internal standard to check the injection amount. The corrected peak height (Peak A/Peak B) has a 5% coefficient

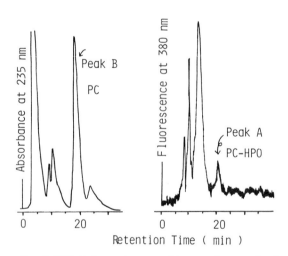

FIG. 3. HPLC chromatograms of human plasma extract.

of variation for 5 replicate samples. The linear relation is obtained between the corrected peak heights and the concentrations of added PC-HPO is in the range of 0–60 pmol/500 μl plasma. The 15 plasma samples gave 20–55 pmol/ml of PC-HPO.

Recently, we obtained better sensitivity, reproducibility, and rapidity by a modification of HPLC conditions as follows. The analytical column is Develosil 60-5 (4.6 i.d. $\times$ 150 mm), the mobile phase is chloroform–methanol–water (20 : 40 : 3, containing 1 mg/10 ml cetyltrimethylammonium bromide and 5 μl/100 ml acetic acid), and the reaction coil length and temperature are 20 m (0.5 mm i.d.) and 47°. Other conditions are the same as described previously.

[14] Phycoerythrin Fluorescence-Based Assay for Reactive Oxygen Species

By Alexander N. Glazer

Introduction

Phycobiliproteins are photosynthetic accessory proteins in cyanobacteria (blue-green algae) and in two groups of eukaryotic algae, the red algae and the cryptomonads.[1] Among these proteins, B- and R-phycoerythrins are the most remarkable with respect both to the stability of their quaternary structure and to their spectroscopic properties. Both these phycoerythrins occur in numerous red algae[2-4] and can be readily purified by conventional procedures. B- and R-phycoerythrins have molecular weights of 240,000 and carry 34 covalently attached tetrapyrrole prosthetic groups.[5-7] Both are multisubunit proteins with the structure

[1] A. N. Glazer, in "The Biochemistry of Plants" (M. D. Harch and N. K. Boardman, eds.), p. 51. Academic Press, London, 1981.

[2] C. OhEocha, in "Physiology and Biochemistry of Algae" (R. A. Lewin, ed.), p. 421. Academic Press, New York, 1962.

[3] A. N. Glazer, J. A. West, and C. A. Chan, Biochem. Syst. Ecol. **10**, 203 (1982).

[4] E. Honsell, V. Kosovel, and L. Talarico, Bot. Mar. **27**, 1 (1984).

[5] D. J. Lundell, A. N. Glazer, R. J. DeLange, and D. M. Brown, J. Biol. Chem. **259**, 5472.

[6] A. V. Klotz and A. N. Glazer, J. Biol. Chem. **260**, 4856.

[7] W. Sidler, B. Kumpf, F. Suter, A. V. Klotz, A. N. Glazer, and H. Zuber, Biol. Chem. Hoppe-Seyler **370**, 115 (1989).

METHODS IN ENZYMOLOGY, VOL. 186

$(\alpha\beta)_6\gamma$.[8,9] The complete amino acid sequences of the α and β subunits of *Porphyridium cruentum* B-phycoerythrin are known,[5,7] and the sequences about the tetrapyrrole attachment sites in the α, β, and γ subunits of *Gastroclonium coulteri* R-phycoerythrin have been determined.[6]

No dissociation of these complex macromolecules is observed even at concentrations as low as 10^{-15} M. At its absorption maximum of 545 nm, B-phycoerythrin has a molar extinction coefficient of 2.41×10^6 M^{-1} cm^{-1} per 240,000,[8] and it fluoresces with a quantum yield of 0.98 with a maximum at 575 nm.[10] R-Phycoerythrin has absorption maxima at 566 (ε_M 1.96 $\times 10^6$ M^{-1} cm^{-1}) and 497 nm (1.54×10^6 M^{-1} cm^{-1} per 240,000)[8] and a fluorescence emission maximum at 578 nm (Q 0.82).[11] Because of these very high extinction coefficients and fluorescence quantum yields, B- and R-phycoerythrin can be readily detected by fluorescence spectroscopy at concentrations as low as 10^{-12} M.

The assay for reactive oxygen species depends on the detection of chemical damage to phycoerythrin through the decrease in its fluorescence emission. The fluorescence of phycobiliproteins is highly sensitive to the conformation and chemical integrity of the protein and of the prosthetic groups. Under appropriate conditions, in the presence of reactive oxygen species, the rate of loss of phycoerythrin fluorescence is an index of free radical damage. The effect of added compounds on the rate of this fluorescence loss is a measure of their ability to protect the protein.

Phycoerythrin Fluorescence-Based Assay for Peroxy–Radical Scavengers[12,13]

Sources and Preparation of Phycoerythrins

The unicellular red alga *Porphyridium cruentum* is a convenient source of phycoerythrin. Laboratory cultures of this organism are easy to maintain,[14,15] and straightforward procedures for the purification of B-phycoerythrin are available.[8,16] R-Phycoerythrin can be isolated from nu-

[8] A. N. Glazer and C. S. Hixson, *J. Biol. Chem.* **252**, 32 (1977).
[9] C. Abad-Zapatero, J. L. Fox, and M. L. Hackert, *Biochem. Biophys. Res. Commun.* **78**, 266 (1977).
[10] J. Grabowski and E. Gantt, *Photochem. Photobiol.* **28**, 39 (1978).
[11] D. J. W. Barber and J. T. Richards, *Photochem. Photobiol.* **25**, 565 (1977).
[12] A. N. Glazer, *FASEB J.* **2**, 2487 (1988).
[13] R. J. DeLange and A. N. Glazer, *Anal. Biochem.* **177**, 300 (1989).
[14] R. F. Jones, H. L. Speer, and W. Kury, *Physiol. Plant.* **16**, 636 (1963).
[15] A. N. Glazer and C. S. Hixson, *J. Biol. Chem.* **250**, 5487 (1975).
[16] R. W. Schoenleber, S.-L. Leung, D. J. Lundell, A. N. Glazer, and H. Rapaport, *J. Am. Chem. Soc.* **105**, 4072 (1973).

merous higher red algae.[3] The seaweed *Gastroclonium coulteri,* found along the Pacific coast of California from Fort Ross to San Diego, is a suitable field material, and the preparation of this phycoerythrin has been described. Both B- and R-phycoerythrin are available from many commercial suppliers (e.g., Molecular Probes, Eugene, OR; Polysciences, Warrington, PA; Sigma, St. Louis, MO). The relevant spectroscopic properties of these proteins are mentioned in the Introduction. The phycoerythrins are very soluble (>10 mg/ml) in the 75 mM sodium phosphate buffer, pH 7.0, used in the assay, and stock solutions can be stored at 4° for months.

Other Reagents

Contaminating metal ions are removed from 75 mM sodium phosphate buffer, pH 7.0, by passage through a column of Chelex 100 resin (200–400 mesh), sodium form (Bio-Rad, Richmond, CA). A 40 mM stock solution of the water-soluble free radical initiator 2,2′-azobis(2-amidinopropane) hydrochloride (AAPH; MW 267; Polysciences) is prepared in the pH 7.0 buffer immediately before use and stored on ice.

A stock 10 mM solution of the water-soluble vitamin E analog 6-hydroxy-2,5,7,8-tetramethylchroman-2-carboxylic acid (Trolox; MW 250.3; 97% pure; Aldrich, Milwaukee, WI) is prepared freshly in 50% (v/v) methanol–water. An aliquot of this solution is used for standardization of the rate of peroxyl radical formation in the reaction mixture.

Principle of Assay

The free radical initiator AAPH undergoes thermal decomposition by the following mechanism,[17] where the structure

$$Cl^- \quad {}^+H_2N{=}C{-}\underset{\underset{CH_3}{|}}{\overset{\overset{CH_3}{|}}{C}}{-}N{=}N{-}\underset{\underset{CH_3}{|}}{\overset{\overset{CH_3}{|}}{C}}{-}C{=}NH_2^+ \quad Cl^-$$

is represented by A—N=N—A:

$$A{-}N{=}N{-}A \rightarrow [A \cdot N_2 \cdot A] \rightarrow 2e\, A \cdot + (1 - e)\, A{-}A + N_2 \tag{1}$$

where e is the efficiency of free radical generation

$$A \cdot + O_2 \rightarrow AO_2 \cdot \tag{2}$$
$$AO_2 \cdot + \text{phycoerythrin} \rightarrow \text{stable products} \tag{3}$$
$$AO_2 \cdot + AO_2 \cdot \rightarrow \text{stable products} \tag{4}$$

---

[17] E. Niki, this volume [3].

Assay Conditions

A convenient assay mixture has the following composition:

Sodium phosphate buffer (75 mM, pH 7.0)	1.78 ml	
Phycoerythrin solution (1.7 × 10⁻⁶ M)	0.02 ml	(final conc. 1.7×10^{-8} M)
AAPH solution (40 mM)	0.20 ml	(final conc. 4×10^{-3} M)

All solutions are made up in the pH 7.0 phosphate buffer. The assay is carried out at 37° in 1-cm rectangular fluorimeter cuvettes, and the AAPH solution is added last. The wavelengths for fluorescence excitation and emission are 540 and 575 nm, respectively. For R-phycoerythrin, 495 nm may be chosen as an alternative excitation wavelength.

Under these conditions, the decay in the fluorescence of phycoerythrin is zero order (i.e., linear with time). At 37°, the rate (R) of peroxyl radical formation from AAPH is 1.1×10^{-6} [AAPH] per second.[18] Since the time required for an assay is about 60 min, the total change in AAPH concentration is less than 1%. The observed zero-order kinetics of fluorescence decay indicate that free radical generation is the rate-limiting step. The rate of change of 575 nm emission is linearly proportional to AAPH concentration up to 4 mM.

Standardization of Assay

Since the rate of peroxyl radical generation is dependent on the precise AAPH concentration and exact temperature, inclusion of a reaction mixture containing a known amount of Trolox with each set of assays is advisable. Trolox reacts with peroxyl radicals much more rapidly than does phycoerythrin. Until the Trolox is consumed, no loss in phycoerythrin fluorescence is observed.[12] The reaction mixture used for standardization has the following composition:

Sodium phosphate buffer (75 mM, pH 7.0)	1.78 ml
Phycoerythrin solution (1.7 × 10⁻⁶ M)	0.02 ml
AAPH solution (40 mM)	0.20 ml

The rate of change of fluorescence at 575 nm is followed for about 12 min, and then 10 μl of Trolox [5 × 10⁻⁴ M in 50% (v/v) methanol–water] is added to give a final Trolox concentration of 2.5×10^{-6} M and the fluorescence at 575 nm monitored for a further 40 to 50 min.

A standardization assay performed as described above is shown in

---

[18] K. Terao and E. Niki, *J. Free Radicals Biol. Med.* **2**, 193 (1986).

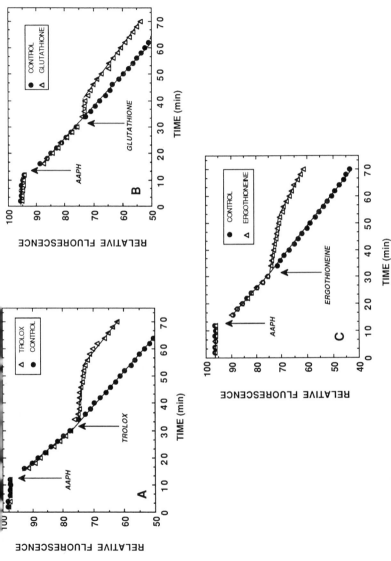

Fig. 1. Effect of Trolox, glutathione, and ergothioneine on the peroxyl radical-dependent loss of phycoerythrin fluorescence. In each instance, at time zero, the assay mixture contained 1.8 ml of 1.9×10^{-8} M R-phycoerythrin solution. At 12 min, 0.2 ml of 40 mM AAPH was added and the reaction followed for 18 min. The following additions were made at that time: A, 10 μl Trolox (5.03×10^{-4} M); B, 10 μl glutathione (4.67×10^{-4} M); C, 10 μl ergothioneine (4.9×10^{-4} M). The final concentrations of the compounds in the assay mixture were 2.5×10^{-6}, 2.3×10^{-6}, and 2.45×10^{-6} M, respectively. An R-phycoerythrin control, with no added free radical scavenger, was included in each case. The temperature was 37°, and the fluorescence excitation and emission wavelengths were 495 and 575 nm, respectively.

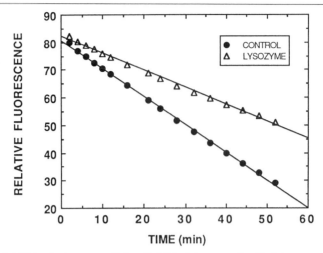

FIG. 2. Inhibition by lysozyme (8.7×10^{-7} M) of the peroxyl radical-dependent loss of B-phycoerythrin (1.65×10^{-8} M) fluorescence. Measurements were performed under the standard conditions described in the text. (Data from Ref. 13.)

Fig. 1A. Each mole of Trolox traps 2 mol of peroxyl radical.[19–21] Consequently, the rate of free radical generation can be calculated from the length of time a known amount of Trolox protects phycoerythrin from damage. In the assay shown in Fig. 1A, this time interval was 22 min. Consequently, the rate of peroxyl radical production was $(2 \times 2.5 \times 10^{-6})/22 = 2.27 \times 10^{-7}$ M min^{-1}.

Applications of Assay

This assay has been applied to the assessment of the free radical scavenging capacity of human plasma, of proteins, DNA, and numerous small molecules.[12,13,22] Molecules that react with peroxyl radicals much more rapidly than does phycoerythrin show effects similar to those shown for Trolox. This is illustrated for glutathione and for ergothioneine[23] in Fig. 1B, C. Molecules that react at rates similar to that of phycoerythrin, when assayed at appropriate concentrations, do not provide complete protection of phycoerythrin. In the presence of such compounds, the rate of phycoerythrin fluorescence decay is seen to be decreased (see Fig. 2).

[19] G. W. Burton, L. Hughes, and C. U. Ingold, *J. Am. Chem. Soc.* **105**, 5950 (1983).
[20] L. R. C. Barclay, J. T. Locke, J. M. MacNeil, J. Van Kessel, G. W. Burton, and K. U. Ingold, *J. Am. Chem. Soc.* **106**, 2479 (1984).
[21] Y. Yamamoto, S. Haga, E. Niki, and Y. Kamiya, *Bull. Chem. Soc. Jpn.* **57**, 1260 (1984).
[22] R. J. DeLange and A. N. Glazer, *Gerontologist* **28** (Oct., Special Issue), 229A (1988).
[23] P. E. Hartman, this volume [32].

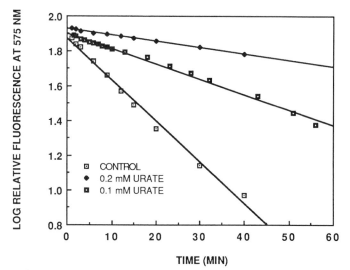

FIG. 3. The decay in R-phycoerythrin emission on incubation with ascorbate (3.4×10^{-4} M) and Cu^{2+} (4.5×10^{-7} M) in the absence and presence of urate at the concentrations indicated. For assay conditions, see text. (Data from Ref. 12.)

Assay of Radical Damage in Cu^{2+}–Ascorbate System

In the presence of ascorbate and Cu^{2+}, hydroxyl radicals are generated at copper-binding sites on macromolecules.[12] Site-specific damage to macromolecules[24–28] results from the reaction

$$\text{Target–}Cu^{2+}\text{–HO} \cdot \rightarrow \text{damaged target} + Cu^{2+} \qquad (5)$$

This assay is particularly useful in screening for compounds that protect against damage by chelating metal ions necessary for the site-specific formation of the radical species.

Assay Conditions[12]

The assay is performed in a manner similar to that described above for peroxyl radicals. The standard assay mixture contains 1.5×10^{-8} M phycoerythrin, 4.5×10^{-7} M $CuSO_4 \cdot 5H_2O$, and 3.6×10^{-4} M ascorbate, all in a total volume of 2.0 ml 75 mM sodium phosphate, pH 7.0, at 37°. The ascorbate solution is prepared immediately before use, and the reaction is

[24] E. K. Hodgson and I. Fridovich, *Biochemistry* **14**, 5294 (1975).
[25] R. L. Levine, *J. Biol. Chem.* **258**, 11828 (1983).
[26] A. Samuni, M. Chevion, and G. Chapski, *J. Biol. Chem.* **256**, 12632 (1981).
[27] D. S. Sigman, *Acc. Chem. Res.* **19**, 180 (1986).
[28] E. R. Stadtman, *J. Gerontol.* **43**, B112 (1988).

initiated by the addition of ascorbate. In this assay, the fluorescence of phycoerythrin is seen to decrease with first-order kinetics. The observed first-order rate constant for fluorescence decay was found to be linearly proportional to the Cu^{2+} concentration from 4×10^{-7} to 24×10^{-7} M. A representative assay is illustrated in Fig. 3.

Acknowledgments

The author is grateful to Amir H. Koushafar for performing the experiments shown in Fig. 1. Work in the author's laboratory was supported by National Science Foundation Grant DMB 88-16727 and National Institutes of Health Grant GM 28994.

[15] Sulfur-Centered Free Radicals

By KLAUS-DIETER ASMUS

Introduction

Biochemical reaction mechanisms often involve free radicals. This has emerged from an increasing number of studies. Among the biologically relevant compounds, those containing sulfur functions are a source of a large variety of free radicals. Most of them are sulfur-centered or at least influenced by the presence of the sulfur atom.

The generation and the properties of sulfur-centered radicals in general resemble many features of their oxygen analogs. There are, however, also some significant differences which can mostly be associated with the relative position of the two elements in the periodic table. Sulfur is much less electronegative and, as the larger atom, electronically considerably softer than oxygen.

The most important classes of sulfur-containing compounds from the biological point of view are thiols and disulfides. They are related to each other in a redox system with thiyl radicals as an intermediate:

$$RSH \xrightarrow{-e^-/-H^+} RS \cdot \to \frac{1}{2} RSSR \tag{1}$$

Evidence, although mostly indirect, for the generation of thiyl radicals in biological systems is manifold. Direct proof of their existence has been provided by spin-trapping experiments.[1]

---

[1] L. S. Harman, C. Mottley, and R. P. Mason, *J. Biol. Chem.* **259**, 5606 (1984).

Thiyl radicals are not the only radical species in the thiol/disulfide system though. $(RSSR)^{-}$ radical anions, $(RSSR)^{+}$ radical cations, perthiyl radicals, thiyl radical adducts, and carbon-centered radicals are equally interesting and important from the chemical point of view, and they may be expected to play more or less significant roles in biological systems as well. A third source of radicals possibly relevant to biological processes are organic monosulfides such as the amino acid methionine.

In this chapter a general survey is given on the formation and essential physicochemical properties of radical species from the three major classes of organic sulfur compounds, namely, thiols, disulfides, and monosulfides. It has to be realized, however, that the enormous amount of data available on this subject precludes any quantitative account within the given format of this chapter.

Experimental Techniques

Experimental techniques to investigate radicals are manifold and are not described here in any detail. It should be mentioned, however, that electron spin resonance (ESR) is often hampered for sulfur-centered radicals owing to the lack of nuclear spin in the main sulfur isotope, ^{32}S. This practically prevents any information on structural aspects. Most of the data available nowadays on sulfur-centered radicals stem from radiation chemical studies, particularly those based on the technique of pulse radiolysis. The latter essentially relies on a time-resolved detection of optical or other physical properties of the radicals. Details of this method have been described previously in this series.[2]

Generation of Sulfur-Centered Radicals

Thiyl radicals can mechanistically be generated via displacement (homolytic bond breakage) as well as one-electron redox processes from thiols and disulfides. An example of the former is the well-known "repair" reaction,[3]

$$-\overset{|}{\underset{|}{C}} \cdot + RSH \rightarrow -\overset{|}{\underset{|}{C}}-H + RS \cdot \tag{2}$$

in which a carbon-centered radical (a "damage" site) abstracts a hydrogen atom from the thiol. Rate constants for this type of reaction are of the

[2] K.-D. Asmus, this series, Vol. 105, p. 167.
[3] G. E. Adams, G. S. McNaughton, and B. D. Michael, *Trans. Faraday Soc.* **64**, 902 (1968).

order of 10^8 M^{-1} sec^{-1}, that is, somewhat below the diffusion limits.[4,5] Energetically, the reaction of the hydroxyl radical,

$$RSH + \cdot OH \rightarrow RS \cdot + H_2O \tag{3}$$

is particularly favorable, in that it requires practically no activation energy. Direct one-electron oxidation is likely to occur in the reaction of thiols, and in particular their anionic form RS$^-$, with electron acceptors such as transition metal ions (Cu^{2+}, Fe^{3+}, etc.)[6] and with many oxidizing organic radicals exhibiting redox potentials of at least +(0.8–1.0) V (for RS$^-$ oxidation) or higher (>+1.3 V for RSH oxidation).

A similarly important pathway to thiyl radicals is provided by displacement and reduction reactions with disulfides. Practically all kinds of organic and inorganic radicals R$'\cdot$ (carbon- or heteroatom-centered) are able to undergo the following general process:

$$R'\cdot + RSSR \rightarrow R'SR + RS \cdot \tag{4}$$

Rate constants for such displacements are, in general, significantly below the limits of diffusion and typically of the order of 10^6 M^{-1} sec^{-1} and lower.[7–9] They, therefore, require relatively high disulfide concentrations in order to compete with other reactions of the R$'\cdot$ radicals.

In some cases reaction (4) does not lead to breakage of the sulfur–sulfur bond but to cleavage of the sulfur–carbon bond with generation of perthiyl radicals:

$$R'\cdot + RSSR \rightarrow RSS \cdot + \text{products} \tag{5}$$

Examples of such a process are, however, limited and for biologically relevant systems have only been claimed to occur with RSSR being penicillamine disulfide.[10] (Energetically, particularly stabilized perthiyl radicals and/or product molecules are required for this type of reaction to occur.)

[4] G. E. Adams, R. C. Armstrong, A. Charlesby, B. D. Michael, and R. L. Willson, *Trans. Faraday Soc.* **65,** 732 (1969).

[5] K.-D. Asmus and M. Bonifačić, *in* "Landolt-Börnstein: Zahlenwerte und Funtionen" (H. Fischer, ed.), New Series, Vol. 13b. Springer-Verlag, Berlin, 1984.

[6] B. C. Gilbert, H. A. H. Laue, R. O. C. Norman, and R. C. Sealy, *J. Chem. Soc., Perkin Trans. 2,* 892 (1975).

[7] M. Bonifačić and K.-D. Asmus, *J. Phys. Chem.* **88,** 6286 (1984).

[8] W. A. Pryor and K. J. Smith, *J. Am. Chem. Soc.* **92,** 2731 (1970).

[9] W. A. Pryor, *in* "Mechanisms of Sulfur Reactions," p. 42. McGraw-Hill, New York, 1962.

[10] A. J. Elliot, R. J. McEachern, and D. A. Armstrong, *J. Phys. Chem.* **85,** 68 (1981); Z. Wu, T. G. Back, R. Ahmad, and D. A. Armstrong, *J. Phys. Chem.* **86,** 4417 (1982).

One-electron reduction of disulfides is also a convenient way to produce thiyl radicals. The reaction mechanism proceeds via an intermediate $(RSSR)^-$ radical anion which generally decays immediately, that is, within 1 μsec, into thiyl radicals and thiolate:

$$RSSR + e^- \rightarrow RSSR^- \cdot \rightarrow RS \cdot + RS^- \tag{6}$$

Hydrated electrons (e_{aq}^-), for example, add to disulfides with rate constants of the order of 10^{10} M^{-1} sec^{-1}.[11,12] Reductions may also be initiated by other reducing species. An efficient reduction of lipoate has, for example, been reported to occur by CO_2^- ($k = 5.5 \times 10^8$ M^{-1} sec^{-1}).[13] According to recent investigations on the $RSSR/(RSSR)^-$ redox potentials, it seems that the reductant must, however, exhibit an E^0 more negative than -1.6 V (versus NHE).[14] Further aspects of the relationship between RS $\cdot$ and $(RSSR)^-$ are discussed in a later section (see below).

One-electron oxidations of organic sulfur compounds occur mostly at the nonbonded electrons of the bivalent sulfur atoms. The radical cations generated this way from organic disulfides[15–17] and monosulfides[18,19] are $(RSSR)^+$ and R_2S^+, respectively. Oxidation of disulfides requires moderately good oxidants with a redox potential greater than $+1.4$ V,[20] for example, Br_2^-, $\cdot$OH, R_2S^+, SO_4^-. Monosulfides are not quite as easy to oxidize as disulfides, and the E^0 of the oxidants should exceed $+1.5$ V.[20] Sufficiently strong oxidants are N_3^-, SO_4^-, Cl_2^-, Ag^{2+}, and CH_3I^+, for example. Hydroxyl radicals also readily oxidize monosulfides to sulfur-centered radical cations. Mechanistically, this process proceeds, however, not directly via an electron transfer but through an $\cdot$OH adduct intermediate.[19]

Rate constant measurements have been conducted for many of these radical-induced oxidation processes and often indicate diffusion- or near-

[11] G. E. Adams, G. S. McNaughton, and D. B. Michael, *in* "Excitation and Ionization" (G. Scholes and G. R. A. Johnson, eds.), p. 281. Taylor & Francis, London, 1967.

[12] M. Anbar, M. Bambenek, and A. B. Ross, *in* "Selected Specific Rates of Transients from Water in Aqueous Solution: 1, Hydrated Electron," NSRDS-NBS 43, U.S. Department of Commerce, Washington D.C., 1973.

[13] R. L. Willson, *Chem. Commun.*, 1425 (1970).

[14] P. S. Surdhar and D. A. Armstrong, *J. Phys. Chem.* **91**, 6532 (1987).

[15] K.-D. Asmus, *in* "Radioprotectors and Anticarcinogens" (O. F. Nygaard and M. G. Simic, eds.), p. 23. Academic Press, New York, 1984.

[16] H. Möckel, M. Bonifačić, and K.-D. Asmus, *J. Phys. Chem.* **78**, 282 (1974).

[17] M. Bonifačić and K.-D. Asmus, *J. Phys. Chem.* **80**, 2426 (1976).

[18] K.-D. Asmus, *Acc. Chem. Res.* **12**, 436 (1979).

[19] M. Bonifačić, H. Möckel, D. Bahnemann, and K.-D. Asmus, *J. Chem. Soc., Perkin Trans. 2*, 675 (1975).

[20] M. Bonifačić and K.-D. Asmus, *J. Chem. Soc., Perkin Trans. 2*, 1805 (1986).

diffusion-controlled reactions.[15-22] The individual rate constants depend, of course, on the nature of the reaction participants, namely, mainly on the redox potentials of the respective radical/molecule couples.

Although sulfur is generally the prime target of a radical attack at organic sulfur compounds, the formation of some non-sulfur-centered radicals also has to be envisaged. This statement is particularly true for the reaction of hydroxyl radicals with organic monosulfides. Owing to their high reactivity, $\cdot$OH radicals are very unselective and attack at almost any site in an organic molecule. Radiation chemical studies have revealed that even in CH_3SCH_3, the simplest sulfide, $\cdot$OH reacts only to 80% at sulfur whereas the rest undergoes hydrogen atom abstraction.[19]

Identification of Sulfur-Centered Radicals

Direct identification of the various sulfur-centered radicals by means of ESR is difficult as has been mentioned already. A comparatively more informative, more sensitive, and complementary method of investigation has proved to be the radiation chemical technique of pulse radiolysis. It relies in particular on the direct observation of optical and conductometric properties of the radical species.[2]

Strong optical absorptions are a characteristic feature of all ionic radicals listed in the previous section. The disulfide radical anions, $(RSSR)^{\overline{\cdot}}$, typically absorb in the 380–450 nm range, that is, in the near-UV extending into the visible.[11,23] Extinction coefficients have been determined as $(6-7) \times 10^3 \ M^{-1} \ cm^{-1}$ for a number of aliphatic and other $(RSSR)^{\overline{\cdot}}$ radicals, including some of biological significance (cystine, cystamine, penicillamine disulfide, glutathione disulfide).

The optical absorptions of disulfide radical anions are directly related to their electronic structures, the essential feature being a three-electron bond between the two sulfur atoms: $(R—S \therefore S—R)^-$. Two of the electrons are bonding in nature, that is, σ electrons, while the third is an antibonding σ^* electron. The optical absorption is essentially a $\sigma-\sigma^*$ transition (in first approximation),[18,23] and the differences between these two energy levels are strongly dependent on the electronic and geometrical structure of the three-electron-bonded species. (The weaker the $S \therefore S$ bond the more red-shifted the absorption, and vice versa.[18,19,23])

Disulfide radical cations, $(RSSR)^{\overline{+}}$, absorb in the same region as the

[21] H. Mohan and K.-D. Asmus, *J. Phys. Chem.* **92**, 118 (1988).

[22] M. Bonifačić, J. Weiss, S. A. Chaudhri, and K.-D. Asmus, *J. Phys. Chem.* **89**, 3910 (1985).

[23] M. Göbl, M. Bonifačić, and K.-D. Asmus, *J. Am. Chem. Soc.* **106**, 5984 (1984).

above radical anions.[15–17] Their extinction coefficients are, however, generally lower by a factor of about three, that is, only in the 2×10^3 M^{-1} cm^{-1} range. [Electronically, the sulfur–sulfur bond in (RSSR)$^{\dot{+}}$ is $2\sigma/2\pi/1\pi^*$.[15]]

Most monosulfide radical cations, R$_2$S$^{\dot{+}}$, exhibit an absorption band in the UV around 300 nm.[18,19,24] These species are usually extremely short-lived and suffer fast reaction with an unattacked sulfur atom located in a second or even the same molecule:

$$R_2S^{+\cdot} + R_2S \rightleftharpoons R_2S \therefore SR_2^+ \qquad (7)$$

The resulting structure is characterized by a three-electron $2\sigma/1\sigma^*$ bond as in the disulfide radical anions. The optical absorptions are broad bands peaking mainly in the visible (400–650 nm range). The actual position of λ_{max} again reflects the S∴S bond strength as mentioned above for the disulfide radical anions.[18,19,23]

Such three-electron bonds may also be established between an oxidized sulfide function and other heteroatoms, for example, oxygen, nitrogen, and halogens. One relevant example is an S∴N bonded intermediate in the oxidation of methionine,

$$(H_3C-\underset{\underline{\qquad\qquad}}{S \therefore NH_2 \diagdown \diagup COOH})^+$$

which absorbs at 400 nm.[25,26]

Optical detection of thiyl radicals is difficult. Most of them seem to absorb in the UV slightly above 300 nm, but only with very low extinction coefficients. Appreciable absorptions have been observed in only a very few cases, for example, for the radical derived from penicillamine, PenS$\cdot$, which absorbs at 330 nm ($\varepsilon = 1.2 \times 10^3$ M^{-1} cm^{-1}).[27,28] Somewhat easier to identify are perthiyl radicals, which generally exhibit reasonably strong absorptions around 380 nm.

Pronounced absorption bands around 270–300 nm with an ε value of 3×10^3 M^{-1} cm^{-1} are typical for α-thioalkyl radicals, $-\overset{\displaystyle\cdot}{\underset{|}{C}}-S-$. As

[24] S. A. Chaudhri, M. Göbl, T. Freyholdt, and K.-D. Asmus, *J. Am. Chem. Soc.* **106**, 5988 (1984).

[25] K.-O. Hiller, B. Masloch, M. Göbl, and K.-D. Asmus, *J. Am. Chem. Soc.* **103**, 2734 (1981).

[26] K.-D. Asmus, M. Göbl, K.-O. Hiller, S. Mahling, and J. Mönig, *J. Chem. Soc, Perkin Trans. 2*, 641 (1985).

[27] J. W. Purdie, H. A. Gillis, and N. V. Klassen, *Chem. Commun.*, 63 (1971); J. W. Purdie, H. A. Gillis, and N. V. Klassen, *Can. J. Chem.* **51**, 3132 (1973).

[28] M. Z. Hoffman and E. Hayon, *J. Am. Chem Soc.* **94**, 7950 (1972); M. Z. Hoffman and E. Hayon, *J. Phys. Chem.* **77**, 990 (1973).

mentioned above, they are indicative for hydrogen abstraction from a C—H bond and/or deprotonation of a R_2S^+ radical cation.[29]

In addition to direct detection based on the measurement of optical (and other physical) parameters, indirect methods also play an important role in the identification of sulfur-centered radicals and their associated chemistry. Product analysis, in particular, has to be mentioned in this context. Over the past decade a number of sensitive chromatographic methods have been developed for the analysis of even very small concentrations of organic sulfur compounds.[30]

Chemical Properties of Sulfur-Centered Radicals

RS · + RS⁻ ⇌ (RSSR)⁻ Equilibrium

Probably one of the most important aspects concerning free radicals in thiol/disulfide systems is an equilibrium between RS · and (RSSR)⁻, controlled by the thiolate concentration[11]:

$$RS \cdot + RS^- \rightleftharpoons (RSSR)^{\bar{}} \tag{8}$$

The establishment of this equilibrium is inherently connected with the electronic structure of the disulfide radical anion, namely, the bond-weakening character of the antibonding σ^* electron.[23] Generally, the strength of the sulfur–sulfur three-electron bond amounts to less than one-half of that for a normal two-electron σ bond. This is, in fact, the rationale for the easy dissociation of the disulfide radical anion.

The equilibrium can be approached from two sides, namely, via association of RS · with RS⁻ or via one-electron reduction of the disulfide. Generally, it lies on the right-hand side, that is, the three-electron-bonded radical anion is thermodynamically the more stable system. Equilibrium constants around $10^3 \ M^{-1}$ have been determined for simple aliphatic systems[11,31,32] and are expected to be of the same order of magnitude for most others as well.

It is very important to realize that equilibrium (8) is essentially controlled by the concentration of the nonradical component, namely, thiolate, which is usually present in excess over the concentrations of the two

[29] K.-O. Hiller and K.-D. Asmus, *Int. J. Radiat. Biol. Relat. Stud. Phys. Chem. Med.* **40**, 597 (1981).

[30] H. J. Möckel, *in* "Advances in Chromatography" (J. C. Giddings, E. Grushka, and P. R. Brown, eds.), Vol. 26, p. 1. Dekker, New York, 1987.

[31] H. P. Schenck and K.-D. Asmus, unpublished results.

[32] W. Karmann, A. Granzow, G. Meissner, and A. Henglein, *Int. J. Radiat. Phys. Chem.* **1**, 395 (1969).

radicals involved. This shows up, for example, in the kinetic stability of the disulfide radical anions. Lifetimes of up to milliseconds and longer are possible in the presence of thiolate. In the absence of thiolate, however, for example, if (RSSR)$^{\cdot-}$ radicals are generated via reduction of the disulfide, the radical anions may decay within less than 1 μsec into the corresponding thiyl radicals. Rate constants for the dissociation of the three-electron bond [reverse reaction in equilibrium(8)] have been determined to 10^6–10^7 sec^{-1}.[31]

It must be recognized, though, that these latter considerations do not apply to backbone-linked disulfide radical anions, which seemingly are more stable and exhibit long lifetimes even in the absence of free thiolate.[13,28,33,34] An example is the radical anion generated on one-electron reduction of lipoic acid, which lives for more than 100 μsec:

$$\tag{9}$$

After dissociation, the thiyl and thiolate components cannot freely diffuse apart, and thus retain a high local concentration which facilitates reassociation to the three-electron-bonded species. This example nicely emphasizes the relative importance of molecular structures.

Equilibrium (8) determines to a great extent the overall chemistry of radical-mediated processes in thiol/disulfide systems. As discussed in the following sections, RS$\cdot$ and (RSSR)$^{\cdot-}$ exhibit very different chemical properties.

(RSSR)$^{\cdot-}$ Radical Anions

Disulfide radical anions are good reductants. RSSR/(RSSR)$^{\cdot-}$ redox potentials (E^0) have been estimated to be approximately -1.6 V for the lipoic acid system, for example, and may be expected to be of similar magnitude for other such couples as well.[14] The most important reaction a disulfide radical anion can undergo is probably the reduction of molecular oxygen to yield superoxide[35]:

$$(RSSR)^{\cdot-} + O_2 \rightarrow RSSR + O_2^{\cdot-} \tag{10}$$

[33] M. Bonifačić and K.-D. Asmus, *Int. J. Radiat. Biol. Relat. Stud. Phys. Chem. Med.* **46,** 35 (1984).

[34] M. Farragi, J. L. Redpath, and Y. Tal, *Radiat. Res.* **64,** 452 (1975).

[35] O. I. Micic, V. M. Markovic, and M. T. Nenadovic, *Bull. Soc. Chim. Beograd* **40,** 277 (1975).

Electron exchange between disulfides should also not be neglected. It often reflects the relative stability of the disulfide radical anions. An example is the equilibrium in the cysteamine (CyaSSCya) and lipoic acid [Lip(SS)] system:

$$(CyaSSCya)^{\cdot} + Lip(SS) \rightleftharpoons CyaSSCya + Lip(SS)^{\cdot} \tag{11}$$

which lies clearly on the right-hand side.[33]

RS · Radicals

Thiyl radicals can undergo a variety of reactions. The most often quoted recombination to the corresponding disulfide,

$$2RS\cdot \rightarrow RSSR \tag{12}$$

is kinetically a fast process and controlled only by diffusion. It has to compete with the many possible radical–molecule reactions the thiyl radical may undergo and which are discussed below.

An important property of thiyl radicals is that they can act as oxidants. Redox potentials of approximately +0.75 and +1.33 V have been estimated for a number of RS·/RS⁻ and RS·/RSH couples, respectively.[14,36] Substrates that have been shown to be oxidized in the general reaction with an electron donor,

$$RS\cdot + D \rightarrow RS^- + D^{\ddagger} \tag{13}$$

(D‡ denotes the oxidized form of the donor irrespective of the actual state of charge) are, for example, ascorbate (vitamin C), α-tocopherol (vitamin E), and various phenothiazine derivatives. Absolute rate constants of the order of $10^8 \ M^{-1} \ sec^{-1}$ and above indicate high efficiencies in this electron-transfer processes.[31,37–39] It should be mentioned that many of the oxidations of phenothiazines seem to be reversible owing to the similarity of the respective redox potentials.

Another reaction where reversibility appears to be involved is the reaction of thiyl radicals with molecular oxygen, that is,

$$RS\cdot + O_2 \rightleftharpoons RSOO\cdot \tag{14}$$

[36] P. S. Surdhar and D. A. Armstrong, *J. Phys. Chem.* **90**, 5915 (1986).

[37] L. G. Forni and R. L. Willson, *in* "Protective Agents in Cancer" (D. C. H. McBrien and T. F. Slater, eds.), p. 159. Academic Press, New York, 1983; L. G. Forni and R. L. Willson, *Biochem. J.* **240**, 896, 905 (1986).

[38] L. G. Forni, J. Mönig, V. O. Mora-Arellano, and R. L. Willson, *J. Chem. Soc., Perkin Trans. 2*, 961 (1983).

[39] J. Mönig, K.-D. Asmus, L. G. Forni, and R. L. Willson, *Int. J. Radiat. Biol. Relat. Stud. Phys. Chem. Med.* **52**, 589 (1987).

Originally, the few available rate constant data had been interpreted in terms of a fast (diffusion-controlled) and irreversible oxygen addition.[40] This view must, however, be changed, judging from a number of recent investigations.[39,41,42] Although there is still some ambiguity about the exact interpretation of these new experimental data the following statement can be made: The experimental set of high rate constants indicating a diffusion-controlled addition of oxygen ($k > 10^9 \ M^{-1} \ \text{sec}^{-1}$)[40,42] can only be correct if reaction (14) is reversible. If oxygen addition is irreversible though, much lower rate constants ($k = 10^7$–$10^8 \ M^{-1} \ \text{sec}^{-1}$),[39] as measured in a different set of experiments, must apply. (These latter rate constants adjust to higher values in case of reversibility.) In any case and irrespective of which possibility applies, the interaction of oxygen with thiyl radicals is considerably less efficient than previously thought.[39,43] The chemistry of the thioperoxyl radicals, RSOO·, has not yet been studied in sufficient detail to characterized them conclusively.

Reversibility must also be considered for the so-called repair reaction. This has become evident from several recent studies. The forward reaction of

$$\text{RSH} + {-}\overset{|}{\underset{|}{C}}{\cdot} \ \rightleftharpoons \ \text{RS}{\cdot} + {-}\overset{|}{\underset{|}{C}}{-}\text{H} \qquad (15)$$

is generally the faster process and for many systems occurs with rate constants of the order of 10^7–$10^9 \ M^{-1} \ \text{sec}^{-1}$.[4,5] However, rate constants of the order of 10^3–$10^7 \ M^{-1} \ \text{sec}^{-1}$ have also been measured for the reverse reaction, namely, for hydrogen abstraction from activated C–H bonds.[44–47] It should be noted in this context that the difference in bond energies for S–H and C–H is relatively small (approximately 80 and 100 kcal mol^{-1}, respectively, in unactivated systems). One important conse-

[40] J. P. Barton and J. E. Packer, *Int. J. Radiat. Phys. Chem.* **2**, 159 (1970).

[41] K. Schäfer, M. Bonifačić, D. Bahnemann, and K.-D. Asmus, *J. Phys. Chem.* **82**, 2777 (1978).

[42] M. Tamba, G. Simone, and M. Quintiliani, *Int. J. Radiat. Biol. Relat. Stud. Phys. Chem. Med.* **50**, 595 (1986).

[43] K.-D. Asmus, M. Lal, J. Mönig, and C. Schöneich, *in* "Oxygen Radicals in Biology and Medicine" (M. G. Simic and K. A. Taylor, eds.), p. 67. Plenum, New York, 1988.

[44] M. S. Akhlaq, H. P. Schuchmann, and C. v. Sonntag, *Int. J. Radiat. Biol. Relat. Stud. Phys. Chem. Med.* **51**, 91 (1987).

[45] W. A. Pryor, G. Gojon, and D. F. Church, *J. Org. Chem.* **43**, 793 (1978).

[46] C. Schöneich, M. Bonifačić, and K.-D. Asmus, *Free Radical Res. Commun.* **6**, 393 (1989).

[47] C. Schöneich, K.-D. Asmus, U. Dillinger, and F. v. Bruchhausen, *Biochem. Biophys. Res. Commun.* **161**, 113 (1989).

quence of equilibrium (15) would be that thiols are not only able to repair a potentially damaging site but indirectly, namely, via their thiyl radicals, can also contribute to biological damage. Equilibrium (15) at least provides a rationale for possible incompleteness of a repair process.

Displacement reactions involving thiyl radicals are not restricted to hydrogen abstraction from C–H bonds. A corresponding process (P–H cleavage) has, for example, been observed for the reaction of various RS · radicals with phosphite.[48] Formally, the exchange of thiyl groups with disulfides may also be regarded as a displacement reaction but has been shown to proceed via an intermediate adduct radical[7,33]:

$$RS \cdot + R'SSR' \rightleftharpoons [R'SS(SR)R'] \cdot \rightleftharpoons R'SSR + R'S \cdot \qquad (16)$$

In fact, the adduct exists in equilibria with the respective thiyl/disulfide couples. Stability constants of the order of 100 M^{-1} and rate constants of about 10^6 M^{-1} sec^{-1} for the thiyl/disulfide reaction have been measured (including the biologically relevant CysS ·/cystine system).[7]

These and many other addition reactions are based on the electrophilicity of the thiyl radical. In other words, thiyl radicals are likely to add to centers of relatively high electron density, particularly to π systems and nonbonded electron pairs at heteroatoms. Rate constants are, however, usually many orders of magnitude below the diffusion limits since the electrophilic character of the thiyl radical is not very pronounced and much lower than that of the analogous oxygen-centered RO · radicals.

Perthiyl Radicals and α-Thioalkyl Radicals

Little is known about the chemistry of perthiyl radicals. Strong evidence for their participation in a reaction mechanism would always be the formation of tri- and tetrathia compounds.

The α-thioalkyl radicals, —Ċ—S—, appear to be moderately good reductants, but there is reason to believe that most reactions proceed via addition. In the presence of oxygen a peroxyl radical adduct, —C(—OO ·)—S—, is formed.[49]

Sulfur-Centered Radical Cations

All sulfur-centered radical cations, that is, $(RSSR)^{\dagger}$, $R_2S^{\dagger}$, $R_2S \therefore SR_2^{\dagger}$, etc., exhibit no measurable reactivity toward molecular oxygen, the ma-

[48] K. Schäfer and K.-D. Asmus, *J. Phys. Chem.* **85**, 852 (1981).

jor reason being the electrophilicity of O_2.[15,49] The pronounced chemical property of these radical cations is their strong oxidative character. Redox potentials of (RSSR)[+]/RSSR couples have been determined to as approximately +1.4 and +1.1 V for the dimethyl disulfide and lipoic acid systems, respectively.[20] Most others can be expected within the range of these values.

Thus, (RSSR)[+] radicals readily oxidize thiolates, phenothiazines, phenol derivatives, vitamin E, and vitamin C, for example. A particularly interesting situation arises when the RS· radical is generated via disulfide radical cations, for example,

$$Lip(SS)^{\ddagger} + CyaS^- \rightarrow CyaS\cdot + Lip(SS) \tag{17}$$

The CyaS· radical would then immediately associate with CyaS⁻ to yield the three-electron-bonded (CyaS∴SCya)⁻ radical anion, finally followed by the electron transfer to another lipoic acid molecule as formulated in Eq. (11). This puts thiols into the interesting role of mediating the conversion of a highly oxidizing to a highly reducing species.[33]

The radical cations derived from organic monosulfides are still better oxidants, and most $R_2S^{\ddagger}$ radicals are able to oxidize disulfides to (RSSR)[+] with diffusion-controlled rate constants.[17,22] The redox potentials of $R_2S^{\ddagger}$/R_2S couples must therefore be significantly more positive than those of the corresponding disulfide couples, namely, above +1.5 V. The oxidizing power of the three-electron-bonded dimers $R_2S\therefore SR_2^+$ are comparatively lower, with E^0 values presumably similar to those of the disulfide couples. The sulfur–nitrogen three-electron-bonded radical cations formed as intermediates on oxidation of methionine-based peptides (see above) are also very good oxidants. An E^0 value of approximately +1.4 V has, for example, been measured for the respective species from glycylmethionine.[50]

Conclusion

The variety of sulfur-centered radicals seems to exceed that of oxygen-centered radicals. Also, the spectrum of their chemical reactions is at least as manifold as for the oxygen analogs. Among the significant aspects emerging from the available data is the pronounced tendency of sulfur-

[49] K.-D. Asmus, D. Bahnemann, M. Bonifačić, and K. Schäfer, in "Oxygen and Oxy-Radicals in Chemistry and Biology" (M. A. J. Rodgers and E. L. Powers, eds.), p. 69. Academic Press, New York, 1981.

[50] W. A. Prütz, J. Butler, E. J. Land, and A. J. Swallow, Int. J. Radiat. Biol. Relat. Stud. Phys. Chem. Med. **55**, 539 (1989); W. A. Prütz, J. Butler, E. J. Land, and A. J. Swallow, Free Radical Res. Commun. **2**, 69 (1986).

centered radicals to coordinate with electron-rich centers. Specific electronic structures such as the $2\sigma/1\sigma^*$ three-electron bonds provide a rationale for many physical and chemical characteristics of the radical species.

Another important finding is the frequent involvement of sulfur-centered radicals in thermodynamic equilibria. This appears to be of particular significance for the role of thiyl radicals since it could explain, for example, some of the insufficiencies in the repair and protection mechanism of thiols. In fact, the equilibria open the possibility that thiyl radicals indirectly could even contribute to biological damage.

Finally, it must be recognized that thiyl radicals can undergo many more reactions than just recombination to their respective disulfides. From product analysis alone this is often difficult to realize since most of the thiyl radical reactions ultimately lead also to either disulfides or thiols. It must clearly be recognized that thiyl radicals are able to undergo a large variety of reactions, and depending on the environmental conditions, that is, local concentrations of possible reaction partners, may thus become imbedded in complex reaction schemes.

[16] Quinoid Compounds: High-Performance Liquid Chromatography with Electrochemical Detection

By Enrique Cadenas and Lars Ernster

Introduction

The carbonyl groups in quinones are found in the same or separate rings and are conjugated with the double bond. The particular arrangement of double bonds inside and outside the six-membered ring of quinones is largely responsible for their chemical properties. The chemistry of quinones is, in many aspects, similar to that of α,β-unsaturated ketones, and most of their redox properties are based on the electrophilic reactivity determined by the carbonyl groups and the reaction of the polarized double bonds with nucleophiles.

The redox features of quinones are essential to the understanding of their overall biological activity, which encompasses functional, toxicological, mutagenic, and antitumor actions. Quinones can participate in redox transitions involving (1) pure electron transfer, such as in the enzymatic reduction of quinones or in the several nonenzymatic sources of semi-

METHODS IN ENZYMOLOGY, VOL. 186

quinones, and (2) nucleophilic addition reactions,[1] implying either oxidation or reduction of the quinone ring. The former is exemplified by the oxidation of the $—C_2=C_3—$ bond to quinone epoxides by O_2 nucleophiles and the latter by the reductive addition via sulfur nucleophiles to yield thioether–hydroquinone adducts.

Since hydroquinones and quinones can be oxidized and reduced electrochemically,[2,3] it appears reasonable to attempt to detect and quantify them by high-performance liquid chromatography (HPLC) with electrochemical methods (HPLC-EC). The sensitivity afforded by electrochemical detection (picomole range) along with the separation methods of HPLC[4] seems ideal for detection and quantification of quinoid compounds in biological systems.[5] HPLC with oxidative electrochemical methods has been successfully applied to the detection of aromatic alcohols and amines, indoles, phenothiazines, purines, as well as other substances, such as ascorbic acid and thiols.[6,7] HO ·, a species often responsible for cellular damage, can also be identified by HPLC-EC by measuring the adduct resulting from the reaction of HO · with 5,5-dimethylpyrroline N-oxide (DMPO)[8] or the oxidation products of salicylate[9,10]; in the latter instances, however, some reservations should be taken into account, for the high oxidation state of certain hemoproteins, e.g., ferrylmyoglobin, can yield a hydroxylation pattern of salicylate similar to that observed with hydroxyl radical.[9,11]

[1] K. T. Finley, in "The Chemistry of Quinonoid Compounds" (S. Patai, ed.), p. 877. Wiley, London, 1974.

[2] J. Q. Chambers, in "The Chemistry of Quinonoid Compounds" (S. Patai, ed.), p. 737. Wiley, London, 1974.

[3] G. Dryhurst, K. M. Kadish, F. Scheller, and R. Renneberg, in "Biological Electrochemistry" (G. Dryhurst et al., eds.), vol. 1, p. 1. Academic Press, New York, 1982.

[4] P. T. Kissinger, K. Bratin, W. P. King, and J. R. Rice, ACS Symp. Ser. 136, 57 (1981).

[5] D. S. Fluck, S. M. Rappaport, D. A. Eastmond, and M. T. Smith, Arch. Biochem. Biophys. 235, 351 (1984).

[6] T. Kurahashi, H. Nishino, S. Parvez, H. Parvez, K. Kojima, and T. Nagatsu, in "Progress in HPLC, Volume 2: Electrochemical Detection in Medicine and Chemistry" (H. Parvez, M. Bastart-Malsot, S. Parvez, T. Nagatsu, and G. Carpentier, eds.), p. 3. VNU Science Press, Utrecht, 1987.

[7] H. Parvez, M. Bastart-Malsot, S. Parvez, T. Nagatsu, and G. Carpentier (eds.), "Progress in HPLC, Volume 2: Electrochemical Detection in Medicine and Chemistry." VNU Science Press, Utrecht, 1987.

[8] R. A. Floyd, C. A. Lewis, and P. K. Wong, this series, Vol. 105, p. 231.

[9] R. A. Floyd, R. Henderson, J. J. Watson, and P. K. Wong, J. Free Radicals Biol. Med. 2, 13 (1986).

[10] D. Mira, U. Brunk, A. Boveris, and E. Cadenas, Free Radical Biol. Med. 5, 155 (1988).

[11] D. Galaris, D. Mira, A. Sevanian, E. Cadenas, and P. Hochstein, Arch. Biochem. Biophys. 262, 221 (1988).

The principle and versatile applications of HPLC-EC have been authoritatively reviewed.[4,6,12] The scope of this chapter encompasses only the amperometric detection of quinones of the *p*-benzo- and 1,4-naphthoquinone series and the effect of substituents on the electrochemical properties of these compounds.

Principle of HPLC with Electrochemical Detection

The principle of HPLC-EC is based on the redox properties of the compounds of interest; hence, the chemical reaction, namely, electron transfer, as well as its dependence on environmental factors (e.g., composition and physicochemical properties of the mobile phase) are essential to the detection process.[4,6,12] HPLC-EC is an amperometric determination that measures the potential difference between the electrode (embedded in one wall of the electrochemical cell) and the bulk of the solution. This determines the reactivity of the electrochemically active compound at the electrode surface. The detector used in this work, a three-pole potentiostat system, was a TL-5A three-electrode glassy carbon flow cell (Bioanalytical Systems Inc., West Lafayette, IN). On the basis of the potential of a reference electrode (Ag/AgCl), constant potential electrolysis is given between the working electrode (glassy carbon) and the auxiliary electrode (stainless steel tube). There are two modes of electrochemical detection: electron transfer from the solute to the electrode surface is known as oxidative electrochemical detection, whereas electron transfer from the electrode surface to the solute is known as reductive electrochemical detection. The former requires a positive applied potential, and the latter requires a negative applied potential (Fig. 1).

The interfacial current measured at a fixed electrode potential is proportional to the concentration of electrochemically active material and is converted to voltage by means of an amplifier current-to-voltage converter (Model LC-4, Bioanalytical Systems). The gain of the converter ($nA\ V^{-1}$) over the recorder sensitivity ($V\ cm^{-1}$) sets the units of the recorder scale ($nA\ cm^{-1}$). Figure 2 shows a chromatogram with reductive applied potential of a mixture of various 1,4-naphthoquinones and naphthoquinone epoxides. Other quinones of biological interest, such as vitamin E quinone, can also be measured by this technique.[13] Table I[14–17]

[12] P. T. Kissinger, K. Bratin, G. C. Davis, and L. A. Pachla, *J. Chromatogr. Sci.* **17**, 137 (1979).

[13] S. Ikenoya, K. Abe, T. Tsuda, Y. Yamano, O. Hiroshima, M. Ohmae, and K. Kawabe, *Chem. Pharm. Bull.* **27**, 1237 (1979).

[14] A. Brunmark, E. Cadenas, J. Segura-Aguilar, C. Lind, and L. Ernster, *Free Radical Biol. Med.* **5**, 133 (1988).

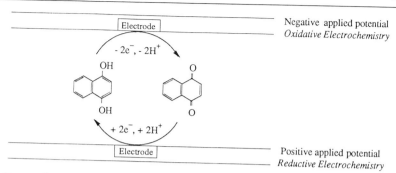

FIG. 1. Schematic representation of electrochemical oxidation and reduction during HPLC with amperometric detection of 1,4-naphthoquinone.

lists the retention times observed for several *p*-benzo- and 1,4-naphtho-quinone derivatives with different mobile phases. The mobile phases were chosen to have a high buffer capacity and a high ionic strength by addition of sodium sulfate. Within the same mobile phase (e.g., phase A, Table I), methyl substitution increases the retention time of the quinone, whereas the quinones substituted with the more polar hydroxyl and methoxyl groups have a shorter retention time. The same applies to naphtho-quinone epoxides, whose methyl-substituted derivatives are retained longer in the column.

Since amperometric detection involves heterogenous electron transfer between an electrochemically active material (solute or analyte) and the electrode surface, the fundamental redox process occurring in the cell will obviously be affected by the composition and pH of the mobile phase as well as the applied potential at the electrode. The selection of the mobile phase is an important factor, allowing the electrode reaction to occur; although the mobile phase should be electrochemically inert at the electrode surface, the amperometric character of the determination requires the presence of electrolytes to convey charge through the electrochemical cell and a solvent with a sufficiently high dielectric constant to permit dissociation of the electrolyte. Therefore, amperometric detection is most satisfactory in mobile phases of high ionic strength (>50 mM) which contain a minimum amount of nonaqueous solvent.[4] The pH of the mobile phase is another important consideration, for the heterogenous electron transfer between the electrochemically active molecule and the electrode surface is affected by pH, as are those chemical reactions linked to the

[15] A. Brunmark and E. Cadenas, *Free Radical Biol. Med.* **3**, 169 (1987).
[16] A. Brunmark and E. Cadenas, *Free Radical Biol. Med.* **6**, 149 (1989).
[17] G. Buffinton, K. Öllinger, A. Brunmark, and E. Cadenas, *Biochem. J.* **257**, 561 (1989).

initial electron-transfer process. Stationary phases or columns for amperometric detection include all type of reversed phase materials, for they are compatible with polar mobile phases containing dissolved electrolytes.[4]

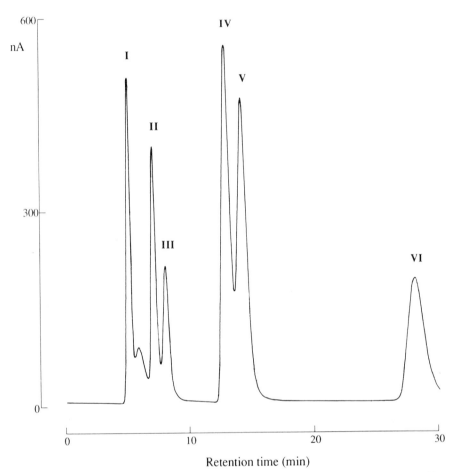

FIG. 2. Analysis of quinones by HPLC with electrochemical detection. Chromatographic conditions: Stationary phase: 15 cm × 3.9 mm Novapak 4 μM C$_{18}$ reversed phase (Waters-Millipore, AB, V. Frölunda, Sweden). Mobile phase: 35% 2-propanol–65% H$_2$O buffered to pH 6.5 with 50 mM sodium phosphate. Mobile phase flow rate: 0.5 ml min^{-1}. Detector: LC-4 with a TL-5A three-electrode glassy carbon flow cell and a Ag/AgCl reference electrode (Bioanalytical Systems). Applied potential: -0.9 V versus Ag/AgCl. Amount injected: 2 nM of each compound. Peaks: I, 2,3-epoxy-1,4-naphthoquinone; II, 1,4-naphthoquinone; III, 2-methyl-2,3-epoxy-1,4-naphthoquinone; IV, 2-methyl-1,4-naphthoquinone; V, 2,3-dimethyl-2,3-epoxy-1,4-naphthoquinone; VI, 2,3-dimethyl-1,4-naphthoquinone.

TABLE I
HPLC with Electrochemical Detection Analysis of
1,4-Naphthoquinone Derivatives

$R_1{}^b$	R_2	R_3	Retention time (min); mobile phase[a]			
			A	B	C	D
H	H	H	6.8			
CH₃	H	H	12.4			
CH₃	CH₃	H	27.4			
OCH₃	OCH₃	H	4.7			
OH	H	H	2.1	4.6		
CH₃	OH	H				2.3
H	H	OH	6.3			
CH₃	H	OH	36.0			
H	SG	H		10.1	16.0	
SG	SG	H			5.4	
NAC	H	H			16.5	
H	SG	OH			16.0	
OH	SG	H		1.8		
CH₃	SG	H			38.0	4.1
CH₃	SG	OH			46.0	
CH₃	NAC	H			26.0	

			A	B	C	D
H	H	H	4.9	17.8		
CH₃	H	H	7.8			13.3
CH₃	CH₃	H	13.8			

[a] (A) 35% 2-propanol–65% H₂O, buffered to pH 6.5 with 50 mM sodium phosphate.[5,14,15] (B) Five milliliters phosphoric acid diluted to 1 liter with H₂O and adjusted to pH 7 with Tris base; 10% of this buffer, 25% methanol, and 65% H₂O containing 50 mM sodium sulfate was used as mobile phase.[16] (C) 35% Methanol–65% H₂O–0.2 mM Tris buffer, 50 mM sodium sulfate, pH 7.0.[17] (D) As mobile phase B, but methanol content increased to 37%.[16]

[b] CH₃, Methyl; OCH₃, methoxyl; OH, hydroxyl; SG, glutathionyl; NAC, N-acetylcysteinyl.

The primary problem in reductive electrochemical detection is the removal of dissolved O_2 in the mobile phase, because of the interference caused by O_2 on reduction to peroxide and, eventually, to water. This problem can be circumvented partly by purging the mobile phase and sample with helium or argon and by using an all-steel liquid chromatograph construction. However, even under these conditions, it is difficult to keep a driftless baseline in a reducing reaction, not only because of the interference of dissolved O_2, but also because of the presence of traces of metal ions leached by the eluent from the stainless steel parts of the chromatograph. Alternative ways to avoid this problem are the redox mode[18] and postcolumn mobile phase changes.[12] The redox mode has been successfully applied to the determination of vitamin K_1 compounds and consists of two sequential generator/detector electrodes arranged in a way such that one is directly upstream of the other. The principle involves the initial reduction of vitamin K_1 to vitamin K_1 hydroquinone ($Q + 2\ e^- + 2\ H^+ \rightarrow QH_2$) by applying a sufficiently negative potential at the generator electrode; the subsequent electrochemical oxidation of vitamin K_1 hydroquinone ($QH_2 \rightarrow Q + 2\ e^- + 2\ H^+$) takes place at the detector electrode, which is set at a positive potential.[18] By this method, pure standards of vitamin K_1 can be detected down to 100 pg with the further advantage that it does not require the removal of O_2 from the system, because while the negative potential at the generator electrode is sufficient to reduce vitamin K_1 and O_2, the positive potential at the detector electrode can oxidize K_1 hydroquinone but not peroxide or water.[13] On the other hand, postcolumn mobile phase changes[12] obtained by using postcolumn mixing to alter the mobile phase pH (1 M $HClO_4$ mixed with the mobile phase prior to detection) allow an optimal determination of vitamin K_3 at lower detector potential and without interference from dissolved O_2.

Multielectrode electrochemical detection[19] offers greater versatility than a three-pole detector, because it helps to overcome the shortcomings inherent in conventional electrochemical detection by allowing the simultaneous determination of oxidized and reduced species. This can be obtained with a detector with three working electrodes, two of which can be set to desired potentials. By this means, reduced and oxidized coenzyme Q was detected simultaneously by setting the potentials of the first and second electrodes at $+0.7$ and -0.3 V, respectively.[9,20]

[18] Y. Haroon, C. A. W. Schubert, and P. V. Hauschka, *J. Chromatogr. Sci.* **22**, 89 (1984).
[19] O. Hiroshima, S. Ikenoya, T. Naitoh, K. Kusube, M. Ohmae, K. Kawabe, S. Ishikawa, H. Hoshida, and T. Kurahashi, *Chem. Pharm. Bull.* **31**, 3571 (1983).
[20] M. Takada, S. Ikenoya, T. Yuzuriha, and K. Katayama, this series, Vol. 105, p. 147. (1984).

The use of amperometric detection combined with HPLC is not restricted to compounds with redox properties. Electrochemically inactive compounds can be derivatized to electrochemically active compounds to be analyzed by HPLC-EC. In this regard, nitrophenyl, aniline, quinoneimine, and phenolate species are ideal candidates for sensitive amperometric methods.

Effect of Substituents on Half-Wave Reduction Potentials of Quinones Determined by HPLC with Electrochemical Detection

There is a substantial body of data on half-wave potentials ($E_{1/2}$) of quinones originating from electrochemical measurements corresponding to the addition of first and second electrons in nonaqueous solvents.[2,3] These measurements are usually devoid of the complications introduced by protonation steps. Values obtained with a standard calomel electrode can be converted to standard hydrogen electrode by adding 241 mV and those obtained with a Ag/AgCl electrode by adding 221 mV. These values cannot be compared directly with those observed in aqueous solvents, as obtained with fast techniques as pulse radiolysis, because of differences in solvation energies and occurrence of junction potentials.[21] The electrochemical reduction of quinones in aqueous or protic media exhibits a complex behavior, which has the appearance of reversible couples[3]; a cyclic voltammetric study of the aqueous electrochemistry of some quinones[22] indicates a strong pH dependence of the redox process.

The half-wave reduction potential ($E_{1/2}$) values of the 1,4-naphthoquinone derivatives in aqueous solutions examined here have been determined by means of hydrodynamic voltamograms against a Ag/AgCl reference electrode. The peak heights are plotted as a function of applied potential for each compound, resulting in the usual sigmoidal voltammetric wave. Although these $E_{1/2}$ values cannot be equated directly with those derived from pulse radiolysis approaches in aqueous media or cyclic voltammetry in protic or aprotic media, they provide a general indication of the redox properties of quinones within the same system, which permits one to predict the likelihood of redox interactions between different molecules.

The effects exerted by different substituents on the $E_{1/2}$ values of p-benzo- and 1,4-naphthoquinones are tabulated in Table II. Figure 3 illustrates the hydrodynamic voltamograms (used to calculate $E_{1/2}$ values) of

[21] A. J. Swallow, in "Function of Quinones in Energy Conserving Systems" (B. L. Trumpower, ed.), p. 59. Academic Press, London, 1982.
[22] S. I. Bailey and I. M. Ritchie, *Electrochim. Acta* **30**, 3 (1985).

TABLE II

HALF-WAVE POTENTIALS ($E_{1/2}$) OF p-BENZO- AND 1,4-NAPHTHOQUINONE DERIVATIVES[a]

Quinone	$E_{1/2}$ ($-$mV) versus Ag/AgCl
p-Benzoquinone series	
Unsubstituted	
p-Benzoquinone	30
Hydroxyl-substituted	
2-Hydroxy-p-benzoquinone	440
Methyl-substituted	
2-Methyl-p-benzoquinone	50
2,6-Dimethyl-p-benzoquinone	120
p-Benzoquinone epoxides	
2,3-Epoxy-1,4-p-benzoquinone	510
2,3-Epoxy-2-methyl-p-benzoquinone	510
2,3-Epoxy-2,6-dimethyl-p-benzoquinone	580
1,4-Naphthoquinone series	
Unsubstituted	
1,4-Naphthoquinone	180
Methyl-substituted	
2-Methyl-1,4-naphthoquinone	225
2,3-Dimethyl-1,4-naphthoquinone	300
Methoxyl-substituted	
2,3-Dimethoxyl-1,4-naphthoquinone	300
Hydroxyl-substituted (quinone ring)	
2-Hydroxy-1,4-naphthoquinone	460
2-Methyl-3-hydroxy-1,4-naphthoquinone	500
Hydroxyl-substituted (benzene ring)	
5-Hydroxy-1,4-naphthoquinone	140
2-Methyl-5-hydroxy-1,4-naphthoquinone	200
Glutathionyl-substituted	
3-Glutathionyl-1,4-naphthoquinone	225
2,3-Diglutathionyl-1,4-naphthoquinone	265
2-N-Acetylcysteinyl-1,4-naphthoquinone	255
2-Methyl-3-glutathionyl-1,4-naphthoquinone	265–300
2-Methyl-3-N-acetylcysteinyl-1,4-naphthoquinone	310
Glutathionyl- and hydroxyl-substituted	
3-Glutathionyl-5-hydroxy-1,4-naphthoquinone	195
2-Methyl-3-glutathionyl-5-hydroxy-1,4-naphthoquinone	220
2-Hydroxy-3-glutathionyl-1,4-naphthoquinone	680
1,4-Naphthoquinone epoxides	
2,3-Epoxy-1,4-naphthoquinone	720
2-Methyl-2,3-epoxy-1,4-naphthoquinone	770–790
2,3-Dimethyl-2,3-epoxy-1,4-naphthoquinone	820

[a] Data from Refs. 14–17.

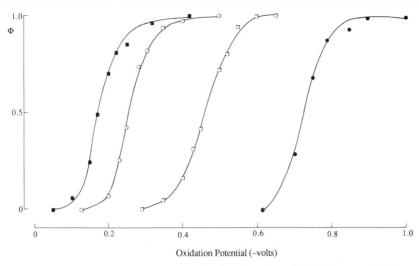

Oxidation Potential (–volts)

FIG. 3. Hydrodynamic voltamograms showing the effect of substituents on half-wave potentials of 1,4-naphthoquinones. Chromatographic conditions: Stationary phase: 15 cm × 3.9 mm Novapak 4 μM C_{18} reversed phase (Waters-Millipore). Mobile phase: 35% 2-pro-panol–65% H_2O buffered to pH 6.5 with 50 mM sodium phosphate. Mobile phase flow rate: 0.5 ml min^{-1}. Detector: LC-4 with a TL-5A three-electrode glassy carbon flow cell and a Ag/AgCl reference electrode (Bioanalytical Systems). Other assay conditions in Refs. 14–17. Hydrodynamic voltamograms: ■, 1,4-naphthoquinone ($E_{1/2} = -180$ mV); ○, glutathionyl-1,4-naphthoquinone ($E_{1/2} = -220$ mV); □, 2-hydroxy-1,4-naphthoquinone ($E_{1/2} = -460$ mV); ●, 2,3-epoxy-1,4-naphthoquinone ($E_{1/2} = -720$ mV).

unsubstituted 1,4-naphthoquinone and derivatives bearing glutathionyl and hydroxyl substituents as well as 1,4-naphthoquinone epoxide. Most quinones can be reduced at working electrode potentials of -300 to -400 mV.[5] The $E_{1/2}$ values for the unsubstituted parent compounds of the *p*-benzo- and 1,4-naphthoquinone series are -30 and -180 mV, respectively. In general, substitution produces discrete changes in the $E_{1/2}$ values (Table II) with exception of hydroxyl substituents in the quinone ring and quinone epoxides, compounds with $E_{1/2}$ values 300–400 mV more negative than the parent unsubstituted quinones.

The introduction of methyl groups in *p*-benzoquinone or the quinone ring of 1,4-naphthoquinone decreases the $E_{1/2}$ values in an additive man-ner.[23] Mono- and dimethyl substitution of 1,4-naphthoquinone decreases the $E_{1/2}$ by about 45 and 120 mV, respectively (Table II).[14,17]

[23] P. Zuman, "Substituent Effects in Organic Polarography." Plenum, New York, 1967.

Hydroxyl Substitutents

Hydroxyl-substituted *p*-benzoquinones can originate from different sources including monooxygenase-mediated metabolism of benzene,[24] enzymatic reduction of *p*-benzoquionone epoxides,[14,25,26] and sulfur nucleophilic addition to quinone epoxides.[16,27] The hydroxyl substituent considerably lowers the $E_{1/2}$ values of *p*-benzoquinones in aqueous solutions,[14,22,28] an effect explained on the basis of mesomeric or inductive effects introduced by the hydroxyl group (similar considerations apply to methoxyl substituents).[28] Reduced hydroxy-*p*-benzoquinones are autoxidized at rates substantially higher than the parent compounds.[14,25]

The effect of hydroxyl substituents on the $E_{1/2}$ values of 1,4-naphthoquinones varies depending on whether the hydroxyl group is situated in the quinone ring (C-2) or in the adjacent benzene ring (C-5). In the former instance, hydroxyl substitution exerts an effect similar to that observed with *p*-benzoquinones, that is, a marked decrease in the $E_{1/2}$ value; in addition, the hydroxyl substituent in the quinone ring raises the pK_a value with regard to the parent quinone.[29] On hydroxyl substitution at C-2, both 1,4-naphthoquinone and 2-methyl-1,4-naphthoquinone decrease their $E_{1/2}$ by 280 and 275 mV, respectively.[14,17] A similar decrease of 1,4-naphthoquinone was observed in the half-wave reduction potential (calculated against a standard calomel electrode) ($\Delta E_{1/2} = -280$ mV)[30] and in the one-electron reduction potential [$E(Q/Q^{-})_7$] (calculated by pulse radiolysis) ($\Delta E_{1/2} = -275$ mV) (data from Ian Wilson quoted in Ref. 31).

Introduction of methoxyl groups causes a similar effect as the hydroxyl substituents vicinal to the carbonyl groups: the $E_{1/2}$ value of 2,3-dimethoxyl-1,4-naphthoquinone is 120 mV more negative than that of the parent compound. A 119 mV more negative $E_{1/2}$ value (against standard calomel electrode) was found with the monomethoxy derivative of 1,4-naphthoquinone in aqueous solutions.[30] Thus, methoxyl substitution seems not to decrease the reduction potential in an additive manner; similar considerations may explain the not very negative one-electron

[24] W. F. Greenlee, J. D. Sun, and J. S. Bus, *Toxicol. Appl. Pharmacol.* **59**, 187 (1981).
[25] A. Brunmark, E. Cadenas, C. Lind, J. Segura-Aguilar, and L. Ernster, *Free Radical Biol. Med.* **3**, 181 (1987).
[26] E. Cadenas, D. Mira, A. Brunmark, J. Segura-Aguilar, C. Lind, and L. Ernster, *Free Radical Biol. Med.* **5**, 71 (1988).
[27] A. Brunmark and E. Cadenas, *Chem.–Biol. Interact.* **68**, 273 (1988).
[28] W. Flaig, H. Beutelspacher, H. Riemer, and E. Kälke, *Liebigs Ann. Chem.* **719**, 96 (1968).
[29] T. Mukherjee, *Radiat. Phys. Chem.* **29**, 455 (1987).
[30] E. M. Hodnett, C. Wongewiechintana, W. J. Dunn, and P. Marrs, *J. Med. Chem.* **26**, 570 (1983).
[31] M. d'Arcy Doherty, A. Rodgers, and G. M. Cohen, *J. Appl. Toxicol.* **7**, 123 (1987).

reduction potential of 2,3-dimethoxyl-1,4-naphthoquinone $[E(Q/Q^{\bar{\cdot}}) = -183$ mV].[32]

When the hydroxyl substituent is in the adjacent benzene ring, it causes an increase in the $E_{1/2}$ value, an effect understood as a polar effect caused by the hydroxyl group in the adjacent benzene ring. At variance with 2-hydroxy-1,4-naphthoquinone, a hydroxyl substituent in the benzene ring lowers the pK_a with respect to the parent compound.[29] Another property of hydroxyquinones is their tendency to undergo strong intramolecular hydrogen bonding, which leads to stabilization of the semiquinone transient species involving displacement toward the left of the disproportionation reaction[33,34]: $2\ Q^{\bar{\cdot}} \rightleftharpoons Q + Q^{2-}$. On hydroxyl substitution at the aromatic ring, the $E_{1/2}$ values versus Ag/AgCl of 1,4-naphthoquinone and 2-methyl-1,4-naphthoquinone increase by 40 and 25 mV, respectively (Table II).[17] This is in agreement with previous reports on the first and second electron reduction potential (calculated by cyclic voltammetry) of 5-hydroxy-1,4-naphthoquinone in apolar media[33] and in aqueous media.[29]

Glutathionyl Substituents

Reduced glutathione (GSH) can be measured in tissue extracts by HPLC-EC with a method based on the separation of GSH from other components by cation-exchange chromatography coupled to the electrochemical oxidation of the thiol to the corresponding disulfide.[35] The oxidation of GSH is accomplished with a graphite paste electrochemical detector with an applied potential of 1.0 V versus Ag/AgCl. A simultaneous detection of thiols and disulfides can be obtained with a dual Hg/Au electrode thin-layer cell to perform both the reduction and detection functions. The electrodes are arranged in series with reduction of oxidized glutathione (GSSG) to GSH at the generator electrode and oxidation at the detector electrode.[36]

Glutathionyl substitution is the result of the 1,4-reductive addition of the thiol across the double bond of the quinone. Since glutathionyl-quinone conjugates retain the redox properties inherent to the quinone moiety, they can be separated and detected by HPLC-EC. The glutathionyl substituent affects both the retention time and the $E_{1/2}$ value of the quinone.[15–17] Because of the relatively weak electron-withdrawing

[32] T. W. Gant and G. M. Cohen, personal communication.
[33] A. Ashnagar, J. M. Bruce, P. L. Dutton, and R. C. Prince, Biochim. Biophys. Acta 801, 351 (1984).
[34] N. F. J. Dodd and T. Mukherjee, Biochem. Pharmacol. 33, 379 (1984).
[35] I. Mefford and R. N. Adams, Life Sci. 23, 1167 (1978).
[36] L. A. Allison, J. Keddington, and R. E. Shoup, J. Liq. Chromatogr. 6, 1785 (1983).

properties of thioether substituents, only minute changes in the quinone reduction potential are expected on glutathionyl substitution. The $E_{1/2}$ values of thioether adducts of p-benzoquinone are slightly more negative than the parent compounds.[37] Despite the minor changes in the $E_{1/2}$ on glutathionyl substitution, significant alterations in the oxidation equilibrium, involving both cross-oxidation and autoxidation reactions, are observed.[27] Electrochemical oxidation of glutathionyl-p-benzohydroquinone shows $E_{1/2}$ values (versus a Ag/AgCl electrode) of +220 mV, 20 mV more positive than those of the unsubstituted compound.[15]

Unsubstituted 1,4-naphthoquinone and derivatives bearing a methyl and/or hydroxyl (in the benzene ring) substituent undergo 1,4-reductive addition with thiols as GSH [reaction (1)]. Glutathionyl substitution of 1,4-naphthoquinone decreases additively the $E_{1/2}$ value from -180 to -225 mV for the monoconjugate and to -265 mV for the diconjugate (Table II). The glutathionyl substitution of 2-methyl-1,4-naphthoquinone decreased the $E_{1/2}$ value by 40 mV. The nucleophilic addition of N-acetyl-L-cysteine to 1,4-naphthoquinone and 2-methyl-1,4-naphthoquinone results in a decrease of the $E_{1/2}$ value more pronounced (75 and 85 mV, respectively) than that observed with GSH.[17] The decrease in $E_{1/2}$ value caused by the thiol addition may be interpreted as a mesomeric effect similar to that produced by hydroxyl and methoxyl substituents. Minor changes in the one-electron reduction potential of 2-methyl-1,4-naphthoquinone [$E(Q/Q^{\overline{\cdot}})_7 = -203$ mV] are observed on GSH addition [$E(GS—Q/GS—Q^{\overline{\cdot}})_7 = -192$ mV].[38] In spite of the small variations in reduction potential, the autoxidation of glutathionylnaphthohydroquinone derivatives, following the reduction of the oxidized counterpart by DT-diaphorase, is 12- to 16-fold higher than that of the parent naphthohydroquinones.[17]

$$\text{(1)}$$

$$R_1 = H, CH_3$$

Hydroxyl and Glutathionyl Substituents

Hydroxyl- and glutathionyl-disubstituted quinones can result from nucleophilic addition of GSH to (1) aromatic-, hydroxyl-substituted naphthoquinones, (2) hydroxy-p-benzoquinones, and (3) p-benzo- or 1,4-

[37] E. R. Brown, K. T. Finley, and R. L. Reeves, *J. Org. Chem.* **36**, 2849 (1971).
[38] I. Wilson, P. Wardman, T.-S. Lin, and A. C. Sartorelli, *J. Med. Chem.* **29**, 1381 (1986).

naphthoquinone epoxides. The primary molecular products resulting from these reductive additions differ in all three instances. In the first case, the product bears a hydroxyl substituent in the benzene ring and a glutathionyl substituent in the quinone ring. In the two latter cases, the hydroxyl and glutathionyl substituents are found in the quinone ring. The occurrence of the new sulfur substituent in these hydroxyquinones controls to a large extent the subsequent redox chemistry in terms of cross-oxidation, autoxidation, and disproportionation reactions.

The nucleophilic addition of GSH to the 5-hydroxyl derivatives of 1,4-naphthoquinone and 2-methyl-1,4-naphthoquinone [reaction (2)] results in compounds with $E_{1/2}$ values close to those of 1,4-naphthoquinone and 2-methyl-1,4-naphthoquinone; the difference amounts to -15 and $+5$ mV, respectively. This may be understood as the counterbalance of the rise in $E_{1/2}$ value caused by the hydroxyl substituent at C-5 and the decrease in $E_{1/2}$ value caused by the glutathionyl substituent.

$$\tag{2}$$

The nucleophilic addition of GSH to either 2-hydroxy-p-benzoquinone or 2,3-epoxy-p-benzoquinone yields the same primary molecular product, 2-hydroxy-5-glutathionyl-p-benzoquinone [reaction (3)], a compound that autoxidizes at rates substantially higher than the parent compound lacking a hydroxyl substituent.[27] The rate of nucleophilic addition is considerably slowed down on methyl substitution.[16] p-Benzohydroquinones with both hydroxyl and glutathionyl substituents autoxidize at rates 350-fold higher than the unsubstituted p-benzohydroquinone.[27] It should be noted that the individual effects of glutathionyl and hydroxyl substituents are not additive when present in the same quinoid molecule, but they seem to potentiate the overall autoxidation.

$$\tag{3}$$

The electron-donating properties of the hydroxyl substituent in the quinone ring increases the electron density at C-3, thereby preventing the nucleophilic addition of GSH to 2-hydroxy-1,4-naphthoquinone. However, a hydroxyglutathionyl adduct of 1,4-naphthoquinone, namely,

2-hydroxy-3-glutathionyl-1,4-naphthoquinone, was obtained during the cleavage of the epoxide ring of 2,3-epoxy-1,4-naphthoquinone on GSH nucleophilic addition[25] [reaction (4)]. A fraction of the product of reaction (4) undergoes oxidative elimination to yield glutathionylnaphthoquinone.[16]

$$\text{(4)}$$

Quinone Epoxides

Vitamin K_1 2,3-epoxide originates from the coupled activity of a two-enzyme system. Vitamin K_1 epoxide is not active in prothrombin synthesis, but it is readily converted back to vitamin K by vitamin K epoxide reductase.[39] Chemical models for the molecular mechanism of vitamin K epoxide reductase emphasize the importance of a primary thiol addition to open the epoxide ring; reaction with a second thiolate results in reductive cleavage of the intermediate, followed by elimination of H_2O to yield the quinone.[40,41]

Although simple quinone epoxides, such as 2,3-epoxy-p-benzoquinone and 2,3-epoxy-1,4-naphthoquinone and their methyl derivatives, are not of direct biological significance, they provide suitable experimental models to understand the electrochemical properties of biologically important quinone epoxides. The formation of quinone epoxides is attained in the addition of O_2 nucleophiles to p-benzoquinones and 1,4-naphthoquinones,[1] as implied in the H_2O_2-dependent oxidation of the 2,3-double bond,[14,15] which is analogous to the reaction of sodium hydroperoxides with naphthoquinones.[42] The reaction is suggested to occur through a nucleophilic addition of HOO^- at the β carbon of the unsaturated system to give an enolate ion as an intermediate. The carbanionic α carbon atom on the enolate ion displaces HO^-, breaking the O—O bond in the hydroperoxyl group and rearranging to an epoxide [reaction (5)]. Quinone epoxides also result from the reaction of $O_2^{\bar{\cdot}}$ with several vitamin K analogs.[43]

[39] J. W. Suttie, *CRC Crit. Rev. Biochem.* **8**, 191 (1980).
[40] R. B. Silverman, *J. Am. Chem. Soc.* **103**, 5939 (1981).
[41] P. C. Preusch and J. W. Suttie, *J. Org. Chem.* **48**, 3301 (1983).
[42] L. F. Fieser, W. P. Campbell, E. M. Fry, and M. D. Gates, *J. Am. Chem. Soc.* **61**, 3216 (1939).
[43] I. Saito, T. Otsuki, and T. Matsura, *Tetrahedron Lett.* **19**, 1693 (1979).

$$\text{(5)}$$

The epoxidation of p-benzo- and 1,4-naphthoquinones is accompanied by changes in their physicochemical properties in terms of electrophilicity, polarity, and reduction potential. The more hydrophilic quinone epoxide is a compound with a weaker electrophilic character than the parent quinone. Because quinone epoxides retain the redox properties of the quinones, they can be analyzed and identified by HPLC-EC.[14–16] The $E_{1/2}$ values for quinone epoxide reduction are about 300–400 mV more negative than those for the parent quinone compounds lacking an epoxide ring (Table II; Fig. 3). The reduction of a quinone epoxide is an irreversible event, which precludes any equilibrium with other redox couples and, consequently, the determination of the one-electron reduction potential by pulse radiolysis approaches.

DT-diaphorase reduces various quinone epoxides at different rates.[14,25] The two-electron reduction of quinone epoxides catalyzed by DT-diaphorase results in epoxide ring opening, yielding a 2-hydroxyhydroquinone as primary product. These hydroxyl derivatives show a higher rate of autoxidation than do the parent hydroquinones lacking a hydroxyl substituent,[14,25] similar to the autoxidation following the 1,4-reductive addition of GSH to quinone epoxides.[16,27]

Concluding Remarks

The development of sensitive and selective electrochemical detectors[4] for use in liquid chromatography has created a new technology applicable to many problems of clinical, pharmacological, and environmental importance. In addition to high sensitivity, the electrochemical detector exhibits good selectivity; it responds only to electrochemically active substances with functional groups or to those compounds derivatized to electrochemically active species. The utility of HPLC combined with electrochemical detection for the investigation of quinone metabolite formation has been previously shown[5] and extended to those derivatives resulting from the addition of oxygen or sulfur nucleophiles.[14–17] This technique can also be used to detect small quantities of physiological quinones, such as vitamin E quinone[13] as well as coenzyme Q[20] and its related changes in mitochondrial membranes on oxidative stress.[44] The constant improvement of amperometric detectors will allow a more efficient evaluation of the cellular redox metabolism of quinones.

The overall process of electron transfer is a complex event influenced by the interaction of the inborn chemical properties of quinoid compounds with environmental factors, which contemplate homogeneous and heterogeneous systems. This applies also to the electrochemical reduction or oxidation of quinones and conveys the requirements for the optimal conditions for amperometric detection, that is, for the heterogenous electron transfer between an electrochemically active compound and the electrode surface.

Although reduction potentials are a critical factor in the metabolism of quinones by one- or two-electron transfer flavoproteins, no statistical correlation was found between the reduction potential of the quinone and its rate of reduction by NADPH–cytochrome-P-450 reductase[45] or DT-diaphorase.[17] Moreover, autoxidation of unsubstituted- or methyl-substituted naphthoquinones, following their two-electron reduction by DT-diaphorase, was significantly different from those naphthoquinone derivatives bearing glutathionyl, hydroxyl (in the benzene ring), or methoxyl substituents. In the latter instances, autoxidation was 6–12 times higher than the unsubstituted 1,4-naphthoquinone, despite small changes in the reduction potential of the substituted naphthoquinones.

Acknowledgments

This research was supported by Grant 4481 from the Swedish Medical Research Council, Grant 2703-B89-01XA from the Swedish Cancer Foundation, and a grant from the Swedish Natural Science Research Council.

[44] M. T. Smith, C. G. Evans, H. Thor, and S. Orrenius, in "Oxidative Stress" (H. Sies, ed.), p. 91. Academic Press, London, 1985.
[45] G. Powis and P. L. Appel, Biochem. Pharmacol. 29, 2567 (1980).

[17] Preparation of Tocopheroxyl Radicals for Detection by Electron Spin Resonance

By ROLF J. MEHLHORN, JÜRGEN FUCHS, SATOSHI SUMIDA, and LESTER PACKER

Introduction

Although the effectiveness of tocopherols as lipophilic antioxidants has long been known, many details about their interactions with biological membranes and with chemical and enzymatic reductants remain to be resolved. Oxidation of tocopherols can produce persistent tocopheroxyl free radicals, which can be observed with electron spin resonance (ESR). In principle, ESR analysis of the perturbation of free radical signals by paramagnetic ions[1] or perturbation of NMR signals of fatty acyl chain residues[2] are tools that can be used for studying the location of the to-copheroxyl radical in the membrane lipid bilayer. ESR studies can also be used to analyze the kinetics of radical formation and decay, and this approach is yielding valuable information about processes that allow to-copherol to act catalytically to scavenge lipid free radicals.[3] Such studies benefit from efficient and specific procedures for producing the to-copheroxyl radical. A variety of procedures may be needed to produce detectable ESR signals of tocopheroxyl radicals in diverse biological systems, which may be adversely affected by some, but not all, oxidation systems.

A general principle governing optimal tocopheroxyl production is that the oxidation system should efficiently produce the one-electron oxidation product of tocopherol but should not involve molecular species that can readily react with the resulting tocopheroxyl radical. Operationally, particularly for studies of the tocopheroxyl radical in liposomes and natural membranes, a suitable oxidations system is one that produces ESR signals of sufficient magnitude to be quantitated and that allows the tocopheroxyl radical to be observed for prolonged periods in the absence of reducing agents or molecular species that can react with it. In solvents, even inefficient oxidation methods or drastic oxidation procedures that rapidly destroy the tocopheroxyl radical as soon as it is formed may be

---

[1] J. M. Herz, R. J. Mehlhorn, and L. Packer, *J. Biol. Chem.* **258,** 9899 (1983).

[2] P. E. Godici, F. R. Landsberger, *Biochemistry* **13,** 362 (1974).

[3] L. Packer, J. J. Maguire, R. J. Mehlhorn, E. Serbinova, and V. E. Kagan, *Biochem. Biophys. Res. Commun.* **159,** 229 (1989).

adequate to produce detectable tocopheroxyl radicals, provided the solvents allow high concentrations (~ 100 mM) of tocopherol to be homogeneously dispersed in them. For this reason experiments of tocopherols in solvent permit much more latitude in the choice of oxidation systems than experiments with aqueous membranes, where far lower tocopherol concentrations are generally attainable. Here, we have compiled oxidation methods that have proved effective in both chemical and biological systems.

Chemical and Physical Oxidation Schemes

For the most part chemical oxidation of tocopherols is useful for solutions of tocopherols in organic solvents, for example, ethanol or chloroform. Chemical oxidation can also be applied to aqueous dispersions of tocopherols in detergents, but, unless carefully controlled to avoid generation of excessive oxidizing potential, it is of limited utility for studies of synthetic and natural membranes, which often contain many oxidant-sensitive sites, for example, thiols and polyunsaturated lipids.

Among the methods that have been successfully employed for oxidizing tocopherol in solvents are pulse radiolysis,[4,5] ultraviolet irradiation,[6] and chemical oxidation by PbO_2,[7] $AlCl_3$,[8] superoxide anion,[9] 2,2-diphenylpicrylhydrazyl,[5,10–12] and a variety of free radical oxidants derived from azo and peroxyl initiators.[11] Tocopherol oxidation with iron–triethylenetetramine has also served to generate tocopheroxyl radicals in liposomes.[13]

Alkaline Oxidation

Alkaline conditions favor the oxidation of phenols and hydroquinones. The application of alkaline oxidation to tocopherols merely requires alka-

[4] J. E. Packer, T. F. Slater, and R. L. Willson, *Nature (London)* **278,** 737 (1979).
[5] D. Jore and C. Ferradini, *FEBS Lett.* **183,** 299 (1985).
[6] W. Boguth and H. Niemann, *Biochem. Biophys. Acta* **248,** 121 (1971).
[7] K. Mukai, N. Tsuzuki, S. Ouchi, and K. Fukusawa, *Chem. Phys. Lipids* **30,** 337 (1982).
[8] K. Mukai, N. Tsuzuki, K. Ishizu, S. Ouchi, and K. Fukuzawa, *Chem. Phys. Lipids* **35,** 199 (1984).
[9] E. J. Nanni, M. D. Stallings, and D. T. Sawyer, *J. Am. Chem. Soc.* **102,** 4481 (1980).
[10] M. Matsuo and S. Matsumoto, *Lipids* **18,** 81 (1983).
[11] J. Tsuchiya, E. Niki, and Y. Kamiya, *Bull. Chem. Soc. Jpn.* **56,** 229 (1983).
[12] C. Rousseau-Richard, C. Richard, and R. Martin, *FEBS Lett.* **233,** 307 (1988).
[13] M. Scarpa, A. Rigo, M. Maiorino, F. Ursini, and C. Gregolin, *Biochim. Biophys. Acta* **801,** 215 (1984).

line treatments that are compatible with the organic solvents. Concentrated ethanolic solutions of tocopherols can be treated directly with aqueous sodium hydroxide to yield slowly decaying tocopheroxyl ESR radical signals. Other strong bases, for example, sodium methoxide, can be added as solids directly to tocopherol solutions in organic solvents if water is to be avoided.

Oxidation by Diphenylpicrylhydrazyl

Diphenylpicrylhydrazyl (DPPH) is a stable organic free radical whose oxidation potential is sufficient to cause it to rapidly oxidize tocopherol to its radical product. The disadvantage of this oxidant is that it has an ESR signal that overlaps with that of the tocopheroxyl radical, which may cloud the interpretation of some experiments. However, for simple solvent systems conditions can be found which cause nearly complete oxidation of tocopherol, without any residual ESR radical signal of the DPPH being apparent.[12] For example, treatment of an ethanolic solution of 300 μM tocopherol with 200 μM DPPH gave exclusively the tocopheroxyl radical signal, whose decay kinetics could be used to determine the bimolecular decay constant for tocopheroxyl in this solvent.[12]

Thermal Initiators

Azo compounds spontaneously decompose into free radical fragments at a rate dependent on the temperature. In principle, these fragments can react directly with tocopherols, but usually, under aerobic conditions, they form peroxyl derivatives that oxidize tocopherols. Azo compounds are available both as water-soluble and hydrophobic derivatives, which allows for comparisons of tocopherol oxidation in hydrophobic environments and in more polar environments like detergent–water interfaces. The rate of free radical production for many available compounds is rather low and may require high concentrations and elevated temperatures for detectability of tocopheroxyls in ESR experiments. For example, at 37°, the water-soluble compound 2,2'-azobis(2-amidinopropane) dihydrochloride undergoes unimolecular decomposition with a rate constant of 1.1×10^{-6} sec^{-1}, and the efficiency of radical release from this decomposition is poor in some solvents.[14] Some azo compounds pose explosive hazards and must be treated with considerable care.

[14] Y. Yamamoto, E. Niki, Y. Kamiya, and H. Shimasaki, *Biochim. Biophys. Acta* **795,** 332 (1984).

Superoxide

Superoxide is an attractive chemical oxidant because it leaves no residue after oxidation has been carried to completion. Potassium superoxide is readily available and convenient to manipulate but, because of its extremely rapid decomposition in water, is suitable only for work with organic solvents. In organic solvents it yields large and persistent tocopheroxyl ESR signals. The reaction products of superoxide with reductants are hydrogen peroxide and a potassium salt. The latter is usually strongly basic, so, in general, treatment with superoxide owes its effects to a combination of alkaline and oxidant effects.

Ferricyanide

The ferricyanide trivalent anion is a remarkably effective tocopherol oxidant considering its midpoint potential is only +0.36 V. Owing to its strongly ionic nature it is most suited for work with aqueous systems, including detergent and liposome dispersions of tocopherols. However, ferricyanide also is effective in ethanol and chloroform, where it can be used for qualitative work only, since the effective ferricyanide concentration involved in the reaction with tocopherol cannot be quantitated.

Photochemical Oxidation Procedures

Ultraviolet Irradiation

The λ_{max} of α-tocopherol is at 294 nm, and UV irradiation in the presence of oxygen or other electron acceptors is an effective means for generating the tocopheroxyl radical ESR signal. Irradiation, in contrast to chemical oxidation schemes, is especially useful for experiments where tocopheroxyl radical decay is being quantitated, since oxidation can be terminated by switching off the light. The interpretation of data obtained with this method must weigh the possible involvement of other UV-absorbing species present in a sample, which may also form free radical products whose reactivity with tocopherol or tocopheroxyl may affect the intensity of the tocopheroxyl radical signal.

Enzymatic Oxidation Procedures

Enzymatic oxidation systems are attractive for work with natural membranes, partly because they can be used to simulate the production of oxidants under realistic physiological conditions. The systems used thus far (peroxidase and lipoxygenase) act on oxygen to produce strong oxidants, which in turn oxidize tocopherols. In the usual glass ESR tube, this type of oxidation system may rapidly become anaerobic, thus losing activity. Therefore, control experiments must be conducted to establish that depletion of oxygen does not play a role in altered ESR signal intensities with these enzymatic methods.

Peroxidases Linked to Enzymatic Hydrogen Peroxide Generators

For studies of natural membranes, an enzyme system for generating phenoxyl radicals has been shown to specifically generate the tocopheroxyl radical, while minimizing the involvement of other free radicals, for example, lipid radical species. The system consists of horse radish peroxidase coupled to a hydrogen peroxide-generating system:

$$\text{Glucose} + O_2 \xrightarrow{\text{GO}} \text{gluconolactone} + H_2O_2$$

$$H_2O_2 + 2 \text{ PhOH} \xrightarrow{\text{HRP}} 2 H_2O + \text{PhO} \cdot$$

$$\text{PhO} \cdot + \alpha\text{-tocOH} \longrightarrow \text{PhOH} + \alpha\text{-tocO} \cdot$$

where GO refers to glucose oxidase; HRP, horseradish peroxidase; PhOH, arbutin (4-hydroxyphenyl-β-D-glucopyranoside); and α-tocOH, α-tocopherol. Since this system consumes oxygen, ESR samples should be placed into gas-permeable tubing to maintain a relatively constant enzyme activity.

Treatment of a variety of tocopherol-containing liposomes (dioleoylphosphatidylcholine, asolectin, and egg lecithin) with this system has been shown to yield the ESR spectrum of the tocopheroxyl radical. Similar spectra, but of smaller magnitudes, can be generated in rat liver submitochondrial and microsomal membranes isolated from animals nutritionally enriched with d-α-tocopherol. In a typical liposome preparation, consisting of dioleoylphosphatidylcholine at 100 mg/ml and d-α-tocopherol at 230 μM, an oxidation system consisting of 5 mM glucose, 0.4 U/ml GO, 100 μM arbutin (a hydrophilic phenol), and 0.2 U/ml HRP produced ESR signals of tocopheroxyl with excellent signal-to-noise characteristics (see Fig. 1).[15]

Subcellular membranes are usually contaminated with catalase, which reduces the hydrogen peroxide concentration available for reaction with the peroxidase. This problem can be overcome with a compensatory increase in horseradish peroxidase. For example, in experiments with submitochondrial particles (SMPs) and microsomes isolated from vitamin E-supplemented rats, useful ESR signals were obtained with the following oxidation system[15]: SMPs at 26.6 mg protein/ml or microsomes at 24 mg/

[15] R. J. Mehlhorn, S. Sumida, and L. Packer, *J. Biol. Chem.* **264**, 13448 (1989).

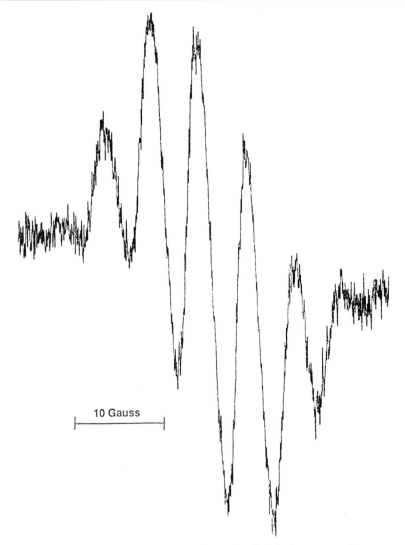

FIG. 1. ESR spectrum of the d-α-tocopheroxyl radical in liposomes using enzymatic oxidation. The reaction mixture contained 1 mM d-α-tocopherol in 100 mg/ml of asolectin treated with 0.4 U/ml glucose oxidase, 10 mM glucose, 40 μM phenol, and 0.2 U/ml horse-radish peroxidase. Instrument conditions: Modulation Amplitude, 2.5 G; gain, 1.25 × 10^6 (Brucker ER 200 D-SCR spectrometer, X-band).

ml treated with 10 mM glucose, 0.4 U/ml GO, 200 μM arbutin, and 5 U/ml HRP.

Peroxidases Linked to Photochemical Peroxide Sources

Many of the ambiguities associated with the multiple effects of UV irradiation, especially in complex biological systems, can be avoided by working with photosensitizers, which absorb light of lower energy. An excellent photosensitizer is flavin mononucleotide (FMN), which absorbs maximally at 450 nm and yields reactive excited species that can abstract electrons from a variety of organic molecules. In recent experiments in our laboratory, a hybrid photochemical–enzymatic system for generating phenoxyl radicals, consisting of FMN, EDTA, and horseradish peroxidase yielded excellent tocopheroxyl radicals in liposomes (Fig. 2). Interestingly, the presence of superoxide dismutase (SOD) proved to increase significantly the tocopheroxyl radical signal, suggesting that superoxide exerted a strong inhibitory effect on the production of oxidant (probably by the peroxidase) in this system.

Lipoxygenase System

Lipoxygenase produces lipid free radicals in the presence of polyunsaturated fatty acids,[16] some of which can be used to oxidize tocopherol in liposomes and natural membranes.[1] The activity of this system is a function of several parameters, including the free fatty acid concentration (in membrane studies, this will depend on how much of the fatty acid partitions into the bilayer; in cellular systems this may be affected by fatty acid metabolism, including phospholipase activity) and the oxygen tension (in glass capillaries commonly used for ESR samples this will decrease with time of incubation), and it may be affected by reducing agents, particularly hydroquinones (which may convert lipoxygenase to the inactive form).[17] The oxygen tension affects the production of radical products in a complex manner.[16] Oxygen is required for sustained activity of the normal enzyme reaction, which produces lipid hydroperoxides but not free radicals. In the absence of oxygen, lipoxygenase will react with hydroperoxides to generate a variety of free radicals. Therefore, in a sealed ESR capillary system, progressive depletion of oxygen as it is incorporated

[16] L. Petersson, S. Slappendel, M. C. Feiters, and F. G. Vliegenthart, *Biochim. Biophys. Acta* **913**, 228 (1987).
[17] P. Louis-Flamberg, R. Krupinski-Olsen, A. L. Shorter, and C. Kemal, *Ann. N.Y. Acad. Sci.* **524**, 382 (1988).

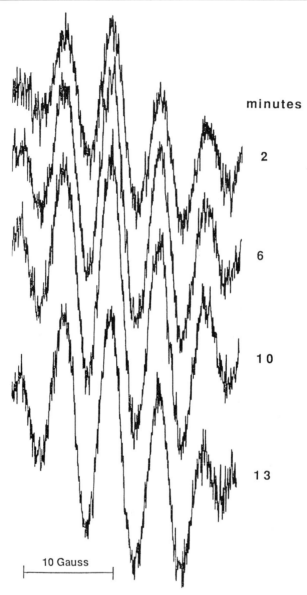

FIG. 2. ESR spectra of tocopheroxyl radicals in liposomes using a mixed photochemical–enzymatic oxidation system. The reaction mixture contained 100 mg/ml dioleoylphosphatidylcholine, 200 μM d-α-tocopherol, 10 μM FMN, 1 mM EDTA, 100 U/ml superoxide dismutase, 0.2 U/ml horseradish peroxidase, and 100 μM arbutin; irradiation was at 450 nm, using an interference filter with a 50 nm bandwidth. Instrument conditions were as for Fig. 1.

into polyunsaturated fatty acids will lead to increasing free radical formation from the accumulated lipid hydroperoxides.

The nonlinear relationship between free radical formation and reaction time places a severe limitation on the use of the lipoxygenase system as a tool for quantitative ESR studies of tocopheroxyl radicals. These difficulties can be overcome to some extent by the use of gas-permeable tubing, but the high enzyme activity that is required for eliciting the ESR signal of tocopheroxyl may lead to oxygen gradients in such tubing, which may be further aggravated in membranes that exhibit other oxygen-consuming activities. For qualitative studies, however, the lipoxygenase system is often suitable for generating easily detectable ESR signals of tocopheroxyl radicals. For example, in a typical liposome preparation consisting of phosphatidylcholine at a concentration of about 5 mg/ml and containing 200 μM of d-α-tocopherol, tocopheroxyl radicals can be observed by ESR with excellent signal-to-noise ratios by treatment with 0.5 mg/ml of lipoxygenase and 1.2 mM arachidonic acid.

Acknowledgments

This research was supported by the National Institutes of Health (AG 04818) through the U.S. Department of Energy under Contract DE-AC03-76SF00098 and the NIH (CA 47597).

Section II

Assay of Enzymes or Substances Involved in Formation or Removal of Oxygen Radicals and Derived Products

A. Proteins
Articles 18 through 30

B. Small Molecules
Articles 31 through 37

[18] Determination of Superoxide Dismutase Activity by Purely Chemical System Based on NAD(P)H Oxidation

By FRANCESCO PAOLETTI and ALESSANDRA MOCALI

Introduction

Most of the currently employed methods for the assay of superoxide dismutase (SOD) activity in tissue extracts are based on the ability of these enzymes to inhibit a superoxide-driven reaction. The extent to which the rate of reaction is reduced can be taken as an indirect measurement of enzyme activity. The generation of superoxide can be achieved by either enzymatic or nonenzymatic systems[1]; subsequently a suitable detection method, whether colorimetric, polarographic, or luminometric according to different approaches,[2] must be devised. Essential requirements for any assay are sensitivity, reliability, and simplicity of the procedure so that it can be easily performed in the laboratory without the aid of expensive equipment and reagents.

The present assay satisfies all these needs and allows the determination of minute amounts of SODs, such as 2 ng, which are far below of the detection limit of most employed methods.[3] Moreover, there is another important aspect worth mentioning: the usual detectors like nitro blue tetrazolium (NBT), cytochrome *c,* or other chromogenic substrates, which are reduced by superoxide, might also be electron acceptors for reducing agents known to occur in biological samples. This fact will explain the difficulties, often encountered with these assays, in attaining saturation levels and reproducible titration curves. Our method, on the contrary, relies on the oxidation of NAD(P)H, and this makes the detection less prone to interferences by aspecific reduction from cellular components. Besides, since NAD(P)H is the detector, measurements of SOD activities can easily be accomplished spectrophotometrically as for the majority of other enzymes.

The method, originally developed in our laboratory,[4] consists of a purely chemical reaction sequence which generates superoxide from mo-

[1] J. M. McCord, M. Crapo, and I. Fridovich, *in* "Superoxide and Superoxide Dismutases" (A. M. Michelson, J. M. McCord, and I. Fridovich, eds.), p. 11. Academic Press, New York, 1977.
[2] L. Flohé and F. Ötting, this series, Vol. 105, p. 93.
[3] W. F. Beyer and I. Fridovich, *Anal. Biochem.* **161,** 559 (1987).
[4] F. Paoletti, D. Aldinucci, A. Mocali, and A. Caparrini, *Anal. Biochem.* **154,** 536 (1986).

lecular oxygen in the presence of EDTA, manganese(II) chloride, and mercaptoethanol. NAD(P)H oxidation is linked to the availability of superoxide anions in the medium. As soon as SOD is added to the assay mixture, it brings about the inhibition of nucleotide oxidation. Therefore, at high concentrations of the enzyme the absorbance at 340 nm remains unchanged, while in the control (no SOD added) absorbance decreases according to predictable kinetics.

Measurements are carried out at physiological pH and consist of a single spectrophotometric step by which SOD activities are reliably and sensitively determined in both crude and pure enzyme preparations. This procedure, involving stable and inexpensive reagents, appears suitable for application in biochemistry, plant physiology, and clinical chemistry.

Procedures

Preparation of Samples

SODs are soluble enzymes and can be extracted from tissues after disaggregation and homogenization in suitable media. For small amounts of material, like cultured cells, leukocytes collected from a few millileters of peripheral blood, or bioptic fragments, treatment with ultrasonics or freezing and thawing might be more appropriate than mechanical homogenization. Simple lysis in hypotonic solutions is adopted for red cells. Tissue solubilization by detergents could also be performed since those commonly used do not interfere directly with the assay. Besides, the interactions of detergents with extract components may affect either NAD(P)H oxidation rates or the stability of some SOD preparations (for further information, see Precautions). Whole tissue extracts should be freed from either subcellular organelles or debris, according to the need, by centrifugation to obtain a clear and homogeneous supernatant for analysis.

Prior to the assay, tissue extracts should always be dialyzed or desalted through a Sephadex G-25 column in order to remove low molecular weight compounds that might have nonenzymatic SOD-like activity. The dialysis buffer or the medium used to develop gel filtration must be kept and used to replace the sample in the control cuvette (see Conditions for Assay). Provided the controls are suitable, the assay seems to be unaffected by most of the buffers and media commonly employed for tissue extraction; namely, Tris-HCl and disodium phosphate buffer (0.1 M, pH 7.4), sodium acetate–acetic acid buffer (0.1 M, pH 5–8), phosphate-buffered saline (pH 7.4), sodium bicarbonate (0.1 M, pH 8.3), and distilled water have been individually used as sample solvents for several SOD preparations without any problem in activity determination, under the conditions given below.

Unless one is dealing with erythrocytes, specimens from laboratory animals should be removed by taking care to avoid trapped blood. Either perfusion or exhaustive wash of the tissue will prevent sample contamination by the red cell SOD and by excess hemoglobin. The latter interferes with the assay by counteracting the inhibitory effect of SOD; therefore, hemoglobin must be eliminated from the extract before SOD determinations. There are three different procedures, which can alternatively be used according to the need. The first two are useful for samples with a slight hemoglobin contamination (i.e., partially hemolysed plasmas, differentiated erythroleukemic cells, not well-perfused tissues).

The first approach is anion-exchange chromatography, using cellulose CM-52 (equilibrated with 8 mM Tris-HCl buffer, pH 6.8). Extracts to be treated must be first dialyzed or equilibrated with the developing buffer. The sample is applied to the column and eluted with Tris-HCl. Several proteins, including hemoglobin, bind to the cellulose while Cu,Zn-SOD and Mn-SOD are completely recovered in the excluded fraction. This step is quite rapid and particularly suitable for small samples (up to 10 ml) which can be processed using a Pasteur pipet filled with approximately 0.5–1 g of wet cellulose.

The second method, gel filtration, takes advantage of differences in molecular weight among Mn-SOD (~86K), hemoglobin (~68K), and Cu,Zn-SOD (~32K). To this purpose it is better to use a gel with a fractionation range of at least 5K–150K, allowing a fine peak resolution, like Sephadex G-100 superfine or Sephacryl S-200.

The third procedure for hemoglobin removal, which applies to crude hemolysates and tissue extracts with high hemoglobin content, was originally developed by Tsuchihashi[5] and further employed by McCord and Fridovich[6] for the purification of Cu,Zn-SOD from erythrocytes. For clarity we report here an example of hemoglobin precipitation and SOD extraction from human erythrocytes (for values of SOD activity see Reference Ranges). The hemolysate (2.5 ml), prewarmed at 37°, is treated with 1 ml of a mixture of ethanol–chloroform (2 : 1, v/v) and mixed thoroughly to obtain a thick precipitate. Add 2 ml of distilled water and mix again with the vortex. Incubate at 37° for about 15 min with occasional stirring and then use a bench centrifuge to spin down the precipitate. The almost

[5] M. Tsuchihashi, *Biochem. Z.* **140**, 65 (1963).
[6] J. M. McCord and I. Fridovich, *J. Biol. Chem.* **244**, 6049 (1969).

colorless supernatant is recovered, dialyzed, and then assayed after suitable dilution. However, because of the sensitivity of the method and high SOD levels in erythrocytes, human samples are usually diluted by such a large factor that dialysis might be omitted. The recovery of a standard of pure Cu,Zn-SOD added to the hemolysate is practically complete.

Chemicals and Equipment

Reduced adenine nucleotides (β-NADH, β-NADPH, disodium salt), $MnCl_2 \cdot 4H_2O$, 2-mercaptoethanol, ethylenediaminetetraacetic acid (EDTA), triethanolamine, and diethanolamine are all reagent grade and are used without further purification. Pure SOD preparations for standard calibration curves are prepared from bovine erythrocytes ($\sim$3,000 unit/mg lyophilizate, according to McCord and Fridovich[6]; Boehringer-Mannheim, West Germany). All solutions are made up with well-aerated distilled or deionized water. Measurements are performed with a spectrophotometer connected to a recorder and equipped with a multicell holder (at least 4 cells). A temperature control set unit may be useful.

Reagents and Solutions

Triethanolamine–diethanolamine (100 mM each)–HCl buffer (TDB): triethanolamine (14.9 g), diethanolamine (10.5 g), and approximately 13.8 ml of concentrated HCl are dissolved in 1 liter of distilled water; the final pH should be around 7.4

NAD(P)H (7.5 mM): dissolve 20 mg of either NADH or NADPH, disodium salt (see Precautions), in 4 ml of water (amount for 100 assays)

EDTA–MnCl$_2$ (100 mM/50 mM): prepare stock solutions of 200 mM EDTA (dissolve 11.69 g EDTA–acid in 200 ml water and adjust the pH to around 7 with 1 M NaOH) and 100 mM MnCl$_2$ (dissolve 3.95 g MnCl$_2 \cdot 4H_2O$ in 200 ml water), combine the EDTA and MnCl$_2$ stock solutions in a ratio of 1:1 (v/v), and adjust the pH of the mixture (Reagent 3) to pH 7 by dropwise addition of 10 M NaOH ($\sim$0.14 ml per 25 ml of reagent)

Mercaptoethanol (10 mM): dilute 50 μl of concentrated thiol (14.2 M) with 71 ml of water

Stability of Solutions. TDB, mercaptoethanol, and both EDTA and MnCl$_2$ stock solutions are quite stable even at room temperature if microbial contamination is avoided. Dissolved NAD(P)H can be kept at 4° and used within 1 week. For longer storage, keep the coenzyme solution at −20°. The EDTA–MnCl$_2$ solution, once made up, is maintained at room temperature and can be used for over 1 month. Solutions of EDTA–

MnCl$_2$ may present some crystals, but these have no influence on the assay.

Conditions for Assay

Each set of assays must include its own control. The control consists of a cuvette where the sample is replaced by an equal volume of the medium used for enzyme solutions. The following solutions are subsequently pipetted into the cuvette (semimicrocuvette; light path 10 mm): 0.8 ml TDB, 40 μl NAD(P)H, 25 μl EDTA–MnCl$_2$, and 0.1 ml sample (or sample solvent for the control). Mix thoroughly and read at 340 nm against air for a stable baseline recorded over a 5-min period. Within this interval the cuvettes will equilibrate at the right temperature (if needed), and the NAD(P)H oxidase activity possibly occurring in samples can be evaluated. Then add 0.1 ml mercaptoethanol. The final volume in the cuvette is 1.065 ml. Mix and monitor the decrease in absorbance for about 20 min to allow full expression of the chain length leading to NAD(P)H oxidation.

A typical assay is reported in Fig. 1. Measurements of the relative

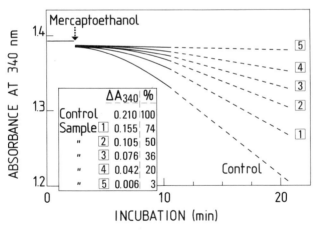

FIG. 1. Kinetics of NADH oxidation at different concentrations of SOD. Rates of NADH oxidation reported concern six assays carried out simultaneously in the absence (Control) and in the presence of increasing amounts of pure Cu,Zn-SOD (sample 1, 12.5 ng; sample 2, 31 ng; sample 3, 53 ng; sample 4, 95 ng; sample 5, 300 ng of lyophilized enzyme, Boehringer-Mannheim). The reaction was started by the addition of mercaptoethanol, and the decrease in absorbance at 340 nm was recorded over 20 min at room temperature. The linear portions of the curves used for calculation are indicated by dashed lines. The rates of NADH oxidation, within an interval of 8 min (ΔA_{340}), are reported in the inset, together with percentages of inhibition relative to the control value.

rates in both the control and sample cuvettes are usually made on the straight portion of the curve, that is, starting from 5–10 min after the addition of mercaptoethanol. The rate of nucleotide oxidation of the control, calculated over an 8-min interval (ΔA_{340}), should be in the range of 0.12–0.35. Values of ΔA_{340} recorded for samples with SOD activity will progressively decrease depending on the amount of enzyme in the assay mixture (see Fig. 1, inset). For the calculation, use (sample rate/control rate) $\times$ 100 = % inhibition.

The results of a calibration curve with pure Cu,Zn-SOD from bovine erythrocytes are shown in Fig. 2. Increasing amounts of enzyme (from 0 to 350 ng; abscissa) induce a proportionate decrease in the rate of NAD(P)H oxidation expressed as percentages of the control value (ordinate). With this enzyme preparation, 50% and almost complete saturation levels (98%) were obtained with about 31 and 350 ng lyophilized protein, respectively.

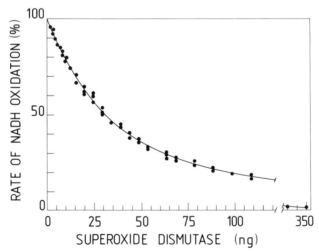

FIG. 2. Calibration curve of pure Cu,Zn-SOD. Increasing amounts (0–350 ng) of pure enzyme from bovine erythrocytes (Boehringer-Mannheim) were assayed for activity under standard conditions as reported in the text, using NADH as the detector. The rates of NADH oxidation at 25° were expressed as percentages of the control. Heat-inactivated SOD up to 350 ng did not affect NADH oxidation at any rate (data not shown). The curve fitting experimental values in the range 10–55 ng was obtained by least-squares regression analysis of SOD amounts transformed into logarithms: $y = 141.2649 - 61.156 \log(x)$, where y is percent inhibition and x is nanograms of enzyme; the correlation coefficient r is -0.9804 and n is 29. The portion of the curve in the range 0–10 ng is linear and represented by the equation $y = 99.974 - 2.003x$, where r is -0.999 and n is 10. The last part of the curve, in the range 55–350 ng, has been traced by hand. With this enzyme preparation, 50% inhibition of NADH oxidation relative to the control was produced by approximately 31 ng of lyophilized protein. Almost complete saturation levels (98%) were obtained with 350 ng.

Calculation

One unit of SOD activity is defined as the amount of enzyme required to inhibit the rate of NAD(P)H oxidation of the control by 50%. However, since calibration curves for all SOD preparations assayed attain saturation levels, the catalytic activity of the enzyme in the system will be equally expressed by either 50% or half-maximal inhibition. This means that the measurement of SOD in a preparation does not require running a full titration curve each time; an estimate of enzyme activity will be easily obtained by means of a single set of assays where serial sample dilutions (at least three) are challenged to reach half-maximal inhibition. Besides, for routine determinations on the same enzyme preparation, it might be convenient to have a calibration curve as a reference. This will allow extrapolation of 50% inhibition values in samples very low in enzyme activity. For practical purposes we have presented the data from the calibration curve of Fig. 2 in Table I, where percentages of inhibition and corresponding units of enzyme are reported together. The conversion

TABLE I
CONVERSION OF PERCENTAGES OF INHIBITION TO UNITS OF ENZYME[a]

%	Units	%	Units	%	Units	%	Units
96	0.064	80	0.323	64	0.590	48	1.078
95	0.080	79	0.335	63	0.613	47	1.119
94	0.096	78	0.348	62	0.636	46	1.162
93	0.112	77	0.362	61	0.661	45	1.207
92	0.128	76	0.375	60	0.686	44	1.253
91	0.144	75	0.390	59	0.712	43	1.301
90	0.160	74	0.405	58	0.740	42	1.350
89	0.176	73	0.421	57	0.768	41	1.403
88	0.192	72	0.437	56	0.797	40	1.457
87	0.208	71	0.453	55	0.828	39	1.513
86	0.225	70	0.471	54	0.860	38	1.571
85	0.241	69	0.489	53	0.893	37	1.631
84	0.257	68	0.508	52	0.927	36	1.694
83	0.273	67	0.527	51	0.963	35	1.759
82	0.289	66	0.547	50	1.000		
81	0.305	65	0.568	49	1.038		

[a] Conversion of percentages of inhibition to units of SOD activity was obtained by solving for x in the following equations: $x = (99.974 - y)/2.003$, in the range 96–81%; and $\log(x) = (141.2649 - y)/61.156$, in the range 80–35%; where y is percent inhibition and x is SOD concentration (ng). Units of SOD activity were calculated by dividing x by 31 ng, which is the amount of pure enzyme yielding 50% inhibition, namely, 1 unit of SOD. These results are also suitable for calculation of SOD activity, with good approximation, in rat liver extracts and in preparations of both Cu,Zn- and Mn-SOD from human leukocytes.

table turns out to be quite useful for quick and accurate calculation of activity; especially in those portions of calibration curve where percentages of inhibition below 50% are difficult to assess.

Proper calculation of enzyme activity should take into account a specific nucleotide oxidation possibly occurring in crude extracts. The contribution of NAD(P)H oxidase to the final rate must be evaluated before mercaptoethanol addition and eventually be subtracted. This precaution is unnecessary when assaying for clear diluted cytosols and pure SOD preparations.

Calibration curves obtained with several pure and crude SOD preparations are practically overlapping. Variations in the slope of the standard curve may result from changes in assay conditions. For a correct estimate of SOD activity in a sample it would be worth referring to a new calibration curve where assay conditions (temperature, sample buffer and volume, reagents, and type of extract) reflect those of routine determinations.

Precautions and Special Conditions for Measurement

Samples to be analyzed should not contain chelators which would alter EDTA/Mn^{2+} ratios required for optimal nucleotide oxidation. Similarly, the presence of endogenous Mn^{2+} will affect the rate of the reaction. Other divalent cations of the second transition series do not interfere directly with NAD(P)H oxidation but could compete with Mn^{2+} for the chelator. Nucleotide oxidation rates are dependent on mercaptoethanol concentration in the assay mixture; therefore, samples should be devoid of free thiols which will increase oxidation rates over control values. To avoid all these problems and those derived from low molecular weight compounds with SOD-like activity, the sample should be dialyzed or desalted before the assay is performed. Several buffers and media (see Preparation of Samples) have been reliably employed as sample solvents.

The concentrations of urea and ethanol (0.6 and 1 M, respectively) in the cuvette are compatible with reliable enzyme measurements. Detergents such as Triton X-100, Brij 35, sodium deoxycholate, Nonidet, digitonin, Tween 20, and SDS, at a final concentration of 0.1%, can be included in the assay mixture without appreciable interferences in the SOD activity determination. However, Triton X-100 has been reported to catalyze NADH oxidation by artificial electron acceptors,[7] and SDS was found to enhance membrane-bound NADPH oxidase activity in cell-free systems.[8] In addition, SDS is also known to inactivate some Mn-SOD

---

[7] U. M. Rao, *Biochem. Int.* **5**, 585 (1982).
[8] Y. Bromberg and E. Pick, *J. Biol. Chem.* **260**, 13539 (1985).

preparations.[9] Therefore, reagents used for either tissue solubilization or enzyme extraction, even without interfering directly with the assay, should be checked in advance to avoid misinterpretation of results.

Both NADH and NADPH are equally susceptible to oxidation in our system. There are crude extracts, however, which contain microsomal or plasma membrane fractions and thus show high levels of NADH or NADPH oxidase activities. In this case it may be worthwhile to run the assay with the nucleotide that is not the favored substrate for sample–oxidase to ensure low rates of aspecific oxidation.

Freshly made solutions of EDTA–MnCl$_2$ show an initial delay in coenzyme oxidation; therefore, as pointed out above, linear kinetics are obtained only 5–10 min after mercaptoethanol addition. Conversely, aged preparations of EDTA–MnCl$_2$ yield almost straight kinetics right from the beginning, thus reducing the time required for the assay. Slight variations in control maximal rates ($\Delta A_{340} \times 8$ min) observed along with storage of the EDTA–MnCl$_2$ reagent are without effect on the calculation of enzyme activity.

The presence of cyanide up to 0.2 mM in the assay mixture is still compatible with the measurement even if it partially slows down the control rate. Conversely, concentrations of cyanide over 1 mM yield almost complete inhibition of nucleotide oxidation and cannot be used in the system. A protocol for differential measurements of SOD activity by cyanide inhibition experiments is described in the following paragraph.

Despite the sensitivity of the method, samples very low in SOD activity (e.g., the levels of both Mn- and Cu,Zn-SOD are greatly reduced in tumor tissues[10]) are difficult to assay. To overcome this problem the sample volume in the assay mixture could appropriately be increased up to 0.8 ml while the buffer (TDB) volume is reduced to 0.1 ml. Naturally, the sample medium should be such as to not substantially affect the final pH in the assay mixture. Whenever samples are diluted in 0.1 M TDB, the TDB in the assay can be omitted, and the procedure is carried out directly on 0.9 ml of sample. The above approach can also be very useful in the determination of enzyme activity in chromatographic fractions when strikingly different levels of SOD isoenzymes are encountered within the same eluate. We have been able to detect and estimate amounts of Mn-SOD as small as 0.1–0.4 U/ml, separated from cell extracts where approximately 97% of the total SOD activity was due to the Cu,Zn isoenzyme.

[9] B. L. Geller and D. R. Winge, *Anal. Biochem.* **128,** 86 (1983).
[10] L. W. Oberley, *in* "Superoxide Dismutase" (L. W. Oberley, ed.), Vol. 2, p. 127. CRC Press, Boca Raton, Florida, 1982.

Enzyme Properties Relevant in Analysis

SODs in fresh tissue extracts are fairly stable, but they loose activity during storage and purification. The Mn-SOD isoenzyme, in particular, shows a varied degree of stability depending on the source from which it is extracted. Rat liver enzyme, for instance, is much more labile than human Mn-SOD.[11] A treatment with organic solvents as suggested for hemoglobin precipitation (ethanol–chloroform, 2 : 1), inactivates Mn-SOD and extracellular (EC) SOD,[12] whereas Cu,Zn-SOD is usually unaffected. This is not a problem when assaying for activity in erythrocytes which contain only the Cu,Zn isoenzyme, but it might be limiting for other cell types. The inactivation of Mn-SOD by SDS has been reported,[9] but in our experience this effect cannot be taken as a general rule for discriminating between Mn- and Cu,Zn-SOD in any extract.

Cyanide is a powerful inhibitor of copper-containing SODs, and this is the basic principle of a widely used procedure for the discrimination between cyanide-sensitive (Cu,Zn and extracellular isoenzyme) and cyanide-insensitive (Mn and Fe isoenzyme) SOD activity in tissue extracts. However, despite the selectivity of the cyanide effect, inhibition experiments do not always provide clear-cut results, owing to the relative amounts of isoenzymes present, the concentration of the inhibitor, and the time of incubation with the extract. There are also differences in both the sensitivity and detection limit as well as in the detection system of several methods available, making it difficult to evaluate the relative contribution of cyanide-sensitive and -insensitive SOD activity in a given sample. We describe herein a procedure which applies to the present method and has proved to be useful for isoenzyme discrimination. The typical experiment requires the incubation of an extract (up to 800 units SOD/ml) at 37° with and without 1 mM NaCN in 0.1 M TDB using stoppered test tubes. After 2 hr samples are assayed for SOD activity, and relative rates of nucleotide oxidation are referred to those of controls containing the same amount of incubation buffers with and without cyanide. The difference in activities between samples with and without the inhibitor corresponds to that arising from the copper-containing SOD, inactivated by the treatment. Inhibition by cyanide of endogenous Cu,Zn-SOD in a rat liver extract and of pure Cu,Zn-SOD added to it as internal standard is over 97%. On the contrary, Mn-SOD activity, estimated following chromatographic separation of the extract, is practically unaffected by cyanide.

[11] F. Paoletti and A. Mocali, unpublished observations (1988).
[12] S. L. Marklund, *Biochem. J.* **220,** 269 (1984).

Comments

The precision of the method has been tested by assaying for 0.5, 1, and 1.5 units of pure Cu,Zn-SOD, yielding a relative standard deviation (RSD) of 1.5, 2, and 2.5%, respectively. The detection limit, ΔA_{340} per 8 min, is about 0.01, corresponding to approximately 2 ng of standard enzyme (Boehringer preparation), and the sensitivity at A_{340} is about 0.005.

Inhibitors of the reaction are Mn^{2+}, uric acid, ascorbic acid, hydrogen peroxide at millimolar levels, other possible superoxide scavengers, and low molecular weight compounds exhibiting SOD-like activity. Since NAD(P)H oxidation is oxygen dependent, the use of deaerated solutions also impairs the efficiency of the assay. Catalase does not interfere with the assay; in fact, it may be included in the mixture to lower hydrogen peroxide levels in the system.

Acid EDTA–Mn solutions increase their activity with time, whereas neutralized preparations, as reported here, yield more stable and constant kinetics during storage. The EDTA–$MnCl_2$ reagent is the real trigger of all the reaction sequences; therefore, variations in EDTA–Mn solutions may lead to differences in NADH oxidation rates. However, for rates within a relatively wide range (see Conditions for Assay) there are no significant changes in the SOD calibration curve.

Most of the inconsistencies in measurement depend on poor dialysis or desalting of samples and on the presence of hemoglobin, as already discussed. The fact that titrations of both crude and pure SOD preparations yield almost overlapping curves confirms the good specificity of the assay and the lack of gross interferences from physiological compounds occurring in tissue extracts. Moreover, the possibility of changing the standard assay conditions and measuring very low SOD activities on a time scale of minutes points out the flexibility of the method. It is also worth recalling that the physiological pH at which the assay is carried out allows a correct estimate of Mn-SOD activity that is not well-expressed in alkaline media as required by other sensitive methods.[13,14]

The assay is based on a purely chemical reaction sequence, and this excludes any possibility of introducing contaminating SOD activity in the system by the use of impure preparations of auxiliary enzymes or cytochrome c. In addition, the sample assayed can be recovered from the cuvette and collected for further analysis after simple dialysis. Both these facts represent clear advantages of the chemical over enzymatic methods.

[13] S. L. Marklund, *J. Biol. Chem.* **251**, 7504 (1976).
[14] H. P. Misra and I. Fridovich, *J. Biol. Chem.* **247**, 3170 (1972).

References Ranges

The specific activities of total SOD from various sources, according to the present method, are tabulated below. These values are intended only to provide references for practical purposes, and a certain degree of variability may be expected depending on the material and treatments employed.

Source	Activity	
Human plasma	0.33	U/mg protein
Human erythrocytes	11.5	U/mg Hb
Human erythrocytes	0.36	U/10^6 cells
Human mononuclear leukocytes	20	U/mg cytosolic protein
Rat liver (perfused)	324	U/mg cytosolic protein
HL-60 (human promyelocytic cells)	37	U/mg cytosolic protein
3T3-BALB (normal fibroblasts)	63	U/mg cytosolic protein
3T3-B77 (transformed fibroblasts)	27	U/mg cytosolic protein
NIH-3T3 KiMSV (transformed fibroblasts)	26	U/mg cytosolic protein
Friend erythroleukemia cells	45	U/mg cytosolic protein
Cu,Zn-SOD (Boehringer; Cat. 837 113)	32,180	U/mg lyophilizate

Acknowledgements

This work was supported by grants from the Ministero della Pubblica Istruzione (60 and 40%) and from the Associazione Italiana per la Ricerca sul Cancro (A.I.R.C.).

[19] Assays for Superoxide Dismutase Based on Autoxidation of Hematoxylin

By JOSEPH P. MARTIN, JR.

Introduction

Superoxide dismutase (SOD) catalyzes the reaction

$$O_2^- + O_2^- + 2 H^+ \rightarrow H_2O_2 + O_2 \tag{1}$$

A variety of spectrophotometric assays for the enzyme have been described. One category of assays combines superoxide-generating systems, such as xanthine oxidase,[1] alkaline dimethyl sulfoxide,[2] and photochemical oxidation of reduced substrates,[3] with an indicating scavenger of superoxide. These scavengers include cytochrome c,[1] hydroxylamine,[4] and nitro blue tetrazolium.[3] In these assays superoxide dismutase activity is quantitated by determining the ability of the enzyme to compete with the scavenger for the available O_2^-. In contrast, autoxidation assays can be performed in which the oxidizing species is both a source of O_2^- and an indicating scavenger for O_2^-. Assays based on the autoxidation of pyrogallol,[5,6] hydroxylamine,[7] 6-hydroxydopamine,[8] and epinephrine[9,10] are simple to perform, involve few reaction components, and are easily monitored in the visible region of the absorption spectrum. One disadvantage of autoxidation assays is that most must be carried out at high pH where the specific activity of manganese and iron SODs is low.[11]

In this chapter, I describe a new autoxidation assay based on the transformation of hematoxylin to its two-electron-oxidized product, hematein. The assay is sensitive and simple. The reaction product, hematein, is relatively stable and has an absorption maximum and extinction coefficient similar to those of the commonly used O_2^- scavenger, cytochrome c. The reaction is inhibited by SOD and can be performed in the physiological pH range of 6.8–7.8 where SOD inhibits the reaction 90–95%. In addition, at pH values above 8.1, the autoxidation reaction is accelerated by added SOD, and the reaction becomes a positive assay for the enzyme, similar in principle to SOD assays based on the photochemical and enzymatic oxidation of dianisidine.[12,13]

Materials and Methods

The activity of purified SOD is determined by the xanthine oxidase–cytochrome c method.[1] Xanthine oxidase is isolated from unpasteurized

[1] J. M. McCord and I. Fridovich, *J. Biol. Chem.* **244**, 6049 (1969).
[2] K. Hyland, E. Voisin, H. Banoun, and C. Auclair, *Anal. Biochem.* **135**, 280 (1983).
[3] C. Beauchamp and I. Fridovich, *Anal. Biochem.* **44**, 276 (1971).
[4] E. Elstner and A. Heupel, *Anal. Biochem.* **70**, 616 (1976).
[5] S. Marklund and G. Marklund, *Eur. J. Biochem.* **47**, 469 (1974).
[6] S. Marklund, *in* "CRC Handbook of Oxygen Radical Research" (R. Greenwald, ed.), p. 243. CRC Press, Boca Raton, Florida, 1985.
[7] Y. Kono, *Arch. Biochem. Biophys.* **186**, 189 (1978).
[8] R. E. Heikkela and F. S. Cabbat, *Anal. Biochem.* **75**, 356 (1976).
[9] H. P. Misra and I. Fridovich, *J. Biol. Chem.* **247**, 3170 (1972).
[10] M. Sun and S. Zigman, *Anal. Biochem.* **90**, 81 (1981).
[11] H. J. Forman and I. Fridovich, *Arch. Biochem. Biophys.* **159**, 396 (1973).
[12] H. P. Misra and I. Fridovich, *Arch. Biochem. Biophys.* **181**, 308 (1977).
[13] H. P. Misra and I. Fridovich, *Anal. Biochem.* **79**, 553 (1977).

cream.[14] General protein is determined by the Bradford assay using bovine serum albumin (BSA) as a standard.[15]

Hematoxylin Assay. Hematoxylin is made up as a 1–5 mM stock solution in 50 mM monobasic potassium phosphate buffer. Reactions are started by adding aliquots of this stock solution to 3 ml of 50 mM potassium phosphate, 0.1 mM ethylenediaminetetraacetic acid (EDTA), at the indicated pH at 25°. Autoxidation is monitored as an increase in the absorbance at 560 nm in an IBM Model 9420 UV–visible spectrophotometer. The rates reported represent those obtained over the first 4 min of reaction. Anaerobic measurements are made in anaerobic quartz cuvettes.[16] The buffer is scrubbed for 25 min with ultrapure nitrogen, and hematoxylin is tipped into the buffer from a sidearm. SOD inhibition studies are carried out by adding aliquots of purified copper zinc superoxide dismutase (Cu,Zn-SOD) solution to the hematoxylin autoxidation assay. The SOD solution is precalibrated using the xanthine oxidase–cytochrome c method.[1]

Results

Hematoxylin Autoxidation. When hematoxylin dissolved in an acidic stock solution is diluted into 50 mM potassium phosphate, 0.1 mM EDTA (pH 7.5, 25°), its autoxidation is detectable as an increase in absorbance between 400 and 670 nm (Fig. 1). The spectra of the oxidized substrate are identical to those of the oxidation product, hematein, and show an absorption maximum at 558 nm. The anaerobic spectrum, measured after 3 hr of incubation, is similar to the aerobic spectrum obtained immediately following dilution. Hematein in 50 mM potassium phosphate, 0.1 mM EDTA (pH 7.5) has an A_{560} extinction coefficient of 27,000 M^{-1} cm^{-1}.

The autoxidation rate, estimated at 560 nm, is first order with respect to hematoxylin concentration and is pH dependent. At pH 6.5 the autoxidation of 66 μM hematoxylin is nearly undetectable, but the rate steadily increases with increasing pH and at pH 8.9 is 0.293 A_{560}/min. Hematoxylin autoxidation at the low end of the pH range is easily observed by increasing the hematoxylin concentration in the assay. For example, at pH 7.2 a rate of 0.022 A_{560}/min is obtained at a hematoxylin concentration of 165 μM, whereas 72 μM gives the same reaction rate at pH 7.5.

Effects of Superoxide Dismutase. SOD inhibits hematoxylin autoxida-

[14] W. A. Waud, F. D. Brady, R. D. Wiley, and K. V. Rajagopalan, *Arch. Biochem. Biophys.* **169,** 695 (1975).

[15] M. Bradford, *Anal. Biochem.* **72,** 248 (1976).

[16] E. K. Hodgson, J. M. McCord, and I. Fridovich, *Anal. Biochem.* **5,** 470 (1973).

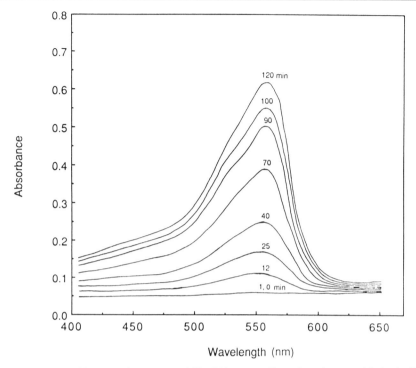

FIG. 1. Repetitive scanning spectra of 22 μM hematoxylin undergoing autoxidation in 50 mM potassium phosphate, 0.1 mM EDTA (pH 7.5, 25°). Scans were initiated at the times indicated. Line 1 represents the spectrum after 3 hr of anaerobic incubation and the aerobic spectrum at zero time. [From J. P. Martin, Jr., M. Dailey, and E. Sugarman, *Arch. Biochem. Biophys.* **255,** 329 (1987).]

tion below pH 8.0. Although initial rates of autoxidation are slightly autocatalytic over the first 4 min of reaction, SOD inhibition can be quantitated by comparing the rates obtained over this interval in the presence of varying concentrations of SOD.

Hematoxylin autoxidation is progressively inhibited by increasing concentrations of Cu,Zn-SOD at pH 7.5 up to a maximum of 92%. When the hematoxylin concentration is adjusted to yield an initial ΔA_{560} of 0.02/min, the reaction is inhibited 50% by the addition of 0.55 μg of Cu,Zn-SOD. This corresponds to 2 units of SOD activity in the xanthine oxidase–cytochrome c assay. Thus, the hematoxylin assay is one-half as sensitive as the cytochrome c method under these conditions.

Transition Metal Effects. Hematoxylin autoxidation is slightly faster in potassium phosphate buffer without EDTA than in the presence of 0.1

mM EDTA. Hematoxylin autoxidation is stimulated by Cu^{2+} and Mn^{3+}, and the rate enhancements caused by these metal cations are not reversed by SOD addition. Cu^{2+}–EDTA and Mn^{3+}–EDTA complexes do not enhance the autoxidation rate. Ferric iron inhibits the autoxidation reaction, but this inhibition is eliminated in buffer containing EDTA. Maximum inhibition by SOD is greater than 90% in the presence of the three metal–EDTA complexes at pH 7.5.

Potential Interfering Factors. Compounds commonly added to SOD assays include cyanide, to inhibit Cu,Zn-SOD and cytochrome oxidase,[17] or sodium azide and hydrogen peroxide, to inhibit Cu,Zn-SOD and iron-SOD.[18,19] Ethanol and ammonium sulfate are often added to SOD assays as extract components during purification of the enzyme.[1,17] Catalase can be added to oxidation assays of pyrogallol, 6-hydroxydopamine, and dianisidine to suppress enzymatic peroxidation of these substrates.[5,6,8,12,13] None of these compounds significantly influences the rate of hematoxylin autoxidation. However, the intracellular reducing agents NADH, ascorbate, and reduced glutathione and the reducing agent dithiothreitol completely inhibit the autoxidation reaction, as do crude undialyzed extracts of *Escherichia coli* B. Most SOD assays are susceptible to interferences by reductants. Biological reductants can be removed from crude cell extracts by dialysis. When extracts of *E. coli* B are dialyzed prior to assay, the inhibition of hematoxylin autoxidation by the extracts is dependent on the amount of SOD activity they contain.

Effects of pH on SOD Inhibition. Although SOD inhibits hematoxylin autoxidation more than 90% below pH 7.9, the observed inhibition diminishes precipitously between pH 7.9 and 8.1 (Fig. 2A). Above pH 8.1 SOD actually enhances the autoxidation rate (Fig. 2B), and this enhancement is maximal at pH 8.3. The degree of maximum rate enhancement is dependent on the level of SOD (Fig. 3). At pH 8.3 approximately 34 cytochrome *c* units of Cu,Zn-SOD are required to double the reaction rate, and a maximum enhancement of 5.3-fold is obtained on addition of 480 SOD units to the reaction. Cu,Zn-SOD, inactivated by boiling, has no effect on the autoxidation rate at pH 8.3. Acceleration of the reaction rate is proportional to SOD concentration up to approximately 200% (Fig. 3, inset). Similar rate enhancements by SOD have been observed during autoxidation of the antitumor antibiotic 9-hydroxyellipticene[20] and during enzymatic[13] and photochemical[12] oxidation of dianisidine.

[17] R. A. Weisiger and I. Fridovich, *J. Biol. Chem.* **248**, 3582 (1973).
[18] E. K. Hodgson and I. Fridovich, *Biochemistry* **14**, 5294 (1975).
[19] H. P. Misra and I. Fridovich, *Arch. Biochem. Biophys.* **189**, 317 (1978).
[20] C. Auclair, K. Hyland, and C. Paoletti, *J. Med. Chem.* **26**, 1438 (1983).

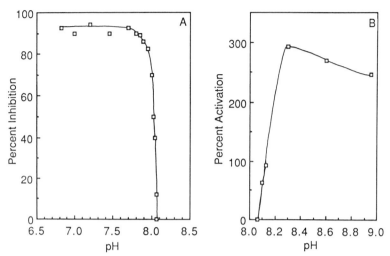

FIG. 2. Effect of Cu,Zn-SOD on the hematoxylin autoxidation rate as a function of pH: (A) inhibition, (B) activation. Eighty cytochrome *c* units of Cu,Zn-SOD was added to reactions containing 66 μM hematoxylin. Other assay conditions were as described in the text. [From J. P. Martin, Jr., M. Dailey, and E. Sugarman, *Arch. Biochem. Biophys.* **255**, 329 (1987).]

Discussion

Hematoxylin autoxidation forms the basis for two simple and convenient assays for SOD activity which complement existing methods. Hematein, the immediate oxidation product, possesses a bright red chromophore with a well-defined absorption maximum. Although hematein eventually decomposes into secondary products, the time course for its disappearance is slow compared to its rate of formation through autoxidation. The sensitivity of the hematoxylin method may be adjusted simply by changing the concentration of hematoxylin in the assay. The negative assay is useful between pH 6.8 and 7.8, and the maximum inhibition that may be obtained with SOD is constant and greater than 90% within this range. Hematoxylin is not toxic. Moreover, it is widely available, chemically stable at room temperature, and inexpensive. Hematoxylin autoxidation is not inhibited by components frequently added to SOD assays in order to distinguish among the three types of SOD. Although the autoxidation is inhibited by reducing agents, this problem is shared by other autoxidation assays and by assay methods which employ ferricytochrome *c* and nitro blue tetrazolium as indicating scavengers. This difficulty may be avoided by dialyzing protein extracts before assay. Hematoxylin autoxidation provides a sensitive negative assay at low pH, and at high pH it

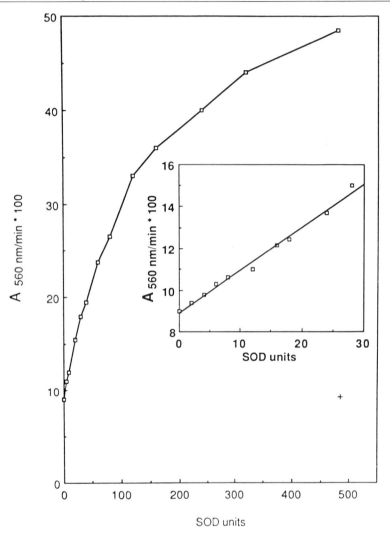

FIG. 3. Enhancement of hematoxylin autoxidation by Cu,Zn-SOD. Cu,Zn-SOD was added to reaction assays containing 66 μM hematoxylin at pH 8.3. Other assay conditions were as described in the text. The reaction rate obtained after the addition of 480 units of SOD inactivated by boiling is indicated by a plus symbol (+). Inset: Activation of hematoxylin autoxidation by up to 28 units of SOD.

can be used as a positive assay for SOD activity in which the autoxidation rate is stimulated in proportion to the level of added SOD. The positive method is an order of magnitude less sensitive than the negative assay, but it may be applied in situations in which the SOD concentration is high, such as during purification of SOD.

Acknowledgements

This work was supported by Public Health Service Research Grant AI-19695 from the National Institute of Allergy and Infectious Diseases; Grant C-900 from the Robert A. Welch Foundation; and a grant from the American Heart Association, Texas Affiliate.

[20] Assay for Superoxide Dismutase Based on Chemiluminescence of Luciferin Analog

By MINORU NAKANO

Principle

The methods for estimating superoxide dismutase (SOD) concentrations in erythrocytes and tissues are based on the inhibitory effect of SOD extracted from these materials. Using 2-methyl-6-(p-methoxyphenyl)-3,7-dihydroimidazo[1,2-a]pyrazin-3-one (MCLA), the chemiluminescence induced by a hypoxanthine–xanthine oxidase system (an O_2^--generating system) is measured and the inhibition determined.

Preparation of 2-Methyl-6-(p-methoxyphenyl)-3,7-dihydroimidazo[1,2-a] pyrazin-3-one Hydrochloride

MCLA can be synthesized from 2-amino-5-(p-methoxyphenyl) pyrazine according to the following procedure.[1] To a solution of 2-amino-5-(p-methoxyphenyl)pyrazine (100 mg) and 40% aqueous methylglyoxal (0.16 ml) in methanol (4.1 ml) is added concentrated hydrochloric acid (0.1 ml). The mixture is heated and stirred under an atmosphere of nitrogen at 60–70° for 1.5 hr. After drying *in vacuo*, the product mixture is partially purified by silica gel chromatography. The crude product which is eluted from the column in 2-propanol is recrystallized from benzene–methanol to give a pale yellow crystalline power (89 mg, 70%), mp 197–201° (dec.). Analysis: Calc. for $C_{14}H_{14}O_2N_3Cl$: C, 57.64; H, 4.84; N,

---

[1] A. Nishida, H. Kimura, M. Nakano, and T. Goto, *Clin. Chim. Acta* **179**, 177 (1989).

14.40%. Found: C, 57.62; H, 4.70; N, 14.39%. UV_{max} (MeOH), nm (ε): 430 (8260), 357 (5460), 261 (20800), 203 (20900); (MeOH/HCl): 348 (5080), 278 (21700), 208 (18600); (MeOH/NaOH): 409 (6000), 348 (5530), 264 (22900), 207 (20900). Mass spec. (EI): m/z 255 (M^+), 240, 226, 186, 171. ^1H NMR (DMSO-d_6 + D_2O), δ: 8.05 and 7.70 (each $1H$, s), 7.69–7.73 ($2H$, $A_2'X_2'$, $J = 9$ Hz), 7.04–7.11 ($2H$, $A_2'X_2'$, $J = 9$ Hz), 3.82 ($3H$, s), 2.36 ($3H$, s). The $\varepsilon_{430\ nm}$ value of 9600 $M^{-1}cm^{-1}$ in water can be used for preparation of the MCLA solution.

Preparation of SOD Samples

Preparation of Erythrocyte Lysates. Blood is drawn into heparinized syringes (20 units/ml) from human volunteers or animals. Samples are prepared according to Oyanagui.[2] Heparinized blood (1–3 ml) is centrifuged at 2500 rpm for 30 min at 0–4° and the plasma carefully separated. Saline is added to the erythrocyte pellet up to the original volume, and 0.4 ml of the cell suspension is transferred to a 50-ml centrifuge tube. After washing the erythrocytes twice with saline, the cells are diluted with water to 4 ml to lyse the erythrocytes. Ethanol (1 ml) and chloroform (0.6 ml) are then added to remove hemoglobin (Tsuchihashi method[3]). The tubes are shaken vigorously for 15 min and centrifuged at 2500 rpm for 10 min at 0–4°. Exactly 0.1 ml of the water–ethanol layer is aspirated and diluted with 0.7 ml of water to make a final concentration of 1/1000 of whole blood cells and 0.25% ethanol. Ethanol (2.5%) is used for further dilution of erythrocyte lysates to obtain a 0.25% solution.

Granulocytes and Lymphocytes from Human Blood. Granulocytes and lymphocytes are obtained by the methods described by Boyum.[4] Granulocytes and lymphocytes are over 95% homogeneous. Washed granulocytes obtained from 1 ml whole blood or washed lymphocytes obtained from 3 ml whole blood are resuspended in 1 ml of Hanks' balanced salt solution and are sonicated over ice in 30 consecutive 0.5-sec bursts at a power setting of 30 W (Branson sonifier). The sonication is repeated after 1 min. The sonicated sample is centrifuged at 78,000 g for 60 min, and the clear supernatant is used for the assay of total SOD. An aliquot (0.4 ml) of the supernatant is further treated with sodium dodecyl sulfate (SDS) to inactivate Mn-SOD activity (see below).

Tissues from Rats. Rats, weighing between 230 and 250 g, are anesthetized with ketamine (60 mg/kg) and subjected to a gradual perfusion with

[2] Y. Oyanagui, *Anal. Biochem.* **142,** 290 (1984).
[3] M. Tsuchihashi, *Biochem. Z.* **140,** 65 (1923).
[4] A. Boyun, *Scand. J. Clin. Lab. Invest.* **21** (Suppl. 97), 77 (1967).

lactate–Ringer solution (pH 7.0) containing heparin (20 units/ml), allowing outflow from the decending aorta through femoral vein, and sacrificed by decapitation. Each organ is immediately removed, weighed, and homogenized in 9 volumes (w/v) of buffered sucrose solution (0.25 M sucrose in 20 mM Tris-HCl buffer containing 1 mM EDTA at pH 7.4). The homogenate is then sonicated over ice in 30 consecutive 0.5-sec bursts at a power setting of 60 W (Branson sonifier). This procedure is repeated 4 times at 1-min intervals. The sonicated sample is centrifuged at 78,000 g for 30 min, and the supernatant is diluted with 50 mM Tris-HCl buffer at pH 7.4 to achieve a 1000- to 5000-fold dilution of the wet tissue. An aliquot of the sample is further treated with SDS to inactivate Mn-SOD.

Differentiation of Cu,Zn-SOD from Mn-SOD. Several methods have been used to distinguish Cu,Zn-SOD activity from Mn-SOD activity in whole tissues or blood cell extracts. These are the Tsuchihashi reagent, N,N'-diethyldithiocarbamate, sodium dodecyl sulfate (SDS), and CN^-. Only the SDS method proved workable with MCLA chemiluminescence.

Inactivation of Mn-SOD activity in the sample by SDS is accomplished as described by Bruce and Dennis.[5] SDS is added to the diluted tissue extract or to a mixture of bovine erythrocyte Cu,Zn-SOD and human liver Mn-SOD in 50–100 mM Tris-HCl buffer at pH 7.4 to obtain a final concentration of 2% (w/v), and the mixture is incubated at 37° for 2 hr to inactivate Mn-SOD. At the end of the incubation, the solution is cooled to 4° and mixed with 1/10 volume of 3 M KCl to precipitate the SDS. After standing for 30 min at 4°, the mixture is centrifuged at 20,000 g for 10 min at 4°. The supernatant obtained is used to assay for Cu,Zn-SOD.

Assay of Xanthine Oxidase Activity

Enzyme activity can be determined in glycine-HCl buffer at pH 8.8 and 37°, using hypoxanthine as substrate.[6] One unit of enzyme is defined as the amount of the enzyme that catalyzes an increase in absorbance of 0.001/30 min at 290 nm.

Reagents

30 μM MCLA in water (prepared from a stock solution of 300 μM MCLA)

1.5 mM hypoxanthine in water

Xanthine oxidase (grade III, buttermilk, Sigma Chemical Co., St. Louis, MO)

[5] L. D. Bruce and R. W. Dennis, *Anal. Biochem.* **123,** 86 (1983).
[6] G. G. Roussos, this series, Vol. 12A, p. 5.

Superoxide dismutase, bovine erythrocyte SOD (3500 units/mg of protein)

0.3 M Tris-HCl buffer containing 0.6 mM EDTA at pH 7.8

Procedure. A standard reaction mixture contains 20 μl of MCLA, 100 μl of hypoxanthine, 50 μl of 6.5 units of xanthine oxidase, 6–200 μl purified bovine erythrocyte SOD or sample (or none), and 500 μl of the Tris-HCl buffer containing EDTA and water, for a total volume of 3.0 ml. The chemiluminescence measurement is initiated by the addition of MCLA to the standard incubation mixture excluding xanthine oxidase, continued for 4 min without additions, and then continued for an additional 4 min after the addition of xanthine oxidase. Chemiluminescence can be measured with a Luminescence Reader (Aloka, BLR 102) at 25° and expressed as counts per minute.

Preparation of the standard curve for MCLA in the standard incubation mixture excluding xanthine oxidase and SOD (control system) is based on the light emitted in the visible region (λ_{max} 465 nm). Such luminescence originates from the autooxidation of MCLA and is a nonspecific chemiluminescence. The addition of xanthine oxidase to the above mixture (experimental system) rapidly enhances the luminescence, which reaches a maximum and then remains constant for at least 3 min. The luminescence in the system containing xanthine oxidase decreases with increasing SOD concentration. However, nonspecific luminescence is almost constant for 10 min after the addition of MCLA and is not significantly influenced by SOD.

The xanthine oxidase-induced luminescence (I_0), expressed in terms of light intensity (counts/min), can be calculated by subtraction of the nonspecific light intensity at 8 min (after the addition of MCLA) from the light intensity at 2 min (after the addition of xanthine oxidase). The same incubation experiments, except that SOD is present both in the experimental and the control systems, are carried out, and the enzyme-induced luminescence (I_i) is calculated in the same manner as described above. The percentage of SOD inhibition dependent on the xanthine oxidase-induced luminescence (I_0) can be calculated from Eq. (1) and is then plotted against the SOD concentration to obtain the standard curve.

$$\% \text{ Inhibition} = (I_0/I_i)/I_0 \times 100 \tag{1}$$

The SOD concentration for 50% inhibition of the xanthine oxidase-induced luminescence has been compared with that obtained by the cytochrome c method.[7] The luminescence method may give 95 times higher sensitivity than the cytochrome c method.

---

[7] J. McCord and I. Fridovich, *J. Biol. Chem.* **244**, 6049 (1969).

TABLE I
SOD CONCENTRATIONS IN HUMAN BLOOD CELLS[a]

Cell	Total SOD	Cu,Zn-SOD	Mn-SOD	Mn-SOD/Total SOD
Erythrocyte	1.43 ± 0.11	1.43 ± 0.11	0	0
Granulocyte	1.22 ± 0.12	1.09 ± 0.16	0.13 ± 0.05	10.7 ± 5.1%
Lymphocyte	5.66 ± 0.42	4.57 ± 0.53	1.10 ± 0.31	19.5 ± 5.5%

[a] Data are expressed as ng/10^5 cells and are means ± standard deviations of five observations on samples measured in duplicate.

TABLE II
SOD CONCENTRATIONS IN RAT TISSUES[a]

Tissue	Total SOD	Cu,Zn-SOD	Mn-SOD	Mn-SOD/Total SOD
Brain	133.4 ± 31.0	94.4 ± 16.5	39.0 ± 26.0	27.7 ± 14.0%
Lung	115.6 ± 23.8	84.8 ± 10.2	30.8 ± 16.7	25.6 ± 8.6%
Heart	191.2 ± 32.0	106.2 ± 20.8	85.0 ± 18.6	44.5 ± 7.0%
Kidney	516.0 ± 43.2	413.0 ± 60.5	103.0 ± 53.2	19.1 ± 10.2%
Liver	819.2 ± 113.2	603.2 ± 158.8	216.0 ± 51.0	27.4 ± 9.9%

[a] Data are expressed as μg/g wet weight and are means ± standard deviations of five observations of samples measured in duplicate.

Assay of SOD in Sample

The standard reaction mixture is used for the assay of SOD in the sample. Cu,Zn-SOD and Mn-SOD activities in human blood cells and rat tissues determined by the MCLA method are shown in Tables I and II. All SOD activities are expressed as nanograms (ng) or micrograms (μg) of bovine erythrocyte SOD based on a specific activity of 3500 units/mg protein. The values for human blood (Cu,Zn-SOD activities in erythrocyte lysate) have been quoted from the previous report.[8]

Cautions Regarding the Assay

The xanthine oxidase-induced luminescence is inhibited by purified bovine serum albumin,[8] probably also by purified human albumin or other protein, at concentrations of more than 5 μg/ml of the reaction mixture. Protein concentrations in samples used for the assay are less than 1.2 μg/ml, which would not interfere with the assay of SOD by the lumi-

[8] H. Kimura and M. Nakano, *FEBS Lett.* **239**, 347 (1988).

nescence method. Glutathione, ascorbate, uric acid, and glucose are also inhibitory of the XOD-induced luminescence at concentrations greater than 4×10^{-6}, 3×10^{-8}, 1×10^{-5}, and 6×10^{-4} M, respectively. However, the concentration of each inhibitor in the reaction mixture corresponds to more than 1 100-fold dilution of the concentration which interferes with the SOD assay. Thus, it is not necessary in the SOD assay to consider the interference by contaminants in the biological samples, provided they are adequately diluted.

MCLA in solution is relatively unstable and autoxidizes with light emission at 465 nm. The stock solution of 300 μM MCLA in water (1 ml) should be stored at $-80°$. The solution is then thawed, diluted with water to obtain 15 μM MCLA, and should be used as soon as possible thereafter.

[21] Automated Assay of Superoxide Dismutase in Blood

By MARY R. L'ABBÉ and PETER W. F. FISCHER

The analysis of copper zinc superoxide dismutase (Cu,Zn-SOD) in erythrocytes and extracellular SOD in plasma must be carried out by an indirect method since the substrate ($O_2^{\cdot -}$) is an unstable free radical. The $O_2^{\cdot -}$ is generated by xanthine plus xanthine oxidase. The SOD activity is calculated from the rate at which the generated $O_2^{\cdot -}$ reduces cytochrome c, a reaction which is followed spectrophotometrically. If the activity is high, more of the generated radicals are dismutated and less cytochrome c is reduced, whereas if the activity is low, the opposite is true.[1] The method is adapted to an Abbott VP Super System bichromatic analyzer, and, since it uses small volumes and allows for a relatively large number of samples, it is suitable for use in both clinical laboratories and research laboratories utilizing small animals.[2]

Apparatus

Abbott VP Super System bichromatic discrete analyzer (Abbott Laboratories Ltd., Diagnostics Division, Mississauga, Ontario, Canada) equipped with a 415/450 nm filter. The Abbott ABA-200 bichromatic analyzer has also been used.[2]

[1] J. M. McCord and I. Fridovich, *J. Biol. Chem.* **224**, 6049 (1969).
[2] M. R. L'Abbé and P. W. F. Fischer, *Clin. Biochem.* **19**, 175 (1986).

Reagents

Sodium carbonate buffer (20 mM, pH 10.0) containing 0.1 mM EDTA is prepared by dissolving 2.12 g Na_2CO_3 in distilled water and bringing the volume to 1 liter (A), and by dissolving 1.68 g $NaHCO_3$ in a similar fashion (B). Solutions A and B are mixed to obtain the buffer. EDTA, 0.372 g (disodium salt, Analar analytical reagent, BDH Chemicals Ltd., Toronto, Ont.), is dissolved in 1 liter of mixed buffer. Ferricytochrome c (50 μM) is prepared by dissolving 123.84 mg (type III from horse heart, Sigma Chemical Co., St. Louis, MO) in sodium carbonate buffer containing EDTA and bringing the volume to 20 ml. Xanthine stock solution (1.0 mM) is prepared by dissolving 15.21 mg of xanthine (purified grade, Fisher Scientific Co., Fairlawn, NJ) in sodium carbonate buffer containing EDTA and bringing the volume to 100 ml.

Xanthine oxidase (~15 mU/ml) is prepared fresh each day by adding approximately 10 μl of xanthine oxidase (EC 1.1.3.22, grade III, from buttermilk, 29.4 U/ml, Sigma) to 20 ml of sodium carbonate buffer containing EDTA. As the activity of the xanthine oxidase varies, sufficient enzyme is used to produce a rate of cytochrome c reduction resulting in a change in A_{415} of 0.025 per minute in the assay system without any added SOD. The xanthine oxidase is kept in an ice bath during the assay.

The reagent mixture is prepared by combining 93 ml of the sodium carbonate buffer, containing EDTA, 2 ml of cytochrome c solution, and 5 ml of xanthine solution. The reagent mixture is stable for several weeks if stored refrigerated. Standards are prepared fresh daily by diluting a stock solution of bovine liver SOD, 15,000 U/ml (Sigma), to a working range of 0.75 to 15 U/ml. One unit is defined as the activity that inhibits the rate of cytochrome c reduction by 50% under the conditions specified for a particular system in a 1 ml reaction volume.[1] SOD extraction solution is prepared by mixing 150 ml of chloroform and 250 ml of ethanol.

Sample Preparation

Erythrocytes. Blood is collected in a heparinized microhematocrit capillary tube (75 × 1.5 mm) and stored in an ice bath. The tube is centrifuged in an Autocrit II centrifuge (Fisher) at 13,000 g for 3 min. The tube is cut above the lower sealant and below the buffy coat layer, and the length of this section is recorded. The section of capillary tube containing the erythrocytes is placed in a 2-ml screw-capped tube containing 1.0 ml of cold isotonic saline. The erythrocytes are washed out of the capillary tube by shaking. The piece of tube is discarded and the cells are centrifuged at 15,600 g in a microcentrifuge (Brinkmann Instruments, Rexdale, Ont.) for 2 min at 4°. The supernatant is discarded, and the cells are lysed by the addition of 1.0 ml of cold distilled water. Cold chloroform–ethanol extraction solution (0.4 ml) is added, and the mixture is agitated using a vortex mixer. After centrifugation at 15,600 g for 3 min in the microcentrifuge, the upper aqueous layer is carefully decanted and retained for assay of the SOD activity. The aqueous phase can be stored frozen at $-70°$ for several months without any change in activity. The erythrocyte samples, prior to lysing, can be stored frozen for several weeks with no apparent loss of activity.[2]

Plasma. Rat plasma SOD activity is determined using the plasma in the microhematocrit tube. The section of the tube containing the plasma is cut above the buffy layer and is placed in a tube containing 0.4 ml of isotonic saline. The plasma is rinsed out of the capillary tube by shaking. Human plasma SOD activity is measured on undiluted samples collected in conventional ways because the activity is lower than the rat activity and the microhematocrit tube does not yield sufficient sample volume. Plasma samples do not require extraction and can be analyzed directly. Plasma samples can be stored frozen for several months without apparent loss of activity.

Assay Procedure

Buffer alone is placed in positions 01 to 04 of the carousel, followed by the standards and the samples. The assay reagent mixture is connected to the instrument reagent line, and the cold xanthine oxidase solution is connected to the auxiliary line. The instrument is programmed as shown in Table I.

Calculations

The reading obtained for position 01 is added to the readings for positions 02–04, so that four values are obtained (that for position 01 plus the three sums). These are the changes in absorbance caused by the reduction of cytochrome c in the absence of SOD, or the baseline ΔA. The mean of the baseline ΔA is calculated (ΔA_i) and should be approximately $0.050A$ per 2 min revolution. The change in absorbance for each sample (ΔA_s) is equal to ΔA_i plus the change in absorbance obtained from the instrument for each sample. This latter value is negative since the instrument automatically subtracts ΔA_i. The percentage of the original rate of cytochrome c reduction (Rate %) is calculated for each sample as follows:

$$\text{Rate \%} = \Delta A_s / \Delta A_i \times 100 \tag{1}$$

Rate % is plotted against the activity of the standards on semilogarithmic graph paper (Fig. 1). The activity of the samples is read directly from

TABLE I
SETTINGS FOR ABBOTT VP FOR ANALYSIS OF SOD
IN ERYTHROCYTES OR PLASMA

Parameter	Setting
Units of measure	Absorbance
Sample volume (μl)[a]	(1.25–2.5)
Reagent volume (μl)	250.0
Filter	415/450
Reaction type	Rate
Direction	Up
Analysis time (min)	2
Number of revolutions	3
Temperature (°)	30
First revolution read	Yes
Auxillary volume (μl)	10.22
Assay factor	Equivalent to filter factor[b]

[a] Sample volume is dependent on the type of sample being analyzed: 1.25 μl for rat and human erythrocyte lysate, 2.5 μl for diluted rat and undiluted human plasma.

[b] The instrument determines the filter factor from the filter used.

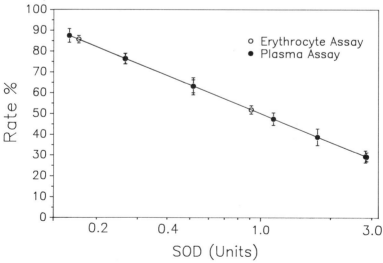

FIG. 1. Typical calibration curve of the automated SOD method. Rate % is equivalent to the rate of cytochrome c reduction observed with samples or standards divided by the rate in the absence of SOD multiplied by 100.

the graph. Alternatively, the activity can be calculated using Eq. (2):

$$\text{Rate \%} = a + b \ln(\text{SOD activity}) \tag{2}$$

where, by definition, Rate % is 50 when SOD activity is 1 unit, so that a equals 50. The slope, b, is calculated from the standard curve. The activity of the unknowns is therefore determined with Eq. (3) or (4):

$$\text{Activity} = e^{(\text{Rate \%}-50)/b} \, df \quad \text{U/ml plasma} \tag{3}$$
$$\text{Activity} = e^{(\text{Rate \%}-50)/b} \, df(1000/vol. \, RBC, \mu l) \quad \text{U/ml packed cells} \tag{4}$$

for plasma or erythrocytes, respectively, where df is the dilution factor calculated by the instrument. Alternatively, the activity can also be expressed as U/mg hemoglobin for erythrocytes and as U/ml or U/mg of protein for plasma.

Comments

The coefficient of variation for SOD in plasma ranged from 3.4 to 3.9% within runs and was 4.5% when a single plasma sample was analyzed over a period of 6 months. The within-run variability for erythrocyte SOD ranged from 3.5 to 5.3%, while variation for analyses performed over 3 days was 5.8%.[2] Recovery of known activities of SOD ranged from 92 to 101% with an overall recovery of 98%.[2] The method is linear for both the erythrocyte assay and the plasma assay from 0.2 to 3.0 units, which corresponds to 0.5 to 10 ng of purified bovine liver SOD (Fig. 1). Typical values for erythrocyte samples from humans, rats, and monkeys are given in Table II.

TABLE II
ERYTHROCYTE SOD IN VARIOUS SPECIES[a]

Species	Sex	n	Activity		U/mg hemoglobin
			U/ml packed cells $\times 10^2$		
			Mean ± SD	Range	
Rat	F	11	331 ± 31	262–367	100 ± 9
Human	M	25	339 ± 44	255–414	107 ± 14
Monkey	M	20	107 ± 20	73–149	37 ± 7
	F	20	100 ± 21	66–144	34 ± 7

[a] Reprinted with permission from M. R. L'Abbé and P. W. F. Fischer, *Clin. Biochem.* **19,** 175 (1986).

It is possible to prepare 40 to 60 erythrocyte samples for analysis per day. The automated method allows for the analysis of 25 samples in 6 min in contrast to time-consuming manual methods that require large sample volumes.[3] In the present method, the pH for the assay was increased to 10 from 7.8,[4] thus increasing the sensitivity of the assay for Cu,Zn-SOD 17 times. The reduction of cytochrome c was followed at the more sensitive 415 nm compared to 550 nm used in most assays.[1,3] With these modifications, the sensitivity of the assay was increased 28-fold. Using this method, a standard preparation of bovine liver SOD was found to have an activity of 90,560 U/mg of protein compared to the reported activity of 3200 U/mg. Thus, the units of activity calculated with the present assay are equivalent to 0.035 U reported by Sigma. This does not affect the use of bovine liver SOD as standards since they are used only to calculate the slope. A unit of SOD activity is arbitrarily defined as that activity which will decrease the rate of cytochrome c reduction by 50% under standard assay conditions.

[3] L. Flohé and F. Ötting, this series, Vol. 105, p. 93.
[4] C. O. Beauchamp and I. Fridovich, *Biochim. Biophys. Acta* **317,** 50 (1973).

[22] Assay of Superoxide Dismutase Applicable to Whole Bacterial Cells

By F. S. Archibald

Since 1968 a wide variety of assays for the quantitation of superoxide dismutase (SOD) activity have been proposed and employed. Most are indirect assays requiring both a generator of a known flux of O_2^- and a chemical detector of the flux. SOD is then quantitated indirectly in a sample by the ability of that sample to reduce the detected O_2^- flux. Such assays have proved suitable for purified SOD and for most crude, dialyzed broken cell preparations. The present alternatives to this sort of indirect SOD assay, for example, direct viewing of O_2^- in the far-UV, electron paramagnetic resonance using free radical spin traps such as DMPO (5,5′-dimethyl-1-pyrroline 1-oxide), ^{19}F NMR, stopped-flow spectroscopy, and chemiluminescence, are not practical with the usual broken cell extracts employed and SOD levels found in biological systems.

Given the need for a controllable, known flux of O_2^- and known amounts of an O_2^- detector in all present assays for SOD in crude cell extracts, it is probably not possible to develop an "intact cell" bacterial

SOD assay because (1) the cytoplasmic membrane (phospholipid bilayer) is impermeable to O_2^- and most macromolecules and (2) SOD is always found within the cytoplasmic membrane. Thus, a usable whole cell assay must permeabilize the cells, allowing entry of O_2^- and the O_2^- detector or egress of SOD.

Unfortunately, many bacteria, especially gram-positive rods and cocci have heavy walls, and are broken quantitatively only by harsh, labor-intensive methods such as the hydraulic (French) press alone or combined with freeze–thaw or ball and glass bead mills. This creates severe problems in quantitating SOD in genetic and other studies that produce large numbers of small samples. It also prevents the screening of large numbers of bacterial colonies or clones on solid media for individuals unusually high or low in SOD. Lacking heavy walls, most animal cells readily lyse under mild physical, osmotic, or chemical insult, reducing the problem, but for many plant and protistan cell types (e.g., algae and fungi) the problem of accessibility also exists.

Assay of Toluenized Whole Bacterial Cells for Superoxide Dismutase

Treatment of whole prokaryotic cells with toluene results in loss of membrane integrity and exit of cytoplasmic proteins and small molecules, although the cells often retain a normal microscopic appearance. Toluene permeabilization has been used for the quantitation of specific cytoplasmic proteins in a number of prokaryotes and yeasts.[1-5] The following is a modification of the light- and riboflavin-driven hydroxylamine–nitrite SOD assay of Whitelam and Codd[5] as used on the toluenized cells of several bacterial species.

Method. Cells are harvested from agar plate or broth cultures, washed with an appropriate buffer (e.g., 25 mM Tris base, 1 mM sodium EDTA, pH 8.0, for neisserial species), and resuspended to a cell protein concentration of 0.1–1.0 mg/ml. To 4 ml of the cell suspension is added 2 ml of toluene. The suspension shaken for 5 min, placed on ice, and the aqueous phase recovered. A 0.2 ml sample of the aqueous toluenized cell phase is added to a tube containing 5 ml of the Tris buffer, 10 μM flavin mononucleotide (FMN), and 5 mM hydroxylamine. The tubes are incubated (25°) under uniform cool white fluorescent light (~80 foot-candles or 860 lux)

1 R. Serrano, H. M. Gancedo, and D. Gancedo, *Eur. J. Biochem.* **34**, 479 (1973).
2 F. R. Tabita, P. Caruso, and W. Whitman, *Anal. Biochem.* **84**, 462 (1978).
3 E. Flores, M. G. Guerrero, and M. Losada, *Arch. Microbiol.* **128**, 137 (1980).
4 F. S. Archibald and M.-N. Duong, *Infect. Immun.* **51**, 631 (1986).
5 G. C. Whitelam and G. A. Codd, *Anal. Biochem.* **121**, 207 (1982).

for 25 min on a white background. The incubation area should be checked with a photometer for uniformity of illumination.

A 1.0 ml sample of each assay mixture is assayed for nitrite formation by the addition of 1.0 ml of 20 mM sulfanilic acid and 1.0 ml of 7 mM α-naphthylamine, both in 4.2 M acetic acid. These reagents are made by dissolving 0.5 g α-naphthylamine in 100 ml boiling water, to which, after cooling to 25°, is added 125 ml acetic acid, and the volume is made up to 500 ml with cold water; the reagent is stored in the dark. The sulfanilic acid (1.7 g) is dissolved in 375 ml of warm water, and 125 ml of acetic acid is added. The assay mixture is incubated for 20 min, and the colored diazo complex is quantitated at 530 nm on a good spectrophotometer. Initial controls should include several concentrations of purified SOD protein, boiled cells, buffer blanks, and a SOD assay on mechanically broken cells to check the efficiency of SOD release by toluene from the cell type being assayed.

Principle. The assay depends on the production of O_2^- by light, FMN, and EDTA and by the O_2^--mediated oxidation of hydroxylamine to nitrite by O_2^-. The oxidation of hydroxylamine occurs in two steps, both of which are sensitive to inhibition by SOD,

$$NH_2OH + O_2^- + H^+ \rightarrow \cdot NHOH + H_2O_2 \tag{1}$$
$$\cdot NHOH + O_2^- \rightarrow NO_2^- + H_2O \tag{2}$$

The nitrite produced is quantified in a conventional assay in which NO_2^- reacts with sulfanilic acid in an acid solution to form a diazo compound. This compound is then reacted with α-naphthylamine to form a red complex readily measured at 530 nm. Care should be exercised in the handling of α-naphthylamine as it has been identified as a carcinogen. The system will detect NO_2^- levels ranging from less than 10 to 180 ng/ml.

Caveats. Although a number of workers have reported the hydroxylamine–NO_2^- reaction to be a useful detector in crude extracts,[4-8] Bielski *et al.*[9] have presented evidence from pulse radiolysis studies that the initial reaction of O_2^- and hydroxylamine is in fact much slower (in a defined buffer–formate–water system) than it appears to be in actual samples. They speculate that this discrepancy is due to modulation of the initial reaction rate by metals, namely, the observed sample inhibition of NO_2^- formation may be dependent on a factor other than the SOD content of the sample. However, comparison of the results with those obtained using other SOD assay methods, demonstration of SOD-inhibitable

[6] E. F. Elstner and A. Heupel, *Anal. Biochem.* **70,** 616 (1977).

[7] Y. Kono, *Arch. Biochem. Biophys.* **186,** 189 (1978).

[8] W. Bors, C. Michel, and M. Saran, *Z. Naturforsch., C: Biosci.* **33,** 891 (1978).

[9] B. H. J. Bielski, R. L. Arudi, D. E. Cabelli, and W. Bors, *Anal. Biochem.* **142,** 207 (1984).

and O_2^--dependent hydroxylamine oxidation in metal-poor Tris-buffered controls, and the absence of hydroxylamine oxidation inhibition in boiled cell assay controls[4-8] all argue for the observed NO_2^- production being dependent on O_2^- and largely or entirely inhibitable by SOD.

Alternative Assays

The NADH, PMS (phenazine methosulfonate), and NBT (nitro blue tetrazolium) O_2^- generator and detector system described by Nishikimi *et al.*[10] can reportedly be substituted for the light-driven FMN–EDTA hydroxylamine system for toluenized cells described above with satisfactory results.[5] In this system NADH reduces PMS which autoxidizes in the presence of O_2 to yield O_2^-. This NADH–PMS-driven reduction of NBT to its blue insoluble monoformazan is 95% inhibitable by SOD, suggesting that 5% of the NBT reduction is due to direct reaction with reduced PMS. In contrast, cytochrome *c* added to reduced PMS is almost entirely reduced by an O_2^--independent mechanism. As with other SOD assays employing nitro blue tetrazolium (NBT) as the O_2^- detector, the analyst should keep in mind that this tetrazolium dye can be oxidized and reduced by a variety of species other than O_2^-, including oxidases such as xanthine oxidase,[11] and appropriate controls should be performed.

Cells are harvested, washed (as described above), and toluenized in 17 mM sodium pyrophosphate buffer, pH 8.3. A 0.2 ml sample of the aqueous toluenized cell phase is added to a tube of the same buffer containing 50 μM NBT and 78 μM NADH (freshly prepared). PMS (5 μM final concentration) is added to initiate the reaction, and the rate of NBT reduction is followed at 560 nm. Boiled sample and excess SOD controls should always be run to ensure that the observed NBT reduction is O_2^- dependent and that its inhibition by the sample is independent of heat-stable species. It should be noted that D-amino-acid oxidases and some diaphorases can also reduce PMS so that if the toluenized cells contain significant quantities of these enzymes the initial flux of O_2^- generated may be elevated.

Measurement of Superoxide Dismutase in Individual Colonies

A method for determining the approximate SOD levels in each of numerous colonies on an agar plate can be of great value in isolating SOD structural and regulatory mutants and distinguishing SOD-rich from SOD-

---

[10] M. Nishikimi, N. A. Rao, and K. Yagi, *Biochem. Biophys. Res. Commun.* **46**, 849 (1972).
[11] C. Beauchamp and I. Fridovich, *Anal. Biochem.* **44**, 276 (1971).

poor species in mixed populations. The only suitable method developed at present is that of Schiavone and Hassan for *Escherichia coli*,[12] an abbreviated version of which follows.

Method. Colonies of *E. coli* are picked, grown overnight on LB agar plates, and blot-transferred to a Whatman #1 filter paper disk with an index mark. Replica plating onto the paper from dilution plates having scattered, well defined colonies can also be done. The paper disks are inverted and placed in 0.5 ml of a 1 mg/ml lysozyme solution for 30 min and then transferred to a desiccator having a chloroform atmosphere for further lysis (30 min). The disks are frozen (10 min) and thawed 3 times, then transferred (colony side up) to plates containing 1% agar, 50 mM phosphate, and 0.1 mM EDTA, which are incubated for 3 hr at room temperature. The incubation allows diffusion of the released cytoplasmic enzymes, including SOD, into the agar. There should be no syneresis (free) water on the plates as this will cause blurring of the zones.

The disks are removed from the plates, and 15 ml of 0.55 mM NBT, 66 μM riboflavin, 0.1% TEMED (N,N,N',N'-tetramethylethylenediamine), 1% agar, 50 mM phosphate, and 0.1 mM EDTA at 55° are overlaid on the plates in low light. After a further 3-hr incubation in the dark, the plates are exposed to even cool-white fluorescent until there is full development of the deep blue monoformazan color from the NBT (5–10 min). Areas containing SOD will block the O_2^--dependent production of color and will remain pale or achromatic. The assay is sufficiently sensitive to detect 0.05 μg pure SOD (about 0.16 units). This protocol, especially the cell lysis steps, may require modification depending on the cell type employed.

$$\text{Riboflavin} \xrightarrow{\text{light}} \text{riboflavin*} \tag{3}$$
$$\text{Riboflavin*} + \text{TEMED} \longrightarrow \text{riboflavin}^{\overline{\cdot}} + \text{TEMED}_{ox} \tag{4}$$
$$\text{Riboflavin}^{\overline{\cdot}} + O_2 \longrightarrow \text{riboflavin} + O_2^- \tag{5}$$
$$O_2^- + \text{NBT (yellow)} \longrightarrow O_2 + \text{monoformazan (deep blue)} \tag{6}$$

Comments. Use 5–50 mM EDTA with the lysozyme in the absence of divalent cations. EDTA permeabilizes the outer membranes of many gram-negative bacteria, improving access of lysozyme to its substrate, the peptidoglycan wall. Placing of the freeze–thaw step before lysozyme treatment may also improve access of the lysozyme to the wall polymers. Since chloroform has been reported to release periplasmic, but not cytoplasmic, enzymes from *E. coli*,[13] and toluene to release cytoplasmic proteins including SOD,[1-5] replacement of chloroform with toluene should improve results.

[12] J. R. Schiavone and H. M. Hassan, *Anal. Biochem.* **168,** 455 (1988).
[13] G. F.-L. Ames, C. Prody, and S. Kustu, *J. Bacteriol.* **160,** 1181 (1984).

For large numbers of small-volume cell suspensions, use of a micro-multiwell filter suction apparatus (e.g., Minifold I of Schleicher and Schuell, Keene, NH) can produce sheets of paper with large numbers of precise spots containing known numbers of washed cells. Use of a 96-well microtiter plate format and dimensions for the paper-bound cells would allow subsequent rapid reading and photometric quantitation in a conventional microtiter plate reader.

Toluene-mediated release of active enzymes from yeasts has also been reported.[1] Thus, unicellular algae and fungi, as well as distinct colonies or clones of cells from plants or animals adherent on a solid surface should be amenable to SOD quantitation by the above methods.

[23] Superoxide Dismutase Mimic Prepared from Desferrioxamine and Manganese Dioxide

By Wayne F. Beyer, Jr., and Irwin Fridovich

Introduction

Superoxide dismutases (SODs) are a family of metalloenzymes which provide a defense against one aspect of the toxicity of dioxygen, by catalyzing the conversion of O_2^- to H_2O_2 plus O_2.[1-3] Low molecular weight mimics of SOD might be very useful, both as antioxidants and as pharmaceutical agents, and have been vigorously sought. Manganese or copper ions, either free or complexed, are efficient catalysts of the dismutation reaction. SOD mimics based on complexed copper suffer from problems which include insolubility in the aqueous buffer systems, instability in the presence of serum proteins, and the possibility that copper, once freed from the complexing agent, might catalyze hydroxyl radical formation. In contrast, manganese complexes are stable in the presence of serum proteins and are unable to catalyze hydroxyl radical formation. We present a brief description of the preparation and properties of a soluble green complex prepared from desferrioxamine and manganese dioxide (MnO_2)[4] which can catalyze the dismutation of O_2^- in vitro and which could pro-

[1] J. V. Bannister, W. H. Bannister, and G. Rotilio, CRC Crit. Rev. Biochem. 22, 111 (1987).
[2] I. Fridovich, Arch. Biochem. Biophys. 247, 1 (1986).
[3] I. Fridovich, Adv. Enzymol. 58, 61 (1986).
[4] D. Darr, K. A. Zarilla, and I. Fridovich, Arch. Biochem. Biophys. 258, 351 (1987).

tect green algae against the toxic effects of paraquat[5] or of sulfite.[6] In addition, the preparation of a more active and more stable pink variant of this complex is described.

Materials and Methods

Water which has been sequentially deionized and glass-distilled is used throughout. Buffers are stirred for 2–4 hr at 25° with 10 g/liter of 200–400 mesh Chelex 100 (Bio-Rad, Richmond, CA) and then filtered and readjusted to the desired pH, under the glass electrode, with H_3PO_4. This is necessary to diminish trace metal contaminants that otherwise impose a syncatalytic inactivation of xanthine oxidase. Phosphate salts are from Mallinckrodt, desferrioxamine methane sulfonate (Desferal) from Ciba Geigy or Sigma (St. Louis, MO), and ultrapure MnO_2 from Aesar (J. Matthey, Inc.). Xanthine oxidase is prepared from unpasteurized cream.[7] SOD-like activities are assayed as previously described but in the absence of EDTA.[8] In some assays 10 mM of freshly distilled CH_3CHO replaces xanthine as the substrate for xanthine oxidase.

Preparation of Green Complex. Desferrioxamine methane sulfonate (328 mg) is dissolved in 10 ml of water, yielding a 50 mM solution at approximately pH 4.0. A 25% molar excess of powdered MnO_2 (56 mg) is added. After stirring at 25° for 2–4 hr the unreacted MnO_2 is removed by centrifugation, and the dark green solution is passed through a 0.22-μm Millex-GV syringe filter and is either used immediately or lyophilized for storage. When needed the lyophilized material is dissolved in water. The SOD-like activity of this complex decreases, concomitant with the formation of a fine white precipitate, if stored in aqueous solution for prolonged periods of time (>7–10 days).

Preparation of Pink Complex. Desferal (328 mg) and ascorbic acid (100 mg) are dissolved in 10 ml of water, and the pH is adjusted to 8.5 with NaOH. MnO_2 (56 mg) is then added, and the mixture is stirred at 25° for 2 hr and then clarified by centrifugation and ultrafiltration. The orange-pink solution is either used immediately or lyophilized for storage. In aerobic solutions, this complex has a limited stability (<24 hr), decomposing to

[5] H. D. Rabinowitch, C. T. Privalle, and I. Fridovich, *Free Radical Biol. Med.* **3,** 125 (1987).

[6] H. D. Rabinowitch, G. M. Rosen, and I. Fridovich, *Free Radical Biol. Med.* **6,** 45 (1989).

[7] W. R. Waud, F. O. Brady, R. D. Wiley, and K. V. Rajagopalan, *Arch Biochem. Biophys.* **169,** 695 (1975).

[8] J. M. McCord and I. Fridovich, *J. Biol. Chem.* **244,** 6049 (1969).

the green complex. It can be stored at 4° in anaerobic, aqueous solutions for 2–3 weeks.

Results

Manganese Content and Extinction Coefficients. The optical spectra of the green and of the pink complexes are shown in Fig. 1A,B. The Mn contents of these complexes are determined by atomic absorption spectrophotometry. The desferrioxamine content of the complexes is based on the assumption that all of the desferrioxamine had been converted to the manganese complexes by the excess of MnO_2 used. The molar ratio of Mn to desferrioxamine in the green complex is 1.13 ± 0.11. This complex exhibits absorption maxima at 640 and 910 nm, with a shoulder at 320 nm. The molar extinction coefficients, based on the manganese content, are $1.09 \pm 0.07 \times 10^2 \ M^{-1} \ cm^{-1}$ at 640 nm and $1.57 \pm 0.1 \times 10^3 \ M^{-1} \ cm^{-1}$ at 320 nm. The extinction of the 910 nm absorption band is approximately 50 $M^{-1} \ cm^{-1}$. The pink complex yields a ratio of Mn to desferrioxamine of 1.09 ± 0.01 and exhibited an absorption maximum at 496 nm ($\varepsilon = 31.2$ $M^{-1} \ cm^{-1}$) and a shoulder at 550 nm ($\varepsilon = 21.8 \ M^{-1} \ cm^{-1}$). The low-energy

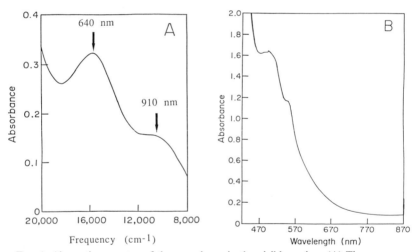

FIG. 1. Absorption spectra of the complexes in the visible region. (A) The green complex, prepared from MnO_2 and Desferal as described in the text, was scanned from 8000 to 20,000 cm^{-1} using a Perkin Elmer spectrophotometer. The complex was present at approximately 3 mM in water. Note that the abscissa is presented as wave numbers (cm^{-1}). (B) The pink complex, prepared from MnO_2, ascorbate, and Desferal as described in the text, was scanned from 900 to 400 nm using a Hitachi 100-80 spectrophotometer. The complex was present at approximately 50 mM in water. Note the absence of the low-energy transition.

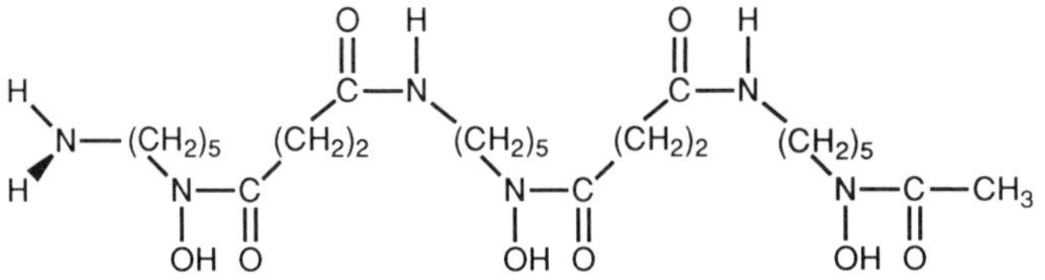

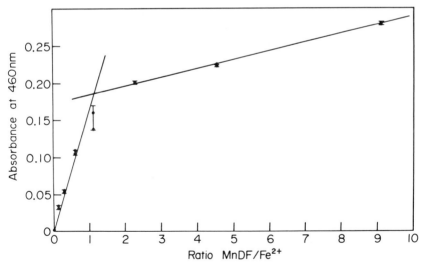

FIG. 2. Structure of the linear trihydroxamic acid desferrioxamine B (Desferal).

band (910 nm) present in the green complex is noticeably absent in the pink form. It thus appears that both the green and the pink compounds are 1:1 complexes of manganese with desferrioxamine, which differ either in the valence of the metal or in the state of oxidation of one of the ligand moieties (Fig. 2).

Valence of Manganese in Green Complex. When the green complex is mixed with desferrioxamine and $FeSO_4$, at pH 6.0, the yellow-orange color of ferrioxamine develops rapidly, indicating oxidation of the ferrous salt by the green complex, followed by reaction of the resultant Fe(III) with desferrioxamine. The manganese in the green complex is evidently an oxidant toward Fe(II). Figure 3 presents the results obtained when the ratio of the green complex to the Fe(II) salt is varied, at a fixed concentration of the iron salt, and in the presence of excess desferrioxamine. It is

FIG. 3. Oxidation of Fe^{2+} by the green complex in the presence of excess Desferal. The final conditions were 0.28 mM Fe^{2+}, 0.35 mM desferal, and 0–1.2 mM green complex. The reaction was performed in 25 mM N-morpholinoethanesulfonic acid (MES) buffer at pH 6.0. MnDF, Manganese complex of desferrioxamine.

clear from the results of this titration that 1.15 mol of the green complex causes the oxidation of 1 mol of Fe(II) to Fe(III). The modest increase in $A_{460 nm}$ with increasing green complex seen after the equivalence point had been passed is due to the absorbance of the green complex itself at 460 nm. It seems likely from this result that the green complex contains trivalent manganese.

The green complex can also be prepared by air oxidation of a mixture of desferrioxamine and Mn(II); no complex is formed if O_2 is excluded. $Mn(OH)_3$, when made by a modification of the Winkler method,[9] rapidly generates the green complex when stirred with desferrioxamine.

Formation of the green complex from desferrioxamine and Mn(II) under aerobic conditions must involve oxidation of Mn(II), which is favored by the relative stabilization of Mn(III) and provided by its much higher affinity for the chelating agent. It follows that addition of desferrioxamine to a salt of Mn(II) should result in consumption of O_2. The data in Fig. 4 show that addition of desferrioxamine to Mn(II), or of Mn(II) to desferrioxamine, at pH 9.15, results in rapid consumption of O_2. Since the subsequent addition of bovine liver catalase caused a return of approximately one-half of the O_2 consumed, it follows that H_2O_2 is accumulating in the reaction mixtures.

The foregoing data are consistent with the view that the green complex contains Mn(III). In that case formation of the complex from MnO_2 plus desferrioxamine must involve reduction of the Mn(IV) by the desferrioxamine. O_2 is not involved in this process, which was seen to proceed anaerobically (data not shown). Hydroxamates are reductants, and it is likely that one of the hydroxamate groups of the desferrioxamine (Fig. 2) served as the reductant for the Mn(IV). Provision of an alternate reductant, such as ascorbate, might spare the hydroxamate group and yield a complex in which all three hydroxamates remain intact. Such, presumably, is the nature of the pink complex.

Effects of EDTA on Green and Pink Complexes. EDTA causes an immediate bleaching of the green complex, which is complete when the EDTA is added in stoichiometric, or greater, amounts. When EDTA is added to a relatively concentrated solution of the green complex ($\sim$10 mM), a faint pink color rapidly develops and then disappears in less than 5 min at 25°. It seems likely that EDTA can rapidly displace the Desferal from the metal, yielding the faintly pink Mn(III)–EDTA complex, which is unstable because the EDTA then acts as a reductant toward the trivalent manganese. The pink complex, in contrast, is not bleached by EDTA, even when the EDTA is added in 20-fold molar excess over the complex.

[9] G. S. Sastry, R. E. Hamm, and K. H. Pool, *Anal. Chem.* **41,** 857 (1969).

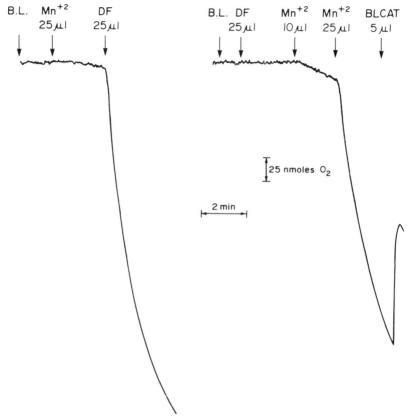

FIG. 4. Stimulation of O$_2$ consumption by MnCl$_2$ and Desferal. The reaction was done in 50 mM 2-(N-cyclohexylamine)ethanesulfonic acid (CHES) (pH 9.15) at 25°. At the arrows, aliquots of either MnCl$_2$ (100 mM) or Desferal (100 mM) were added, and the change in pO$_2$ was recorded using a Clarke-type oxygen electrode. Bovine liver catalase (BLCAT) stock solution was 6.02 mg/ml. The final reaction volume was 2.0 ml. B.L., Baseline.

The pink complex is thus stable to EDTA. Nevertheless, the abilities of both the pink and the green complexes to catalyze the dismutation of O$_2^-$ are eliminated by EDTA. This may be explained by a valence change at the Mn center, during the catalytic cycle, which allows EDTA to displace Desferal from the pink complex.

SOD-Like Activities of the Complexes. Figure 5 compares the abilities of the green and the pink complexes, with that of Mn(II), to inhibit the reduction of cytochrome c by an enzymic source of O$_2^-$. Mn(II) can be distinguished because its inhibition is biphasic whereas that due to the complexes is not. Thus, addition of Mn(II) causes an immediate and

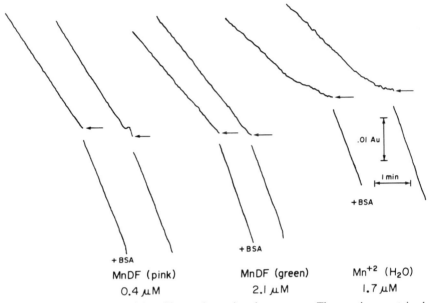

FIG. 5. SOD mimetic activity of free and complexed manganese. The reactions contained 50 mM potassium phosphate buffer (pH 7.8), 10 μM cytochrome c^{3+}, 50 μM xanthine, and approximately 8–10 nM xanthine oxidase in a final volume of 3.0 ml. The absorbance change at 550 nm owing to the reduction of cytochrome c^{3+} by O_2^- was recorded in the absence and presence of the indicated manganese species, which were added at the arrows.

strong inhibition, which then diminishes until a lesser, but stable, inhibition is achieved. In contrast, the green and the pink complexes immediately impose a degree of inhibition which is stable with time. When compared on the basis of stable inhibitions the SOD-like activities of Mn(II), pink, and green complexes on a per manganese basis fall in the ratio 1.0 : 1.7 : 0.3. The pink complex is thus between 5 and 6 times more active than the green complex.

Discussion

Both the green and the pink complexes of manganese with Desferal contain one atom of Mn(III) per molecule of Desferal. Since the green complex can be prepared by reacting MnO_2 with Desferal it follows that some moiety of the Desferal must have been oxidized in providing the electron which reduced Mn(IV) to Mn(III). The pink complex was obtained when ascorbate was present during the reaction of MnO_2 with Desferal, and this suggests that the ascorbate then provided the electron.

In this view the difference between the green and the pink complexes is that the former contains oxidized Desferal while the latter contains intact Desferal. Oxidation of one of its hydroxamate ligand groups should decrease the affinity of the Desferal for Mn(III), and, as expected, the green complex was less stable to EDTA than was the pink complex.

Desferal is known to act as an electron donor, and the nitroxide product of the univalent oxidation of one of its hydroxamate groups has been studied by electron paramagnetic resonance.[10] Since the pink complex was prepared by reacting 1.0 mol ascorbate per mole of Desferal with a 25% molar excess of MnO_2, it must be presumed to contain dehydroascorbate. Dehydroascorbate at 250 μM dd not inhibit the reduction of cytochrome c by the flux of O_2^- generated by the action of xanthine oxidase on 10 mM CH_3CHO. It follows that this concentration of dehydroascorbate did not exhibit any detectable ability to intercept O_2^- and that the very much lower concentration (4 μM) present during assays of the pink complex would not interfere with assays of the activity of that complex.

The greater stability and catalytic activity of the pink complex suggest that it may be a more useful mimic of SOD activity than the green complex. This remains to be explored *in vivo*.

Acknowledgments

This work was supported by research grants from the National Science Foundation, the Council for Tobacco Research—USA, Inc., and the National Institutes of Health.

[10] K. M. Morehouse, W. D. Flitter, and R. P. Mason, *FEBS Lett.* **222,** 246 (1987).

[24] Purification of Exocellular Superoxide Dismutases

By KENNETH D. MUNKRES

Introduction

Although superoxide dismutases (SODs, EC 1.15.1.1) have been intensively investigated for many years, the existence of exocellular forms of the enzyme was recognized only in the 1980s. The existence of exocellular forms of the enzyme (EC-SODs) in bovine milk was discovered almost simultaneously in three laboratories around 1975.[1] Subsequently,

[1] M. Korycka-Dahl, T. Richardson, and C. L. Hicks, *J. Food Prot.* **42,** 867 (1979).

Marklund and associates[2-5] reported a series of investigations of EC-SODs from human and animal tissues.

Marklund[2] purified human lung EC-SOD on a large scale with low yield by a lengthy series of ion-exchange and affinity chromatographic steps. Subsequently, the human EC-SOD structural gene was cloned, and the enzyme was synthesized in large quantities by genetic engineering techniques.[4,5] The complete amino acid sequence of the polypeptide subunit and signal peptide of the enzyme was deduced from the nucleotide sequence of the gene. Comparison of sequence similarities indicated that the exo- and endocellular SOD isozymes arose from a common ancestral enzyme before the evolution of fungi and plants; hence, one may infer a broad phylogenetic distribution of the exoenzyme. The results of this investigation support that inference. Facile procedures for large-scale purification and crystallization were invented and applied to bovine milk and blood serum, wheat germ, yeast, and bacteria. The enzyme was also partially purified from *Neurospora*.[6]

The widespread phylogenetic distribution of the enzyme indicates that it plays an essential but yet unknown biological role. Knowledge of its biological role may be significant for application of the engineered human enzyme in proposed clinical therapy of human diseases mediated by free radicals.

Our laboratory is investigating the genetic regulation and biological function of EC-SOD. EC-SOD-deficient mutants have been isolated in *Neurospora* and yeast.[6] The aim of this study was to develop the technology to conduct comparative EC-SOD therapy of the physiological "diseases" of the mutants. Human EC-SOD is a tetrameric glycoprotein whose molecular weight is 135,000.[2] The present purification procedures are designed to select proteins of this nature.

Materials

Defatted, pasteurized dry milk (Carnation Co., Los Angeles, CA), natural (untoasted) wheat germ (Hodgson Mill Enterprise, Gainsville, MO), and Red Star Quick-Rise dried bakers' yeast *in vacuo* (Universal Foods Corp., Milwaukee, WI) were obtained from local grocers. Lyophi-

[2] S. L. Marklund, *Proc. Natl. Acad. Sci. U.S.A.* **79,** 7634 (1982).
[3] S. L. Marklund, *Biochem. J.* (*London*) **222,** 649 (1984).
[4] K. Hjalmarsson, S. L. Marklund, Å. Engström, and T. Edlund, *Proc. Natl. Acad. Sci. U.S.A.* **84,** 6340 (1987).
[5] L. Tibell, K. Hjalmarsson, T. Edlund, G. Skogman, Å. Engström, and S. L. Marklund, *Proc. Natl. Acad. Sci. U.S.A.* **84,** 6634 (1987).
[6] K. D. Munkres, unpublished (1988).

lized cells of *Bacillus subtilis* (ATCC 6633) and *Escherichia coli* (strain W) and concanavalin A-agarose were obtained from Sigma Chemical Co. (St. Louis, MO). Sterile calf serum was from HyClone Laboratories (Logan, UT).

Methods and Results

SOD Assay. The spectrophotometric method of Misra and Fridovich[7] is used. The method is based on the inhibition by the enzyme of the spontaneous, alkali-catalyzed autoxidation of epinephrine to norepinephrine.

Enzyme in 10–100 μl of distilled water diluent is added to 1 ml of the reaction mixture in a semimicrocuvette at 36°. The reaction rate is monitored in a Gilford recording spectrophotometer at 480 nm for 6 min. After a lag period of 1 to 3 min, the linear rate during the next 2 to 3 min is graphically determined. Plots of percent inhibition versus protein concentration are constructed with two or three points, and the concentration producing 50% inhibition (1 unit) is determined by interpolation. Usually the plots are linear from 20 to 80% inhibition. Rarely some samples exhibit linear plots that plateau at 40–60% inhibition. In those instances, a unit of activity is defined as that concentration producing half-maximal inhibition.

Employing commercial bovine red blood cell SOD, we confirmed the observation of Misra[7,8] that the present assay method is about 2-fold more sensitive than the xanthine oxidase method of McCord and Fridovich. Ethanol mimics SOD: at least 1% inhibits the assay. Ethanol extracts should contain sufficient enzyme to permit its assay after dilution in water to noninhibitory ethanol concentrations. Tetraborate does not interfere with the assay.

Protein Assay. The spectrophotometric method of Murphy and Kies[9] is used.

Tetraborate Assay. The course of dialysis to remove tetraborate is monitored by a colorimetric assay. A sample of the dialysate (0.9 ml) is mixed with 0.1 ml of 1% silver nitrate and incubated at room temperature about 10 min or until the absorbancy at 500 nm becomes constant. The extinction coefficient of tetraborate (E_{mM}, cm^{-1}) is 0.255. The assay detects as little as 10–100 μg sodium tetraborate per milliliter.

Extraction of Exocellular SOD. Survey of 16 strains of bakers', beer, and wine yeast indicated that Red Star Quick-Rise Bakers' yeast is an

[7] H. P. Misra and I. Fridovich, *J. Biol. Chem.* **247**, 3410 (1972).

[8] H. P. Misra, *in* "CRC Handbook of Methods for Oxygen Radical Research" (R. A. Greenwald, ed.), p. 237. CRC Press, Boca Raton, Florida, 1985.

[9] J. B. Murphy and M. W. Kies, *Biochim. Biophys. Acta* **45**, 382 (1960).

extraordinarily rich source of EC-SOD.[6] The EC-SOD activity of the yeast *in vivo* is determined by assay with washed cells diluted in water. The specific activity is about 0.89 units/mg dry cells.

Two methods efficiently extract the enzyme, as indicated by either units extracted or units remaining with the cell. Method A uses 0.3 M KBr in 5 mM acetate buffer (pH 5.5); method B uses 80% ethanol. One gram of dry yeast is stirred 10–20 min with 10–20 volumes of extraction solvent at room temperature. Cells are removed by centrifugation at 10,000 g for 10 min. Two extractions yield at least 95% of the enzyme units initially present. Method A is preferred if we wish to subsequently break the cells and measure endocellular SOD activities. The exoenzyme may be concentrated and purified from the extract on a small scale by adjusting the extract to 80% ethanol followed by precipitation with tetraborate; however, the volumes of ethanol required for large-scale preparations are prohibitively large; therefore, method B is used for the latter.

A preliminary trial indicates that method C may be a suitable compromise for small-scale analysis. After extraction of endocellular enzymes in a neutral buffer, the debri of unbroken cells and cell walls is collected by centrifugation at 10,000 g for 10 min and extracted with 80% ethanol.

KBr, unlike ethanol, interferes with spectrophotometric protein assay at 215/225 nm; therefore, the Warburg–Christian[10] method is used for method A.

Large-Scale Purification Procedures. Kilogram quantities of dry milk, wheat germ, and dry yeast are used in large-scale purifications (Tables I–III). Since most of the procedures are essentially the same for these materials, a general method is described.

Step 1: Extraction. The dry material is suspended in 2–4 volumes of ethanol per weight in the stainless steel cup of a commercial Waring blendor and homogenized 2–4 min at full speed. The ensuing increase of homogenate temperature from about 20 to 40° is not detrimental to enzyme activity and may actually facilitate purification by denaturing unglycosylated proteins. The homogenate is slowly poured into a 30-cm-diameter bench-type Büchner funnel containing Whatman No. 1 paper while collecting the filtrate with the aid of an oil vacuum pump. The extraction is repeated, and filtrates are pooled. The filtrate may require clarification by additional filtration with Celite (diatomaceous earth). The Celite is washed on the filter with water and ethanol before use.

Step 2: Tetraborate crystallization. Dry milk and wheat germ are extracted with 95% (w/v) ethanol and diluted to 80%. Na$_2$B$_4$O$_7 \cdot$ 10H$_2$O and sodium acetate are slowly dissolved in the solution while stirring to 1 and 0.5% (w/v), respectively. Eighty percent ethanol is more efficient than

[10] O. Warburg and W. Christian, *Biochem. Z.* **310**, 384 (1942).

TABLE I
PURIFICATION OF BOVINE MILK SOD

Step	Protein (mg)	SOD Activity Units × 10⁻⁶	SOD Activity Units/mg	Purification factor	Recovery (-fold)
1. Ethanol extraction[a]	1,020	6.1	6.0×10^3	1^b	
2. Borate	102	6.5	6.4×10^4	10.7	1.1
3. Recrystallization (1×)	—	4,350	2×10^7	3.3×10^3	713
4. Recrystallization (4×)	92	5,600	6×10^7	9.9×10^3	918
5. Dialysis	2.8	12,000	4.3×10^9	7×10^5	2,000
6. Crystallization[c]					

[a] Four liters of 95% ethanol extract recovered from extraction of 1.8 kg dry milk.
[b] 390-fold with respect to total milk protein.
[c] After dialyses 3 times at 20°, the solution containing about 0.1 mg protein/ml was stored at 5°.

TABLE II
PURIFICATION OF WHEAT GERM EXOCELLULAR SOD

Step	Volume (ml)	Protein (mg)	SOD Activity Units × 10⁻⁷	SOD Activity Units/mg	Purification factor	Recovery (-fold)
1. Ethanol extract[a]	2,500	118,000	3.1	2.6×10^2	1^b	
2. Borate	350	—	—			
3. Ethanol	1,750	470	10.9	2.3×10^5	8.8×10^2	3.5
4. Borate	85	138	3.2	2.3×10^5	8.8×10^2	1.0
5. Dialysis	100	2.6	5.0	1.9×10^7	7.3×10^4	1.6
6. Centrifugation	10	0.4	3.5	8.8×10^7	3.4×10^5	1.1

[a] From 1 kg.
[b] About 10-fold with respect to weight of wheat germ.

TABLE III
PURIFICATION OF BAKER'S YEAST EXOCELLULAR SOD

Step	Volume (ml)	Protein (mg)	SOD Activity Units × 10⁻⁶	SOD Activity Units/mg	Purification factor	Recovery (-fold)
1. Ethanol extract[a]	1,600	2,500	1.6	6.4×10^2	1^b	
2. Borate		—	—			
3. Ethanol		—				
4. Borate	60	3.56	11.7	3.3×10^6	5.2×10^3	7.3
5. Dialysis	80	0.15	3.4	2.3×10^7	3.6×10^4	2.1

[a] From 500 g dry yeast.
[b] Purification was 213-fold with respect to cell dry weight.

95% for yeast SOD extraction. Solid salts are slowly dissolved in the extract as described above. The tetraborate step is preferably carried out with a glass vessel because the crystals subsequently formed tenaciously stick to glass, a fact that facilitates their collection from the mother liquor.

The ethanol–borate solution is allowed to sit overnight at 10°. The mother liquor is decanted and stored another day for the collection of a smaller crop of crystals. The crystals are collected in a minimum volume of 5 mM acetate buffer (pH 5.5) at room temperature. A glass rod or spatula is used to dislodge them from the glass vessel. Dissolution of the crystals is facilitated by magnetically stirring the solution on a hot plate while raising the temperature to 39–40°.

At this stage of the process, either one of two tactics provide additional purification: recrystallization from buffer (Tables I, VI) or repetition of the ethanol–borate steps (Tables II–IV), another mode of recrystallization. The former tactic is more economical in time, effort, and material.

Step 3: Recrystallization from buffer. After dissolving the crystals in buffer at 39°, the rate of subsequent crystal growth and the size of crystals are primarily regulated by the rate of temperature reduction. Addition of a few seed crystals or scratching the wall of the glass container with a glass rod sometimes hastens crystal growth. Large crystals are grown by storage of the solution several hours or overnight while decreasing the temperature slowly in increments from 20 to 10 to 5°. Rapid formation of small, birefringent crystals occurs if the solution is immediately set in an ice bath. Sham experiments indicate that tetraborate alone crystallizes under these conditions. Comparative analysis of the dry weight and protein content of the crystals indicates that the majority of the crystalline mass is sodium tetraborate.

Step 4: Dialysis. Regardless of the method of recrystallization, subsequent dialysis enhances specific activity with respect to protein 10- to 100-fold and removes excess borate salt (Tables I–IV). We suspect that low molecular weight glycopeptides are removed by dialysis. The amount of borate associated with the enzyme after dialysis has not been determined. Although tetraborate can be measured with the silver assay, the reactivity of proteins with silver precludes assay of protein-bound tetraborate.

The crystals are dissolved as described and dialyzed against 200–400 volumes of distilled water at 20–25° with continuous stirring for 1 day. One or two additional dialyses are performed. The second but not the first dialysate is free of measurable tetraborate. The dialyzed samples are usually free of particulate matter. Dilute protein solutions are opaque and faintly blue.

Step 5a: Crystallization from water. The ethanol and water solubilities of the enzyme indicate that it is an amphiphilic glycoprotein. After exhaustive dialysis against water at 20°, dilute enzyme protein solutions are opaque, unlike dilute or concentrated ethanolic solutions. That observation indicates a relatively low solubility in water. Although it is not known if tetraborate remains bound to the enzyme and influences its solubility, the situation indicates that it may also crystallize in a form at least relatively free of tetraborate. Indeed, after overnight storage at 5° the dialyzed milk enzyme (Table I) and some dialyzed yeast enzyme preparations formed crystals that subsequently grew to several millimeters in length. The crystalline milk SOD remains active in water at 5° for 2 years.

Step 5b: Centrifugation. The dialyzed wheat germ enzyme preparation was exceptionally opaque although it contained only 26 μg protein/ml (Table II). It was centrifuged at 4° for 30 min at 30,000 g. The precipitate was dissolved in 10 ml of 0.2 M acetate buffer (pH 5.5). The solution contained 41 μg protein/ml and was highly opaque and slightly blue. The step recovered 70% of the activity and enriched specific activity about 4-fold (Table II). The absorbancy ratios at 215/225, 280/260, and 280/560 nm were 1.27, 0.69, and 4.5, respectively. The 215/225 nm protein assay indicated 0.27 mg/ml; the 280/260 assay indicated 0.81 mg/ml. The latter probably overestimates concentration because of the exceptionally low 280/260 ratio, a general feature of the enzyme from several sources (Table VII).

Affinity Chromatography. Highly purified yeast or milk SOD is bound to a column of concanavalin A-agarose in 10 mM acetate buffer (pH 5.5) containing 0.1 mM CaCl$_2$ and eluted with the buffer containing 0.1 M α-methylmannoside. The binding and elution are nearly perfectly efficient as indicated by both protein and enzyme assays. The column must be washed thoroughly to ensure that the sample does not become contaminated with dissociable concanavalin. Without calcium chloride, the enzyme is not retained. This procedure might prove useful for the removal of possible trace amounts of bound tetraborate after dialysis. Affinity chromatography is not used for purification with respect to protein because it may not be sufficiently specific: most if not all exocellular proteins are probably glycoproteins.

Purification of Calf Serum SOD. Serum presents special problems because it is extraordinarily rich in nonglycosylated protein. Three rather than one or two ethanol–borate cycles are necessary to increase the specific activity to a level comparable to that of the highly purified enzymes from other sources (Table IV). The gigantic mass of protein formed on adjusting serum to 80% (w/v) ethanol occludes a substantial amount of the solution volume after centrifugation. This problem is largely circum-

TABLE IV
PURIFICATION OF BOVINE CALF SERUM EXOCELLULAR SOD

Step	Volume (ml)	Protein (mg)	SOD Activity Units × 10⁻⁴	SOD Activity Units/mg	Purification factor	Recovery (-fold)
Serum	250	185,000	6.2	0.33	1	
1. Ethanol extract	—	—				
2. Borate	—	—				
3. Ethanol	150	320	7.2	2.4×10^2	7.3×10^2	1.2
4. Borate	30	86	30	3.5×10^3	1.1×10^4	4.8
5. Ethanol	—	—				
6. Borate	50	1.1	27	2.5×10^5	8.3×10^5	4.4
7. Dialysis	50	0.20	300	1.5×10^7	4.5×10^7	48

vented by a preliminary fractionation with 50% (w/v) ethanol. The subsequent 50–80% ethanol step also must be clarified by centrifugation.

Purification of Bacterial Exocellular SOD. Tables V and VI summarize purifications of the enzymes from dried cells and culture filtrates of *Escherichia coli* and *Bacillus subtilis*. It is not clear whether *E. coli* has genuine cell-bound ethanol-soluble EC-SOD. The attempt to precipitate it with borate failed, the only failure encountered among the various materials examined. Conversely, *B. subtilus* appears to possess a genuine cell-bound enzyme. The results of purification of the enzyme from culture filtrates are relatively unequivocal (Table VI). Comparative analysis indicates that *B. subtilis* is a richer source (Tables V, VI).

TABLE V
PURIFICATION OF EXOCELLULAR SOD FROM LYOPHILIZED BACTERIA

Step	Cells	Volume (ml)	Protein[c] (mg/g)	SOD Activity[c] Units × 10⁻⁴	SOD Activity[c] Units/mg	Purification factor	Recovery (%)
1. Ethanol extract	*E. coli*	7.3[a]	4.46	3.6	8.1×10^3	1	
	B. subtilis	17[b]	19	47	2.5×10^4	1[d]	
2. Borate	*E. coli*	5	0.52	0.1	1.9×10^3	0.2	2.7
	B. subtilis	10	0.47	230	4.9×10^6	196	489

[a] One gram extracted with 10 ml of 85% ethanol; 7.3 ml recovered.
[b] One gram extracted with 20 ml of 95% ethanol; 17.3 ml recovered.
[c] Corrected for loss of extract solvent occluded with cells.
[d] Purification was 50-fold with respect to cell weight.

TABLE VI
PURIFICATION OF BACTERIAL EXOCELLULAR SOD FROM CULTURE FILTRATES

| | | | | SOD Activity | | | |
Step	Cells[a]	Volume (ml)	Protein (mg)	Units $\times 10^{-4}$	Units/mg	Purification factor	Recovery (-fold)
1. Culture	E	50	375	0.65	17.3	1	
filtrate[a]	B	50	1,600	21.4	134	1	
2. Ethanol			—		—		
3. Borate	E	50	1.5	4.38	2.9×10^4	1.7×10^3	6.7
	B	50	—				
4. Recrystalliza-	E	5	0.38	3.1	8.2×10^4	4.8×10^3	4.8
tion (1X)	B	50	1.0	36.1	3.6×10^5	2.7×10^3	1.7
5. Recrystalliza-	B	10	0.145	38	2.6×10^6	2×10^4	1.8
tion (2X)							

[a] E. coli (E) and B. subtilis (B) were grown overnight at 36° in shake cultures containing 50 ml of medium with 2% glucose. The absorbancies at 600 nm of the cultures were 0.465 and 0.720 for E and B, respectively. After removal of cells by centrifugation, the supernatant was clarified by Millipore filtration. The culture medium was Vogel's minimal (R. H. Davis and F. J. de Serres, this series, Vol. 17A, p. 79).

Concanavalin A Inhibition of Yeast Exocellular SOD in Vivo and in Vitro. Inhibition titrations of either highly purified or *in vivo* yeast EC-SOD with concanavalin A (Con A) were performed.[6] Calcium chloride (0.1 mM) is essential for inhibition. The range of Con A concentrations for *in vivo* inhibition is much less than that which causes the cells to flocculate.

Discussion

Purification methods and some enzyme properties briefly discussed in the preceding section are discussed at greater length here. The purification procedures are based on the assumption that all EC-SODs are large, highly glycosylated proteins; the previously demonstrated prototype is the human enzyme.[2] The literature on highly glycosylated proteins indicates that their properties resemble that of polysaccharides and sugars. Solubility and resistance to denaturation in organic solvents is one feature of highly glycosylated proteins exhibited by EC-SODs. They are soluble and remain active in 80–95% ethanol. The yeast enzyme is also soluble and fully active after extraction in 80% phenol.[6] Furthermore, unlike ordinary protein, EC-SODs are soluble in aqueous saturated ammonium sulfate.[6]

Sugars or glycoproteins form diol–charge complexes with borate.[11,12] EC-SODs are relatively insoluble in the presence of tetraborate. The enzyme–tetraborate complex apparently cocrystallizes with free tetraborate salt from 80% ethanol or acidic acetate buffer.

Since human and animal tissue,[3] bovine milk, and yeast EC-SODs bind to the plant lectin concanavalin A and are eluted with α-methylmannoside, at least a portion of their polysaccharide probably consists of mannose, for which Con A is primarily specific.

The human EC-SOD polypeptide subunit contains a hydrophobic amino-terminal amino acid sequence to residue 95 and a hydrophilic carboxy-terminal sequence at residues 194–222.[4] Assuming phylogenetic sequence homology, those features and bound polysaccharide probably account for the amphiphilic nature of EC-SODs observed here. The efficient extraction of the enzyme from cells with the chaotropic agents KBr or ethanol indicates that the hydrophobic region of the protein may bind to cell wall or plasmalemma. The inhibition of yeast EC-SOD *in vivo* by Con A indicates that it is bound to the cell wall surface. Inhibition of the isolated yeast enzyme by Con A may indicate that the lectin induces a conformational change.

Yeast and calf serum EC-SOD exhibit fluorescence excitation and emission spectra characteristic of tryptophan-containing proteins[6]; however, the relatively low 280/260 absorbancy ratios (Table VII) indicate a relatively low tryptophan content, a feature that probably accounts for the consistent discrepancies between the two spectrophotometric methods of measurement of protein concentration.

The specific activity of bovine red blood cell (RBC) SOD (endocellular) in the assay is about 6,000 units/mg protein. Conversely, the specific activities of the exocellular enzymes from various sources approach or slightly exceed that of RBC SOD in relatively impure samples and are 10^4- to 10^6-fold more active in highly purified preparations (Tables I–VI). Marklund[2] compared the specific activities of human EC-SOD and the endocellular Cu,Zn-SOD and concluded that they are not appreciably different; however, the assay was at pH 7 and our results are at 10.8.

The kinetic features are further clouded by our observations that the enzyme is frequently activated during the course of purification. Marklund[2] did not observe activation of the human enzyme during purification. The activation was sufficiently frequent and of sufficient magnitude (Tables I, III, IV, V) to indicate that it is not a spurious error. Removal of a naturally occurring inhibitor may account for the activation

[11] J. X. Khym, this series, Vol. 12A, p. 93.
[12] J. R. Cann, this series, Vol. 25, p. 157.

TABLE VII
ABSORBANCE RATIOS OF HIGHLY PURIFIED
EXOCELLULAR SODs

Enzyme source	Absorbance ratio, wavelength	
	280/260 nm	280/560 nm
Milk	0.78	2.8
Calf serum	0.80	3.0[a]
Wheat germ	0.60	4.7
Yeast	0.67	5.6[a]

[a] λ_{max} 560 nm in visible light.

effect. Even if the activation effect is discounted, the data indicate that the procedure provides excellent recoveries of activity and that specific activities are enriched many orders of magnitude with respect to ethanol-soluble protein. Moreover, the initial extraction alone enriches specific activity 10- to 400-fold with respect to biomass (Tables I–III, V).

Although activity recoveries are high, protein yields are low, ranging from 0.15 to 2.8 mg (Tables I–IV); hence, either larger scale preparations or microanalytical techniques will be required for chemical analyses. The ready tendency of the enzymes to crystallize from dilute solutions, with or without the addition of borate, indicates that crystallographic analysis of their structure may be feasible.

The highly purified enzyme solutions are faintly blue and exhibit significant 280/560 absorbancy ratios (Table VII). The visible absorbancy spectra of the calf and yeast enzymes exhibit a peak at 560 nm.[6] The yeast and calf enzymes are inhibited by 1 mM cyanide.[6] Collectively, those observations indicate that the enzymes contain copper, as proved for the human enzyme.[2]

An attempt by a commercial analytical laboratory to determine the isoelectric points and molecular weights of the calf and yeast enzyme subunits by the O'Farrell method was unsuccessful, apparently because the proteins failed to stain with Coomassie blue.[6] Specific glycoprotein stains may be required.

The long-range research goals motivating this study, noted in the Introduction, are now approachable. Minute quantities of milk or yeast EC-SODs are active in therapy of physiological and biochemical abnormalities of EC-SOD-deficient mutants of *Neurospora* and yeast.[6] These findings have permitted development of quantitative bioassays of the enzymes and discovery of the probable biological function of the enzyme.[6]

Acknowledgments

This research was supported by the University of Wisconsin College of Agriculture and Life Sciences and Graduate School and by a National Institutes of Health Biomedical Research grant administered by the Graduate School. Contribution 3111 from the Department of Genetics.

[25] Analysis of Extracellular Superoxide Dismutase in Tissue Homogenates and Extracellular Fluids

By Stefan L. Marklund

Introduction

Extracellular superoxide dismutase (EC 1.15.1.1, EC-SOD) is a secretory, tetrameric, copper- and zinc-containing glycoprotein with a subunit molecular weight of about 30,000.[1,2] EC-SOD is the major SOD isoenzyme in extracellular fluids, such as plasma, lymph,[3] and synovial fluid.[4] Although EC-SOD is the least predominant SOD isoenzyme in tissues, 90–99% of the EC-SOD in the body of mammals is located in the extravascular space of tissues.[5,6]

A prominent feature of EC-SOD is its affinity for heparin. On chromatography on heparin-Sepharose, plasma EC-SOD from man,[7] pig, cat, mouse, guinea pig, and rabbit[8] can be divided into at least three fractions: (A) a fraction without weak heparin affinity, (B) a fraction with weak heparin affinity, and (C) a fraction which elutes relatively late in a NaCl gradient. EC-SOD from tissues is mainly composed of forms with high heparin affinity (S. L. Marklund, unpublished data). In rat plasma, however, only fractions A and B can be demonstrated.[8] The binding to heparin is of electrostatic nature.[9] Since EC-SOD carries a net negative charge at neutral pH, the binding to the strongly negatively charged heparin mole-

[1] S. L. Marklund, Proc. Natl. Acad. Sci. U.S.A. 79, 7634 (1982).
[2] L. Tibell, K. Hjalmarsson, T. Edlund, G. Skogman, Å. Engström, and S. L. Marklund, Proc. Natl. Acad. Sci. U.S.A. 84, 6634 (1987).
[3] S. L. Marklund, E. Holme, and L. Hellner, Clin. Chim. Acta 126, 41 (1982).
[4] S. L. Marklund, A. Bjelle, and L.-G. Elmqvist, Ann. Rheum. Dis. 45, 847 (1986).
[5] S. L. Marklund, J. Clin. Invest. 74, 1398 (1984).
[6] S. L. Marklund, Biochem. J. 222, 649 (1984).
[7] K. Karlsson and S. L. Marklund, Biochem. J. 242, 55 (1987).
[8] K. Karlsson and S. L. Marklund, Biochem. J. 255, 223 (1988).
[9] K. Karlsson, U. Lindahl, and S. L. Marklund, Biochem. J. 256, 29 (1988).

cule must be mediated by a cluster of positively charged amino acid residues in the enzyme. Such a cluster occurs in the very hydrophilic carboxy-terminal end of EC-SOD C, which contains three lysines and six arginine residues among the last 20 amino acids.[10] The differences between EC-SOD A, B, and C probably reside in this region. The *in vivo* correlate of the heparin affinity is apparently binding to heparan sulfate proteoglycan in the glycocalyx of cell surfaces.[7-9,11]

The middle portion of the EC-SOD amino acid sequence shows a strong similarity with that part of the Cu,Zn-SOD sequence which defines the active site.[10] The degree of homology in this portion between human EC-SOD and Cu,Zn-SODs from man, pig, cow, horse, swordfish, fruit fly, spinach, and bakers' yeast does not differ significantly. Only the Cu,Zn-SOD from *Photobacterium leiognathi* was clearly less homologous.[10] The data indicate that the EC-SODs have evolved from the Cu,Zn-SODs or that they have a common ancestry. The divergence may have occurred before the evolution of plants and fungi. The EC-SODs may thus be widely distributed among higher phyla. So far EC-SODs have only been sought and found in mammals,[3,5,6,8] birds, and fish (S. L. Marklund, unpublished data). For continued research on the functional role of the EC-SODs, their distribution in the body, as well as their phylogenetic occurrence, methods allowing their specific analysis are of the essence.

The present chapter deals with methods aiding in distinguishing between EC-SOD and other SOD isoenzymes. EC-SOD is, like the CuZn SODs but unlike Mn-SODs and Fe-SODs, very sensitive to inhibition by cyanide. The problem is thus reduced to distinguishing between the cyanide-sensitive isoenzymes EC-SOD and Cu,Zn-SOD. We have so far not found any distinguishing inhibitor. In accord with the similarities in active site structure,[10] both isoenzymes are inhibited by cyanide, azide, H_2O_2, diethyl dithiocarbamate,[12] and the arginine-specific reagent phenylglyoxal.[13] Availability of polyclonal or monoclonal antibodies toward EC-SOD and/or Cu,Zn-SOD solves the problem of distinction within a certain species, since we have not found any antigenic cross-reactivity between the isoenzymes.[12] However, our antihuman EC-SOD antibodies produced so far have reacted poorly or not at all with EC-SOD from other mammals. More generally applicable procedures for distinction must be based on physical differences between the isoenzymes. Below are described methods employing differences in size and affinity for lectins and heparin.

[10] K. Hjalmarsson, S. L. Marklund, Å. Engström, and T. Edlund, *Proc. Natl. Acad. Sci. U.S.A.* **84**, 6340 (1987).

[11] K. Karlsson and S. L. Marklund, *J. Clin. Invest.* **82**, 762 (1988).

[12] S. L. Marklund, *Biochem. J.* **220**, 269 (1984).

[13] T. Adachi and S. L. Marklund, *J. Biol. Chem.* **264**, 8537 (1989).

Experimental Procedures

Preparation of Samples

Tissues are homogenized with an Ultra-Turrax (or similar apparatus) in 10 volumes of ice-cold 50 mM potassium phosphate, pH 7.4, with 0.3 M KBr (chaotropic salt that increases extraction of EC-SOD severalfold[5]), 3 mM diethylenetriaminepentaacetic acid, 100 kIU/ml aprotinin, and 0.5 mM phenylmethylsulfonyl fluoride (the latter three additions inhibit proteases). The tissue extract may then be subjected to ultrasonication followed by centrifugation (20,000 g, 20 min). The supernatants are then used for the separations.

For subsequent separation on Con A-Sepharose or with gel chromatography, samples can be processed fresh or thawed after storage below $-70°$. The affinity for heparin is, however, very sensitive to treatment of the samples and is easily lost, probably owing to proteolysis of the carboxy-terminal end, which is responsible for heparin binding. To avoid this, tissues should preferably be processed fresh. If the tissues are allowed to thaw after freezing, we have noted a loss of high heparin affinity. We interpret the effect to be due to proteolytic enzymes released by freezing-induced cellular and subcellular rupture. Frozen tissues can, however, be mechanically pulverized at $-196°$ (liquid N_2) followed by addition of ice-cold extraction buffer and subsequent sonication without loss of heparin affinity. In the final extract, the antiproteolytic measures seem to prevent further degradation. The extracts can be kept at $-70°$.

The stability problem is apparently not so great in extracellular fluids. The plasma EC-SOD heparin affinity pattern was not influenced by freezing and thawing or by a 3-day storage in a refrigerator.[7] Apparently the large amounts of protein and the numerous antiproteases in plasma confer protection. EDTA (or citrate) but not heparin should be used as anticoagulant, since heparin might interfere with the heparin-Sepharose procedure and also induces a large increase in the apparent molecular mass of EC-SOD C.[8]

Analysis of SOD Activity

The EC-SOD activity is low both in extracellular fluids and in tissue extracts. The amount of EC-SOD in the fractions collected in the suggested procedures is consequently very low and varies between about 4 and 100 ng/ml. Concentrations this small cannot be determined with most common SOD assays. Highly sensitive SOD assays are necessary. We only use the direct spectrophotometric assay with O_2^- obtained from

KO_2.[14,15] A unit corresponds to 4–9 ng Cu,Zn-SOD and EC-SOD, and the assay is 40 times more sensitive than the xanthine oxidase–cytochrome c procedure.[16]

Separation by Gel Chromatography

The principle behind gel chromatography is that most EC-SODs have an apparent molecular weight of around 150,000 on gel chromatography, whereas Cu,Zn-SOD elutes at a position corresponding to 30,000. EC-SOD in plasma from man, cat, pig, sheep, mouse, rabbit, guinea pig, and rat was very easily distinguished from the small amounts of Cu,Zn-SOD that occur.[3,8] The rat plasma EC-SOD elutes at an apparent molecular weight of 90,000.[8] Whether this is due to, for example, smaller subunits or to a dimeric state is not yet known. Such atypical molecular weights might also occur in the EC-SODs from other phyla. In tissue extracts EC-SOD is a minor isoenzyme, but it is still usually possible to assess the amount of EC-SOD from the gel chromatography pattern.[5,6] This was not possible in rat tissue homogenates, however, where the EC-SOD peak was hidden by the large Cu,Zn-SOD and Mn-SOD peaks.[6]

Procedure. Any gel chromatography system with good resolution in the range 150–30 kDa can probably be used. We use the following procedure. The sample (1–5 ml) is applied to a column (1.6 × 90 cm) of Sephacryl S-300 (Pharmacia LKB Biotechnology, Bromma, Sweden), eluted at 20 ml/hr with 10 mM potassium phosphate (pH 7.4), 0.15 M NaCl, and collected in 3-ml fractions. The absorbance at 280 nm and the SOD activity are determined in the collected fractions.

Separation on Con A-Sepharose

EC-SOD is, unlike the other SOD isoenzymes, a glycoprotein and has been found to bind to lectins such as concanavalin A, lentil lectin, and wheat germ lectin.[1,2] Chromatography of samples on concanavalin A-substituted Sepharose (Con A-Sepharose, Pharmacia) has proved to be a useful procedure for distinguishing EC-SOD from the other SOD isoenzymes. The EC-SOD is bound and can then be eluted with α-methylmannoside. The recovery of EC-SOD in the suggested procedure varies mostly between 70 and 90%.

Procedure. The chromatography is executed manually in a stepwise fashion. The tissue extract (1–2 ml) is applied to a 1-ml Con A-Sepharose

[14] S. L. Marklund, *J. Biol. Chem.* **251**, 7504 (1976).

[15] S. L. Marklund, *in* "Handbook of Methods for Oxygen Radical Research" (R. Greenwald, ed.), p. 249. CRC Press, Boca Raton, Florida, 1985.

[16] J. M. McCord and I. Fridovich, *J. Biol. Chem.* **244**, 6049 (1969).

column equilibrated with 50 mM Na–HEPES (pH 7.0), 0.25 M NaCl. The sample is applied in 0.5-ml portions with 5-min intervals between applications. After 5 min, 3 ml equilibration buffer is added. The eluting fluid from the tissue extract and buffer additions is collected and contains the SOD activity which lacks Con A affinity. The column is then washed with 10 ml equilibration buffer. EC-SOD is finally eluted with 5 ml of 0.5 M α-methylmannoside dissolved in equilibration buffer added in 1-ml portions with 5-min intervals. The column is regenerated with 5 ml of 0.5 M α-methylmannoside followed by 10 ml of equilibration buffer.

Chromatography on Heparin-Sepharose

Procedure. The chromatography is carried out on a 2-ml Heparin-Sepharose column equilibrated with 15 mM sodium cacodylate (pH 6.5), 50 mM NaCl, eluted at 5 ml/hr. The tissue extract (1–10 ml) or plasma (up to 2 ml) is applied. Many plasma proteins bind to heparin, and if more than 2 ml is applied there is a risk of column saturation.[7] The samples should be dialyzed against the elution buffer. After application of samples, the column is eluted with 15 ml of the buffer. Thereafter, bound proteins are eluted with a linear gradient of NaCl in the buffer (0–1 M, total volume 50 ml). The eluent is collected in 1.5-ml fractions, and the SOD activity and absorbance at 280 nm are determined.

By definition EC-SOD A is the fraction that elutes without binding, B is the fraction that elutes early in the gradient, and C is the fraction that elutes relatively late. In man, pig, cat, mouse, rabbit, and guinea pig plasma the B fractions eluted between 0.17 and 0.30 M NaCl and the C fractions between 0.42 and 0.62 M NaCl.[7,8] In all these species Cu,Zn-SOD and Mn-SOD eluted without binding, together with EC-SOD A. To distinguish between EC-SOD A and the other isoenzymes in the nonbinding fraction, the Con A-Sepharose procedure can be used. All SOD activity in the gradient in these species was given by EC-SOD. Note, however, that it cannot be taken for granted that all SOD activity in the gradient is given by EC-SOD. The strongly negatively charged heparin-Sepharose gel will also function as a cation-exchange chromatography column. The net charge of Cu,Zn-SOD and Mn-SOD from other taxa might be such that they bind to the heparin-Sepharose.

General Comments

Demonstration and quantification of EC-SOD using the procedures outlined here should be possible in samples from most mammalian species. Analysis in other phyla is more uncertain, since we have as yet no

generally applicable knowledge concerning molecular weights, glycosylation, and heparin affinity of the secretory EC-SODs. Combination of the
procedures should aid in identifying a SOD fraction as an EC-SOD. For
example, a cyanide-sensitive SOD fraction with high heparin affinity that
binds to concanavalin A and/or has a high molecular weight is likely to be
an EC-SOD. Better procedures might emerge as EC-SODs from a wider
phylogenetic spectrum are isolated and characterized.

[26] Oxidative Reactions of Hemoglobin

By CHRISTINE C. WINTERBOURN

Introduction

Hemoglobin readily undergoes one-electron oxidations and reductions, and it can act as a source or sink of free radicals. Autoxidation of
the heme groups produces O_2^- and, indirectly, H_2O_2.[1,2] Hemoglobin also
interacts with redox-active xenobiotics and metabolites, forming the
xenobiotic radical and initiating a series of reactions that generate other
radicals and oxidant species and often result in oxidative denaturation of
the hemoglobin.[3] In the red cell, the outcome can be hemoglobin precipitation as Heinz bodies and oxidative damage to other cellular constituents. Production of O_2^- and other radicals can be detected by conventional techniques (e.g., superoxide dismutase-inhibitable cytochrome c
reduction or electron spin resonance with or without the use of a spin
trap). Care must be taken with interpretation, however, because intermediates of hemoglobin oxidation can sometimes participate in the detection
reaction.

This chapter focuses on the measurement of oxidant-mediated
changes to hemoglobin. First, I shall define the different oxidation products of hemoglobin and briefly describe general mechanisms for their
production. Oxidation of oxyhemoglobin (oxyHb)[4] gives O_2^- and methemoglobin (metHb). If the globin structure is destabilized, metHb can convert to hemichrome, in which either the distal histidine or an external

[1] H. P. Misra and I. Fridovich, *J. Biol. Chem.* **247**, 6960 (1972).
[2] C. C. Winterbourn, B. M. McGrath, and R. W. Carrell, *Biochem. J.* **155**, 493 (1976).
[3] C. C. Winterbourn, *Environ. Health Perspect.* **64**, 32 (1985).
[4] Abbreviations: oxyHb, oxyhemoglobin; metHb, methemoglobin; ferrylHb, ferrylhemoglobin; DTNP, 2,2'-dithiobis(5-nitropyridine); SDS–PAGE, polyacrylamide gel electrophoresis in the presence of sodium dodecyl sulfate.

ligand occupies the sixth coordination position of the ferric heme. Hemichromes precipitate readily and are the main constituent of Heinz bodies.[5] Choleglobin is a term used to describe denatured hemoglobin in which the porphyrin ring has been hydroxylated or broken open. Ferrylhemoglobin (ferrylHb or $Hb^{2+}H_2O_2$) is an Fe(IV) complex formed from ferrous hemoglobin and H_2O_2.[6] The reaction of metHb with H_2O_2 produces short-lived ferryl radicals, which are Fe(IV) species with the additional oxidizing equivalent localized on the globin.[7-9]

Reactions with Superoxide. Superoxide can oxidize oxyHb [reaction (1)] and reduce metHb [reaction (2)].[2,10] Both reactions are relatively slow[10]; at pH 7.8 $k_1 = 4 \times 10^3 M^{-1} sec^{-1}$ and $k_2 = 6 \times 10^3 M^{-1} sec^{-1}$. Thus, in concentrated hemoglobin solutions O_2^- reacts predominantly with hemoglobin, but in more dilute solutions most spontaneously dismutates. Since k_1 and k_2 are almost the same, O_2^- cannot oxidize or reduce hemoglobin completely. Provided H_2O_2 is removed with catalase, continued exposure to O_2^- produces a steady-state mixture containing approximately 40% metHb.

$$2 H^+ + O_2^- + oxyHb \rightarrow metHb + O_2 + H_2O_2 \qquad (1)$$
$$O_2^- + metHb \rightarrow oxyHb \qquad (2)$$

Reactions can be followed spectrally, but because O_2^- is usually produced in combination with other species that react with hemoglobin, interpretation can be complicated. O_2^- involvement can be probed with superoxide dismutase. However, superoxide dismutase can also inhibit reactions of other radicals that exist in equilibrium with O_2^-.[11,12] Thus, for O_2^- to be responsible for hemoglobin oxidation or reduction, the following criteria should apply: (1) The reaction should occur in the presence of catalase. (2) It should not occur in the absence of O_2. (3) It should be inhibited by low concentrations (<5 $\mu g/ml$ with 40 μM hemoglobin) of superoxide dismutase. (4) Oxidation of oxyHb should not proceed beyond 40%, nor reduction of metHb beyond 60%. (These percentages vary, but not substantially, with pH.) Otherwise, species formed reversibly from O_2^- may be responsible for the reaction.

[5] E. A. Rachmilewitz, J. Peisach, and W. E. Blumberg, *J. Biol. Chem.* **246**, 3356 (1971).
[6] K. D. Whitburn, *in* "Oxygen Radicals in Chemistry and Biology" (W. Bors, M. Saran, and D. Tait, eds.), p. 447. de Gruyter, Berlin, 1984.
[7] P. George and D. H. Irvine, *Biochem. J.* **52**, 511 (1952).
[8] J. F. Gibson, D. J. E. Ingram, and P. Nicholls, *Nature (London)* **181**, 1398 (1958).
[9] N. K. King, F. D. Looney, and M. E. Winfield, *Biochim. Biophys. Acta* **113**, 65 (1976).
[10] H. C. Sutton, P. B. Roberts, and C. C. Winterbourn, *Biochem. J.* **155**, 503 (1976).
[11] C. C. Winterbourn, J. K. French, and R. F. C. Claridge, *FEBS Lett.* **94**, 269 (1978).
[12] C. C. Winterbourn, J. K. French, and R. F. C. Claridge, *Biochem. J.* **179**, 665 (1979).

Reactions with H_2O_2. The reaction of excess H_2O_2 with oxyHb or oxymyoglobin forms the ferryl species.[6] With excess oxyHb, ferrylHb is formed as an intermediate, but the final product is metHb.[6,13] FerrylHb can oxidize a variety of electron donors.[7] It does not accumulate during the interaction between hemoglobin and redox drugs, unless excess H_2O_2 is produced, as with fava bean pyrimidines.[14] The ferryl radical formed from H_2O_2 and metHb oxidizes further metHb to ferrylHb. It can also undergo internal reactions leading to irreversibly oxidized heme derivatives (choleglobin) or modifications to globin residues.[8,9] It is also a strong oxidant of exogenous electron donors.[15]

Reactions with Xenobiotics. Hemoglobin reacts with many reducing agents, which are oxidized by heme-bound oxygen,[3,13,16–19]

$$Hb^{2+}O_2 + RH_2 + H^+ \rightarrow Hb^{3+} + H_2O_2 + RH \cdot \qquad (3)$$

and oxidizing agents, which are reduced by the heme iron,[3,12,19]

$$Hb^{2+}(O_2) + R + H^+ \rightarrow Hb^{3+} + RH \cdot (+O_2) \qquad (4)$$

In either case the xenobiotic radical is formed initially, and secondary reactions of it, O_2^-, and H_2O_2 can follow. The oxidation products of some compounds (e.g., phenylhydrazines) bind to the hemoglobin to form hemichromes.[20] Other compounds react only slowly with hemoglobin directly but cause heme oxidation through generation of H_2O_2.[13,14] Reactions can be followed spectrally, and, with the provisos listed above, involvement of O_2^- and H_2O_2 can be probed using superoxide dismutase and catalase.

Hemoglobin Preparation

The hemoglobin must be freed from other red cell proteins, particularly, superoxide dismutase and catalase. A satisfactory one-step procedure is column chromatography of hemolysates on DEAE-Sephadex. Up to 3 ml of a 10 g/dl hemolysate (equilibrated against start buffer, 50 mM Tris, 0.1 mM EDTA, pH 8.3) is applied to a column of DEAE-Sephadex A-50 (25 cm $\times$ 1.5 cm i.d.) equilibrated with start buffer. The hemoglobin

[13] P. Eyer, H. Hertle, M. Kiese, and G. Klein, *Mol. Pharmacol.* **11**, 326 (1975).
[14] C. C. Winterbourn, U. Benatti, and A. de Flora, *Biochem. Pharmacol.* **35**, 2009 (1986).
[15] D. Keilin and E. F. Hartree, *Biochem. J.* **60**, 310 (1955).
[16] J. K. French, C. C. Winterbourn, and R. W. Carrell, *Biochem. J.* **173**, 19 (1978).
[17] S. Kawanishi and W. S. Caughey, *J. Biol. Chem.* **260**, 4622 (1985).
[18] B. Goldberg, A. Stern, and J. Peisach, *J. Biol. Chem.* **251**, 3045 (1976).
[19] B. Goldberg and A. Stern, *Biochim. Biophys. Acta* **437**, 628 (1976).
[20] H. A. Itano, K. Hirota, and T. S. Vedvick, *Proc. Natl. Acad. Sci. U.S.A.* **74**, 2556 (1977).

is eluted at a flow rate of 0.5 ml/min using a linear pH gradient consisting of 350 ml start buffer, pH 8.3, in the first chamber and 350 ml of buffer adjusted to pH 7.0 with HCl in the second. The major oxyHb peak from the column can be concentrated and used directly.

MetHb is prepared by oxidation of oxyHb with a slight excess of potassium ferricyanide and passage down a column of Sephadex G-25. Commercial hemoglobin preparations should be checked for superoxide dismutase and catalase activity. Note also that the superoxide dismutase inhibitor diethyl dithiocarbamate oxidizes hemoglobin, so hemoglobin oxidation or reduction should not be studied in its presence.

An oxyHb or metHb concentration of 40 μM (expressed per heme group), or 0.7 mg/ml, is convenient for spectral monitoring of radical reactions.

Measurement of Spectral Changes

Reaction of oxyHb or metHb with radical- or oxidant-generating systems can be followed spectrally, either with purified solutions or with red cells after appropriate dilution.[2,12,16] The spectra of oxyHb, metHb, hemichrome, and ferrylHb are shown in Fig. 1; choleglobin absorbance is almost flat over this range. It should be noted that the spectrum of methemoglobin is pH dependent. The spectrum shown and the data given below are all for pH 7.4. For studies at other pH values, the correct extinction coefficients must be substituted.[21]

Initially, when studying hemoglobin oxidation, it is advisable to record spectral changes with time over the 500–700 nm range and attempt to recognize features that are characteristic of different products. The following points should be noted: (1) MetHb gives a characteristic peak or shoulder at 630 nm. (2) Hemichrome in oxyHb–metHb mixtures is indicated by a shallower trough at 560 nm.[16] (3) Choleglobin is indicated by an increase in absorbance at 700 nm.[16] (4) FerrylHb is distinguished from MetHb by its lack of a shoulder at 630 nm. The distinction between ferrylHb and hemichrome is more subtle, although the flat absorbance of ferrylHb in the 580–600 nm range can be characteristic. Adding a few crystals of dithionite to hemichrome gives a hemochrome spectrum with a large peak at approximately 558 nm and a smaller one at 530 nm.[5] This does not occur with ferrylHb.

Solutions Containing Oxyhemoglobin and Methemoglobin. If only oxyHb–metHb transformations are involved, the reaction can be followed by monitoring absorbances at 577 and 630 nm. From these, and the

[21] C. C. Winterbourn and R. W. Carrell, *J. Clin. Invest.* **54,** 678 (1974).

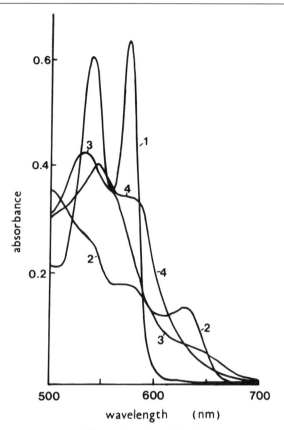

FIG. 1. Absorption spectra of (1) oxyHb, (2) metHb, (3) hemichrome, and (4) ferrylHb. All solutions are approximately 40 μM in phosphate buffer (pH 7.4).

extinction coefficients given in Table I, both concentrations (μM) can be calculated:

$$[\text{Oxyhemoglobin}] = 66A_{577} - 80A_{630} \qquad (5)$$
$$[\text{Methemoglobin}] = 279A_{630} - 3.0A_{577} \qquad (6)$$

The sum of these concentrations should remain constant during the course of the reaction. If not, the reaction is more complex, and more wavelengths must be monitored, as described below. Once it is established that only oxyHb and metHb are present, a simpler procedure can be used (although it is less accurate since the total hemoglobin concentration is not monitored). The absorbance change at 577 nm is measured and Eq. (7) is used for the calculation.

TABLE I

MILLIMOLAR EXTINCTION COEFFICIENTS OF
HEMOGLOBIN DERIVATIVES[a]

Derivative	Wavelength (nm)		
	560	577	630
Oxyhemoglobin	8.6	15.0	0.17
Methemoglobin	4.30	4.45	3.63
Hemichrome	8.6	6.8	0.92

[a] All values are expressed per heme group and are from C. C. Winterbourn, B. M. McGrath, and R. W. Carrell, *Biochem. J.* **155**, 493 (1976) and A. Riggs, this series, Vol. 76, p. 22.

$$\Delta[\text{Oxyhemoglobin}] = \Delta A_{577}/0.0110 \qquad (7)$$

Solutions Also Containing Hemichrome and Choleglobin. If hemichrome and choleglobin are also formed in the reaction, precipitated oxidized hemoglobin derivatives are removed by centrifugation, the contribution of choleglobin is subtracted, and the oxyHb, metHb, and hemichrome concentrations are measured, using the following procedure: (1) If the solution is turbid, centrifuge for 1 min in a minifuge to sediment the precipitate. (2) Measure the absorbance of the solution at 560, 577, 630, and 700 nm. (3) Subtract $A_{700} - 0.005$, as the contribution of choleglobin, from each of the other absorbances. (The absorbance of choleglobin is essentially the same at each wavelength. A 40 μM oxyhemoglobin, methemoglobin, or hemichrome solution has an A_{700} of approximately 0.005. With different concentrations, this figure should be adjusted accordingly.) (4) Calculate the micromolar concentrations of the three species using Eqs. (8)–(10), which are derived from the extinction coefficients given in Table I. In the absence of choleglobin formation or precipitation, the total concentration should remain approximately constant; otherwise it will decrease as these components increase.

$$[\text{Oxyhemoglobin}] = 119A_{577} - 39A_{630} - 89A_{560} \qquad (8)$$
$$[\text{Methemoglobin}] = 28A_{577} + 307A_{630} - 55A_{560} \qquad (9)$$
$$[\text{Hemichrome}] = -133A_{577} - 114A_{630} + 233A_{560} \qquad (10)$$

Solutions Containing Ferrylhemoglobin. Approximate concentrations can be calculated for solutions containing oxyHb, metHb, and ferrylHb by measuring absorbances at 700, 630, 577, and 560 nm and using the millimolar extinction coefficients for ferrylmyoglobin determined by

Whitburn et al.[22] (ε_{560} 14.1, ε_{577} 3.9, ε_{630} 3.0) along with those in Table I, to solve the simultaneous equations generated.

Hemoglobin Thiol Oxidation

The hemoglobin tetramer contains six cysteine residues. Four are buried and poorly accessible to oxidants, but those at position 93 of the β chains are exposed. Oxidation to give either mixed disulfides (e.g., with glutathione) or interchain disulfide bonds can occur during exposure of hemoglobin to redox-active compounds[23] or during precipitation of oxidized hemoglobin.[21] Thiol oxidation can be measured using Ellman's regent or by a modification using 2,2'-dithiobis(5-nitropyridine) (DTNP), which has the advantage that its product has an absorption maximum that does not overlap with the Soret peak of hemoglobin.[24] Interchain disulfides can be detected using SDS–PAGE under nonreducing conditions.[21]

Free Thiol Analysis. To measure only the β-93 thiols, dilute hemoglobin solutions to a concentration of 5–15 μM (heme groups) in pH 8.0 phosphate buffer. To 1 ml of solution, add 10 μl of 10 mM DTNP (3.1 mg/ml in ethanol), and, after 20 min, measure A_{386} against a blank consisting of hemoglobin at the same concentration. Also subtract a blank corresponding to the diluted DTNP, then calculate the thiol concentration using an ε_{386} (mM) value of 14.0.[24] To measure the total thiol groups, use the same procedure but with 1% SDS added to the pH 8 buffer.

The thiol concentration is related to the hemoglobin concentration, measured using Drabkin's reagent (100 mg NaCN and 300 mg K_3FeCN_6 dissolved in 1 liter water). Dilute the hemoglobin solution (to 0.5–1 mg/ml) and read the A_{450} after 5 min against a Drabkin's solution blank. Calculate the hemoglobin concentration using an ε_{540} (mM) value of 11.5 (per heme).

Detection of Interchain Disulfides. Hemoglobin monomers (M_r 17,000) are separated from disulfide bonded dimers and polymers by SDS–PAGE using 10% gels and 0.1 M Tris–citrate buffer, pH 6.8 (10.3 g citric acid, 6.2 g Tris, 1 g SDS, 97.5 ml N NaOH to 1 liter). Protein bands can be stained with Coomassie blue or stained for heme groups using 3,3',5,5'-tetramethylbenzidine (0.1 g in 25% aqueous acetic acid) followed by a few drops of 30% H_2O_2. The slightly acidic electrophoresis buffer is used to minimize thiol–disulfide exchange reactions.

[22] K. D. Whitburn, J. J. Shieh, R. M. Sellers, M. Z. Hoffman, and I. A. Taub, *J. Biol. Chem.* **257**, 1860 (1982).

[23] W. H. Reinhart, L. A. Sung, and S. Chien, *Blood* **68**, 1376 (1986).

[24] D. R. Grassetti and J. F. Murray, *J. Chromatogr.* **41**, 121 (1969).

Blockage of β-93 Thiol Groups with Iodoacetamide. A number of compounds can form thioether derivatives with the β-93 thiol groups of hemoglobin. These include menadione and other quinones with free α-methylene groups.[12] This reaction generates the semiquinone radical and can initiate subsequent radical reactions. When it is required to study interaction with the heme groups without this complication, or if the involvement of the thiol groups in other oxidative processes is to be probed, thiol groups can be blocked with iodoacetamide. To a solution of hemoglobin (1 mM heme is convenient) in phosphate buffer, pH 7.4, add a 40-fold molar excess of iodoacetamide and allow to react for 1 hr at 37° in the dark. Remove the excess iodoacetamide by passing the solution through a column of Sephadex G-25.

Heinz Body Formation in Intact Red Cells

Intracellular precipitation of hemoglobin as Heinz bodies can be assessed by determining the number and size of inclusions in a fixed number of red cells, stained with methyl violet and viewed by optical microscopy. This is tedious, however, and at best semiquantitative. A simple quantitative assessment of relative amounts of Heinz body formation can be made by hypotonically lysing the red cells and measuring the turbidity of the resultant solution.[25]

Method. Lyse red cells by the addition of 5 mM phosphate buffer, pH 7.4. Addition of 0.1 ml of a 10% cell suspension to 1 ml buffer is an appropriate dilution. Mix and let stand for approximately 15 min, then read the A_{700} of the suspension against a buffer blank, before and after centrifugation for 1 min in a minifuge.

Membranes from lysed control cells give an absorbance difference (~0.01). Values in excess of this represent hemoglobin precipitation as Heinz bodies. The assay remains linear up to an A_{700} of at least 1.5. Comparison with microscopic examination indicates that an A_{700} reading of 0.1 corresponds very approximately to 1–4 small Heinz bodies per cell. Note, however, that even when cells contain several large Heinz bodies, this represents only a very small fraction of the total hemoglobin content.

[25] C. C. Winterbourn, *Br. J. Haematol.* **41**, 245 (1979).

[27] Assays for Cytochrome *P*-450 Peroxygenase Activity

By MINOR J. COON, ROBERT C. BLAKE II, RONALD E. WHITE, and
GERALD D. NORDBLOM

Cytochrome *P*-450 is an extraordinarily versatile enzyme that catalyzes numerous types of chemical reactions with a variety of substrates, including hydroxylation of xenobiotics such as drugs, pesticides, dyes, organic solvents, anesthetics, and carcinogens, as well as naturally occurring lipids, including steroids, fatty acids, and prostaglandins.[1] The versatility extends to the oxygen donor. Molecular oxygen serves as the natural donor when electrons are supplied to cytochrome *P*-450 by NADPH (or, in some instances, NADH) via proteins that serve as electron carriers. In the case of liver microsomes or the reconstituted enzyme system, NADPH–cytochrome-*P*-450 reductase serves as the carrier, and in the overall reaction equimolar amounts of substrate, O_2, and NADPH are consumed and equimolar amounts of hydroxylated substrate, H_2O, and NADP are formed.[2,3] Molecular oxygen, however, can be replaced by a variety of hydroperoxides and related artificial donors. Following the discovery of the ability of cytochrome *P*-450-containing microsomal suspensions to promote the oxygenation of an organic substrate at the expense of an alkyl hydroperoxide,[4] other oxidants such as peroxy acids,[5] periodate,[6] iodosobenzene,[7] iodobenzene diacetate,[8] and *N*-oxides[9,10] were found to function in a similar capacity.

$$RH + R'OOH \rightarrow ROH + R'OH \qquad (1)$$

[1] S. D. Black and M. J. Coon, *Adv. Enzymol. Relat. Areas Mol. Biol.* **60**, 35 (1987).

[2] M. J. Coon, T. A. van der Hoeven, S. B. Dahl, and D. A. Haugen, this series, Vol. 52, p. 109.

[3] M. J. Coon, this series, Vol. 52, p. 200.

[4] F. F. Kadlubar, K. C. Morton, and D. M. Ziegler, *Biochem. Biophys. Res. Commun.* **54**, 1255 (1973).

[5] G. D. Nordblom, R. E. White, and M. J. Coon, *Arch. Biochem. Biophys.* **175**, 524 (1976).

[6] E. G. Hrycay, J.-Å. Gustafsson, M. Ingelman-Sundberg, and L. Ernster, *Biochem. Biophys. Res. Commun.* **66**, 209 (1975).

[7] R. Lichtenberger, W. Nastainszyk, and V. Ullrich, *Biochem. Biophys. Res. Commun.* **70**, 939 (1976).

[8] J.-Å. Gustafsson, L. Rondahl, and J. Bergman, *Biochemistry* **18**, 865 (1979).

[9] S. G. Sligar, K. A. Kennedy, and D. C. Pearson, *in* "Oxidases and Related Redox Systems" (T. E. King, H. S. Mason, and M. Morrison, eds.), p. 837. Pergamon, Elmsford, New York, 1982.

[10] D. C. Heimbrook, R. I. Murray, K. D. Egeberg, S. G. Sligar, M. W. Nee, and T. C. Bruice, *J. Am. Chem. Soc.* **106**, 1514 (1984).

METHODS IN ENZYMOLOGY, VOL. 186

Equation (1) indicates the stoichiometry of the reaction in which a substrate, RH, undergoes hydroxylation and a hydroperoxide, R′OOH, is converted to the corresponding alcohol, as established with purified cytochrome P-450 in a reconstituted enzyme system.[5] Full activity is retained under anaerobic conditions and in the absence of NADPH and the reductase, but not when phosphatidylcholine is omitted. Since the oxygen atom in the hydroxylated substrate is derived from the hydroperoxide,[5] cytochrome P-450 functions in such reactions as a peroxygenase rather than as a typical peroxidase.[11]

Assay in Reconstituted System for N-Methylaniline Demethylation by Cumyl Hydroperoxide

The determination of the cytochrome P-450-catalyzed hydroxylation of numerous substrates by artificial oxidants in the reconstituted system in the absence of NADPH and the reductase has been described.[5] For example, cumyl hydroperoxide, p-nitro- or m-chloroperbenzoic acid, p-menthyl hydroperoxide, hydrogen peroxide, and sodium chlorite have been used for the demethylation of benzphetamine, and cumyl hydroperoxide has been used for the hydroxylation of cyclohexane and the demethylation of N,N-dimethylaniline, N-methylaniline, N-methylbenzylamine, methyl cumyl ether, N-methyl-n-butylamine, aminopyrine, and methyl octanoate. The usual assays for product formation may be used provided that the substrate and the product are stable to the particular artificial oxidant used. Another potential problem is that the heme of cytochrome P-450 may be destroyed by certain oxidants and that the extent of destruction is influenced as well by the particular substrate present. Cytochrome P-450 form 2 is sufficiently stable with cumyl hydroperoxide and N-methylaniline for the reaction to be linear with time for 3 min at 30° or with hydrogen peroxide and benzphetamine for 5 min.

N-Methylaniline + cumyl hydroperoxide → aniline +

formaldehyde + cumyl alcohol   (2)

The conditions for the assay of N-methylaniline demethylation by cumyl hydroperoxide, according to reaction (2), are as follows. The components of the reaction mixture are added in the order shown and are at the final concentrations indicated, in a total volume of 1.5 ml: purified rabbit liver microsomal cytochrome P-450 form 2[2] (1.0 μM), sonicated suspension of dilauroylglyceryl-3-phosphorylcholine (0.15 mM when N-methylaniline demethylation is to be assayed, but a different phospholipid concentration may be optimal with other substrates), potassium phos-

[11] R. E. White, S. G. Sligar, and M. J. Coon, J. Biol. Chem. **255,** 11108 (1980).

phate buffer, pH 7.4 (0.1 *M*), *N*-methylaniline (6.7 m*M*), and cumyl hydroperoxide (previously purified by extraction with alkali[12]) (1.0 m*M*) to initiate the reaction. The mixture is incubated at 30°, and at 3 min the reaction is terminated; the formaldehyde formed is measured by the method of Nash[13] as modified by Cochin and Axelrod,[14] except that color development is carried out for 35 min at 25°, under which conditions formaldehyde is completely stable in the presence of the hydroperoxide. Controls are included in which the cytochrome *P*-450 is omitted. Under these conditions the turnover number for the demethylation reaction is 56 nmol of formaldehyde formed per minute per nanomole of cytochrome *P*-450.

Assay in Reconstituted System for Toluene Hydroxylation by Cumyl Hydroperoxide

Toluene, which yields benzyl alcohol in the cytochrome *P*-450-catalyzed peroxygenase reaction [Eq. (3)], is a useful substrate in mechanistic studies.

$$\text{Toluene } + \text{ cumyl hydroperoxide} \rightarrow \text{benzyl alcohol } + \text{ cumyl alcohol} \tag{3}$$

To determine benzyl alcohol formation at fixed time intervals, a gas chromatographic assay is employed.[15] The components of the reaction mixture are added in the following order and are at the final concentrations indicated, in a total volume of 2.0 ml: cytochrome *P*-450 form 2 (2.4 μ*M*), sonicated suspension of dilauroylglyceryl-3-phosphoryl choline (0.1 m*M*), phosphate buffer, pH 7.0 (50 m*M*), toluene (4 m*M*), and cumyl hydroperoxide (0.1 m*M*). The reaction, which is carried out at 10° to stabilize the cytochrome *P*-450, is initiated by addition of the hydroperoxide and terminated at 15 to 60 sec by the addition of 0.2 ml of 2 *M* NaOH. The mixture is then treated with 0.2 mg of sodium bisulfite to reduce the unreacted hydroperoxide, and, as an internal standard, 20 nmol of *p*-methylbenzyl alcohol is added in 10 μl of methylene chloride. The resulting mixture is extracted with 1.0 ml of methylene chloride, and the extract is dried with anhydrous sodium sulfate and analyzed by gas chromatography with use of a glass column (2 mm × 6 ft) of 3% Carbowax 20M on 100/120 Supelcoport (Varian) with a nitrogen flow rate of 20 ml/min. The temperature of the column is maintained at 120°, and that of the injection port and detector is maintained at 150°. The turnover number for the

[12] H. Hoch and S. Lang, *Chem. Ber.* **77**, 257 (1944).
[13] T. Nash, *Biochem. J.* **55**, 416 (1953).
[14] J. Cochin and J. Axelrod, *J. Pharmacol. Exp. Ther.* **125**, 105 (1959).
[15] R. C. Blake II and M. J. Coon, *J. Biol. Chem.* **256**, 5755 (1981).

hydroxylation reaction under these conditions is 12 nmol of benzyl alcohol formed per minute per nanomole of cytochrome P-450.

Alternatively, to obtain initial rate measurements, benzyl alcohol formation is determined at 10° at 365 nm by the reduction of 3-acetylpyridine adenine dinucleotide (3-AcPyAD)[16] in the presence of alcohol dehydrogenase according to reaction (4),[15]

$$\text{Benzyl alcohol} + \text{3-AcPyAD}^+ \rightarrow \text{benzaldehyde} + \text{3-AcPyADH} + \text{H}^+ \qquad (4)$$

The assay is initiated by rapidly mixing in a stopped-flow spectrophotometer (or in a cuvette in a conventional spectrophotometer) an aliquot of a solution containing horse liver alcohol dehydrogenase (10 mg/ml), dilauroylglyceryl-3-phosphorylcholine (0.2 mM), and cytochrome P-450 form 2 (4.8 μM) with an equal volume of a solution containing 3-AcPyAd (10 mM), cumyl hydroperoxide or one of its derivatives (1.8 mM), and toluene or one of its derivatives (4 mM). The absorption coefficient for reduction of this NAD analog is 9,100 M^{-1} cm^{-1}.[17] The alcohol dehydrogenase preparation is purified before use according to the procedure of Bernhard et al.[18] to remove ethanol.

Two aspects of this coupled enzyme assay limit the range of experimental conditions within which it may be used. First, high levels of 3-AcPyADH and cumyl hydroperoxide interfere with the reactions under study, but this is avoided by limiting the experimental observations to cumyl hydroperoxide concentrations of less than 0.3 mM and to time intervals when less than 0.2 nM 3-AcPyADH has accumulated. Second, the equilibrium constant for the reaction catalyzed by alcohol dehydrogenase is sensitive to the electronic nature of substituents on benzyl alcohol.[19] The stronger the electron-withdrawing character of the substituent, the more unfavorable is the free energy change for alcohol oxidation. When p-nitrotoluene or p-cyanotoluene is the substrate being hydroxylated, this dependence significantly reduces the linear portion of the absorbance change observed at 365 nm in the coupled enzyme assay. This effect is minimized with the high concentration of alcohol dehydrogenase specified above, which ensures that the initial lag time required to reach a linear reaction velocity is less than 10 sec.[20,21]

[16] E. J. Marco and R. Marco, Anal. Biochem. 62, 472 (1974).
[17] A. M. Stein, N. O. Kaplan, and M. M. Ciotti, J. Biol. Chem. 234, 979 (1959).
[18] S. A. Bernhard, M. F. Dunn, P. L. Luisi, and P. Schack, Biochemistry 9, 185 (1970).
[19] M. F. Dunn, S. A. Bernhard, D. Anderson, A. Copeland, R. G. Morris, and J.-P. Roque, Biochemistry 18, 2346 (1979).
[20] W. R. McClure, Biochemistry 8, 2782 (1969).
[21] F. B. Rudolph, B. W. Baugher, and R. S. Beissner, this series, Vol. 63, p. 22.

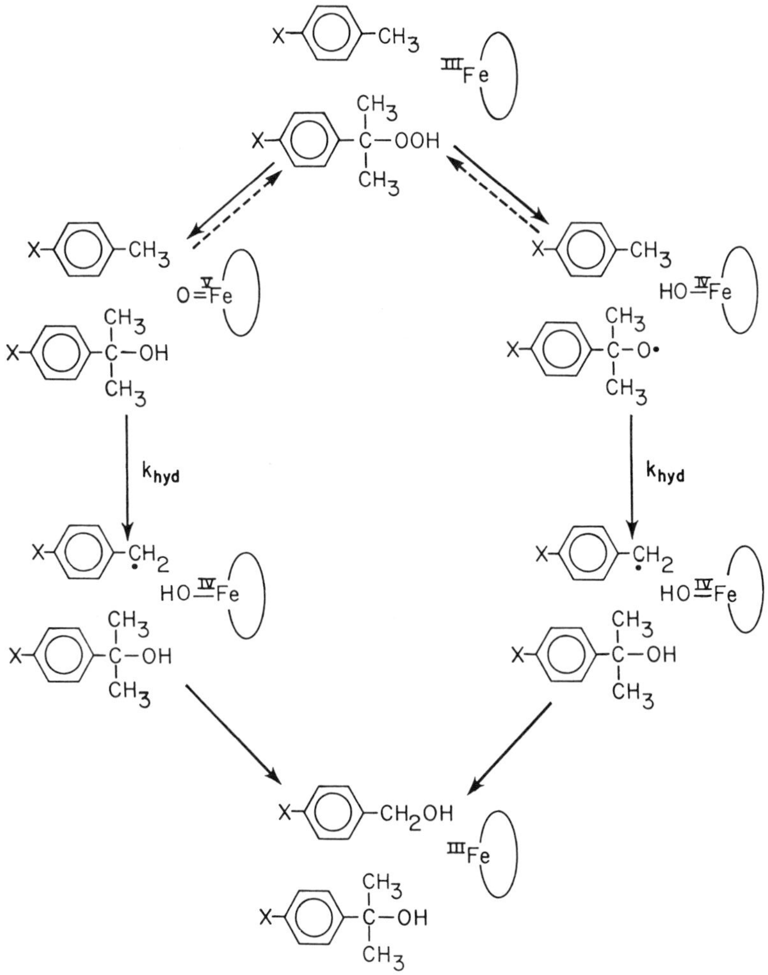

FIG. 1. Alternative schemes for homolytic and heterolytic oxygen–oxygen bond cleavage of heme cumyl hydroperoxide as a step in toluene hydroxylation. Fe represents the iron atom at the catalytic center of cytochrome *P*-450. (Reproduced from Blake and Coon,[22] with permission of the *Journal of Biological Chemistry*.)

Role of Oxygen Radicals and Alkoxy Radicals

The broad substrate specificity of cytochrome *P*-450 form 2 both for substrates and for hydroperoxides as oxygen donors has been established by assay procedures of the kind described above. The versatility with respect to substituted toluenes and cumyl hydroperoxides has been exploited to determine the effect of small structural and electronic perturbations on the formation of spectral intermediates or on the hydroxylation rate constants.[22] The following compounds with various electron-donating and withdrawing substituents were examined: toluene and its p-methyl, p-fluoro, p-chloro, p-bromo, p-cyano, p-nitro, m-methyl, m-fluoro, m-chloro, m-bromo, m-cyano, and m-nitro derivatives, and cumyl hydroperoxide and its p-methyl, p-fluoro, p-chloro, p-bromo, p-cyano, and p-nitro derivatives. The hydroxylation rate constant was found to be sensitive to alterations in both reactants. These results are not compatible with a heterolytic mechanism of oxygen–oxygen bond cleavage leading to a common iron–oxo intermediate, as shown on the left-hand side of Fig. 1. Instead, the findings support a homolytic mechanism, as shown on the right-hand side of Fig. 1, with the hydroperoxide yielding a cumyloxy radical and HO–Fe, which represents a hydroxyl radical equivalent complexed to the heme iron atom. The cumyloxy radical then abstracts a hydrogen atom from toluene to yield cumyl alcohol and benzyl radical, and the latter combines with the hydroxyl radical to yield benzyl alcohol. It may be noted that an alkoxy radical is also believed to be an intermediate in the reductive rearrangement of cumyl hydroperoxide or fatty acid hydroperoxides to produce carbonyl compounds and alkanes.[23]

[22] R. C. Blake II and M. J. Coon, J. Biol. Chem. 256, 12127 (1981).
[23] A. D. N. Vaz and M. J. Coon, this volume [28].

[28] Reductive Cleavage of Hydroperoxides by Cytochrome P-450

By Alfin D. N. Vaz and Minor J. Coon

Various forms of liver microsomal cytochrome P-450 (cytochrome P-450_{LM}) are known to catalyze numerous chemical reactions, including the hydroxylation of fatty acids, steroids, and a variety of foreign compounds, such as drugs and carcinogens.[1] These monooxygenation reactions require aerobic conditions and NADPH as the electron donor in microsomal suspensions or in the reconstituted system containing purified cytochrome P-450_{LM}, purified NADPH–cytochrome-P-450 reductase,

[1] S. D. Black and M. J. Coon, Adv. Enzymol. Relat. Areas Mol. Biol. 60, 35 (1987).

and phosphatidylcholine.[2] On the other hand, molecular oxygen, NADPH, and the reductase are not required when organic hydroperoxides or related compounds serve as the oxygen donor in cytochrome $P\text{-}450_{LM}$-catalyzed substrate hydroxylations.[3,4] In such reactions the hydroperoxides are converted to the corresponding alcohols, so that cumyl hydroperoxide, for example, is converted to cumyl alcohol.

During the course of mechanistic studies in which cumene hydroperoxide was added to the complete reconstituted system in the presence of NADPH under anaerobic conditions, acetophenone was shown to be produced instead of cumyl alcohol,[5] and the other product was identified by gas chromatography–mass spectrometry (GC–MS) as methane[6] [Eq. (1)].

$$C_6H_5(CH_3)_2C\text{—}OOH + NADPH + H^+ \rightarrow C_6H_5COCH_3 + CH_4 + H_2O + NADP^+ \quad (1)$$
$$CH_3(CH_2)_4CH(OOH)CH\text{=}CH\text{—}CH\text{=}CH(CH_2)_7COOH + NADPH + H^+ \rightarrow$$
$$CHO\text{—}CH\text{=}CH\text{—}CH\text{=}CH\text{—}(CH_2)_7COOH + CH_3(CH_2)_3CH_3 + H_2O + NADP^+ \quad (2)$$

The mechanism of the rearrangement reaction is of biological interest because organic peroxides and hydroperoxides are used industrially,[7] and some of these compounds are known to produce malignant tumors[8,9] or to damage cytochrome $P\text{-}450$ *in vitro*.[3] Furthermore, lipid hydroperoxides are naturally occurring substrates for the reaction, as shown in Eq. (2) for the hydroperoxide derived from linoleic acid as the substrate.[6] Following the report that ADP-activated lipid peroxidation, as measured by malonaldehyde formation, is linked to the NADPH oxidase system of microsomes,[10] numerous papers have dealt with this subject. NADPH–cytochrome-$P\text{-}450$ reductase may serve as the primary source of hydroxyl[11] and superoxide radicals[12] for initiation of lipid peroxidation, and

[2] M. J. Coon, this series, Vol. 52, p. 200.

[3] G. D. Nordblom, R. E. White, and M. J. Coon, *Arch. Biochem. Biophys.* **175**, 524 (1976).

[4] M. J. Coon, R. C. Blake II, R. E. White, and G. D. Nordblom, this volume [27].

[5] A. D. N. Vaz and M. J. Coon, *in* "Cytochrome $P\text{-}450$: Biochemistry, Biophysics and Induction" (L. Vereczkey and K. Magyar, eds.), p. 545. Elsevier, Amsterdam, and Akademiai Kiado, Budapest, 1985.

[6] A. D. N. Vaz and M. J. Coon, *Proc. Natl. Acad. Sci. U.S.A.* **84**, 1172 (1987).

[7] C. S. Sheppard and O. L. Mageli, *in* "Encyclopedia of Chemical Technology (M. Grayson, ed.), Vol. 17, p. 27. Wiley (Interscience), New York, 1982.

[8] M. Sittig, "Handbook of Toxic and Hazardous Chemicals." Noyes, Park Ridge, New Jersey, 1981.

[9] "Commerical Organic Peroxide Toxicological Data," Society of the Plastics Industry Bulletin. Society of the Plastics Industry, New York, 1982.

[10] P. Hochstein and L. Ernster, *Biochem. Biophys. Res. Commun.* **12**, 388 (1963).

[11] C.-S. Lai, T. A. Grover, and L. H. Piette, *Arch. Biochem. Biophys.* **193**, 373 (1979).

[12] S. D. Aust, D. L. Roerig, and T. C. Pederson, *Biochem. Biophys. Res. Commun.* **47**, 1133 (1972).

cytochrome *P*-450 has also been implicated in the initiation process.[13–15] Various products are known to arise from lipid peroxidation, including alkanes, and hydrocarbon exhalation is believed to be a measure of this pathophysiological process *in vivo*.[16]

Reconstitution of Enzyme System and Determination of Reactants and Products

The reaction with cumyl hydroperoxide is carried out as described earlier.[6] Stock solutions of cytochrome *P*-450 form 2[17] (40–95 μM) and the reductase (14–22 μM) are mixed to give a 1:1 molar ratio of the two proteins. An appropriate volume of a freshly sonicated, aqueous dispersion of dilauroyl-glyceryl-3-phosphorylcholine (1.0 mg/ml) is added in an amount so that in the final reaction mixture, 3.0 ml in volume, the concentration of each enzyme is 2.0 μM and that of the phospholipid is 102 μg/ml. The concentrated system is allowed to stand at 4° for 1–2 hr before use. Potassium phosphate buffer, pH 7.4, containing EDTA (to give final concentrations of 50 and 10 mM, respectively) is added to a cuvette with a 1.0-cm optical path length designed for anaerobic work and equipped with two sidearms, and the cuvette is made anaerobic by purging with oxygen-free, water-saturated nitrogen gas. The reconstituted enzyme system is then added from one sidearm, followed by a solution containing 600 nmol of NADPH from the other sidearm. A constant absorbance at 340 nm during a 15-min equilibration at 25° provides assurance that the mixture is anaerobic. The reaction is initiated by the injection of a 30-μl sample of an anaerobic aqueous solution of cumyl hydroperoxide (100 nmol), and absorbance changes are monitored at the same temperature for 2–3 min. The initial rate and final extent of NADPH oxidation are determined with use of an absorption coefficient of 6.22 mM^{-1} cm^{-1}.

Cumyl hydroperoxide and acetophenone are determined by HPLC with use of a data integrator. An IBM C$_{18}$ reversed-phase column is used at room temperature with an isocratic solvent system of 30% acetonitrile in water at a flow rate of 1 ml/min. The extraction of acetophenone from the reaction mixtures by 1.0 ml of petroleum ether is quantitative, whereas that of cumyl hydroperoxide gives a recovery of 56%. The areas

[13] E. G. Hrycay and P. J. O'Brien, *Arch. Biochem. Biophys.* **147,** 14 (1971).
[14] B. A. Svingen, J. A. Buege, F. O. O'Neal, and S. D. Aust, *J. Biol. Chem.* **254,** 5892 (1979).
[15] G. Ekström and M. Ingelman-Sundberg, *Biochem. Pharmacol.* **33,** 2521 (1984).
[16] A. Wendel and E. E. Dumelin, this series, Vol. 77, p. 10.
[17] M. J. Coon, T. A. van der Hoeven, S. B. Dahl, and D. A. Haugen, this series, Vol. 52, p. 109.

of the peaks are found to be linear with respect to known amounts of the standards in the range of 5 to 1000 pmol for acetophenone and 2 to 150 nmol for cumyl hydroperoxide. Remaining hydroperoxide in reaction mixtures is determined iodometrically. Gas chromatographic analysis of methane in the head space is done on a 2-ft Supelco Carbosieve G glass column at 100° with a 7 ml/min flow rate of N_2 as carrier gas.

The reaction with 13-hydroperoxy-9,11-octadecadienoic acid is carried out under similar conditions. The gas chromatographic analysis for pentane is the same as for methane, except that the reaction mixture is heated for 20 min at 60° before the determination, and a 2-ft 5% Carbowax 20M glass column is used. The other product has been identified as 13-oxo-9,11-tridecadienoic acid.[18]

Substrate Specificity

Other model hydroperoxides in addition to cumyl serve as substrates in the reductive cleavage reaction.[6] α-Methylbenzyl hydroperoxide yields acetophenone as well as benzaldehyde since either a hydrogen atom or a methyl group can be lost, benzyl hydroperoxide yields benzaldehyde, and *tert*-butyl hydroperoxide yields acetone and methane. With limiting amounts of cytochrome P-450 form 2 present, the turnover numbers, expressed as nanomoles of NADPH oxidized per minute per nanomole of the cytochrome, are 60, 30, 24, and 6 for *tert*-butyl, cumyl, α-methylbenzyl, and benzyl hydroperoxides, respectively. Pentane has been identified as a cleavage product of the 15-hydroperoxide derived from arachidonic acid as well as from the 13-hydroperoxide derived from linoleic acid.

Cytochrome P-450 Specificity

Of the six forms of rabbit liver microsomal P-450 examined, alcohol-inducible form 3a[19] is the most active as determined by the turnover numbers for NADPH oxidation at 25°, namely, 57 and 84 for cumyl and *tert*-butyl hydroperoxides, respectively. Phenobarbital-inducible form 2[17] has intermediate activity with values of 30 and 60, respectively, and forms 3b, 3c, 4, and 6 have low activity with values of 17 to 23, and 5 to 9 for the respective hydroperoxides.[20] The yield of carbonyl compounds from the model hydroperoxides with cytochrome P-450 forms 2 and 3a was as predicted by the stoichiometry of Eq. (1), but with the less active cyto-

---

[18] A. D. N. Vaz, E. S. Roberts, and M. J. Coon, *in* "Microsomes and Drug Oxidations, Proceedings of the 7th International Symposium" (J. O. Miners, D. J. Birkett, R. Drew, B. K. May, and M. E. McManus, eds.), p. 184. Taylor & Francis, London, 1988.

[19] D. R. Koop, E. T. Morgan, G. E. Tarr, and M. J. Coon, *J. Biol. Chem.* **257**, 8472 (1982).

[20] A. D. N. Vaz and M. J. Coon, *Fed. Proc, Fed. Am. Soc. Exp. Biol.* **46**, 1956 (1987).

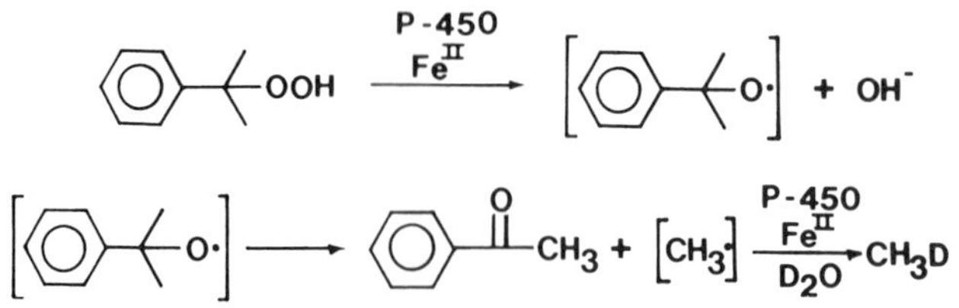

FIG. 1. Proposed mechanism of cytochrome P-450-catalyzed reductive cleavage of hydroperoxides, with cumyl hydroperoxide as the example. (Reproduced from Coon and Vaz,[21] with permission of Alan R. Liss, Inc.)

chromes somewhat less of the carbonyl compounds was formed than predicted.

Role of Oxygen Radicals and Alkoxy Radicals

The accompanying scheme (Fig. 1)[21] shows the proposed reaction mechanism as applied to cumyl hydroperoxide. This substrate is believed to undergo homolytic cleavage of the oxygen–oxygen bond with generation of a cumyloxy radical and input of the first electron from ferrous cytochrome P-450 to yield a hydroxide ion. Rearrangement of the cumyloxy radical by β-scission would yield acetophenone and a transient methyl radical, which by acceptance of the second electron from ferrous cytochrome P-450 to yield a hydroxide ion. Rearrangement of the cumyloxy radical by β-scission would yield acetophenone and a transient demonstrated by the incorporation of one deuteron into the methane formed when the reaction was carried out in 70% deuterium oxide as solvent.[6] Ethane was not detected as a product, thus indicating that radical dimerization does not occur.

[21] M. J. Coon and A. D. N. Vaz, in "Oxidases and Related Redox Systems, Proceedings of the Fourth International Symposium" (T. E. King, H. S. Mason, and M. Morrison, eds.), p. 497. Alan R. Liss, New York, 1988.

[29] Prostaglandin H Synthase

By JEFF A. BOYD

Introduction

Prostaglandin H synthase (PHS) (EC 1.14.99.1) is a membrane-bound glycoprotein complex, consisting of two identical heme-containing subunits of M_r 70,000.[1] Two distinct activities, a cyclooxygenase and a peroxidase, are present in the single polypeptide chain, and heme is absolutely required for both activities.[2] The physiologic role of PHS is to initiate the conversion of arachidonic acid to the biologically active eicosanoids, including prostaglandins, thromboxane, and prostacyclin.[3] This reaction is accomplished through the cyclooxygenase-catalyzed bisdioxygenation of arachidonic acid to the unstable hydroperoxy endoperoxide PGG_2, followed by the peroxidase-catalyzed reduction of PGG_2 to the hydroxy endoperoxide PGH_2 (Fig. 1).[4–6] The cyclooxygenase reaction proceeds through a chain mechanism of carbon- and oxygen-centered radicals,[7–9] whereas the hydroperoxidase reaction generates radical species from fatty acid hydroperoxide, reducing cofactors, and the enzyme itself.[10–12]

The peroxidase reaction of PHS is typical of heme-containing peroxidases, in that a cycle of native enzyme, compound I, and compound II may be observed.[13] During this reaction, both the cyclooxygenase and

[1] F. J. Van der Ouderaa, M. Buytenhek, F. J. Slikkerveer, and D. A. Van Dorp, *Biochem. Biophys. Acta* **572**, 29 (1979).

[2] T. Miyamoto, N. Ogino, S. Yamamoto, and O. Hayaishi, *J. Biol. Chem.* **251**, 2629 (1976).

[3] B. Samuelsson, M. Goldyne, E. Granstrom, M. Hamberg, S. Hammerstrom, and C. Malmsten, *Annu. Rev. Biochem.* **47**, 997 (1978).

[4] S. Ohki, N. Ogino, S. Yamamoto, and O. Hayaishi, *J. Biol. Chem.* **254**, 829 (1979).

[5] M. Hamberg, J. Svensson, T. Wakabayashi, and B. Samuelsson, *Proc. Natl. Acad. Sci. U.S.A.* **71**, 345 (1974).

[6] D. H. Nugteren and E. Hazelhof, *Biochem. Biophys. Res. Commun.* **326**, 448 (1973).

[7] M. E. Hemler and W. E. M. Lands, *J. Biol. Chem.* **255**, 6253 (1980).

[8] N. A. Porter, *in* "Free Radicals in Biology" (W. A. Pryor, ed.), Vol. 4, p. 261. Academic Press, New York, 1980.

[9] R. P. Mason, B. Kalyanaraman, B. E. Tainer, and T. E. Eling, *J. Biol. Chem.* **255**, 5019 (1980).

[10] B. Kalyanaraman, R. P. Mason, B. Tainer, and T. E. Eling, *J. Biol. Chem.* **257**, 4764 (1982).

[11] R. W. Egan, P. H. Gale, E. M. Baptista, K. L. Kennicot, W. J. A. Vanden-Heuvel, R. W. Walker, P. E. Fagerness, and F. A. Kuehl, Jr., *J. Biol. Chem.* **256** 7352 (1981).

[12] T. A. Dix and L. J. Marnett, *J. Am. Chem. Soc.* **103**, 6744 (1981).

[13] A.-M. Lambeir, C. M. Markey, H. B. Dunford, and L. J. Marnett, *J. Biol. Chem.* **260**, 14894 (1985).

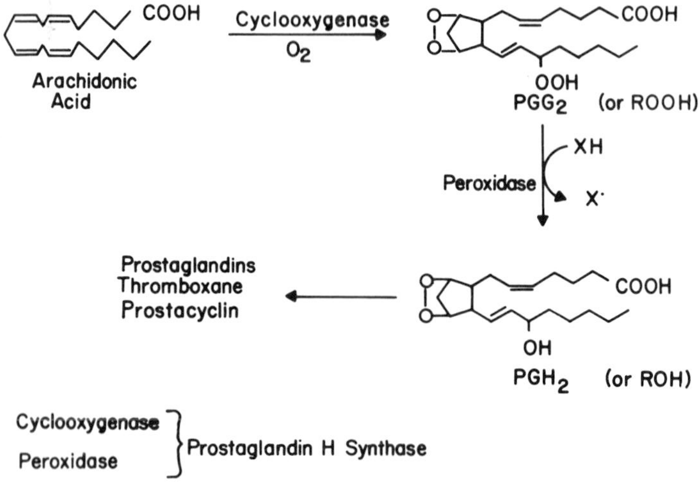

FIG. 1. Role of prostaglandin H synthase in the conversion of arachidonic acid to eicosanoids. The enzyme complex consists of two activities, both of which generate radical products.

the peroxidase undergo irreversible substrate-dependent self-deactivation,[14,15] as the result of an attack on the enzyme by a free radical generated during the peroxidase reaction.[16] This free radical derives from the enzyme itself and may be formed by the oxidation of an amino acid located near the iron of the heme group.[10] Observations that certain organic compounds such as phenol, tryptophan, and serotonin stimulate both enzymatic activities[15,17,18] are now understood in the context of their protecting the enzyme from self-deactivation by donating an electron and undergoing oxidation, hence serving as reducing cofactors for the peroxidase reaction.[10,11]

Numerous xenobiotics are among the several classes of compounds which may serve as reducing cofactors for the PHS peroxidase reaction.[19] A small number of other compounds are oxygenated during prostaglandin biosynthesis but do not serve as reducing cosubstrates; the oxidant in this

[14] W. L. Smith and W. E. M. Lands, *Biochemistry* **11**, 3276 (1972).

[15] R. W. Egan, J. Paxton, and F. A. Kuehl, Jr., *J. Biol. Chem.* **251**, 7329 (1976).

[16] R. W. Egan, P. H. Gale, and F. A. Kuehl, Jr., *J. Biol. Chem.* **254**, 3295 (1979).

[17] T. Miyamoto, S. Kamamoto, and O. Hayaishi, *Proc. Natl. Acad. Sci. U.S.A.* **71**, 3645 (1974).

[18] C. J. Sih, C. Takeguchi, and P. Foss, *J. Am. Chem. Soc.* **92**, 6670 (1970).

[19] L. J. Marnett and T. E. Eling, *in* "Reviews in Biochemical Toxicology" (E. Hodgson, J. R. Bend, and R. M. Philpot, eds.), Vol. 5, p. 135. Elsevier/North Holland, New York, 1983.

system is a fatty acid peroxyl radical,[12] and the source of oxygen is molecular O_2.[20] Thus, there are two major components of the PHS-catalyzed reaction, each involving radical formation, for which convenient assays exist: (1) the cyclooxygenase-dependent oxygenation of arachidonic acid and (2) the peroxidase-dependent oxidation of a reducing cosubstrate, in which the oxidant is a higher oxidation state of the enzyme. Assays for these reactions are presented below; an assay for the oxygenation of nonreducing cosubstrates was presented in an earlier volume of this series.[21]

Arachidonic Acid Oxygenation

Principle. Oxygenation of arachidonic acid (AA) to PGG_2 by PHS cyclooxygenase is the rate-limiting step in prostaglandin biosynthesis. An initial hydrogen abstraction, resulting in a carbon-centered AA radical, leads to a series of sequential reactions in which two molecules of molecular oxygen are incorporated to give PGG_2.[9] The substrate requirement of the cyclooxygenase is quite specific, namely, a methylene-interrupted trienoic fatty acid in which the last double bond is located six carbon atoms from the methyl terminus.[22] Various PHS enzyme preparations may be tested using this assay, in which the uptake of molecular oxygen is monitored. This assay is also useful in preliminary assessment of cyclooxygenase inhibitors and peroxidase reducing cofactors: addition of the former results in decreased oxygen consumption, whereas addition of the latter results in increased oxygen consumption owing to protection of PHS from self-deactivation.

Materials

Potassium phosphate buffer, 1 M, pH 7.8
PHS enzyme preparation[23] (e.g., microsomes from ram seminal vesicle or rabbit kidney medulla, 10 mg/ml)
Arachidonic acid, 20 mM in ethanol

Protocol. Reaction mixtures consist of 0.1 M phosphate buffer and 1 mg/ml of microsomal enzyme preparation at 37°. The reaction is initiated by addition of arachidonic acid to 100 μM and can be monitored by following the utilization of molecular oxygen using a Clark-type oxygen electrode and strip chart recorder. Oxygenation is relatively rapid, becoming maximal during the first 20–30 sec and nearing completion in less

[20] L. J. Marnett, M. J. Bienkowski, and W. R. Pagels, *J. Biol. Chem.* **254**, 5077 (1979).
[21] P. H. Siedlik and L. J. Marnett, this series, Vol. 105, p. 412.
[22] S. Bergstrom, H. Danielson, and B. Samuelsson, *Biochem. Biophys. Acta* **90**, 207 (1964).
[23] J. A. Boyd and T. E. Eling, *J. Pharmacol. Exp. Ther.* **219**, 659 (1981).

than 2 min. The effect of cyclooxygenase inhibitors (e.g., indomethacin) or peroxidase reducing cofactors (e.g., phenol) may be assayed by adding the inhibitors to the reaction mixture 2–3 min prior to the addition of arachidonic acid.

Comment. Purified PHS (>50,000 units/mg protein) is now commercially available, and may be assayed with a similar protocol. Generally, 1–2 × 10³ units/ml of enzyme are used. A reducing cofactor such as phenol, 100 μM, may also be added prior to the addition of arachidonic acid.

Reducing Cofactor Oxidation

Principle. The peroxidase activity of PHS is much less substrate specific than that of the cyclooxygenase. In addition to PGG$_2$, a wide variety of organic hydroperoxides may serve as substrates, there being a preference for alkyl hydroperoxides.[4] Following the reduction of peroxide to alcohol (PGG$_2$ to PGH$_2$), two molecules of suitable cofactor may donate one electron each in the sequential reduction of the peroxidase back to its native state.[13] The resultant free radical metabolites may then react with molecular oxygen, producing a measurable oxygen uptake response as described above for the cyclooxygenase reaction. The antiinflammatory drug phenylbutazone is one such substrate,[24] commonly employed in studies of PHS peroxidase.

Materials

Potassium phosphate buffer, 1 M, pH 7.0
PHS microsomal enzyme preparation, 10 mg/ml
H$_2$O$_2$, 20 mM in water, or 15-hydroperoxy-5,8,11,13-eicosatetraenoic acid (15-HPETE)[25]
Phenylbutazone, 50 mM in ethanol (new stock solutions should be made before each experiment)

Protocol. Reaction mixtures consist of 0.1 M phosphate buffer, 0.4 mg/ml of PHS microsomal enzyme preparation, and 500 μM phenylbutazone. The reaction is initiated by the addition of hydroperoxide to 100 μM, and it may be monitored as above by following the consumption of molecular oxygen. A rapid time course is essentially complete after approximately 60 sec.

Comment. Other compounds may serve as reducing cofactors for PHS peroxidase, but their one-electron oxidation products do not trap molecu-

---

[24] L. J. Marnett. T. A. Dix, R. J. Sachs, and P. H. Siedlik, *in* "Advances in Prostaglandin, Thromboxane, and Leukotriene Research" (B. Samuelsson, R. Paoletti, and P. Ramwell, eds.), Vol. 11, p. 79. Raven, New York, 1983.

[25] M. O. Funk, R. Isaac, and N. A. Porter, *Lipids* **11,** 113 (1976).

lar oxygen to any appreciable degree; one example is 2-aminofluorene.[26] Such compounds may be identified as potential reducing cofactors for the PHS peroxidase by their inhibition of phenylbutazone oxygenation. The compound to be studied is placed in the reaction mixture with phenylbutazone prior to the addition of peroxide, and its effect on oxygen consumption is monitored.

It should also be noted that for PHS peroxidase-dependent oxidations of reducing cofactors, classical Michaelis–Menton kinetics are inapplicable. A linear phase of reaction is not observed during oxygen uptake measurements; initial rates must therefore be extrapolated from points early in the reaction curves.

Acknowledgments

The author is indebted to Dr. Thomas E. Eling of the Laboratory of Molecular Biophysics, National Institute of Environmental Health Sciences.

[26] J. A. Boyd and T. E. Eling, *J. Biol. Chem.* **259,** 13885 (1984).

[30] DT-Diaphorase: Purification, Properties, and Function

By Christina Lind, Enrique Cadenas, Paul Hochstein, and Lars Ernster

In 1958, Ernster and Navazio[1,2] reported the occurrence of a highly active diaphorase in the soluble fraction of rat liver homogenates, which catalyzed the oxidation of NADH and NADPH at equal rates. A partial purification and some properties of the enzyme were described in 1960 by Ernster *et al.*[3] They named the enzyme DT-diaphorase because of its reactivity with both NADH and NADPH (at that time DPNH and TPNH). The same authors subsequently published a detailed report on DT-diaphorase, which included its purification, assay conditions, data regarding kinetics, electron acceptors, activators, and inhibitors, as well as a comparison of the enzyme with various diaphorases and quinone reductases earlier described in the literature.[4]

[1] L. Ernster and F. Navazio, *Acta Chem. Scand.* **12,** 595 (1958).
[2] L. Ernster, *Fed. Proc., Fed. Am. Soc. Exp. Biol.* **17,** 216 (1958).
[3] L. Ernster, M. Ljunggren, and L. Danielson, *Biochem. Biophys. Res. Commun.* **2,** 88 (1960).
[4] L. Ernster, L. Danielson, and M. Ljunggren, *Biochim. Biophys. Acta* **58,** 171 (1962).

Over the past three decades DT-diaphorase has been extensively studied from various points of view, including its structure and reaction mechanism, its biosynthesis and induction, its role as a protective device against the cytotoxicity and mutagenicity of quinone-derived oxygen radicals, and its involvement in vitamin K-dependent protein carboxylation. These aspects of DT-diaphorase have been reviewed on several occasions,[5-10] including a chapter in a previous volume of this series.[6] A collection of papers summarizing more recent information is found in a symposium volume[11] published in 1987.

Occurrence and Intracellular Distribution

The flavoprotein DT-diaphorase [NAD(P)H : quinone oxidoreductase; EC 1.6.99.2, NAD(P)H dehydrogenase (quinone)] catalyzes the two-electron reduction of quinoid compounds to hydroquinones:

$$Q + NAD(P)H + H^+ \rightarrow QH_2 + NAD(P)^+$$

where Q is a quinone. It also reacts with a number of other electron acceptors, for example, various dyes and nitro compounds (see below). DT-Diaphorase is widely distributed in the animal kingdom, although its activity varies greatly from one species to another. Diaphorases reacting with both NADH and NADPH have been described in plants and bacteria, but their relationship to the animal DT-diaphorase remains unclear.[12]

Among animal tissues, liver is one of the richest sources of the enzyme, but other tissues, including brain, heart, lung, kidney, small intestine, skeletal muscle, and mammary gland, exhibit varying DT-diaphorase activities.[12] On fractionation of liver homogenates, the bulk (>90%) of

[5] L. Ernster, *in* "Biological Structure and Function" (T. W. Goodwin and O. Lindberg, eds.), Vol. 2, p. 139. Academic Press, New York, 1961.

[6] L. Ernster, this series, Vol. 10, p. 309.

[7] J. M. Hall, C. Lind, M. P. Golvano, B. Rase, and L. Ernster, *in* "Structure and Function of Oxidation Reduction Enzymes" (Å. Åkeson and A. Ehrenberg, eds.), p. 433. Pergamon, Oxford, 1972.

[8] C. Lind, P. Hochstein, and L. Ernster, *in* "Symposium on Oxidases and Related Redox Systems" (T. E. King, H. S. Mason, and M. Morrison, eds.), p. 321. Pergamon, New York, 1982.

[9] P. Hochstein, A. S. Atallah, and L. Ernster, *in* "Cellular Antioxidant Defense Mechanisms" (C. K. Chow, ed.), Vol. 2, p. 123. CRC Press, Boca Raton, Florida, 1988.

[10] L. Ernster, *Chem. Scr.* **27A**, 1 (1987).

[11] L. Ernster, R. W. Estabrook, P. Hochstein, and S. Orrenius (eds.), "DT-Diaphorase: A Quinone Reductase with Special Functions in Cell Metabolism and Detoxication," *Chem. Scr.* **27A** (1987).

[12] C. Martius, *in* "The Enzymes" (P. D. Boyer, H. Lardy, and K. Myrbäck, eds.), Vol. 7, p. 517. Academic Press, New York, 1963.

DT-diaphorase is recovered in the cytosolic fraction, whereas minor portions of the enzyme are associated with mitochondria and microsomes.[1-3,13] Appreciable DT-diaphorase activities are found in different tumors[14-25] and in preneoplastic liver nodules[16-19,26-33] as well as in various cell cultures.[34-41] In the latter case, a nitroreductase enzyme has been isolated from Walker 256 rat carcinoma cells which can convert 5-(aziridin-1-yl)-2,4-dinitrobenzamide to a cytotoxic DNA interstrand cross-

[13] L. Danielson, L. Ernster, and M. Ljunggren, *Acta Chem. Scand.* **14**, 1837 (1960).

[14] R. E. Beyer, J. E. Segura-Aguilar, and L. Ernster, *Anticancer Res.* **8**, 233 (1988).

[15] E. E. Gordon, L. Ernster, and G. Dallner, *Cancer Res.* **27**, 1371 (1967).

[16] N. A. Schor, B. F. Rice, and R. E. Huseby, *Proc. Soc. Exp. Biol. Med.* **151**, 418 (1976).

[17] N. A. Schor and H. P. Morris, *Cancer Biochem. Biophys.* **2**, 5 (1977).

[18] N. A. Schor and C. J. Cornelisse, *Cancer Res.* **43**, 4850 (1983).

[19] N. A. Schor, K. Ogawa, G. Lee, and E. Farber, *Cancer Lett.* **5**, 167 (1978).

[20] F. F. Becker and D. L. Stout, *Carcinogenesis* **5**, 785 (1984).

[21] J. Koudstaal, B. Makkink, and S. H. Overdiep, *Eur. J. Cancer* **2**, 111 (1975).

[22] G. Batist, K. H. Cowan, G. Curt, A. G. Katki, and C. E. Myers, *Proc. Amer. Assoc. Cancer. Res.* **26**, 345 (1985).

[23] S. R. Keyes, P. M. Fracasso, D. C. Heimbrok, S. Rockwell, S. G. Sligar, and A. C. Sartorelli, *Cancer Res.* **44**, 5638 (1984).

[24] R. E. Talcott, M. Rosenblum, and V. A. Levin, *Biochem. Biophys. Res. Commun.* **111**, 346 (1983).

[25] M. S. Berger, R. E. Talcott, M. L. Rosenblum, M. Silva, R. Aliosman, and M. T. Smith, *J. Toxicol. Environ. Health* **16**, 713 (1985).

[26] R. Cameron, G. D. Sweeney, K. Jones, G. Lee, and E. Farber, *Cancer Res.* **36**, 3888 (1976).

[27] K. Okita, K. Noda, Y. Fukumoto, and T. Takemoto, *Gann* **67**, 899 (1976).

[28] M. J. Griffin and D. E. Kizer, *Cancer Res.* **38**, 1137 (1976).

[29] W. Levin, A. Y. H. Lu, P. E. Thomas, D. Ryan, D. Kizer, and M. J. Griffin, *Proc. Natl. Acad. Sci. U.S.A.* **75**, 3240 (1978).

[30] P. Bentley, F. Waechter, F. Oesch, and W. Stäubli, *Biochem. Biophys. Res. Commun.* **91**, 1101 (1979).

[31] R. N. Sharma, H. L. Gurtoo, E. Farber, R. K. Murray, and R. G. Cameron, *Cancer Res.* **41**, 3311 (1981).

[32] A. Kitahara, K. Satoh, and K. Sato, *Biochem. Biophys. Res. Commun.* **112**, 20 (1983).

[33] A. Åström, J. W. DePierre, and L. Eriksson, *Carcinogenesis* **4**, 577 (1983).

[34] K. Kumaki, N. M. Jensen, J. G. M. Shire, and D. W. Nebert, *J. Biol. Chem.* **252**, 157 (1977).

[35] R. M. Robertson, D. W. Nebert, and O. Hankinson, *Chem. Scr.* **27A**, 83 (1987).

[36] D. Galaris and J. Rydström, *Biochem. Biophys. Res. Commun.* **110**, 364 (1983).

[37] D. Galaris, A. Georgellis, and J. Rydström, *Biochem. Pharmacol.* **34**, 989 (1985).

[38] H. J. Prochaska, M. J. De Long, and P. Talalay, *Proc. Natl. Acad. Sci. U.S.A.* **82**, 8232 (1985).

[39] A. S. Atallah, J. R. Landolph, L. Ernster, and P. Hochstein, *Biochem. Pharmacol.* **37**, 2451 (1988).

[40] M. J. De Long, H. J. Prochaska, and P. Talalay, *Proc. Natl. Acad. Sci. U.S.A.* **83**, 787 (1986).

[41] R. E. Beyer, J. E. Segura-Aguilar, C. Lind, and V. Castro, *Chem. Scr.* **27A**, 145 (1987).

linking agent; the enzyme has been recently identified as a form of NAD(P)H : quinone reductase or DT-diaphorase.[42]

Purification

Most of our present knowledge about DT-diaphorase has been obtained in studies of the cytosolic enzyme, which was first purified from both rat[9,10,43] and beef liver[44] by employing conventional methods. The introduction of affinity chromatography[45–47] for the purification of DT-diaphorase reduced the number of purification steps considerably. In the present method for the purification of DT-diaphorase,[47] the key isolation step is the biospecific adsorption of the enzyme to immobilized dicoumarol, which is a potent competitive inhibitor of the enzyme with respect to NAD(P)H.[4] A gel with both a high affinity and a high binding capacity for DT-diaphorase is obtained by coupling dicoumarol through an azo linkage to divinyl sulfonate (DVS)-activated Sepharose 6B as follows.

Sepharose 6B (100 g wet weight) in 100 ml of 0.5 M Na$_2$CO$_3$ (pH 11) is allowed to react with 6 ml DVS for 2 hr at room temperature. The cross-linked and activated gel is thoroughly washed with water prior to coupling 4-aminobenzoic acid hydrazide to its free vinyl groups. The coupling reaction is initiated by the addition of 0.5 g of 4-aminobenzoic acid hydrazide (dissolved in 50 ml of 0.5 M Na$_2$CO$_3$, pH 11) to the sedimented gel, and the reaction mixture is kept overnight at room temperature. Unreacted vinyl groups are inactivated by the addition of 1 ml of 2-mercaptoethanol for 90 min. The derivatized gel is washed extensively with water, ethanol, and acetone. The sedimented, slightly brownish gel is diazotized by addition of 1 M HCl and 50 ml of 1 M NaNO$_2$. The reaction is allowed to proceed for 10 min at 4°. Dicoumarol (2 g, freshly dissolved in 20 ml of 1 M NaOH and 20 ml of 0.5 M Na$_2$CO$_3$, pH 11) is added to the diazotized gel slurry, and the mixture is rapidly adjusted to pH 10–11. The coupling of dicoumarol to the diazotized gel is terminated after 2 hr at 4° by extensive washing of the affinity gel with water until neutral pH is obtained. Prior to use, the affinity gel is washed with 0.5 M Na$_2$CO$_3$ (pH 11) to remove unbound dicoumarol. The affinity gel contains approximately 0.4 μmol bound ligand per ml gel and has a capacity to bind 1.5–2

[42] R. J. Knox, M. P. Boland, F. Friedlos, B. Coles, C. Southan, and J. J. Roberts, *Biochem. Pharmacol.* **37**, 4671 (1988).

[43] S. Hosoda, W. Nakamura, and K. Hayashi, *J. Biol. Chem.* **249**, 6416 (1974).

[44] F. Märki and C. Martius, *Biochem. Z.* **333**, 111 (1960).

[45] B. Rase, T. Bartfai, and L. Ernster, *Arch. Biochem. Biophys.* **172**, 380 (1976).

[46] R. Wallin, D. Gebhardt, and H. Prydz, *Biochem. J.* **169**, 95 (1978).

[47] B. Höjeberg, K. Blomberg, S. Stenberg, and C. Lind, *Arch. Biochem. Biophys.* **207**, 205 (1981).

mg enzyme per ml gel. It is not affected by repeated usage for at least 3 months at 4°, and it can be stored at −20° in the presence of 30% (v/v) ethylene glycol for at least 6 months with no change in binding capacity.

A homogeneous preparation of DT-diaphorase is obtained in high yield from rat liver cytosol by affinity chromatography and subsequent gel filtration as summarized in Table I. The detailed procedure is given below.

All purification steps are performed at 4°. Livers from 10–20 rats are homogenized in 0.25 M sucrose, and the homogenate (10% wet weight by volume) is centrifuged at 105,000 g for 1 hr. The supernatant is filtered through cheesecloth and applied at a flow rate of 1.5 ml/min to an azodi-coumarol-Sepharose 6B column (2 × 3.5 cm) equilibrated with 0.25 M sucrose in 50 mM Tris-HCl, pH 7.5 (buffer A). After application of the sample, the gel is washed first with buffer A until no protein is detected in the effluent, then with 2 M KCl in 0.25 M sucrose–50 mM Tris-HCl, pH 8, to remove nonspecifically adsorbed proteins. Prior to elution of the enzyme, buffer A is applied to the column in the opposite direction, and the flow rate is decreased to 0.5 ml/min. Elution of DT-diaphorase is achieved by a pulse of 10 ml of 20 mM NADH in buffer A. The fractions containing DT-diaphorase activity are pooled and concentrated by ultrafiltration in a Diaflo cell (Amicon Corp., Lexington, MA) equipped with a PM10 filter. The concentrated sample is applied at a flow rate of 0.5 ml/min to a Sephacryl S-200 column (3.2 × 55 cm) equilibrated with buffer A. Fractions exhibiting a constant ratio of protein to flavin (Fig. 1) are pooled.

The purified enzyme can be stored at −20° for several months without appreciable loss in activity. It should be noted that the presence of 0.25 M

TABLE I

PURIFICATION OF HEPATIC DT-DIAPHORASE FROM
3-METHYLCHOLANTHRENE-TREATED RATS[a]

Purification step	Total protein (mg)	Total activity (μmol/min)		Specific activity (μmol/min/mg)		Yield (%)
		DCPIP	K_3	DCPIP	K_3	
105,000 g supernatant	855.0	8640	17390	10	20	100
Azodicoumarol-Sepharose 6B	19.7	6420	13040	326	662	75
Sephacryl S-200	3.9	4728	8886	1212	2452	55

[a] DCPIP, 2,6-Dichlorophenol-indophenol; K_3, 2-methyl-1,4-naphthoquinone or menadione.

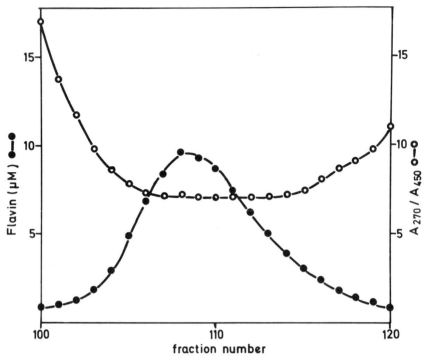

FIG. 1. Elution profile of DT-diaphorase from Sephacryl S-200. The various fractions were analyzed by absorption spectrometry between 600 and 240 nm. The flavin content was calculated from the absorbance at 450 nm employing an extinction coefficient of 11.3 mM^{-1} cm^{-1}. (From Ref. 47.)

sucrose in all buffers is essential for stabilizing DT-diaphorase throughout the purification as well as during storage at $-20°$. Liver mitochondrial and microsomal DT-diaphorase can also be purified by this method,[48] after solubilization of the membranes with deoxycholate in the presence of glycerol and dithioerythritol.

Assay

DT-diaphorase activity is measured routinely with NADH or NADPH as the electron donor and 2,6-dichlorophenol-indophenol (DCPIP) or 2-methyl-1,4-naphthoquinone (menadione) as the electron acceptor. The standard assay system contains 50 mM Tris-HCl, pH 7.5, 0.08% Triton X-100, 0.5 mM NADH or NADPH, and 40 μM DCPIP or 10 μM menadione. When menadione is used as electron acceptor, 77 μM cytochrome c is

[48] C. Lind and B. Höjeberg, *Arch. Biochem. Biophys.* **207,** 217 (1981).

included in the assay system in order to reoxidize continuously the menadiol formed. The reaction is started by the addition of the enzyme. The reduction of DCPIP is followed at $\lambda_{600 \text{ nm}}$ ($\varepsilon = 21 \text{ m}M^{-1} \text{ cm}^{-1}$) and that of cytochrome c at $\lambda_{550 \text{ nm}}$ ($\varepsilon = 18.5 \text{ m}M^{-1} \text{ cm}^{-1}$).

In nonpurified preparations, such as cell fractions, DT-diaphorase can be suitably assayed with NADH or NADPH as the electron donor and menadione plus cytochrome c as the electron acceptor (which results in a linear reaction velocity for a period longer than that found with other electron acceptors). A fairly selective measure of the DT-diaphorase activity is that portion of the reaction rate which is inhibited by $10^{-5} M$ dicoumarol; this rate should be equal with either electron donor, NADH or NADPH.

Properties

Cytosolic DT-diaphorase has a molecular weight of about 55,000 and contains two equally sized subunits and two molecules of FAD. The primary structure of DT-diaphorase has been deduced from cDNA sequences.[49–51] Further, essential tyrosine and lysine residues may be present in the active site of the enzyme in hydrophobic and hydrophilic domains, respectively.[52] The enzyme has also been obtained from murine liver in a crystalline form suitable for X-ray analysis.[53]

DT-diaphorase reacts with NADH or NADPH as electron donors with equal maximal velocities;[4] the K_m value for NADH is somewhat higher than that for NADPH (45 and 85 μM, respectively). As electron acceptors, DCPIP and certain benzo- and naphthoquinones are most active, and cytochrome b_5 and c are practically inactive. Among the quinones, those without a side chain are the most active electron acceptors, and the activity decreases with increasing length of the side chain. The naturally occurring forms of vitamin K react with the enzyme only at a very slow rate, but this can be enhanced by incorporating the vitamin into liposomes.[54] Ubiquinones with a long side chain (e.g., coenzyme Q_{10}) are virtually inactive as electron acceptors for DT-diaphorase.[4] Benzo[a]-pyrene quinones have been shown to act as electron acceptors for

[49] J. A. Robertson, H.-C. Chen, and D. W. Nebert, *J. Biol. Chem.* **261**, 15794 (1986).
[50] R. M. Bayney, J. A. Rodkey, C. D. Bennett, A. Y. H. Lu, and C. B. Pickett, *J. Biol. Chem.* **262**, 572 (1987).
[51] A. K. Jaiswal, O. W. McBride, M. Adesnik, and D. W. Nebert, *J. Biol. Chem.* **263**, 13572 (1988).
[52] M. Haniu, H. Yuan, S. Chen, T. Iyanagi, T. D. Lee, and J. E. Shively, *Biochemistry* **27**, 6877 (1988).
[53] H. P. Prochaska and P. Talalay, *Chem. Scr.* **27A**, 43 (1987).
[54] C. Martius, R. Ganser, and A. Viviani, *FEBS Lett.* **59**, 13 (1975).

DT-diaphorase in the course of their conjugation with glucuronate.[55–58] Anthracycline antibiotics and nitro compounds,[59–62] azo dyes,[63–65] hexavalent chromium compounds,[66] p-benzo- and naphthoquinone epoxides,[67,68] naphthoquinone–glutathione conjugates,[69] quinoneimines,[70] and dopamine quinone[71] have also been shown to serve as electron acceptors for DT-diaphorase. It is noteworthy that a poor substrate for DT-diaphorase in terms of V_{max} or K_{cat} does not necessarily reflect the affinity of the enzyme for that substrate.[55] It has been shown that DT-diaphorase reduces benzo[a]pyrene quinones at very modest rates, but its affinity for these quinones is very high. For example, benzo[a]pyrene quinones compete effectively with menadione, thereby inhibiting the reduction of the latter by DT-diaphorase.[55] Table II provides a list of quinoid- and nonquinoid electron acceptors for DT-diaphorase.

A prominent feature of cytosolic DT-diaphorase, which is especially pronounced on purification, is its requirement for a nonpolar milieu. Phospholipids, nonionic detergents, polyols (such as sucrose), and serum albumin are activators of the enzyme; this feature might be of physiological significance, particularly because a fraction of DT-diaphorase is bound to the membranes of mitochondria and endoplasmic reticulum.

Dicoumarol, an anticoagulant, is the strongest known inhibitor of DT-diaphorase. The inhibition is competitive with respect to the electron donors NADH and NADPH.[4] The K_i values are in the range 10^{-8} to 10^{-10}

[55] C. Lind, Ph.D. thesis, University of Stockholm, 1978.
[56] C. Lind, H. Vadi, and L. Ernster, Arch. Biochem. Biophys. 190, 97 (1978).
[57] C. Lind, Biochem. Pharmacol. 34, 895 (1985).
[58] C. Lind, Arch. Biochem. Biophys. 240, 226 (1985).
[59] T. Sugimura, K. Okabe, and M. Nagao, Cancer Res. 26, 1717 (1966).
[60] M. Nagao and T. Sugimura, Adv. Cancer Res. 23, 131 (1966).
[61] K. Tatsumi, N. Koga, S. Kitamura, H. Yoshimura, P. Wardman, and Y. Kato, Biochim. Biophys. Acta 567, 75 (1979).
[62] H. Tsuda, D. Yoshida, and S. Mizusaki, Carcinogenesis 5, 331 (1984).
[63] M.-T. Huang, G. T. Miwa, and A. Y. H. Lu, J. Biol. Chem. 254, 3930 (1979).
[64] M.-T. Huang, G. T. Miwa, N. Cronheim, and A. Y. H. Lu, J. Biol. Chem. 254, 11223 (1979).
[65] J. G. Dent, M. E. Graichen, R. W. Roth, and J. R. Heys, Anal. Biochem. 112, 299 (1981).
[66] S. De Flora, A. Morelli, C. Basso, M. Ramano, D. Serra, and A. De Flora, Cancer Res. 45, 3188 (1985).
[67] A. Brunmark, E. Cadenas, C. Lind, J. Segura-Aguilar, and L. Ernster, Free Radical Biol. Med. 3, 181 (1987).
[68] A. Brunmark, E. Cadenas, J. Segura-Aguilar, C. Lind, and L. Ernster, Free Radical Biol. Med. 5, 133 (1988).
[69] G. D. Buffinton, K. Öllinger, A. Brunmark, and E. Cadenas, Biochem. J. 257, 561 (1988).
[70] G. Powis, K. L. See, K. S. Santone, D. C. Melder, and E. M. Hodnett, Biochem. Pharmacol. 36, 2473 (1987).
[71] J. Segura-Aguilar and C. Lind, Chem.-Biol. Interact. 72, 309 (1989).

TABLE II
Electron Acceptors for DT-Diaphorase[a]

Acceptor	K_{cat} (nmol/min/μg enzyme)
Quinoid compounds	
I. p-Benzoquinone series	
Unsubstituted	
p-Benzoquinone	3345
Methyl-substituted	
2-Methyl-p-benzoquinone	2578
2,6-Dimethyl-p-benzoquinone	2648
Methyl- and methoxyl-substituted	
2-Methyl-5,6-dimethoxy-1,4-p-benzoquinone (CoQ$_0$)	2370
Methyl-, methoxyl-, and prenoyl-substituted	
2-Methyl-3-diprenoyl-5,6-dimethoxy-p-benzoquinone (CoQ$_2$)	465
2-Methyl-3-decaprenoyl-5,6-dimethoxy-p-benzo-quinone (CoQ$_{10}$)	3.38
p-Benzoquinone epoxides	
2,3-Epoxy-p-benzoquinone	10.40
2-Methyl-2,3-epoxy-p-benzoquinone	4.20
2,6-Dimethyl-2,3-epoxy-p-benzoquinone	0.36
II. Naphthoquinone series	
Unsubstituted	
1,4-Naphthoquinone	1252
1,2-Naphthoquinone	1069
Methyl-substituted	
2-Methyl-1,4-naphthoquinone (vitamin K$_3$)	2323
2,3-Dimethyl-1,4-naphthoquinone	1252
Methyl- and phytyl-substituted	
2-Methyl-3-phytyl-1,4-naphthoquinone (vitamin K$_1$)	3
Methoxyl-substituted	
2,3-Dimethoxy-1,4-naphthoquinone	1075
Hydroxyl-substituted (quinoid ring)	
2-Hydroxyl-1,4-naphthoquinone	13
Hydroxyl-substituted (benzene ring)	
5-Hydroxy-1,4-naphthoquinone	896
2-Methyl-5-hydroxy-1,4-naphthoquinone	1312
5,8-Dihydroxy-1,4-naphthoquinone	1301
Glutathionyl-substituted	
3-Glutathionyl-1,4-naphthoquinone	907
2,3-Diglutathionyl-1,4-naphthoquinone	155
2-Methyl-3-glutathionyl-1,4-naphthoquinone	1208
3-Glutathionyl-5-hydroxy-1,4-naphthoquinone	310
2-Methyl-3-glutathionyl-5-hydroxy-1,4-naphthoquinone	619
3-Glutathionyl-5,8-dihydroxy-1,4-naphthoquinone	256
Naphthoquinone epoxides	
2,3-Epoxy-1,4-naphthoquinone	0.27
2,3-Epoxy-2-methyl-1,4-naphthoquinone	0.18
2,3-Epoxy-2,3-dimethyl-1,4-naphthoquinone	0.02

(continued)

TABLE II (*continued*)

Acceptor	K_{cat} (nmol/min/μg enzyme)
III. Quinoneimines	
N-Acetyl-p-benzoquinoneimine	238
2-Amino-1,4-naphthoquinoneimine	36
N,N-Dimethylindoaniline	14
2-Acetamido-N,N-dimethylindoaniline	13
IV. Other quinones	
Benzo[a]pyrene-3,6-quinone	11
Cyclized dopamine o-quinone	23
9,10-Phenanthrenequinone	242
Nonquinone compounds	
I. Organic compounds	
2,6-Dichlorophenol-indophenol	997
Methylene blue	459
Cytochrome c	0.1
II. Inorganic compounds	
Ferricyanide	747

[a] Data from Refs. 10, 55–58, 67–71 and from H. Thor, M. T. Smith, P. Hartzell, G. Bellomo, S. A. Jewell, and S. Orrenius, *J. Biol. Chem.* **257,** 12419 (1982); J. Segura-Aguilar and J. Llopis (unpublished results).

M. Related coumarin derivatives, for example, warfarin, as well as derivatives of inandione, another group of anticoagulants, also inhibit DT-diaphorase in a competitive fashion with respect to NAD(P)H,[7,72–74] but the K_i values are considerably higher. It is noteworthy that the inhibitions by coumarin and inandione derivatives are not additive but synergistic. Other, less effective inhibitors of DT-diaphorase include flavins and flavin antagonists, thyroxine and related compounds, and p-chloromercuribenzoate.[4,6]

A unique property of DT-diaphorase is that the enzyme transfers two electrons,[75] at variance with other NAD(P)H-linked quinone reductases. As discussed below, this feature is of crucial importance for the role of DT-diaphorase as a quinone reductase in connection with quinone detoxication, that is, with conjugation reactions.

Like many flavoenzymes, DT-diaphorase reacts with its electron donor and acceptor according to a binary complex or Ping-Pong mecha-

[72] P. M. Hollander, Ph.D. thesis, University of Stockholm, 1975.
[73] L. Ernster, J. M. Hall, C. Lind, B. Rase, and M. P. Golvano, *in* "Atti del Seminario di Studi Biologici" (E. Quagliariello, ed.), Vol. 5, p. 217. Adriatica Editrice, Bari, 1973.
[74] P. M. Hollander and L. Ernster, *Arch. Biochem. Biophys.* **169,** 560 (1975).
[75] T. Iyanagi and I. Yamazaki, *Biochim. Biophys. Acta* **216,** 282 (1970).

nism.[7,73] A detailed study[72] on the reaction mechanism of DT-diaphorase suggested the existence of two kinetically distinct species of the enzyme. Immunological studies, together with the finding that purified DT-diaphorase contains two different amino-terminal amino acids, had indicated the existence of multiple forms of the enzyme. Indeed, recently several forms of DT-diaphorase differing in isoelectric points and substrate specificity, possibly representing isozymes, have been isolated.[76] The finding that the proportion between the various forms of DT-diaphorase is different in control and 3-methylcholanthrene-treated rats might have an interesting implication regarding the mechanism of regulation.

DT-Diaphorase is induced by a large number of xenobiotics, both in various tissues of rats and mice *in vivo* and in cell cultures. The inducers include various polycyclic hydrocarbons and azo dyes, polychlorinated dioxins and diphenyls, 2(3)-*tert*-butyl-4-hydroxyanisole and related antioxidants, mercapto- and thionosulfur compounds, *trans*-stilbene oxide, β-naphthoflavone, 2-acetylaminofluorene, coumarin, quercitin, ethoxyquin, *tert*-butylhydroquinone, daunorubicin, and phenobarbital. For a comprehensive list of inducers of DT-diaphorase, see Ref. 77. In all cases investigated, the induction has been shown to involve a net increase in newly synthesized DT-diaphorase. The induction of DT-diaphorase is controlled by an interaction of two genes, one of which is identical with the *Ah* (aryl hydrocarbon hydroxylase) gene.[49–51] On the other hand, evidence has been brought forward that the *Ah* gene is involved only in those cases when the inducer requires oxygenation by the aryl hydrocarbon hydroxylase system in order to activate the DT-diaphorase gene, whereas compounds which already are redox-labile are capable of inducing DT-diaphorase even in *Ah*-negative mutants.[77] In dioxin-treated human hepatoblastoma cells, DT-diaphorase has been shown to be encoded by a single gene located on chromosome 16. The human DT-diaphorase cDNA and protein are 83 and 85% similar to rat liver cytosolic DT-diaphorase cDNA and protein, respectively.[51]

Biological Significance of DT-Diaphorase

Since the 1950s, the biological significance of DT-diaphorase has been outlined from different points of view, including (1) its involvement in electron transfer and oxidative phosphorylation in the mitochondrial respiratory chain, (2) its involvement in the vitamin K-dependent protein carboxylation, and (3) its role as a quinone reductase in connection with

[76] J. Segura-Aguilar and C. Lind, *Chem. Scr.* **27A,** 37 (1987).
[77] P. Talalay and H. J. Prochaska, *Chem. Scr.* **27A,** 61 (1987).

the detoxication of quinones. The early proposal[78] of the involvement of DT-diaphorase in electron transfer and oxidative phosphorylation in the NADH–cytochrome *b* region of the respiratory chain proved to be unrealistic.[79] DT-Diaphorase is not a member of the respiratory chain and can even constitute a bypass over the normal, amytal- and rotenone-sensitive, phosphorylating electron transfer between NADH and cytochrome *b*, provided that an artificial electron mediator, such as menadione, is added.[79]

The high sensitivity of DT-diaphorase to dicoumarol and other anticoagulant drugs has drawn attention to its possible involvement in the vitamin K-dependent biosynthesis of prothrombin and other proteins that undergo posttranslational carboxylation. In the course of prothrombin biosynthesis vitamin K undergoes an oxidation–reduction cycle, with the intermediate formation of its 2,3-epoxide; it has been proposed that DT-diaphorase may serve as a vitamin K reductase.[8,80] Evidence has been brought forward that there exists a vitamin K epoxide reductase, which was proposed to operate through a thiol–disulfide cycle[81] and to constitute an alternative pathway to that proceeding via DT-diaphorase.[82] DT-diaphorase itself is an active quinone epoxide reductase, capable of converting a quinone epoxide to a hydroxyhydroquinone.[67,68]

The suggestion that DT-diaphorase may play a role as a quinone reductase in connection with the detoxication of quinones is based mainly on two early findings. One is the predominant feature of the enzyme to catalyze two-electron transfers to several quinoid compounds with formation of relatively stable hydroquinones.[75] The other originates from the observation that rats and mice treated with small doses of various polycyclic hydrocarbons and azo dyes developed an increased resistance to the cytotoxic and carcinogenic effects of these compounds and, concomitantly, such a treatment caused an elevation in the cellular levels of DT-diaphorase.[83] Conversely, it has been shown that dicoumarol enhances quinone toxicity *in vivo*,[9] in cell cultures,[36,37,39] and in isolated cells.[84]

The proposed protective effect of DT-diaphorase against carcinogenicity, mutagenicity and other toxicities caused by quinones would reflect the two-electron reduction catalyzed by the enzyme, which would com-

[78] C. Martius, *in* "Ciba Foundation Symposium on Cell Metabolism" (G. E. W. Wolstenholme and C. M. O'Connor, eds.), p. 194. Churchill, London, 1959.

[79] T. E. Conover, L. Danielson, and L. Ernster, *Biochim. Biophys. Acta* **67**, 254 (1963).

[80] J. W. Suttie, *CRC Crit. Rev. Biochem.* **8**, 191 (1980).

[81] J. J. Lee and M. J. Fasco, *Biochemistry* **23**, 2246 (1984).

[82] R. Wallin, S. R. Rannels, and L. F. Martin, *Chem. Scr.* **27A**, 193 (1987).

[83] C. B. Huggins, E. Ford, and E. V. Jensen, *Science* **147**, 1153 (1965).

[84] H. Thor, M. T. Smith, P. Hartzell, G. Bellomo, S. A. Jewell, and S. Orrenius, *J. Biol. Chem.* **257**, 12419 (1982).

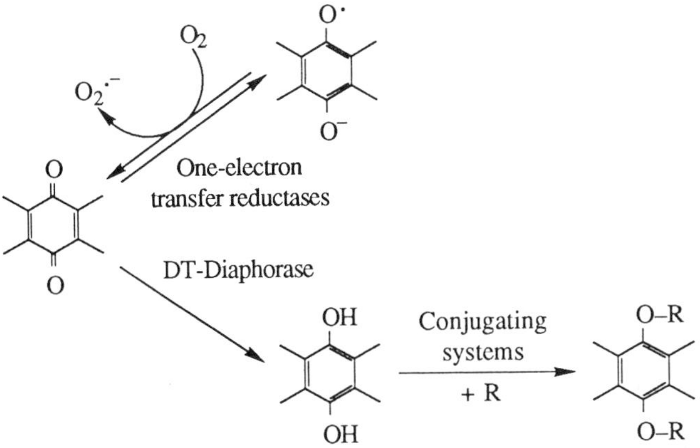

FIG. 2. Proposed role of DT-diaphorase in promoting hydroquinone formation and hydroquinone conjugation. R is UDPglucuronic acid or phosphoadenosine phosphosulfate.

pete with the formation of reactive oxygen species that may be generated as a result of one-electron transfer processes, such as those catalyzed, among others, by NADPH–cytochrome-P-450 reductase and NADPH–cytochrome b_5 reductase (Fig. 2). The stability of hydroquinones against autoxidation during DT-diaphorase catalysis can be understood in terms of their protonation state and the very positive reduction potential of the $Q^{2-}/Q^{\cdot-}$ couple. At the pH of the biological milieu, most two-electron reduced quinones are protonated, and, as such, they do not participate readily in electron transfer reactions. Although the reduction potential at the intermediate step ($Q^{2-} \rightleftharpoons Q^{\cdot-}$) varies for every quinone, in general it is more positive than that of the $O_2/O_2^{\cdot-}$ couple ($E = -155$ mV) and the autoxidation reaction ($Q^{2-} + O_2 \rightleftharpoons Q^{\cdot-} + O_2^{\cdot-}$), thus thermodynamically restricted, proceeds at very slow rates. DT-diaphorase catalyzes the formation of stable (against autoxidation) hydroquinones of the 1,4-naphthoquinone series without substituents or with methyl substituents[85]; however, the introduction of methoxyl, glutathionyl, or hydroxyl substituents (the latter independent of whether these are located in the quinone or the benzene ring) dramatically increases the reactivity of the hydroquinones formed during DT-diaphorase catalysis. For example, 1,4-naphthoquinone or its methyl derivatives are autoxidized poorly, whereas the glutathionyl- or hydroxyl-substituted compounds are autoxidized at rates 15- to 50-fold higher.[69] Similar considerations apply to the autoxidation of the hydroxyhydroquinones of the p-benzo- and 1,4-naphthoquinone series, which are formed subsequent to the reduction by DT-diaphorase of the corresponding quinone epoxides.[67,68]

[85] C. Lind, P. Hochstein, and L. Ernster, *Arch. Biochem. Biophys.* **216,** 175 (1982).

Superoxide dismutase has been shown to enhance the autoxidation of semiquinones of the 1,4-naphthoquinone series formed during NADPH–cytochrome-P-450 reductase catalysis,[86] an effect understood as a displacement of the equilibrium of the autoxidation reaction ($Q^{\cdot-} + O_2 \rightleftharpoons Q + O_2^{\cdot-}$) on removal of $O_2^{\cdot-}$ by superoxide dismutase. On the other hand, Cu,Zn-superoxide dismutase has been shown to inhibit the autoxidation of several hydroquinones of the p-benzo-,[87] o-benzo-,[71] and 1,4-naphthoquinone[86,87] series as well as that of dialuric acid[88] and 6-hydroxydopamine.[89] The inhibition of hydroquinone autoxidation by superoxide dismutase has been interpreted in terms of (1) a displacement of the equilibrium of the autoxidation reaction involving changes in the steady-state concentration of $O_2^{\cdot-}$ and in the reaction propagated by this species[88,89] and (2) a direct catalytic interaction of the free radical products with superoxide dismutase[86,87] involving semiquinone reduction at expense of $O_2^{\cdot-}$. The former possibility has been discussed assuming the $O_2^{\cdot-}$ is the propagating species in a free radical chain: it is formed during semiquinone autoxidation and consumed during hydroquinone autoxidation.[88] The latter possibility implies a mixed function of superoxide dismutase, which requires the simultaneous breakdown of the organic compound into a semiquinone intermediate and $O_2^{\cdot-}$. The thermodynamic restrictions imposed on this reaction could be overcome by the formation of a ternary complex ($^{\cdot-}Q$—Cu^{2+}—$O_2^{\cdot-}$), as proposed for the inhibition of adrenaline autoxidation by superoxide dismutase.[90] Independent of the molecular mechanism(s) by which superoxide dismutase inhibits the autoxidation of several hydroquinones, it is clear that this activity, connected to the two-electron transfer catalyzed by DT-diaphorase or, in some instances, to the reduced glutathione nucleophilic addition,[91] represents an efficient means of detoxication against quinone toxicity.

The increased activity of DT-diaphorase in several cancer cell lines is constitutive, not transiently induced,[92] and its role remains still unclear.

[86] K. Öllinger, G. Buffinton, L. Ernster, and E. Cadenas, *Chem.–Biol. Interact.* **73,** 53 (1990).
[87] E. Cadenas, D. Mira, A. Brunmark, C. Lind, J. Segura-Aguilar, and L. Ernster, *Free Radical Biol. Med.* **5,** 71 (1988).
[88] C. C. Winterbourn, W. B. Cowden, and H. C. Sutton, *Biochem. Pharmacol.* **38,** 611 (1989).
[89] P. Gee and A. Davison, *Free Radical Biol. Med.* **6,** 271 (1989).
[90] L. M. Shubotz, M. Younes, and U. Weser, *in* "Chemical and Biochemical Aspects of Superoxide and Superoxide Dismutase" (J. V. Bannister and H. A. O. Hill, eds.), p. 328. Elsevier/North Holland, Amsterdam, 1980.
[91] A. Brunmark and E. Cadenas, *Chem.–Biol. Interact.* **68,** 273 (1988).
[92] E. Farber, *Chem. Scr.* **27A,** 131 (1987).

This enhanced activity of DT-diaphorase has been proposed to play a role in the resistance of cancer cells against certain toxins and to be involved in the metabolic activation of carcinogens.[24,59,60] Several recent studies[67–69,86] indicate that the two-electron reduction of quinoid compounds by DT-diaphorase cannot be generalized as a detoxication process per se and that the subsequent redox transitions of the hydroquinone formed are a function of the physicochemical properties of the quinoid compound, which also influences the prospective role of superoxide dismutase on autoxidation.

[31] Antioxidant Activities of Bile Pigments: Biliverdin and Bilirubin

By ROLAND STOCKER, ANTONY F. MCDONAGH, ALEXANDER N. GLAZER, and BRUCE N. AMES

Introduction

Bile pigments are open-chain tetrapyrroles formed *in vivo* by the oxidative cleavage of ferriprotoporphyrin IX (heme), whereby biliverdin is produced as the primary catabolic product.[1] The pigments are present in almost all organisms ranging from primitive life forms to mammals.[2] In photosynthetic blue-green bacteria, algae, and in higher plants, biliverdin serves as the biosynthetic precursor of a family of bilins. These bilins are covalently attached to a group of proteins, the phycobiliproteins, which function as photosynthetic accessory pigments in cyanobacteria, red algae, and the cryptomonads.[3] In eukaryotic algae and higher plants, phytochromobilin, a bilin derived from biliverdin, is covalently attached to the photomorphogenetic protein phytochrome. Thus, bile pigments have a vital role in the physiology and development of these organisms. In mammals and some species of fish, biliverdin is reduced to bilirubin in a highly specific, NADPH-requiring reaction that is catalyzed by the cytosolic enzyme biliverdin reductase [bilirubin : NAD(P)$^+$ oxidoreductase, EC 1.3.1.24].[1]

[1] R. Schmid and A. F. McDonagh, *in* "The Metabolic Basis of Inherited Disease" (J. B. Stanbury and J. B. Wyngaarden, eds.), p. 1221. McGraw-Hill, New York, 1978.

[2] A. F. McDonagh, *in* "The Porphyrins" (D. Dolphin, ed.), Vol. 6, p. 293. Academic Press, New York, 1979.

[3] A. N. Glazer, *Annu. Rev. Biochem.* **52**, 125 (1983).

In adult humans about 300 mg of bilirubin is produced daily by reticuloendothelial cells. Because of its intramolecular hydrogen bonding, bilirubin is sparingly soluble in aqueous solutions at physiological pH and ionic strength[2] and in order to be transported in the circulation is tightly bound to albumin. Under physiological conditions, plasma bilirubin concentrations range from 5 to 20 μM, practically all of which is unconjugated pigment bound to albumin.[1] Bilirubin is removed from the circulation through uptake by hepatocytes, where it is bound to glutathione S-transferases before being transformed into a family of water-soluble derivatives by conjugation of one or both of its propionyl groups with glucuronic acid, glucose, or xylose. In human liver, conjugated bilirubin is present at concentrations between 20 and 40 μM[4] and becomes further concentrated (0.35 to 4 mM) after excretion into bile. Conjugated bilirubin finally reaches the intestine, where it is transformed into urobilinogens and urobilins, which are then excreted. From the foregoing it is evident that bilirubin and its derivatives have a wide variation in their physicochemical properties as well as in their location.

Biliverdin and bilirubin have been generally regarded as waste products of heme catabolism in higher animals. However, we have proposed recently that these bile pigments are biological antioxidants of potential importance[5] since, under conditions of physiological relevance, they scavenge peroxyl radicals with high efficiency. Some earlier work, which received little notice, also suggested that bile pigments might play a role as natural antioxidants.[6,7] Indeed, biliverdin and bilirubin are potent scavengers of singlet oxygen,[8,9] and a protective function of the biliverdin-related bilins in the photoreceptors of algae and plants seems possible, in analogy to the role of carotenoids in plants.[10] Bilirubin does react with superoxide anion to some extent[11,12] and serves as a substrate for peroxidases in the presence of hydrogen peroxide or organic hydroperoxides.[13,14]

[4] W. H. M. Peters and P. L. M. Jansen, *J. Hepatol.* **2**, 182 (1986).
[5] R. Stocker, Y. Yamamoto, A. F. McDonagh, A. N. Glazer, and B. N. Ames, *Science* **235**, 1043 (1987).
[6] K. Bernhard, G. Ritzel, and K. U. Steiner, *Helv. Chim. Acta* **37**, 306 (1954).
[7] H. P. Kaufmann and H. Garloff, *Fette, Seifen, Anstrichm.* **63**, 334 (1961).
[8] A. F. McDonagh, *Biochem. Biophys. Res. Commun.* **48**, 408 (1972).
[9] B. Stevens and R. D. Small, Jr., *Photochem. Photobiol.* **23**, 33 (1976).
[10] B. Halliwell, *Chem. Phys. Lipids* **44**, 327 (1987).
[11] R. Kaul, H. K. Kaul, P. C. Bajpai, and C. R. K. Murti, *J. Biosci.* **1**, 377 (1979).
[12] P. Robertson, Jr., and I. Fridovich, *Arch. Biochem. Biophys.* **213**, 353 (1982).
[13] R. Brodersen and P. Bartels, *Eur. J. Biochem.* **10**, 468 (1969).
[14] J. Jacobsen and O. Fedders, *Scand. J. Clin. Lab. Invest.* **26**, 237 (1970).

The methods described here apply to the measurement of the peroxyl radical scavenging activity of biliverdin and the various biological forms of bilirubin. Most of the methods, however, are generally applicable to testing the antioxidant activity of other compounds (e.g., see also this volume [3]).

General Principle for Measuring Antioxidant Activity

The method is based on the thermal decomposition of an azo compound (A—N=N—A)[15] under aerobic conditions whereby peroxyl radicals are produced at a known and constant rate [Eqs. (1) and (2)]:

$$A—N=N—A \rightarrow [A \cdot N=N \cdot A] \rightarrow 2E\,A \cdot + (1 - E)\,A—A + N_2 \qquad (1)$$
$$A \cdot + O_2 \rightarrow AO_2 \cdot \qquad (2)$$

where E is the efficiency of free radical generation. In the presence of phospholipids containing polyunsaturated fatty acids (LH), the peroxyl radical $(AO_2 \cdot)$ will abstract a hydrogen atom to give rise to a lipid radical, thus initiating the chain reaction of lipid peroxidation [Eqs. (3)–(5)]:

$$AO_2 \cdot + LH \rightarrow AOOH + L \cdot \qquad (3)$$
$$L \cdot + O_2 \rightarrow LOO \cdot \qquad (4)$$
$$LOO \cdot + LH \rightarrow LOOH + L \cdot \qquad (5)$$

Because the peroxyl radical-mediated oxidation of polyunsaturated fatty acids initially results in quantitative formation of the corresponding lipid hydroperoxides (LOOH),[16] the extent of oxidation can be followed simply by measuring the accumulation of hydroperoxides. Any compound possessing peroxyl radical scavenging activity will compete with LH for the lipid peroxyl radical [LOO · in Eq. (5)], resulting in inhibition of the chain reaction and hence inhibition of formation of lipid hydroperoxides.

Materials

Azo Compounds

Water-soluble 2,2'-azobis(2-amidinopropane) dihydrochloride (AAPH)[15,17] and lipid-soluble 2,2'-azobis(2,4-dimethylvaleronitrile) (AMVN)[16,17] (Polysciences, Warrington, PA) can be used as radical generators to test the peroxyl radical scavenging activity of hydrophilic and

[15] L. R. C. Barclay, S. L. Locke, J. M. MacNeil, J. VanKessel, G. W. Burton, and K. U. Ingold, *J. Am. Chem. Soc.* **106,** 2479 (1984).
[16] Y. Yamamoto, E. Niki, and Y. Kamiya, *Bull. Chem. Soc. Jpn.* **55,** 1548 (1982).
[17] Y. Yamamoto, S. Haga, E. Niki, and Y. Kamiya, *Bull. Chem. Soc. Jpn.* **57,** 1260 (1984).

lipophilic compounds, respectively. AMVN should be recrystallized from hot methanol before use.

Lipids

Commercially available lipids containing polyunsaturated fatty acids need to be purified before use to remove endogenous antioxidants and contaminating hydroperoxides.

Linoleic Acid. Five grams of linoleic acid (Sigma, St. Louis, MO, Cat. No. L-1376) is dissolved in 100 ml of hexane and applied onto a low pressure Econo-column (15 i.d. × 300 mm, Bio-Rad, Richmond, CA) containing silica (Sigma, 100–200 mesh) previously equilibrated with hexane. During loading of the lipid the flow rate is adjusted to about 1–2 drops/sec with the aid of a Teflon Luer-lock three-way stopcock. After all the lipid has been applied, the column is washed with 100–200 ml of hexane to remove endogenous antioxidants. The purified linoleic acid is then eluted with hexane–ethyl ether (95 : 5, v/v), and its presence is verified by spotting on a silica TLC plate and exposing the plate to iodine vapor. The eluant containing purified linoleic acid is collected, the solvent evaporated under reduced pressure, and the amount of lipid recovered determined by weighing. Finally, the purified linoleic acid is dissolved in hexane at a known concentration and stored in the dark at −20° until used.

Soybean Phosphatidylcholine. Five hundred milligrams of soybean phosphatidylcholine (Sigma) is added to 50 ml of chloroform and applied onto an Econo-column (15 i.d. × 600 mm, Bio-Rad) containing silica (Sigma, 100–200 mesh) previously equilibrated with chloroform. The loading of the lipid is performed using a very slow flow rate (~1 drop/sec) and normally takes about 40–50 min. After loading, the column is washed with 500 ml of chloroform using a maximal flow rate. To remove some of the endogenous lipid hydroperoxides the column is then washed with 150 ml of chloroform–methanol (90 : 10, v/v) followed by a second wash with chloroform–methanol (60 : 40, v/v). During the two washing steps, which are performed using a flow rate of 1 drop/sec, the column material turns turbid, and a slight yellowish band (oxidized lipid?) may run at the solvent front of the second wash. After this front has eluted, an additional 50 ml of chloroform–methanol (60 : 40, v/v) is passed through the column before the purified phosphatidylcholine is eluted with 1 liter of the same solvent. The eluate is collected, the solvent removed under reduced pressure, and the purified lipid dissolved in benzene at 50 mg/ml and stored in the dark at −20°. Recovery is normally around 70–80%.

Lipid Hydroperoxides. Standards of linoleic acid and phosphatidylcholine hydroperoxide for the quantitation of lipid hydroperoxide by high-performance liquid chromatography (HPLC) are prepared as described by Yamamoto *et al.*[18] (see also this volume [38]).

Bile Pigments

Bilirubin. Commercial preparations of bilirubin normally contain bilirubin isomers and nonbilirubin contaminants. To remove these impurities, bilirubin is purified according to the method described by McDonagh and Assisi.[19] A mixture of 400 mg of bilirubin IXα (Sigma, from bovine gallstones) and 459 ml of chloroform is stirred and heated until the solvent boils vigorously. The mixture is then cooled to room temperature, washed in a separatory funnel with 0.1 M NaHCO$_3$ solution (3 times, 100 ml each, or until the washings are colorless), dried over anhydrous NaSO$_4$ (10 g), and filtered. The filtrate is heated until the solvent boils, and about one-third of the chloroform is distilled off. Methanol is then added to the boiling solution in small portions until the solution becomes just perceptibly turbid. The mixture is allowed to cool to 4° (overnight), and the crystalline precipitate is collected by filtration. The precipitate is washed with chloroform–methanol (1 : 1, v/v) and dried under high vacuum at 65° for at least 12 hr. The entire procedure should be done under dim, diffuse light. Recovery of crystalline product is generally about 80%.

Conjugated Bilirubin and Biliverdin. Bilirubin ditaurate can be used as a useful model compound for conjugated bilirubin. Like biliverdin IX dihydrochloride, it can be obtained from Porphyrin Products (Logan, UT) and used without further purification. Alternatively, biliverdin free acid can be readily synthesized from bilirubin.[2]

Methods

To test the peroxyl radical scavenging activities of bile pigments, several test systems have been developed to make allowance for the physicochemical differences between biliverdin and the various biological forms of bilirubin. All experiments are carried out in reaction tubes covered with aluminum foil to prevent exposure of the bile pigments to light. Although experiments are normally performed under air, the use of reaction mixtures in sealed tubes that have been previously equilibrated with a gas mixture of known composition allows the examination of the effect of oxygen concentration on the antioxidant activity of the compound to be

[18] Y. Yamamoto, M. H. Brodsky, J. C. Baker, and B. N. Ames, *Anal. Biochem.* **160,** 7 (1987).
[19] A. F. McDonagh and F. Assisi, *Biochem. J.* **129,** 797 (1972).

tested. As the absolute amounts of lipid hydroperoxides formed do vary between different sets of experiments, an internal control sample (no antioxidant) is always included.

Free Bilirubin and Biliverdin

Owing to the low solubility of bilirubin in aqueous solutions at physiological pH and ionic strength, the peroxyl radical scavenging activity of this pigment is best tested in a homogeneous system using an organic solvent and the lipid-soluble radical initiator.[5] In a typical set of experiments, chloroform (or benzene) solutions with linoleic acid (158 mM), AMVN (1.25 mM), and different micromolar amounts of bilirubin are prepared in the cold. The production of peroxyl radicals is initiated by placing the reaction tube in a shaking water bath at 37°. Immediately after this (time zero) and at various time points thereafter, aliquots of the reaction mixture are removed and analyzed directly by HPLC for either lipid hydroperoxides or bile pigments. Similar conditions can also be used to test the antioxidant activity of biliverdin if methanol is used as the solvent.

Membrane-Bound Bilirubin

Liposomes containing bilirubin are prepared by a method analogous to that described by Yamamoto et al.[20] Phosphatidylcholine and oil-soluble additives such as AMVN and bilirubin are dissolved in chloroform or benzene. The solution is mixed vigorously, and the solvents are removed under reduced pressure to obtain a thin film. An appropriate amount of 0.15 M NaCl or phosphate-buffered saline (136 mM Nacl, 2.6 mM KCl, 1.4 mM KH$_2$PO$_4$, 8 mM NaHPO$_4$, pH 7.2) is then added, and the liposomes are prepared by slowly peeling off the film by agitation. The resulting liposomes are multilamellar vesicles. In a typical experiment the final concentrations are 20 mM phosphatidylcholine and 0.2 mM AMVN in the absence and presence of low micromolar concentrations of bilirubin. Standard incubations are carried out under air in a water bath at 50°. The temperature is raised for the liposome experiments because of the high viscosity of the bilayer which results in a low E value [Eq. (1)]. As the phase transition temperature of soybean phosphatidylcholine is below 0°, the liposome structure is not expected to be significantly different between 37 and 50°. At various time points aliquots of the reaction mixture are removed and analyzed directly for phosphatidylcholine hydroperoxide or bilirubin.

[20] Y. Yamamoto, E. Niki, Y. Kamiya, and H. Shimasaki, Biochim. Biophys. Acta 795, 332 (1984).

Albumin-Bound Bilirubin

Recrystallized bilirubin is dissolved in a few drops of degassed 50 mM NaOH immediately before it is added to a phosphate-buffered solution of essentially fatty acid-free human albumin (Sigma). To this optically clear solution purified linoleic acid is added as an aqueous dispersion, and the mixture is stirred at 4° until the solution is again clear (this normally takes about 20 min). Since the affinity of the primary binding site for bilirubin to human albumin is not changed by cobinding of up to four molecules of fatty acids and since bilirubin dianion binds to the primary binding site on albumin when the pigment is added in a sodium hydroxide solution, this experimental model system is representative of the *in vivo* form of bilirubin bound to the primary binding site on human albumin.[21] Cold AAPH is then added to the reaction mixture, and radical production is initiated by placing the reaction vessel in a water bath equilibrated at 37°. Typical final concentrations are as follows: albumin 500 μM, linoleic acid 2 mM, AAPH 50 mM, and bilirubin at low micromolar concentrations.

To analyze the extent of oxidation of linoleic acid, aliquots are removed at various time points, and the fatty acids are extracted from the albumin by the addition of 10 volumes of chloroform to 1 volume of reaction mixture. The two phases are separated by a 2-min centrifugation at 11,000 g (Eppendorf centrifuge), and a known volume of organic phase is removed, evaporated to dryness under a stream of nitrogen, and redissolved in methanol for HPLC analysis. Evaporation to dryness of linoleic acid hydroperoxide in the presence of bilirubin does not result in any significant loss of the hydroperoxide.[22] For the analysis of bilirubin and its oxidation products, aliquots are removed at various time points, and the bile pigments are extracted by the addition of 4 volumes of cold methanol to 1 volume of reaction mixture. The protein is pelleted, the supernatant removed, and an aliquot is then analyzed. Quantitation of linoleic acid hydroperoxide and bile pigments by HPLC is done as described below.

Conjugated Bilirubin and Biliverdin

The nonhomogeneous liposome system described above for the determination of the peroxyl radical scavenging activity of membrane-bound bilirubin can easily be adapted to test the antioxidant activity of water-soluble bile pigments. Liposomes are prepared in the absence of AMVN and bilirubin, and, instead, AAPH (10 mM) and bilirubin ditaurate or free biliverdin are included in the aqueous phase. Biliverdin is dissolved in a

[21] R. Brodersen, *CRC Crit. Rev. Clin. Lab. Invest.* **11**, 305 (1979).

[22] R. Stocker, A. N. Glazer, and B. N. Ames, *Proc. Natl. Acad. Sci. U.S.A.* **84**, 5918 (1987).

few drops of 50 mM NaOH immediately before the experiment. To mimic biliary conditions, the peroxyl radical-mediated oxidation of phosphatidylcholine can also be performed in a micellar system by including taurocholate (50 mM) in the reaction mixture.

Analysis of Lipid Hydroperoxides

Most of the methods used are based on the HPLC separation of lipid hydroperoxides described by Yamamato et al.[18] (see also this volume [38]). Under the experimental conditions chosen, the amounts of hydroperoxides formed lie in the high micromolar range, and UV absorption at 234 nm can be used for detection. Lipid hydroperoxides are quantitated by comparison of the peak areas with that of the corresponding standard.

Linoleic Acid Hydroperoxide

Separation and quantitation of linoleic acid hydroperoxide by HPLC can be done on an analytical LC-NH$_2$ column (4.6 × 250 mm, 5 μm particle size; Supelco, Bellefonte, PA) using methanol–40 mM NaH$_2$PO$_4$ (9 : 1, v/v) (1 ml/min) as the mobile phase. The same chromatographic conditions can also be used for the analysis of phosphatidylcholine hydroperoxide. Alternatively, linoleic acid hydroperoxide may be analyzed with an LC-18 column (4.6 × 250 mm, 5 μm particle size; Supelco) using methanol (1 ml/min) as the eluant. Under these conditions, linoleic acid hydroperoxide elutes close to the solvent front. However, the contribution of the solvent front relative to the absorbance of linoleic acid hydroperoxide at 234 nm is normally small and can be eliminated by subtraction.

Phosphatidylcholine Hydroperoxide

Separation and quantitation of phosphatidylcholine hydroperoxide by HPLC can be achieved using an LC-18DB column (4.6 × 250 mm, 5 μm particle size; Supelco) with methanol containing 0.02% triethylamine (v/v) (Fisher Scientific, Fairlawn, NJ) (1 ml/min) as the mobile phase. For repetitive analysis of multiple samples in a short time, phosphatidylcholine hydroperoxide is conveniently analyzed after separation on a silica gel column (4.6 × 150 mm, 5 μm particle size; Supelco) with methanol–water (85 : 15, v/v) (1.8 ml/min) as the eluant. Under these conditions, the lipid hydroperoxide elutes as a single broad peak at around 3.5 min, well after the solvent front.

Analysis of Bile Pigments

Isocratic HPLC analysis of free bilirubin and biliverdin is achieved using an LC-18 column (4.6 × 250 mm, 5 μm particle size) with 0.1 M di-n-octylamine acetate in methanol–water (96 : 4, v/v) (1 ml/min) as the mobile phase and monitoring at 460 and 650 nm, respectively. The pH of the mobile phase is adjusted to 7.7 by the addition of the appropriate amount of acetic acid to the di-n-octylamine solution in methanol. Under these conditions the retention times for biliverdin and bilirubin are 5.4 and 12.5 min, respectively. These chromatographic conditions are also suitable for the separation and quantitation of the three configurational photoisomers of the naturally occurring (4Z,15Z)-bilirubin IXα.[23] By increasing the water content of the mobile phase to 15 vol % and the flow rate to 1.5 ml/min, this method is also suitable for the quantitation of bilirubin ditaurate.

Conclusions

We have presented general approaches for quantitation of the peroxyl radical scavenging activities of biliverdin and the biologically relevant forms of bilirubin. The results obtained using these systems,[5,22,24] in conjunction with our recent observations that both biliverdin and conjugated bilirubin efficiently scavenge hypochlorous acid[25] and also act synergistically with vitamin E in protecting lipid membranes from peroxidation initiated within the lipid phase,[26] add further credit to the notion that the two bile pigments function as important physiological antioxidants.

Acknowledgments

We thank Dr. Y. Yamamoto for advice and helpful suggestions. Experimental studies were supported by the National Cancer Institute Outstanding Investigator Grant CA39910 to B.N.A., by the National Institute of Environmental Health Sciences Center Grant ES01896 (B.N.A.), by National Institutes of Health Grants DK26307 and HD20551 to A.F.McD., and by National Science Foundation grant DMB8518066 to A.N.G. In addition, we thank Dr. E. Peterhans for interest and support.

[23] A. F. McDonagh, L. A. Palma, F. R. Trull, and D. A. Lightner, *J. Am. Chem. Soc.* **104,** 6865 (1982).

[24] R. Stocker and B. N. Ames, *Proc. Natl. Acad. Sci. U.S.A.* **84,** 8130 (1987).

[25] R. Stocker and E. Peterhans, *Free Radical Res. Commun.* **6,** 57 (1989).

[26] R. Stocker and E. Peterhans, *Biochim. Biophys. Acta* **1002,** 238 (1989).

[32] Ergothioneine as Antioxidant

By PHILIP E. HARTMAN

In 1909 the ergot fungus, *Claviceps purpurea,* was reported to contain about 4 mM of a sulfur-bearing compound that was isolated, crystallized, and named ergothioneine.[1] The compound was quickly identified as the betaine of 2-thiolhistidine[2] and was eventually shown to be the L-isomer by synthesis.[3] Synonyms appearing in the early literature are "thiasine," "thioneine," and "sympectothion."

$$(1)$$

Ergothioneine was a popular focus of investigation in the 1920s through the 1950s following the discovery of its occurrence in mammalian erythrocytes. This work has been thoroughly reviewed[4–6]; only a few selected aspects are reiterated here. Although the molecule is a very interesting antioxidant of biological origin, ergothioneine has received little general notice among biologists and biochemists.[6,7]

Some Properties of Ergothioneine

Ergothioneine is depicted in reaction (1) above (left-hand side) as a thiol. It forms disulfides, including mixed disulfides, on catalysis by Cu^{2+} in acidic solutions; the disulfides, however, are unstable in aqueous solu-

[1] C. Tanret, *C. R. Acad. Sci.* **149,** 222 (1909).
[2] G. Barger and A. J. Ewins, *J. Chem. Soc.* **1911,** 2336 (1911).
[3] H. Heath, A. Lawson, and C. Rimington, *J. Chem. Soc.,* 2215 (1951).
[4] D. J. Bell, *Annu. Rep. Prog. Chem.* **52,** 285 (1955).
[5] D. B. Melville, *Vitam. Horm. (N.Y.)* **17,** 155 (1958).
[6] E. C. Stowell, *in* "Organic Sulfur Compounds" (N. Kharasch, ed.), Vol. 1, p. 462. Pergamon, New York, 1961.
[7] Z. Hartman and P. E. Hartman, *Environ. Mol. Mutagen.* **10,** 3 (1987).

tions.[5,8,9] One important attribute is that the standard redox potential (E_0') at pH 7 of the thiol–disulfide couple of ergothioneine is -0.06 V as opposed to -0.20 to -0.32 V for other widely occurring natural thiols.[10,11] The disulfide of ergothioneine is reduced by cysteine with the production of cystine.[8] Halogens readily remove the sulfur of ergothioneine, whereas the other thiols are stable.[12] In contrast, the sulfur atom of ergothioneine is remarkably stable to alkali.[5,6] The pK_{SH}, earlier given as 10.8,[13] is 10.4 to 10.5[9,14]; λ_{max} is 258 nm and ε_{max} is 14,500 up to pH 9.[6,8] Ergothioneine dissolves in water at 25° to a level of almost 0.9 M, with greater solubility as the temperature is raised. Solubility is much lower in hot methanol, ethanol, and acetone with almost no solubility in ether, chloroform, or benzene.[1,15]

A second important attribute of ergothioneine is its presence in aqueous solutions predominantly as the thione [reaction (1), right-hand side] rather than as the thiol; the thiolate ion is present in basic solutions.[5,8,9,14] Because of its redox potential and propensity to tautomerize to the thione, ergothioneine shares only some properties of representative thiols and thiones. It reacts with the traditional sulfhydryl reagents iodoacetamide,[16] p-chloromercuribenzoate,[17] nitroprusside,[5,6,8] and with bromobimanes.[18–20] In contrast, ergothioneine fails to react with N-ethylmaleimide,[6,17] N-methyl-N'-nitro-N-nitrosoguanidine,[7,21,22] 4-nitroquinoline

[8] H. Heath and G. Toennies, *Biochem. J.* **68,** 204 (1958).

[9] N. Motohashi, I. Mori, and Y. Sugiura, *Chem. Pharm. Bull.* **24,** 1737 (1976).

[10] M. Calvin, in "Glutathione" (S. Colowick, ed.), p. 20. Academic Press, New York, 1956.

[11] P. C. Jocelyn, "Biochemistry of the SH Group." Academic Press, New York, 1972.

[12] O. Touster, *J. Biol. Chem.* **188,** 371 (1951).

[13] B. Stanovnik and M. Tisler, *Anal. Biochem.* **9,** 68 (1964).

[14] H. Sakuri and S. Takeshima, *Talanta* **24,** 531 (1977).

[15] "The Merck Index, 10th Edition" (M. Windholz and S. Budavari, eds.), p. 529. Merck & Co., Rahway, New Jersey, 1983.

[16] J. Carlsson, M. P. J. Kierstan, and K. Brocklehurst, *Biochem. J.* **139,** 221 (1974).

[17] T. Hama, T. Konishi, N. Tamaki, F. Tunemori, and H. Okumura, *Vitamins* **46,** 121 (1972).

[18] R. C. Fahey, G. L. Newton, R. Dorian, and E. M. Kosower, *Anal. Biochem.* **111,** 357 (1981).

[19] R. C. Fahey, R. Dorian, G. L. Newton, and J. Utley, in "Radioprotectors and Anticarcinogens" (O. F. Nygaard and M. G. Simic, eds.), p. 103. Academic Press, New York, 1983.

[20] R. C. Fahey and G. L. Newton, in "Functions of Glutathione: Biochemical, Physiological, Toxicological, and Clinical Aspects" (A. Larsson, S. Orrenius, A. Holmgren, and B. Mannervik, eds.), p. 251. Raven, New York, 1983.

[21] R. A. Owens and P. E. Hartman, *Environ. Mutagen.* **8,** 659 (1986).

[22] J. Y. H. Chan, M. Ruchirawat, J.-N. Lapeyre, and F. F. Becker, *Carcinogenesis* **4,** 1097 (1983).

N-oxide,[7] metmyoglobin,[23] and 5,5'-dithiobis(nitrobenzoic acid).[24] It also fails to react with the Grote reagent for detection of thiones such as thiourea and substituted thioureas.[8]

Ergothioneine as Defensive Molecule

A wide array of functions have been proposed for ergothioneine through the years, and this conjecture continues.[25] We endorse the proposition[5] that a primary function of ergothioneine, both in its native species and in organisms assimilating it, is its service as an interceptor of H_2O_2, some radical species, and some toxic electrophilic organic molecules.

Ergothioneine is oxidized and catalyzes the oxidation of reduced glutathione to the disulfide in the presence of H_2O_2.[6,8,26] Ergothioneine and catalase protect germination of ergot conidia from H_2O_2; H_2O_2 is an agent that may be used by plants as a defense against fungal infection.[27] Ram semen exposed to H_2O_2 or to Cu^{2+} is protected by exogenous ergothioneine.[28] Ergothioneine and thiourea at 0.1 mM protect bacteriophages T4 and P22 from inactivation by γ-irradiation; inactivation in this lipid-free system is a process dependent on H_2O_2 and superoxide or on hydroxyl radical formation generated from these reduced oxygen species.[29] Mutagenesis of *Salmonella* bacteria by *tert*-butyl hydroperoxide is decreased by the presence of ergothioneine[7]; in an analogous situation, hemolysis induced by *tert*-butyl hydroperoxide is decreased in the presence of a series of synthetic 2-imidazolethiones structurally related to ergothioneine.[30] The 2-imidazolethiones examined react more readily with the stable free radical 1,1-diphenyl-2-picrylhydrazyl than do analogous 2-imidazolones such as uric acid,[30] which is also an antioxidant of note.[31,32]

Lipid peroxidation in rat liver by endogenous autoxidation or administration of ethionine *in vivo* is significantly blocked in animals administered ergothioneine.[33] Addition of ergothioneine to mouse liver homogenates

[23] N. Motohashi and I. Mori, *Chem. Pharm. Bull.* **31**, 1702 (1983).
[24] E. Turner, R. Klevit, P. B. Hopkins, and B. M. Shapiro, *J. Biol. Chem.* **261**, 13056 (1986).
[25] M. C. Brummel, *Med. Hypotheses* **18**, 351 (1985).
[26] N. W. Pirie, *Biochem. J.* **27**, 1181 (1933).
[27] A. S. Garay, *Nature (London)* **177**, 91 (1956).
[28] T. Mann and E. Leone, *Biochem. J.* **53**, 140 (1953).
[29] P. E. Hartman, Z. Hartman, and M. J. Citardi, *Radiat. Res.* **114**, 319 (1988).
[30] R. C. Smith, J. C. Reeves, R. C. Dage, and R. A. Schnettler, *Biochem. Pharmacol.* **36**, 1457 (1987).
[31] B. N. Ames, R. Cathcart, E. Schwiers, and P. Hochstein, *Proc. Natl. Acad. Sci. U.S.A.* **78**, 6858 (1981).
[32] P. Hochstein, L. Hatch, and A. Sevanian, this series, Vol. 105, p. 162.
[33] H. Kawano, K. Cho, Y. Haruna, Y. Kawai, T. Mayumi, and T. Hama, *Chem. Pharm. Bull.* **31**, 1676 (1983).

decreases lipid peroxidation and enhances glutathione peroxidase and Mn^{2+} superoxide dismutase activities.[34] Ergothioneine in the diet of rabbits[35] and rats[36] significantly blocks nitrite-induced methemoglobin formation from hemoglobin *in vivo;* erythrocyte ergothioneine is more effective than erythrocyte glutathione. *In vitro* alkylation of RNA by tris(2-chloroethyl)amine is decreased in the presence of ergothioneine,[37] as is mutagenicity for *Salmonella* of directly acting mutagens produced by the nitrosation of spermidine.[7] Ergothioneine appreciably decreases singlet oxygen generation by illuminated Rose Bengal solutions, possibly by quenching the excited state dye molecules.[38] In contrast, ergothioneine fails to react significantly with and intercept singlet oxygen[38] and superoxide[39] in aqueous solutions. Ergothioneine quenches singlet oxygen produced in D_2O solutions, and at a rate greater than that found for glutathione,[40] but it is doubtful if this quenching is of significance in biological systems near neutral pH.

Ergothioneine injected a few minutes prior to irradiation does not protect mice from the early lethal effects of high doses of whole-body irradiation,[41,42] nor do intraperitoneal (i.p.) injections of ergothioneine given over a 6-day period prior to irradiation increase survival time at high X-ray doses (cited in Ref. 5). Such "first-week mortality" of high X-ray doses is caused by the "gastrointestinal syndrome" arising from killing of stem cells in the intestinal crypts.[43,43a] Ergothioneine is as yet untested as a protective molecule in animals at lower X-ray doses where effects on bone marrow predominate in lethal effects. It has been suggested that ergothioneine may be involved in effects of ultraviolet light since 2-thioimidazoles have molar extinction coefficients of around 15,000 at 255–265 nm.[44]

[34] H. Kawano, H. Murata, S. Iriguchi, T. Mayumi, and T. Hama, *Chem. Pharm. Bull.* **31,** 1682 (1983).

[35] S. S. Spicer, J. G. Wooley, and V. Kessler, *Proc. Soc. Exp. Biol. Med.* **77,** 418 (1951).

[36] R. A. Mortensen, *Arch. Biochem. Biophys.* **46,** 241 (1953).

[37] L. Szinicz, G. J. Albrecht, and N. Weger, *Arzneim.-Forsch./Drug Res.* **31,** 1713 (1981).

[38] T. A. Dahl, W. R. Midden, and P. E. Hartman, *Photochem. Photobiol.* **47,** 357 (1988).

[39] P. E. Hartman, W. J. Dixon, T. A. Dahl, and M. E. Daub, *Photochem. Photobiol.* **47,** 699 (1988).

[40] M. Rougee, R. V. Bensasson, E. J. Land, and R. Pariente, *Photochem. Photobiol.* **47,** 485 (1988).

[41] Z. M. Bacq and A. Herve, *Strahlentherapie* **95,** 215 (1954).

[42] P. Alexander, Z. M. Bacq, S. F. Cousens, M. Fox, A. Herve, and J. Lazar, *Radiat. Res.* **2,** 392 (1955).

[43] T. L. Phillips, G. Y. Ross, and L. S. Goldstein, *Int. J. Radiat. Oncol. Biol. Phys.* **5,** 1441 (1979).

[43a] J. V. Moore, *Cancer Treat. Rep.* **68,** 1005 (1984).

[44] M. C. Brummel, H. G. Claycamp, and L. D. Stegink, *Photochem. Photobiol. Suppl.* **47,** 34S (1988).

Ergothioneine as Chelating Agent

Metal ions, predominantly iron but also possibly copper, appear to be involved in multiple phases of oxidative damage in critical situations.[45–48] These metal ions can facilitate formation of the hydroxyl radical from H_2O_2, generate H_2O_2 and superoxide via the nonenzymatic oxidation of endogenous biomolecules such as reduced glutathione, and facilitate decomposition of lipid peroxides. The impact on the oxidation of biomolecules by iron complexes can differ considerably from the activity of free iron salts.[49] Generally, agents that strongly chelate iron, such as desferroxamine, exhibit protection and may have important clinical applications.[45–48]

Ergothioneine is an avid chelator of divalent metal ions, forming complexes of 2 mol of ergothioneine to 1 mol of metal.[9,50,51] Formation constants range up to 10^{18} for metal ions, in the relative order $Cu^{2+} > Hg^{2+} > Zn^{2+} > Cd^{2+} > Co^{2+} \cong Ni^{2+}$ (rate constant 10^8).[50,51] Intraperitoneally administered ergothioneine protects mice against Cd^{2+} teratogenicity, possibly via chelating ability.[52] The author has been unable to find any studies with regard to chelation of Fe^{2+} by ergothioneine; however, Fe^{2+} exhibits strong interference in the Pauly reaction for imidazoles, as does Cu^{2+}.[53] A related thione, thiourea, is an excellent hydroxyl radical scavenger and is characterized as weakly binding iron.[49] The chelation proficiency of ergothioneine has been suggested as predisposing mammals to diabetes by creating a metal deficiency.[54] This proposition appears to be without foundation with regard to animals on mineral-sufficient diets.

Distribution of Ergothioneine

Ergothioneine is synthesized exclusively by fungi and mycobacteria.[5,6,11,17–20] Both plant pathogenic and saprophytic fungi, such as *Neurospora crassa*, synthesize ergothioneine, and the biochemistry of its syn-

[45] J. M. McCord, *N. Engl. J. Med.* **312,** 159 (1985).
[46] S. D. Aust and B. C. White, *Adv. Free Radical Biol. Med.* **1,** 1 (1985).
[47] B. Halliwell and J. M. C. Gutteridge, *Arch. Biochem. Biophys.* **246,** 501 (1986).
[48] B. Halliwell and J. M. C. Gutteridge, this volume [1].
[49] J. M. C. Gutteridge, *Biochem. J.* **224,** 761 (1984).
[50] D. P. Hanlon, *J. Med. Chem.* **14,** 1084 (1971).
[51] N. Motohashi, I. Mori, Y. Suriura, and H. Tanaka, *Chem. Pharm. Bull.* **22,** 654 (1974).
[52] T. Mayumi, K. Okamoto, K. Yoshida, Y. Kawai, H. Kawano, T. Hama, and K. Tanaka, *Chem. Pharm. Bull.* **30,** 2141 (1982).
[53] D. B. Melville and R. Lubschez, *J. Biol. Chem.* **200,** 275 (1953).
[54] R. M. Epand, *Med. Hypothesis* **9,** 207 (1982).

thesis has been most carefully studied in the latter species.[11,55] Fungal localization is principally in critical cells, the conidia.[17-20,56] No mutants have been obtained lacking in ergothioneine biosynthesis, so its role in fungal life remains conjectural. Plants assimilate ergothioneine via the roots.[5,57]

Mammals ingest ergothioneine and avidly conserve it with minimal metabolism[6,58]; in rats the whole-body half-life of ingested ergothioneine is about 1 month.[58] Storage is in the free form in critical tissues subject to oxidative stress such as erythrocytes, seminal fluid, liver, and kidney, where it can reach near millimolar levels following continuous ingestion.[5,6,11,17,58,59] Millimolar levels also are found in the human lens, especially the cataract-free lens.[60] The content of ergothioneine varies greatly among tissues; in a particular tissue, the ergothioneine concentration is strongly dependent on its dietary level as well as the animal species; superimposed on these factors is an important individual variation within a single population.[5,6] Entry into seminal fluid and most other tissues is rapid, but erythrocyte ergothioneine enters only during erythropoiesis with no exchange in mature red blood cells; thus, the ergothioneine content slowly declines with erythrocyte age.[5,12,61] Micromolar concentrations of ergothioneine are found in the brain.[62] The compound crosses the placental barrier and is found in milk of animals ingesting it.[5] Egg white can contain ergothioneine for delivery to embryos of the next generation.[63] In addition to mammals and birds, accumulation also occurs in animals as diverse as horseshoe crabs and frogs.[5,6] It thus appears that most animals in native environments are to some degree exposed to ergothioneine, and they ingest and accumulate it. Human populations in modern times ingesting less mold-contaminated foods are certainly exposed to fewer fungal toxins than heretofore; however, these populations also may be consuming less of what seems to be a universally treasured molecule. We simply have no knowledge of the sources of ergothioneine consumption by human populations, a matter worthy of investigation.

[55] Y. Ishikawa, S. E. Israel, and D. B. Melville, *J. Biol. Chem.* **249**, 4420 (1974).

[56] H. Heath and J. Wildy, *Biochem. J.* **64**, 612 (1956).

[57] B. G. Audley and C. H. Tan, *Phytochemistry* **7**, 1999 (1968).

[58] T. Mayumi, H. Kawano, Y. Sakamoto, E. Suehisa, Y. Kawai, and T. Hama, *Chem. Pharm. Bull.* **26**, 3772 (1978).

[59] H. Kawano, M. Otani, K. Takeyama, Y. Kawai, T. Mayumi, and T. Hama, *Chem. Pharm. Bull.* **30**, 1760 (1982).

[60] Y. Shukla, O. P. Kulshrestha, and K. P. Khuteta, *Indian J. Med. Res.* **73**, 472 (1981).

[61] P. W. Kuchel and B. E. Chapman, *Biomed. Biochim. Acta* **42**, 1143 (1983).

[62] I. Briggs, *J. Neurochem.* **19**, 27 (1972).

[63] T. Konishi, N. Tamaki, F. Tunemori, N. Masumitu, H. Okumura, and T. Hama, *Vitamins* **46**, 127 (1972).

Pharmacology

Numerous experiments cited above have given no indications of deleterious effects of ergothioneine ingestion by rodents or man. Neither have several abbreviated pharmacological studies shown deleterious effects at physiological ergothioneine levels.[64,65] Careful, modern studies are called for; these must take into consideration the pharmacodynamics of ergothioneine in mammals described to date. The route of administration may be important because there appears to be far greater breakdown of ergothioneine when it is administered i.p.[66] than when it is consumed in the diet.[5,6,58]

Potential Applications

Ergothioneine is a protective and relatively nontoxic naturally occurring molecule that is not readily autoxidizable and one that reaches appreciable (micromolar to millimolar) concentrations in certain mammalian tissues. Its stability and pharmacodynamics suggest several potential applications, albeit limited to aqueous systems:

1. Already used as a constituent in the long-term culture of human cells,[67,68] ergothioneine could possibly decrease H_2O_2-mediated toxicity of tissue culture media that contain riboflavin and are exposed to light.[69] Cell toxicity owing to H_2O_2 generation from oxidation of readily autoxidizable reducing agents such as cysteine and cysteamine[70–72] is also a problem that may be alleviated by ergothioneine addition.

2. The growth of aerohypersensitive microorganisms *in vitro*[73–77] may be facilitated by addition of ergothioneine to the culture media. For exam-

[64] M. L. Tainter, *Proc. Soc. Exp. Biol. Med.* **24**, 621 (1926).
[65] E. Trabucchi, *Boll.-Soc. Ital. Biol. Sper.* **11**, 117 (1936).
[66] G. Wolf, J. G. Bergan, and D. E. Sunko, *Biochim. Biophys. Acta* **54**, 287 (1961).
[67] R. Holmes, *J. Biophys. Biochem. Cytol.* **6**, 535 (1959).
[68] R. Holmes, J. Helms, and G. Mercer, *J. Cell Biol.* **42**, 262 (1969).
[69] J. S. Zigler, Jr., J. L. Lepe-Zuniga, B. Vistica, and I. Gery, *In Vitro Cell. Dev. Biol.* **21**, 282 (1985).
[70] G. Saez, P. J. Thornalley, H. A. O. Hill, R. Hems, and J. V. Bannister, *Biochim. Biophys. Acta* **719**, 24 (1982).
[71] Y. Takagi, M. Shikita, T. Terasima, and S. Akaboshi, *Radiat. Res.* **60**, 292 (1974).
[72] J. E. Biaglow, R. W. Issels, L. E. Gerweck, M. E. Varnes, B. Jacobson, J. B. Mitchell, and A. Russo, *Radiat. Res.* **100**, 298 (1984).
[73] J. Carlsson, G. Nyberg, and J. Wrethen, *Appl. Environ. Microbiol.* **36**, 223 (1978).
[74] F. J. Bolton and D. Coates, *J. Appl. Bacteriol.* **54**, 115 (1983).
[75] B. Steiner, G. H. W. Wong, and S. Graves, *Br. J. Vener. Dis.* **60**, 14 (1984).
[76] N. R. Krieg and P. S. Hoffman, *Annu. Rev. Microbiol.* **40**, 107 (1986).
[77] F. S. Archibald and M.-N. Duong, *Infect. Immunol.* **51**, 631 (1986).

ple, freshly isolated *Lactobacillus* strains produce H_2O_2 but contain little or no catalase.

3. Because ergothioneine is uniquely produced by many filamentous fungi, ergothioneine content may be a useful indicator for mold contamination of processed foods such as tomato puree, catsup, and grains. Current methods[78] for quantitating fungal contamination appear to be laborious and erratic.

4. As a nonvolatile and relatively nonautoxidizable agent, ergothioneine may have utility in enzyme purification and stabilization, both as an antioxidant and as a chelator of deleterious metal ions.[79]

5. Many disorders of erythrocytes (including sickle cell anemia, thalassemia, deficiencies in glucose-6-phosphate dehydrogenase, vitamin E, reduced glutathione, and malarial infections) are thought to sensitize red blood cells to oxidative stress.[11,80–82] Such conditions would seem to be amenable to modulation by the high and nonexchangeable content of ergothioneine that is built into erythrocytes of animals ingesting it.

6. Slow replenishment of cellular glutathione owing to impaired cysteine transport may partially underlie the human disorder ataxia telangiectasia.[83] At least in rat liver *in vivo,* accumulation of ingested ergothioneine has no effect on residual glutathione levels[33]; namely, it can serve as an adjunct reducing agent in those tissues where it is concentrated. Of two mechanisms responsible for killing of hepatocytes by H_2O_2,[84] stored ergothioneine may act as a supplementary defense to mitigate the class of cell death that is accompanied by lipid peroxidation.

7. A tissue-damaging oxidative burst often follows temporary deprivation of the blood supply ("reperfusion injury") as can occur during tissue transplantation or following a stroke.[45,48] Particularly in the former instance, ergothioneine may have utility because of its inability to be rapidly autoxidized and its ability to chelate divalent metal ions.

8. Some deleterious aspects of Down's syndrome are probably due to an excess of superoxide dismutase.[85,86] Ergothioneine supplementation may possibly alleviate some of these effects.

[78] M. A. Cousin, C. S. Zeidler, and P. E. Nelson, *J. Food Sci.* **49,** 439 (1984).

[79] D. H. Hug and D. Roth, *Biochim. Biophys. Acta* **293,** 497 (1973).

[80] D. Chiu, B. Lubin, and S. B. Shohet, *in* "Free Radicals in Biology" (W. A. Pryor, ed.), Vol. 5, p. 115. Academic Press, New York, 1982.

[81] J. Anastasi, *Med. Hypothesis* **14,** 311 (1984).

[82] L. Schacter, J. A. Warth, E. M. Gordon, A. Prasad, and B. L. Klein, *FASEB J.* **2,** 237 (1988).

[83] M. J. Meredith and M. L. Dodson, *Cancer Res.* **47,** 4576 (1987).

[84] P. E. Starke and J. L. Farber, *J. Biol. Chem.* **260,** 86 (1985).

[85] P. M. Sinet, *Ann. N.Y. Acad. Sci.* **396,** 83 (1982).

[86] J. Kedziora and G. Bartosz, *Free Radical Biol Med.* **4,** 317 (1988).

9. Ergothioneine generates a steady-state level of superoxide from illuminated 4,9-dihydroxyperylene-3,10-quinones[39,87] and anthraquinones.[88] Although other active oxygen species are most likely also generated, quinone–ergothioneine mixtures may be versatile systems in the presence of suitable quencher and interceptor molecules; steady-state superoxide production is nonenzymatic and may be readily manipulated with regard to reactant concentrations as well as by light intensity.

Ergothioneine is clearly indicated as a molecule worthy of further study both in model and in more complex biological systems.

Assays

Early assays for ergothioneine are summarized and discussed elsewhere.[5,6,11,25] Paper chromatographic separation of Pauly-positive imidazoles has been widely utilized.[6,89] Recently, spectrophotometric,[16] TLC,[90] and HPLC[18–20,58] techniques have been developed. It is likely that NMR techniques[61] will find increasing use for *in vivo* studies. In the future, HPLC combined with electrochemical measurements may be used for quantitation *in vitro*, although the latter refinement would not seem necessary in order to answer many critical questions concerning the status of ergothioneine in living systems.

[87] P. E. Hartman, C. K. Suzuki, and M. E. Stack, *Appl. Environ. Microbiol.* **55**, 7 (1989).
[88] P. E. Hartman and M. A. Goldstein, *Environ. Mol. Mutagen.* **14**, 42 (1989).
[89] B. N. Ames and H. K. Mitchell, *J. Am. Chem. Soc.* **74**, 252 (1952).
[90] I. Kaneko, Y. Takeuchi, Y. Yamaoka, Y. Tanaka, T. Fukuda, Y. Fukumori, T. Mayumi, and T. Hama, *Chem. Pharm. Bull.* **28**, 3093 (1980).

[33] Thiyl Free Radical Metabolites of Thiol Drugs, Glutathione, and Proteins

By Ronald P. Mason and D. N. Ramakrishna Rao

The oxidation of the thiol group of cysteine is involved in several functions of glutathione (GSH) and proteins. The one-electron oxidation of the thiol group of cysteine catalyzed by a metal ion or a radiolytic process results in the formation of a thiyl radical. The thiyl radical can undergo dimerization to the corresponding disulfide or react with oxygen or hydrogen peroxide (H_2O_2) to produce the oxygen-containing products [Eq. (1)]. Although the thiol oxidation of cysteine and related compounds by metal ions and radiolytic processes has been studied for several years, the enzymatic oxidation of GSH has been only recently reported.[1,2]

$$RSH \xrightarrow{-e^-} \tfrac{1}{2} RS \cdot \overset{RSSR}{\underset{\uparrow}{\xrightarrow{-e^-}}} RSOH \xrightarrow{-e^-}$$

$$RSO \cdot \xrightarrow{-e^-} R\overset{O}{\underset{}{\overset{\|}{S}}OH \xrightarrow{-e^-} R\overset{O}{\underset{O}{\overset{\|}{S}}} \cdot \xrightarrow{-e^-} R\overset{O}{\underset{O}{\overset{\|}{S}}}\text{—OH} \quad (10)$$

Thiyl Radicals of Cysteine and Glutathione Generated by Peroxidases

The thiyl radical and products derived therefrom are responsible for the bactericidal effect of cysteine[3] and the radioprotection afforded by GSH and related thiol compounds.[4] The glutathionyl radical generated from glutathione thionitrite has been shown to cause mutagenicity in *Salmonella typhimurium* TA100.[5] However, Stark *et al.* suggest that the thiyl radical metabolite leads to the formation of H_2O_2 which may be responsible for its mutagenicity.[6]

One of the mechanisms by which the thiyl radical of cysteine or GSH may be formed *in vivo* is by reaction with a peroxidase enzyme.[1,2] This has been demonstrated for cysteine and GSH using isolated enzymes such as horseradish peroxidase (HRP),[7,8] lactoperoxidase,[9] and prostaglandin H synthase.[10] Since HRP is a plant enzyme, there is some criticism of its use in studies of the oxidation of drugs and toxic chemicals because it may not have any pharmacological relevance, but it certainly serves as a good model for the peroxidase reaction in biological systems.

Thiyl free radical metabolites formed *in vivo,* or in the presence of

[1] R. P. Mason, in "Biological Reactive Intermediates III" (J. J. Kocsis, D. J. Jollow, C. M. Witmer, J. O. Nelson, and R. Snyder, eds.), p. 493. Plenum, New York, 1986.

[2] R. P. Mason, L. S. Harman, C. Mottley, and T. E. Eling, in "Oxygen and Sulfur Radicals in Chemistry and Medicine" (A. Breccia, M. A. J. Rodgers, and G. Semerano, eds.), p. 103. Edizioni Scientifiche (Lo Scarabeo), Bologna, 1986.

[3] G. K. Nyberg, G. P. D. Granberg, and J. Carlsson, *Appl. Environ. Microbiol.* **38**, 29 (1979).

[4] M. Z. Baker, R. Badiello, M. Tamba, M. Quintiliani, and G. Gorin, *Int. J. Radiat. Biol. Relat. Stud. Phys. Chem. Med.* **41**, 595 (1982).

[5] M. H. Carter and P. D. Josephy, *Biochem. Pharmacol.* **35**, 3847 (1986).

[6] A.-A. Stark, E. Zeiger, and D. A. Pagano, *Carcinogenesis* **9**, 771 (1988).

[7] L. S. Harman, C. Mottley, and R. P. Mason, *J. Biol. Chem.* **259**, 5606 (1984).

[8] L. S. Harman, D. K. Carver, J. Schreiber, and R. P. Mason, *J. Biol. Chem.* **261**, 1642 (1986).

[9] C. Mottley, K. Toy, and R. P. Mason, *Mol. Pharmacol.* **31**, 417 (1987).

[10] T. E. Eling, J. F. Curtis, L. S. Harman, and R. P. Mason, *J. Biol. Chem.* **261**, 5023 (1986).

isolated enzymes, can be detected by electron spin resonance (ESR) spectroscopy using a spin trap. Spin trapping is a technique where a diamagnetic molecule (the spin trap) reacts with a free radical to produce a more stable radical (or spin adduct) that can be detected by ESR spectroscopy. Spin adducts are substituted nitroxide free radicals; 5,5-dimethyl-1-pyrroline N-oxide (DMPO) is a commonly used spin trap for the detection of the thiyl radical. Spin trapping offers an excellent alternative to other conventional methods of detecting unstable radicals formed in biological systems. The conventional methods are flow experiments that require large quantities of enzymes, which limit applicability, and the fast-freeze method, which involves ESR measurements at very low temperatures where most of the structural information on the radical is lost. The thiyl radical is extremely reactive; hence, this radical is usually detected by spin trapping.

The ESR spectrum of the DMPO–glutathionyl radical formed in a system of GSH, HRP, and H_2O_2 is shown in Fig. 1A. This radical adduct has a distinctive ESR spectrum consisting of four relatively broad lines with coupling constants of $a^N = 15.4$ G and $a_\beta^H = 16.2$ G. These values are similar to those reported by Josephy et al. for the adduct in a chemical system.[11] The formation of this radical adduct requires H_2O_2 (Fig. 1B,E) and HRP (Fig. 1C,D). There is no effect of superoxide dismutase (SOD) on the ESR spectrum (Fig. 1F), suggesting that the superoxide radical is not involved in generation of the DMPO–glutathionyl radical adduct. In the absence of diethylenetriaminepentaacetic acid (DTPA) in the assay system, the radical decay was much faster (data not shown), presumably because of the reduction of the DMPO–glutathionyl adduct by GSH, which is catalyzed by trace transition metals.

The oxidation of a wide variety of aromatic amines, phenols, chlorpromazine, and related compounds has been studied using HRP and H_2O_2. The previously known evidence for thiyl radical formation in the oxidation of thiol compounds was the formation of compound II [Eqs. (2)–(4)]. Compound II, the second peroxidase intermediate, is formed from compound I by one-electron reduction. This requires a one-electron oxidation of the thiol compound to the thiyl radical.

$$\text{HRP} + H_2O_2 \rightarrow \text{HRP–compound I (green)} \qquad (2)$$
$$\text{HRP–compound I} + \text{RSH (RS}^-) \rightarrow \text{HRP–compound II (red)} + \text{RS} \cdot \qquad (3)$$
$$\text{HRP–compound II} + \text{RSH (RS}^-) \rightarrow \text{HRP} + \text{RS} \cdot \qquad (4)$$

The ESR evidence presented from our laboratory is the first direct evidence for the formation of the thiyl radical.[7] The rate of formation of the thiyl radical in the peroxidase reaction can also be measured from

[11] P. D. Josephy, D. Rehorek, and E. G. Janzen, *Tetrahedron Lett.* **25**, 1685 (1984).

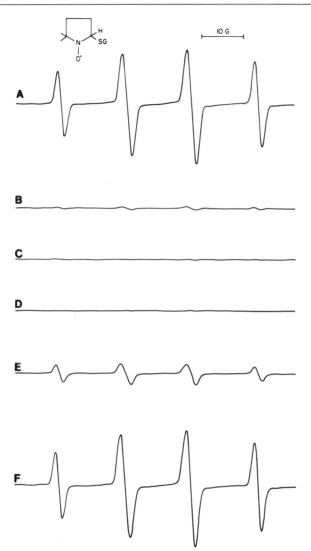

FIG. 1. ESR spectra of the DMPO–glutathione thiyl radical adduct produced in a system of HRP and H_2O_2 under aerobic conditions. (A) Incubation containing 10 mM GSH, 50 μM H_2O_2, 90 mM DMPO, and 0.1 mg/ml HRP in 0.1 M Tris-HCl buffer containing 1 mM DTPA, pH 8.0. (B) Same as in (A), but in the absence of H_2O_2. (C) Same as in (A), but in the absence of HRP. (D) Same as in (A), but in the presence of heat-denatured HRP. (E) Same as in (A), but in the presence of catalase (1000 units/ml). (F) Same as in (A), but in the presence of SOD (40 μg/ml). (From Ref. 8, with permission.)

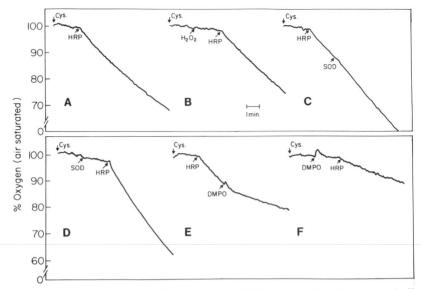

FIG. 2. Oxygen uptake curves for the cysteine–HRP system in 0.1 M phosphate buffer containing 1 mM DTPA, pH 7.5. Additions were made as indicated, and the final concentrations were as follows: 10 mM cysteine, 100 μM H_2O_2, 0.1 mg/ml HRP, 40 μg/ml SOD, 270 mM DMPO. (From Ref. 7, with permission.)

oxygen consumption experiments. The oxygen consumption by solutions containing cysteine and HRP (Fig. 2A) shows that cysteine (1 mM) supports oxygen consumption in the presence of HRP. Addition of H_2O_2 (Fig. 2B) does not affect oxygen consumption, because H_2O_2 already exists in the system owing to low level autoxidation of cysteine.[7] SOD increases the oxygen consumption (Fig. 2C,D), probably because of the formation of a complex between copper at the active site and the thiolate anion that can undergo oxidation using molecular oxygen.[12] The rate of oxygen consumption decreased with the addition of DMPO (Fig. 2E,F). These results are consistent with the cysteine thiyl free radical being formed and then trapped by DMPO before further reactions of the cysteinyl radical can lead to oxygen consumption.

The generation of H_2O_2 and the consumption of oxygen by the HRP-catalyzed oxidation of cysteine can be explained based on the known reactions [Eqs. (5)–(7)] of the cysteine thiyl radical. The thiyl free radical reacts with the thiol anion to form the cysteine disulfide radical anion which is air oxidized to cystine, forming superoxide. Superoxide dispro-

[12] F. J. Davis, B. C. Gilbert, R. O. C. Norman, and M. C. R. Symons, *J. Chem. Soc., Perkin Trans. 2*, 1763 (1983).

$$RS\cdot + RSH\ (RS^-) \rightarrow RSSR^{\overline{\cdot}} + H^+ \tag{5}$$
$$RSSR^{\overline{\cdot}} + O_2 \rightarrow RSSR + O_2^{\overline{\cdot}} \tag{6}$$
$$2\ O_2^{\overline{\cdot}} + 2\ H^+ \rightarrow H_2O_2 + O_2 \tag{7}$$

portionates rapidly to form H_2O_2, which will drive the HRP-catalyzed reaction. Catalase does not affect the oxygen consumption. The absence of a catalase effect may be due to the fact that peroxides other than H_2O_2 are supporting the cysteine oxidation in the presence of air.

Addition of molecular oxygen or H_2O_2 to the thiyl radical ultimately leads to the production of cysteine sulfinic acid (RSO_2H). Cysteine thioperoxide (RSOH) is an unstable intermediate that has not been isolated, but the subsequent oxidation steps have been studied [Eq. (1)]. The oxidation of cysteine sulfinic acid by HRP and H_2O_2 using *tert*-nitrosobutane (tNB) produced the $R\dot{S}O_2$ adduct.[7] Subsequently, the sulfur trioxide radical adduct was obtained. It is well known that the cysteine radical can undergo fragmentation to produce sulfur dioxide and a carbon-centered radical [Eqs. (8)–(10)]. The sulfur trioxide dianion is known to be oxidized by the HRP–H_2O_2 system to the sulfur trioxide radical anion.[13] This ex-

$$RS\cdot + O_2 \rightarrow RSOO\cdot \rightarrow RSO\cdot \rightarrow R\overset{\displaystyle O}{\underset{\displaystyle O}{\overset{\|}{\underset{\|}{S}}}}\cdot \tag{8}$$

$$R\overset{\displaystyle O}{\underset{\displaystyle O}{\overset{\|}{\underset{\|}{S}}}}\cdot \rightarrow R\cdot + SO_2 \xrightarrow{\ H_2O\ } SO_3^{2-} \tag{9}$$

$$SO_3^{2-} \xrightarrow[H_2O_2]{\ HRP\ } \cdot SO_3^- + e^- \tag{10}$$

plains the disappearance of the first radical adduct followed by the appearance of the sulfur trioxide radical adduct. Sevilla and co-workers[14] have demonstrated the formation of thioperoxyl radicals (RSO ·) in irradiated GSH in low-temperature matrices. They propose that at low temperature the peroxysulfenyl radical decomposes to give the thioperoxyl radical.

Glutathione Peroxidase

Glutathione peroxidase (GSHP) is known to reduce H_2O_2 and organic hydroperoxides. In this reaction GSH is oxidized to glutathione disulfide,

---

[13] C. Mottley and R. P. Mason, *Arch. Biochem. Biophys.* **267**, 681 (1988).

[14] D. Becker, S. Swarts, M. Champagne, and M. D. Sevilla, *Int. J. Radiat. Biol.* **53**, 767 (1988).

and it is thought that the one-electron oxidation product, namely, the thiyl radical, is not formed. GSHP does not support oxygen consumption in the presence of H_2O_2. It also inhibits the oxygen consumption stimulated by HRP after an initial enhancement of oxygen consumption.[8]

Prostaglandin H Synthase

Purified prostaglandin H synthase (PHS) in the presence of either H_2O_2 or 15-hydroperoxy-5,8,11,13-eicosatetraenoic acid (15-HPETE) oxidizes GSH to its thiyl radical.[10] Glutathione also inhibited cyclooxygenase activity as determined by measuring oxygen incorporation into arachidonic acid. Reversed-phase high performance liquid chromatography analysis of the arachidonic acid metabolites indicated that the presence of glutathione in an incubation altered the metabolite profile. In the absence of the cofactor, the metabolites were PGD_2, PGE_2, and 15-hydroperoxy-PGE_2 (where PG indicates prostaglandin), whereas in the presence of glutathione, the only metabolite was PGE_2. These results indicate that glutathione not only serves as a cofactor for prostaglandin E isomerase but is also a reducing cofactor for prostaglandin H hydroperoxidase. Assuming that glutathione thiyl free radical observed in the trapping experiments is a result of the enzymatic reduction of 15-hydroperoxy-5,8,11,13-eicosatetraenoic acid to 15-hydroxy-5,8,11,13-eicosatetraenoic acid, then a one-electron donation from glutathione to prostaglandin hydroperoxidase is indicated.

Lactoperoxidase

Several thiol compounds of pharmacological relevence such as captopril, cysteine, cysteamine, N-acetylcysteine, GSH, and penicillamine were studied by lactoperoxidase–H_2O_2 oxidation.[9] In all cases the corresponding thiyl radical was formed. The reaction was strongly peroxide dependent with either H_2O_2- or thiol-derived peroxides, but in some cases the addition of exogenous hydrogen peroxide was unnecessary.

Thyroid Peroxidase

Methylmercaptoimidazole, an antithyroid compound, reacts with hog thyroid peroxidase compound I and compound II. One of the mechanisms for the antithyroid action of methylmercaptoimidazole is proposed to involve thiol oxidation by peroxidase compound I.[15]

[15] S. Ohtaki, H. Nakagawa, M. Nakamura, and I. Yamazaki, *J. Biol. Chem.* **257,** 761 (1982).

Myeloperoxidase–Oxidase

Svensson and Lindvall have shown[16,17] by oxygen consumption measurements that myeloperoxidase–oxidase (MPO) can oxidize compounds such as cysteamine and methyl and ethyl esters of cysteine to their corresponding thiyl radicals at pH 7.0. Penicillamine and GSH were not oxidized by this system, and cysteine was a poor substrate compared to its esters. Oxidation of cysteamine by MPO is shown to involve H_2O_2 as an intermediate.[17]

Reaction of Thiol Compounds with Nitro Compounds

Nitrofurantoin, the urinary antiseptic, can be reduced to the nitrofurantoin anion radical by thiol compounds such as cysteine and related thiol compounds in a slightly alkaline solution (pH 8.0) by a reaction that presumably forms the thiyl radical.[18]

Some Reactions of Glutathione Thiyl Radical

The GSH thiyl radical generated by HRP or PHS has been shown to form GSH conjugates. The thiyl radical reacts with styrene to yield a carbon-centered radical that subsequently reacts with molecular oxygen to give the styrene–GSH conjugate.[19] Under anaerobic conditions additional GSH conjugates were formed, one of which was S-(2-phenyl) ethylglutathione [Eq. (11)].

$$GS \cdot + C_6H_5CH{=}CH_2 \rightarrow C_6H_5\dot{C}HCH_2SG \xrightarrow{O_2} C_6H_5\overset{OO\cdot}{\underset{|}{C}}HCH_2SG \rightarrow\rightarrow C_6H_5\overset{OH}{\underset{|}{C}}HCH_2SG$$

$$(11)$$

Thiol Pumping

Thiol pumping may be defined as the oxidation of a thiol compound by another free radical metabolite that results in formation of the parent molecule and the thiyl free radical, or a radical derived therefrom. This is a futile metabolism. The first report of thiol pumping was from the laboratory of Yamazaki and co-workers,[20] who showed that GSH reacts with

[16] B. E. Svensson and S. Lindvall, *Biochem. J.* **249,** 521 (1988).

[17] B. E. Svensson, *Biochem. J.* **253,** 441 (1988).

[18] D. N. Ramakrishna Rao and R. P. Mason, unpublished observations (1988).

[19] B. H. Stock, J. Schreiber, C. Guenat, R. P. Mason, J. R. Bend, and T. E. Eling, *J. Biol. Chem.* **261,** 15915 (1986).

[20] T. Ohnishi, H. Yamazaki, T. Iyanagi, T. Nakamura, and I. Yamazaki, *Biochem. Biophys. Acta* **172,** 357 (1969).

chlorpromazine radical cation to produce the GSH thiyl radical. The literature on thiol pumping has been reviewed recently.[21]

The GSH thiyl radical is also shown to be formed in the aminopyrine-catalyzed reaction.[22] Aminopyrine is oxidized by prostaglandin H synthase hydroperoxidase to the aminopyrine radical cation, which then disproportionates to the iminium cation. The iminium cation is further hydrolyzed to the demethylated amine and formaldehyde; in the presence of GSH, however, the aminopyrine radical cation is reduced, resulting in the formation of aminopyrine and the GSH thiyl radical (Fig. 3). The phenoxyl radical of diethylstilbestrol, which is proposed as a possible determinant of the genotoxicity of this compound, is also shown to react with GSH to generate the gluthionyl radical.[23]

Moldeus and co-workers[24] have done extensive studies on the fate of GSH during the HRP- and PHS-catalyzed oxidations of p-phenetidine and acetaminophen. They have shown that the free radical metabolites of both p-phenetidine and acetaminophen react readily with GSH to form the parent molecules and the glutathionyl radical, the latter being trapped by DMPO. In the absence of DMPO, the thiyl radical of GSH reacts with oxygen or thiolate anion [Eqs. (5)–(7)].

Recent pulse-radiolysis experiments failed to detect any reaction between the acetaminophen phenoxyl radical and cysteine.[25] The reduc-

$$\cdot OC_6H_5NHAC + RSH \rightleftharpoons RS\cdot + HOC_6H_5NHAC \qquad (12)$$

tion potential of a thiol group is $+0.83$ V,[26] which is about 0.1 V higher than that for the acetaminophen phenoxyl radical; hence, the reverse reaction [Eq. (12)], the reduction of the thiyl radical by acetaminophen, is thermodynamically favored. Wardman and co-workers[27] have studied a similar phenomenon involving the reaction between the aminopyrine radical cation (AP$^+$) and GSH [Eq. (13)]. They have shown that, thermo-

$$AP^+ + GSH \rightleftharpoons AP (+H^+) + GS\cdot \qquad (13)$$

dynamically, the equilibrium is on the left-hand side of the reaction. The reaction proceeds to the right only because the thiyl radical reacts with oxygen and/or thiolate, ultimately forming GSSG. Reaction (13) is driven

[21] D. Ross, *Pharmacol. Ther.* **37,** 231 (1988).

[22] T. E. Eling, R. P. Mason, and K. Sivarajah, *J. Biol. Chem.* **260,** 1601 (1985).

[23] D. Ross, R. J. Mehlhorn, P. Moldeus, and M. T. Smith, *J. Biol. Chem.* **260,** 16210 (1985).

[24] D. Ross, R. Larsson, B. Andersson, U. Nilsson, T. Lindquist, B. Lindeke, and P. Moldéus, *Biochem. Pharmacol.* **34,** 343 (1985).

[25] R. H. Bisby and N. Tabassum, *Biochem. Pharmacol.* **37,** 2731 (1988).

[26] P. S. Surdhar and D. A. Armstrong, *J. Phys. Chem.* **90,** 5915 (1986).

[27] I. Wilson, P. Wardman, G. M. Cohen, and M. D. Doherty, *Biochem. Pharmacol.* **35,** 21 (1986).

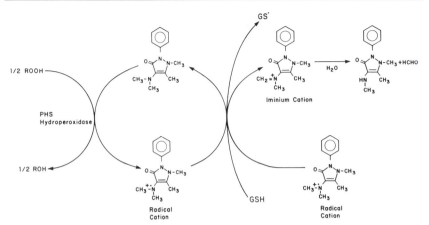

FIG. 3. Proposed mechanism for the oxidation of aminopyrine by PHS. ROOH is 15-HPETE.

to the right kinetically because of the thermodynamic stability of GSSG. We have studied reaction (12) by the direct detection of radicals with ESR spectroscopy in flow experiments. At a high concentration of GSH (100 mM), the acetaminophen phenoxyl radical oxidizes GSH to the thiyl radical, which reacts with thiolate anion to form the disulfide radical anion.[28] In the presence of oxygen, the thiyl radical is known to produce the thioperoxyl radical (RSO ·).[14]

The thiyl radical of rat hemoglobin is another example of xenobiotic-induced radical formation. The reaction of oxyhemoglobin with phenylhydrazine and hydrazine-based drugs within red blood cells induces a series of processes which lead to destruction of the cell and result in hemolytic anemia. Considerable evidence obtained from *in vitro* ESR investigations implicates free radicals in the events contributing to red blood cell hemolysis.

An immobilized radical adduct (a_{zz}^{N} = 31.8 G and a_{zz}^{H} = 9.5 G) is formed in the blood of rats which received an intraperitoneal injection of DMPO followed by an intragastric dose of phenylhydrazine.[29] This immobilized radical adduct is detected when phenylhydrazine is administered at a dose of only 1 mg/kg. Hydrazine itself gives a weaker spectrum of the same species. The immobilized radical adduct cochromatographs with oxyhemoglobin and can be detected *in vitro* using purified rat hemoglobin, phenylhydrazine, and DMPO. The sulfhydryl reagents iodoacetamide, maleimide, and N-ethylmaleimide all inhibit phenylhydrazine-dependent

[28] D. N. R. Rao, V. Fischer, and R. P. Mason, *J. Biol. Chem.* **265**, 844 (1990).

[29] K. R. Maples, S. J. Jordan, and R. P. Mason, *Mol. Pharmacol.* **33**, 344 (1988).

radical adduct formation when whole rat blood is treated *in vitro*. This sulfhydryl-dependent radical adduct has been assigned to a DMPO–hemoglobin thiyl radical adduct. This is the first report of free radical formation from a biological macromolecule formed as a consequence of free radical metabolism.

The hydrazine-based drugs hydralazine, iproniazid, isoniazid, and phenelzine were also examined.[30] Of the four, only iproniazid and phenelzine were able to induce DMPO–hemoglobin thiyl radical adduct formation *in vivo*. Only hydralazine and phenelzine were able to form this adduct *in vitro*. The *in vivo* iproniazid-induced radical adduct formation was decreased by pretreating the rats with bis-*p*-nitrophenyl phosphate, an arylamidase inhibitor. These results support the argument that iproniazid is hydrolyzed in the liver to a more reactive metabolite, most likely isopropylhydrazine, which is subsequently released into the bloodstream. In contrast, phenylhydrazine and phenelzine react directly with red blood cells to yield the DMPO–hemoglobin thiyl radical adduct. As hydralazine did not yield this adduct *in vivo*, we proposed that hydralazine is metabolized *in vivo* into a less reactive compound, possibly via acetylation.

In summary, a DMPO–hemoglobin thiyl radical adduct has been detected *in vivo*. This species is formed by the reaction of phenylhydrazine[29,30] and some hydrazine-based drugs[30] with oxyhemoglobin [Eqs. (14)–(20)]. The free radical R · could be any of the free radicals produced in these reactions [Eqs. (14)–(20)], but the phenyl radical and other carbon-centered free radicals probably play a significant role in the generation of the hemoglobin thiyl radical.

$$C_6H_5NHNH_2 + HbFe(II)-O_2 \rightarrow C_6H_5\dot{N}NH_2 + HbFe(III)-O_2H \qquad (14)$$
$$2\ C_6H_5\dot{N}NH_2 \rightarrow C_6H_5N{=}NH + C_6H_5NHNH_2 \qquad (15)$$
$$C_6H_5N{=}NH + O_2 \rightarrow C_6H_5N{=}N \cdot + (H^+ + O_2 \overline{\cdot}) \qquad (16)$$
$$C_6H_5N{=}N \cdot \rightarrow \cdot C_6H_5 + N_2 \qquad (17)$$
$$\cdot C_6H_5 + DMPO \rightarrow DMPO/\cdot C_6H_5 \qquad (18)$$
$$HbSH + R \cdot \rightarrow HbS \cdot + RH \qquad (19)$$
$$HbS \cdot + DMPO \rightarrow DMPO/HbS \cdot \qquad (20)$$

Role of Thiyl Radicals in Enzymatic Reactions

The thiol group of lipoamide dehydrogenase is thought to involve thiyl radical intermediates in its reaction mechanism.[31] Thiyl free radical intermediates are also involved in the reaction of pyruvate synthase (pyruvate : ferredoxin oxidoreductase) of *Tritrichomonas foetus* hydrogenosomes. This reaction involves oxidative decarboxylation of pyruvate

[30] K. R. Maples, S. J. Jordan, and R. P. Mason, *Drug. Metab. Dispos.* **16,** 799 (1988).
[31] Y. M. Torchinsky, in "Sulfur in Proteins," p. 164. Pergamon, New York, 1981.

to acetyl-CoA.[32] In the first step, pyruvate undergoes decarboxylation, binding thiamin pyrophosphate. This is followed by dehydrogenation to form a C_3 pyruvate radical which is bound to the enzyme complex. This complex subsequently interacts with coenzyme A, resulting in the formation of the CoA thiyl radical, which has been trapped by DMPO.[32]

Role of Thiyl Radicals in Radiation Damage

The radical repair mechanism proposed by Howard-Flanders[33] suggests that thiols compete with oxygen for transiently damaged biological targets (T) through two competing reactions: the damage repair by hydrogen donation from sulfhydryls *versus* the damage of target molecules (T) by peroxidation [Eq. (21)]. In order to explain the increased radiosensitiv-

$$TH \rightarrow T \cdot \xrightarrow{O_2} TO_2 \cdot \text{ (damage)}$$
$$\Big\downarrow RSH \tag{21}$$
$$TH + RS \cdot \text{ (damage repair)}$$

ity of GSH-depleted cells in the presence of oxygen, Biaglow *et al.*[34] have suggested [Eqs. (22) and (23)] that peroxyl radicals are first reduced to hydroperoxides and then to alcohols, the latter by the action of glutathione peroxidase enzyme. However, Tamba *et al.*[35] suggest that the

$$TOO \cdot + GSH \rightarrow TOOH + GS \cdot \tag{22}$$
$$TOOH + 2\,GSH \rightarrow TOH + H_2O + GSSG \tag{23}$$

peroxysulfenyl radical (RSOO·) is involved in the radiation damage [Eq. (8)].

Conclusion

The thiyl radical of GSH can be produced either by a peroxidase reaction or by reaction with a free radical metabolite of a drug or toxic chemical. The thiyl radical formed by the above reactions can undergo further reactions to produce oxysulfur free radicals and ultimately diamagnetic products. Thus, it can participate in important detoxification reactions in a biological system. However, the toxicological significance of the thiyl radicals themselves is unknown.

[32] R. Docampo, S. N. J. Moreno, and R. P. Mason, *J. Biol. Chem.* **262,** 12417 (1987).

[33] P. Howard-Flanders, *Nature (London)* **186,** 485 (1960).

[34] J. E. Biaglow, M. E. Varnes, E. R. Epp, E. P. Clark, and M. Astor, *in* "Oxygen and Sulfur Radicals in Chemistry and Medicine" (A. Breccia, M. A. J. Rodgers, and G. Semerano, eds.), p. 89. Edizioni Scientifiche (Lo Scarabeo), Bologna, 1986.

[35] M. Tamba, G. Simone, and M. Quintiliani, *in* "Anticarcinogenesis and Radiatino Protection" (P. A. Cerutti, O. F. Nygaard, and M. G. Simic, eds.), p. 25. Plenum, New York, 1987.

[34] Estrogens as Antioxidants

By Etsuo Niki and Minoru Nakano

Background

Chain-breaking antioxidants such as vitamin E and vitamin C protect biological membranes and tissues from oxygen toxicity and free radical attack by rapidly scavenging the oxygen radicals to terminate free radical chain oxidations. Recently, it has been found that estrogens, especially catechol estrogens, function as radical scavengers and suppress the peroxidations both *in vitro* and *in vivo*.[1-4]

Polyunsaturated fatty acids and their esters having two or more double bonds are readily oxidized by molecular oxygen by a free radical chain mechanism. Estradiol, estrone, and estriol suppress the peroxidation of methyl linoleate under UV irradiation,[1] that is, they reduce the rate of formation of conjugated diene and thiobarbituric acid reactive substances (TBARS). Estrone, estradiol, and estriol also suppress the oxidations of phospholipids of rat liver microsomes induced by $Fe^{3+}-ADP-NADPH$,[1] $Fe^{3+}-ADP-Adriamycin$,[2] or $Fe^{3+}-ADP-ascorbate$.[2] 2-Hydroxyestradiol and 2-hydroxyestrone, the major metabolites of estradiol and estrone, exhibit more profound antioxidant effects as measured by oxygen uptake and fatty acid decrease.[3] Furthermore, it is reported that estradiol and 2-hydroxyestradiol administered intraperitoneally to mice decrease the serum and liver lipid peroxide levels.[4]

Estimation of Antioxidant Activity

The free radical-mediated chain oxidations of lipids are accompanied by oxygen uptake, loss of polyunsaturated lipids, formation of lipid hydroperoxides and their decomposition products, denaturation of proteins by, for example, cross-linking and cleavage, evolution of low molecular weight gases, chemiluminescence, and loss of antioxidants. The extent of oxidation can be followed by measuring any of the above changes, and the activities of antioxidants can be estimated from their effects on the extent of oxidations.

[1] K. Yagi and S. Komura, *Biochem. Int.* **13**, 1051 (1986).
[2] K. Sugioka, Y. Shimosegawa, and M. Nakano, *FEBS Lett.* **210**, 37 (1987).
[3] M. Nakano, K. Sugioka, I. Naito, S. Takekoshi, and E. Niki, *Biochem. Biophys. Res. Commun.* **142**, 919 (1987).
[4] K. Yoshino, S. Komura, I. Watanabe, Y. Nakagawa, and K. Yagi, *J. Clin. Biochem. Nutr.* **3**, 233 (1987).

Oxidation of linoleic acid, its esters, and dilinoleoylphosphatidylcholine gives conjugated diene hydroperoxides quantitatively.[5,6] Accordingly, when oxidation is induced by a constant flux of free radicals, constant rates of oxygen uptake, substrate disappearance, and formation of conjugated diene hydroperoxides are observed. In fact, it has been observed that the amounts of oxygen uptake, substrate loss, and conjugated diene formed agree with each other in the oxidation of methyl linoleate and dilinoleoylphosphatidylcholine, both in solution and in aqueous dispersion.[7] Thus, the direct quantitation of peroxidation would be achieved from measurement of loss of substrate, uptake of oxygen, or formation of conjugated diene and hydroperoxide.

When a strong chain-breaking antioxidant such as α-tocopherol and 2-hydroxyestradiol is added to the reaction mixture, the oxidation is suppressed and gives a clear induction period. The antioxidant is consumed at a constant rate, and when it is depleted, the induction period is over and the oxidation proceeds at the same rate as that without antioxidant. Under these conditions, the oxidation of lipids proceeds according to the following scheme:

Chain initiation:

$$LH \rightarrow LO_2 \cdot \tag{1}$$

Chain propagation:

$$LO_2 \cdot + LH \xrightarrow{k_p} LOOH + L \cdot \tag{2}$$
$$L \cdot + O_2 \longrightarrow LO_2 \cdot \tag{3}$$

Chain termination:

$$LO_2 \cdot + IH \xrightarrow{k_{inh}} LOOH + I \cdot \tag{4}$$
$$(n - 1) LO_2 \cdot + I \cdot \longrightarrow \text{stable product(s)} \tag{5}$$

where LH, $L \cdot$, $LO_2 \cdot$, IH, and $I \cdot$ are lipid, lipid radical, lipid peroxyl radical, antioxidant, and the radical derived from the antioxidant, respectively, and k_p and k_{inh} are the rate constants for the hydrogen atom abstractions by lipid peroxyl radical from the lipid and antioxidant, respectively. The constant n is the stoichiometric number of peroxyl radicals trapped by each antioxidant. When the initiating radicals are generated at a constant rate, for example, by the use of a radical initiator (see [3], this

[5] N. A. Porter, *Acc. Chem. Res.* **19,** 252 (1986); N. A. Porter, *Adv. Free Radical Biol. Med.* **2,** 283 (1986).

[6] Y. Yamamoto, E. Niki, and Y. Kamiya, *Bull. Chem. Soc. Jpn.* **55,** 1548 (1982); Y. Yamamoto, E. Niki, and Y. Kamiya, *Lipids* **17,** 870 (1982).

[7] Y. Yamamoto, E. Niki, Y. Kamiya, and H. Shimasaki, *Biochim. Biophys. Acta* **795,** 332 (1984).

volume), the length of the induction period, t_{inh}, and the rate of oxidation during the induction period, R_{inh}, are given by

$$t_{inh} = \frac{n[IH]}{R_i} \tag{6}$$

$$R_{inh} = \frac{k_p[LH]R_i}{nk_{inh}[IH]} \tag{7}$$

where R_i is the rate of chain initiation.

Equations (6) and (7) can be combined to give Eq. (8). The ratio of the rate constants, k_{inh}/k_p, determines the ratio of the rate of inhibition to that

$$\frac{k_{inh}}{k_p} = \frac{[LH]}{R_{inh}t_{inh}} \tag{8}$$

of chain propagation; in other words, it determines the efficacy of the antioxidant for scavenging the chain carrying peroxyl radical before the peroxyl radical continues the chain propagation. Therefore, the ratio k_{inh}/k_p is a quantitative measure of the antioxidant activity. Since the rate of oxidation during the induction period and the length of induction period can be measured accurately, the ratio k_{ihh}/k_p can be calculated from Eq. (8). Furthermore, the length of the induction period and the ratio of the rate of the inhibited oxidation to that of uninhibited oxidation show how long and how efficiently the antioxidant suppresses the oxidation, and they are also good measures of the antioxidant activity.

Table I summarizes the effect of 2-hydroxyestradiol in the oxidation of methyl linoleate emulsions in aqueous dispersions. The oxidations were

TABLE I

INHIBITION BY 2-HYDROXYESTRADIOL OF OXIDATION OF METHYL LINOLEATE MICELLES[a]

[MeLH] (mM)	[AAPH] (mM)	[AMVN] (mM)	[IH] (μM)	t_{inh} (sec)	$10^8 R_{inh}$ (M/sec)	$10^8 R_p$ (M/sec)	$10^{-3} k_{inh}/k_p$[b]
72.9	2.02		0	0		7.26	
72.9	2.02		3.03	1220	2.28	5.34	2.62
72.9	2.02		4.03	1700	2.11	4.65	2.04
72.9	2.02		5.02	2170	1.66	2.87	2.02
211		1.01	2.04	1420	5.41	7.90	2.74
211		0.99	3.15	1480	5.92	8.16	2.40
211		0.99	4.08	2240	4.07	5.28	2.31
211		1.01	5.03	2890	2.81	5.21	2.60

[a] In 10 mM Triton X-100 aqueous dispersions induced by AAPH or AMVN in air at 37°.
[b] See text.

initiated either by lipophilic 2,2′-azobis(2,4-dimethylvaleronitrile (AMVN) or hydrophilic 2,2′-azobis(2-amidinopropane) dihydrochloride (AAPH). It gives a k_{inh}/k_p ratio of 2.4×10^3. This value shows that 2-hydroxyestradiol is over 1000 times more reactive than methyl linoleate. If the ratio of the concentration of 2-hydroxyestradiol to that of methyl linoleate is 1 to 1000, the ratio of the rate of chain inhibition to that of chain propagation is

$$\frac{\text{Chain inhibition}}{\text{Chain propagation}} = \frac{k_{inh}[\text{LO}_2 \cdot][\text{IH}]}{k_p[\text{LO}_2 \cdot][\text{LH}]} = 2.4 \tag{9}$$

suggesting that 70% of the peroxyl radicals are scavenged by 2-hydroxy-estradiol before they react with methyl linoleate to continue the chain oxidation.

Thus, the antioxidant parameters can be obtained quantitatively from the simple model systems when the substrate and reaction conditions are carefully controlled. Obviously, the estimate for antioxidant activity is more difficult and less quantitative as the substrate and reaction medium become more complicated, since the direct quantitation of their peroxida-tion becomes more difficult. For example, in biological systems, various substrates are oxidized, and there are other oxygen-utilizing processes such as respiration. The hydroperoxides formed as primary products un-dergo both enzymatic and nonenzymatic decomposition to give alcohols, aldehydes, and volatile hydrocarbon gases, none of which are quantita-tive, and the stoichiometries are not the same.

In conclusion, it has been found that estrogens, especially 2-hydroxy-estradiol, function as chain-breaking antioxidants. The antioxidant ac-tivity has been determined quantitatively in simple model systems and semi-quantitatively in more complex, biologically related systems by the methods described above.

[35] Characterization of _o_-Semiquinone Radicals in Biological Systems

By B. KALYANARAMAN

Introduction

_o_-Semiquinone radicals are intermediates in the one-electron oxida-tion of catechols, catecholamines, and catechol estrogens. For example, autoxidation and enzymatic oxidation of these materials to give free radi-

METHODS IN ENZYMOLOGY, VOL. 186

cals are well known.[1,2] The toxicity of catechols and catecholamines generally is felt to be related to production of semiquinones and toxic molecular products such as o-quinones.[3-6] Except at high pH, o-semiquinone radicals are transient in neutral and acid solution,[7] decaying rapidly via disproportionation to give the catechol and o-quinone. At low pH, where the semiquinone is present as the neutral species, the rate constant for second-order recombination $(2k)$ is typically about 10^9 M^{-1} sec^{-1}.[7,8] In neutral solution where the anion predominates (pK_a values reported for o-semiquinones range from 3.6 to 5.2),[7-9] a rate constant of around 10^8 M^{-1} sec^{-1} has been reported for the o-semiquinone from epinephrine.[8] (Rate constants reported for p-semiquinones[9] are quite similar.)

Thus, in a system where rates of radical production are low, steady-state radical concentrations can be below the level for detection by electron spin resonance (ESR) (~0.1 μM). For this reason, most ESR studies of semiquinone radical production in enzymatic systems[10] have utilized high enzyme concentrations and a rapid flow system in order to have high rates of radical production and to minimize the associated rapid depletion of starting materials.

Methods

Principle of Spin Stabilization. Spin stabilization is a technique by which o-semiquinones are stabilized by chelation through the use of diamagnetic di- or trivalent ions,[11-14]

[1] M. Adams, M. S. Blois, Jr., and R. H. Sands, *J. Chem. Phys.* **28,** 774 (1958).
[2] H. S. Mason, E. Spencer, and I. Yamazaki, *Biochem. Biophys. Res. Commun.* **4,** 236 (1961).
[3] P. Hochstein and G. Cohen, *Ann. N.Y. Acad. Sci.* **100,** 876 (1963).
[4] E. Dybing, S. D. Nelson, J. R. Mitchell, H. A. Sasame, and J. R. Gillette, *Mol. Pharmacol.* **12,** 911 (1976).
[5] A. Rotman, J. W. Daly, and C. R. Creveling, *Mol. Pharmacol.* **12,** 887 (1976).
[6] D. G. Graham, S. M. Tiffany, and F. S. Vogel, *J. Invest. Dermatol.* **70,** 113 (1978).
[7] H. W. Richter, *J. Phys. Chem.* **83,** 1123 (1979).
[8] M. Gohn, N. Getoff, and E. Bjerbakke, *J. Chem. Soc., Faraday Trans. 2* **73,** 406 (1977).
[9] P. S. Rao and E. Heyon, *J. Phys. Chem.* **77,** 2274 (1978).
[10] I. Yamazaki, *in* "Free Radicals in Biology" (W. A. Pryor, ed.), Vol. 3, p. 183. Academic Press, New York, 1977.
[11] C. C. Felix and R. C. Sealy, *J. Am. Chem. Soc.* **103,** 2831 (1981).
[12] C. C. Felix and R. C. Sealy, *J. Am. Chem. Soc.* **104,** 1555 (1982).
[13] B. Kalyanaraman, C. C. Felix, and R. C. Sealy, *Environ. Health Perspect.* **64,** 185 (1985).
[14] D. R. Eaton, *Inorg. Chem.* **3,** 1268 (1964).

Although several metal ions (e.g., Al^{3+}, Y^{3+}, Cd^{2+}, Ca^{2+}, Mg^{2+}, or Zn^{2+}) have been employed to stabilize *o*-semiquinones in aqueous[13] and non-aqueous media,[15] we feel that either Mg^{2+} or Zn^{2+} is more likely to be useful in biological systems, and consequently only Zn^{2+}- and Mg^{2+}-complexed *o*-semiquinones in aqueous media are discussed in this chapter.

We illustrate here the principle of kinetic stabilization of *o*-semiquinone using the catecholamine dopa as an example. Ultraviolet photolysis of neutral solutions of dopa gave the steady-state ESR spectrum shown in Fig. 1a which is typical of dopa semiquinone.[11] Time-resolved experiments (Fig. 1b) show that the radical is transient, decaying with second-order kinetics under these conditions. The radical decay corresponds to a second-order rate constant of 2.5×10^8 M^{-1} sec^{-1}, close to that reported for the semiquinone from epinephrine.[16]

The results obtained are markedly different when complexing metal ions are included in the reaction mixture. Using Zn^{2+} ions at pH 5.0, the only radical detected (Fig. 1c) is the semiquinone complexed with Zn^{2+}.[11] (The ESR parameters of the complex are modified from those of the uncomplexed species because of the redistribution of spin density that occurs in the presence of the metal ion.[12]) The complexed radicals are much less transient than the uncomplexed ones. Decay (Fig. 1d) remains second order, but the radical lifetime is now several seconds for a steady state concentration of approximately 10^{-5} M. The calculated second order rate constant is 1.1×10^4 M^{-1} sec^{-1} for these conditions.

Chelation (or complexation) is therefore extremely effective in decreasing the rate of radical termination. The uncomplexed *o*-semiquinone at neutral pH has a rate constant over 10,000-fold greater. Thus, the complexed radical can be detected at rates of radical formation 10,000 times lower than are necessary to detect the uncomplexed *o*-semiquinone. This allows the use of static rather than flow systems.

Preparation of Buffers. Spin stabilization of *o*-semiquinones is usually carried out in acetate and Tris buffers.[17–21] Acetate buffer is prepared by

[15] H. B. Stegmann, H. V. Bergler, and K. Scheffler, *Angew. Chem. Int. Ed.* **20**, 389 (1981).
[16] B. Kalyanaraman, C. C. Felix, and R. C. Sealy, *J. Biol. Chem.* **259**, 7584 (1984).
[17] B. Kalyanaraman and R. C. Sealy, *Biochem. Biophys. Res. Commun.* **106**, 1119 (1982).
[18] B. Kalyanaraman, C. C. Felix, and R. C. Sealy, *J. Biol. Chem.* **259**, 354 (1984).

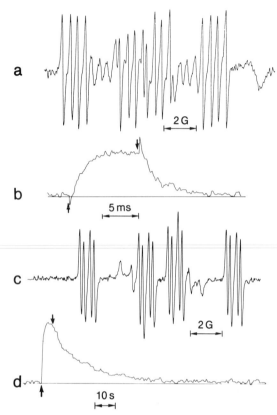

FIG. 1. ESR spectra (a and c) and kinetic profiles (b and d) for free and spin-stabilized *o*-semiquinones produced by the UV photolysis of dopa. (From Kalyanaraman *et al.*,[16] with permission.)

the addition of zinc acetate to 0.2 *M* acetic acid solution. The buffer is brought to the desired pH by dropwise addition of either 10 *M* NaOH or glacial acetic acid. Anhydrous magnesium chloride is dissolved in Tris buffer (50 m*M*, pH 7.5) and the solution filtered to remove insoluble impurities. Experiments in acetate buffer are carried out over a range of pH from 3.0 to 6.0, and those in Tris buffer are carried out over pH 7.5 to 9.5. At higher pH values, however, the Tris buffer is stirred more vigor-

[19] B. Kalyanaraman, R. C. Sealy, and K. Sivarajah, *J. Biol. Chem.* **259**, 14018 (1984).
[20] B. Kalyanaraman, P. Hitz, and R. C. Sealy, *Fed. Proc., Fed. Am. Soc. Exp. Biol.* **45**, 2477 (1986).
[21] B. Kalyanaraman, P. I. Premovic, and R. C. Sealy, *J. Biol. Chem.* **262**, 11080 (1987).

ously to dissolve Mg^{2+}. Final concentrations of Zn^{2+} and Mg^{2+} in most cases are 0.45 and 0.5 M, respectively. For experiments designed to study the effects of deuterium exchange, buffers are prepared at appropriate pD values, taking pD to equal pH plus 0.5.

Generation and Detection of o-Semiquinones. *o*-Semiquinones are generated from the oxidation of catechols by photooxidation,[11,12,22] autoxidation/chemical oxidation,[13,23] and enzymatic oxidation.[17] The enzyme activity in the presence of Zn^{2+} and Mg^{2+} is generally checked by independent assays.[17] ESR measurements are carried out at ambient temperature on solutions contained in a quartz aqueous flat cell, using a Varian E-109 spectrometer operating at 9.5 GHz and employing 100 kHz field modulation.

Spectral Parameters of Complexed and Uncomplexed o-Semiquinones. Let us consider, for example, the simple case where R = H in radicals **I** and **II**. The ESR parameters of the uncomplexed and Zn^{2+}-

I   II

complexed radicals **I** and **II** are

$$a_3^H = a_6^H = 0.76 \text{ G}; a_5^H = a_4^H = 3.66 \text{ G} \quad \text{(radical I)}$$
$$a_3^H = a_6^H = 0.5 \text{ G}; a_4^H = a_5^H = 3.9 \text{ G} \quad \text{(radical II)}$$

In general, the magnetic parameters of the complexed species are modified from those of the uncomplexed species. Typically, for the Zn^{2+}- and Mg^{2+}-complexed species, a_3^H and a_6^H decrease by about 0.28 G in going to the complex, whereas a_5^H increases by about the same amount.[11,12]

The methylene proton couplings in substituted metal-complexed *o*-semiquinones (R = CH_2X in radical **II**) decrease as the bulk of the substituent on the carbon atom bearing the methylene protons is increased.[22] For example, this particular hyperfine coupling (a_H^β) varies from 4.8 G (R = CH_3, radical **II**) to about 2.2 G [R = $CH_2C(NH_3^+)(CH_3)CO_2^-$, radical **II**].

The pK_a of the side-chain amino group in catecholamines is usually around 9. Since the pK_a is probably not very different in the derived radical, one should obtain ESR spectra from both protonated and unpro-

[22] C. C. Felix and R. C. Sealy, *Photochem. Photobiol.* **34**, 423 (1981).
[23] R. C. Sealy, W. Puzyna, B. Kalyanaraman, and C. C. Felix, *Biochim. Biophys. Acta* **800**, 269 (1984).

tonated species, which will definitely complicate the interpretation of spectra.[22,23] However, under conditions used in spin-stabilization experiments, the amino group present in o-semiquinones of catecholamines is protonated, and, consequently, only the spectrum of protonated species is usually observed. The only enzymatic system in which clear evidence for a pH dependence at around pH 5–6 was found is that of dopa. This material in the presence of Zn^{2+} in a horseradish peroxidase (HRP)–H_2O_2 system gives a fully protonated species at pH 4.5 but a mixture at pH 6.0.[17,22]

Identification of Secondary o-Semiquinones. During enzymatic oxidation of catecholamines, two types of free radicals have been identified: primary "open chain" semiquinones, formed by one-electron oxidation of the parent catecholamines, and secondary semiquinones, formed subsequent to cyclization reactions. For example, addition of HRP to an incubation mixture containing epinephrine, H_2O_2, and Zn^{2+} in acetate buffer initially gives an intense spectrum that is assigned to radical **II**, where R = $CH(OH)CH_2\overset{+}{N}H_2CH_3$. With prolonged incubation, a spectrum of a secondary radical (**III**; R_1 = H and R_2 = CH_3) appears. Radical

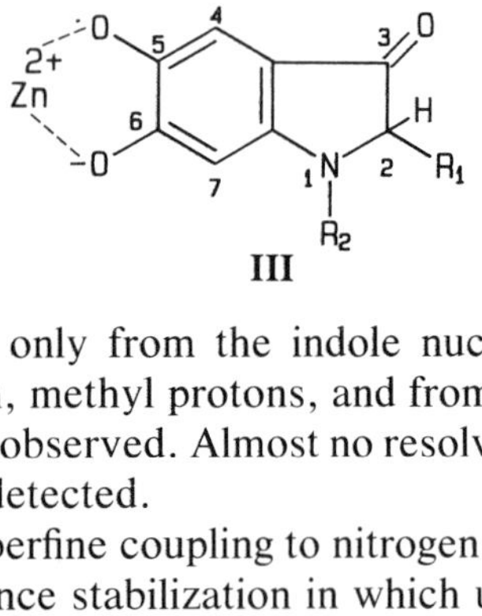

III

III exhibits couplings only from the indole nucleus; a large hyperfine coupling from nitrogen, methyl protons, and from two equivalent hydrogens (at C-2) is usually observed. Almost no resolvable couplings from the aromatic protons are detected.

The fairly large hyperfine coupling to nitrogen for radicals of this kind is indicative of resonance stabilization in which unpaired spin density is delocalized onto nitrogen (**IV↔V**).[18,24,25] Radicals of this type (R_2 = H in

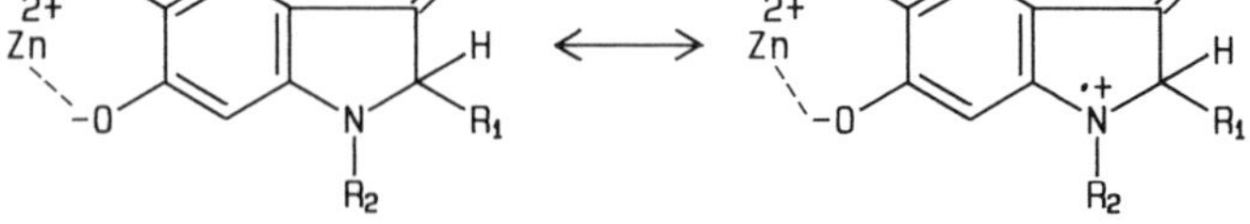

IV   V

[24] B. S. Prabhananda, B. Kalyanaraman, and R. C. Sealy, *Biochim. Biophys. Acta* **840**, 21 (1985).
[25] B. S. Prabhananda, C. C. Felix, B. Kalyanaraman, and R. C. Sealy, *J. Magn. Reson.* **76**, 264 (1988).

radical **III**) are sensitive to a change of medium from H_2O to D_2O since the NH proton is readily exchangeable. A $1:2:2:1$ pattern is visible, characteristic of equal hyperfine couplings from a nitrogen and a hydrogen. Radical **III** from norepinephrine and 3,4-dihydroxynorephedrine exhibits this pattern.

Characterization of o-Semiquinones from Quinone–Amino Acid, Quinone–Peptide, and Quinone–Protein Adducts. Spin stabilization has been used to detect *o*-semiquinones formed from addition of nucleophiles to *o*-quinones, generated by enzymatic oxidation of catechols.[21] Two types of radicals are formed, namely **VI** and **VII**. Oxidation of 4-methylcatechol and tyrosinase in the presence of excess glycine and Mg^{2+} forms

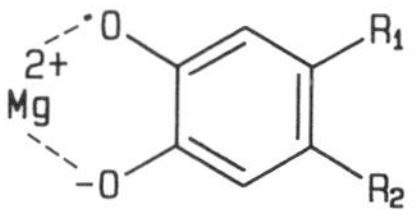

VI 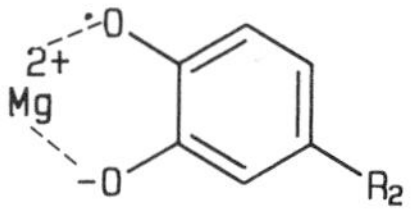   VII

a well-resolved ESR spectrum attributable to **VI** ($R_1 = CH_3$; $R_2 = NHCH_2COO^-$). The spectrum shows major couplings to side-chain methyl protons (CH_3), methylene protons (CH_2), nitrogen, and an NH proton and minor couplings attributable to two aromatic protons (Fig. 2). The spectral assignments were made through the use of specifically deuterated amino acids, $D_2NCH_2COO^-$, $H_2NCD_2COO^-$, and $D_2NCD_2COO^-$ (Fig. 2A–D). Reaction with other amino acids (alanine, phenylalanine, histidine, lysine, and proline) also gives related spectra attributable to the corresponding adducts.

Oxidation of catechol in the presence of tyrosinase, Mg^{2+}, and amino acids gives overlapping spectra owing to formation of a mixture of adducts. However, with hydroxyproline, a single ESR spectrum attributable to **VII**:

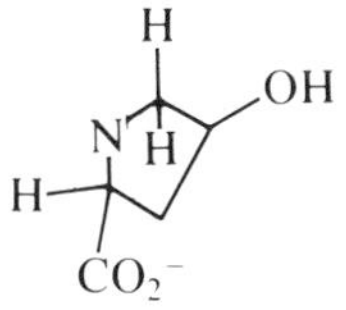

is obtained (Fig. 3a). The structure was confirmed through the use of fully deuterated phenol (Fig. 3b). *o*-Semiquinone radicals (formed from the addition of glycyl peptides to 4-methyl-*o*-quinone) have also been detected using Mg^{2+}. Spectra show a decreasing overall spectral width with increasing peptide chain lengths (Fig. 4). Semiquinones formed from addition of the protamine salmine to *o*-quinone have also been characterized.

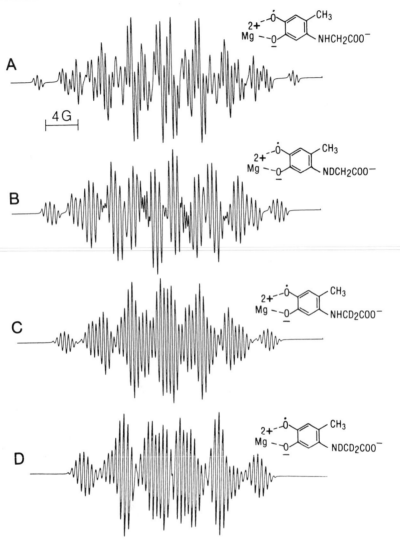

FIG. 2. ESR spectra of magnesium-complexed secondary *o*-semiquinones from 4-methylcatechol–glycine adducts. (From Kalyanaraman *et al.*,[21] with permission.)

The resulting spectrum is somewhat broader owing to slower tumbling of the radical adduct. Semiquinones arising from addition of aromatic amines and thiols to 4-methyl-*o*-quinone have been characterized. In general, peptides and proteins (with exposed amino acid residues) form ad-

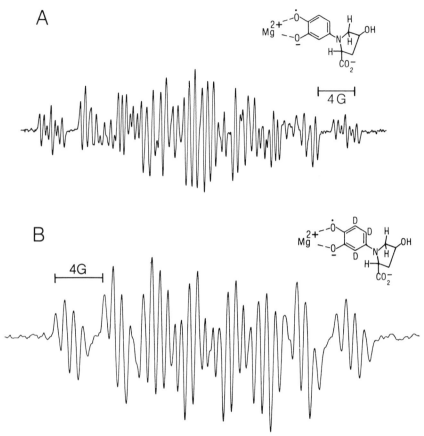

FIG. 3. ESR spectra of magnesium-complexed secondary *o*-semiquinones formed from (a) oxidation of catechol, hydroxyproline, and tyrosinase or (b) oxidation of phenol-d_6, hydroxyproline, and tyrosinase. (From Kalyanaraman *et al.*,[21] with permission.)

ducts with *o*-quinones more easily because of the lower pK_a of amino groups.

Spin Stabilization of o-Semiquinones of Catechol Estrogens. The spin-stabilization approach has also been used to detect semiquinones produced during enzymatic oxidation of catechol estrogens. Two types of semiquinones have been detected: radical **VIII** formed from 2-hydroxyestradiol and radical **IX** from 4-hydroxyestradiol. Using either Zn^{2+} or Mg^{2+} as the complexing agents, semiquinones **VIII** (R = H), **VIII** (R = OH), **IX** (R = H), and **IX** (R = OH) have been detected. The ESR specra of **VIII** (R = H)–Mg^{2+} and **VIII** (R = H)–Zn^{2+} complexes are characterized by three large hyperfine couplings to β-alicyclic protons (at C-6 and C-9).

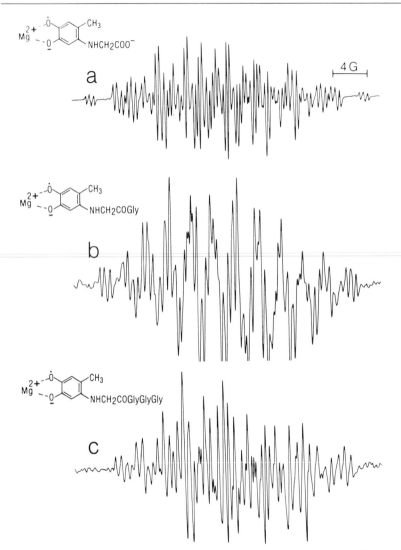

FIG. 4. ESR spectra of magnesium-complexed secondary o-semiquinones from 4-methylcatechol–glycyl, –diglycyl, and –tetraglycyl adducts. (From Kalyanaraman *et al.*,[21] with permission.)

Species **VIII** (R = OH)–Mg^{2+} and **VIII** (R = OH)–Zn^{2+} exhibit only two large hyperfine couplings. Species **IX** (R = OH)–Mg^{2+} and **IX** (R = H)–Zn^{2+} are characterized by only one large hyperfine coupling to an alicyclic β-proton (C-9) and a significant coupling to an aromatic proton at C-1.

VIII   IX

Using the spin-stabilization technique, one can also obtain the kinetic parameters (V_{max}, K_m, etc.) in enzymatic systems.[16,26]

Acknowledgments

This work was supported by National Institutes of Health Grants GM29035 and RR01008.

[26] W. Korytowski, T. Sarna, B. Kalyanaraman, and R. C. Sealy, *Biochim. Biophys. Acta* **924**, 383 (1987).

[36] Flavonoids as Antioxidants: Determination of Radical-Scavenging Efficiencies

By WOLF BORS, WERNER HELLER, CHRISTA MICHEL, and MANFRED SARAN

Introduction

Flavonoids are phenolic compounds widely distributed in plants, with over 4000 individual substances known and the list constantly expanding.[1-3] This multiplicity is not surprising in view of the structural diversity of flavonoids (Scheme I). Based on just a few backbone structures (chalcones, flavanols, flavanones, flavones, flavonols, flavylium salts or anthocyanidins, isoflavonoids, neo-, and biflavanoids), various hydroxylation, methoxylation, sulfation, and/or glycosylation patterns exist. Consider-

[1] J. Kühnau, *World Rev. Nutr. Diet.* **24**, 117 (1976).
[2] J. B. Harborne, *in* "Plant Flavonoids in Biology and Medicine" (V. Cody, E. Middleton, and J. B. Harborne, eds.), p. 15. Alan R. Liss, New York, 1986.
[3] J. B. Harborne (ed.), "Flavonoids: Advances in Research." Chapman & Hall, London, 1988.

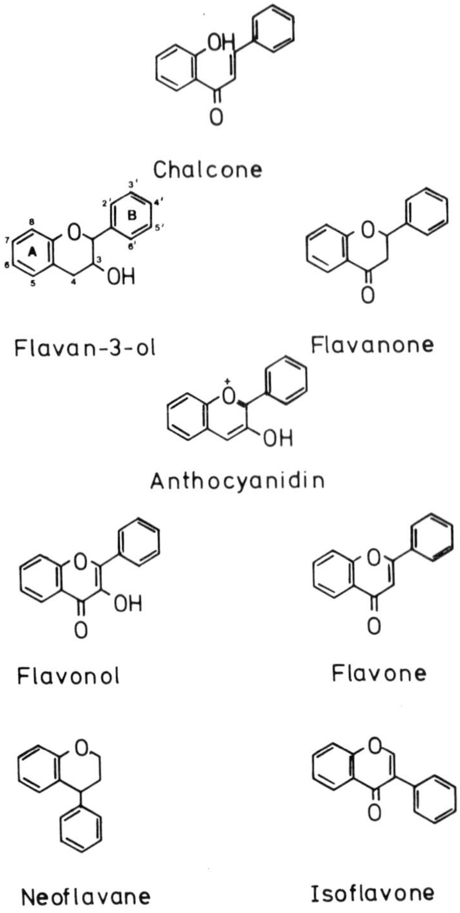

Chalcone

Flavan-3-ol   Flavanone

Anthocyanidin

Flavonol   Flavone

Neoflavane   Isoflavone

SCHEME I. Structures of flavonoids.

ing the biosynthetic pathways of these substances[4] and their evolutionary development, a major function for the intensely colored flavonols and anthocyanidins in flower petals seems to be a visual signal for pollinating insects (or hummingbirds), whereas catechins and other flavanols, owing to their astringent character, probably evolved as feeding repellants.[5] Isoflavonoid derivatives constitute one of the most important groups of plant-protective phytoalexins.[6]

[4] W. Heller and G. Forkmann, in "Flavonoids: Advances in Research" (J. B. Harborne, ed.), p. 399. Chapman & Hall, London, 1988.
[5] J. B. Harborne, Biochem. Syst. Ecol. 5, 7 (1977).
[6] D. A. Smith and S. W. Banks, Phytochemistry 25, 979 (1986).

The antioxidative effect of flavonoids has been of interest for a considerable time. In the early 1930s, the discovery of their vitamin C-sparing activity[7] led to the short-lived proposal of "vitamin P."[8] After a hiatus of more than 10 years, a flurry of activity lasted throughout the 1950s and into the 1960s.[1,9,10] In these studies, probable mechanisms and structure–activity relationships were put forward as an explanation for the role of flavonoids as food protectants. Owing to the fact that flavonoids are rapidly degraded in the digestive system to inactive compounds,[1,11] interest in their *in vivo* antioxidative effects in mammals ebbed. Only after the recognition that the antioxidative and lipid peroxidation-inhibiting potential predominantly resides in the radical-scavenging capacity rather than the chelation of metals[12,13] did the problem of the radical chemistry of flavonoid aglycones become a topic of interest.

While a general capability of flavonoids to scavenge superoxide was proposed repeatedly,[14-18] the specific scavenging of hydroxyl,[19] superoxide,[20] and peroxyl[21] radicals may have been an overinterpretation of the results. In all cases, the suggestions were derived from inhibition studies after more or less nonspecific radical generation in lipid peroxidation, enzymatic or chemical systems, with additional problems arising from the nonspecific assay methods which were used in these studies. As shown below, flavonoids do react rapidly with hydroxyl radicals, which is, however, of no surprise because of the generally high reactivity of this radical with aromatic compounds. For O_2^-, in contrast, even for the very efficient flavonol radical scavengers kaempferol and quercetin, only very low rate constants were found.

[7] C. A. B. Clementon and L. Andersen, *Ann. N.Y. Acad. Sci.* **136,** 339 (1966).

[8] A. Bentsath, S. Rusznyak, and A. Szent-Györgyi, *Nature (London)* **139,** 326 (1937).

[9] R. E. Hughes and H. K. Wilson, *Med. Chem.* **14,** 285 (1977).

[10] L. R. Dugan, *in* "Autoxidation in Food and Biological Systems" (M. G. Simic and M. Karel, eds.), p. 261. Plenum, New York, 1980.

[11] A. M. Hackett, *in* "Plant Flavonoids in Biology and Medicine" (V. Cody, E. Middleton, and J. B. Harborne, eds.), p. 177. Alan R. Liss, New York, 1986.

[12] C. G. Fraga, V. S. Martino, G. E. Ferraro, J. F. Coussio, and A. Boveris, *Biochem. Pharmacol.* **36,** 717 (1987).

[13] A. K. Ratty and N. P. Das, *Biochem. Med. Metab. Biol.* **39,** 69 (1988).

[14] J. Baumann, G. Wurm, and F. von Bruchhausen, *Arch. Pharm. (Weinheim, Ger.)* **313,** 330 (1980).

[15] I. Ueno, M. Kohno, K. Haraikawa, and I. Hirono, *J. Pharmacobio-Dyn.* **7,** 798 (1984).

[16] U. Takahama, *Photochem. Photobiol.* **42,** 89 (1985).

[17] M. Damon, F. Michel, C. Le Doucen, and A. Crastes de Paulet, *Bull. Liaison Groupe Polyphenols* **13,** 569 (1986).

[18] U. Takahama, *Plant Cell Physiol.* **28,** 953 (1987).

[19] S. R. Husain, J. Cillard, and P. Cillard, *Phytochemistry* **26,** 2489 (1987).

[20] J. Robak and R. J. Gryglewski, *Biochem. Pharmacol.* **37,** 837 (1988).

[21] J. Torel, J. Cillard, and P. Cillard, *Phytochemistry* **25,** 383 (1986).

We limited our present study on the reactivity of flavonoids with selectively generated radicals to a number of commercially available hydroxylated and methoxylated flavonoid aglycones.[22] Sulfate ester derivatives are a recently recognized subgroup,[23] and the multitudinous C- and O-glycosylated sugar moieties have no bearing on the reactivities of the remaining phenolic hydroxy groups. The hydroxyl groups are considered to be of prime importance for the radical-scavenging properties.

To generate specific types of radicals, pulse radiolysis is a uniquely suited method.[24,25] Data evaluation after fast kinetic spectroscopy of the buildup and/or decay of transient absorption leads to kinetic parameters for the primary radical attack, the stability of the secondarily formed radicals, as well as for the redox potentials of these radicals.[26] It is the radical-scavenging property which is the focus of this chapter, whereas the radical-generating function as studied by Pardini and co-workers,[27] pharmacological aspects, interactions with various enzymes, and cytotoxic and mutagenic/antimutagenic activities[28,29] are not covered.

Methods

Radiolytic and Photolytic Principles of Radical Generation

To test whether a certain substance acts as radical scavenger, the clearest evidence comes from the determination of reaction rate constants with a set of different specifically produced radicals. In our investigation we used hydroxyl ($\cdot OH$), azide ($N_3 \cdot$), superoxide (O_2^-), linoleic acid peroxyl ($LOO \cdot$), $tert$-butoxyl (t-$BuO \cdot$), and sulfite radicals ($SO_3^{\cdot -}$). Formylmethyl radical ($\cdot CH_2$—CH=O) was reacted with some flavonoids in an earlier study,[30] and it was assumed that at pH 13.5 this radical at least partially exists as the ethylene oxy radical (CH_2=$CHO \cdot$).[31] All

[22] E. Wollenweber and V. H. Dietz, *Phytochemistry* **20**, 896 (1981).

[23] D. Barron, L. Varin, R. K. Ibrahim, J. B. Harborne, and C. A. Williams, *Phytochemistry* **27**, 2375 (1988).

[24] K.-D. Asmus, this series, Vol. 105, p. 167.

[25] W. Bors, M. Saran, C. Michel, and D. Tait, *in* "Advances on Oxygen Radicals and Radioprotectors" (A. Breccia, C. L. Greenstock, and M. Tamba, eds.), p. 13. Lo Scarabeo, Bologna, 1984.

[26] L. G. Forni and R. L. Willson, this series, Vol. 105, p. 179.

[27] W. F. Hodnick, E. B. Milosavljevic, J. H. Nelson, and R. S. Pardini, *Biochem. Pharmacol.* **37**, 2607 (1988).

[28] B. Havsteen, *Biochem. Pharmacol.* **32**, 1141 (1983).

[29] E. Middleton, *Trends Pharmacol. Sci.* **5**, 335 (1984).

[30] S. Steenken and P. Neta, *J. Phys. Chem.* **86**, 3661 (1982).

[31] S. Steenken, *J. Phys. Chem.* **83**, 595 (1979).

these radicals are oxidizing species and are assumed to form aroxyl radicals with phenolic compounds.[32] Hydrated electrons (e_{aq}^-) and formate radicals (CO_2^-) are reducing radicals and react only with a few flavonoid structures.

Spectral observation of radical reactions with flavonoids is generally assisted by the strong absorption characteristics of both the parent compounds and their respective aroxyl radicals. The 15-photomultiplier array used at our institute,[33] which allows analysis of spectral changes within a 130 nm spectral region at a time resolution of 500 nsec after just one pulse, could be optimally applied to determine whether different absorption bands show identical kinetic behavior. These data are especially important for establishing structure–activity relationships. The aroxyl radicals are preferentially generated with $N_3 \cdot$ radicals rather than $\cdot OH$ radicals, because it cannot be ruled out that after attack of the latter radical highly unstable hydroxycyclohexadienyl radical derivatives are formed, which may exhibit different spectral and kinetic properties.[34] The strong pH dependence of the absorption spectra of flavonoids also has to be considered. Owing to the various dissociable phenolic hydroxyl groups,[35–38] both the transient spectra of the radicals and the kinetics of their formation (e.g., $N_3 \cdot$ radicals react with phenolate anions rather than with undissociated phenols)[39] can change drastically.

To determine whether substances act as antioxidants, the stability of the radical formed after scavenging should be known. A very reactive secondary radical would propagate rather than interrupt a chain reaction! Furthermore, effective antioxidants have been shown to react in a 1 : 2 stoichiometry,[40] that is, one antioxidant molecule reacts with two radical species, the second reaction being a radical–radical recombination process. This type of reaction has thus far been observed only for aliphatic peroxyl radicals reacting with phenolic and arylamine antioxidants[40] and

[32] M. Erben-Russ, W. Bors, and M. Saran, *Int. J. Radiat. Biol. Relat. Stud. Phys. Chem. Med.* **52,** 393 (1987).
[33] M. Saran, G. Vetter, M. Erben-Russ, R. Winter, A. Kruse, C. Michel, and W. Bors, *Rev. Sci. Instrum.* **58,** 363 (1987).
[34] G. E. Adams and B. D. Michael, *Trans. Faraday Soc.* **63,** 1171 (1967).
[35] N. P. Slabbert, *Tetrahedron* **33,** 821 (1977).
[36] P. K. Agrawal and H. J. Schneider, *Tetrahedron Lett.* **24,** 177 (1983).
[37] J. A. Kennedy, M. H. G. Munro, H. K. J. Powell, L. J. Porter, and L. Y. Foo, *Aust. J. Chem.* **37,** 885 (1984).
[38] O. S. Wolfbeis, M. Leiner, P. Hochmuth, and H. Geiger, *Ber. Bunsenges. Phys. Chem.* **88,** 759 (1984).
[39] Z. B. Alfassi and R. H. Schuler, *J. Phys. Chem.* **89,** 3359 (1985).
[40] C. E. Boozer, G. S. Hammond, C. E. Hamilton, and J. N. Sen, *J. Am. Chem. Soc.* **77,** 3233 (1955).

with α-tocopherol.[41,42] Recently, we obtained kinetic evidence that recombination of kaempferol and quercetin aroxyl radicals with linoleic acid peroxyl radicals takes place with rate constants exceeding $10^8 \ M^{-1}$ sec^{-1}.[32]

It is easily understood that with increasing stability of an antioxidant-derived aroxyl radical, a recombination reaction becomes more and more likely:

$$\text{PhOH} + \text{ROO} \cdot \rightarrow \text{PhO} \cdot + \text{ROOH} \tag{1}$$
$$\text{PhO} \cdot + \text{ROO} \cdot \rightarrow \text{ROO—Ph}(=\text{O}) \tag{2}$$

Consequently, to classify a certain substance as an antioxidant, the following points have to be known: (i) rate constants with different types of radicals; (ii) stability (and decay kinetics) of the "antioxidant radical"; and (iii) stoichiometry of the radical-scavenging process. Pulse radiolysis yields data for the first two points, while unequivocal evidence of the stoichiometry can only be obtained from product identification.

Rate constants of reaction with the photolytically produced t-BuO $\cdot$ radical were, with one exception,[43] determined in the so-called crocin assay (see below).[44] The $\cdot$ OH radicals simultaneously formed are scavenged by the presence of 20% $tert$-butanol, which in addition helps to dissolve the flavonoid aglycones. Only substances not absorbing in the wavelength region of crocin (λ_{max} 440 nm), such as flavanols and flavanones, can be measured by this method.

Another problem with flavonoid aglycones is their rather poor solubility in water. This can be overcome by dissolving the substances in alkaline solutions and titrating back to the desired pH value, taking care to avoid autoxidation of the substances[27] by keeping the solution constantly under nitrogen.

Determination of Reaction Rate Constants

Pseudo-First-Order Reactions. If a considerable excess of substrate concentration over radical concentration (pulse dose) can be achieved, a concentration-dependent plot of the first-order rate constant should yield a straight line intersecting the origin. At the same time, a value for the second-order rate constant is obtained by division with the substrate concentration. Rate constants with highly reactive radicals ($\cdot$ OH, $N_3 \cdot$, e_{aq}^-) can usually be obtained by this method.

[41] J. Tsuchiya, E. Niki, and Y. Kamiya, *Bull. Chem. Soc. Jpn.* **56**, 229 (1983).
[42] J. Winterle, D. Dulin, and T. Mill, *J. Org. Chem.* **49**, 491 (1984).
[43] M. Erben-Russ, C. Michel, W. Bors, and M. Saran, *J. Phys. Chem.* **91**, 2362 (1987).
[44] W. Bors, C. Michel, and M. Saran, *Biochim. Biophys. Acta* **796**, 312 (1984).

Competitive Radical Decay Processes. When the radical concentration approaches that of a poorly soluble substrate, direct kinetic evaluation of the data is no longer feasible. In that case, competitive bimolecular decay of the radicals dominates. To obtain reaction rate constants under such conditions, kinetic modeling of parallel and sequential reactions, as they occur for the various radical-generating systems, is required. Combinations of differential equations are run, and optimized rate constants are obtained as variables after iterative approximation of the experimental results with the theoretical model. Rate constants with O_2^- (C. Michel, unpublished results), t-BuO·,[43] SO_3^-,[45] and LOO·[32] have been determined by this procedure. The kinetic model for LOO· generation and reactions turned out to be the most complex one.

Competitive Radical Attack. Evaluation of the parallel attack of radicals at a substrate and a reference compound (competitor) is usually applied, if one of the substances is optically transparent at the respective wavelength.[46,47] The relative rate constants with t-BuO· in the crocin assay are always obtained from competition plots by observing the bleaching of the strong absorption of crocin with a molar absorptivity of $133.500 \ M^{-1} \ cm^{-1}$.[44] Using $3 \times 10^9 \ M^{-1} \ sec^{-1}$,[43] the absolute rate constant of crocin with t-BuO·, as the reference value, relative rate constants can be converted to absolute ones.

Radical Chemistry of Flavonoids

Rate Constants of Flavonoid Aroxyl Radical Formation and Decay

Owing to the low solubility of flavonoids, conditions for pseudo-first-order reactions with radical species can rarely be achieved for flavonoid aglycones. Therefore, kinetic modeling calculations have to be employed.

Table I gives a compilation of all presently known rate constants of flavonoid aglycones with three different types of radicals. Both ·OH and N_3· are highly electrophilic radicals and consequently show little difference in their diffusion-controlled rate constants. The values for t-BuO· are 20- to 80-fold lower, with the only exception of the flavonol quercetin, whose rate constant was obtained directly by pulse radiolysis.[43] This

[45] M. Erben-Russ, C. Michel, W. Bors, and M. Saran, *Radiat. Environ. Biophys.* **26,** 289 (1987).

[46] G. E. Adams, J. W. Boag, J. Currant, and B. D. Michael, *in* "Pulse Radiolysis" (M. Ebert, J. P. Keene, A. J. Swallow, and J. H. Baxendale, eds.), p. 131. Academic Press, New York, 1965.

[47] W. Bors, C. Michel, and M. Saran, *in* "CRC Handbook of Methods for Oxygen Radical Research" (R. A. Greenwald, ed.), p. 181. CRC Press, Boca Raton, Florida, 1985.

TABLE I
REACTION RATE CONSTANTS OF FLAVONOIDS WITH STRONGLY OXIDIZING RADICALS

Flavonoid			Rate constant $(\times 10^8 \ M^{-1} \ \text{sec}^{-1})$		
	Substitution pattern				
Trivial name	OH	OCH$_3$	$\cdot$OH[a]	N$_3\cdot$[b]	t-BuO$\cdot$[c]
Dihydrochalcones					
Phloretin	4,2',4',6'	—	—	—	0.4
Flavanols					
(+)-(2R,3S)-Catechin	3,5,7,3',4'	—	66	50	1.35
(−)-(2R,3R)-Epicatechin	3,5,7',3',4'	—	64	51	—
Flavanones					
Dihydrofisetin	3,7,3',4'	—	45	56	—
Eriodictyol	5,7,3',4'	—	—	31	—
Dihydroquercetin	3,5,7,3',4'	—	—	24	1.0
Hesperetin	5,7,3'	4'	—	58	0.7
Flavylium salts (anthocyanidins)					
Pelargonidin chloride	3,5,7,4'	—	45	62	—
Cyanidin chloride	3,5,7,3',4'	—	—	30	—
Flavones					
Apigenin	5,7,4'	—	—	48	—
Luteolin	5,7,3',4'	—	—	41	—
Acacetin	5,7	4'	—	28	1.3
Flavonols					
Fisetin	3,7,3',4'	—	—	52	—
Kaempferol	3,5,7,4'	—	46	88	—
Quercetin	3,5,7,3',4'	—	43	66	25.0[d]
Morin	3,5,7,2',4'	—	—	73	—
Kaempferid	3,5,7	4'	—	65	—

[a] Determined pulse-radiolytically from the buildup of aroxyl radical absorption [M. Erben-Russ, W. Bors, and M. Saran, *Int. J. Radiat. Biol.* **52**, 393 (1987)]; the values for dihydrofisetin and pelargonidin chloride have not previously been published.

[b] Determined pulse-radiolytically from the buildup of aroxyl radical absorption [W. Bors and M. Saran, *Free Radical Res. Commun.* **2**, 289 (1987)].

[c] Data obtained in the crocin assay [W. Bors, C. Michel, and M. Saran, *Biochim. Biophys. Acta* **796**, 312 (1984)] and recalculated using the reference value for crocin of $3 \times 10^9 \ M^{-1} \ \text{sec}^{-1}$ [M. Erben-Russ, C. Michel, W. Bors, and M. Saran, *J. Phys. Chem.* **91**, 2362 (1987)].

[d] Determined pulse-radiolytically from the buildup of aroxyl radical absorption [M. Erben-Russ, C. Michel, W. Bors, and M. Saran, *J. Phys. Chem.* **91**, 2362 (1987)].

points to flavonols having structural features which may facilitate the attack of more discriminately reacting radicals such as t-BuO$\cdot$. The investigation was therefore expanded to include reactions with some other oxidizing radicals and the two flavonols kaempferol and quercetin (Table II).

TABLE II
REACTION RATE CONSTANTS OF FLAVONOLS WITH OTHER
OXIDIZING RADICALS

	Rate constant ($\times 10^8 \ M^{-1} \ sec^{-1}$)			
Flavonol	$(SCN)_2 \cdot^{-a}$	$SO_3 \cdot^{-b}$	$LOO \cdot^c$	$O_2 \cdot^{-d}$
Kaempferol	8.5	4.0	0.34, 0.42	0.0055
Quercetin[e]	4.0	2.5	0.18, 0.15	0.0009

[a] Determined pulse-radiolytically from the kinetics of $(SCN)_2 \cdot^-$ decay, substrate bleaching, as well as the buildup of aroxyl radical absorption [M. Erben-Russ, W. Bors, and M. Saran, *Int. J. Radiat. Biol.* **52,** 393 (1987)].

[b] Determined pulse-radiolytically from the buildup of aroxyl radical absorption [M. Erben-Russ, C. Michel, W. Bors, and M. Saran, *Radiat. Environ. Biophys.* **26,** 289 (1987)].

[c] Value dependent on method of radical generation, with the first one for a mixture of $LOO \cdot$ radical isomers and the second for 13-$LOO \cdot$ radicals specifically [M. Erben-Russ, W. Bors, and M. Saran, *Int. J. Radiat. Biol.* **52,** 393 (1987)].

[d] Determined pulse-radiolytically from kinetic modeling of the O_2^- decay at pH 7.5; at higher pH no reaction at all was observed (C. Michel, unpublished results).

[e] Exceptionally high values of $3.1 \times 10^9 \ M^{-1} \ sec^{-1}$ (and $1.8 \times 10^9 \ M^{-1} \ sec^{-1}$ for catechin) were determined for the reaction with $CH_2{=}CHO \cdot / \cdot CH_2{-}CH{=}O$ radicals at pH 13.5 [S. Steenken and P. Neta, *J. Phys. Chem.* **86,** 3661 (1982)].

Table III compiles the decay rate constants of flavonoid aroxyl radicals after generation by $N_3 \cdot$ radicals at pH 11.5, which in all cases represent second-order dismutation processes. The most striking feature is the wide range of decay rates, and this should be viewed in context with the radical-scavenging versus antioxidative capacity. While both kaempferol and quercetin are highly efficient radical scavengers, only the quercetin aroxyl radical decays slowly enough to make this flavonol a potent antioxidant. Looking at the structures of the most stable aroxyl radicals, those with decay rates of 10^5 to $10^6 \ M^{-1} \ sec^{-1}$, it is evident that without exception all contain the 3′,4′-catechol structure. All other phenolic compounds form far less stable aroxyl radicals.

Transient Spectra of Radical Intermediates

The wide variety of flavonoid structures (see Scheme I) and hydroxylation patterns offers a unique opportunity to study the influence of substrate substitution pattern on shape and intensity of transient spectra.

TABLE III
SECOND-ORDER DECAY RATE CONSTANTS AND SPECTRAL PARAMETERS
OF FLAVONOID AROXYL RADICALS[a]

Substance[b]	Rate constant $2k$ ($\times$ 10^6 M^{-1} sec^{-1})	Molar absorptivity (M^{-1} cm^{-1})	Absorption maximum (nm)
(+)-Catechin	0.6	10,000	310
Eriodictyol	0.4	8,900	315
Dihydroquercetin	0.1	7,600	315
Pelargonidin chloride	210	19,500	685
Apigenin	170	8,100	360
Luteolin	0.2	8,100	475
Acacetin	500	6,600	325
Fisetin	0.3	4,100	600
Kaempferol	140	24,000	550
Quercetin	3.4	15,600	530
Morin	63	17,600	525
Kaempferid	370	8,000	480

[a] Aroxyl radicals were generated by attack of $N_3 \cdot$ radicals at pH 11.5 [W. Bors and M. Saran, *Free Radical Res. Commun.* **2**, 289 (1987)].
[b] For substitution pattern of substances see Table I.

This was best exemplified in the comparison of aroxyl radical spectra generated by $N_3 \cdot$ attack.[48] The study revealed three structural principles governing the formation of transient spectra (with one exceptional case): flavan-3-ols and flavanones, which possess a saturated 2,3-bond, show only the semiquinone radical of the B ring; flavones, with either the 3-hydroxyl group (the flavonol fisetin) or the 5-hydroxyl group (e.g., the flavone luteolin), show transient spectra of intermediate strength and similar shape; flavonols which also possess a 5-hydroxyl group show the strongest absorption of the aroxyl radical, in wavelength regions remote from the parent phenol absorption. These substances are thus best suited for competition experiments. The exception is pelargonidin chloride, a flavylium salt. Cyanidine chloride, the corresponding B-ring catechol analog, forms a weakly absorbing water adduct,[1] a pseudo-base, where the conjugation to the heterocyclic ring is interrupted (Scheme II). As expected, the aroxyl radical exhibits only the semiquinone structure of the B ring.

[48] W. Bors and M. Saran, *Free Radical Res. Commun.* **2**, 289 (1987).

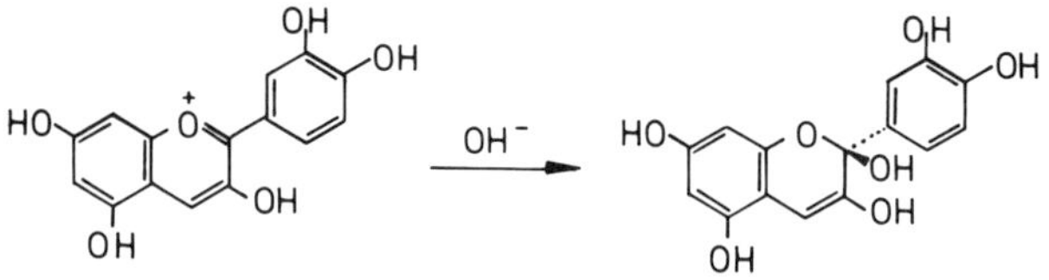

SCHEME II. Reaction of flavylium salt to form a water adduct.

Structural Principles for Effective Radical Scavenging by Flavonoids

On the basis of the presently available spectral and kinetic evidence on the formation and decay of the flavonoid aroxyl radicals, a structure–activity relationship can be derived. As shown in Scheme III, three structural groups are important determinants for radical scavenging and/or antioxidative potential: (a) the *o*-dihydroxy (catechol) structure in the B ring, which is the obvious radical target site for all flavonoids with a saturated 2,3-bond (flavan-3-ols, flavanones, cyanidine chloride) [whether semiquinone formation itself occurs with flavonols cannot be determined because of spectral overlaps and would require electron paramagnetic resonance (EPR) investigations; however, the catechol moiety does confer a higher stability to aroxyl radicals and obviously participates in elec-

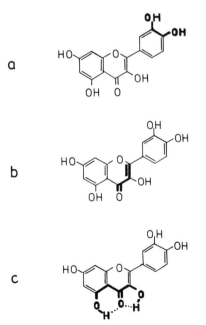

SCHEME III. Structural groups for radical scavenging.

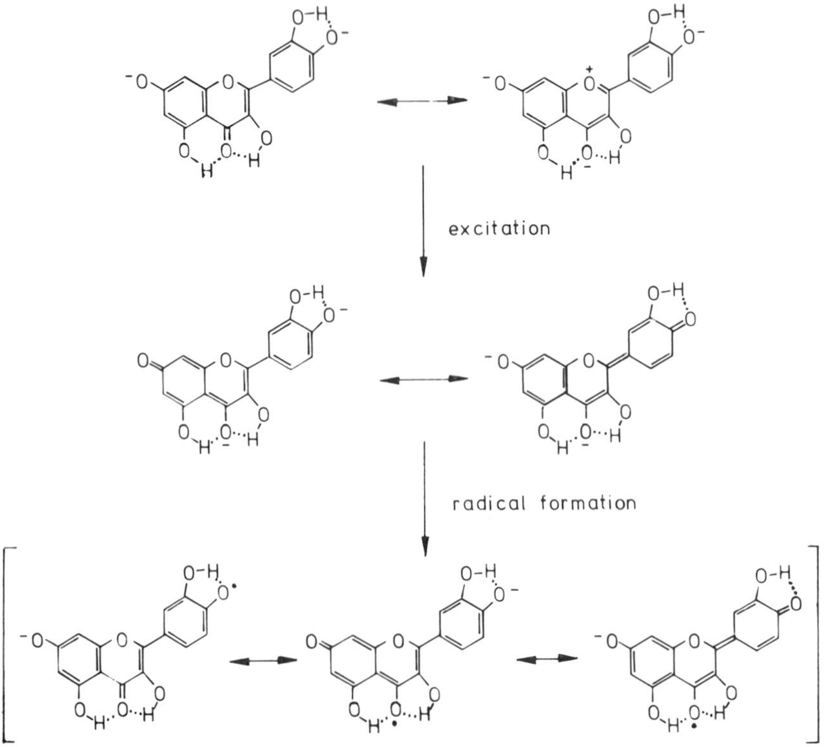

SCHEME IV. Reactions of quercetin showing principal radical structures involving three ring systems.

tron delocalization]; (b) the 2,3-double bond in conjugation with a 4-oxo function, which are responsible for electron delocalization from the B ring; and (c) the additional presence of both 3- and 5-hydroxyl groups for maximal radical-scavenging potential and strongest radical absorption (from a kinetic standpoint, the 3- and 5-hydroxyl groups are equivalent owing to their hydrogen bonds with the keto group).[35]

In Scheme IV, using quercetin as an example, we attempted to draw the principal mesomeric structures for the flavonol in its doubly-dissociated state which, at pH 11.5, is to be expected based on the dissociation sequence of 7—OH > 4'—OH > 5—OH.[49] For the aroxyl radicals formed by electron transfer to $N_3 \cdot$ we assume that attack at the most likely site is kinetically indistinguishable. Dipolar structures and espe-

[49] T. J. Mabry, K. R. Markham, and M. B. Thomas (eds.), "The Systematic Identification of Flavonoids," Part 2. Springer-Verlag, Berlin, 1970.

cially the involvement of the 7- and 4'-hydroxyl groups in the excited state (see top part of Scheme IV), have already been noted.[38] Taking these mesomeric structures as guide, we present three principal radical structures involving all three ring systems.

Conclusions

The radical chemistry of flavonoids not only is of interest from a kinetic or mechanistic point of view but also offers considerable insight into structural relationships of these highly evolved plant components. First of all, the consistently high rate constants for attack by different types of radicals (see Tables I and II) demonstrate the effective radical-scavenging capabilities of most flavonoids. Second, as shown in Scheme IV, owing to extensive electron delocalization as a prerequisite for radical stabilization, multiple mesomeric structures exist for aroxyl radical species of flavonoids. From Scheme IV it can also be predicted which structural principles are necessary for optimal antioxidative capacity. By the method of pulse radiolysis, as described here, it was possible to obtain the data necessary to confirm the hypothesis. This might allow construction of even better flavonoid antioxidants, were it not for nature itself which has already optimized the biosynthesis along these lines.[4]

Acknowledgments

The dedicated assistance during the pulse-radiolytic experiments by Alf Kruse is greatly appreciated. Michael Erben-Russ determined a number of the quoted data (as reported in his Ph.D. thesis). Matthias Born and Armin Furch were very helpful during discussion of the manuscript.

[37] Fluorescence Measurements of Incorporation and Hydrolysis of Tocopherol and Tocopheryl Esters in Biomembranes

By VALERIAN E. KAGAN, RUMYANA A. BAKALOVA, ELENA E. SERBINOVA, and TSANKO S. STOYTCHEV

In qualitative terms, the stabilizing action of vitamin E (α-tocopherol) in biomembranes can be ascribed to four effects, namely, (1) interaction with lipid peroxide radicals, (2) specific rearrangement in the membrane lipid bilayer (i.e., restriction of molecular mobility), (3) protection of membranes from the damaging action of phospholipases, and (4) quenching of singlet molecular oxygen.[1-4] Quantitative evaluation of the role of each of these particular mechanisms should be based on information on the real concentration of tocopherols in the lipid bilayer of the membrane. Estimation of the real concentration of tocopherols in the lipid bilayer of the membranes (localization in membrane monolayers, incorporation and distribution within the hydrophobic zone) on the basis of its total concentration is not possible, because of its nonuniform (nonrandom) distribution in the membrane: (1) tocopherols exogenously added to membrane or liposome suspensions could be but partly bound to the membranes, (2) membrane-bound tocopherols could be but partly incorporated within the hydrophobic core of the membrane, (3) incorporated tocopherols could associate to form clusters within the membrane, thus preventing monomeric distribution of molecules, and (4) tocopherol molecules incorporated and distributed in the outer leaflets of the membranes could be unaccessible to the inner leaflets of the membrane owing to low rate of transbilayer migration.[5,6] In addition to these simple physicochemical mechanisms, there are also metabolic complications arising from the different rates of hydrolysis of tocopherol esters (tocopheryl acetate is a standard form of vitamin E)[7,8] by membrane-bound deesterases and from the different activities of chromanoxyl radical-regenerating enzymes and of tocopherol-binding proteins.[9-13]

Some of these problems can be solved by studying the intrinsic fluorescence of tocopherol molecules within liposomes or natural mem-

[1] C. De Duve and O. Hayaishi (eds.), "Tocopherol, Oxygen and Biomembranes." Elsevier/North-Holland, Amsterdam, 1978.

[2] L. J. Machlin (ed.), "Vitamin E: A Comprehensive Treatise." Dekker, New York, 1980.

[3] A. L. Tappel, *Ann. N.Y. Acad. Sci.* **205**, 12 (1972).

[4] G. W. Burton and K. U. Ingold, *Acc. Chem. Res.* **19**, 194 (1986).

[5] E. Niki, Y. Yamamoto, and Y. Kamiya, *Life Chem. Rep.* **3**, 35 (1985).

[6] V. A. Tyurin, V. E. Kagan, E. A. Serbinova, N. V. Gorbunov, A. N. Erin, L. L. Prilipko, and Ts. S. Stoytchev, *Bull. Exp. Biol. Med. USSR* **12**, 689 (1986).

[7] J. Whitaker, E. G. Fort, S. Vimokesan, and J. S. Dinning, *Am. J. Clin. Nutr.* **20**, 783 (1967).

[8] M. L. Scott, *in* "Handbook of Lipid Research" (H. F. De Luka, ed.), p. 133. Plenum, New York, 1978.

[9] Y. Takahashi, K. Urano, and S. Kimura, *J. Nutr. Sci. Vitaminol. (Tokyo)* **23**, 201 (1977).

[10] W. A. Behrens and R. Madere, *Fed. Proc., Fed. Am. Soc. Exp. Biol.* **42**, 813 (1983).

[11] L. Packer, J. J. Maguire, R. J. Melhorn, E. Serbinova, and V. E. Kagan, *Biochem. Biophys. Res. Commun.* **159**, 229 (1989).

[12] M. Scarpa, A. Rigo, M. Majorino, F. Ursini, and C. Gregolin, *Biochim. Biophys. Acta* **801**, 215 (1984).

[13] H. H. Draper, *in* "Fat Soluble Vitamins" (R. A. Morton, ed.), p. 333. Pergamon, Oxford, 1970.

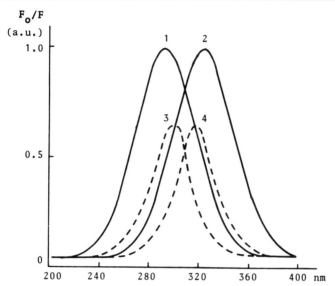

FIG. 1. Excitation (1, 3) and emission (2, 4) spectra of α-tocopherol (10 μM) in solvents of different polarity: ethanol (1, 2) and n-pentanol (3, 4).

branes. Figure 1 shows excitation and emission spectra of α-tocopherol (α-T) in ethanol and n-pentanol. The positions of the maxima in the spectra and the fluorescence intensity depend on the polarity of the solvent. In aqueous solutions α-T forms micelles, and the intensity of its fluorescence is very low owing to fluorescence self-quenching.[14-16] Addition of lipids and subsequent incorporation of α-T molecules into liposomes allow for the distribution of α-T molecules between phospholipid molecules, thus providing conditions for a decrease of the self-quenching effect and an increase of fluorescence intensity (Fig. 2). It should also be mentioned that neither α-T esters nor α-T oxidation products possess this characteristic fluorescence.[17]

This chapter shows the possibilities of fluorescence measurements in the estimation of binding, incorporation, and distribution of α-T and its synthetic homologs with different chain lengths in membranes as well as the hydrolysis of esterified forms of these compounds (Fig. 3).

[14] V. E. Kagan and P. Quinn, *Eur. J. Biochem.* **171**, 661 (1988).

[15] D. Schmidt, H. Steffen, and C. Von Planta, *Biochim. Biophys. Acta* **443**, 1 (1976).

[16] J. B. Massey, H. D. She, and H. S. Pownall, *Biochem. Biophys. Res. Commun.* **106**, 842 (1982).

[17] S. Hou and Z. Zhu, *Fenxi Huaxue* **10**, 535 (1982).

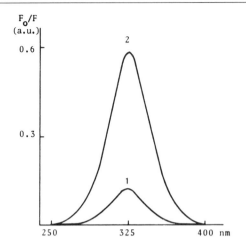

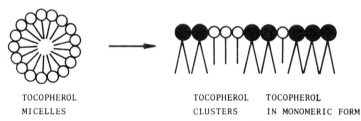

TOCOPHEROL   TOCOPHEROL   TOCOPHEROL
MICELLES   CLUSTERS   IN MONOMERIC FORM

FIG. 2. Fluorescence spectra of α-tocopherol dispersed in 0.1 M potassium, sodium phosphate buffer (pH 7.4) in the absence (1) and presence (2) of multilayer liposomes prepared from rat liver microsomal lipids. α-Tocopherol (20 μM) in ethanol solution was added to the buffer or lipid dispersion (concentration of lipids 0.5 mg/ml, ethanol concentration 0.1%). Incubation was for 20 min at 20°; the excitation wavelength was 298 nm, and the slit width was 4 nm.

Hydrolysis of Esterified Chromanols in Membrane Suspensions

Hydrolysis of nonfluorescent esterified chromanols can be studied by the appearance and increase of the intensity of fluorescence emitted by chromanols produced in the course of deesterification reactions (Fig. 4). Excitation at 303 rather than 292 nm should be used to minimize the interference of protein fluorescence. The sensitivity of the assay is determined by the minimum amounts of chromanols producing a distinct fluorescence increment in the presence of a given concentration of membrane protein. Preliminary calibration by addition of standard amounts of chromanols to membrane suspensions allows us to obtain the rates of hydroly-

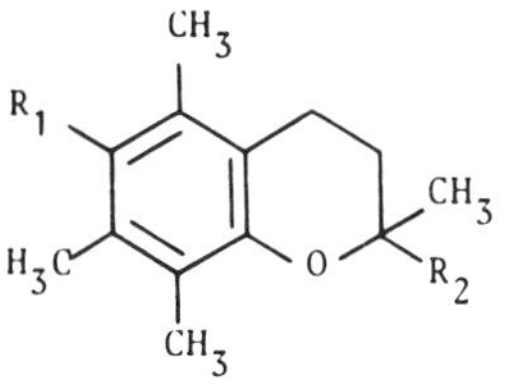

$R_1 = -OH;\ R_2 = -(CH_2CH_2CH_2CH-)_3CH_3$ CHROMANOL C_{16} (α-TOCOPHEROL)
$\qquad\qquad\qquad\qquad\ CH_3$

$R_1 = -OH;\ R_2 = -(CH_2CH_2CH_2CH-)_2CH_3$ CHROMANOL C_{11}
$\qquad\qquad\qquad\qquad\ CH_3$

$R_1 = -OH;\ R_2 = -CH_2CH_2CH_2CH-CH_3$ CHROMANOL C_6
$\qquad\qquad\qquad\qquad\quad\ CH_3$

$R_1 = -OH;\ R_2 = -CH_3$ CHROMANOL C_1

$R_1 = CH_3COO-\ ;\ R_2 = -(CH_2CH_2CH_2CH-)_3CH_3$ α-TOCOPHERYL ACETATE
$\qquad\qquad\qquad\qquad\qquad\ CH_3$

$R_1 = CH_3-(CH_2CH=CH)_3(CH_2)_7COO-\ ;$ α-TOCOPHERYL LINOLEATE

$R_2 = -(CH_2CH_2CH_2CH-)_3CH_3$
$\qquad\qquad\quad\ CH_3$

$R_1 = CH_3(CH_2)_{16}COO-\ ;\ R_2 = -(CH_2CH_2CH_2CH-)_3CH_3$ α-TOCOPHERYL STEARATE
$\qquad\qquad\qquad\qquad\qquad\qquad\ CH_3$

$\qquad ONa$
$\qquad\ |$
$R_1 = \ O=P-O-\ ;\ R_2 = -(CH_2CH_2CH_2CH-)_3CH_3$ α-TOCOPHERYL PHOSPHATE
$\qquad\ |$
$\qquad ONa \qquad\qquad\qquad\qquad\ CH_3$

$R_1 = CH_3COO-\ ;\ R_2 = -CH_3$ CHROMANE C_1 ACETATE

FIG. 3. Structural formulas of α-tocopherol, its esters, and homologs.

sis of esterified chromanols. Tables I and II summarize the data on the hydrolysis rates of α-T acetate, α-stearate, linoleate, phosphate, and 2,2,5,7,8-pentamethylchromane acetate (C_1 acetate) in suspensions of rat liver mitochondria and microsomes at three different temperatures (20, 34, and 37°). The deesterification rates depend on the chemical nature of the ester and the incubation temperature, and they differ significantly in microsomes, mitochondria, and, particularly, in blood serum (Fig. 4).

Binding of Chromanols to Membrane Suspensions

Binding of chromanols to membranes after their addition to and equilibration in membrane suspensions can be estimated by the amount of

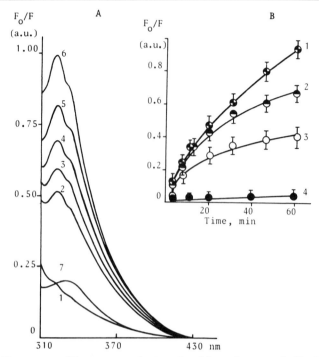

FIG. 4. Time course of fluorescence of α-tocopherol formed as a result of hydrolysis of α-tocopherol acetate. (A) Fluorescence spectra of α-tocopherol formed during incubation of α-tocopherol acetate (5 μM) with rat liver microsomal suspension (0.2 mg protein/ml). Incubation was in 0.1 M potassium, sodium phosphate buffer (pH 7.4) at 37°. Spectra 1 to 6 were recorded at 20, 25, 30, 40, 50, and 60 min, respectively. Spectrum 1, α-tocopherol acetate in buffer; spectrum 7, membrane suspension before addition of α-tocopherol acetate. (B) Kinetic curves of α-tocopherol fluorescence formed during incubation of α-tocopherol acetate (5 μM) with rat liver microsomes (1), rat liver mitochondria (2), rat blood plasma (3), as well as a multilamellar liposomal dispersion (rat liver lipids) (4). Incubation conditions were the same as in (A). The lipid concentration in the liposomal dispersion was 0.5 mg/ml. F_0/F, Fluorescence intensity (arbitrary units); λ_{ex} = 303 nm; λ_{em} = 325 nm; F_0 and F, fluorescence intensity before and after addition of preparations.

chromanols remaining in the aqueous phase and the amount of chromanols adsorbed by and incorporated into membranes. The chromanol concentrations in both aqueous solutions and membranes can be estimated by fluorescence. To avoid interference with protein-dependent fluorescence, chromanols are extracted from both the aqueous phase and membranes using aliquots of *n*-pentanol (vigorous shaking in a vortex mixer for 3 min). The concentrations of chromanols in the pellet and

TABLE I

RATES OF HYDROLYSIS OF α-TOCOPHEROL
DERIVATIVES IN RAT LIVER MICROSOMAL AND
MITOCHONDRIAL SUSPENSIONS[a]

α-Tocopherol derivative	Rate of hydrolysis [nmol α-T (C_1)/60 min]	
	Microsomes	Mitochondria
α-T acetate	5.00 ± 0.02	2.94 ± 0.03
α-T phosphate	0.95 ± 0.03	0.80 ± 0.03
α-T stearate	0.50 ± 0.01	0
α-T linoleate	1.13 ± 0.02	0.50 ± 0.02
C_1 acetate	5.00 ± 0.04	3.80 ± 0.03

[a] Initial concentration of α-T esters or C_1 acetate in suspensions was 5 nmol/mg protein. Incubation conditions were as in Fig. 4.

TABLE II

RATES OF HYDROLYSIS OF α-TOCOPHEROL ESTERS
IN RAT LIVER MICROSOMAL SUSPENSIONS
AT DIFFERENT TEMPERATURES[a]

α-Tocopherol ester	Rate of hydrolysis (nmol α-T/60 min)		
	20°	34°	37°
α-T acetate	1.53 ± 0.03	2.40 ± 0.04	5.00 ± 0.02
α-T phosphate	0.52 ± 0.01	0.80 ± 0.02	0.95 ± 0.03
α-T stearate	0	0	0.50 ± 0.01

[a] Initial concentration of α-T esters in the incubation medium was 5 nmol. Incubation conditions were as in Fig. 4.

supernatant are estimated by comparing the fluorescence intensities of the samples with the calibration curve. The fluorescence of endogenous tocopherol should also be taken into account. The results in Table III illustrate the binding of α-T and homologs with different hydrocarbon chain lengths to rat liver microsomes. α-Tocopherol and homologs with hydrocarbon chains partition predominantly into the membranous phase, whereas 2,2,5,7,8-pentamethyl-6-hydroxychromane (C_1), a homolog devoid of a hydrocarbon chain, is localized mainly in the aqueous phase.

TABLE III
BINDING OF α-TOCOPHEROL AND HOMOLOGS TO
RAT LIVER MICROSOMAL MEMBRANES[a]

α-Tocopherol homolog	Amount of α-T homolog in membrane pellet (P) and supernatant (S), %		
	P	S	S/P
C_1	14	86	6.14
C_6	90	10	0.11
C_{11}	71	29	0.40
α-T	84	16	0.19

[a] Concentration of α-T homologs was 20 μM; the concentration of microsomal protein in suspension was 0.2 mg/ml.

Evaluation of Chromanol Incorporation into Hydrophobic Core of Membranes and Formation of Clusters

The results on chromanol binding to membranes provide no information on whether the compounds are adsorbed by the membrane surface or incorporated into the hydrophobic zone of the membrane and, if incorporated, whether they are uniformly distributed in the hydrophobic core or form clusters. Figure 5 shows that an increase of the "membrane concentration" in suspensions (rat liver microsomes and rat brain synaptosomes) containing definite amounts of chromanols results in an enhancement of the characteristic chromanol fluorescence (measured after a 20-min preincubation). This effect can be explained by the elimination of fluorescence self-quenching due to chromanols in the course of their incorporation and distribution in the hydrophobic zone of the membrane. The increase of fluorescence intensity follows the saturation pattern for α-T and homologs with hydrocarbon chains, and there is no saturation for C_1. Obviously this difference is due to the almost complete binding of α-T and homologs with hydrocarbon chains to membranes and to the predominant localization of C_1 in the aqueous phase. The increase in the amount of the "membrane solvent" allows for a further partitioning of additional C_1 into the membranes. α-Tocopherol, C_6, and C_{11} are completely bound to the membrane at protein concentrations as high as 0.3–0.4 mg/ml.

The fluorescence intensity of chromanols even at saturating concentrations of membrane protein is not maximal (Fig. 5). Addition of detergent to the incubation medium causes a sharp increase in fluorescence

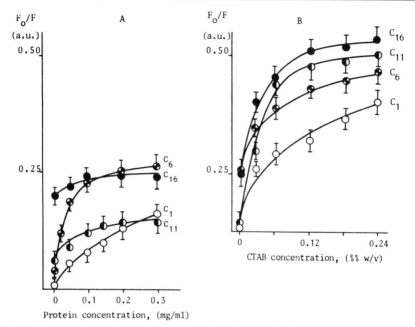

FIG. 5. Dependence of fluorescence intensity of α-tocopherol homologs on the concentration of microsomal protein and the detergent cetyl-3-methylammonium bromide (CTAB). Microsomal suspensions (0.2 mg protein/ml) were treated with detergent after the addition of α-tocopherol homologs (20 μM). The incubation time was 10 min at 20°. F_0/F, Fluorescence intensities (λ_{ex} = 298 nm, λ_{em} = 325 nm) before (F_0) and after (F) addition of microsomes or detergent.

intensity, which depends on the detergent concentrations [below the critical micelle concentration (cmc)] and reaches the maximum at concentrations exceeding the cmc. The finding suggests that in the presence of detergent chromanols are uniformly distributed in mixed micelles. Thus, one can evaluate the amounts of chromanols uniformly distributed in the membrane in the form of monomers and/or clusters. These estimations should take into consideration the partition coefficients of a given chromanol between the aqueous phase and the membranes. Table IV summarizes the results on cluster formation by exogenous chromanols in rat liver microsomes. Almost one-half of the added α-T or homologs with hydrocarbon chains are localized within the microsomal membrane in the form of clusters, whereas C_1 exists predominantly in the monomeric form.

In a similar way incorporation and uniform or nonuniform distribution of chromanols added to liposomal suspensions can be evaluated. Two examples of such measurements are presented on Fig. 6A,B. α-Tocopherol in ethanol solution is added to a multilamellar liposomal suspen-

TABLE IV

CONTENT OF CHROMANOLS IN MONOMERIC FORM OR IN CLUSTERS IN PRESENCE OR ABSENCE OF MICROSOMAL MEMBRANES[a]

α-Tocopherol homolog	F_0/F				Chromanol clusters in microsomal suspension, %	Chromanol in monomeric form in microsomal membranes, %	Chromanol clusters in aqueous phase, %	Chromanol clusters in microsomal membranes, %
	−Microsomes		+Microsomes					
	−CTAB	+CTAB	−CTAB	+CTAB				
C_1	0	48	14	44	69.6	30.4	59.9	9.7
C_6	4	50	28	51	47.8	52.2	4.8	43.0
C_{11}	7	56	16	56	81.6	18.4	23.7	57.9
α-T	22	61	28	59	84.3	15.7	13.5	70.8

[a] Incubation conditions: 0.1 M potassium, sodium phosphate buffer (pH 7.4), 20°C, 0.2 mg of protein/ml, 20 μM α-T or homolog, 0.24% (w/v) detergent (CTAB). F_0/F, Fluorescence intensities in arbitrary units (λ_{ex} = 298 nm, λ_{em} = 325 nm) before (F_0) and after (F) addition of microsomes or detergent.

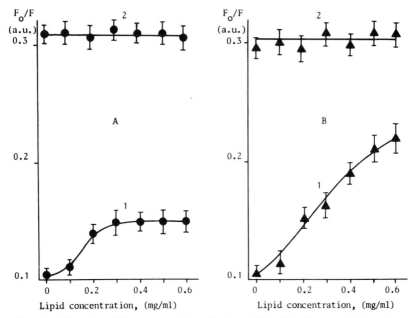

FIG. 6. α-Tocopherol fluorescence intensity in multilamellar liposomal dispersions at different concentrations of lipid before (1) and after (2) addition of detergent (sodium deoxycholate, 25 mM). (A) α-Tocopherol dissolved in ethanol was added to liposomes from rat liver microsomal lipids to a final concentration of 20 μM (ethanol concentration 0.1%). Incubation was in 0.1 M potassium, sodium phosphate buffer (pH 7.4) at 20° for 40 min. (B) α-Tocopherol and rat liver microsomal lipids were dissolved in chloroform, evaporated to dryness, and the mixture thus obtained was vigorously shaken (10 min) to obtain liposomal dispersion. Fluorescence was recorded as indicated in Fig. 5.

sion already prepared from liver microsomal lipids (Fig. 6A), or α-T and microsomal lipids are dissolved in chloroform, evaporated to dryness, after which multilayer liposomes are obtained by adding buffer and shaking vigorously (Fig. 6B). The α-T fluorescence intensity reaches saturation at lipid concentrations of 0.3–0.4 mg/ml in the case presented in Fig. 6A, whereas not in the case shown in Fig. 6B. Addition of detergent (sodium deoxycholate) in concentrations exceeding the cmc causes an increase of fluorescence intensity, reaching the same level in both cases. This suggests that when α-T is introduced into suspensions of multilamellar liposomes its binding, incorporation, and distribution occur mainly in the outer layer, thus causing saturation of the fluorescence intensity.

It should be mentioned that the rate of transbilayer migration of α-T is extremely low (see below).[5,6] Hence, exogenous α-T can hardly be distributed between the layers of multilamellar liposomes. The more homogeneous incorporation and distribution of α-T mixed with lipids in chloroform solution are probably responsible for the higher fluorescence intensity level. In the presence of detergent α-T in monomeric form is randomly dispersed in mixed detergent–lipid micelles, giving the maximum fluorescence intensity.

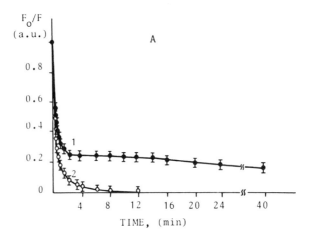

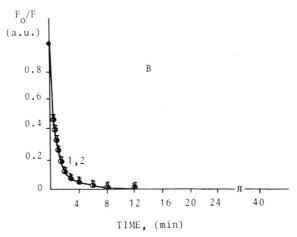

FIG. 7. Kinetic curves of fluorescence quenching of α-tocopherol (A) and C_1 (B) in unilamellar egg-yolk lecithin liposome suspensions after addition of ferricyanide. (1) α-Tocopherol or C_1 was incorporated into both monolayers; (2) α-tocopherol or C_1 was incorporated into the outer monolayer. Incubation was in 0.1 M potassium, sodium phosphate buffer (pH 7.4) at 20°, and the concentrations of reactants were 10 μM α-tocopherol or C_1, 100 μM ferricyanide, and 0.5 mg lipids/ml. F_0/F, Fluorescence intensity (λ_{ex} = 303 nm, λ_{em} = 325 nm) before (F_0) or after (F) addition of ferricyanide.

Estimation of Topography and Mobility of Chromanols in Lipid Bilayer

Because the oxidation products of chromanols do not possess the characteristic fluorescence, a simple procedure for the estimation of the mobility and content of chromanols in outer and inner monolayers of membranes, based on their accessibility to nonpermeable oxidizers, can be suggested. In the experiments described below ferricyanide is used to oxidize chromanols. Figure 7 shows that α-T, incorporated into both outer and inner layers of monolamellar egg lecithin liposomes using a sonication procedure, appears to be only partly accessible to ferricyanide. About 30% of the α-T presumably localized in the inner liposomal monolayer did not interact with ferricyanide when it was added to the liposomal suspension after sonication, but α-T was oxidized when sonication was performed in the presence of ferricyanide. α-Tocopherol incorporated only into the outer monolayer of unilamellar liposomes (addition of α-T without subsequent sonication) is fully accessible to ferricyanide. In contrast, C_1 is completely oxidized by ferricyanide added either before or after sonication. These data suggest that C_1 possesses a high transmembrane mobility in the lipid bilayer, whereas α-T undergoes no transbilayer migration within tens of minutes. Obviously, this effect, suggesting a very low rate of α-T transmembrane mobility, can be used for the determination of tocopherol asymmetry in liposomes and natural membranes.[18]

[18] V. A. Tyurin, V. E. Kagan, N. F. Avrova, and M. P. Prozorovskaya, *Bull. Exp. Biol. Med. USSR* **6,** 667 (1988).

Section III

Assay and Repair of Biological Damage Due to Oxygen and Oxygen-Derived Species

A. Lipid Peroxidation
Articles 38 through 48

B. Protein Oxidation
Articles 49 through 51

C. Modification of Nucleic Acids and Chromosomes
Articles 52 through 58

[38] Assay of Lipid Hydroperoxides Using High-Performance Liquid Chromatography with Isoluminal Chemiluminescence Detection

By Yorihiro Yamamoto, Balz Frei, and Bruce N. Ames

Introduction

Lipid peroxidation has been implicated in several diseases such as atherosclerosis, cancer, and rheumatoid arthritis, as well as in drug-associated toxicity, postischemic reoxygenation injury, and aging. Lipid peroxidation proceeds by a free radical chain mechanism and yields lipid hydroperoxides as major initial reaction products. Therefore, it would be desirable to detect and identify lipid hydroperoxides in biological tissues. The thiobarbituric acid assay, which measures an aldehydic breakdown product (malondialdehyde) of lipid hydroperoxides, has served as the most commonly used method for measuring lipid peroxidation. However, the application of the assay to biological samples is very problematic, since a variety of biological compounds such as deoxyribose, many other carbohydrates, sialic acid, prostaglandins, and thromboxanes also react with thiobarbituric acid and interfere with the assay.[1] There are methods for the detection of lipid hydroperoxides themselves[2,3] rather than their breakdown products, but these methods, too, cannot be applied unreservedly to biological samples, because of interference by antioxidants present in biological tissues and/or lack of sensitivity. Warso and Lands[4] have developed a method for the detection of lipid hydroperoxides based on activation of prostaglandin H synthase by hydroperoxides. However, this method, like most other methods for measuring lipid peroxidation, is unable to distinguish between different classes of lipid hydroperoxides. Furthermore, the specificity of this assay remains to be established.

The luminol chemiluminescence assay for the detection of hydrogen peroxide is well known for its picomole-level sensitivity.[5–7] Microperox-

[1] J. M. C. Gutterdige, *Free Radical Res. Commun.* **1**, 173 (1986).
[2] T. F. Slater, this series, Vol. 105, p. 283, and references therein.
[3] R. Cathcart, E. Schwiers, and B. N. Ames, *Anal. Biochem.* **134**, 111 (1983).
[4] M. A. Warso and W. E. M. Lands, *J. Clin. Invest.* **75**, 667 (1985).
[5] D. T. Bostick and D. M. Hercules, *Anal. Chem.* **47**, 447 (1975).
[6] W. R. Seitz, this series, Vol. 57, p. 445.
[7] B. Olsson, *Anal. Chim. Acta* **136**, 113 (1982).

METHODS IN ENZYMOLOGY, VOL. 186

idase (a heme fragment of cytochrome c) has been shown to be the most effective catalyst for this assay.[7,8] We have adapted the luminol–microperoxidase assay to organic hydroperoxides and obtained a sensitivity similar to that for hydrogen peroxide.[9] Since isoluminol gave better results than luminol in terms of signal-to-noise ratio, we used isoluminol as the substrate in our assay.

The reaction sequence leading to the emission of light from isoluminol in the presence of lipid hydroperoxides (LOOH) and microperoxidase is assumed to be as follows:

$$\text{LOOH} \xrightarrow{\text{microperoxidase}} \text{LO} \cdot \tag{1}$$
$$\text{LO} \cdot + \text{isoluminol (QH}^-) \rightarrow \text{LOH} + \text{semiquinone radical (Q}^{\overline{\cdot}}) \tag{2}$$
$$\text{Q}^{\overline{\cdot}} + \text{O}_2 \rightarrow \text{quinone (Q)} + \text{O}_2^{\overline{\cdot}} \tag{3}$$
$$\text{Q}^{\overline{\cdot}} + \text{O}_2^{\overline{\cdot}} \rightarrow \text{isoluminol endoperoxide} \rightarrow \text{light} \qquad (\lambda_{max} \ 430\text{nm}) \tag{4}$$

Reactions (1)[10] and (2) are not well established, but it is known that production of the semiquinone radical is essential for the emission of light from isoluminol. Reactions (3) and (4) have been well studied.[11,12]

Analysis of biological samples with the isoluminol assay can be hampered by antioxidants, which quench the production of chemiluminescence by scavenging intermediary radicals in the above reaction sequence [Eqs. (1)–(4)]. We have eliminated this problem by separating biological antioxidants from lipid hydroperoxides by high-performance liquid chromatography (HPLC) and using the isoluminol assay as an on-line postcolumn chemiluminescence detection system.[9]

HPLC–Isoluminol Chemiluminescence Assay

Equipment is arranged as outlined in Fig. 1. Antioxidants and hydroperoxides in the sample are separated by HPLC, using one of the chromatographic conditions described below, and then mixed with a reaction solution containing isoluminol (6-amino-2,3-dihydro-1,4-phthalazinedione; Sigma, St. Louis, MO) and microperoxidase (Type MP-11, Sigma) in a special mixer (Model 2500-0322, Kratos, Westwood, NJ). The reactions leading to the emission of light take place in a mixing coil made from

[8] H. R. Schroeder, R. C. Boguslaski, R. J. Carrico, and R. T. Buckler, this series, Vol. 57, p. 424.

[9] Y. Yamamoto, M. H. Brodsky, J. C. Baker, and B. N. Ames, *Anal. Biochem.* **160**, 7 (1987).

[10] T. A. Dix and L. J. Marnett, *J. Biol. Chem.* **260**, 5351 (1986).

[11] J. Lind, G. Merényi, and T. E. Eriksen, *J. Am. Chem. Soc.* **105**, 7655 (1983).

[12] G. Merényi, J. Lind, and T. E. Eriksen, *J. Phys. Chem.* **88**, 2320 (1984).

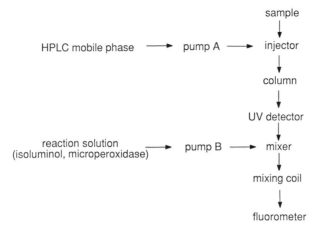

Fig. 1. Schematic diagram of the HPLC–isoluminol chemiluminescence assay.

a piece of HPLC tubing. The emitted light is measured in a fluorometer (Model FS 970 equipped with a 10-μl flow cell; Schoeffel Instrument Corporation, Westwood, NJ) used as a photon detector with the excitation source turned off and in the absence of an emission filter. The time constant and the supplying voltage for the fluorometer are set at 4 sec and 1160 V (58% of full range), respectively. The flow rates of pump A and B (Fig. 1) are 1.0 and 1.5 ml/min, respectively. The length of the mixing coil is 45 cm (inner volume ~92 μl).

The reaction solution is prepared as follows: 100 mM aqueous borate buffer (38.14 g of sodium tetraborate decahydrate per liter) is prepared, and the pH is adjusted to 10 with sodium hydroxide. Isoluminol (177.2 mg, final concentration 1 mM) is dissolved in 300 ml of methanol and 700 ml of the above borate buffer, and then 25 mg of microperoxidase is added. For the analysis of the hexane phase of plasma (see below), the reaction solution is composed of isoluminol and microperoxidase in 700 ml of methanol and 300 ml of borate buffer instead. All solutions as well as the HPLC mobile phases are kept in brown bottles in order to minimize light-induced generation of chemiluminescence-producing material.

Several useful HPLC conditions for the separation of various classes of lipid hydroperoxides and biological antioxidants are summarized in Table I. For example, condition III is useful for the separation of neutral lipid hydroperoxides such as hydroperoxides of cholesterol, triglycerides, and cholesteryl esters. Condition V can be used for the separation and quantitation of hydrogen peroxide, free fatty acid hydroperoxides, and phospholipid hydroperoxides. Under all chromatographic conditions the coefficients of variation (standard deviation/mean value) for the chemi-

TABLE I

HPLC CONDITIONS AND RETENTION TIMES OF HYDROPEROXIDES AND ANTIOXIDANTS[a]

HPLC condition:	I	II	III	IV	V
Guard column/column[b]:	LC18	LC18DB[c]	LC18DB[c]	LCSi	LCNH$_2$
Mobile phase:	d	e	f	g	h
Compound[i]		Retention times as monitored by chemiluminescence (min)			
Hydrogen peroxide	3.0	3.1	3.0	—	3.9
tert-Butyl hydroperoxide	3.4	3.2	3.1	2.2	3.6
Linoleic acid hydroperoxide	2.5	2.2	3.1	1.5	4.8
Cholesterol hydroperoxide	6.2	7.1	4.0	2.3	3.5
PC18:2-OOH	—	6.3	—	14.2	7.4
PE18:2-OOH	—	4.6	—	2.6	5.1
Trilinolein hydroperoxide	—	—	5.9	2.1	ND
Ch18:2-OOH	—	—	8.0	2.1	ND
		Retention times as monitored by UV (min)			
α-Tocopherol	10.4	12.9	4.8	2.0	3.3
γ-Tocopherol	9.2	11.2	4.5	2.0	3.3
Ascorbate	2.1	2.0	2.3	1.4	9.9
Urate	1.9	1.8	2.0	1.2	8.7

[a] Flow rate = 1.0 ml/min; — , not eluted within 30 min; ND, not determined. Adapted from Yamamoto et al.[9]

[b] The 5-μm analytical columns (25 cm × 4.6 mm i.d., except LCSi: 3 μm, 15 cm × 4.6 mm i.d.) were purchased from Supelco (Bellefonte, PA), with the corresponding guard columns (5 μm, 2 cm × 4.6 mm i.d.).

[c] DB, Deactivated for basic compounds.

[d] Methanol.

[e] Methanol containing 0.01% triethylamine.

[f] Methanol–tert-butanol (50 : 50 by volume).

[g] Acetonitrile–tert-butanol–water (55 : 30 : 15 by volume).

[h] Methanol–40 mM sodium phosphate, monobasic (95 : 5 by volume). The combination of LCNH$_2$ and LCSi (both 5 μm, 25 cm × 4.6 mm i.d.) in series using methanol–40 mM sodium phosphate, monobasic (90 : 10 by volume) gave similar and more reproducible retention times.

[i] PC18:2-OOH, Dilinoleylphosphatidylcholine hydroperoxide; PE18:2-OOH, dilinoleyl-phosphatidylethanolamine hydroperoxide; Ch18:2-OOH, cholesteryl linoleate hydroperoxide.

luminescence peak areas of the hydroperoxides listed in Table I are less than 3% ($n \geq 3$). Table II summarizes the sensitivity of the assay to various hydroperoxides relative to the sensitivity to linoleic acid hydroperoxide under chromatographic conditions II–V. The assay is less sensitive to hydrogen peroxide and tert-butyl hydroperoxide than to other hydroperoxides.

TABLE II

SENSITIVITY OF HPLC–ISOLUMINOL CHEMILUMINESCENCE
ASSAY TO VARIOUS HYDROPEROXIDES RELATIVE TO
SENSITIVITY TO LINOLEIC ACID HYDROPEROXIDE[a]

Compound[b]	HPLC condition[c]			
	II	III	IV	V
Hydrogen peroxide	0.57	0.39	—	0.53
tert-Butyl hydroperoxide	0.57	0.39	0.80	0.53
Linoleic acid hydroperoxide	1.00	1.00	1.00	1.00
Cholesterol hydroperoxide	0.84	0.52	0.91	0.82
PC18:2-OOH	1.44	—	1.13	0.92
PE18:2-OOH	1.04	—	1.16	0.83
Trilinolein hydroperoxide	—	1.18	1.04	—
Ch18:2-OOH	—	0.75	1.54	—

[a] Sensitivity is defined as the ratio of peak area to amount of hydroperoxide. From Yamamoto *et al.*[9]
[b] For abbreviations see footnote *i* of Table I.
[c] See Table I.

The chemiluminescence response relative to the amount of linoleic acid hydroperoxide is linear over a range of 1 to 1000 pmol. Other hydroperoxides such as hydrogen peroxide and *tert*-butyl hydroperoxide also give good linear relationships. The detection limit of the assay for linoleic acid hydroperoxide is about 1 pmol.

Application of Assay to Analysis of Human Blood Plasma

In human blood plasma, the major lipids are unesterified fatty acids, phospholipids, cholesterol, triglycerides, and cholesteryl esters.[13] As evident from Table I, it is very difficult to separate all hydroperoxides of these lipids from the plasma antioxidants ascorbate, urate, and α-tocopherol in one chromatographic step. Therefore, plasma is separated into two phases by extraction with methanol and hexane prior to analysis with the HPLC–isoluminol chemiluminescence assay.

Analytical Procedures. For extraction, 0.5 ml of heparinized plasma is added to 2 ml of HPLC-grade methanol and mixed vigorously. This is followed by the addition of 10 ml of hexane (HPLC grade; Aldrich, Milwaukee, WI). (Before use the hexane should be washed with water in order to remove trace amounts of hydroperoxides. About 1 volume of HPLC-grade water is added to 10 volumes of hexane in a brown bottle

[13] B. Frei, R. Stocker, and B. N. Ames, *Proc. Natl. Acad. Sci. U.S.A.* **85,** 9748 (1988).

TABLE III
RECOVERY OF LIPID HYDROPEROXIDES ADDED TO PLASMA PRIOR
TO EXTRACTION WITH METHANOL AND HEXANE[a]

| | Recovery (%) | |
Compound[b]	Aqueous phase	Hexane phase
Linoleic acid hydroperoxide	72	0
PC18:2-OOH	58	0
Cholesterol hydroperoxide[c]	10	14
Trilinolein hydroperoxide	0	56
Ch18:2-OOH	0	56

[a] Lipid hydroperoxides were added at 10 μM 1.0 min prior to
   extraction. Adapted from Frei et al.[15]
[b] For abbreviations see footnote i of Table I.
[c] Cholesterol hydroperoxide is degraded rapidly in plasma [B.
   Frei, R. Stocker, and B. N. Ames, Proc. Natl. Acad. Sci.
   U.S.A. 85, 9748 (1988)].

and stirred vigorously overnight. The water is allowed to settle for several
hours before the upper-phase hexane is used for extraction.) The methanol–hexane extract of plasma is vortexed for 1 min and then spun at 1000
g for 10 min at 4°. Nine milliliters of the upper, hexane phase is collected
and dried under vacuum in a rotary evaporator. Dried hexane extracts are
dissolved in 0.45 ml of methanol–*tert*-butanol (50 : 50 by volume). One
hundred microliters of this solution (corresponding to 100 μl of plasma) is
subjected to HPLC using condition III (see Table I). Under this chromatographic condition hydroperoxides of cholesterol, triglycerides, and
cholesteryl esters, and the antioxidants α- and γ-tocopherol, all of which
are extracted into the hexane phase of plasma, are well separated.[14,15]

The aqueous methanol phase of extracted plasma is passed through a
0.2-μm filter (Type ARCO LC 13, Gelman Science, Ann Arbor, MI), and
20 μl of this solution (corresponding to 4 μl of plasma) is subjected to
HPLC using condition V (see Table I). This chromatographic condition is
used since it allows good separation of hydrogen peroxide, free fatty acid
hydroperoxides, phospholipid hydroperoxides, ascorbate, and urate, all
of which are extracted into the aqueous methanol phase of plasma.[14,15]

The recoveries of lipid hydroperoxides added to plasma at a concentration of 10 μM prior to biphasic extraction with methanol and hexane
and analysis as described above are given in Table III. The detection
limits for lipid hydroperoxides in plasma extracting into the aqueous

[14] Y. Yamamoto and B. N. Ames, Free Radicals Biol. Med. 3, 359 (1987).
[15] B. Frei, Y. Yamamoto, D. Niclas, and B. N. Ames, Anal. Biochem. 175, 120 (1988).

methanol phase and the hexane phase are about 0.03 and 0.01 μM, respectively.[15]

In order to assess the purity of all reagents and materials used, for each set of analyses a blank (0.5 ml of water) should be extracted and then carried through the entire analytical procedure as described above. After each set of HPLC runs, a calibration is performed under the respective chromatographic condition, using linoleic acid hydroperoxide as a standard. This calibration is necessary since the sensitivity of the assay varies slightly, depending mainly on the freshness of the microperoxidase–isoluminol solution.

Figures 2A and 2B show typical chemiluminescence chromatograms

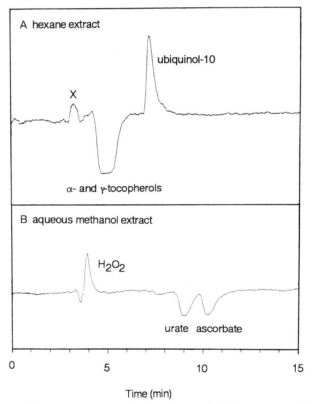

FIG. 2. Typical chemiluminescence chromatograms of the hexane extract (A) and aqueous methanol extract (B) of human blood plasma from a healthy subject. The evaporated hexane phase was dissolved in methanol–*tert*-butanol (50 : 50 by volume) and analyzed under chromatographic condition III (Table I). The aqueous methanol phase was analyzed on an LCNH₂ column using condition V (Table I). Chemiluminescence was recorded at 0.02 μA.

of the hexane phase and the aqueous methanol phase, respectively, of extracted human blood plasma from a healthy subject.[14,15] No chemiluminescence-producing compounds are observed that comigrate with spiked standards of hydroperoxides of linoleic acid, dilinoleylphosphatidylcholine, cholesterol, trilinolein, or cholesteryl linoleate (see Table I). However, the hexane extract (Fig. 2A) contains two chemiluminescence-producing compounds, one of which has been identified as ubiquinol-10 (see also below). In the methanol extract (Fig. 2B) a chemiluminescence-producing compound comigrating with hydrogen peroxide is observed (see below). Both extracts contain plasma antioxidants which produce negative peaks in the chemiluminescence chromatograms owing to quenching of background chemiluminescence.

Figure 3 illustrates the successful application of the assay to the analysis of human blood plasma after it has been incubated with activated polymorphonuclear leukocytes (PMNs). Besides a decrease in the plasma level of ubiquinol-10, such incubation led to formation of detectable amounts of triglyceride hydroperoxides, cholesteryl ester hydroperoxides (Fig. 3), and phospholipid hydroperoxides.[13]

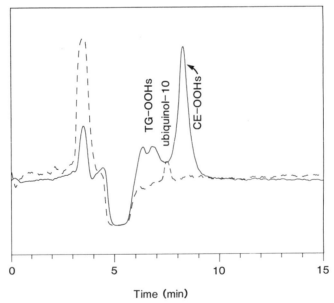

Time (min)

FIG. 3. Chemiluminescence chromatogram of the hexane phase of plasma incubated with activated PMNs with (dashed line) and without (solid line) subsequent treatment with sodium borohydride. Chemiluminescence was recorded at 0.02 μA. TG-OOHs, Triglyceride hydroperoxides; CE-OOHs, cholesteryl ester hydroperoxides.

Interference

Hydrogen peroxide can be produced by autoxidation of ascorbate during workup and analysis.[15] Thus, the assay is not suitable for the quantitation of hydrogen peroxide in biological samples containing ascorbate.

The assay is not sensitive to endoperoxides, hydroxyalkenals, alkenals, quinones, aldehydes, and alcohols.[15] However, ubiquinols produce chemiluminescence in the assay (see Fig. 2A).[15] Using chromatographic condition III, the assay yields a chemiluminescence signal for ubiquinol-10 that is 47% of the signal for an equimolar amount of linoleic acid hydroperoxide; ubiquinols-6,-9, and -10 give chemiluminescence peaks with retention times of 4.2, 6.1, and 7.2 min, respectively.[15] Naturally occurring ubiquinols-7 and -8, too, produce chemiluminescence in the assay, as do other hydroquinones. Human plasma contains about 0.7 μM ubiquinol-10 (Fig. 2A).[15] We also occasionally observed trace amounts of ubiquinol-9 (see also Ref. 16).

Chemiluminescence signals produced by ubiquinols and other hydroquinones cannot be eliminated by treatment with reducing agents, in contrast to the chemiluminescence signals produced by hydroperoxides. Therefore, in order to distinguish between lipid hydroperoxides and ubiquinols, the evaporated hexane extract of plasma is dissolved in 0.5 ml of methanol–tert-butanol (50 : 50 by volume) and incubated with 0.5 ml of a freshly prepared solution containing 10 mg/ml of sodium borohydride, triphenylphosphine, or stannous chloride in methanol. The sample is incubated for 60 min at 4° in the dark, and then 1 ml of methanol and 0.5 ml of water are added, followed by 10 ml of hexane. After extraction and centrifugation of the sample, 9 ml of the hexane phase is collected and analyzed as described above. Figure 3 demonstrates the usefulness of this procedure for the identification of chemiluminescence-producing compounds generated in plasma by incubation with stimulated PMNs: the chemiluminescence peaks of triglyceride hydroperoxides and cholesteryl ester hydroperoxides are eliminated, in contrast to the chemiluminescence peak produced by ubiquinol-10.

Besides ubiquinol-10, a second compound producing chemiluminescence appears in the hexane phase of plasma (compound X in Fig. 2A). This compound elutes from the LC18DB column near the solvent front. It is absent in unextracted plasma,[15] and thus seems to be produced artifactually during the analytical procedure. The nature of compound X as well as the reason for the increase of its chemiluminescence signal following sodium borohydride treatment (Fig. 3) are unknown.

[16] S. Vadhanavikit, N. Sakamoto, N. Ashida, T. Kishi, and K. Folkers, *Anal. Biochem.* **142,** 155 (1984).

As mentioned earlier and illustrated in Fig. 2, antioxidants such as ascorbate, urate, and tocopherols give negative peaks since they quench background chemiluminescence. The background chemiluminescence most probably arises from trace amounts of hydroperoxides present in the HPLC mobile phase. The negative peaks provide only qualitative, but not quantitative, information, since this type of response is dependent on the background chemiluminescence.

Conclusion

Detection of microperoxidase–isoluminol-dependent chemiluminescence on-line to high-performance chromatographic separation of oxidants and antioxidants is a rapid, sensitive, and selective assay for lipid hydroperoxides in biological samples. Possible interference by naturally occurring ubiquinols and other hydroquinones can be disclosed and corrected for by reductive treatment, for instance, with sodium borohydride. The assay has been successfully applied to the analysis of normal human blood plasma,[15] pulmonary edema fluid from patients with adult respiratory distress syndrome,[17] and traumatic spinal cord tissue of rats.[18] In addition, the assay has proved useful in *in vitro* studies on bile[19] and human blood plasma.[13,20] When used with the appropriate precautions, the assay allows for the detection, identification, and quantitation of lipid hydroperoxides in body fluids and tissues and thus should contribute to our understanding of diseases associated with oxidative stress.

[17] C. E. Cross, T. Forte, R. Stocker, S. Louie, Y. Yamamoto, B. N. Ames, and B. Frei, *J. Lab. Clin. Med.* (in press).
[18] M. Lemke, B. Frei, B. N. Ames, and A. I. Faden, *Neurosci. Lett.* **108**, 201 (1990).
[19] R. Stocker and B. N. Ames, *Proc. Natl. Acad. Sci. U.S.A.* **84**, 8130 (1987).
[20] B. Frei, L. England, and B. N. Ames, *Proc. Natl. Acad. Sci. U.S.A.* **86**, 6377 (1989).

[39] Determination of Methyl Linoleate Hydroperoxides by ^{13}C Nuclear Magnetic Resonance Spectroscopy

By E. N. Frankel, W. E. Neff, and D. Weisleder

Much work has been published on the hydroperoxides of unsaturated lipids. Significant progress during the last decade in understanding the mechanism of lipid oxidation can be attributed to a large extent to the development of new analytical tools such as the combination of gas chromatography and mass spectrometry (GC–MS), high-performance liquid chromatography (HPLC), and ^{13}C-NMR spectroscopy. Methods in GC–MS and HPLC have been used extensively to elucidate the course of oxidation of various unsaturated fatty esters. The structural information afforded by these techniques has led to new, modified, and refined mechanisms for the stereochemistry of hydroperoxide formation by free radical and singlet oxidation.[1–3]

In the past 20 years ^{13}C NMR has developed into an important technique for instrumental analysis in organic chemistry. Several investigators have reported on applications of ^{13}C NMR to elucidate configurations of unsaturated long-chain fatty acids.[4–8] In our previous studies,[9] ^{13}C-NMR spectroscopy afforded a very useful method for the study of the stereochemistry of methyl oleate autoxidation. We have developed methods to determine quantitatively the cis/trans isomeric ratio of hydroxy derivatives from hydroperoxides of methyl oleate by measuring peak heights and integrals of resonances for cis C-OH (67.5–67.8 ppm), cis C-OOH (81.1 ppm), trans C-OH (73.1 ppm), and trans C-OOH (86.9 ppm).

This chapter reports developments extending this analytical technique to the study of stereochemistry of methyl linoleate autoxidation. We have used the combination of HPLC and NMR techniques to establish the structures and to determine quantitatively the positional and stereoisomers of methyl linoleate hydroperoxides.

Preparative HPLC Separation of Dienols from
Linoleate Hydroperoxides

Methyl linoleate gives by free radical autoxidation a mixture of four cis,trans- and trans,trans-conjugated diene hydroperoxides that can be

[1] E. N. Frankel, *in* "Fatty Acids" (E. H. Pryde, ed.), p. 353. American Oil Chemists' Society, Champaign, Illinois, 1979.
[2] E. N. Frankel, *in* "Autoxidation in Food and Biological Systems" (M. G. Simic and M. Karel, eds.), p. 141. Plenum, New York, 1980.
[3] E. N. Frankel, *Prog. Lipid Res.* **23**, 197 (1985).
[4] J. Bus and D. J. Frost, *Recl. Trav. Chim. Pays-Bas* **93**, 213 (1974).
[5] J. Bus and D. J. Frost, *in* "Lipids" (R. Paoletti, G. Jacini, and R. Porcellati, eds.), p. 343. Raven, New York, 1976.
[6] J. Bus, I. Sies, and M. S. F. Lie Ken Jie, *Chem. Phys. Lipids* **17**, 501 (1976); J. Bus, I. Sies, and M. S. F. Lie Ken Jie, *Chem. Phys. Lipids* **18**, 130 (1977).
[7] F. D. Gunstone, M. R. Pollard, C. M. Scrimgeour, and H. S. Vedanayagam, *Chem. Phys. Lipids* **18**, 115 (1977).
[8] A. P. Tulloch and M. Mazurek, *Lipids* **11**, 228 (1976).
[9] E. N. Frankel, R. F. Garwood, B. P. S. Khambay, G. Moss, and B. C. L. Weedon, *J. Chem. Soc., Perkin Trans. 1*, 2233 (1984).

completely separated by HPLC.[10-13] Hydroperoxide isomers isolated by HPLC are subject to decomposition; therefore, Chan and Levett[10] relied on molar absorbances based on the peroxide content of isolated fractions by iodometric titration rather than by conventional gravimetric techniques. Analyses obtained by this approach can at best only be considered of relative significance. The more stable dienol derivatives of linoleate hydroperoxides are more easily separated by HPLC and are reported to have higher molar absorptivities for the cis,trans (e_{236} 27,200–28,300) and trans,trans (e_{233} 30,500–31,600) than the corresponding hydroperoxide isomers.[10]

We avoided the difficulties inherent in an ultraviolet detector by using a preparative HPLC system. An ultraviolet detector provides data which for quantitation require calculations based on the different absorptivities of each dienol isomer. However, preparative HPLC provides pure dienol isomers which allow direct quantitation based on weight. Further, the pure dienol isomers from preparative HPLC are used as ^{13}C-NMR standards to develop a method for quantitation by ^{13}C NMR of the intact autoxidation mixtures. This gravimetric HPLC technique thus provides a suitable method for the quantitative determination of a complex mixture of hydroperoxide derivatives having different ultraviolet absorption maxima and absorptivities.

Hydroperoxide mixtures from samples of methyl linoleate, autoxidized at different temperatures with pure oxygen, are reduced with sodium borohydride in methanol to the corresponding dienol derivatives.[14] The dienol mixtures are separated from unoxidized methyl linoleate by preparative TLC using plates coated with 0.25 mm silica gel treated with a UV marker. Plates are developed with diethyl ether–hexane (6:4, v/v), and UV-active materials are eluted from plates with diethyl ether. Then, the dienol isomers are separated on a 25.0 × 2.12 cm Zorbax Sil 6 μm column, operated at 27°, (Du Pont Analytical Instruments Division, Wilmington, DE), using a solvent mixture of ethanol–hexane (0.5:99.5, v/v), at a flow rate of 20 ml/min. The separation achieved by such preparative HPLC of dienol isomers from a 31.0-mg sample isolated from methyl linoleate autoxidized at 50° and reduced with sodium borohydride is shown in Fig. 1. The dienol components are identified on the basis of published relative retention volumes for each isomer,[10] and identities are

[10] H. W.-S. Chan and G. Levett, *Lipids* **12,** 99 (1977).
[11] N. A. Porter, J. Logan, and V. Kontoyiannidon, *J. Org. Chem.* **44,** 3177 (1979).
[12] N. A. Porter, B. A. Weber, H. Weenen, and J. A. Khan, *J. Am. Chem. Soc.* **102,** 5597 (1980).
[13] J. P. Koskas, J. Cillard, and P. Cillard, *J. Chromatogr.* **258,** 280 (1983).
[14] E. N. Frankel, W. E. Neff, W. K. Rohwedder, B. P. S. Khambay, R. F. Garwood, and B. C. L. Weedon, *Lipids* **12,** 1055 (1977).

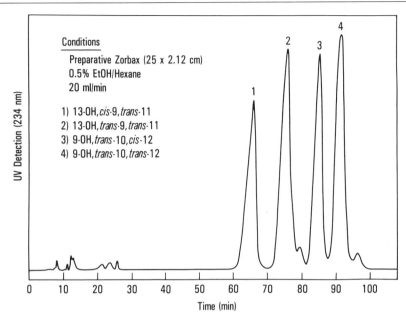

FIG. 1. Preparative high-performance liquid chromatographic separation of isomeric dienols prepared by sodium borohydride reduction of hydroperoxides from autoxidized methyl linoleate.

confirmed by ^{13}C NMR as described in the following section. The dienol weight percent composition is based on the weight of each fraction collected.

^{13}C-NMR Spectrometry

Spectra are recorded in CDCl$_3$ with a Brucker NMR spectrometer at 75.469 MHz operating in the pulse mode with Fourier transform at 37° and proton noise decoupling. The deuterium signal from CDCl$_3$ is used for the field frequency lock. ^{13}C-NMR signals are accumulated 300 to 500 times with a 10-μsec pulse width and a 3.0-sec pulse repetition time using 16.384 data points. The dienol mixtures isolated by preparative TLC are analyzed by both ^{13}C NMR and gravimetric HPLC.

^{13}C-NMR studies of the four geometric dienol isomers isolated by preparative HPLC from autoxidized methyl linoleate showed characteristic differences in resonance for the olefinic carbons and for the methylene carbon atoms α to the diene systems (Table I). Assignments were based on published data for similar carbons in other unsaturated long-chain fatty acids.[4–8] The position of the unsaturated carbons was established in the

TABLE I
13C-NMR BANDS OF GEOMETRIC DIENOL ISOMERS FROM
AUTOXIDIZED METHYL LINOLEATE[a]

Assignment	13-OH, c-9,t-11	13-OH, t-9,t-11	9-OH, t-10,c-12	9-OH, t-10,t-12
CH₃		51.4		
C-1		174.2		
C-2		34.2		
C-3		25.0–25.4		
C-8	27.8	32.6	37.6	37.5
C-9	132.4	135.0	72.9	72.9
C-10	127.9	129.6	125.8	131.0
C-11	136.1	133.9	136.1	133.8
C-12	125.5	130.7	128.0	129.6
C-13	72.9	72.9	132.7	135.9
C-14	37.4	37.4	27.8	32.7
C-16		31.5–31.9		
—CH₂—		29.0–29.5		
C-17		22.6		
C-18		14.0		

[a] c, cis; t, trans. Data in ppm.

cis,trans and trans,trans isomers on the basis of lanthanide shift reagent experiments.

The ^{13}C-NMR spectrum of methyl 13-hydroxy-*cis*-9,*trans*-11-octadecadienoate (Peak 1, Fig. 1) shows four unique signals for the olefinic carbons, in parts per million (ppm) as follows: C-9, 132.4; C-10, 127.9; C-11, 136.1; and C-12, 125.5 (Fig. 2). The corresponding methyl 13-hydroxy-*trans*-9,*trans*-11-octadecadienoate (Peake 2, Fig. 1) shows four different signals for the olefinic carbons, in ppm as follows: C-9, 135.0; C-10, 129.6; C-11, 133.9; and C-12, 130.7 (Fig. 3).

The ^{13}C-NMR spectrum of methyl 9-hydroxy-*trans*-10,*cis*-12-octadecadienoate (Peak 3, Fig. 1) shows four unique signals for the olefinic carbons, in ppm as follows: C-10, 125.7; C-11, 135.9; C-12, 127.8; and C-13, 132.7. The corresponding methyl 9-hydroxy-*trans*-10,*trans*-12-octadecadienoate (Peak 4, Fig. 1) shows four different signals for the olefinic carbons, in ppm as follows: C-10, 130.8; C-11, 133.7; C-12, 129.5; and C-13, 135.0.

The ^{13}C-NMR spectrum obtained with an unfractionated sample of dienol isomers from autoxidized methyl linoleate shows 16 lines for the olefinic carbons corresponding to the four geometric isomers listed in Fig. 1 (Fig. 4). These signals were identified on the basis of pure isomers (Figs.

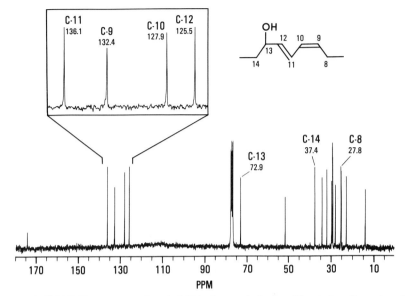

FIG. 2. ¹³C-NMR spectrum of methyl 13-hydroxy-*cis*-9,*trans*-11-octadecadienoate, with enlarged olefinic region between 125 and 140 ppm.

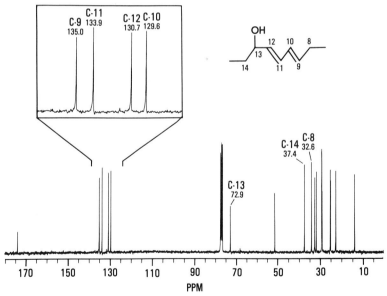

FIG. 3. ¹³C-NMR spectrum of methyl 13-hydroxy-*trans*-9,*trans*-11-octadecadienoate, with enlarged olefinic region between 130 and 135 ppm.

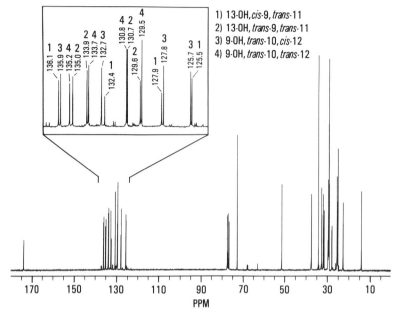

FIG. 4. ¹³C-NMR spectrum of isomeric dienols from hydroperoxides of autoxidized methyl linoleate, with enlarged olefinic region between 125 and 140 ppm.

1–3). The integrals of these signals were used as a basis for quantitative analyses of the dienol isomeric composition in autoxidized methyl linoleate.

In Table II, we compare the quantitative ¹³C-NMR analyses based on the integrals of the olefinic carbons with the gravimetric HPLC analyses on the same samples of autoxidized methyl linoleate. The analysis of isomeric composition determined by ¹³C NMR were in good agreement ($r = .97$) with those determined by HPLC on samples of methyl linoleate autoxidized at three temperatures. Differences between ¹³C-NMR and HPLC analyses averaged 1.1 and 2.6% at 65 and 25°, respectively. Standard deviations varied between 0.02 and 1.3% (averaging 0.40%) for the ¹³C-NMR analyses and between 0.1 and 2.0% (averaging 1.0%) for the HPLC analyses. The ratios of cis,trans- to trans,trans-dienol isomers decreased at a rate of 0.029 units per 1° with temperatures of autoxidation from 1.7 at 25° to 0.56 at 65°.

In summary, the ¹³C-NMR procedure described in this chapter provides an alternate approach to HPLC for the quantitative analysis of positional and geometric isomers of linoleate hydroperoxides formed by autoxidation. This direct analytical technique has the additional powerful

TABLE II
^{13}C-NMR AND PREPARATIVE HPLC ANALYSES OF DIENOL
ISOMERS FROM HYDROPEROXIDES OF METHYL LINOLEATE
AUTOXIDIZED AT DIFFERENT TEMPERATURES[a]

Temperature and method[b]	13-OH, c-9,t-11	13-OH, t-9,t-11	9-OH, t-10,c-12	9-OH, t-10,t-12	Ratio (c,t/t,t)
25°					
NMR (rel %)	31.6	18.9	31.7	17.8	1.72
S.D. (±)	(0.2)	(0.2)	(0.2)	(0.1)	
HPLC (wt %)	30.5	21.4	27.6	20.5	1.39
S.D. (±)	(0.9)	(1.0)	(0.8)	(1.4)	
50°					
NMR (rel %)	18.2	28.1	23.5	30.2	0.72
S.D. (±)	(0.1)	(0.7)	(0.5)	(0.1)	
HPLC (wt %)	20.8	27.8	22.3	29.1	0.76
S.D. (±)	(1.0)	(0.9)	(0.0)	(0.1)	
65°					
NMR (rel %)	17.7	34.0	18.1	30.2	0.56
S.D. (±)	(0.02)	(0.7)	(0.7)	(1.3)	
HPLC (wt %)	19.5	33.0	16.9	30.6	0.57
S.D. (±)	(1.7)	(1.2)	(1.0)	(2.0)	

[a] c, cis; t, trans.
[b] S.D., Standard deviations of triplicate NMR analyses and duplicate HPLC analyses. See Fig. 1 for HPLC separation of dienols.

advantage of providing detailed structural information on any new lipid hydroperoxides under investigation. The effective use of this method to analyze the allylic hydroperoxides of methyl oleate[9] and of linoleate in this study indicates that it should also be applicable to the allylic hydroperoxides from other similar substrates.

[40] Gas Chromatography–Mass Spectrometry Assays for Lipid Peroxides

By Frederik J. G. M. van Kuijk, David W. Thomas, Robert J. Stephens, and Edward A. Dratz

Introduction

The purpose of this chapter is to summarize recently developed gas chromatography–mass spectrometry (GC–MS) techniques for the measurement of lipid peroxidation in tissues. GC–MS methods for identification of peroxidation products have not been applied to phospholipids with one exception. Hughes *et al.*[1,2] used a GC–MS method to measure phospholipid oxidation products in a study of acute carbon tetrachloride toxicity in mouse liver. Their methodology required an enzymatic incubation step to liberate fatty acids from phospholipids or triglycerides, formation of methyl esters using diazomethane, and several purification steps before analyses.

The key to the methods described here is a mild transesterification procedure that allows simple and direct chemical identification and semiquantitative measurement of phospholipid or other glyceroester peroxide products by GC–MS.[3,4] The method can be simply extended to full quantitation with the use of isotopically labeled internal standards.[5] The method uses sodium borohydride reduction of the hydroperoxides and trimethylsilyl (TMS) derivatives of hydroxy fatty acid esters. Phospholipid hydroperoxide standards were synthesized and characterized to test the method.[3] The approach presented is applicable to the study of peroxidation of triglyceride storage depots as well as membrane phospholipids.[6] The transesterification procedure does not detect free fatty acid peroxides which are usually formed by enzymatic prostanoid metabolism.

[1] H. Hughes, C. V. Smith, J. O. Tsokos-Kuhn, and J. R. Mitchel, *Anal. Biochem.* **130,** 431 (1983).

[2] H. Hughes, C. V. Smith, J. O. Tsokos-Kuhn, and J. R. Mitchel, *Anal. Biochem.* **152,** 107 (1986).

[3] F. J. G. M. van Kuijk, D. W. Thomas, R. J. Stephens, and E. A. Dratz, *J. Free Radicals Biol. Med.* **1,** 215 (1985).

[4] F. J. G. M. van Kuijk, D. W. Thomas, R. J. Stephens, and E. A. Dratz, *J. Free Radicals Biol. Med.* **1,** 387 (1985).

[5] F. J. G. M. van Kuijk, D. W. Thomas, R. J. Stephens and E. A. Dratz, this volume [41].

[6] F. J. G. M. van Kuijk, D. W. Thomas, R. J. Stephens, and E. A. Dratz, *in* "Lipid Peroxidation in Biological Systems" (A. Sevanian, ed.), p. 117. American Oil Chemists' Society, Champaign, Illinois, 1988.

METHODS IN ENZYMOLOGY, VOL. 186

Methods

Sources of Reagents

Methanolic (*m*-trifluoromethylphenyl)trimethylammonium hydroxide (0.2 *N*) was obtained from Applied Sciences Laboratories, Inc. (Bellefonte, PA). *N*,*O*-Bis(trimethylsilyl)trifluoroacetamide (BSTFA) plus 1% trimethylchlorosilane (TMCS) was obtained from Regis Chemical Co. (Morton Grove, IL). Pentafluorobenzyl alcohol and sodium *tert*-pentoxide were purchased from Aldrich Chemical Company (Milwaukee, WI). All solvents were HPLC grade from Fisher Scientific (Fairlawn, NJ).

Extraction of Lipid Peroxides from Tissues

Tissue samples or phospholipid hydroperoxides obtained by photooxidation[3] are extracted by the method of Bligh and Dyer,[7] which is modified to use dichloromethane instead of chloroform.[3] The solvent ratios used during the extraction are as follows: 1 ml dichloromethane, 1 ml methanol containing 50 μg/ml butylated hydroxytoluene (BHT), and 0.25 ml aqueous buffer (2 m*M* EDTA, pH 7.0) are added to 5–50 mg tissue. This single-phase solvent mixture is homogenized in the first step in the extraction. One part dichloromethane (1 ml) is added in the second step, followed by vortex mixing for 60 sec. In the third step 0.5 ml water is added followed by vortex mixing, which creates two phases for extraction. The samples are centrifuged for 2 min at 1000 *g*, the dichloromethane lower phase is collected, and extraction of the aqueous phase with dichloromethane is repeated. Fractions of dichloromethane containing phospholipid and/or triglyceride hydroperoxides are pooled, dried over sodium sulfate, and evaporated under nitrogen.

The aqueous component of the single-phase organic solvent used for lipid peroxide extraction contains EDTA in order to inhibit iron-promoted breakdown of lipid peroxides. BHT is added to the organic component to inhibit formation of additional lipid peroxides during the workup. Desferal (desferroxamine mesylate, desferrioxamine) has been recommended as a substitute for EDTA as a powerful and perhaps even more effective chelating agent.[8]

The importance of substitution of dichloromethane for chloroform in the extraction procedure should be emphasized. Chloroform is used in the organic phase for total lipid extraction by most workers, following the widely used methods of Folch *et al.*[9] or Bligh and Dyer.[7] However, we

[7] E. C. Bligh and W. J. Dyer, *Can. J. Biochem. Physiol.* **37**, 911 (1959).
[8] J. M. C. Gutteridge, R. Richmond, and B. Halliwell, *Biochem. J.* **184**, 469 (1979).
[9] J. Folch, M. Lees, and G. H. Sloane Stanley, *J. Biol. Chem.* **226**, 497 (1957).

found that phospholipid hydroperoxides in chloroform tended to undergo decomposition, which may be mediated by formation of the trichloromethyl radical.[10] We found that dichloromethane greatly stabilized lipid peroxides, and, therefore, the extraction method of Bligh and Dyer[7] was modified by substituting dichloromethane for chloroform. The modifications require a substantial change in solvent ratios as described above. Dichloromethane is less toxic to laboratory workers than chloroform and more convenient to remove by evaporation. Colorimetric analysis established that phospholipid hydroperoxides are stable for months in HPLC grade dichloromethane at $-20°$.

Derivatization Reactions for GC–MS Analysis of Lipid Peroxides

Simplified GC–MS methods were recently published by our laboratory, based on a one-step transesterification reaction at room temperature, to convert fatty acid esters to either their corresponding methyl esters or pentafluorobenzyl (PFB) esters.[3,4] A scheme of the derivatization procedures for analysis of oxidized lipids by GC–MS using PFB esters is shown in Fig. 1.

Transesterification to Form Methyl Esters

The lipids of interest are collected in dichloromethane, and the dichloromethane is evaporated with a stream of dry nitrogen. The hydroperoxides are chemically reduced with 10 mg sodium borohydride in 1 ml methanol at $4°$ for 1 hr. After incubation, 1 ml water is added to decompose the sodium borohydride, and the samples are reextracted into dichloromethane. The extract is dried over sodium sulfate for 1 hr, and the liquid phase is transferred to a clean vial and evaporated under nitrogen to dryness. Subsequently, 50 μl of dichloromethane and 20 μl of transesterification reagent (0.2 M m-trifluoromethylphenyltrimethylammonium hydroxide in methanol) are added, and samples are shaken on a vortex mixer and incubated at room temperature for 30 min. This procedure provides quantitative conversion of phospholipids and triglycerides to fatty acid methyl esters as shown by TLC.[4,11] After incubation, 150 μl methanol and 200 μl water are added, and the samples are extracted into 1 ml hexane. The methanol–water phase is washed 2 times with 1 ml hexane, and the pooled hexane fractions are evaporated to dryness. Dry

[10] R. O. Recknagel, E. A. Clende, and A. M. Hruszkewycz, *In* "Free Radicals in Biology" (W. A. Pryor, ed.), Vol. 3, p. 97. Academic Press, New York, 1977.
[11] F. J. G. M. van Kuijk, D. W. Thomas, J. P. Konopelski, and E. A. Dratz, *J. Lipid Res.* **27,** 452 (1986).

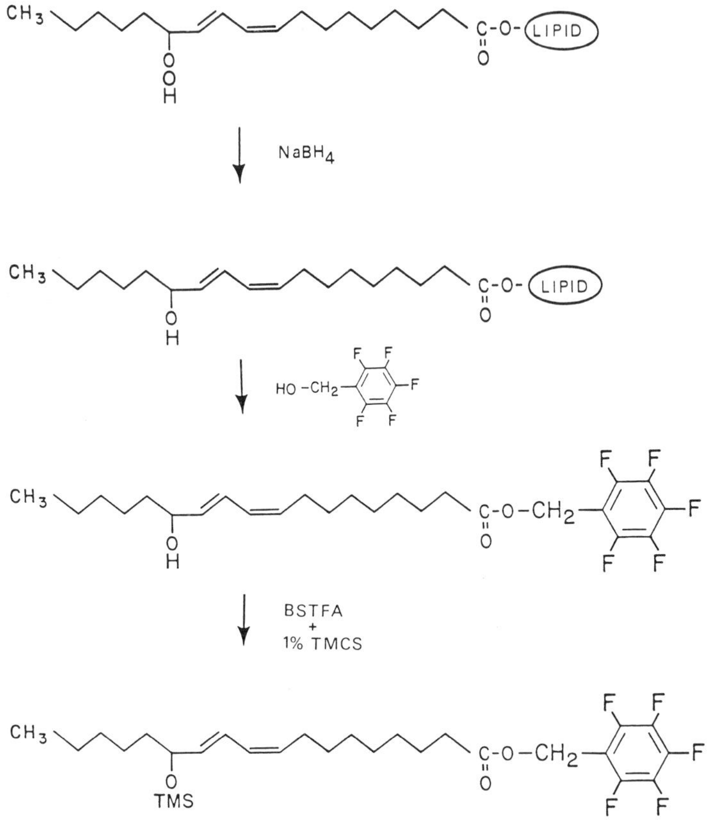

FIG. 1. Derivatization procedure for analysis of oxidized phospholipids by GC–MS. The top structure shows the fatty acid hydroperoxide esterified in a lipid. First, a reduction step with sodium borohydride is employed to form a hydroxy derivative (still esterified to the lipid). Second, transesterification is performed in the presence of pentafluorobenzyl alcohol to yield PFB esters. Methyl esters are produced if methanol is used instead of pentafluorobenzyl alcohol. The third step shows the conversion of the hydroxyl groups to TMS ethers.

pyridine (25 μl) and BSTFA containing 1% TMCS (25 μl) are added to convert the alcohols to the corresponding TMS derivatives.

Transesterification to Form Pentafluorobenzyl Esters

Phospholipid and/or triglyceride hydroperoxides are extracted and reduced to alcohols as described above and shown in Fig. 1. Transesterification to form PFB esters is carried out with a reagent containing 20% (v/v) pentafluorobenzyl alcohol and 3% (w/v) sodium *tert*-pentoxide in dichloromethane[4,11] (Fig. 1). A 20-μl aliquot of this solution is added to a sample of phospholipids or triglycerides in 50 μl dichloromethane. The vial is closed with a Teflon-lined screw cap, shaken on a vortex mixer, and incubated for 30 min at room temperature (phospholipids) or at 60° (triglycerides). After incubation dichloromethane is first evaporated, and 1 ml hexane, 200 μl water, and 200 μl methanol are added in this order. Further derivatization is carried out as described for methyl esters.

It is recommended that the completeness of the transesterification reaction be monitored by TLC on representative samples.[4,11] Sodium *tert*-pentoxide is hygroscopic and may be inactivated to the corresponding alcohol by traces of water, which leads to incomplete transesterification. This problem can be avoided by preparing the transesterification reagent immediately before use. Furthermore, traces of methanol react rapidly to form fatty acid methyl esters instead of PFB esters and therefore should be carefully excluded. Methyl esters are not formed if hexane and water are added first after transesterification.

Gas Chromatography–Mass Spectrometry

GC–MS analysis is carried out using a Ribermag R10-10 C GC–MS system. Chromatography is carried out on a 5-m DB-5 capillary column (J + W Scientific, Rancho Cordova, CA) with a temperature program from 150 to 250° at 10°/min for fatty acid methyl esters or from 180 to 280° at the same rate for PFB esters. A Ros glass falling needle injector is used at 270°. Mass spectra are obtained by either electron ionization (EI) at 70 eV for methyl esters or by negative ion chemical ionization (NICI) at 60 eV using ammonia reagent gas for PFB esters.

Results and Discussion

Methyl ester derivatives with EI detection proved to be useful for obtaining structural information on lipid peroxides. EI allowed detection of the different positional isomers from photooxidized unsaturated fatty acid chains.[3] It was found that the nonconjugated isomers, which are specific for singlet oxygen-mediated photooxidation reactions, could be easily separated and distinguished from the conjugated isomers.[3,4] The conjugated isomers are formed both by autoxidation and by photooxidation.

The pentafluorobenzyl (PFB) esters only yield one ion (M − H) in the mass spectrometer if detected by NICI.[4,12] The lack of fragmentation and the higher thermal stability of the PFB esters allow detection at the picogram level.

---

[12] R. J. Strife and R. C. Murphy, *Prostaglandins, Leukotrienes Med.* **13,** 1 (1984).

TABLE I

NICI FRAGMENTS OF PENTAFLUOROBENZYL ESTERS FROM
NONOXIDIZED FATTY ACIDS AND SINGLE- AND
DOUBLE-PHOTOOXIDIZED FATTY ACIDS, AFTER
REDUCTION OF HYDROPEROXIDES AND FORMATION OF
O-TMS DERIVATIVES

| Fatty acid | $M - C_7F_5H_2{}^a$ | Oxidized fatty acid $(M - C_7F_5H_2)^a$ | |
		Single	Double
16 : 1	253	341	
16 : 0	255		
17 : 0	269		
18 : 3	277	$(365)^b$	
18 : 2	279	367	
18 : 1	281	369	
18 : 0	283		
20 : 5	301	$(389)^b$	
20 : 4	303	391	479
20 : 0	311		
22 : 6	327	415	503
22 : 5	329	417	
22 : 4	332	419	

[a] On chemical ionization in the mass spectrometer, the $C_7F_5H_2$ group is cleaved from the PFB esters and the carboxylate anion is generated.

[b] The oxidized fatty acids and mass values of carboxylic ions shown in parentheses were not detected during this study.

Table I shows the m/e values for the major negative ions of each of the PFB esters from the most common naturally occurring fatty acids in biological systems, and values for the products of single oxidations and double oxidations of these unsaturated fatty acids. The double-oxidized species are observed in small amounts in these experiments with the characteristic masses shown in Table I and are discussed below. As can be seen in Table I, all these species can be distinguished by the unique masses of their negative carboxylate ion formed from the PFB esters in the mass spectrometer.

Photooxidized Rat Retinal Lipids

A NICI total ion chromatogram of total rat retina lipids, photooxidized *in vitro* and analyzed as PFB esters, is shown in Fig. 2a. About 2 ng of a

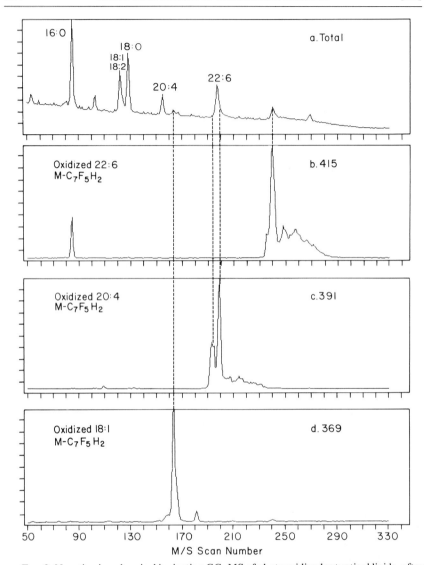

FIG. 2. Negative ion chemical ionization GC–MS of photooxidized rat retinal lipids, after derivatization according to the PFB ester method. The mass spectrometer performed 1 scan per 1.5 sec. (a) Total ion chromatogram. (b) Specific ion monitoring for the carboxylate anion of 22:6-O-TMS pentafluorobenzyl esters minus $C_7F_5H_2$ at m/e 415. (c) Specific ion monitoring for the carboxylate anion of 20:4-O-TMS pentafluorobenzyl esters minus $C_7F_5H_2$ at m/e 391. (d) Specific ion monitoring for the carboxylate anions of 18:1-O-TMS pentafluorobenzyl esters minus $C_7F_5H_2$ at m/e 369.

mixture of pentafluorobenzyl ester derivatives was injected on the GC–MS system. Figure 2b–d shows chromatograms that were obtained by specific ion monitoring for the carboxylate anions of the products of single oxidations of docosahexaenoic acid, arachidonic acid, and oleic acid (molecular weights are listed in Table I). The oxidation was limited to about 10–15% of the total double bonds by limiting the amount of oxygen available.[3] Under these conditions it was possible to detect only small amounts of a double-oxidized arachidonic acid di-TMS derivative (m/e 479) near scan number 225 in Fig. 2 and double-oxidized docosahexaenoic acid di-TMS derivative (m/e 503) near scan number 275 (data not shown). These double-oxidized derivatives were about one-tenth as abundant as the single-oxidized products.

The nonconjugated and conjugated derivatives from oxidized arachidonic acid are separated (Fig. 2c), but separation was not obtained between conjugated and nonconjugated isomers from oxidized docosahexaenoic acid (Fig. 2b). The unresolved hump at longer retention times than the sharp, large oxidized docosahexaenoic acid peak monitored at mass 415 contained predominantly conjugated species with some double-oxidized species on the trailing edge. We purposely used short capillary columns with modest chromatographic resolution to maximize the speed of the assays and the recovery of the products.[3,4,6]

No common impurities such as the plasticizer dioctyl phthalate were observed in the NICI total ion chromatograms.[4] Owing to the enhanced NICI sensitivity obtained on analysis of PFB esters, the samples can often be diluted to such an extent that impurities are not detected. The traces in Figures 2b–d show that, with specific ion monitoring, small amounts of oxidation products can easily be detected in the presence of larger amounts of nonoxidized fatty acids.

Evaluation of Pentafluorobenzyl Ester Method

NICI Mass Spectrometry. In the NICI system used in this study, more than 90% of the total ionization was found in the carboxylate anion.[4] Under optimal EI conditions ionization is spread over many more fragments, and there is typically at most 0.5% ionization in any structurally significant ion. Strife and Murphy[12] calculated an advantage factor of 20–100 using NICI conditions with PFB esters relative to optimal EI conditions with PFB esters. When methyl esters are used, there is little difference in sensitivity between EI and NICI.

Transesterification. Methods previously available for forming PFB esters required free fatty acids.[12] Therefore, a lipase or saponification step would be required to use these methods for analysis of phospholipids or triglycerides. Both the lipase and the saponification conditions are relatively vigorous and can lead to formation and/or decomposition of peroxidized polyunsaturated phospholipids or glycerides depending on the detailed conditions. Furthermore, the manipulation required impairs the recovery of small amounts of oxidized products.

Direct transesterification is much simpler and, with the use of a sterically hindered quaternary ammonium hydroxide catalyst in methanol to form methyl esters, appears to be significantly milder than the enzymatic approaches mentioned above. McCreary et al.[13] showed that this transesterification method produced derivatization of triglycerides to fatty acid methyl esters at room temperature. Their method was applied successfully for derivatization of phospholipid peroxides that were first reduced to the corresponding hydroxy derivatives.[3] The one-step transesterification reaction to form PFB esters, rather than using enzymatic release and derivatization of free fatty acids in two steps, was desirable.

Analysis of several authentic synthetic phospholipids such as $(16:0)(18:2)$-phosphatidylcholine and $(16:0)(22:6)$-phosphatidylcholine by the PFB ester method gave the expected ratio of fatty acids. These results indicate that quantitative conversion of the fatty acid esters in phospholipids to PFB esters was obtained with saturated as well as with highly polyunsaturated fatty acids. Analysis of the total fatty acid composition of the whole rat retina by flame ionization detection using PFB ester derivatives produced a fatty acid distribution identical to that reported by Farnsworth et al.,[14] who used conventional methanol–BF$_3$-catalyzed transesterification to form methyl esters. Derivatization on photooxidized model phospholipids, after reduction of the hydroperoxides to the corresponding hydroxy derivatives, showed that PFB esters were also efficiently produced from hydroxy fatty acids. This indicates that these functional groups are preserved during the transesterification step with the base catalyst employed.

Ultrasensitive Detection. The PFB esters are clearly superior to the methyl esters for analysis of small amounts of lipid peroxidation products by GC–MS. The combined advantages of thermal stability in the gas chromatograph and the enhanced NICI sensitivity in the mass spectrometer provide an at least 1000 times lower detection level for the PFB ester derivatives. As little as 1–10 pg (2.5–25 fmol) of oxidation product can be detected. The greater stability of the PFB ester derivatives of the phospholipid oxidation products on the GC column also allowed detection of oxidation products with two (or perhaps more) oxidation sites. When

[13] D. K. McCreary, W. C. Kossa, S. Ramachandran, and R. R. Kurtz, *J. Chromatogr. Sci.* **16,** 329 (1978).
[14] C. C. Farnsworth, W. L. Stone, and E. A. Dratz, *Biochim. Biophys. Acta* **552,** 281 (1979).

TABLE II
COMPARISON OF METHYL ESTER AND PFB ESTER METHODS[a]

Parameter	Methyl ester	PFB ester
Substrate specificity	Fatty acid hydroperoxide esters Hydroxy fatty acid esters	
Detection limit	10 ng	1–10 pg
MS detection	EI	NICI
Main ion	Fragments	M − H
Application	Adipose	All tissues
GC detection	FID	ECD

[a] EI, Electron ionization; NICI, negative ion chemical ionization; FID, flame ionization detection; and ECD, electron capture detection.

oxidation products were analyzed as methyl esters,[3] only the single oxidation products were detected, probably because of lower stability of the products. The differences between the methyl ester and PFB ester methods are summarized in Table II.

Quantitation. Analysis as PFB esters by GC–MS under NICI conditions provides a relatively simple approach to detect lipid peroxidation products with as yet unsurpassed sensitivity and selectivity. However, this method yields qualitative information only, since recoveries may vary considerably between different tissues and different analyses. Hence, it would be highly desirable if absolute levels could be determined by including suitable internal standards.

Hughes *et al.*[2] reported a method for quantitation of lipid peroxidation products in total hepatic lipid based on the use of methyl 15-hydroxy-arachidonate as internal standard. Three isomers, 11-, 12-, and 15-hydroxy fatty acids of arachidonate, were quantitated as the methyl ester *O*-TMS ether derivatives. A disadvantage of this approach for quantitation results from differences in retention time for the internal standard and the compounds of interest. The conditions in the mass spectrometer may change rapidly in time, especially with the very sensitive NICI method; therefore, it is much more desirable to use stable isotope internal standards. We attempted to use a rare saturated fatty acid (20:0) as an internal standard during the present study and could not obtain reliable results with NICI methods.

Quantitative GC–MS analyses are typically carried out, using compounds labeled with deuterium (D), carbon-13, or oxygen-18.[2,12,15,16] Since

[15] W. K. Rohwedder, *Prog. Lipid Res.* **24,** 1 (1985).
[16] H. Frank, M. Wiegand, M. Strecker, and T. Dietmar, *Lipids* **22,** 689 (1987).

these types of internal standards elute very close to (D) or coelute (^{13}C and ^{18}O) with the authentic compounds, errors arising from changes in mass spectrometer conditions are minimized or eliminated. Also, peak identification is more facile in the presence of the higher backgrounds usually encountered when tissue extracts are analyzed. An example of this approach is presented elsewhere in this volume, where a quantitative assay of cleavage products of lipid peroxides is developed. In this assay deuterium-labeled internal standards are used which nearly coelute with the compounds measured. Quantification of the methods developed here for analysis of hydroxy fatty acids awaits future application of stable isotope internal standards. An example of such an assay was recently reported by Frank et al.,[16] who employed ^{18}O-labeled standards.

Conclusions

Transesterification to form PFB esters and NICI GC–MS provide several advantages compared to analyses with methyl esters. Lipid peroxides are detected not only in vitamin E deficiency, but also in normal nutritional states. The PFB NICI method can detect and identify lipid peroxides in animals raised on lab chow diets. The EI GC–MS methods allow identification of the contribution of singlet oxygen-mediated reactions during damage to tissues. Future work requires quantitation with stable isotope internal standards[5] in order to determine the differences in lipid peroxide contents between groups of animals raised on different diets, animals exposed to different conditions that produce pathology, and in investigation of human pathology.

The methods developed can also be used in a simple GC system without a mass spectrometer. In case of PFB derivatives, very sensitive detection (1 pg) can be obtained with electron capture detection (ECD). A major difference in the GC assay is the requirement for only one internal standard which must be chromatographically separated from commonly occurring oxidized fatty acids. A commercially available synthetic (16 : 0)(16 : 1)-phospholipid can serve as an internal standard for GC after photooxidation, since the palmitoleic acid content of most biological tissues is relatively low.

[41] Gas Chromatography–Mass Spectrometry of 4-Hydroxynonenal in Tissues

By Frederik J. G. M. van Kuijk, David W. Thomas, Robert J. Stephens, and Edward A. Dratz

Introduction

Lipid peroxides may undergo a variety of secondary reactions including cleavage to form 4-hydroxyalkenals. The 4-hydroxyalkenals are of particular interest because they have been reported to elicit a variety of powerful biological activities,[1] such as inhibition of enzymes,[2] inhibition of calcium sequestration by microsomes,[3] exhibition of chemotactic activity toward neutrophils,[4] and inhibition of protein synthesis.[5] The 4-hydroxyalkenals are thought to exert these effects because of their high reactivity toward molecules with sulfhydryl groups.[4,6]

4-Hydroxynonenal has been reported as a specific indicator for lipid peroxidation in rat liver microsomes[7] and to be the major aldehyde formed during peroxidation of linoleic acid (18 : 2ω6) and arachidonic (20 : 4ω6) acid.[6] 4-Hydroxynonenal is derived from oxidized ω6 fatty acids,[6,8] but the mechanism by which it is generated is not yet fully understood. Winker *et al.* postulated a mechanism for the formation of 4-hydroxynonenal from the 13-hydroperoxide of linoleic acid.[8] Another mechanism indicates the formation of 4-hydroxynonenal in the luciferin reaction.[9] A closely related compound, 4-hydroxyhexenal, was also produced during peroxidation of rat liver microsomes, and it is thought to be derived from oxidized ω3 fatty acids. Therefore, generation of 4-hydroxy-

[1] E. Schauenstein, H. Esterbauer, and H. Zollner, "Aldehydes in Biological Systems: Their Natural Occurrence and Biological Activities." Pion Press, London, 1977.

[2] E. Schauenstein, *J. Lipid Res.* **8**, 417 (1967).

[3] A. Benedetti, R. Fulceri, and M. Comporti, *Biochim. Biophys. Acta* **793**, 489 (1984).

[4] M. Curzio, H. Esterbauer, C. Di Mauro, G. Cecchini, and M. U. Dianzani, *Biol. Chem. Hoppe-Seyler* **367**, 321 (1986).

[5] A. Benedetti, L. Barbieri, M. Ferrali, A. F. Casisni, R. Fulceri, and M. Comporti, *Chem.–Biol. Interact.* **35**, 331 (1981).

[6] H. Esterbauer, *in* "Free Radicals, Lipid Peroxidation, and Cancer" (D. C. H. McBrien and T. F. Slater, eds.), p. 101. Academic Press, New York, 1982.

[7] A. Bennedetti, M. Comporti, and H. Esterbauer, *Biochim. Biophys. Acta* **620**, 281 (1980).

[8] P. Winker, W. Lindner, H. Esterbauer, E. Schauenstein, R. J. Schauer, and G. A. Khoschsorur, *Biochim. Biophys. Acta* **796**, 232 (1984).

[9] F. McCapra, *Q. Rev., Chem. Soc.* **20**, 485 (1966).

hexenal might be expected on peroxidation of lipids rich in ω3 fatty acids as are present in synaptic vesicles of the retina and brain.

In most of the previous studies aldehyde products were detected in biological systems through the formation of 2,4-dinitrophenylhydrazone derivatives, followed by analysis with a combination of thin-layer chromatography (TLC) and high-performance liquid chromatography (HPLC).[6] Lang et al.[10] published a HPLC method for quantitative determination of free 4-hydroxynonenal down to 2 ng in biological samples, based on extraction of the aldehyde with dichloromethane and trapping on an Extralut column. This method relies on extraction of free aldehydes, which are difficult to extract from tissues since they are highly reactive and become covalently bound to amino groups of proteins and lipids as Schiff bases[11] and since the 4-hydroxyalkenals also bind strongly to sulfhydryl groups.[4,6]

The method described here improves the extraction efficiency for 4-hydroxyalkenals in biological systems by formation of an oxime derivative, analogous to the methods used for determination of retinals.[11] An excess of hydroxylamine or O-ethylhydroxylamine has been used to liberate vitamin A aldehydes from Schiff base binding sites.[11] The specific procedure developed here for the 4-hydroxyalkenals is based on the use of O-(pentafluorobenzyl)hydroxylamine hydrochloride (PFBHA-HCl). This reagent forms the O-pentafluorobenzyl oxime derivatives of 4-hydroxyalkenals, from either the free aldehyde or from aldehydes bound to amino groups by Schiff base linkages. The volatility of the 4-hydroxyalkenal oxime is increased by formation of trimethylsilyl (TMS) ethers of the hydroxyl functions. These derivatives can be detected with high sensitivity by GC–MS with negative ion chemical ionization (NICI).[12] This approach was inspired by the successful use of PFB ester TMS ether derivatives of hydroxy fatty acids,[13] also described elsewhere in this volume [40]. Isotopically labeled dideuterated analogs of 4-hydroxyalkenals were used as internal standards, which allowed 4-hydroxynonenal analyses in quantitative measurements. The method is illustrated by data derived from retinas of rats.

Methods: Preparation of Standards and Samples

Source of Standards and Reagents

Chemically synthesized 4-hydroxynonenal and 4-hydroxyhexenal were generous gifts from Prof. H. Esterbauer (Department of Biochemis-

[10] J. Lang, C. Celotto, and H. Esterbauer, Anal. Biochem. 150, 369 (1985).
[11] G. W. T. Groenendijk, W. J. de Grip, and F. J. M. Daemen, Biochim. Biophys. Acta 617, 430 (1980).

try, University of Graz, Graz, Austria). Two triple-bonded precursors, hydroxynonynal diethylacetal and hydroxyhexynal diethylacetal, were also gifts from Prof. Esterbauer and were used to synthesize deuterated internal standards as described below. *O*-Pentafluorobenzylhydroxylamine hydrochloride was purchased from Aldrich Chemical Co. (Milwaukee, WI). PIPES buffer was obtained from Sigma (St. Louis, MO). *N,O*-Bis(trimethylsilyl)trifluoroacetamide (BSTFA) plus 1% trimethylchlorosilane (TMCS) and pyridine were obtained from Regis Chemical Co. (Morton Grove, IL). Solvents were electron capture detection (ECD) grade from EM Science (Cherry Hill, NJ).

Synthesis of Deuterated Internal Standards

Deuteration of the triple-bonded precursor of 4-hydroxyalkenals is accomplished by reduction with lithium aluminum deuteride (LiAlD$_4$) to form vinyl perdeuterated 2,3-trans double bonds. The method follows Esterbauer and Weger[14] and Esterbauer[15] who used lithium aluminum hydride (LiAlH$_4$) to form the normal (protonated) analogs of 4-hydroxy-2,3-*trans*-alkenals. Specific formation of the trans isomer is necessary, since the cis isomer is unstable.[1,14,15] The cis isomer immediately forms cyclic acetals (dihydrofuran derivatives), which cannot be converted back to the desired 4-hydroxyalkenals.

Preparation of Samples for Gas Chromatography–Mass Spectrometry

Standard solutions of 4-hydroxyalkenals are stored in dichloromethane containing 50 μg/ml butylated hydroxytoluene (BHT) at −90°. For derivatization, 200 μl methanol is added and the dichloromethane is evaporated under a stream of nitrogen. If dichloromethane is evaporated directly without added methanol it is possible to lose 4-hydroxyalkenals at this step, because of their relatively high volatility. A 200-μl aliquot of 0.1 *M* PIPES (disodium salt) buffer containing 50 m*M* PFBHA-HCl (pH 6.5) is added to the 200 μl methanol containing up to 0.1 mg 4-hydroxyalkenal. The mixture is vortexed and incubated for 5 min at room temperature to allow formation of the *O*-PFB oxime derivatives. The optimum pH for the formation of the oxime derivatives is 6.5.

[12] F. J. G. M. van Kuijk, D. W. Thomas, R. J. Stephens, and E. A. Dratz, *Biochem. Biophys. Res. Commun.* **139**, 144 (1986).

[13] F. J. G. M. van Kuijk, D. W. Thomas, R. J. Stephens, and E. A. Dratz, *J. Free Radicals Biol. Med.* **1**, 387 (1985).

[14] H. Esterbauer and W. Weger, *Monatsh. Chem.* **98**, 1994 (1967).

[15] H. Esterbauer, *Monatsh. Chem.* **102**, 824 (1971).

After incubation to form the oxime, 1 ml hexane is added, the mixture is vortexed for 1 min, and complete phase separation is accomplished by centrifugation at 1000 g for 1 min. The oxime partitions into the hexane upper layer which is collected. The hexane extraction is repeated twice to recover the oxime in the small amounts of upper layer that could not be collected without contamination by the lower layer. The pooled hexane fractions are stored at $-20°$ until analysis by GC–MS. Prior to analysis, vials are warmed to room temperature and the hexane is evaporated under a stream of nitrogen. The samples are dissolved in 50 μl silylation-grade pyridine, and 50 μl of BSTFA reagent is added. The samples are incubated for 5 min at 80° to form TMS ethers of the hydroxyl groups and analyzed immediately.

Extraction of 4-Hydroxyalkenals from Tissues

Tissue samples varying in amount from 3 to 50 mg are homogenized in a mixture of 400 μl methanol containing 50 μg/ml BHT, 200 μl of a 2 mM EDTA buffer (pH 7.0), and 200 μl of 0.1 M PIPES containing 50 mM PFBHA-HCl (pH 6.5). Samples are incubated at room temperature for 5 min, extracted with hexane, and processed as described above. If tissues rich in triglycerides are extracted, a different procedure is followed. The pooled hexane fractions are evaporated, dissolved in 0.5 ml acetonitrile, and vortexed for 1 min. This procedure extracts the derivatives into the acetonitrile, whereas the triglycerides stay bound to the walls of vial. The liquid phase is collected and can be stored at $-20°$ for up to several months, until analysis by GC–MS. Just prior to GC–MS the samples are evaporated under nitrogen, the pyridine and BSTFA are added to form TMS derivatives, and the analysis is carried out immediately. For quantitative analysis, deuterated internal standards are added to the tissues before homogenization.

Gas Chromatography–Mass Spectrometry

GC–MS analysis is performed on a Ribermag R10-10 C instrument using a Ros falling needle injector at 270°. Chromatography is carried out on a 5-m DB-5 capillary column at 150°. Alternatively, a VG-Analytical 7070 instrument equipped with a Varian 3700 split/splitless capillary injector is used with a 5-m DB-5 capillary column and a temperature gradient from 50 to 200° at 10°/min. With both instruments mass spectra are obtained by NICI with specific ion monitoring at 60 eV. Ammonia reagent gas is used on the Ribermag, whereas methane reagent gas is used on the VG instrument.

Results and Discussion

Analysis of Synthetic 4-Hydroxyalkenals

Complete derivatization of 4-hydroxynonenal for GC–MS gave the *O*-PFB oxime TMS ether derivative as shown in Fig. 1. Under the conditions used in the mass spectrometer, the most abundant fragment found at 152 atomic mass units (amu) is the parent carbon chain arising from loss of both the PFB group (shown in Fig. 1A by a dashed line) and the silanol residue (HOTMS). The loss of the silanol residue indicates that the aldehyde has a hydroxyl function. On the nonpolar GC column employed, each aldehyde compound is recovered as two peaks (the anti and syn isomers). The paired anti/syn peaks occur with characteristic intensity ratios, and the peak pairs are useful to confirm the presence of aldehydes when tissues are analyzed. Separation of the anti/syn isomers can be avoided by using more polar columns if desired. This increases the sensitivity of the assay, but then the anti/syn ratio is lost as an additional parameter for identification.

Synthetic 4-hydroxyalkenals were also analyzed as *tert*-butyldimethylsilyl (TBDMS) derivatives. The TBDMS derivatives are generally more stable than the TMS derivatives and usually yield an abundant ion at M −

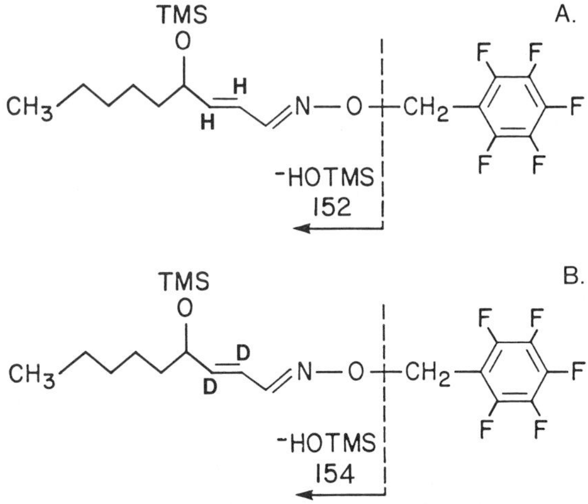

FIG. 1. Structures of the *O*-PFB oxime TMS derivatives of 4-hydroxynonenal: (a) protonated (normal) 4-hydroxynonenal and (b) dideuterated 4-hydroxynonenal. The dashed line indicates where the molecules break to form the most abundant specific fragments, 152 and 154 amu for the protonated and deuterated compound, respectively. The fragmentation also involves loss of the silanol (HOTMS) group.

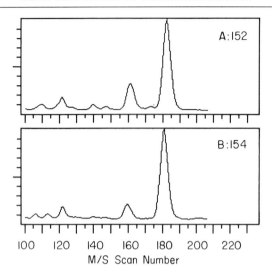

FIG. 2. Negative ion chemical ionization GC–MS of O-PFB oxime TMS ether derivatives of 4-hydroxynonenal and its 2,3-dideutero internal standard in the rat retina. The mass spectrometer performed 1 scan per 0.44 sec. (A) Specific ion monitoring at 152 amu for 4-hydroxynonenal occurring in the rat retina. (B) Specific ion monitoring at 154 amu for dideuterated internal standard.

57, which is due to loss of the TBDMS group.[16] However, the O-PFB oxime TBDMS ether derivatives of 4-hydroxyalkenals gave the same fragmentation patterns as the TMS ethers under NICI conditions (data not shown). Since the TBDMS ethers are more difficult to form than TMS ethers,[17] these derivatives were not used for routine analysis.

Several rat tissues were analyzed after the method was shown to work satisfactorily on different synthetic 4-hydroxyalkenals. A typical example of quantitative analysis for 4-hydroxynonenal in rat retina using the dideuterated internal standard is shown in Fig. 2. Figure 2A shows specific ion monitoring at 152 amu, monitoring native 4-hydroxynonenal. The deuterated internal standard yields the most abundant fragment at 154 amu (Fig. 1B), which is monitored in Fig. 2B. Typically, the deuterated internal standard elutes about 1 sec earlier (scan number 181) than the protonated authentic analog (scan number 182). Quantitation is carried out by comparing the peak area ratios of the second (syn) peak of the

[16] J. T. Watson and B. J. Sweetman, *Org. Mass Spectrom.* **9**, 39 (1974).
[17] F. J. G. M. van Kuijk, D. W. Thomas, R. J. Stephens, and E. A. Dratz, unpublished results (1986).

unknown (152 amu) to the internal standard (154 amu). An absolute amount of unknown can be determined since a known amount of deuterated 4-hydroxynonenal is added.[18]

Using this methodology, 4-hydroxyalkenals could be detected down to 10 pg. 4-Hydroxyhexenal and 4-hydroxydecenal could be analyzed by the same method as well, with monitoring for the specific fragments of 110 and 166 amu, respectively. The molecular weights of all the specific fragments from 4-hydroxyhexenal are 42 amu lower, owing to the shorter alkyl chain, whereas the weights of specific fragments from 4-hydroxydecenal are 14 amu higher. It appears that the most abundant fragments of other 4-hydroxyalkenals can simply be calculated by adding or subtracting the appropriate number of CH_2 groups.

Quantitation of 4-Hydroxyalkenals

Low levels of protonated 4-hydroxynonenal were often found in reagent blanks containing the deuterated standard. We have obtained evidence that about 10–15% of the deuterated internal standard might be converted to the protonated analog by D/H exchange with the solvent during sample preparation and GC–MS. In addition, it has been found that typical HPLC-grade solvents are contaminated with traces of lipids, which oxidize and decompose (H. Esterbauer, personal communication). As a precaution, reagent blanks should be run daily for every set of analyses, and any small amount of protonated 4-hydroxynonenal in the deuterated blank should be subtracted from the unknown in the quantitative calculations. Using the deuterated standard, quantitation is not fully reliable because of the D/H exchange. We are now synthesizing carbon-13 (^{13}C) labeled standards to overcome this problem.

Binding of 4-Hydroxyalkenals to Proteins

Esterbauer[6] reported that 4-hydroxyalkenals react rapidly with SH groups when plasma is incubated with 4-hydroxynonenal. Curzio *et al.*[4] reported that 4-hydroxyalkenals react with SH groups of bovine serum albumin (BSA) to form the S-alkylated BSA. Curzio *et al.*[4] provided evidence for an equilibrium reaction which is established in 75–110 min. Under the assay conditions used here, the *O*-PFB oxime of the thioethers is certainly formed; however, the thioether formation is not expected to be reversed, which limits the quantitative assay presented here.

[18] M. L. Selley, M. R. Barlett, J. A. McGuiness, A. J. Hapel, N. G. Ardlie, and M. J. Lacey, *J. Chromatogr.* **488,** 329 (1989).

Semiquantitative analysis of 4-hydroxynonenal in retina tissues indicated enhanced levels in the vitamin E-deficient tissues.[19] The occurrence of these compounds in the retina is the most direct proof to date for lipid peroxidation and aldehydic products of lipid peroxidation in retina tissues of vitamin E-deficient and -supplemented animals. The differences in 4-hydroxyalkenal levels between vitamin E-deficient and -supplemented animal tissues are not fully consistent, and this may be because the aldehydes are largely bound as thioethers which are not yet recovered by the present form of the assay as discussed above. In addition, ^{13}C labeled internal standards are necessary for full confidence in the analytical quantitation.

Conclusions

A new GC–MS-based method was developed for analyses of secondary lipid peroxidation products based on formation of PFB oxime derivatives from 4-hydroxyalkenals. These aldehydes were found in tissues of rats.

[19] E. A. Dratz, C. C. Earnsworth, E. C. Loew, R. J. Stephens, D. W. Thomas, and F. J. G. M. van Kuijk, in "Vitamin E: Biochemistry and Health Implications" (A. T. Diplock, L. J. Machlin, L. Packer, and W. A. Pryor, eds.), Vol. 570, p. 46. New York Academy of Sciences, New York, 1989.

[42] Determination of Aldehydic Lipid Peroxidation Products: Malonaldehyde and 4-Hydroxynonenal

By Hermann Esterbauer *and* Kevin H. Cheeseman

Introduction

Aldehydes are always produced when lipid hydroperoxides break down in biological systems,[1–3] and it is of interest to identify and measure these compounds both as an index of the extent of lipid peroxidation and as an aid to elucidate the role of aldehydes as causative agents in certain pathological conditions.[2–4] We deal here with current analytical methods used for the qualitative and quantitative determination of aldehydes in biological systems, and we pay particular attention to 4-hydroxynonenal (HNE) and malondialdehyde (MDA).

4-Hydroxynonenal is produced as a major product of the peroxidative decomposition of $\omega 6$ polyunsaturated fatty acids (PUFA) and possesses cytotoxic, hepatotoxic, mutagenic, and genotoxic properties.[2,4,5] Increased levels of HNE were found in plasma and various organs under conditions of oxidative stress (for review, see Refs. 6 and 7). In addition to HNE, lipid peroxidation generates many other aldehydes (alkanals, 2-alkenals, 2,4-alkadienals, protein- and phospholipid-bound aldehydes) which may also be of toxicological significance.[2,4,6,8]

Malondialdehyde is in many instances the most abundant individual aldehyde resulting from lipid peroxidation, and its determination by thiobarbituric acid (TBA) is one of the most common assays in lipid peroxidation studies. *In vitro* MDA can alter proteins, DNA, RNA, and many other biomolecules.[8] Recently, it has been demonstrated with

[1] H. Esterbauer, *in* "Free Radicals, Lipid Peroxidation and Cancer" (D. C. H. McBrien and T. F. Slater, eds.), p. 101. Academic Press, London, 1982.

[2] H. Esterbauer, *in* "Free Radicals in Liver Injury" (G. Poli, K. H. Cheeseman, M. U. Dianzani, and T. F. Slater, eds.), p. 29. IRL Press, Oxford, 1985.

[3] W. Grosch, *in* "Autoxidation of Unsaturated Lipids" (H. W. S. Chan, ed.), p. 95. Academic Press, London, New York, 1987.

[4] M. Comporti, *Lab. Invest.* **53**, 599 (1985).

[5] H. Esterbauer, H. Zollner, and R. J. Schaur, *ISI Atlas Sci. Biochem.* **1**, 311 (1988).

[6] H. Esterbauer, H. Zollner, and R. J. Schaur, *in* "Membrane Lipid Oxidation" (C. Vigo-Pelfrey, ed.), Vol. 1, p. 239. CRC Press, Boca Raton, Florida, 1990.

[7] H. Esterbauer and H. Zollner, *Free Radical Biol. Med.* **7**, 197 (1989).

[8] E. Schauenstein, H. Esterbauer, and H. Zollner, "Aldehydes in Biological Systems: Their Natural Occurrence and Biological Activities." Pion Press, London, 1977.

monoclonal antibodies that malonaldehyde-altered protein occurs in atheroma of hyperlipidemic rabbits.[9]

Standard Determination of Malonaldehyde with Thiobarbituric Acid

In the TBA test reaction one molecule of MDA reacts with two molecules of TBA with the production of a pink pigment having an absorption maximum at 532–535 nm. The reaction should be performed at pH 2–3 at 90–100° for 10–15 min. Typically, the tissue sample (e.g., a liver microsomal suspension) is mixed with 2 volumes of cold 10% (w/v) trichloroacetic acid (TCA) to precipitate protein. The precipitate is pelleted by centrifugation, and an aliquot of the supernatant is reacted with an equal volume of 0.67% (w/v) TBA in a boiling water bath for 10 min. After cooling the absorbance is read at 532 nm and the concentration of MDA calculated based on an ε value of 153,000. This value is the average of several slightly differing figures reported in the literature.[10] The crystalline MDA–TBA adduct in water shows an absorption maximum at 532 nm (ε 159,200).[11]

The TBA reagent should be prepared as an aqueous solution and requires heating to dissolve the TBA solid. A standard curve can be prepared using malonaldehyde bisdimethyl- or bisdiethylacetal as the source of MDA. A 10 mM stock solution is prepared by adding 1 mmol of the acetal to 100 ml of 1% (v/v) sulfuric acid and leaving the mixture at room temperature for 2 hr to achieve complete hydrolysis. For the preparation of the standard curve the MDA stock is further diluted to about 1–10 μM and then reacted with TBA as above. The concentration of the MDA solution can be checked by measuring the UV spectrum. In 1% H_2SO_4 the absorption maximum is at 245 nm (ε 13,700).[12] In alkaline solution (10 mM Na_3PO_4) the maximum is at 267 nm (ε 31,500).[13]

Many factors influence the results obtained with the TBA test, as discussed previously in this series.[12,14] Briefly, conditions and procedures to be avoided if free MDA is to be measured are as follows: preparation of the TBA reagent in strong acid solutions, high concentrations of metals, such as iron, high concentrations of sugars, such as sucrose, and use of the whole tissue sample in the assay. To ensure that no lipid oxidation

[9] M. E. Haberland, D. Fong, and L. Cheng, *Science* **241**, 215 (1988).
[10] H. Esterbauer, K. H. Cheeseman, M. U. Dianzani, G. Poli, and T. F. Slater, *Biochem. J.* **208**, 129 (1982).
[11] V. Nair and G. A. Turner, *Lipids* **19**, 804 (1984).
[12] H. Esterbauer, J. Lang, S. Zadravec, and T. F. Slater, this series, Vol. 105, p. 319.
[13] T. W. Kwon and B. M. Watts, *J. Food Sci.* **28**, 627 (1963).
[14] R. P. Bird and H. H. Draper, this series, Vol. 105, p. 299.

TABLE I
TBA REACTION WITH DIFFERENT COMPOUNDS[a]

Compound	Conditions[b]	ε value
Malonaldehyde	A	153,000
Alkanals	A, B, C	0
2-Alkenals	A without Fe	14–66
	A with Fe	30–90
	B	100–200
	C	130–160
2,4-Alkadienals	A without Fe	48–160
	A with Fe	184–280
	B	1100–2600
	C	4500
4-Hydroxyalkenals	A without Fe	12–119
	A with Fe	38–124
	C	320
Amino acids preincubated with 0.9 mM Fe	D	50–620
Sugars preincubated with 0.9 mM Fe	D	90–2700
Monohydroperoxides from arachidonic acid	E without Fe	3200–8100
	E with Fe	12400–34100

[a] Absorbance at 530–535 nm is expressed per mole of compound.

[b] A, 5% TCA, 10 min, 100°, in the presence or absence of 3 μM FeSO$_4$ (Ref. 10); B, water, 60 min, 95° (Ref. 15); C, 1 N glacial acetic acid, 120 min, 100° (Ref. 16); D, glacial acetic acid, 30 min, 100° (Ref. 17); and E, 10% TCA, 30 min, 100°, in the presence or absence of 2.5 mM FeSO$_4$ (Ref. 18).

occurs during the assay, butylated hydroxytoluene (0.01 vol % of a 2% BHT solution in ethanol) and EDTA (1 mM final concentration) can be added to the sample prior to TCA precipitation.

Determinations from TBA Test Measurements

It is well documented that the TBA test is not specific for MDA. A great variety of substances other than MDA under appropriate conditions also form pink TBA complexes; moreover, MDA or MDA-like substances can arise during the assay from acid-catalyzed or thermal decomposition of precursors (other aldehydes, MDA bound to proteins, oxidized lipids, amino acids, sialic acid) (Table I).[15–18] It would seem, however, that using

[15] R. Marcuse and L. Johansson, *J. Am. Oil Chem. Soc.* **50**, 387 (1973).

[16] G. Witz, N. J. Lawrie, A. Zaccaria, H. E. Ferran, Jr., and B. D. Goldstein, *J. Free Radicals Biol. Med.* **2**, 33 (1986).

[17] J. M. C. Gutteridge, *FEBS Lett.* **128**, 343 (1981).

[18] J. Terao and S. Matsushita, *Lipids* **16**, 98 (1981).

the protocol described above for peroxidized tissue samples, e.g., microsomes, there is little artifactual production of MDA or interference with other TBA-positive substances. This is not merely conjecture but has been demonstrated in practice.[10,12,19,20] In liver microsomal suspensions in which lipid peroxidation has been stimulated by ADP–iron, CCl_4, or ascorbate–iron, the direct determination of free MDA by the HPLC method described below gave precisely the same value as did the TBA test, indicating that in those systems the standard TBA test measures only free MDA and not MDA-like substances. Also, in oxidized low density lipoprotein 80% of the TBA-reactive substances (TBARS) were free MDA.[21]

This does not contradict the low specificity of the TBA test but can be explained as follows. First, in the standard procedure most of the potential MDA precursors, e.g., protein–MDA complexes or oxidized lipids, are removed by TCA precipitation in the cold prior to the actual assay. Second, other TBA-positive compounds that could be present in the deproteinized supernatant, such as aldehydes, amino acids, sugars, and fatty acid hydroperoxides, give only a very weak color in the standard TBA assay. On a molar basis, the absorption at 530–535 nm produced by such compounds is several orders of magnitude lower than the absorption produced by MDA (Table I). The TBA-positive compounds would therefore have to be present in the sample in extremely high concentrations to interfere significantly with the standard determination of MDA. Suspensions of peroxidized rat liver microsomes (ADP–iron, 30 min) contain, e.g., 58 nmol free MDA and 95 nmol of the other aldehydes listed in Table I.[10] A rough estimate shows that 99.7% of the absorbance at 535 nm which would be found in the standard TBA assay results from MDA (ε 153,000), and only 0.3% or less is due to all other aldehydes (assumed average ε 300).

The situation may, however, be completely different if the standard reaction conditions are significantly altered, e.g., heating in the presence of the complete tissue fraction, prolonged reaction times, the use of other acids, and supplementation of the reaction mixture with iron. There can be no doubt that such modified TBA tests are much less specific, and it seems appropriate to refer in such cases to the measurement of TBA-positive substances, TBARS, or simply the TBA value rather than specifying MDA.

Frequently used modifications of the TBA test employ the whole acid-

[19] H. Esterbauer and T. F. Slater, *IRCS Med. Sci.* **9,** 749 (1981)

[20] J. Lang, P. Heckenast, H. Esterbauer, and T. F. Slater, *in* "Oxygen Radicals in Chemistry and Biology" (W. Bors, M. Saran, and D. Tait, eds.), p. 351. de Gruyter, Berlin, New York, 1984.

[21] H. Esterbauer, G. Jürgens, O. Quehenberger, and E. Koller, *J. Lipid Res.* **28,** 495 (1987).

ified sample. Typically[22,23] 1 volume of tissue sample, e.g., 10% (w/v) homogenate, is mixed with 6 volumes of 1% phosphoric acid and 2 volumes 0.6% aqueous TBA solution and heated for 45–60 min in a boiling water bath. After cooling, 8 volumes of *n*-butanol are added and mixed vigorously. The butanol phase which contains the colored TBA reaction products is separated by centrifugation and its absorbance measured at 532 nm. The method with phosphoric acid is basically developed to measure MDA and/or TBARS bound to proteins; it seems that under the assay conditions the binding is, at least in part, reversible. With this method significantly elevated levels of TBARS were found in various fresh tissues of rats fed a vitamin E-deficient diet, e.g., 570 nmol/g liver of vitamin E-deficient rats compared to 76 nmol in controls.[23] In another procedure, 0.2 ml of 35% TCA and 1 ml of 0.5% TBA are added to 1 ml of sample, and the mixture is heated at 60° for 90 min, after which dispersed lipids are extracted with 3 ml CH_2Cl_2 and the clear aqueous phase measured at 532 nm.[24–26] With this method 0.22–0.40 nmol TBARS/ml were found in control plasma. Aust[27] reported a method where 1 ml of sample is mixed with 2 ml of a TCA–TBA–HCl reagent [15% (w/v) TCA, 0.375% (w/v) TBA, 0.25 *N* HCl], the complete mixture is heated on a boiling water bath for 15 min, and after centrifugation, the absorbance is measured at 535 nm.

Numerous other variations have been introduced, but heating the complete assay sample with TBA in acidic solution is common to all. These methods can be subject to various sources of errors if only the absorption at 530–535 nm is measured. For example, yellow compounds with maxima at 450–490 nm and significant absorption at 530–535 nm are often formed and would lead to an overestimation of TBARS. It is therefore strongly recommended that the spectrum in the 430–600 nm range be recorded to prove the existence of the 532 nm maximum typical for the MDA–TBA complex and to correct for possible background absorption by interpolation. The exact amount of the MDA–TBA complex in the reaction mixture can also be determined by HPLC.[14,28] The neutralized reaction mixture or a butanol extract is separated on an ODS column with methanol–water (85 : 15) and detected at 530–535 nm. In a TCA extract of fresh pork liver 30 nmol TBARS was found by the spectrometric method, 36% of which was MDA as determined by HPLC.[28]

[22] M. Uchiyama and M. Mihara, *Anal. Biochem.* **86**, 271 (1978).
[23] M. Mihara, M. Uchiyama, and K. Fukuzawa, *Biochem. Med.* **23**, 302 (1980).
[24] K.-L. Fong, P. B. McCay, and J. L. Poyer, *J. Biol. Chem.* **248**, 7792 (1973).
[25] D. M. Lee, *Biochem. Biophys. Res. Commun.* **95**, 1663 (1980).
[26] F. Bernheim, M. L. Bernheim, and K. M. Wilber, *J. Biol. Chem.* **174**, 257 (1948).
[27] S. D. Aust, this series, Vol. 52, p. 302.
[28] R. P. Bird, S. S. O. Hung, M. Hadley, and H. H. Draper, *Anal. Biochem.* **128**, 240 (1983).

Other modifications of the TBA assay were designed to analyze lipid hydroperoxides. As can be seen from Table I, including iron in the TBA reaction mixture significantly increases the yield of pink products from various substances. A particularly high absorption is obtained with monohydroperoxides from arachidonic acid, and this is the basis of a TBA test for microdetermination of lipid peroxides. The recommended procedure[18,29] is briefly as follows. One milliliter of methanol solution containing the oxidized lipid is mixed with 2 ml of 20% TCA containing 20 μmol ferrous sulfate and 1 ml 0.67% TBA. The mixture is heated in a boiling water bath for 30 min, and after cooling 2 ml CHCl$_3$ is added (to extract turbid material). The mixture is centrifuged, and the absorbance of the clear supernatant is measured. In the case of methyl arachidonate monohydroperoxide isomers (MeHPETE), a linear relationship is found between the absorption at 532 nm and the concentration of MeHPETE. Although each isomer gave a different response, the lowest amount of MDA was found from the 8-OOH isomer (0.081 mol MDA/mol MeHPETE); the highest amount gave the 5-OOH and 15-OOH isomers (0.22 mol MDA/mol MeHPETE). To prevent additional lipid oxidation during the color development, 0.01 vol% of 2% BHT in ethanol can be added to the TBA reagent just prior to use.

Another TBA test to measure preferentially lipid peroxides was developed by Yagi for the analysis of serum or blood.[30,31] Here, the proteins and lipids are first precipitated with phosphotungstic acid. The sediment (equivalent to 20 μl serum) is then suspended in 4 ml water, 0.5 ml glacial acetic acid, and 0.5 ml 0.33% aqueous TBA solution. The mixture is heated for 60 min at 95° and after cooling extracted with 5 ml n-butanol. The concentration of the TBA–MDA complex in the butanol extract is determined fluorimetrically at 553 nm with excitation at 515 nm. The amount of TBARS (assumed to be lipid peroxides) found by this test in normal subjects was in the range of 1.8–3.9 nmol/ml serum. This is about 10 times the amount found with another modified TBA test[25] and clearly shows that different procedures yield very different results. Direct comparison of data reported by different investigators is often not possible.

In conclusion, what is measured by the TBA assay is strongly influenced by the reaction conditions. In assays where the whole sample is heated in an acidic TBA solution, the resulting absorption at 530–535 nm (or fluorescence at 553 nm) can come from all preexisting MDA, protein-bound MDA, and lipid peroxides, as well as any other substances that give rise to MDA or TBARS in the hot acid. Tests using only the TCA-

[29] T. Asakawa and S. Matsushita, *Lipids* **15,** 137 (1980).

[30] K. Yagi, this series, Vol. 105, p. 328.

[31] K. Yagi, *in* "Lipid Peroxides in Biology and Medicine" (K. Yagi, ed.), p. 223. Academic Press, London, New York, 1982.

soluble fraction of the sample are more specific for free MDA. Here again, however, interference can be caused by other TCA-soluble compounds, in particular, if free MDA is low, such as in fresh tissue samples. In any case, additional analyses should be performed to elucidate the nature and the source of the pink color. Such analyses include the demonstration of the TBA–MDA complex by HPLC[14,28] and the direct detection of free MDA as described below.

Direct Determination of Malondialdehyde by HPLC

An HPLC method for the determination of free MDA in biological samples has been described at length previously in this series,[12] and we only outline it here. The principle of the method is that an aqueous sample containing MDA at pH 6.5–8.0 is separated by HPLC using an amino-phase column with acetonitrile–30 mM Tris buffer, pH 7.4 (1 : 9, v/v), as the mobile phase. The effluent is monitored at 267 nm, the absorption maximum of the enolate anion form of free MDA (ε 31,500 at this pH and wavelength). The system is calibrated and the sample MDA peak identified by comparison with a solution of MDA. Using an injection volume of 20 μl, the smallest concentration of MDA in the original solution that can be quantified this way is 0.25 μM. To protect the column it is best to deproteinize the sample, and we find that addition of an equal volume of acetonitrile followed by centrifugation is satisfactory. A typical estimation would be as follows. An aliquot of sample (e.g., liver microsomal suspension) is mixed with an equal volume of acetonitrile and the protein precipitate pelleted at 3000 g for 10 min. A sample (20 μl) of the supernatant is injected into the HPLC and separated on an S-5 Spherisorb-NH$_2$ column (Phase Separations Ltd.) at 1 ml/min.

With this method it was shown[10,12,19,20] that what is measured with the standard TBA test (use of TCA supernatant) in peroxidized microsomes or mitochondria is exclusively free MDA. In addition, this method was used[6] to compare the amount of free MDA and TBARS formed during oxidation of various PUFA (Fig. 1). At the stage when the PUFA were completely oxidized, the yields of free MDA on a molar basis were 0.5% for linoleic acid, 4.5% for linolenic acid, 4.9% for γ-linolenic acid, 4.7% for arachidonic acid, and 7.6% for docosahexaenoic acid. The corresponding yields of TBARS were slightly higher, namely, 0.55, 4.9, 5.1, 6.1, and 8.6%. The very low yield of MDA from linoleic acid agrees with the proposal[32] that MDA is only formed from PUFAs with three or more double bonds.

We have also used this method in a system completely unrelated to

[32] W. A. Pryor, J. P. Stanley, and E. Blair, *Lipids* **11**, 370 (1976).

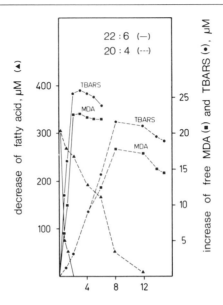

FIG. 1. Free malonaldehyde (MDA) and MDA-like substances (TBARS) formed during autoxidation of arachidonic acid (20 : 4) and docosahexaenoic acid (22 : 6). The fatty acids (0.1 mg/ml) were incubated in Tris buffer, pH 7.4, with ascorbate–iron (10 mM–0.4 mM) at 37°. Consumption of the fatty acids was measured by GC, free MDA by HPLC, and TBARS by the standard TBA assay as described in the text.

lipid peroxidation: the production of MDA from deoxyribose degradation by OH · attack.[33] Further, a modification of this method has been reported in which the proportion of acetonitrile is increased to 80% in order to achieve better separation from interfering substances when measuring plasma.[34]

Two other HPLC methods for measuring free MDA directly have been published. One method[35] uses an ODS column with acetonitrile–water (14 : 86), 50 mM myristyltrimethylammonium bromide, 1 mM phosphate buffer, pH 6.8, as mobile phase at 1 ml/min, with detection at 267 nm. The basis of the separation is ion-pairing chromatography. A reasonably good equivalence between this direct determination and the TBA test was found when measuring peroxidized microsomes. The other method[36] uses a size-exclusion column (Spherisorb TSK G 1000 PW, Phase Separation

---

[33] K. H. Cheeseman, A. Beavis, and H. Esterbauer, *Biochem. J.* **252**, 649 (1988).

[34] C. Largilliere and S. B. Melancon, *Anal. Biochem.* **170**, 123 (1988).

[35] A. W. Bull and J. Marnett, *Anal. Biochem.* **149**, 284 (1985).

[36] A. S. Csallany, M. D. Guan, J. D. Manwaring, and P. B. Addis, *Anal. Biochem.* **142**, 277 (1984).

Ltd.) with 0.1 M phosphate buffer, pH 8.0, and detection at 267 nm. A poor equivalence was found by this method when measuring MDA in beef or pork muscle or rat liver, e.g., 43 nmol (TBA) versus 11 nmol (HPLC) per 1 g of rat liver.

Several other chromatographic methods for the detection of MDA were reported. In one procedure,[37] developed for vegetable oil, the sample (0.1 g) is reacted with dansylhydrazine in hydrochloric acid containing FeCl$_3$. The formed dansylpyrazole is separated by HPLC with fluorimetric detection. In another method,[38] developed for investigation of the formation of MDA from lipid peroxidation products, the oxidized lipid (20–25 mg) is treated for 18 hr at ambient temperature with 1 ml of 5% anhydrous HCl in methanol and 1 ml trimethyl orthoformate. The amount of MDA–tetramethylacetal formed is determined by gas chromatography. Both methods certainly do not measure free MDA but rather the amount of MDA that can be formed from precursors by acid-catalyzed decomposition.

Although in the systems we have studied the TBA test is demonstrated as measuring free MDA,[12,19,20,33] this will not be true in all systems. If the investigator is concerned in knowing whether MDA is the only TBA-reactive product in the test system, then the measurement should be validated with a direct measurement of free MDA by HPLC. If the two determinations are equivalent, the investigator can use the more convenient TBA test.

Determination of Aldehydes via Dinitrophenylhydrazone Formation

The methods most frequently used for determination of aldehydes in biological tissues are based on treatment of the sample with 2,4-dinitrophenylhydrazine. Aldehydes react with dinitrophenylhydrazine to form the corresponding dinitrophenylhydrazone (DNPH) derivatives. In contrast to most free aldehydes the hydrazone derivatives are stable and not volatile, greatly facilitating the subsequent workup procedure. Moreover the DNPH derivatives have a strong yellow color (λ_{max} 360–380 nm, ε 25000–28000 M^{-1} cm^{-1}) which is of great help in detecting the compounds on TLC plates or by HPLC.

An outline of the procedure we routinely use is as follows. The sample is mixed with dinitrophenylhydrazine reagent and allowed to react. The DNPH derivatives are extracted into an organic solvent, concentrated, and preseparated by TLC to yield DNPH classes of different polarity (here termed zones I, II, and III from their positions on the TLC plate).

[37] T. Hirayama, N. Yamada, M. Nohara, and S. Fukui, *J. Sci. Food Agric.* **35**, 338 (1984).
[38] E. N. Frankel and W. E. Neff, *Biochim. Biophys. Acta* **754**, 264 (1983).

The individual classes are recovered and separated by HPLC for identification of their constituent individual aldehydes. The importance of the preliminary separation by TLC should be stressed as it performs several important functions. First, it enables the removal of excess dinitrophenylhydrazine reagent. Second, it enables certain contaminating carbonyls to be eliminated; the DNPH forms of formaldehyde, acetone, and acetaldehyde are always found at this stage even in the reagent blank. Apparently these carbonyls are always present in laboratory air and standard solutions. Finally, analysis of the hydrazones in each zone (I, II, and III) greatly facilitates clear separation of the individual compounds and provides more confident identification of the peaks in the HPLC chromatogram. For example, zone III can only contain alkanals, 2-alkenals, and 2,4-alkadienals and cannot contain the more polar 4-hydroxyalkenals that are restricted to zone I. It is possible to apply all of the DNPH derivatives directly in HPLC without preliminary TLC, e.g., by using gradient programs; however, the resulting chromatograms are complicated, and it is extremely difficult to make definite peak identifications.

A typical determination of aldehydes produced during lipid peroxidation in liver microsomes,[10,39] hepatocytes,[39] or low density lipoproteins,[21] as examples for other biological samples, is as follows. To 1 ml of the sample, e.g., microsomes at 1 mg protein/ml, add 0.1 ml of 1% EDTA, 10 μl of 2% BHT, and 0.5 ml freshly prepared DNPH reagent (2,4-dinitrophenylhydrazine recrystallized from butanol dissolved in 1 N HCl at a concentration of 0.35 mg/ml). Mix vigorously and keep in the dark for 2 hr at ambient temperature and then for 1 hr at 4°. The reaction mixture is extracted with CH_2Cl_2 (2 times 5 ml each); phase separation can be achieved by centrifugation. The pooled extract is left in a freezer for at least 2 hr and then rapidly filtered through a folded filter to remove ice crystals. The extract is brought to dryness on a rotary evaporator ($\leq 35°$) and redissolved in a minimum volume of CH_2Cl_2 (about 0.1–0.5 ml) for application to the TLC plate (silica gel 60 precoated, 20 × 20 cm, Merck, Darmstadt, FRG). The extract is applied across the plate as a band 3–5 cm long; DNPH standards (see Fig. 2) are also applied as a separate spot. The plate is developed first in CH_2Cl_2 (5 cm) and then in benzene (about 15 cm). In Fig. 2, nominal zones I, II, and III are indicated on the developed plate.

The zones are scraped off the TLC plate and eluted with methanol (2 times, 1 ml each). The methanol extracts are dried in a small conical vial with nitrogen, and the residue is finally redissolved in 0.1 ml methanol. Samples (20 μl) are separated by HPLC on an ODS column (5 μm Spheri-

[39] G. Poli, M. U. Dianzani, K. H. Cheeseman, T. F. Slater, J. Lang, and H. Esterbauer, *Biochem. J.* **227**, 629 (1985).

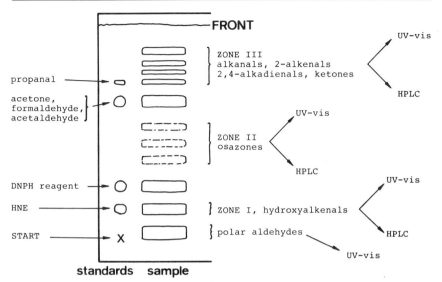

FIG. 2. Scheme showing the determination of aldehydes by the DNPH method.

sorb ODS, 4.6×250 mm) with methanol–water $(31:9, v/v)$ at 1.0 ml/min and detected at a wavelength between 365 and 378 nm. Peak assignment and quantification are made with reference to chromatograms of standard hydrazones. Additionally, the peak material can be collected to determine the assigned structure by mass spectroscopy. Alkanals, 2-alkenals, 2,4-alkadienals, or ketones are commercially available (e.g., Aldrich, Merck). We prepare the corresponding hydrazones as follows. The compound (about 10 mmol) is dissolved in a small volume of ethanol and added to 82 ml of DNPH reagent (2.4 g 2,4-dinitrophenylhydrazine in 100 ml 30% $HClO_4$). The precipitate is filtered, washed acid-free with water, and recrystallized from ethanol or ethanol–water mixtures. Various syntheses for 4-hydroxyalkenals are described (for review, see Refs. 5 and 6), and their hydrazones can be prepared as above.

The compounds identified in zone I include 4-hydroxynonenal and 4-hydroxyhexenal. In addition, this zone can contain 4,5-dihydroxy-decenal[40] and two aldehydes that are probably (based on their mass spectra) 4-hydroxy-4,5-nonadienal and 5-hydroxyoctanal.[41] Zone III contains propanal, butanal, pentanal, hexanal, nonanal, 2-propenal, 2-pentenal, 2-hexenal, 2-heptenal, 2-octenal, 2-nonenal, 2,4-heptadienal, 2,4-deca-

[40] A. Benedetti, M. Comporti, R. Fulceri, and H. Esterbauer, *Biochim. Biophys. Acta* **792,** 172 (1984).
[41] P. Heckenast, Thesis, University of Graz, Austria, 1983.

dienal, butanone, 2-pentanone, 3-pentanone, and 2-octanone. Long-chain aldehydes such as hexadecanal and octadecanal are also present in zone III. They are unrelated to lipid peroxidation but result from plasmalogens. Zone II contains mostly osazones, none of which have been identified.

In multiple analyses of the same sample the reproducibility of the overall procedure is in the range of ±10–15%. In the literature[10,21,39,42–51] the amount of individual aldehyde present in the analyzed sample is usually reported as the figure obtained in the analysis described above, i.e., without considering recovery loss. This seems sufficient for many *in vitro* studies where only the qualitative aldehyde pattern and the relative changes compared to a control sample are needed. For determining the absolute amount, however, recovery losses must be taken into account.

The recovery of aldehydes from biological samples is rather variable.[6] The addition of 4-hydroxynonenal to fresh liver microsomes yields a recovery figure of 40 ± 6%; for hexanal and 2-heptenal the figures were 29 ± 4.2 and 72 ± 8%, respectively. Based on these recovery figures we find in peroxidized rat liver microsomes (ADP–iron, 30 min) the following amounts of aldehydes per milligram of microsomal protein: 4-hydroxynonenal, 20 nmol; hexanal; 40 nmol; propanal, 24 nmol; 2-propenal, 1.6 nmol; and 2-octenal, 2.0 nmol.

The DNPH method has broad applicability and is reasonably selective and sensitive. The detection limit for a single aldehyde is about 1 pmol per 20 μl of injected sample.

The methanol solutions of zones I, II, and III remaining after HPLC

[42] P. Winkler, W. Lindner, H. Esterbauer, E. Schauenstein, R. J. Schaur, and G. A. Khoschsorur, *Biochim. Biophys. Acta* **796**, 232 (1984).

[43] W. E. Turner, R. H. Hill, W. H. Hannon, J. T. Bernert, E. M. Kilbourne, and D. D. Bayse, *Arch. Environ. Contam. Toxicol.* **14**, 261 (1985).

[44] G. Poli, U. Ramenghi, O. David, F. Biasi, G. Cecchini, R. Carini, E. Chiarpotto, and M. U. Dianzani, in "Free Radicals, Cell Damage and Disease" (C. Rice-Evans, ed.), p. 187. Richilieu Press, London, 1986.

[45] G. Poli, G. Cecchini, F. Biasi, E. Chiarpotto, R. A. Canuto, M. E. Biocca, G. Muzio, H. Esterbauer, and M. U. Dianzani, *Biochim. Biophys. Acta* **883**, 207 (1986).

[46] H. Esterbauer, A. Benedetti, J. Lang, R. Fulceri, G. Fauler, and M. Comporti, *Biochim. Biophys. Acta* **876**, 154 (1986).

[47] M. U. Dianzani, G. Poli, R. A. Canuto, M. A. Rossi, M. E. Biocca, F. Biasi, G. Cecchini, G. Muzio, M. Ferro, and H. Esterbauer, *Toxicol. Pathol.* **14**, 404 (1986).

[48] A. Benedetti, A. Pompella, R. Fulceri, A. Romani, and M. Comporti, *Biochim. Biophys. Acta* **876**, 658 (1986).

[49] M. Curzio, H. Esterbauer, G. Poli, F. Biasi, G. Cecchini, C. Di Mauro, N. Cappello, and M. U. Dianzani, *Int. J. Tissue React.* **9**, 295 (1987).

[50] G. D. Buffinton, N. H. Hunt, W. B. Cowden, and I. A. Clark, *Biochem. J.* **249**, 63 (1988).

[51] A. Pompella, A. Romani, R. Fulceri, A. Benedetti, and M. Comporti, *Biochim. Biophys. Acta* **961**, 293 (1988).

separation may be used to determine the total aldehyde concentration in each of the fractions.[39,44,45,47] For that determination the samples are diluted about 20-fold with methanol, and the UV–VIS spectrum is recorded in the range of 200–600 nm. The concentration is calculated from the absorption maximum using an average value of 27,000 for zones I and III (λ_{max} between 360 and 375 nm) and 44,000 for zone II (λ_{max} around 400–430 nm).[10]

An alternative HPLC method has been described for the detection of aldehydic lipid peroxidation products in plasma and liver homogenates.[52–54] This method is based on the reaction of aldehydes with 1,3-cyclohexanedione to yield fluorescent dehydroacridine derivatives that can be separated by HPLC. The sample (0.5 ml) is mixed with an equal volume of methanol and centrifuged. The clear supernatant (0.5 ml) is reacted with 1 ml of 1,3-cyclohexanedione (CHD) reagent at 60° for 1 hr. The CHD reagent is prepared by dissolving ammonium sulfate (10 g), glacial acetic acid (5 ml), and CHD (0.25 g) in 95 ml water. The CHD reaction mixture (1 ml) is poured onto a Sep-Pak C_{18} cartridge for clean up and eluted with 2 ml methanol. This methanol solution is separated by HPLC on a 5-μm ODS column (6 × 100 nm) with fluorescence detection at 445 nm and excitation at 380 nm (flow rate 1 ml/min). The mobile phases for elution are the following: 0–18 min, MeOH–H_2O, 3 : 7; 18–32 min, tetrahydrofuran (THF)–H_2O, 26 : 74; 32–42 min, THF–H_2O, 4 : 6; and 42–50 min, THF. The peaks are assigned using a reference chromatogram; peak quantification is based on the use of 5-hydroxypentanal as internal standard.

The method seems to be rather sensitive and allows detection of about 100 fmol per 100 μl injected aldehydes. The disadvantage is that the chromatogram is rather complex and shows very large peaks resulting from the reagents. With the CHD method various aldehydes including 4-hydroxynonenal in plasma and liver of rats were detected (about 1 nmol/ml plasma or 1 nmol/g liver); rats fed a vitamin E-deficient diet or treated with CCl_4 had significantly increased aldehyde levels.

Direct Determination of 4-Hydroxynonenal by HPLC or GC–MS

The importance of HNE as a cytotoxic lipid peroxidation product has led to the development of two independent analyses specifically for this

[52] K. Yoshino, T. Matsuura, M. Sano, S.-I. Saito, and I. Tomita, *Chem. Pharm. Bull.* **34,** 1694 (1986).

[53] K. Yoshino, M. Sano, M. Fujita, and I. Tomita, *Chem. Pharm. Bull.* **34,** 5184 (1986).

[54] I. Tomita, K. Yoshino, and M. Sano, *in* "Clinical and Nutritional Aspects of Vitamin E" (O. Hayashi and M. Mino, eds.), p. 277. Elsevier, Amsterdam, 1987.

compound.[55,56] Free HNE can easily be detected by HPLC with an UV detector at 220–223 nm owing to its high molar absorptivity (λ_{max} 222 nm, ε 13100, in methanol). A typical analysis of HNE in liver microsomal suspensions is as follows. To 20 ml of microsomal suspension (1 mg protein/ml), 20 μl BHT (10 mg/ml ethanol) is added as well as 200 μl desferrioxamine (10 mg/ml water) to prevent further oxidation during sample workup. The suspension is extracted with CH_2Cl_2 (2 times 20 ml each), acetate buffer (2 ml, 0.1 M, pH 3.0) is added to the pooled extract, and the CH_2Cl_2 is removed on a rotary evaporator ($\leq 20°$). The residual buffer solution containing HNE is quantitatively applied to a disposable C_{18} solid-phase column (Bond-Elut, C_{18}, 3 cm^3 size; Analytichem International, Harbor City, CA) that has been preconditioned with 3 ml methanol and equilibrated with water. The Bond-Elut column is first eluted with 2 ml hexane to remove unwanted nonpolar materials; HNE is then eluted with 2 ml methanol–water (8 : 2) into a 2-ml volumetric flask. Residual hexane in the eluate is removed by nitrogen gassing, and the volume is brought to 2 ml with water.

A volume (20 μl) of the cleaned sample is separated by HPLC on an S-5 Spherisorb ODS column (4.5 × 250 mm) with acetonitrile–water (4 : 6) or methanol–water (6.5 : 3.5) as mobile phase at 1 ml/min, with detection at 220 nm. Peak identification and quantification are done with reference chromatograms of standard solutions of HNE. The lowest amount detectable by this method is about 2 pmol per 20-μl injection. The precision is good and shows a coefficient of variation between 1.0 and 3.2%. With microsomes the recovery as estimated with ^{14}C-labeled HNE is 73%. The HNE value found by this method in peroxidized microsomes (ADP–iron, 30 min) is 4.6 ± 0.67 nmol/mg protein.

If the sample contains higher concentrations of HNE (>2 nmol/ml) the method can be simplified.[57] In such cases the sample is mixed with an equal volume of acetonitrile–acetic acid (97 : 3), which precipitates most of the protein and extracts HNE simultaneously. After centrifugation the clear supernatant is separated by HPLC as described above.

For GC–MS[56] the HNE is first converted under mild conditions to its pentafluorobenzyl oxime and then silylated. The GC–MS analysis is performed in the negative chemical ion mode with specific ion monitoring. Recently this method has been used with deuterated HNE as an internal standard for the quantification of HNE in platelets, monocytes, and

[55] J. Lang, C. Celotto, and H. Esterbauer, *Anal. Biochem.* **150,** 369 (1985).
[56] F. J. G. M. Van Kuijk, D. W. Thomas, R. J. Stephens, and E. A. Dratz, *Biochem. Biophys. Res. Commun.* **139,** 144 (1986).
[57] H. Esterbauer, H. Zollner, and J. Lang, *Biochem. J.* **228,** 363 (1985).

plasma.[58] The details of this GC–MS method are descrit ~~422~~
this volume [40, 41].

Concluding Remarks

The choice of which method for aldehyde analysis sh   ~~ used
depends on the particular interest of the investigator. Is an overall picture
of the complete spectrum of aldehydes required, or is there an interest in a
specific compound such as MDA or HNE? If only MDA is to be determined, the classic TBA test remains a useful method, providing it has
been validated by an HPLC measurement for the particular system under
study. If only HNE is to be determined, the method of choice is direct
HPLC or GC–MS. The latter method is more sensitive, but the resources
required are more expensive.

If the whole spectrum of aldehydes must be measured, then the DNPH
method described is probably more reliable than the current cyclohexane-
dione method, which separates all aldehydes in one run. As the number of
aldehydes present in peroxidized biological samples may exceed 30 and
their relative proportions vary greatly, complex chromatograms are produced and definite peak identification is difficult. The DNPH method is
less sensitive but gives more confidence in peak identification.

Acknowledgments

The authors' work has been supported by the Association for International Cancer
Research (U.K.) and by the Austrian Science Foundation (to H.E., Project P6176B).

[58] M. L. Selley, M. R. Bartlett, J. A. McGuiness, A. J. Hapel, and N. G. Ardlie, *J. Chromatogr.* **488**, 329 (1989).

[43] Malondialdehyde Determination as Index of Lipid Peroxidation

By H. H. Draper and M. Hadley

Introduction

The determination of malondialdehyde (MDA) has attracted widespread interest because it appears to offer a facile means of assessing lipid
peroxidation in biological materials. However, the validity of MDA as an
index of lipid peroxidation has been clouded by controversy regarding its
formation as an artifact of analysis and as a product of enzyme reactions,

_ts occurrence in various bound forms, and the specificity of methods used for its measurement. Some investigators, on the basis of evidence that MDA fails to reflect extensive peroxidation in certain biological materials, have questioned its validity as an index of peroxidation in all such materials. Other investigators have found that MDA is a reliable indicator of peroxidation in many samples, and that difficulties in its determination can be resolved by appropriate modifications in methodology.

Malondialdehyde occurs in biological materials in the free state and in various covalently bound forms. In the only materials so far extensively examined (food and urine), it has been found predominantly in bound forms. Most of the MDA in foods of animal origin (~80%) appears to be bound in Schiff base linkages to the ε-amino groups of the lysine residues of food proteins, from which it is released during digestion as N-ε-propenallysine.[1] A portion of the remaining MDA is present in the form of other minor derivatives. N-ε-Propenallysine and its N-α-acetylated derivative are the main forms in which endogenous MDA is excreted in rat and human urine.[2,3] Urine also contains small amounts of MDA adducts with guanine, the phospholipid bases serine and ethanolamine, and other unidentified reactants. Free MDA is a minor and variable excretory product.

It is apparent from the occurrence of these derivatives in urine that MDA forms adducts with proteins, nucleic acids, and other substances *in vivo,* and this compromises the assessment of lipid peroxidation in the tissues based on the determination of free MDA. For example, although no free MDA is detectable in human plasma,[4] it is found using procedures that involve hydrolysis of bound forms.[5] Free MDA reacts extensively with the serum albumin present in cell culture media,[6] and may do the same in plasma. It may also react with proteins *in vitro,* particularly over long incubation periods.

MDA can be released from its bound forms by hot acid or alkali digestion, but the conditions required for hydrolysis can lead to other complications. The pH required for maximum yield of MDA varies among biological materials depending on the nature of the derivatives present. MDA may be generated during hydrolysis by oxidation of polyunsatu-

[1] L. A. Piché, P. D. Cole, M. Hadley, R. van den Bergh, and H. H. Draper, *Carcinogenesis* **9,** 473 (1988).
[2] L. G. McGirr, M. Hadley, and H. H. Draper, *J. Biol. Chem.* **260,** 15427 (1985).
[3] H. H. Draper, M. Hadley, L. Lissemore, N. M. Laing, and P. D. Cole, *Lipids* **23,** 626 (1988).
[4] C. Largillière and S. B. Mélancon, *Anal. Biochem.* **170,** 123 (1988).
[5] M. A. Warso and W. E. M. Lands, *Clin. Physiol. Biochem.* **2,** 70 (1984).
[6] R. P. Bird and H. H. Draper, *Lipids* **17,** 519 (1982).

rated fatty acids (PUFA) in the sample and by degradation of preexisting oxidation products. Pigments present in the sample, or generated during hydrolysis, also can interfere in the colorimetric assessment of MDA. These problems, and possibilities for their resolution, are discussed in the following sections.

Absorptiometric Methods

Although absorptiometry is the method which has been most frequently used to quantify MDA, many different procedures have been used for its extraction from the sample and for its purification, depending on the nature of the material being analyzed and whether free or total MDA is estimated.

Free MDA. Below its pK_a of 4.65, MDA exists primarily in a cyclic form with a UV_{max} at 245 nm (E_{mol} 1.37 × 10⁴).[7] Raising the pH above 4.65 results in increasing proportions of the enolate ion, which has a UV_{max} at 267 nm (E_{mol} 3.42 × 10⁴).[8] At neutral pH the enolate anion is the predominant form.

Esterbauer and Slater[7] determined free MDA in incubates of peroxidizing hepatic microsomes by measuring its UV absorbance after isolation using high-performance liquid chromatography (HPLC). Aliquots of homogenate were applied to a column designed for carbohydrate analysis, and MDA was measured in the eluates at 270 nm using an absorbance detector. The method gave values similar to those obtained by an unspecified 2-thiobarbituric acid (TBA) procedure, indicating that the MDA formed was still in the free state at the end of the 30-min incubation period and that there was no interference by other aldehydes in the TBA method for MDA determination.

This procedure is short (reported time of analysis <10 min), sensitive (5 ng on the chromatogram), and apparently specific for MDA. It should be useful in monitoring lipid peroxidation *in vitro* over short time periods, and in studying such phenomena as radiation-induced hydroxyl radical degradation of sugars,[9] but it is not suitable for the analysis of complex biological materials, such as food and urine, in which most of the MDA exists in the form of derivatives. For the same reason, it probably has limited application to animal and human tissues.

Largillière and Mélancon[4] applied a modification of the procedure to the determination of MDA in deproteinized human blood plasma. They

---

[7] H. Esterbauer and T. F. Slater, *IRCS Med. Sci.* **9**, 749 (1981).

[8] L. J. Marnett, M. J. Bienkowski, M. Raban, and M. A. Tuttle, *Anal. Biochem.* **99**, 458 (1979).

[9] K. H. Cheeseman, A. Beavis, and H. Esterbauer, *Biochem. J.* **252**, 649 (1988).

failed to detect any free MDA and concluded that "the classical thiobarbituric acid test" is not suitable for determination of MDA in plasma. However, if (as the authors speculate) MDA in plasma is present in complexes with proteins and amino acids, from which it is released by acid hydrolysis under the conditions of the TBA reaction, the converse conclusion may be drawn (i.e., that procedures for the determination of free MDA are not suitable for the analysis of plasma).

Lee and Csallany estimated free MDA in an ultrafiltrate of rat liver homogenate from its absorbance at 267 nm after purification by HPLC on a size-exclusion column.[10] The levels found in the liver of vitamin E-deficient animals were about 15 times those found in control livers. Bound MDA was estimated by subtracting values for free MDA from those obtained after hydrolysis at pH 13 for 30 min at 60°. The concentration of bound MDA was increased only 2-fold in deficient liver, and its concentration in normal liver exceeded that of free MDA. Although the results suggest that free MDA accumulates in the liver in vitamin E deficiency, it is possible that more free MDA was formed in the deficient tissue during the analytical procedure, which did not include the use of a synthetic antioxidant. Analysis for TBA-reactive substances, carried out on a butanol extract of the reaction mixture using a conventional spectrophotometric procedure,[11] yielded values that were 1.4–1.8 times higher than those for total MDA.

Bull and Marnett[12] used myristyltrimethylammonium bromide to prevent polymerization of MDA during HPLC. Chromatography on a reverse-phase column using a mixture of acetonitrile and this ion-pairing reagent enabled MDA to be selectively eluted as the conjugate base and quantified from its absorbance at 267 nm. The procedure was used successfully to estimate free MDA production in peroxidizing liver microsomes and to demonstrate that MDA in urine is present in the form of metabolites. Values similar to those obtained using the colorimetric TBA procedure[11] were observed at high concentrations of MDA in the microsomes, but lower values were found at low concentrations, indicating that the TBA method was nonspecific for MDA (or that MDA was released from bound forms during the TBA procedure).

MDA Derivatives. MDA reacts with a variety of compounds to form derivatives which can be estimated from their absorption in the visible region. These include aniline, 4-hexylresorcinol, *N*-methylpyrrole, indole, 4-aminoacetophenone, ethyl *p*-aminobenzoate, 4,4-sulfonyldiani-

[10] H. Lee and A. S. Csallany, *Lipids* **22,** 104 (1987).
[11] R. O. Sinnhuber, I. C. Yu, and T. C. Yu, *Food Res.* **23,** 620 (1958).
[12] A. W. Bull and L. J. Marnett, *Anal. Biochem.* **149,** 284 (1985).

line, p-nitroaniline, and azulene.[13] Other aldehydes also react with most of these compounds to form yellow or orange complexes which can interfere in the estimation of MDA if they are present in the sample and are not separated from the MDA complex.

The most widely employed method for the determination of MDA in biological materials is based on its reaction with TBA to form a pink complex with an absorption maximum at 532–535 nm (E_{mol} 14.9 × 10⁴).[7] Heating the sample at a pH of 3 or below is necessary for complex formation and for release of MDA from bound forms. The classic procedure of Sinnhuber et al.[11] (often referred to in the literature as "the TBA procedure") involves heating a trichloroacetic acid (TCA) extract of the sample with TBA and measuring the absorbance of the crude mixture in a spectrophotometer.

This simple procedure is subject to several sources of error. The apparent MDA content of many materials of plant and animal origin is inflated by pigments which absorb in the same region as the TBA–MDA complex. On the other hand, MDA content may be underestimated as a result of adsorption of the TBA complex onto the protein precipitate. Other aldehydes, if present, can react with TBA to produce a colored complex. Also, MDA may be formed during the procedure by oxidation of PUFA and by decomposition of oxidized lipids in the sample.

Oxidation of PUFA can be extensively reduced, if not eliminated, by adding butylated hydroxyanisole (BHA) or butylated hydroxytoluene (BHT) to the reaction mixture before processing,[14] and this precaution should be an integral part of the TBA procedure. Chelators may either inhibit or increase metal catalysis of lipid peroxidation. For example, EDTA may promote the formation of hydroxyl radicals, which catalyze lipid peroxidation, by maintaining Fe^{3+} in a soluble form, whereas desferrioxamine may inhibit their formation by blocking the reduction of Fe^{3+} to Fe^{2+}.[15] However, in a study on the MDA content of fish meal, EDTA ($\geq 25\ \mu M$) added to the TBA reaction mixture reduced the value obtained by 22% and desferrioxamine (100 μM) by 48% (H. H. Draper, M. Hadley, and A. J. Ninacs, unpublished).

Formation of MDA by decomposition of preformed lipoxides during the procedure is a more intractable problem, the significance of which depends on the nature of the sample and the purpose of the analysis (i.e., whether assessment of lipid peroxidation or of MDA specifically is the objective). In some materials, such as urine, the quantity of oxidizable

[13] E. Sawicki, T. W. Stanley, and H. Johnson, Anal. Chem. **35**, 199 (1963).
[14] J. Pikul, D. E. Leszczynski, and F. A. Kummerow, J. Agric. Food Chem. **31**, 1338 (1983).
[15] B. Halliwell and J. M. C. Gutteridge, Arch. Biochem. Biophys. **246**, 501 (1986).

lipids present is negligible, whereas in food and tissue samples they may contribute significantly to the amount of MDA determined as the TBA complex. In thermally oxidized oils there may be little relationship between the amount of MDA found using the colorimetric procedure[11] and that found using a gas chromatography–mass spectrometry method specific for MDA.[16] The main precursors of MDA generated in such oils under the conditions required for TBA–MDA complex formation are five-membered hydroperoxy epidioxides and 1,3-dihydroperoxides.[16]

The contribution of oxidized lipids in food and tissue samples to MDA determined as the TBA derivative cannot be determined using current methodology and undoubtedly is highly variable. When analyzing such samples for MDA using a TBA procedure, any interference by pigments and other TBA-reactive substances should be removed and lipid peroxidation reported in terms of MDA equivalents. For most such samples, the error arising from decomposition of oxidized lipids is likely to be much smaller than the error arising from failure to release MDA from its bound forms. Further, inclusion of MDA generated by decomposition of oxidized lipids during the TBA procedure may provide a better assessment of lipid peroxidation in the sample.

MDA can be formed from some sugars, including sucrose and 2-deoxyribose, by hydroxyl radical-generating procedures such as γ-radiolysis of water and incubation with a high concentration of ferrous ions.[9] However, these conditions are unlikely to occur in biological materials. It has been found, for example, that when 0.25 M sucrose was used to prepare mitochondria, the absorbance of the sample blank was not significantly different from that of a water blank.[17]

Interference by other TBA reactants and by pigments can be overcome by isolating the TBA–MDA complex using HPLC prior to determining its absorbance.[18] In the case of urine, in which such interference (as well as the concentration of oxidized lipids) is negligible, the amount of TBA–MDA complex found using an HPLC purification procedure[19] is similar to that found by measuring the absorbance of the crude reaction mixture at 532 nm.[11] In the case of tissue and food samples, which normally contain oxidized lipids, separation of the TBA–MDA complex usually results in lower values.[18]

The fact that aldehydic compounds other than MDA can react with TBA to form a complex that absorbs in the 532–535 nm region has led to use of the term thiobarbituric acid-reactive substances (TBA-RS), ex-

[16] E. N. Frankel and W. E. Neff, *Biochim. Biophys. Acta* **754**, 264 (1983).
[17] G. M. Siu and H. H. Draper, *Lipids* **17**, 349 (1982).
[18] R. P. Bird, S. S. O. Hung, M. Hadley, and H. H. Draper, *Anal. Biochem.* **128**, 240 (1983).
[19] H. H. Draper, L. Polensek, M. Hadley, and L. G. McGirr, *Lipids* **19**, 836 (1984).

pressed in MDA equivalents, to characterize the products of the TBA reaction quantified using the conventional spectrophotometric method.[11] Considering the frequency of references to such derivatives in the literature, there is a surprising paucity of evidence for their actual existence in biological materials. The higher MDA values for tissue samples obtained using the colorimetric TBA procedure,[11] as opposed to a procedure for free MDA, are often attributed to TBA reactions with other aldehydes, but they are more likely to be due to release of bound MDA by acid hydrolysis and/or the presence of pigments in the sample. In any event, isolation of the TBA–MDA complex should be a standard part of the TBA method for MDA determination unless it has been found to be unnecessary in the case of specific samples.

In the TCA extracts of samples high in protein, such as animal tissues, the TBA–MDA complex may adsorb onto the protein precipitate, imparting to it a pink color. This can be largely avoided by removing the protein precipitate before carrying out the TBA reaction. However, some TBA–MDA complex may be bound to peptides soluble in TCA and may be lost in procedures used to purify the complex prior to its estimation. For example, purification of the crude TBA reaction mixture on a Sep-Pak C_{18} cartridge[18] leaves a reddish residue on the cartridge after elution of the free TBA–MDA complex with methanol. Investigation of this so-called TBA-RS material has revealed that it is, in fact, TBA–MDA complex bound to residual protein in the TCA extract. Removal of this bound complex accounts in large part for the lower MDA values obtained by isolating the free TBA–MDA complex[18] and those obtained by spectrophotometric measurement of the free and bound complex in the crude reaction mixture.[11] No evidence of TBA-RS, other than TBA–MDA could be found in human blood serum, pig liver, or fish meal (H. H. Draper, M. Hadley, and A. J. Ninacs, unpublished).

MDA can be determined unambiguously as the TBA–MDA complex using HPLC, and it is therefore not appropriate to report values obtained by this procedure as TBA-RS. It should be recognized, however, that the value obtained by this method may reflect some combination of free and bound MDA originally present in the sample plus MDA formed from oxidized lipids during the procedure. Hence, the method provides a measure of lipid peroxidation in the sample in terms of MDA equivalents. Whether MDA formed by decomposition of oxidized lipids during the procedure is an "artifact" is largely a matter of perception. Since all MDA in a sample (except that of enzymatic origin) is a secondary product of lipid peroxidation, any distinction between MDA formed from oxidized lipids prior to analysis and MDA formed from oxidized lipids during analysis may be fallacious.

HPLC Procedures for Total MDA

The following procedures have been found satisfactory for the determination of total MDA in food samples, animal tissues, and urine.

HPLC Conditions. A 0.39 × 30 cm μBondapak C_{18} stainless steel analytical column attached to a 3 × 22 mm guard column packed with C_{18}/Corasil (Waters, Milford, MA) is used. The absorbance detector is equipped with a 546 nm interference filter attached to a data module for integration of peaks and printout of the elution profile. Sensitivity is set at 0.005 AUFS. The mobile phase is 18% HPLC-grade methanol in distilled water degassed by filtering through a 0.45-μm filter under vacuum with constant stirring. A flow rate of 2.0 ml/min is controlled by a two-phase solvent pump.

Standard Curve. A standard curve is prepared using TBA–MDA complex which has been checked for purity by HPLC, NMR, and elemental analysis. Instrument response is plotted against the molar equivalent weight of MDA in the complex injected. Instrument sensitivity is 1 ng of TBA–MDA complex.

Determination of MDA in Foods and Tissue Samples. The sample (0.5–1.0 g) is homogenized in 5 ml of 5% aqueous TCA plus 0.5 ml of methanolic BHT (0.5 g/liter) and heated in a capped tube for 30 min in a boiling water bath to release protein-bound MDA. To avoid adsorption of the TBA–MDA complex onto insoluble protein, any solid particulate material observed after cooling to room temperature is removed by centrifugation at 1000 *g* for 10 min. A 1-ml aliquot of the supernatant (in duplicate) plus 1 ml of a saturated solution of TBA reagent are heated in boiling water at pH 1.5 for 30 min. After cooling, an acceptable estimate of the MDA content of some samples, including urine, can be obtained by measuring the absorbance of the crude mixture at 532 nm using a spectrophotometer. Two duplicated blanks are used: a reagent blank to zero the instrument and a sample blank that is subtracted from the reading for the sample.

To separate TBA–MDA from other possible TBA–RS, the cooled sample is loaded onto a Sep-Pak C_{18} cartridge (Waters) that has been prewashed with 15 ml of methanol followed by 15 ml of distilled water. The column is developed with 4 ml of distilled water, which is discarded, then with 2 ml of methanol, which is collected in a 4-ml vial. The methanol is evaporated at 70° on a sand bath using a stream of air. The residue is dissolved in water (usually 1 ml) and an aliquot (usually 10–20 μl) is injected onto the HPLC column. Two reagent blanks in duplicate are used. Sample blanks have been found unnecessary. In the case of samples high in protein, this procedure may underestimate MDA content as a result of adsorption of protein-bound TBA–MDA complex onto the resin (see above). For such samples, the simple spectrophotometric procedure[11] may provide a better indication of lipid peroxidation in the sample. However, if a purification procedure is not employed, the reaction mixture should be scanned to determine whether there are pigments present that adsorb in the 532 nm region.

Determination of MDA in Urine. One milliliter of rat or human urine is heated at pH 3.0 ± 0.1 with 4 ml of saturated TBA reagent in a boiling water bath for 30 min. For most purposes, MDA can be satisfactorily estimated from the absorbance of TBA–MDA complex in the cooled sample at 532 nm using a spectrophotometer as described above.

To determine MDA specifically, the cooled sample is subjected to purification on a Sep-Pak cartridge in the manner described for food and tissue samples. After removing the methanol, the residue is dissolved in 2.0 ml of double-distilled water, and an aliquot (10–200 μl) is injected onto the HPLC column. The concentration of MDA in rat urine is typically about 5 times that in human urine. A sample blank and a TBA blank are carried through the procedure in duplicate.

Notes Regarding Procedures

1. The methods described have been found to be specific for MDA and to have good reproducibility. The method for food and tissue samples generally yields lower values than the colorimetric TBA procedure,[11] whereas for urine the values are generally similar.

2. Sep-Pak cartridges can be reused 3–10 times by subjecting them to the pretreatment procedure between samples.

3. In the case of samples that contain large amounts of MDA, more than 2 ml of methanol may be necessary to elute the TBA–MDA complex from the Sep-Pak. Elution of the pink complex can be monitored visually.

4. The TBA–MDA complex should be applied to the HPLC column in water. When applied in methanol, the complex often yields two peaks, one at the solvent front and one at the R_t found with water. The solvent front peak reverts to the proper R_t when it is reinjected in water.

5. The procedure for foods and tissues probably can be applied, with minor modifications, to plasma, cell fractions, *in vitro* incubation mixtures, and other tissue samples.

6. The HPLC column should be rigorously cleaned after 30–40 samples have been injected.

7. Some urine samples form a precipitate, to which the TBA–MDA complex adsorbs, during the procedure. This can be prevented by

heating the urine in a capped tube for 30 min in a boiling water bath at pH 3.0 ± 0.1 in the absence of TBA reagent, then removing the precipitate by centrifugation and using the supernatant fraction for MDA analysis.

8. Variations may be observed in the R_t of the TBA–MDA complex for the same sample. These variations usually can be eliminated by buffering the mobile phase.

9. During HPLC of some samples a "matrix effect" is observed, i.e., the TBA–MDA complex in the sample has a different R_t from that of the TBA–MDA standard. The peak for the sample can be identified by spiking with the standard, which migrates to the R_t of the complex in the sample.

10. In urinalysis for MDA, time can be saved by preparing the TBA reagent in 1.3 M phosphate buffer (pH 2.8), thereby avoiding the necessity of adjusting the pH of each sample. Otherwise, in the analysis of some samples, the pH may rise during the TBA reaction because ammonia is released by hydrolysis of urea, thereby inhibiting TBA–MDA complex formation. Complex formation decreases markedly when the pH exceeds 3.0. Four milliliters of TBA reagent is necessary for maximum complex formation. MDA does not react with the amino groups of urea under the conditions of the TBA reaction.

Gas Chromatographic Methods

Gas chromatography (GC) procedures have been developed for determining free MDA in oils and fats after its conversion to appropriate derivatives. These procedures have been developed mainly to avoid the risk of MDA generation under the hot, acidic conditions required for formation of the TBA derivative. Free MDA in photoirradiated PUFA, corn oil, and beef fat has been determined by reacting it with methylhydrazine to form 1-methylpyrazole, which was measured using a nitrogen–phosphorus-specific detector and a fused silica capillary column.[20] A similar detector has been used to determine MDA in working solutions and in urine after derivitization using 2-hydrazinobenzothiazole.[21] MDA formed by decomposition of PUFA methyl esters has been determined by GC after conversion to a stable 1,3-dioxane derivative,[22] and MDA formed by decomposition of lipoxides has been measured after conversion to a stable tetramethylacetal derivative.[16]

[20] K. Umano, K. J. Dennis, and T. Shibamoto, *Lipids* **23**, 811 (1988).

[21] M. Beljean-Leymarie and E. Bruna, *Anal. Biochem.* **173**, 174 (1988).

[22] G. Lakshminarayana and D. G. Cornwell, *Lipids* **21**, 175 (1986).

So far, these methods have been confined to the determination of free MDA in lipids. They have limited applicability to complex biological materials, in which MDA is present mainly in bound forms.

Other Methods

Fluorometric procedures for determining MDA have been reported in the older literature, but they have not found frequent use. Various compounds, including 4,4-sulfonyldianiline, ethyl p-aminobenzoate, p-aminobenzoic acid, and 4-aminoacetophenone, form fluorescent complexes with MDA. The TBA–MDA complex fluoresces with an emission maximum at 553 nm, but biological samples contain natural fluorescent compounds, as well as compounds other than MDA, that form fluorescent derivatives. Kikugawa et al.[23] have described a fluorometric method for the determination of free and acid-labile MDA in oxidized lipids as the reaction product 1,4-dimethyl-1,4-dihydropyridine-3,5-dicarbaldehyde. These investigators found that their method yielded values lower than those obtained by the measurement of absorbance at 532 nm using the crude TBA reaction mixture, but they concluded that the latter method may provide a better index of lipid oxidation.

A polarographic method for determining MDA in 2 M HCl solution, with reported applicability to biological fluids, has been proposed.[24] This procedure, which measures free MDA (and possibly MDA present in some easily hydrolyzable derivatives), has not been extensively evaluated.

---

[23] K. Kikugawa, T. Kato, and A. Iwata, *Anal. Biochem.* **174**, 512 (1988).
[24] A. M. Bond, P. P. Deprez, R. D. Jones, G. G. Wallace, and M. H. Briggs, *Anal. Chem.* **52**, 2211 (1980).

---

[44] Cyclooxygenase Initiation Assay for Hydroperoxides

By Richard J. Kulmacz, James F. Miller, Jr., Robert B. Pendleton, and William E. M. Lands

Evaluation of the pathophysiological role of hydroperoxides requires a sensitive and specific assay for these compounds. The assay described here exploits the ability of low concentrations of hydroperoxides to initiate the cyclooxygenase reaction catalyzed by prostaglandin endoperoxide synthase. Because cyclooxygenase initiation requires only about 10^{-8} M

lipid hydroperoxide or 10^{-5} M H_2O_2, the assay is quite sensitive. Monitoring the cyclooxygenase reaction requires common laboratory apparatus and a commercially available and relatively inexpensive polarographic oxygen electrode monitor.

Materials

Rose Bengal, soybean lipoxygenase, glutathione, N-ethyl maleimide, phospholipase A_2 from *Naja mocambique mocambique*, esterase from *Pseudomonas*, cholesterol linoleate, and egg yolk phosphatidylcholine were obtained from Sigma (St. Louis, MO). Glutathione peroxidase was purchased from Calbiochem (San Diego, CA) and from Sigma. Trilinolein, 9,12-octadecadienoic (linoleic) acid, and 5,8,11,14-eicosatetraenoic (arachidonic) acid were purchased from NuChek Preps, Inc. (Elysian, MN). Cholesterol [1-^{14}C]oleate, tri [1-^{14}C]olein, 1-stearoyl-2-[1-^{14}C]arachidonyl-L-3-phosphatidylcholine, and [1-^{14}C]arachidonic were obtained from Amersham (Arlington Heights, IL). [1-^{14}C]Linoleic acid and [1-^{14}C]arachidonic acid were purchased from Dupont NEN Products (Boston, MA). Solvents of HPLC grade or better were purchased from American Burdick and Jackson (Muskegon, MI). Other reagents were of analytical grade or better.

Isolation of Prostaglandin Endoperoxide Synthase

Prostaglandin endoperoxide synthase was purified as described previously.[1] Briefly, frozen ram seminal vesicles (250 g) were homogenized in a Waring blender with 350 ml of 50 mM Tris HCl (pH 8.0)/5 mM EDTA/5 mM diethyldithiocarbamate/1 mM phenol. The homogenate was centrifuged in a Sorvall GSA rotor at 8000 rpm for 10 min at 4°. The supernatant liquid was filtered through several layers of cheesecloth and centrifuged at 50,000 rpm in a Beckman Ti60 rotor for 50 min at 4°. The resulting microsomal pellet was resuspended in 200 ml of 50 mM Tris HCl (pH 8.0)/0.1 mM EDTA/1 mM phenol at 4° before the synthase was solubilized by the dropwise addition of 20 ml of 10% Tween-20 to the stirred suspension. Undissolved microsomal material was removed by centrifugation in a Beckman Ti60 rotor at 50,000 rpm for 1 hr at 4°. The solubilized synthase was concentrated by ultrafiltration to about 30 ml, and gel-filtered at 4° on a 5 × 60 cm column of AcA34 (agarose–acrylamide matrix) with 50 mM

[1] R. J. Kulmacz and W. E. M. Lands, *in* "Prostaglandins and Related Substances: A Practical Approach" (C. Benedetto, R. G. McDonald–Gibson, S. Nigram, and T. F. Slater, eds.), p. 209. IRL Press, Washington, D. C., 1987.

Tris HCl (pH 8.0)/0.1 mM EDTA/0.2 mM phenol/0.01% NaN$_3$ as the elution buffer. The active fractions were pooled, washed, and concentrated to a volume of about 5 ml in an ultrafiltration cell. This material was loaded on a horizontal preparative isoelectric focusing bed consisting of a slurry of Ultrodex (refined dextran), ampholytes in the range of pH 3–10, glycerol, octyl glucoside, and water. The apparatus was maintained at 0–2° and focusing was carried out at a power level of 6 W overnight. The gel was fractionated, and protein was eluted from the matrix with ice-cold 10% glycerol/0.1% octyl glucoside. Active fractions, centered around pH 6.5, were pooled and concentrated to about 5 ml by ultrafiltration. This material was gel-filtered on a 1.6 × 75 cm column of AcA34 at 4° with 50 mM Tris HCl (pH 8.0)/0.1 M NaCl/0.1% octyl glucoside/0.1 mM EDTA/ 0.01% NaN$_3$ as the elution buffer. Active fractions were pooled, concentrated by ultrafiltration, and mixed with 0.5 volume of ice-cold glycerol before storage at −70°. Cyclooxygenase activity was assayed as described.[1] One unit of activity gave an optimal velocity of 1 nmol O$_2$/min at 30°.

Synthesis of Hydroperoxides

Several types of hydroperoxides were used to calibrate the assay response. 15-Hydroperoxyeicosatetraenoic acid (15-HPETE) was the primary standard hydroperoxide; it was prepared from arachidonic acid using soybean lipoxygenase.[2]

Mixtures of hydroperoxide isomers from cholesterol linoleate, trilinolein, egg yolk phosphatidylcholine, linoleic acid, and arachidonic acid were synthesized using a modification of the photooxidation procedure of van Kuijk et al.[3] The starting lipid (3–12 mg) with an appropriate radiolabeled lipid tracer (6 μCi) were dissolved in 7.2 ml of methanol containing 150 mg of Rose Bengal. The solutions were placed in glass vials, sealed with Teflon-lined caps, and incubated 4–6 hr at a distance of about 5 cm from the tungsten light of a spectrophotometer. Each solution was then mixed with 4 ml of dichloromethane and 6 ml of water and shaken vigorously. The phases were separated by centrifugation (1 min at 200 g, 25°). The dichloromethane (lower) layer was removed and washed twice with 12 ml of methanol–water (1:1, v/v), and then dried over anhydrous sodium sulfate for 15 min. The resulting crude hydroperoxide solutions were stored at −20°.

[2] G. Graff, this series, Vol. 86, p. 386.
[3] F. J. G. M. van Kuijk, D. W. Thomas, R. J. Stephens, and E. A. Dratz, J. Free Rad. Biol. Med. **1**, 215 (1985).

A modification of the HPLC procedure of Porter *et al.*[4] was used to separate the positional isomers of the hydroperoxides of arachidonic acid prepared by photooxidation. The crude hydroperoxide mixture (556 nmol) was dissolved in 100 μl of the eluting solvent (99.1% hexane, 0.8% acetic acid, and 0.1% isopropyl alcohol) and applied to a 15 × 0.75 cm, 5 μm, "Resolve" silica column (Waters Chromatography, Milford, MA) with a 2 cm guard column (LC-Si silica, Supelco, Inc., Bellefonte, PA). Isocratic elution was at 1 ml/min for 30 min, followed by a linear increase in flow rate to 4 ml/min over the next 120 min. The eluate was monitored for absorbance at 237 nm (conjugated diene). The pattern of eluted peaks was essentially the same as that reported by Porter *et al.*[4] and the structural assignments were based on their results.

A portion of the hydroperoxides of lipid esters were treated to enzymatically cleave ester bonds. Mixtures of hydroperoxides of cholesterol linoleate or trilinolein were suspended in 1 ml of 0.1 M potassium phosphate (pH 7.2) containing 1.5% sodium taurocholate and 40 units of esterase. Crude phosphatidylcholine hydroperoxide was dissolved in 1 ml of 0.1 M potassium phosphate (pH 7.2) containing 10 mM calcium chloride and 12 units phospholipase A_2. These mixtures were incubated 20 min at 25°, after which the lipids were extracted by the method of Bligh and Dyer.[5] An aliquot of each lipid extract was analyzed by TLC on silica in hexane–diethyl ether–acetic acid (80:20:1, v/v) at room temperature with no paper curtains in the tank. The location of radiolabeled lipids was determined with a radioscanner (Model 400, Bioscan, Washington, D.C.); the integrated area under the peaks determined the relative amount of radioactivity in each band and thus the extent of hydrolysis of each lipid.

Iodometric Assay for Hydroperoxide

The hydroperoxide concentrations of the solutions of esterified and nonesterified lipid hydroperoxides were determined using a modification of the iodometric assay for peroxide of Buege and Aust[6] as follows. Acetic acid–chloroform (3:2, v/v) and water were chilled on ice and purged with N_2 gas. The tube containing the acetic acid–chloroform was then sealed and warmed to room temperature. The nitrogen-saturated water (5 ml) was mixed with 6 g of KI to prepare the stock solution of KI that was kept on ice. Potassium iodate (KIO_3) was used as the primary standard. Aliquots of hydroperoxide samples and of KIO_3 standard (0.1–10 nmol)

[4] N. A. Porter, J. Logan, and V. Kontoyiannidou, *J. Org. Chem.* **44**, 3177 (1979).
[5] E. G. Bligh and W. J. Dyer, *Can. J. Biochem. Physiol.* **37**, 911 (1959).
[6] J. A. Buege and S. D. Aust, this series, Vol. 52, p. 302.

were evaporated to dryness under N_2 in glass test tubes and redissolved in 50 μl of ethanol. Under reducing lighting, 0.35 ml of the acetic acid–chloroform and 15 μl of the KI solutions were added to the tubes and mixed. The tubes were incubated for 3 min at room temperature before the reaction was stopped by mixing in 1 ml of cadmium chloride (5 g/liter in water). The phases were separated by centrifugation at 170 g for 5 min at room temperature and the absorbance of the aqueous phase at 353 nm was measured. For the standard curve, A_{353} was plotted against the amount of KIO_3 added, and a linear least-squares regression line calculated. This equation was then used to determine the peroxide content of the individual samples, taking into account the fact that KIO_3 reacts with I^- to produce 3 mol of the chromophore triiodide (I_3^-), whereas hydroperoxide produces only one mol of I_3^-.

Assay of Hydroperoxide by Activation of the Cyclooxygenase Activity of Prostaglandin Endoperoxide Synthase

Acceleration of the cyclooxygenase reaction is very sensitive to hydroperoxide; therefore, it is essential that the arachidonic acid used as substrate be free of hydroperoxide contamination. A stock solution of peroxide-free arachidonic acid was prepared by treatment with $NaBH_4$ as follows. One gram of arachidonic acid was dissolved in 3 ml of toluene, and 50 mg of $NaBH_4$ was added. The mixture was incubated for 1 hr at room temperature with agitation every 5 min. To quench the reaction, 9 ml of water was added, followed by 3 ml of 1 M citric acid added dropwise with constant mixing. The two phases were separated by centrifugation at 200 g for 5 min at room temperature, and the toluene (lower) layer was washed 4 times with 3 ml of water. Butylated hydroxytoluene; (0.5 μmol) was added as an antioxidant, and the solution was dried over anhydrous sodium sulfate for 1 hr at 0° and stored over anhydrous sodium sulfate in an aluminum foil-wrapped tube at $-20°$.

Each day, a working solution of arachidonic acid was prepared by evaporating the toluene from 15 μmol of arachidonic acid with a stream of N_2 gas. The fatty acid was dissolved in 60 μl of ethanol and mixed with 14.6 ml of 0.1 M Tris HCl (pH 8.5), 150 μl of 100 mM glutathione, and 100 μl of 0.2 unit/ul glutathione peroxidase. The solution was shielded from light with aluminum foil and allowed to incubate for 5 min at 25°. It was then stored on ice until used in the cyclooxygenase reaction.

A clean, stirred oxygen electrode cuvette, thermostatted at 25°, was used for each assay reaction. To each cuvette, 2.42 ml of 0.1 M potassium phosphate (pH 7.2) was added. An automatic dispenser (Model 450, Tecan US Ltd., Chapel Hill, NC) was then used to add 30 μl of 100 mM

phenol in 0.1 M potassium phosphate (pH 7.2), 30 μl of 1.0 M sodium cyanide, 70 μl of 100 mM N-ethyl maleimide (NEM), 150 μl of ethanol, and 300 μl of the peroxidase-treated aqueous arachidonic acid solution. The 23-fold molar excess of NEM reacts with the glutathione present in the aqueous arachidonate solution, preventing further glutathione peroxidase activity. The mixture was allowed to stir in the cuvette for about 15 sec; then 0–300 pmol of 15-HPETE standard was added in 50 μl of ethanol. The incubation was continued for 2 min to allow temperature equilibration before the oxygen electrode probe was placed in the cuvette, and then the oxygen concentration was monitored by a Yellow Springs Instruments Model 53 oxygen monitor (Yellow Springs Instruments Co., Inc., Yellow Springs, OH) until a stable baseline was achieved (about 2 min).

The assay reaction was initiated by the injection of 140 units of the synthase holoenzyme in 9 μl of 30% glycerol/20 mM potassium phosphate (pH 7.2). A 1 μl "plug" of 30% glycerol was loaded into the syringe before the enzyme to completely flush the enzyme from the syringe upon injection. It is essential that the amount of enzyme delivered to each cuvette be precise because the rate and amount of oxygen consumed is directly proportional to the amount of enzyme added. The output of the oxygen monitor was recorded on a strip chart recorder. For each reaction, the time required to reach optimal cyclooxygenase velocity (lag time) was determined from the strip chart tracing[7] and plotted as a function of the amount of 15-HPETE added (see Fig. 1). The lag time approximated a linear function of the amount of 15-HPETE,[8] so a linear least-squares regression routine was used to calculate the standard curve. A separate standard curve was obtained each day.

Individual samples were substituted for the ethanol and/or some of the buffer in the cuvette, maintaining a final volume of 3.0 ml containing 200 μl of ethanol. The lag time for each individual sample was measured in the manner described for the 15-HPETE standards, and the standard curve for 15-HPETE was then used to calculate the amount of hydroperoxide in each sample. Samples were typically assayed in triplicate or quadruplicate. This method has been used to measure the amount of hydroperoxide in the supernatant liquid from cells in culture[9] and in plasma after the precipitation of protein with 2 volumes of ethanol.[10]

After each reaction, the electrode probe was washed with 70% ethanol, and then water, to prevent the transfer of any enzyme or hydroperoxide into subsequent cuvettes. The needle of the syringe used to inject the

[7] P. J. Marshall, M. A. Warso, and W. E. M. Lands, *Anal. Biochem.* **145**, 192 (1985).
[8] J. F. Miller Jr., Ph. D. Thesis, Univ. of Illinois at Chicago (1990).
[9] P. J. Marshall and W. E. M. Lands, *J. Lab. Clin. Med.* **108**, 525 (1986).
[10] M. A. Warso and W. E. M. Lands, *J. Clin. Invest.* **75**, 667 (1985).

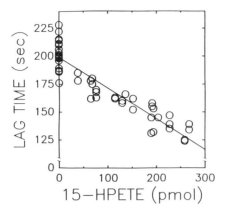

FIG. 1. Standard cuve for the initiation of the cyclooxygenase by 15-HPETE. The line was fitted to the points by a least squares routine. The slope of the fitted line was -274 ± 16 sec/nmol (from Ref. 8).

enzyme was also rinsed with ethanol, and the syringe barrel was flushed several times with 30% glycerol between injections.

The cyanide in the assay cocktail has been found to alter the response characteristics of the oxygen electrode, with a stabilized response time obtained after 3–5 reactions have been performed. Thus, we routinely do about six reactions each day to precondition the electrode.

The results obtained with several different positional isomers of eicosatetraenoic acid hydroperoxide purified by HPLC showed that their effectiveness as activators declined almost linearly with the distance of the hydroperoxide from the methyl end of the fatty acid. The least effective preparation, a mixture of 5-HPETE and 6-HPETE, was only 15% as effective as 15-HPETE in activating cyclooxygenase.[11] This result emphasizes the sensitivity of the activation process to the structure of the added hydroperoxide. An extreme case of this is the several hundred-fold lower response of the cyclooxygenase to hydrogen peroxide relative to 15-HPETE.[10] The sensitivity of the assay to the structure of the hydroperoxide also means that if the structure of the hydroperoxide in a sample is not known, the assay results will only be in terms of an equivalent of the standard, 15-HPETE, instead of a molar concentration of the hydroperoxide. The choice of an appropriate standard for the expected hydroperoxide is thus important.

Results obtained with hydroperoxides of esterified lipids, and of these lipids after esterase treatment, showed that esterified lipid hydroperox-

[11] R. B. Pendleton, Ph. D. Thesis, Univ. of Illinois (1990).

ides are very poor activators of the cyclooxygenase in contradiction to a preliminary observation reported earlier.[7] For example, the esterified hydroperoxide in triglyceride and phosphatidylcholine were 1–4% as effective as the nonesterified 15-HPETE in activating cyclooxygenase. However, after nearly complete enzymatic hydrolysis (determined by TLC analysis), the preparations were 72–76% as effective as 15-HPETE.[11] Any sample likely to contain esterified material should thus be treated with esterases before assay with the cyclooxygenase method, so as to obtain a more valid assessment of the total hydroperoxide content.

We find the assay described above useful in quantitating fatty acid hydroperoxide in amounts ranging from 50–200 pmol (~100–300 μl sample per assay). However, materials other than hydroperoxide may activate cyclooxygenase (e.g., ethanol, acetone); added controls are valuable. Treating samples with glutathione peroxidase selectively reduces the activating fatty acid hydroperoxide and permits measurement of any nonhydroperoxide activator. Comparing the amounts of glutathione peroxidase-reducible and glutathione peroxidase-resistant activator is essential in determining how much of the activator in a sample is hydroperoxide.

Acknowledgments

This work was supported by grants GM30509 and HL34422 from the United States Public Health Service.

[45] Cholesterol Epoxides: Formation and Measurement

By G. A. S. Ansari and Leland L. Smith

The isomeric cholesterol 5,6-epoxides 5,6α-epoxy-5α-cholestan-3β-ol (cholesterol α-oxide, **1**) and 5,6β-epoxy-5β-cholestan-3β-ol (cholesterol β-oxide, **2**) are found in human tissues and foods and on bioassay exhibit diverse toxic effects *in vitro*.[1,2] Both are formed together from cholesterol (cholest-5-en-3β-ol, **3**) by many defined oxidants, including air oxidation,[1,3] and by the actions of soybean lipoxygenases and of liver microsomal lipid peroxidation systems *in vitro*.[4] Both are formed by an un-

[1] L. L. Smith, "Cholesterol Autoxidation." Plenum, New York, 1981.
[2] L. L. Smith, *Chem. Phys. Lipids* **44**, 87 (1987).
[3] J. Gumulka, J. S. Pyrek, and L. L. Smith, *Lipids* **17**, 197 (1982).
[4] L. Aringer and P. Eneroth, *J. Lipid Res.* **15**, 389 (1974).

characterized *in vivo* epoxidation system in mammals,[5] and the $5\alpha,6\alpha$-epoxide (**1**) is formed by a liver microsomal enzyme *in vitro*.[6]

Of all the recognized oxidation products of cholesterol formed by attack on cholesterol of defined active oxygen species, the 5,6-epoxides (**1** and **2**) are the most broadly formed, being formed by ground state dioxygen ($^3O_2$), electronically excited (singlet) dioxygen ($^1O_2$), peroxide (O_2^{2+}), ozone (O_3), dioxygen cation (O_2^+), and hydroxyl radical (HO $\cdot$). The ratio of product 5,6-epoxides (**1:2**) favors the $5\beta,6\beta$-epoxide (**2**) in most cases, the $5\alpha,6\alpha$-epoxide (**1**) predominating only in oxidations involving the specific $5\alpha,6\alpha$-epoxidase, HO $\cdot$, and peracids.[3] In new experimental systems generating the 5,6-epoxides it is of importance that the products be properly identified and the 5,6-epoxide ratio (**1:2**) ascertained. The $5\alpha,6\alpha$-epoxide (**1**) alone has been posed as the 5,6-epoxide formed in biological systems and as the active agent in numerous bioassays but without adequate supporting evidence.

Sources

The $5\alpha,6\alpha$-epoxide (**1**) is commercially available from several sources, including Sigma Chemical Co. (St. Louis, MO), Research Plus Inc. (Bayonne, NJ), and Steraloids Inc. (Wilton, NH); the $5\beta,6\beta$-epoxide (**2**) is also available from Research Plus Inc. The $5\alpha,6\alpha$-epoxide (**1**) is readily prepared from cholesterol by peracid oxidation according to simple directions.[7] Synthesis of the $5\beta,6\beta$-epoxide (**2**) from cholesterol[8] and from 5α-cholestane-$3\beta,5,6\beta$-triol triacetate[9] also provides ready access.

The limited commercial availability of the $5\beta,6\beta$-epoxide (**2**) is probably responsible for the lack of attention given this isomer in prior studies. Commercial and synthetic samples may contain significant amounts of the isomeric 5,6-epoxide, and the identity and purity of all reference samples

[5] A. Sevanian, J. F. Mead, and R. A. Stein, *Lipids* **14,** 634 (1979).
[6] T. Watabe and T. Sawahata, *J. Biol. Chem.* **254,** 3854 (1979).
[7] L. F. Fieser and M. Fieser, "Reagents for Organic Synthesis," Vol. 1, p. 136. Wiley, New York, 1967.
[8] S. G. Levine and M. E. Wall, *J. Am. Chem. Soc.* **81,** 2826 (1959).
[9] A. T. Rowland and H. R. Nace, *J. Am. Chem. Soc.* **82,** 2833 (1960).

should be determined by appropriate chromatography and ready reliable means, generally including melting point, optical rotation, and proton nuclear magnetic resonance (NMR) spectra. The literature gives data in ranges: 5α,6α-epoxide (1), mp 139–148°, $[\alpha]_D$ −40 to −48.5°, δ_H 2.90 (d, J = 3.3–4.1 Hz, 6β-proton); 5β,6β-epoxide (2), mp 130–136°, $[\alpha]_D$ +8 to +11.5°, δ_H 3.05 (d, J = 2.1–3.0 Hz, 6α-proton).[1] The isomers exhibit minor differences in infrared and mass spectra that are useful for confirming the identity of pure samples, but they are not useful for analysis of mixtures.[10–12] Characterization of the 5,6-epoxides as the 3β-acetate, 3β-benzoate, or 3β-trimethylsilyl ethers should also be considered in special cases.

Recovery of 5,6-Epoxides

Typically, total lipids from tissues, biological fluids, *in vitro* incubations, etc. may be extracted with chloroform–methanol (2:1, v/v) using the classic methods of Folch *et al.*[13] or Bligh and Dyer[14] or with hexane–2-propanol (3:2, v/v) according to Hara and Radin.[15] For foods and tissues containing phospholipids, preliminary removal of phospholipids from neutral lipids is necessary; for this purpose chromatography on silica gel using hexane–diethyl ether (9:1, v/v) may be used to elute hydrocarbons, triacylglycerols, and other neutral lipid esters, followed by hexane–diethyl ether (5:3, v/v) to elute sterols and oxysterols. Polar phospholipids are retained on the column. For oxysterols recovered from *in vitro* incubations in which massive amounts of triacylglycerols and phospholipids are not involved, direct thin-layer chromatography (TLC) of the recovered total lipid fraction may provide satisfactorily resolved 5,6-epoxide preparations for analysis.

The column fractionation should be monitored by TLC even though TLC does not resolve the isomeric 5,6-epoxides. Silica gel-coated chromatoplates irrigated with benzene–ethyl acetate (3:1, v/v) in three ascending irrigations or with cyclohexane–diethyl ether (9:1, v/v) resolve the mixed 5,6-epoxides from other oxidized cholesterol derivatives. The 5,6-epoxides are detected with 50% aqueous sulfuric acid as spray, with heating to color development (yellow to tan to brown) and to charring.

In some cases these methods may not resolve the 5,6-epoxides from other common companion oxysterols such as 3β-hydroxycholest-5-en-7-one (7-ketocholesterol) or the cholesterol 7-hydroperoxides. The pres-

[10] E. Chicoye, W. D. Powrie, and O. Fennema, *Lipids* **3**, 335 (1968).

[11] L. S. Tsai and C. A. Hudson, *J. Food Sci.* **49**, 1245 (1984).

[12] J. Nourooz-Zadeh and L.-A. Appleqvist, *J. Food Sci.* **52**, 57 (1987).

[13] J. Folch, M. Lees, and G. H. Sloane-Stanley, *J. Biol. Chem.* **226**, 497 (1957).

[14] E. G. Bligh and W. J. Dyer, *Can. J. Biochem. Physiol.* **31**, 911 (1959).

[15] A. Hara and S. N. Radin, *Anal. Biochem.* **90**, 420 (1978).

ence of the 7-ketone is revealed by its UV-absorbing properties on chromatoplates containing an UV-fluorescing phosphor; hydroperoxide contamination is revealed by N,N-dimethyl-p-phenylenediamine spray applied before the sulfuric acid spray.[16] The 7-ketone and sterol hydroperoxide contaminations do not interfere in the subsequent HPLC analysis of 5,6-epoxides.

As the 5,6-epoxides are subject to hydration, for some studies it is expedient to include their common hydration product, 5α-cholestane-3β,5,6β-triol, among analytes examined. Moreover, the 5β,6β-epoxide (2) is less stable than the 5α,6α-epoxide (1) toward other decomposition, and care must be exercised to avoid alteration of the 5,6-epoxides during recovery and analysis.

Analysis of 5,6-Epoxides

Three major means have been successfully used for analysis of mixtures of the 5,6-epoxides: high-performance liquid chromatography (HPLC), proton NMR spectroscopy, and chemical reduction to cholestanediols.

Chromatography. The 5,6-epoxides are not resolved by thin-layer or packed-column gas chromatography, but their 3β-acetates and 3β-trimethylsilyl ethers may be.[3,4,6,10] Generally the 5α,6α-epoxide (2) is more mobile. Preferred operations are conducted with HPLC and gas chromatography using fused silica capillary columns. Adsorption (silica) HPLC columns with binary solvents, e.g., hexane–2-propanol (24:1, 49:1, or 100:3, all v/v), used isocratically[3,11,17,18] or in a curvilinear gradient[19] are very suitable, the 5α,6α-epoxide (1) being the more mobile. Resolution is also achieved with reversed-phase partition columns using acetonitrile–water (9:1, v/v) or methanol–water (9:1, v/v), the 5β,6β-epoxide (2) being the more mobile.[17] Detection of the 5,6-epoxides in column effluent by UV absorption is not suitable, but detection by differential refractive index or hydrogen flame ionization monitoring is effective.

5,6-Epoxide 3β-benzoates and 3β-p-nitrobenzoates are also resolved by HPLC using adsorption columns with isooctane–2-propanol (499:1, v/v) or reversed-phase columns with acetonitrile–water (19:1, v/v) or methanol–water (9:1, v/v) isocratically[17] and with linear gradients of acetonitrile–2-propanol.[20] Esters of the 5β,6β-epoxide (2) are the more mobile and are readily detected by UV absorption.

Gas chromatographic resolution of the isomeric 5,6-epoxides and their

[16] L. L. Smith and F. L. Hill, *J. Chromatogr.* **66**, 101 (1972).

[17] G. A. S. Ansari and L. L. Smith, *J. Chromatogr.* **175**, 307 (1979).

[18] L. S. Tsai, K. Ichi, C. A. Hudson, and J. J. Meehan, *Lipids* **15**, 124 (1980).

[19] G. Maerker, E. H. Nungesser, and I. M. Zulak, *J. Agric. Food Chem.* **36**, 61 (1988).

[20] K. Sugino, J. Terao, H. Murakami, and S. Matsushita, *J. Agric Food Chem.* **34**, 36 (1986).

3β-trimethylsilyl ethers is conveniently achieved using fused silica capillary columns coated with silicone polymers such as SE-30, SE-54, DB-1, or DB-5, with gas effluent monitoring by the flame ionization method. The $5\beta,6\beta$-epoxide (2) and its 3β-trimethylsilyl ether are the more mobile.[3,21–24] Detailed protocols using these various methods for recovery and analysis of the 5,6-epoxides from human body fluids[24,25] and cholesterol-rich foods are available.[11,12,19,21,26,27]

Proton NMR Spectroscopy. The proton NMR signals of the 6-proton of the isomeric 5,6-epoxides are significantly distinct from one another so as to constitute a ready means of analysis of mixtures. The 6β-proton signal (in deuteriochloroform) of the $5\alpha,6\alpha$-epoxide (1) appears as a doublet at δ_H 2.90 ($J = 3.3–4.1$ Hz); the 6α-proton of the isomeric $5\beta,6\beta$-epoxide (2) is also a doublet found at δ_H 3.05 ($J = 2.1–3.0$ Hz). These key signals serve to identify the individual 5,6-epoxides, and integration of the signals affords ready estimation of relative amounts of each component. Except where the key signals are obscured by extraneous signals from other components of the sample, the proton NMR method is preferred for 5,6-epoxide estimation.

Chemical Reduction Analysis. In the absence of an adequate HPLC or gas chromatography apparatus and NMR spectrometer, chemical reduction analysis may be used. The chemical reduction of mixtures of isomeric 5,6-epoxides yields mixtures of reduction products resolved by TLC with benzene–ethyl acetate (3:2, v/v). The $5\alpha,6\alpha$-epoxide (1) yields 5α-cholestane-$3\beta,5$-diol (4, R_f 0.30) as a single product; the isomeric $5\beta,6\beta$-epoxide (2) yields 5α-cholestane-$3\beta,6\beta$-diol (5, R_f 0.19) and 5β-cholestane-$3\beta,5$-diol (6, R_f 0.53). All three diols give a red color with 50% sulfuric acid, and

4　　　　　　5　　　　　　6

[21] G. Maerker and J. Unruh, *J. Am. Oil Chem. Soc.* **63,** 767 (1986).

[22] S. R. Missler, B. A. Wasilchuk, and C. Merritt, *J. Food Sci.* **50,** 595 (1985).

[23] S. W. Park and P. B. Addis, *J. Agric. Food Chem.* **34,** 653 (1986).

[24] L. D. Gruenke, J. C. Craig, N. L. Petrakis, and M. B. Lyon, *Biomed. Environ. Mass Spectrom.* **14,** 335 (1987).

[25] I. Björkhem, O. Breuer, B. Angelin, and S.-Å Wikström, *J. Lipid Res.* **29,** 1031 (1988).

[26] L. S. Tsai and C. A. Hudson, *J. Food Sci.* **49,** 1245 (1984).

[27] S. W. Park and P. B. Addis, *J. Food Sci.* **42,** 1500 (1987).

following heating to char the components the amounts of each may be estimated using densitometric scanning of the chromatoplate, eluted and weighed,[28] or estimated by gas chromatography.[5]

[28] L. L. Smith and M. J. Kulig, *Cancer Biochem. Biophys.* **1,** 79 (1975).

[46] Oxystat Technique in Study of Reactive Oxygen Species

By HERBERT DE GROOT

Introduction

There are two principal approaches to maintain steady-state oxygen partial pressures (P_{O_2}) in suspensions of respiring biological material. One approach is to deliver O_2 via diffusion from a gas phase to the liquid phase. The other is to infuse O_2-saturated aqueous medium into the suspension of respiring biological material. Based on the second principle we have developed an oxystat technique where O_2 supply is maintained by injecting O_2-saturated aqueous medium using a computer-supported feedback control system (Fig. 1).[1,2] Major advantages of this technique are (1) it is capable of maintaining steady-state P_{O_2} at very low levels, (2) it rapidly responds to alterations in the O_2 uptake rate, and (3) it allows calculations of O_2 uptake from the amounts of O_2-saturated medium added.

Oxystat Technique

Apparatus

The oxystat system (Fig. 1) consists of an incubation chamber, an O_2 sensor, a pump, and a computer. The water-jacketed incubation chamber is made of plexiglass. It is equipped with a magnetic stirrer and a port for the O_2 sensor. At the top two stainless steel needles serve as entrances. The inlet for the O_2-saturated infusion medium ends directly above the magnetic stirring bar. The second needle, which does not project into the

[1] T. Noll, H. de Groot, and P. Wissemann, *Biochem. J.* **236,** 765 (1986).
[2] H. de Groot, A. Littauer, D. Hugo-Wissemann, P. Wissemann, and T. Noll, *Arch. Biochem. Biophys.* **264,** 591 (1988).

METHODS IN ENZYMOLOGY, VOL. 186

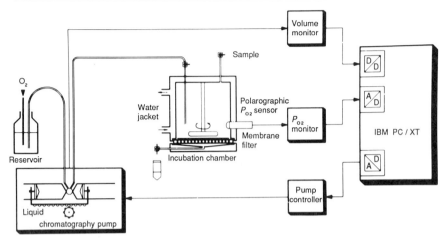

FIG. 1. Diagram of the oxystat system.[2]

chamber, is used as an injection port for the addition of reagents and for taking samples of the incubation mixture. The bottom of the incubation chamber is equipped with a membrane holder covered by a filter membrane (Schleicher & Schüll, Dassel, West Germany; pore size 6–12 μm). The actual P_{O_2} in the incubation chamber is determined by a polarographic O_2 sensor (MT-1-AC, Eschweiler, Kiel, West Germany).

A liquid chromatography pump is used to add the O_2-saturated infusion medium to the incubation mixture. The infusion medium is taken from a reservoir equilibrated with either 100% O_2 or 95% O_2/5% CO_2, depending on the incubation buffer used. The computer (IBM PC/XT) is equipped with a 12-bit analog-to-digital converter to read the P_{O_2} signal, a 12-bit digital-to-analog converter to activate the pump, and a 24-bit digital input to read the volume of O_2-saturated medium infused. The computer reads the O_2 sensor signal 3 times per second, compares it with a preselected value, and activates the pump to deliver O_2-saturated medium to the incubation chamber when required. Programs providing process control (PI-control) and processing of data were developed in Pascal.

Calculations

Oxygen uptake is continuously calculated by the computer from the amounts of O_2-saturated medium added by using the following equation:

$$NO_2(t) = ([O_2]_b - [O_2]_i)V(t)/V_iC_{res}$$

where $NO_2(t)$ is the specific O_2 uptake of the respiring biological material (nmol mg protein^{-1} or nmol 10^6 cells^{-1}), $[O_2]_b$ is the O_2 concentration of the infusion medium (nmol ml^{-1}), $[O_2]_i$ is the O_2 concentration in the incubation mixture (nmol ml^{-1}), V_i is the volume of the incubation chamber (ml), $V(t)$ is the amount of O_2-saturated infusion medium added (ml), and C_{res} is the concentration of the O_2-consuming biological particles (mg protein ml^{-1} or 10^6 cells ml^{-1}). $[O_2]_b$ and $[O_2]_i$ are calculated from the P_{O_2} assuming that 1 mmHg equals 1.4 μM O_2 at 37° or 1.7 μM O_2 at 20°.[3] During incubations of isolated cells C_{res} is constant since the filter membrane prevents dilution of the cells. However, subcellular particles, such as isolated mitochondria and microsomes, easily pass through the filter. Thus, in these cases the biological material is continuously diluted, and C_{res} is a function of time $[C_{res}(t)]$. $C_{res}(t)$ can be calculated as follows:

$$C_{res}(t) = C_{res}(0) \, e^{-[V(t)/V_i]}$$

where $C_{res}(0)$ is the concentration of the biological material at time zero of the incubation.

Comments

With isolated cells, incubation times up to days may be feasible. In experiments with subcellular fractions the maximum incubation time is limited to about 30 min because of the continuous dilution of biological material (see above). The polarographic O_2 sensor is the factor which determines the lower limiting P_{O_2} of 0.1 mmHg of the operation range of the oxystat system. By using an alternative O_2 sensor, namely, the O_2-dependent luminescence of the photobacterium *Vibrio fischeri*,[4] the operating range of the oxystat system can be extended to 0.01 mmHg.[1] The upper value of about 300 mmHg for the operating range of the oxystat system is given by the fact that the O_2-enriched medium can be saturated with O_2 only up to 1.2 mM (20°). When such a medium is used to maintain steady-state P_{O_2} above 300 mmHg (0.42 mM O_2, 37°), the difference between the P_{O_2} of the O_2-saturated medium and the P_{O_2} of the incubation mixture, that is, the effective O_2 concentration added, becomes increasingly smaller. Therefore, at such high P_{O_2} values O_2 should not be added physically dissolved but bound to O_2 carriers or in the form of H_2O_2. The latter would require the presence of catalase in the incubation mixture.

Application of Oxystat Technique to Study of Reactive Oxygen Species

The oxystat system allows samples to be collected from the filtrate as well as directly from the incubation chamber (Fig. 1). Thus, the formation of products of oxidative reactions in biological material, such as malon-

---

[3] H. Forstner and E. Gnaiger, in "Polarographic Oxygen Sensors (E. Gnaiger and H. Forstner, eds.), p. 321. Springer-Verlag, Berlin, 1983.

[4] D. Lloyd, K. James, J. Williams, and N. Williams, *Anal. Biochem.* **116**, 17 (1981).

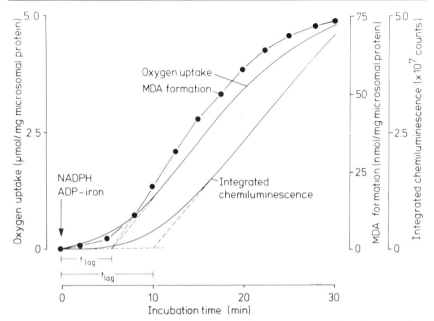

FIG. 2. Time courses of oxygen uptake, malondialdehyde (MDA) formation, and low-level chemiluminescence during NADPH/ADP–iron-induced lipid peroxidation in rat liver microsomes at a steady-state P_{O_2} of 30 mmHg.[6]

dialdehyde,[5] can easily be determined during the course of an incubation (Fig. 2).[6] Furthermore, by putting the incubation chamber directly in front of the photomultiplier of a single-photon counting apparatus, low-level chemiluminescence emitted by electronically excited species such as singlet molecular oxygen ($^1O_2$) and triplet carbonyls (^3RO) can be detected.[7] Finally, the stimulation of O_2 uptake may serve as an additional indicator for the formation of reactive oxygen species.

We have used these methods in the study of ADP–iron- and haloalkane-induced lipid peroxidation in isolated rat liver microsomes and isolated rat liver cells under steady-state O_2 conditions. As an example, the time course of ADP–iron-induced lipid peroxidation in microsomes at a P_{O_2} of 30 mmHg is depicted in Fig. 2. Oxygen uptake and malondialdehyde formation increased simultaneously after a lag time (t_{lag}) of 5 min, whereas low-level chemiluminescence started to rise significantly later.[6] Varying the steady-state P_{O_2} revealed that the rate of O_2 uptake during the lag phase of ADP–iron-induced lipid peroxidation became dependent on

[5] H. Esterbauer and K. H. Cheeseman, this volume [42].

[6] T. Noll, H. de Groot, and H. Sies, Arch. Biochem. Biophys. **252,** 284 (1987).

[7] E. Cadenas and H. Sies, this series, Vol. 105, p. 221.

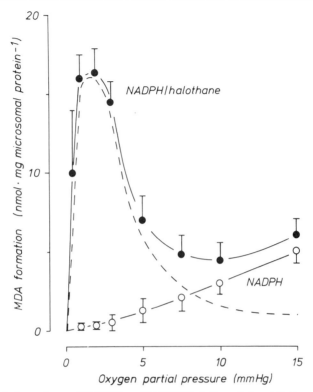

FIG. 3. O_2 dependence of NADPH- and NADPH/halothane-induced lipid peroxidation in rat liver microsomes. NADPH-reduced rat liver microsomes from phenobarbital-pretreated rats were incubated in the absence or presence of halothane under varyting state-state P_{O_2} levels for 30 min.[9] Lipid peroxidation is expressed as malondialdehyde (MDA) formation. The dashed line represents the difference between NADPH- and NADPH/halothane-induced formation of malondialdehyde.

O_2 at P_{O_2} values below 200 mmHg, whereas the rate of O_2 uptake during the propagation phase of lipid peroxidation became limited by O_2 only at P_{O_2} levels below 30 mmHg.[8]

In contrast, haloalkane-induced lipid peroxidation, which occurred at a maximum rate just after addition of the haloalkane,[6] showed a distinct optimum at P_{O_2} values between 1 and 10 mmHg, as exemplified for halothane ($CF_3CHBrCl$) in Fig. 3.[2,8–11] Both examples demonstrate that the actual P_{O_2} can be of major importance for the study of reactive O_2 species.

[8] H. de Groot and T. Noll, *Chem. Phys. Lipids* **44,** 209 (1987).
[9] H. de Groot and T. Noll, *Biochem. Biophys. Res. Commun.* **119,** 139 (1984).
[10] T. Noll and H. de Groot, *Biochim. Biophys. Acta* **795,** 356 (1984).
[11] H. de Groot and T. Noll, *Biochem. Pharmacol.* **35,** 15 (1986).

This conclusion is especially true in view of the fact that the physiological P_{O_2} in liver ranges from 1 to 60 mmHg.[12] Further instances of the importance of the actual P_{O_2} in the formation of reactive O_2 species are Adriamycin-induced lipid peroxidation, where recently an O_2 dependence of lipid peroxidation similar to that described for haloalkane-dependent lipid peroxidation has been reported,[13] and the formation of superoxide anion radical (O_2^{-}) and hydrogen peroxide (H_2O_2) by redox cycling of menadione, where at physiological P_{O_2} the electron transfer from the reduced forms of this quinoid compound to the mitochondrial respiratory chain appears to dominate over the reduction of molecular oxygen.[14]

Acknowledgments

Stimulating discussions with Prof. H. Sies are gratefully acknowledged. Research was supported by Ministerium für Wissenschaft und Forschung des Landes Nordrhein-Westfalen.

[12] M. Kessler, J. Höper, D. K. Harrison, K. Skolasinska, W. P. Klövekorn, F. Sebening, H. J. Volkholz, I. Beier, C. Kernbach, V. Rettig, and H. Richter, *in* "Oxygen Transport to Tissue" (D. W. Lübbers, H. Acker, E. Leniger-Follert, and T. K. Goldstick, eds.), Vol. 5, p. 69. Plenum, New York, 1984.
[13] G. F. Vile and C. C. Winterbourn, *Biochem. Pharmacol.* **37**, 2893 (1988).
[14] H. de Groot, T. Noll, and H. Sies, *Arch. Biochem. Biophys.* **243**, 556 (1985).

[47] Phospholipid Hydroperoxide Glutathione Peroxidase

By MATILDE MAIORINO, CARLO GREGOLIN, and FULVIO URSINI

Introduction

Phospholipid hydroperoxide glutathione peroxidase (PHGPX) is the second selenoenzyme discovered in mammals.[1] The enzyme was first identified and purified by following its antiperoxidant activity, and for this reason it was originally named peroxidation-inhibiting protein (PIP).[2] This enzyme was purified to homogeneity from pig liver,[2] heart,[3] and brain[4] and rat liver,[2] and it has been identified in all tissues in which it was

[1] F. Ursini, M. Maiorino, and C. Gregolin, *Biochim. Biophys. Acta* **839**, 62 (1985).
[2] F. Ursini, M. Maiorino, M. Valente, and C. Gregolin, *Biochim. Biophys. Acta* **710**, 197 (1982).
[3] M. Maiorino, F. Ursini, M. Leonelli, N. Finato, and C. Gregolin, *Biochem. Int.* **5**, 575 (1982).

searched for namely, rat kidney, heart, lung, muscle, and brain,[5,6] dog liver brain and kidney, bull spermatozoa, and fish liver (unpublished). A high activity has been recently observed in rat testis (unpublished). The chromatographic behavior of the protein and purification procedures were identical for all tissues from which it was purified.

Because PHGPX is a soluble enzyme, which has been purified from cell sap it can be classified as a cytosolic enzyme. However, a substantial activity, which can be partially recovered by high ionic strength extraction,[5] is present in membranes of subcellular organelles. Based on present knowledge, therefore, this enzyme must be regarded as a cytosolic enzyme that is active on membranes to which it is bound to some extent. Immunohistochemical localization of the enzyme will clarify this issue.

(PHGPX) is a monomeric enzyme (MW ~20,000 on SDS–PAGE) that contains one selenium atom at the active site. The selenoprotein nature of PHGPX was assessed by proton-induced X-ray fluorescence of the purified protein. The presence of a selenocysteinwas first suggested by inhibition kinetics in the presence of iodoacetate and thiols, and it was subsequently demonstrated by amino acid analysis following carboxymethylation and acidic digestion (L. Flohé, personal communication).

PHGPX reduces the hydroperoxy derivatives of phospholipids to alcohol derivatives, and this activity is absent both in Se-dependent tetrameric glutathione peroxidase and in glutathione transferase B (the so-called Se-independent glutathione peroxidase). PHGPX is therefore unique in that it reduces the hydroperoxides of the major components of membranes. The reduction of phospholipid hydroperoxides was demonstrated by HPLC and mass spectroscopic analysis of intact phospholipid substrate and product[7] and by HPLC analysis of the fatty acid derivatives released by enzymatic hydrolysis.[8]

The specificity for the peroxide substrate is broad, and the enzyme is active on all phospholipid hydroperoxides, as well as fatty acid hydroperoxides, cumene hydroperoxide, *tert*-butyl hydroperoxide, and hydrogen peroxide. The wide specificity for hydroperoxides is also suggested by

[4] F. Ursini, M. Maiorino, L. Bonaldo, and C. Gregolin, *in* "Oxygen Radicals in Chemistry and Biology" (W. Bors, M. Saran, and D. Tait, eds.), p. 713. de Gruyter, Berlin and New York, 1984.

[5] F. Ursini, M. Maiorino, and C. Gregolin, *in* "Oxy Radicals and Their Scavenger System" (R. A. Greenwald and G. Cohen, eds.), Vol. 2, p. 224. Elsevier, New York, 1983.

[6] L. Zhang, M. Maiorino, A. Roveri, and F. Ursini, *Biochim. Biophys. Acta* **1006,** 140 (1989).

[7] S. Daolio, P. Traldi, F. Ursini, M. Maiorino, and C. Gregolin, *Biomed. Mass Spectrom.* **10,** 499 (1983).

[8] F. Ursini, L. Bonaldo, M. Maiorino, and Gregolin, *J. Chromatogr.* **270,** 301 (1983).

evidence that all titratable hydroperoxide groups generated in microsomal membranes during lipid peroxidation are reduced by this enzyme.[9] Furthermore, it has recently been observed directly that PHGPX also reduces cholesterol hydroperoxides.[10] The physiological reducing substrate is glutathione, although inhibition experiments in the presence of iodoacetate suggest that the oxidized active site can also be reduced by mercaptoethanol.[1]

The kinetic mechanism of PHGPX appears identical to that of glutathione peroxidase. Kinetic analysis, indeed, was compatible with a Ping-Pong mechanism without the formation of ternary complexes.[1] In this mechanism, as in the case of glutathione peroxidase, the kinetic parameters for one substrate (V_{max} and K_m) are not defined, being a function of the other substrate. In the suggested reaction scheme, the selenol moiety of PHGPX is first oxidized by hydroperoxides and then reduced back in two steps by two glutathione molecules. This mechanism exactly fits the Dalziel equation and the integrated rate equation describing the Tert-Uni Ping-Pong mechanism that have been applied to glutathione peroxidase[11,12] From the plots obtained by computer processing of the progression curves of the activity, it is possible to calculate the pseudo-first-order rate constant for the first reaction of the catalytic cycle, i.e., the reduction of the hydroperoxide and the oxidation of the selenol. This parameter was measured on different substrates, and the results were compared with those obtained using tetrameric glutathione peroxidase.[13] The results showed that glutathione peroxidase is much more active than PHGPX on hydrogen peroxide, that both enzymes react almost equally well on linoleic acid hydroperoxide, and that only PHGPX is active on phospholipid hydroperoxides (Table I). These kinetic data indicate that an identical kinetic mechanism has been adopted by nature to control the hydroperoxide concentration in different environments: one enzyme is active on small hydrophilic substrates and the other on lipophilic substrates in membranes.

The identical kinetic mechanisms of the peroxidase reactions, the selenium content, and the similarity of molecular weight between PHGPX

[9] M. Maiorino, A. Roveri, F. Ursini, and C. Gregolin, *J. Free Radicals Biol. Med.* **1,** 203 (1985).

[10] J. P. Thomas, M. Maiorino, F. Ursini, and A. W. Girotti, *J. Biol. Chem.* **454,** in press (1990).

[11] L. Flohé, G. Loeschen, W. A. Gunzler, and E. Eichele, *Hoppe-Seyler's Z. Physiol. Chem.* **353,** 987 (1972).

[12] J. Chaudière and A. L. Tappel, *Arch. Biochem. Biophys.* **226,** 448 (1983).

[13] F. Ursini and A. Bindoli, *Chem. Phys. Lipids* **44,** 255 (1987).

TABLE I

APPARENT SECOND-ORDER RATE CONSTANTS FOR REACTION
BETWEEN PHGPX OR GPX AND HYDROPEROXE SUBSTRATES[a]

Substrate	K (mM^{-1} min^{-1})	
	PHGPX	GPX
Hydrogen peroxide	1.9×10^5	2.9×10^6
Cumene hydroperoxide	1.3×10^5	1.0×10^6
tert-Butyl hydroperoxide	7.1×10^4	7.1×10^5
Linoleic acid hydroperoxide	1.8×10^6	2.3×10^6
Phosphatidylcholine hydroperoxide	7.0×10^5	—

[a] From Refs. 1 and 12.

and the monomer of glutathione peroxidase might seem to suggest a structural relationship between these enzymes. However, the identification of PHGPX with the monomer of GPX seems to be excluded for the following reasons: (1) the amino acid composition is different (even though the amino acid composition of pig heart glutathione peroxidase is not available, and the comparison was made between bovine red blood cell glutathione peroxidase and pig heart PHGPX); (2) the monomer of glutathione peroxidase is catalytically inactive, independent of the substrate; and (3) there is a completely different susceptibility to inhibitors.[14]

PHGPX was identified and subsequently purified as a cytosolic protein that inhibits microsomal lipid peroxidation in the presence of glutathione. The first apparent role of this enzyme in cell biology, therefore, seems to be related to the protection of biomembranes against oxidative damage.[2] During microsomal lipid peroxidation, lipid hydroperoxides are produced. The decomposition of lipid hydroperoxides generates the extremely reactive alkoxyl radical which, in turn, generates new hydroperoxides. The enzymatic reduction of the hydroperoxides prevents this free-radical multiplication. However, the catalytic activity of PHGPX cannot account per se for the almost complete protection observed in microsomes undergoing peroxidation. The inhibition of lipid peroxidation, induced by iron and reducing equivalents (ascorbate or NADPH), by PHGPX and glutathione requires a physiological amount of vitamin E in membranes, suggesting a tendem mechanism for the antioxidant activities

[14] M. Maiorino, A. Roveri, C. Gregolin, and F. Ursini, Arch. Biochem. Biophys. **251,** 600 (1986).

of PHGPX and vitamin E.[15] Reacting vitamin E with lipid hydroperoxyl radicals prevents propagation of peroxidation but generates lipid hydroperoxides. If not promptly reduced by PHGPX, the lipid hydroperoxides undergo decomposition, leading to alkoxy radicals[16] that are extremely reactive against vitamin E and lipids as well.[17] In the absence of PHGPX, vitamin E is just cooxidized with lipids, whereas PHGPX without vitamin E can only slow down the free-radical multiplication rate, and its capacity is rapidly saturated.

Although protection against lipid peroxidation is the most prominent function of the enzyme, it is plausible that PHGPX would also affect the physiological control of hydroperoxides enzymatically generated, and that these hydroperoxides would affect cellular functions or activation. In this case a new, exciting chapter on the physiology of PHGPX could begin.

Substrate Preparation and Activity Measurement

Taking advantage of its peroxidation-inhibiting activity, PHGPX was discovered and purified in a simple lipid peroxidation test. The peroxidase activity, which was first identified on a partially purified preparation of the enzyme, was in fact measurable in crude fractions with some difficulties. Thereafter, the introduction, as peroxide substrate, of mixed micelles of phospholipid hydroperoxides and Triton X-100 greatly simplified activity measurements. PHGPX activity, although measurable on liposomes or membranes containing hydroperoxides, is higher and more linear when the substrate is in micellar form. The test we describe can be used throughout all purification steps as well as in tissue homogenates.

Phospholipid hydroperoxides can be prepared either by spontaneous autoxidation or by enzymatic hydroperoxidation of soybean phosphatidylcholine liposomes or dispersions. The enzymatic procedure, which is more practical and reproducible, is described. Polyunsaturated fatty acids of phosphatidylcholine are oxygenated by soybean lipoxidase[18] in the presence of bile salts. At the end of the reaction lipids must be extracted and separated from bile salts, which are strong inhibitors of PHGPX activity.[14]

[15] M. Maiorino, M. Coassin, A. Roveri, and F. Ursini, *Lipids* **24,** 721 (1989).
[16] S. D. Aust and B. A. Svingen, *in* "Free Radicals in Biology" (W. A. Pryor, ed.), Vol. 5, p. 1. Academic Press, New York, 1982.
[17] M. Erben-Russ, C. Michel, W. Bors, and M. Saran, *J. Phys. Chem.* **91,** 2362 (1987).
[18] J. Eskola and S. Laasho, *Biochim. Biophys. Acta* **751,** 305 (1983).

Enzymatic Hydroperoxidation of Soybean Phosphatidylcholine and Extraction of Products

The reaction is carried out in an oxygraph vial, and the oxygen consumption is recorded. The reaction mixture contains, in 2.5 ml, 0.2 M sodium borate (pH 9), 3 mM sodium deoxycholate, and 0.3 mM soybean phosphatidylcholine. The phospholipids are obtained from Sigma (St. Louis, MO; Type III/S) in chloroform solution. The required amount is dried under an argon stream and first dispersed with 10 mM deoxycholate. This dispersion is then diluted with the buffer. The reaction starts by the addition of 0.1 mg of soybean lipoxyganase (EC 1.13.11.12) from Sigma (Type IV). After approximately 10 min, when 50% of the oxygen has been consumed, the mixture is applied to a Sep-Pak C_{18} cartridge (Water Associates, Milford, MA) equilibrated with water. Deoxycholate is washed off with 10 bed volumes of water, and phospholipids are eluted with 2 ml of methanol. This procedure can be scaled up or repeated several times. Methanolic extracts are pooled, and this solution, which contains phospholipid hydroperoxides and native phospholipids, is concentrated to a small volume. It is stable for approximately 1 month at $-20°$.

Using this procedure the recovery of hydroperoxides accounts for approximately 60% of the oxygen consumed. The amount of hydroperoxide groups can be evaluated from the test described below if partially purified PHGPX is available, or by titration. This preparation of phospholipid hydroperoxides is suitable for the test of PHGPX activity. If necessary, peroxidized phospholipid can be purified by HPLC. A detailed description of the HPLC procedures is not within the scope of this chapter; however, we suggest the procedures described by Crawford et al.[19] or Ursini et al.[8]

Test of Activity

PHGPX activity is measured spectrophotometrically; glutathione oxidation is recorded at 340 nm in the presence of an excess of glutathione reductase and a saturating amount of NADPH. The peroxide substrates are phospholipid hydroperoxides prepared as described above and dispersed in mixed micellar form in the presence of Triton X-100. Since Triton X-100 might contain some hydroperoxides, it is important to use high-quality, peroxide-free Triton X-100 (Boehringer-Mannheim, West Germany). Moreover, it is convenient to run a blank (without adding lipid hydroperoxides) to search for these Triton X-100 hydroperoxides.

[19] C. G. Crawford, R. D. Plattner, D. J. Sessa, and J. J. Rackis, *Lipids* **15,** 91 (1979).

Reagents

Buffer and reagent solution: 0.5 M Tris-HCl (pH 7.4), 25 mM EDTA, 0.5 mM NADPH, 5 mM NaN$_3$, 15 mM reduced glutathione

Glutathione reductase, 5 mg/ml, specific activity 120 U/mg (Boehringer-Mannheim)

Triton X-100 (peroxide free), 20% (v/v) in water

Phospholipid containing hydroperoxide groups in methanol ~15 mM containing 0.24 μmol of hydroperoxide groups per micromole of phospholipid)

Assay. To a spectrophotometer cuvette equipped with a magnetic stirrer and a temperature control set at 30°–37° are added 0.5 ml of buffer and reagent mixture, 15 μl of Triton X-100 solution, 5 μl of glutathione reductase, and the sample containing PHGPX activity. The final volume is 2.5 ml. After 5 min for temperature equilibration, complete reduction of glutathione, and PHGPX activation, a baseline is recorded. The reaction is then started by the addition of phospholipid hydroperoxides in methanol. The amount added is usually between 20 and 50 nmol of peroxide groups. The glutathione concentration can range from 0.5 to 5 mM. The suggested concentration of 3 mM gives a good compromise between activity and unspecific oxidation.

The rate becomes linear in less than 10 sec, suggesting that the time required for the formation of mixed micelles is not a limiting factor under these conditions (identical results are obtained using preformed mixed micelles). The NADPH oxidation is usually followed until all the peroxide substrate is consumed and the baseline slope reaches the original value, before the addition of hydroperoxides. To test that the system is functioning correctly, it is useful to check it by adding a second aliquot of hydroperoxides.

This assay allows for the evaluation of a blank due to nonspecific NADPH and glutathione oxidation (the rate before the addition of hydroperoxides and after their consumption). Under our experimental conditions, glutathione and NADPH oxidation in the presence of hydroperoxides, i.e., the blank without enzyme, is negligible.

If necessary, the reaction can be started by adding PHGPX, but in this case two artifacts are possible. If the enzyme is stored in a solution that does not contain thiols, a progressive increase of the rate of the enzyme reaction is observed, possibly owing to activation of the enzyme (as in the case of glutathione peroxidase). If the solution of PHGPX contains thiols, activation is not required, but some disulfides, which are always present, might lead to an overestimation of the activity. If it is necessary to start the reaction by adding the enzyme, the enzyme should be preincubated with glutathione, glutathione reductase, and NADPH, to avoid these artifacts.

The spectrophotometric test can be utilized for routine purification of the enzyme and for evaluation of activity in tissues. In these cases, to get reproducible results, it is extremely important to check the phospholipid hydroperoxide substrate for hydrolysis. Fatty acid hydroperoxides are indeed substrates for other peroxidases that, if present, lead to an overestimation of the PHGPX activity. It is therefore useful to rule out the hydrolysis of phospholipid hydroperoxides using glutathione peroxidase. Since fatty acid hydroperoxides, but not phospholipid hydroperoxides, are reduced by glutathione peroxidase, if any activity is detected, the substrate is not suitable for PHGPX activity measurement.

PHGPX Activity in Tissues

To search for PHGPX activity in tissues, if activity is low and precise measurements are not possible using homogenates, we use two preparations, neither of them giving the true "total activity" present in the tissue. Nevertheless, measurements obtained by these approaches are more reproducible and useful for comparisons between different samples.

In the first preparation, the tissue is broken in a Polytron homogenizer for 5 min in 3 volumes of 0.1 M Tris-HCl (pH 7.4), 0.3 M KCl and centrifuged for 15 min at 15,000 g and 45 min at 100,000 g. The ionic strength of the supernatant is then decreased either by dialysis against 50 mM Tris-HCl (pH 7.4) or by chromatography on Sephadex G-25, since we observed that the activity is partially inhibited by the presence of high ionic strength in the test. This preparation contains the cytosolic PHGPX and the enzyme released from all cellular membranes.

In the second preparation, the tissue is homogenized in 3 volumes of 0.25 M sucrose, 20 mM Tris-HCl (pH 7.4). Following centrifugation for 15 min at 10,000 g, Triton X-100 (final concentration of 0.2% v/v) is added to the supernatant to solubilize microsomes. This fraction contains PHGPX from the cytosol and microsomal fraction.

The activity recovered by the latter procedure is approximately 30% higher than that recovered by the first. However, there are more interfering activities, and the optical density of the sample is higher. By the latter procedure the specific activity or rat liver is 6.2 nmol/min/mg protein at 37°. The amount of protein that can be used in this test is approximately 1 mg.

Purification

Purification of PHGPX from pig heart is described; however, the same procedure can also be applied to other organs, e.g., liver or brain. More-

TABLE II
PURIFICATION OF PHGPX FROM PIG HEART

Step	Specific Activity[a]	Units[b]	Recovery (%)
Homogenization, centrifugation, and dialysis	0.002	190	100
DEAE-Sepharose	0.455	130	69
BSP Affinity Chromatography	5.15	90	47
Sephadex G-50	101.23	36	19
HPLC (TSK CM)	310.15	19	10

[a] Specific activity is measured as μmol/min/mg protein at 37°.
[b] One unit of activity catalyzes the reduction of 1 μmol/min at 37°.

over, the procedure can be easily scaled down if only small samples are available. The purification scheme is reported in Table II.

Approximately 500 g of ventricular muscle is prepared from pig heart (from the local slaughterhouse). The tissue is minced and thoroughly homogenized in a Polytron mixer, set at maximum power, for 5 min in approximately 3 volumes of ice-cold 0.1 M Tris-HCl (pH 7.4), 5 mM 2-mercaptoethanol. The soluble fraction is prepared by two centrifugations of the homogenate: 30 min at 10,000 g and 45 min at 100,000 g. Before the second centrifugation filter through cheesecloth to eliminate fluffy material. The supernatant is dialyzed exhaustively against 10 mM potassium phosphate buffer (pH 7), 5 mM mercaptoethanol. During dialysis some proteins precipitate and are eliminated by centrifugation.

The supernatant is applied to a DEAE-Sepharose 6B column (13 × 5.5 cm) equilibrated with 10 mM potassium phosphate buffer (pH 7), 5 mM mercaptoethanol. The column is washed with 2 liters of the equilibration buffer and then eluted with 0.5 liters of 0.2 M potassium phosphate buffer (pH 7), 5 mM mercaptoethanol. The active fractions are collected and pooled, and 10% (v/v) (final concentration) glycerol is added. This fraction is loaded on a Sepharose–bromosulfophthalein–glutathione (BSP) affinity column, prepared by linking the bromosulfophthalein–glutathione adduct[20] to CNBr-activated Sepharose.[21] The column is equilibrated with 25 mM Tris-HCl (pH 7.2), 5 mM mercaptoethanol, 10% (v/v) glycerol. After the sample is loaded, the column is washed with 300 ml of 25 mM Tris-HCl (pH 7.2), 100 mM KSCN, 5 mM mercaptoethanol, 10% (v/v) glycerol. The elution is carried out with 25 mM Tris-HCl (pH 7.6), 300 mM KSCN, 5 mM mercaptoethanol, 10% (v/v) glycerol. Active fractions

[20] G. Whelan, J. Hoch, and B. Combes, *J. Lab. Clin. Med.* **75,** 542 (1970).
[21] S. C. March, I. Parikh, and P. Cuatrecasas, *Anal. Biochem.* **60,** 149, (1974).

are pooled and concentrated to 5–10 ml using an Amicon ultrafiltration apparatus with an YM10 membrane. This fraction is applied on a Sephadex G-50 column (140 × 5.5 cm) equilibrated with 25 mM Tris-HCl (pH 7.4), 300 mM KSCN, 5 mM mercaptoethanol, 10% (v/v) glycerol. The active portion is eluted 160 ml after the void volume of the column. The active fractions are pooled and concentrated with the same ultrafiltration apparatus as before. This fraction is dialyzed against 10 mM potassium phosphate (pH 6.5), 0.1 M KCl, 5 mM mercaptoethanol. Some proteins precipitate during dialysis and are eliminated by centrifugation.

At this purification stage, PHGPX accounts for 40–70% of the proteins. A preparation at this level of purification is useful for routine purposes such as measuring phospholipid hydroperoxides. If PHGPX is stored at this purification step, 10% (v/v) glycerol should be added to the last dialysis buffer. This preparation is very stable (several months at −20°).

The final purification step is carried out by HPLC using either gel-permeation or ion-exchange columns, e.g., TSK SW 2000 (gel permeation), TSK DEAE (weak anion exchanger), Mono Q (strong anion exchanger), and TSK CM (weak cation exchanger). We describe here the chromatographic conditions for a TSK CM column. Buffer A: 10 mM potassium phosphate, 100 mM KCl, 5 mM mercaptoethanol (pH 6.5); buffer B: 10 mM potassium phosphate, 300 mM KCl, 5 mM mercaptoethanol (pH 6.5). Flow rate: 1 ml/min. The gradient from 0 to 100% buffer B is developed in 25 min after 3 min in isocratic conditions. Detection is at 280 nm; injection volume is less than 0.2 ml. PHGPX is eluted as a single peak when KCl is approximately 200 mM. Peaks isolated from several runs are pooled, and 10% (v/v) glycerol is added. The preparation is then concentrated by ultrafiltration to a final protein concentration no higher than 0.3 mg/ml to avoid aggregation.

[48] Iron Redox Reactions and Lipid Peroxidation

By Steven D. Aust, Dennis M. Miller, and
Victor M. Samokyszyn

Introduction

Iron-catalyzed lipid peroxidation has been studied in many *in vitro* model systems. While the mechanism of iron-catalyzed lipid peroxidation is not completely understood, it is well established that the redox chemis-

METHODS IN ENZYMOLOGY, VOL. 186

try of iron influences both the occurrence and the rate of lipid peroxidation. For example, iron-catalyzed lipid peroxidation in systems comprised initially of Fe(II) and phospholipid liposomes requires some Fe(II) oxidation.[1] Conversely, iron-catalyzed lipid peroxidation in systems containing Fe(III) and phospholipid liposomes requires some Fe(III) reduction.[2] However, complete Fe(II) oxidation or complete Fe(III) reduction results in conditions that do not promote lipid peroxidation. Thus, measurements of the rate and extent of Fe(II) oxidation or Fe(III) reduction aid the interpretation of experimental lipid peroxidation data. While several methods are available, only methods pertinent to lipid peroxidation are detailed here.

Two important factors which influence both the rates and extents of Fe(II) oxidation or Fe(III) reduction are chelation and pH. Chelators which preferentially bind Fe(II) tend to prevent or slow the rate of Fe(II) autoxidation as well as the rates of Fe(II) oxidation by peroxides. Chelators which bind Fe(III) with greater affinity than Fe(II), however, have the opposite effect, that is, they increase the rate of Fe(II) autoxidation as well as Fe(II) oxidation by peroxides. In addition, the rate of Fe(II) autoxidation, particularly unchelated Fe(II), increases with increasing pH.

Fe(III) Reduction

Enzymatic Fe(III) Reduction

Numerous electron-transport enzymes are able to catalyze the reduction of Fe(III) to Fe(II). Perhaps the easiest method for determining the rate of enzymatic Fe(III) reduction is to measure the rate of Fe(II) formation spectrophotometrically using one of the Fe(II) chelators listed in Table I.

Continuous Fe(III) Reduction Assay

Reaction mixtures should contain the enzyme of interest, appropriate reducing cosubstrate, Fe(III) chelated to the compound(s) of interest [owing to the strong absorbances of some of the Fe(II) chelators in Table I, the total iron concentration should be less than 100 μM], and one of the Fe(II) chelators in Table I at a concentration at least 3-fold greater than the total iron concentration [because these chelators form a tris complex with Fe(II)]. If the assay is conducted in the presence of microsomes, liposomes, or fatty acid micelles, an antioxidant should be included [e.g.,

[1] G. Minotti and S. D. Aust, *J. Biol. Chem.* **262**, 1098 (1986).
[2] M. Tien, J. R. Bucher, and S. D. Aust, *Biochem. Biophys. Res. Commun.* **107**, 279 (1982).

TABLE I
Fe(II) Chelators

Chelator	λ_{max} (nm)	ε (M^{-1} cm^{-1})
4,7-Diphenyl-1,10-phenanthroline[a] (bathophenanthroline)	534	22,200
4,7-Diphenyl-1,10-phenanthroline disulfonate	534	22,100
1,10-Phenanthroline (o-phenanthroline)	510	11,100
3-(2-Pyridyl)-5,6-bis(4-phenylsulfonic acid)-1,2,4-triazine (ferrozine)	564	27,900

[a] Bathophenanthroline is insoluble in water but soluble in most organic solvents.

butylated hydroxytoluene, 0.03% (w/v) final concentration] to prevent lipid peroxidation.

Comments. There are several complications and artifacts that may hamper accurate determination of Fe(III) reduction. For example, the rate of Fe(III) reduction is often directly proportional to total iron concentration. In addition, the presence of other redox-active transition metals may interfere with chromophore formation or unpredictably effect iron reduction. Since most reagents are contaminated with iron and other redox-active transition metals, Chelex 100 treatment of these reagents, to remove trace metal contaminants, is essential. Chelex 100 treatment can be performed as a batch method or, preferably, by column chromatography. We routinely use a glass column (2.5 × 40 cm, flow rate ~10 ml/min) of Chelex 100 (pH approximately 6–8) to remove trace metal contaminants from saline solutions.

A simple test for redox-active metal ion contamination, based on the rate of metal-catalyzed ascorbate oxidation, should be used to determine the level of metal contamination.[3] Briefly, the reagent of interest is incubated with 100 μM ascorbate, and the decrease in absorbance at 265 nm is monitored for 15 min. Based on the extinction coefficient (ε 14,500 M^{-1} cm^{-1}), a loss of ascorbate concentration of greater than 0.5% within 15 min indicates significant metal contamination. Note that there is a fairly common misconception that ascorbate autoxidizes; however, this is incorrect as its oxidation actually results from metal reduction followed by autoxidation of the reduced metals. Typically it is sufficient to treat the assay buffer with Chelex 100. However, some buffers, such as phosphate, are inherently resistant to metal removal and may, in fact, absorb addi-

[3] G. R. Buettner, *J. Biochem. Biophys. Methods* **16,** 27 (1988).

tional metals from the Chelex 100 resin. Thus, this test should be performed both before and after Chelex 100 treatment of reagents.

Additionally, many frequently used buffers, such as phosphate, or any of the Good buffers (e.g., HEPES) catalyze rapid autoxidation of Fe(II).[4] These buffers, apparently by chelating or ligating to iron, affect iron redox chemistry and hence the sensitivity of the Fe(III) reduction assay. Careful selection of buffer systems can reduce complications in the assay, or, if possible, saline solutions of desired ionic strength should be used to avoid buffer artifacts altogether. In systems which rely on superoxide (O_2^{-}) as the Fe(III) reductant (e.g., xanthine oxidase), the addition of catalase to remove H_2O_2 produced by O_2^{-} dismutation or by other sources increases the sensitivity of the technique.

Many electron transport enzymes use reducing substrates which have characteristic absorbances, such as the pyridine nucleotides. The rate of reducing substrate oxidation should always be used as an additional source of information to estimate the rate of Fe(III) reduction, providing that the stoichiometry of reducing substrate and Fe(III) is known. If this is unknown, this method coupled with the method described above may be used to determine the stoichiometry.

Discontinuous Fe(III) Reduction Assays

Iron chelators which preferentially bind Fe(III) (e.g., EDTA) often make accurate determination of the rate of Fe(III) reduction difficult owing to rapid Fe(II) autoxidation catalyzed by these chelators. One possible solution for this problem is to conduct the Fe(III) reduction assay under anaerobic conditions as follows. All reactants (except the enzyme) are purged of dioxygen by exhaustive argon bubbling of the solution (argon is preferred over nitrogen because it has less dioxygen contamination and is denser than dioxygen). The reaction is started with the addition of the enzyme, and anaerobiosis is maintained during the time course of the assay by continuously purging the head space of the reaction vessel (e.g., a 13 × 100 mm test tube) of dioxygen with argon. At regular time intervals, 0.5-ml aliquots of the reaction mixture are removed and mixed with 1 ml of 15 mM 1,10-phenanthroline and 0.5 ml of 30% trichloroacetic acid, and the phenanthroline–Fe(II) complex is extracted with 2 ml of *n*-amyl alcohol. Since this procedure requires the use of solvent extraction, it may be necessary to compare the amount of Fe(II) formed to a standard curve of $FeCl_3$ and excess thioglycolic acid subjected to identical assay conditions.

[4] D. O. Lambeth, G. R. Ericson, M. A. Yorek, and P. D. Ray, *Biochim. Biophys. Acta* **719**, 501 (1982).

Comment. Obviously, some enzymes, such as xanthine oxidase, require dioxygen for Fe(III) reduction, and thus cannot be assayed anaerobically. However, this method may be used aerobically to assay for Fe(III) reduction, providing rapid Fe(II) autoxidation is not a problem.

Chemical Fe(III) Reduction

Determination of the rate of Fe(III) reduction by chemical reductants can be performed using the continuous reduction assay, essentially as described in the previous section. Fe(III), complexed to the chelator(s) of interest, is incubated with the reductant in the presence of any of the Fe(II) chelators in Table I. In addition, it is often possible to measure the rate of Fe(III) reduction by monitoring the rate of oxidation of the reductant. For instance, as mentioned above, ascorbate exhibits a characteristic absorbance at 265 nm; thus, by following the decrease in absorbance at 265 nm, an indication of the rate of Fe(III) reduction can be obtained. Other reductants, such as cysteine or glutathione, while lacking characteristic UV–visible absorbances, react with 5,5′-dithiobis-2-nitrobenzoic acid to yield the corresponding 5-thio-2-nitrobenzoate derivatives which absorb at 412 nm.[4] Thus, by following the decrease in 5,5′-dithiobis-2-nitrobenzoic acid-detectable thiols over time, an indication of the rate of thiol oxidation, and hence Fe(III) reduction, is obtained.

Fe(II) Oxidation

Depending on the system under study, a variety of oxidants are present in many *in vitro* lipid peroxidation systems that can oxidize Fe(II) to Fe(III). These include dioxygen, O_2^{-}, H_2O_2, and other peroxides. The oxidation of Fe(II) by dioxygen [Eq. (1)], termed autoxidation, results in the generation of O_2^{-}. Superoxide is itself an oxidant of many Fe(II)

$$Fe(II) + O_2 \rightarrow Fe(III) + O_2^{-} \tag{1}$$

complexes [e.g., Fe(II)–citrate, Fe(II)–EDTA] [Eq. (2)].[5,6] In addition,

$$Fe(II) + O_2^{-} + 2\,H^+ \rightarrow Fe(III) + H_2O_2 \tag{2}$$

various peroxides including H_2O_2 [Eq. (3)] and lipid (LOOH) or other alkyl (ROOH) hydroperoxides [Eq. (4)] are able to oxidize Fe(II) to Fe(III).

$$Fe(II) + H_2O_2 \rightarrow Fe(III) + \cdot OH + OH^- \tag{3}$$
$$Fe(II) + L(R)OOH \rightarrow Fe(III) + L(R)O\cdot + OH^- \tag{4}$$

[5] G. Minotti and S. D. Aust, *J. Free Radicals Biol. Med.* **3,** 379 (1987).
[6] J. Butler and B. Halliwell, *Arch. Biochem. Biophys.* **218,** 174 (1982).

The Fenton Reaction [Eq. (3)] generates the hydroxyl radical ($\cdot OH$), which also is an oxidant of Fe(II) [Eq. (5)].

$$Fe(II) + \cdot OH \rightarrow Fe(III) + OH^- \tag{5}$$

As with the redox chemistry of Fe(III), chelation can greatly affect the rates of Fe(II) oxidation. For example, unchelated Fe(II) does not readily autoxidize in saline solution at neutral pH and ambient O_2 tension; however, incubation of Fe(II) with metal chelators results in variable rates of Fe(II) autoxidation, depending on the nature of the chelator. Similarly, as stated in the previous section, many common buffers can influence the rates of unchelated Fe(II) autoxidation. Chelation of iron also can influence its reactivity with the other oxidants. For example, Fe(II)–EDTA reacts much more readily with hydroperoxides compared with ADP-chelated Fe(II).[7] To measure the rates of Fe(II) autoxidation a continuous method involving measurement of rates of dioxygen consumption is used. In addition, we have developed a discontinuous assay which is applicable for determining rates of Fe(II) oxidation in complex reaction mixtures.

Continuous Assay for Fe(II) Autoxidation

The continuous assay method involves polarographic measurement of dioxygen consumption using a Clark-type electrode. Typically, chelated or unchelated Fe(II), prepared anaerobically, is injected into a water-jacketed reaction chamber maintained at a constant temperature. The initial rates of dioxygen consumption are used to determine the rate of Fe(II) autoxidation. It should be noted that the concentration of dissolved dioxygen decreases with increased temperature or increased ionic strength. Thermographs are available to calculate the dissolved dioxygen concentration under the conditions employed.[8]

Comments. The kinetics of Fe(II) autoxidation are very complex, often dependent on the total Fe(II) and dioxygen concentrations. In addition, many competing side reactions often make accurate determination of the rate of Fe(II) autoxidation difficult. For example, O_2^-, generated during Fe(II) autoxidation [Eq. (1)], can undergo several fates including oxidation of the Fe(II) chelate [Eq. (2)], reduction of the Fe(III) chelate generated by autoxidation, and dismutation yielding dioxygen at a rate which is second order with respect to O_2^- concentration [Eq. (6)]. Fur-

$$2 O_2^- + 2 H^+ \rightarrow H_2O_2 + O_2 \tag{6}$$

thermore, the H_2O_2 generated by the dismutation reaction may also func-

---

[7] B. A. Svingen, J. A. Buege, F. O. O'Neal, and S. D. Aust, *J. Biol. Chem.* **254**, 5892 (1979).
[8] M. J. Green and H. A. O. Hill, this series, Vol. 105, p. 3.

tion as an Fe(II) oxidant. Thus, depending on the rate constants of the competing reactions, the initial rates of dioxygen consumption may not represent the true rate of Fe(II) autoxidation.

Discontinuous Fe(II) Oxidation Assay

Alternatively, the rate of Fe(II) oxidation can be measured by sampling an aliquot of the reaction mixture at regular time intervals, mixing with one of the Fe(II) chelators in Table I, and measuring the absorbance of the corresponding Fe(II) chromophore. This assay has the advantages of being sensitive and applicable to complex reaction systems where the continuous assay described above is unsuitable because of rapid dioxygen depletion owing to other processes (e.g., lipid peroxidation, dioxygen-utilizing enzymes).

At regular time intervals, a 0.5-ml aliquots of the reaction mixture are mixed with 0.5 ml of the desired Fe(II) chelator at a stock concentration of 15 mM. To obtain accurate absorbance values, dilution may be necessary depending on the total Fe(II) concentration in the sample [typically this should be less than 100 μM Fe(II) concentration]. The absorbance of the samples should be determined as soon as possible or the samples stored in the dark until the absorbances can be determined because fluorescent lights cause artifactually higher absorbances. The turbidity inherent in reaction mixtures containing phospholipid can be eliminated by filtering the samples through 0.22-μm filters. Turbidity associated with samples containing fatty acids can be eliminated by dispersion of the fatty acids with Chelex 100-treated detergent (e.g., Tween 20, 1 mM final concentration).

[49] Determination of Carbonyl Content in Oxidatively Modified Proteins

By Rodney L. Levine, Donita Garland, Cynthia N. Oliver, Adolfo Amici, Isabel Climent, Anke-G. Lenz, Bong-Whan Ahn, Shmuel Shaltiel, and Earl R. Stadtman

Introduction

Metal-catalyzed oxidation[1] has been identified as a posttranslational covalent modification of proteins which may be important in several physiological and pathological processes.[2-4] These include the aging process, intracellular protein turnover, arthritis, and pulmonary diseases. Introduction of carbonyl groups into amino acid residues of proteins is a hallmark for oxidative modification. Reaction of these groups with carbonyl-specific reagents provides methods for detecting and quantitating metal-catalyzed oxidation. Other changes in proteins are induced by metal-catalyzed oxidation and can also be utilized as assays of the oxidative modification. These include loss of catalytic activity,[5-7] loss of histidine residues,[7] changes in surface hydrophobicity,[8] and changes in the ultraviolet spectrum of the protein.[5,9] Except for inactivation, these alternative assays are useful only with purified proteins. In any case, the relationship of these indirect assays to the metal-catalyzed oxidation must be established for the protein(s) under study. Consequently, assay for the carbonyl content of proteins is currently the definitive method for assessing metal-catalyzed oxidation. The carbonyl-bearing residues have not been completely identified, but γ-glutamyl semialdehyde appears to be the major residue.

[1] The process is correctly referred to as a mixed-function oxidation because it requires both molecular oxygen and reducing equivalents. However, mixed-function oxidation has been confused with the class of enzymes known as mixed-function oxidases. To avoid confusion, we adopt the alternative nomenclature, metal-catalyzed oxidation.

[2] E. R. Stadtman, *Trends Biol. Sci.* **11**, 11 (1986).

[3] K. J. A. Davies, *J. Free Radicals Biol. Med.* **2**, 155 (1986).

[4] C. N. Oliver, R. L. Levine, and E. R. Stadtman, *J. Am. Geriatr. Soc.* **35**, 947 (1987).

[5] R. L. Levine, C. N. Oliver, R. M. Fulks, and E. R. Stadtman, *Proc. Natl. Acad. Sci. U.S.A.* **78**, 2120 (1981).

[6] L. Fucci, C. N. Oliver, M. J. Coon, and E. R. Stadtman, *Proc. Natl. Acad. Sci. U.S.A.* **80**, 1521 (1983).

[7] R. L. Levine, *J. Biol. Chem.* **258**, 11823 (1983).

[8] J. Cervera and R. L. Levine, *FASEB J.* **2**, 2591 (1988).

[9] A. J. Rivett and R. L. Levine, *Arch. Biochem. Biophys.*, in press (1990).

Of course, carbonyl groups may be present elsewhere in the protein, such as in enzymatically glycosylated proteins or as a result of nonenzymatic glycosylation.[10] For example, glucose may react with protein-bound amino groups to form a Schiff base (an aldimine) which may then undergo the Amadori rearrangement to form a relatively stable keto-amine. Both the Schiff base and the ketoamine would react in the borotritide assay to yield labeled protein. The Schiff base would not yield a protein-bound hydrazone when treated with 2,4-dinitrophenylhydrazine, while the ketoamine would be expected to do so, albeit slowly.[11] Although the procedures described below have been used to determine the carbonyl content of both purified proteins and crude extracts,[4,7,12–16] one must be aware that carbonyl assays are not entirely specific for metal-catalyzed oxidation, especially for crude extracts or homogenates. Specificity is increased by acid hydrolysis of the labeled protein followed by chromatography of the hydrolyzate (see below and Ref. 17). Technical problems may compromise the accurate assessment of the carbonyl content of proteins in cruder extracts. As discussed below, nucleic acids will interfere and must be removed. In addition, oxidatively modified proteins are particularly susceptible to proteolytic degradation.[18] Thus, recovery of such proteins may be enhanced by the inclusion of a cocktail of protease inhibitors in the homogenizing buffer.

The mechanism of metal-catalyzed oxidation of proteins has been considered in detail.[19] In brief, a cation capable of redox cycling (e.g., Fe^{2+}/Fe^{3+}) binds to a divalent cation binding site on the protein. Reaction with O_2 or H_2O_2 generates an active oxygen species which oxidizes amino acid residues at or near that cation binding site. As mentioned, this reaction usually inactivates enzymes, presumably by destroying the essential cation binding site. While the oxidative modification may be mediated by

[10] A. Cerami, *J. Am. Geriatr. Soc.* **33,** 626 (1985).

[11] N. Mori and J. M. Manning, *Anal. Biochem.* **152,** 396 (1986).

[12] A. Benedetti, H. Esterbauer, M. Ferrali, R. Fulceri, and M. Comparti, *Biochim. Biophys. Acta* **711,** 345 (1982).

[13] C. N. Oliver, *Arch. Biochem. Biophys.* **253,** 62 (1987).

[14] C. N. Oliver, B. W. Ahn, E. J. Moerman, S. Goldstein, and E. R. Stadtman, *J. Biol. Chem.* **262,** 5488 (1987).

[15] P. E. Starke, C. N. Oliver, and E. R. Stadtman, *FASEB J.* **1,** 36 (1987).

[16] D. L. Garland, P. Russell, and J. S. Zigler, Jr., *in* "Oxygen Radicals in Biology and Medicine" (M. G. Simic, K. A. Taylor, J. F. Ward, and C. von Sonntag, eds.), p. 347. Plenum, New York, 1989.

[17] A. Amici, R. L. Levine, L. Tsai, and E. R. Stadtman, *J. Biol. Chem.* **264,** 3341 (1989).

[18] A. J. Rivett, *Curr. Top. Cell. Regul.* **28,** 291 (1986).

[19] E. R. Stadtman, *in* "Medical, Biochemical, and Chemical Aspects of Free Radicals" (O. Hayaishi, E. Niki, M. Kondo, and T. Yoshikawa, eds.), p. 11. Elsevier, Amsterdam, 1989.

a variety of enzymatic or nonenzymatic systems,[4] it is a remarkably site-specific free-radical reaction due to the existence of the cation binding site on the protein. A nonenzymatic, model system consisting of ascorbate, oxygen, and iron has been extensively used to study oxidative modification of proteins, particularly the glutamine synthetase from *Escherichia coli*.[5,7,20] Timed sampling of proteins during oxidation by the ascorbate system generates a series of proteins of increasing oxidative modification.[20] Preparation of such oxidized proteins was described earlier in this series.[21]

The methods described below are (1) reduction of the carbonyl group to an alcohol with tritiated borohydride; (2) reaction of the carbonyl group with 2,4-dinitrophenylhydrazine to form the 2,4-dinitrophenylhydrazone; (3) reaction of the carbonyl with fluorescein thiosemicarbazide to form the thiosemicarbazone; and (4) reaction of the carbonyl group with fluorescein amine to form a Schiff base followed by reduction to the secondary amine with cyanoborohydride. Van Poelje and Snell[22] have also quantitated protein-bound pyruvoyl groups through formation of a Schiff base with *p*-aminobenzoic acid followed by reduction with cyanoborohydride. Although a systematic investigation has not appeared, this method should also be useful in detecting other protein-bound carbonyl groups. Carbonyl content of proteins is expressed as moles carbonyl/mole subunit for purified proteins of known molecular weight. For extracts, the results may be given as nanomoles carbonyl/milligram protein. For a protein having a molecular weight of 50,000, a carbonyl content of 1 mol carbonyl/mol protein corresponds to 20 nmol carbonyl/mg protein.

Sample Preparation: Removal of Nucleic Acids

Nucleic acids also contain carbonyl groups and will react with the carbonyl reagents. If not removed before or after reaction, contaminating nucleic acids may thus cause an erroneously high estimate of protein-bound carbonyl.[23] Cellular extracts should be treated to decrease nucleic acid contamination. Precipitation of the nucleic acids with 1% streptomycin sulfate[24] is usually effective in minimizing interference,[25] but this

[20] R. L. Levine, *J. Biol. Chem.* **258**, 11828 (1983).
[21] R. L. Levine, this series, Vol. 107, p. 370.
[22] P. D. van Poelje and E. E. Snell, *Anal. Biochem.* **161**, 420 (1987).
[23] In addition, contaminating nucleic acids slow the rate of oxidation of proteins by metal-catalyzed oxidation systems such as the ascorbate–oxygen–iron system.[21] Thus, nucleic acids should also be removed when subjecting extracts to metal-catalyzed oxidation.
[24] N. S. Oxenburgh and A. M. Snoswell, *Nature (London)* **207**, 1416 (1965).
[25] B. Ahn, S. G. Rhee, and E. R. Stadtman, *Anal. Biochem.* **161**, 245 (1987).

should be confirmed for specific samples. Higher concentrations of streptomycin sulfate are occasionally required. Removal from extracts of bacteria carrying multicopy plasmids may be more efficient with two separate additions of 1% streptomycin, with a centrifugation between additions. In general, interference will be minimized when the nucleic acid is reduced to the point at which the ratio of absorbance at 280 nm to that at 260 nm is greater than 1.

Polyethyleneimine precipitation of nucleic acids[26] from bacterial extracts was not successful because recovery of protein was low. Treatment with nucleases would likely eliminate interferences. This should be avoided, however, because oxidatively modified proteins are distinctly more susceptible to proteolytic degradation[18] and would likely be degraded during incubation with the nucleases.

Reagents

   Streptomycin sulfate, 10%, in 50 mM HEPES (pH 7.2)
   Trichloroacetic acid, 20%

Procedure

Prepare the extract of bacterial or mammalian cells as desired; the final protein concentration should be no greater than 5 mg/ml. Procedures which minimize breakage of nuclei will decrease release of DNA. Centrifuge to remove debris. Add 1 volume streptomycin solution to 9 volumes of supernatant and allow to stand for 15 min. Centrifuge at 11,000 g for 10 min and discard the pellet. Use the supernatant for assay of protein-bound carbonyl groups.

If utilizing the borotritide or 2,4-dinitrophenylhydrazine methods, it is convenient to concentrate the protein at this point by precipitation with trichloroacetic acid. Add an equal volume of 20% trichloroacetic acid, centrifuge, and discard the supernatant. The residual trichloroacetic acid should be removed by careful draining, particularly for the borotritide assay. As an alternative to acid precipitation, simply dry the sample in a vacuum centrifuge (Savant).

Reaction with Tritiated Borohydride

The most sensitive method currently available to measure carbonyls is reaction with tritiated borohydride.[27] Borohydride will reduce all carbonyl

[26] A. Atkinson and C. W. Jack, *Biochim. Biophys. Acta* **308,** 41 (1973).
[27] A. G. Lenz, U. Costabel, S. Shaltiel, and R. L. Levine, *Anal. Biochem.* **177,** 419 (1989).

groups (i.e., aldehydes, ketones, and keto acids) to alcohols. It will also reduce Schiff bases to the amines so that this method will detect carbonyl groups which have formed a Schiff base with the ε-amino group of lysine or the α-amino group of the amino-terminal residue. In either case, a stable tritium label will be introduced if the borohydride is tritiated:

$$\text{Protein}—\text{C}{=}\text{O} \xrightarrow{\text{NaB}^3\text{H}_4} \text{protein}—^3\text{HCH}—\text{OH} \qquad (1)$$

$$\text{Protein}—\text{C}{=}\text{N}—\text{Protein} \xrightarrow{\text{NaB}^3\text{H}_4} \text{protein}—^3\text{HCH}—\text{NH}—\text{protein} \qquad (2)$$

The method has been utilized extensively in studies with purified proteins, with homopolymers, and with extracts of cells and tissues. Total protein content in the assay should not exceed 1 mg as incorporation of tritium may not be proportional to protein content beyond 1 mg, especially with extracts. Note that tritiated borohydride will release tritium gas. All procedures must be conducted in a hood equipped to handle such material. Derivatizations are carried out in 1.5-ml screw-top tubes closed with a cap containing an O ring (Sarstedt, Princeton, NJ).

Reagents

Sodium hydroxide, 100 mM

Tris, 1 M, with 10 mM EDTA, adjusted to pH 8.5 with HCl

Trichloroacetic acid, 10% (w/v)

Guanidine, 6 M, with 20 mM potassium phosphate, adjusted to pH 2.3 with trifluoroacetic acid (we find this solution is a better protein solvent than pH 6–7 guanidine; alternatively, one may simply use 100 mM NaOH to redissolve precipitated protein)

Sodium borohydride, 1 M, in 100 mM NaOH (this solution may appear slightly opalescent)

Sodium borotritide, 100 mM, in NaOH, specific activity ~100 mCi/mmol: Simply add the required volumes of 1 M NaBH$_4$ and 100 mM NaOH to the NaB^{3}H$_4$. We typically purchase 25 mCi to generate 2.5 ml of stock solution, which is divided into 0.5-ml aliquots in 1.5-ml plastic screw-top tubes (Sarstedt), closed with a cap containing an O ring, and stored at $-20°$. We have not conducted a systematic study of the stability of the stock solutions, but reproducible results have been obtained with solutions stored for several months. The specific activity may be adjusted to increase or decrease the labeling, as desired. The 100 mCi/mmol will provide about 10,000 cpm for 1 nmol carbonyl.

Procedure

Redissolve or resuspend[28] the dried protein-containing sample (10–1000 μg) in 50 μl water, then add 6 μl of 1 M Tris-HCl, 10 mM EDTA and 14 μl of 100 mM NaB^{3}H$_4$. After incubation at 37° for 30 min, precipitate the protein with 1 ml of 10% trichloroacetic acid, let stand 5 min in the hood, cap, and centrifuge for 3 min in a tabletop microcentrifuge (11,000 g). Discard the supernatant and wash the pellet twice with 1 ml of 10% trichloroacetic acid. Then redissolve the precipitate by incubation for 15 min at 37° in 0.6–0.8 ml of the guanidine solution. Determine the radioactivity of an aliquot of the protein solution by liquid scintillation counting. If desired, protein recovery may be determined at this point, either chemically or spectrophotometrically. The simplest method is multicomponent analysis[29] of the UV spectrum obtained in a diode-array spectrophotometer.

Hydrolysis

To increase the specificity of the method, one may perform rapid acid hydrolysis,[30] followed by chromatography on a small Dowex-50 column or by formal amino acid analysis.[17,27] The technique for hydrolysis is not complicated and can be performed using a benchtop block heater.

To prepare for hydrolysis, redissolve the trichloroacetic acid precipitate in 70% (v/v) formic acid instead of the acidic guanidine solution. For complete solution, the precipitate may require incubation for 30–40 min at 37°. Transfer the solution to glass vials (Wheaton, Millville, NJ, #224882), remove the formic acid in a vacuum centrifuge (Savant Speed-Vac), and add 100–500 μl constant boiling HCl. Flush each vial with a gentle stream of argon or nitrogen for 30 sec, then cap using a Teflon-faced silicone liner (Wheaton #240583 or Pierce, Rockford, IL, #12712) in an open top 13/425 screw cap (Wheaton #240508 or Pierce #13215). The cap should be snug, but avoid overtightening as this may crack the neck of the vial. Place the vial in a benchtop block heater at 155° for 45 min,[30] dry the hydrolyzed sample, and redissolve in 500 μl of 10 mM HCl. Use an aliquot of 50 μl for amino acid determination with *o*-phthalaldehyde and for liquid scintillation counting of radioactivity (see below). To separate any label not present in amino acids, load the remaining aliquot of 450 μl onto a Dowex 50W-X8 column. This may be packed in a Pasteur

---

[28] Limited experience indicates that even insoluble proteins are effectively labeled by this method.

[29] R. L. Levine and M. M. Federici, *Biochemistry* **21**, 2600 (1982).

[30] P. E. Hare, this series, Vol. 47, p. 3.

pipet or a small, disposable column. Wash the column with water and then elute with 2 M NH$_4$OH. Determine the radioactivity in both the water wash and the NH$_4$OH eluate. The tritium label found in the water wash could presumably be in carbohydrates or nucleic acids but not in amino acids nor amino sugars. The latter bind to the column and elute with the NH$_4$OH. Total recovery of loaded radioactivity is excellent, but a minor fraction is always irreversibly bound. In the case of samples taken from pulmonary washes, 18% was recovered in the water wash and 70% in the NH$_4$OH eluate, leaving 12% unrecovered.[27]

The Dowex procedure is a convenient step and is easy to perform even with a large number of samples. However, more detailed information is available from the pattern of labeling obtained by counting fractions from a full amino acid analysis.[17] Moreover, one can assure virtual specificity for carbonyl groups arising via metal-catalyzed oxidation by quantitation of 5-hydroxy-2-aminovaleric acid, derived on reduction of protein-bound γ-glutamyl semialdehyde.[17]

Determination of Total Amino Acids in Hydrolyzed Samples

Measurement of total amino acid content provides a convenient method of determining recovery of protein after acid hydrolysis. We utilize the method of Church et al.[31] which employs o-phthalaldehyde. As noted in the section above, simply determine the protein concentration in the 50-μl aliquot of the acid hydrolyzate. After measuring the absorbance at 340 nm, count the sample in the liquid scintillation spectrometer to determine the total tritium present in the hydrolyzate.

Tryptophan is destroyed by acid hydrolysis; cystine and proline are not detected by the o-phthalaldehyde assay. Thus, the determined value can be increased by 9.4%, the average content of tryptophan, cystine/cysteine, and proline in proteins.[32] One can then express the carbonyl content as nmol carbonyl/mg protein or as mol carbonyl/mol protein.

Comments

We have used several lots of NaB^{3}H$_4$ in studies in our laboratory, usually from Amersham (Arlington Heights, IL). Among these, we have encountered two separate lots of NaB^{3}H$_4$ which gave spuriously high results with all proteins, even those which had been prereduced with

[31] F. C. Church, D. H. Porter, G. L. Catigani, and H. E. Swaisgood, Anal. Biochem. **146**, 343 (1985).
[32] M. O. Dayhoff, L. T. Hunt, and S. Hurst-Calderone, in "Atlas of Protein Sequence and Structure" (M. O. Dayhoff, ed.), Vol. 5, Suppl. 3, p. 363. National Biomedical Research Foundation, Washington, D.C., 1979.

unlabeled NaBH$_4$ (I. Climent, personal communication). Neither we nor the supplier have determined the cause of this artifact. However, the spurious counts are removed by the acid hydrolysis/Dowex chromatography procedure described later. It is prudent to test each new lot of NaB^{3}H$_4$ by analysis of a prereduced protein. If the resultant "background" is high, then one should utilize the acid hydrolysis/Dowex chromatography step.

EDTA is included in the reaction buffer to prevent exchange of tritium into aromatic residues, which would cause an artifactually high determination for carbonyl content. The exchange is mediated by Mn^{2+} and completely prevented by the inclusion of EDTA.[17] Also, guanidine should not be present in the incubation mixture because it causes very high, nonspecific labeling of proteins.[27] The artifactual nature of this labeling was confirmed by testing proteins that had been pretreated with unlabeled NaBH$_4$. In the presence of guanidine, these prereduced proteins had very high levels of tritium incorporation. This is an example of the utility of prereduced proteins in screening for artifactual labeling. They should be included as a control if the assay conditions are changed from those described here.

Trichloroacetic acid is used to destroy excess NaB^{3}H$_4$. One cannot substitute a scavenger such as acetone because its addition also causes an artifactual increase in protein labeling, presumably owing to the formation of Schiff bases between the acetone and the protein. However, if desired, excess reagent can be separated from the protein by chromatographic means. We have successfully utilized HPLC with gel filtration (Pharmacia, Piscataway, NJ, Superose-12 column) and hydrophobic chromatography[33] (BioRad, Richmond, CA, ToyaSoda PW5 phenyl column). If other columns are employed, they must be tested for their ability to separate protein from excess reagent and protein and also for the possibility that tritium exchange into protein may occur during chromatography. A simple way to evaluate both separation and exchange is to use a prereduced protein in the assay.

Reaction with 2,4-Dinitrophenylhydrazine

2,4-Dinitrophenylhydrazine is the classic carbonyl reagent, and it can be used successfully with proteins.[7,34,35] Considerably more sample is required than for the borotritide method since the hydrazone is quantitated spectrophotometrically. The carbonyl contents determined by the

[33] S. Shaltiel, this series, Vol. 104, p. 69.
[34] O. L. Brady and C. V. Elsmie, *Analyst* **52,** 77 (1926).
[35] R. Fields and H. B. F. Dixon, *Biochem. J.* **121,** 587 (1971).

two methods are equal.[27] Thus, when one has sufficient sample (preferably >0.5 mg protein) the 2,4-dinitrophenylhydrazine technique provides a nonradiochemical assay for carbonyl groups in protein. Assuming that Schiff bases react under the conditions employed, the reactions are as follows:

$$\text{Protein}-\text{C}{=}\text{O} \xrightarrow{\text{H}_2\text{N}-\text{NH}-2,4\text{-DNP}} \text{protein}{=}\text{N}-\text{NH}-2,4\text{-DNP} + \text{H}_2\text{O} \qquad (3)$$

$$\text{Protein}-\text{C}{=}\text{N}-\text{Protein} \xrightarrow{\text{H}_2\text{N}-\text{NH}-2,4\text{-DNP}} \text{protein}{=}\text{N}-\text{NH}-2,4\text{-DNP} + \text{H}_2\text{O} \qquad (4)$$

2,4-Dinitrophenylhydrazine was the reagent originally used to demonstrate the appearance of carbonyl groups in proteins subjected to metal-catalyzed oxidation.[7] Because of the possibility that extraction of the excess 2,4-dinitrophenylhydrazine might also cause loss of the protein-bound hydrazone, the assay was first performed without extraction.[7,21,35] This rendered the method rather insensitive, and subsequent investigations demonstrated that the reagent could be extracted without adversely affecting the results. As noted above, the borotritide and 2,4-dinitrophenylhydrazine methods yield the same value for carbonyl content of a series of oxidized samples of glutamine synthetase (glutamate–ammonia ligase).[27] Derivatization can be carried out in any desired size tube, although we find the use of 1.5-ml plastic centrifuge tubes quite convenient.

Reagents

HCl, 2 M
2,4-dinitrophenylhydrazine, 10 mM, in 2 M HCl
Trichloroacetic acid, 20% (w/v)
Guanidine, 6 M, with 20 mM potassium phosphate, adjusted to pH 2.3 with trifluoroacetic acid

Procedure

The amount of protein assayed may be adjusted in accordance with the carbonyl content. One milligram of protein will give excellent results if the carbonyl content is approximately 0.2–1.0 mol/mol protein, but reasonable results may be obtained with as little as 0.1 mg protein. Purified proteins generally do not require reagent blanks to be run in parallel; cruder mixtures should have a blank prepared by treatment with 2 M HCl instead of 2,4-dinitrophenylhydrazine in HCl. Pipet the protein solution into 1.5-ml centrifuge tubes (e.g., Sarstedt or Eppendorf) and either dry in a vacuum centrifuge or precipitate with trichloroacetic acid.[36] To each

[36] It is possible to omit the drying or precipitation provided that the protein concentration will be at least 0.5 mg/ml during the derivatization. Also, at least 4 volumes of 2,4-

tube, add 500 μl of 10 mM 2,4-dinitrophenylhydrazine in 2 M HCl and allow to stand at room temperature for 1 hr, with vortexing every 10–15 min. Then add 500 μl of 20% trichloroacetic acid, centrifuge the tubes in a tabletop microcentrifuge (11,000 g) for 3 min, and discard the supernatant. Wash the pellets 3 times with 1 ml ethanol–ethyl acetate (1:1) to remove free reagent, allowing the sample to stand 10 min before centrifugation and discarding the supernatant each time. Redissolve the precipitated protein in 0.6 ml guanidine solution. Proteins usually redissolve within 15 min at 37°. Remove any insoluble material by centrifugation in the microcentrifuge for 3 min. Obtain the spectrum, read against the complementary blank in the case of cruder samples or against water in the case of purified proteins. Calculate the carbonyl content from the maximum absorbance (360–390 nm) using a molar absorption coefficient[37] of 22,000 M^{-1} cm^{-1}. As with the borotritide method, protein recovery is generally excellent, but it may be checked as desired. It is easiest to do so spectrophotometrically, particularly by multicomponent analysis.[29]

Filter Paper Extraction Procedure

Extraction of excess 2,4-dinitrophenylhydrazine by ethanol–ethyl acetate can also be performed on filter paper using a vacuum manifold. The method works well with crude extracts but not with some purified proteins because their precipitates do not adhere to the filter paper. Recovery of protein from crude extracts is 80–90% when the sample contains at least 0.1 mg protein.

Whatman 3MM filter paper disks (2.3 cm) have been used routinely, but Whatman glass fiber GFC filters work as well. Filters should be washed to remove contaminants that might contribute to the final spectrum. Place the filters in the guanidine solution and rotate gently for 30 min. Then wash twice in water for 15 min. Finally, transfer the filters to 20% trichloroacetic acid and place them in the vacuum manifold. After derivatization by 2,4-dinitrophenylhydrazine as above, transfer the precipitated sample to the center of the filter paper under low vacuum; an Eppendorf-type pipet tip works well for the transfer. Then rinse the tube with 1 ml 10% trichloroacetic acid and add this to the filter. Wash the precipitate on the filter 3 times with 5 ml ethanol–ethyl acetate. Let the filters dry under vacuum and transfer them to a 30-ml beaker. Add 2 ml guanidine solution, cover the beaker, and rotate gently for 1.5 hr. Remove

---

dinitrophenylhydrazine solution should be added in order to maintain a concentration of 2 mM. When concluding the reaction, add an equal volume of 20% trichloroacetic acid to provide a final concentration of 10%.

[37] G. D. Johnson, *J. Am. Chem. Soc.* **75,** 2720 (1953).

the solution, centrifuge to remove any insoluble material, and determine the carbonyl content as for the tube extraction method.

Comments

Derivatized proteins from a homogenate or other mixture of proteins may be separated by typical techniques, such as reversed-phase HPLC[21] or gel chromatography.[38] While we have reasonable experience with the reversed-phase HPLC method of separating proteins, there are no published studies on stability and recovery of the protein-bound hydrazones. It would be prudent to assess recovery before embarking on a quantitative study utilizing chromatographic separation. A few proteins, such as bovine serum albumin, have some solubility in the ethanol–ethyl acetate extracting solvent; recovery should be checked when indicated. Recovery of protein by trichloroacetic acid precipitation can be quite good. For example, recovery of oxidized glutamine synthetase was over 90% for samples containing as little as 10 μg protein.[27]

Antidinitrophenyl antibodies, which are commercially available, might be useful in selectively removing derivatized, oxidized proteins from a mixture of oxidized and nonoxidized proteins. Again, such studies have not yet been reported.

Any protein-bound chromophore that absorbs in the region of the dinitrophenylhydrazone could be a source of interference in the method. If the chromophore is reproducibly recovered in the blank, then the potential error is avoided. However, it has proved difficult to accomplish this when samples contain significant amounts of heme, especially in hemoglobin. It may be possible to extract the heme before derivatization with 2,4-dinitrophenylhydrazine, but no systematic study is available. Extraction of heme from hemoglobin is easily performed with 2-butanone.[39] However, the hemoglobin must first be converted to methemoglobin by, for example, treatment with ferricyanide. One would need to assess the effect of these treatments on the measurement of carbonyl groups by 2,4-dinitrophenylhydrazine. Alternatively, hemoglobin might be removed by antihemoglobin antibodies.

Reaction with Fluorescein Thiosemicarbazide for Gel Electrophoresis

Fluorescein thiosemicarbazide and fluorescein adipic hydrazide[25] react with carbonyl groups to yield fluorescent thiosemicarbazones or hydrazones analogous to the chromophoric 2,4-dinitrophenylhydrazones.

---

[38] K. Nakamura, C. Oliver, and E. R. Stadtman, *Arch. Biochem. Biophys.* **240**, 319 (1985).
[39] F. W. J. Teale, *Biochim. Biophys. Acta* **35**, 543 (1959).

Proteins labeled with fluorescein thiosemicarbazide may be separated by lithium dodecyl sulfate gel electrophoresis, providing a very sensitive method for visualizing oxidized proteins. While potentially useful as a screening method, the technique does not appear well suited for quantitative measurements on the gels because of differences in stability of the derivatives and of variability in quantum yield.[25]

Reagents and Equipment

  Fluorescein thiosemicarbazide, 0.25%, in dimethylformamide or dimethyl sulfoxide
  Ethanol–ethyl acetate (1 : 1)
  Trichloroacetic acid, 20% and 10% (w/v)
  Hydrochloric acid, concentrated
  Tris, 1 M, adjusted to pH 8.5 with HCl
  Urea, 8 M
  Gel electrophoresis apparatus
  Gelatin filters, Wratten 47B and 12
  Light sources and Polaroid camera system

Procedure

Derivatization has been carried out in 12- to 15-ml conical glass centrifuge tubes. Oxygen should be excluded during the derivatization; flush the tube with argon or nitrogen and stopper during incubation. Protect the reagents and the incubation tubes from light by wrapping with aluminum foil.

As mentioned in the Introduction, if nucleic acids are present they are removed by treatment with 1% streptomycin sulfate. Then add to the sample (1–2 mg protein) an equal volume of cold 20% trichloroacetic acid to precipitate the protein. Wash the precipitate twice with 10% trichloroacetic acid. Resuspend the pellet in 1 ml fluorescein thiosemicarbazide plus 1 drop concentrated HCl. Hold at room temperature for 30 min with frequent stirring. Then precipitate the protein by addition of 1 ml 20% trichloroacetic acid. Centrifuge and drain the supernatant by standing the tubes upside down. Free dye will still be present, but it will be separated from protein during electrophoresis.

The derivative is stable at 0 but not at 25°, so care should be taken to keep the sample cold after derivatization. In preparation for electrophoresis, neutralize the precipitate by adding cold Tris (~50 μl), then redissolve by adding 1 ml of 8 M urea. Electrophoresis is carried out according to Laemmli[40] except for these changes: Prepare the sample and run the

---

[40] U. K. Laemmli, *Nature (London)* **227**, 680 (1970).

electrophoresis in a cold room or box to keep the temperature around 4°. Use lithium dodecyl sulfate because of the greater solubility of the lithium over the sodium salt. Use a gel buffer at pH 8.5 instead of 8.8 because the fluorescent derivative is more stable at the lower pH. Limited experience with the Pharmacia PhastGel System is promising. This automated system was run at 15° because it uses sodium dodecyl sulfate, but the entire electrophoresis required less than 1 hr so that loss of label may be minimized.

The gels should be photographed promptly. They may be stained for protein detection after photography. Place the gel between two glass plates and illuminate at about a 45° angle with bright light through a Wratten 47B gelatin filter (Eastman Kodak, Rochester, NY). For illumination, one can simply use two slide projectors, one on each side. Quartz copy stand lamps might be more convenient if available. The second filter (Wratten 12) is placed in front of the camera lens. Polaroid "instant" film is essential for good results because optimal exposure must be determined by trial. A 30-sec exposure is a good starting point. Lengthen the exposure if the bands are too light; shorten the exposure if they are too dark. The fluorescence is rarely visible to the eye, but good results are obtained with the camera.

Comments

Fluorescein and its derivatives may bind nonspecifically to proteins. Indeed, compounds such as eosin (tetrabromofluorescein) have long been employed as histochemical stains. Consequently, one must take care to assure that labeled proteins are actually derivatized through carbonyl groups. It is known that metal-catalyzed oxidation of proteins may alter their surface hydrophobicity.[8] Such changes could facilitate nonspecific binding of the fluorescein thiosemicarbazide. One should therefore include a control sample which has been reduced with sodium borohydride as noted in the section on Reaction with Tritiated Borohydride. Antibodies against fluorescein are also commercially available and might be utilized to selectively remove labeled proteins from solution.

Reaction with Fluorescein Amine and Cyanoborohydride

The fluorescein amine–cyanoborohydride approach is a relatively new method which has thus far been applied only to purified proteins.[41] It is based on the ability of primary aromatic amines to form Schiff bases with carbonyl groups at or below neutral pH. At this pH, cyanoborohydride

---

[41] I. Climent, L. Tsai, and R. L. Levine, *Anal. Biochem.* **182,** 226 (1989).

will reduce the Schiff base to a secondary amine but will not reduce carbonyl groups.[42] Thus, fluorescein amine and sodium cyanoborohydride can be present simultaneously. Again, assuming that fluorescein amine (H_2N—Fl) would displace any protein–protein Schiff base, the reactions are as follows:

$$\text{Protein—C=O} \xrightarrow{H_2N—Fl} \text{protein=N—Fl} + H_2O \tag{5}$$

$$\text{Protein=N—Fl} \xrightarrow{NaCNBH_3} \text{protein—NH—Fl} \tag{6}$$

The derivative is fluorescent, but the quantum yield is low and varies with solvent conditions. However, the molar absorptivity is high (~86,000 at 490 nm) so that the label provides a very useful chromophoric tag for carbonyl groups. Again, antibodies against fluorescein are available and could be used to selectively remove labeled proteins from a mixture.

Fluorescein amine does not react with all protein-bound carbonyl groups. In the case of oxidized glutamine synthetase (glutamate–ammonia ligase), it appeared to react exclusively with γ-glutamyl semialdehyde, which was shown to constitute 65% of the carbonyl content.[41] The remaining 35% has not yet been identified. Use of the fluorescein amine method on one sample and either the borotritide or 2,4-dinitrophenylhydrazine method on a duplicate would provide quantitation of both γ-glutamyl semialdehyde content and total carbonyl content.

One significant advantage of this method is that it provides an exceptionally stable label. Indeed, after acid hydrolysis one can isolate the labeled amino acid.[41] Quantitation of the labeled amino acid provides a very specific technique for identification of metal-catalyzed oxidation of proteins because it measures γ-glutamyl semialdehyde in protein. The method may thus prove particularly useful with complex or crude samples. However, to date the method has only been applied to purified proteins.

Reagents

MES, 0.25 M, pH 6.0 (MES is 4-morpholinoethanesulfonic acid)
MES, 0.25 M, pH 6.0, with 1% sodium dodecyl sulfate (SDS)
Fluorescein amine,[43] 0.25 M, in 0.52 M NaOH
Sodium cyanoborohydride, 0.4 M, in 0.25 M MES, pH 6.0 (this solution should be freshly prepared each day)

[42] R. F. Borch, M. D. Bernstein, and H. D. Durst, *J. Am. Chem. Soc.* **93**, 2897 (1971).
[43] Specifically, fluorescein amine isomer II, which is 6-aminofluorescein. Care should be taken to obtain isomer II and not isomer I or a mixture of the isomers. The *Chemical Abstracts* registry number is 51649-83-3, and the formal name is 6-amino-3',6'-dihydroxyspiro[isobenzofuran-1(3*H*),9'-[9*H*]xanthen]-3-one. Do not confuse fluorescein amine with fluorescamine.

Trichloroacetic acid, 10% (w/v)
Ethanol–ethyl acetate (1 : 1) plus 10 mM HCl
Sodium hydroxide, 0.1 M

Procedure

Samples are dried by vacuum centrifugation in 1.5-ml plastic centrifuge tubes. Use the following volumes for samples containing up to 250 μg protein; for 250–1000 μg, simply double the volumes. Redissolve the sample in 50 μl MES–SDS and heat at 100° for 1 min. Adjust the volume by adding 18.6 μl MES buffer without SDS, 6.4 μl fluorescein amine, and finally 5 μl NaCNBH$_3$. The reaction mixture will thus be 80 μl and will be 20 mM fluorescein amine and 25 mM cyanoborohydride.

Incubate at 37° for 1 hr and then precipitate the protein by adding 1 ml trichloroacetic acid. Let stand 5 min in the hood, cap the tube, and centrifuge for 5 min in a tabletop microcentrifuge at 11,000 g. Discard the supernatant and wash the pellet 3 times with 1 ml of the acidified ethanol–ethyl acetate to remove residual dye. Redissolve the protein by incubating for 15 min at 37° in 600 μl of 0.1 M sodium hydroxide. Centrifuge for 3 min to remove any insoluble material, then obtain the spectrum of the sample. Carbonyl content is calculated from the absorbance maximum at 490 nm using an ε_M for fluorescein amine in 0.1 M NaOH of 86,800 M^{-1} cm^{-1}. Protein concentration may be determined as desired. If a spectrophotometric method is used, one may have to correct for the contribution from the fluorescein moiety.

[50] Protein That Prevents Mercaptan-Mediated Protein Oxidation

By Sue Goo Rhee, Kanghwa Kim, Il Han Kim, and Earl R. Stadtman

Several enzymes have been shown to lose catalytic activity on incubation in air with mercaptans such as dithiothreitol (DTT) and 2-mercaptoethanol and trace amounts of iron (or copper) salts.[1] At present, proteins which are inactivated by a mercaptan–Fe^{3+}–O$_2$ mixed-function oxidase (MFO) system include glutamine-dependent carbamoyl-phosphate syn-

---

[1] K. Kim, S. G. Rhee, and E. R. Stadtman, *J. Biol. Chem.* **260**, 15394 (1985).

thase from *Escherichia coli*,[2] rhodanase from bovine liver,[3,4] phosphoenolpyruvate carboxykinase from rat liver,[5] glutamine synthetase (glutamate–ammonia ligase) from *E. coli* and yeast,[1] adenylyltransferase from *E. coli*,[1] pyruvate kinase[1] and enolase[6] from rabbit muscle, and low density lipoprotein.[7] The mercaptan-dependent MFO inactivation is accompanied by the oxidation of unidentified amino acids as evidenced by the increase of reactive carbonyl moieties on the protein,[8] fragmentation, and formation of higher molecular weight aggregates through radical-mediated cross-linking reactions.[1]

It has been shown that the autoxidation of mercaptans in the presence of iron (or copper) generates reactive oxygen species such as superoxide anion, hydrogen peroxide, and hydroxyl radical, as well as thioyl radical.[9,10] It is generally believed that the following reactions take place:

$$RSH + Fe^{3+} \rightarrow RS \cdot + Fe^{2+} + H^+ \tag{1}$$
$$Fe^{2+} + O_2 \rightarrow Fe^{3+} + O_2^- \tag{2}$$
$$O_2^- + O_2^- + 2 H^+ \rightarrow H_2O_2 + O_2 \tag{3}$$
$$Fe^{2+} + H_2O_2 \rightarrow Fe^{3+} + HO \cdot + OH^- \tag{4}$$
$$O_2^- + Fe^{3+} \rightarrow O_2 + Fe^{2+} \tag{5}$$

Thus, Fe^{3+} or Cu^{2+} initiates mercaptan oxidation to produce thioyl radical as in reaction (1). The resulting reduced metal ion can be utilized to produce superoxide [reaction (2)]. The superoxide can undergo the dismutation reaction to form H_2O_2 [reaction (3)]. Hydrogen peroxide in turn can react with reduced metal ion via the Fenton reaction [Eq. (4)], generating hydroxyl radical. The reduced iron can be generated via reaction (5).

Hydroxyl radicals, powerful oxidants, are known to react with many cellular components at rates approaching diffusion control, and, in so doing, they initiate the peroxidation of membrane lipids and oxidation of proteins. When Fe^{2+} bound to an enzyme is involved in the Fenton reaction, the locally generated HO · will react with amino acid residues at the Fe^{2+}-binding site and thus impair enzyme activity. Therefore, any system

[2] P. P. Trotta, L. M. Pinkus, and A. Meister, *J. Biol. Chem.* **249**, 1915 (1974).
[3] M. Costa, L. Pecci, B. Pensa, and C. Cannella, *Biochem. Biophys. Res. Commun.* **78**, 596 (1977).
[4] P. M. Horowitz and S. Bowman, *J. Biol. Chem.* **262**, 8728 (1987).
[5] N. S. Punekar and H. A. Lardy, *J. Biol. Chem.* **262**, 6714 (1987).
[6] K. Kim, I. H. Kim, K. Y. Lee, S. G. Rhee, and E. R. Stadtman, *J. Biol. Chem.* **263**, 4704 (1988).
[7] J. W. Heinecke, H. Rosen, L. A. Suzuki, and A. Chait, *J. Biol. Chem.* **262**, 10098 (1987).
[8] R. L. Levine, D. Garland, C. N. Oliver, A. Amici, I. Climent, A. G. Lenz, B.-W. Ahn, S. Shaltiel, and E. R. Stadtman, this volume [49].
[9] H. P. Misra, *J. Biol. Chem.* **249**, 2151 (1974).
[10] G. Saez, P. J. Thornalley, H. A. O. Hill, R. Hems, and J. V. Bannister, *Biochim. Biophys. Acta* **719**, 24 (1982).

capable of reducing Fe^{3+} to Fe^{2+} can produce $HO \cdot$ in the presence of Fe^{3+} and O_2 and inflict damage to protein. For example, in addition to various mercaptans, ascorbate[11] (which reduces Fe^{3+}), the NADH oxidase/ NADH system[12] (which generates H_2O_2 and reduces Fe^{3+}), and the xanthine oxidase/hypoxanthine system[13] (which generates $O_2^{\overline{}}$ and H_2O_2) can inactivate various enzymes. It follows that any system which removes either Fe^{3+} or H_2O_2, which are essential for $HO \cdot$ production, will inhibit the inactivation reaction. Indeed, EDTA, catalase, and glutathione peroxidase protected glutamine synthetases against the Fe^{3+}/DTT system. Although the Fe^{3+}–EDTA complex can be more readily reduced, the presence of EDTA inhibited the inactivation reaction completely, probably because the inactivation reaction requires iron bound to glutamine synthetase.

All aerobic organisms are equipped with several antioxidant enzymes, which include superoxide dismutases, catalases, and peroxidases. In addition, eukaryotes, represented by yeast and mammals, contain a 27-kDa protein that provides several enzymes with protection against the MFO system containing a mercaptan, but not against the MFO system containing the nonsulfhydryl reducing equivalent. The activity of this mercaptan-specific antioxidant protein is assessed by its ability to inhibit the mercaptan/Fe^{3+}/O_2-mediated inactivation of glutamine synthetase.

Reagents

> Inactivation substrate solution: 1 mg/ml glutamine synthetase (glutamate–ammonia ligase) purified from *E. coli* (a simple purification procedure was described previously[14])
>
> Buffer solution: 0.2 *M* imidazole-HCl, pH 7.0
>
> Inactivation solution: 100 m*M* dithiothreitol (or 200 m*M* 2-mercaptoethanol), 50 μM $FeCl_3$ (this solution should be freshly prepared just before use)
>
> Glutamine synthetase assay mixture: γ-Glutamyltransferase assay mixture containing 0.4 m*M* ADP, 150 m*M* glutamine, 10 m*M* potassium arsenate (K-ASO_4), 20 m*M* NH_2OH, 0.4 m*M* $MnCl_2$, and 100 m*M* HEPES (pH 7.4)
>
> Stop mixture: One liter of solution containing 55 g of $FeCl_3 \cdot 6H_2O$, 20 g of trichloroacetic acid, and 21 ml of concentrated HCl

[11] R. L. Levine, *J. Biol. Chem.* **258,** 11828 (1983).

[12] L. Fucci, C. N. Oliver, M. J. Coon, and E. R. Stadtman, *Proc. Natl. Acad. Sci. U.S.A.* **80,** 1521 (1983).

[13] E. R. Stadtman and M. E. Wittenberger, *Arch. Biochem. Biophys.* **239,** 379 (1985).

[14] S. G. Rhee, P. B. Chock, and E. R. Stadtman, this series, Vol. 113, p. 213.

Assay Procedure

Assays are preformed in a 50-μl reaction mixture containing 0.5 μg of glutamine synthetase, 10 mM DTT (or 20 mM 2-mercaptoethanol), 3 μM FeCl$_3$, 50 mM imidazole-HCl (pH 7.0) and a source of antioxidant. The inactivation reaction is initiated by adding 20 μl of a freshly prepared mixture of FeCl$_3$ and DTT to 30 μl of solution containing glutamine synthetase and antioxidant protein. After 10 min at 30°, the remaining activity of glutamine synthetase is measured by adding 2 ml of γ-glutamyltransferase assay mixture. The glutamine synthetase reaction is incubated at 30° for 5 min, then terminated by the addition of 1 ml of stop mixture. The absorbance resulting from the γ-glutamyl hydroxamate–Fe^{3+} complex is measured at 540 nm.[15] Using this assay procedure, the mercaptan-specific antioxidant protein may be purified from yeast to homogeneity.

Purification Procedure

Step 1: Yeast Culture and Extraction. *Saccharomyces cerevisiae* BJ926 is grown on medium containing, per liter, 3.4 g Difco yeast nitrogen base without amino acids, 3 g (NH$_4$)$_2$SO$_4$, 20 g dextrose, and 1 g glutamate. Cultures are harvested at mid-logarithmic phase and frozen at $-20°$. Frozen cells (400 g) are suspended in 1.5 liters of water and centrifuged at 4000 g for 10 min. The cell pellet and 300 g of 0.5-mm-diameter glass beads are placed in a 450-ml stainless steel chamber for a bead beater (Biospec Products, Bartleville, OK). The chamber is filled with 20 mM HEPES buffer (pH 7.0) containing 2 mM phenylmethylsulfonyl fluoride. The slurry is vigorously blended for six 2-min intervals interspersed with periods of cooling on an ice–NaCl bath. The broken extracts are decanted, and the slurry is washed several times with a total volume of 500 ml of HEPES buffer plus phenylmethylsulfonyl fluoride. The combined cell extracts are centrifuged for 30 min at 9,000 g.

Step 2: Streptomycin Precipitation. The supernatant from Step 1 is brought to 1% streptomycin sulfate by slow addition of 10% streptomycin sulfate solution. After 15 min at 4°, the suspension is centrifuged at 45,000 g for 30 min to remove nucleic acid precipitates.

Step 3: Polyethylene Glycol Precipitation. To the supernatant from Step 2, polyethylene glycol [40% (w/v) polyethylene glycol, M_r 8,000] is slowly added with stirring to a final concentration of 8%. Stirring is continued for 15 min at 4°, and the suspension is centrifuged at 4,000 g for 10 min. The pellet is resuspended in 150 ml of 50 mM Tris-HCl buffer (pH

---

[15] B. M. Shapiro and E. R. Stadtman, this series, Vol. 17A, p. 910.

7.6). Insoluble material remaining after resuspension is removed by centrifugation at 45,000 g for 40 min.

Step 4: DEAE-Cellulose Chromatography. The clear supernatant from Step 3 is loaded onto a DEAE column (5 × 30 cm) previously equilibrated with 50 mM Tris-HCl buffer (pH 7.6). The column is washed with 400 ml of equilibrium buffer and eluted with a linear KCl gradient (0–300 mM) in a total volume of 4 liters Tris-HCl buffer (pH 7.6). Fractions of 20 ml are collected. Two milliliters from every tube is concentrated separately on Centricon 30 microconcentrators (Amicon, Danvers, MA), washed with 20 mM HEPES buffer (pH 7.6), and concentrated again on the same concentrators. The final volume of each concentrated and washed sample is adjusted to 0.2 ml by adding 20 mM HEPES buffer (pH 7.6), before assaying for the protector activity.

A broad peak of protector activity elutes at a KCl concentration of 120–150 mM. The peak fractions are pooled, and the proteins are precipitated with 70% ammonium sulfate. The precipitate is dissolved in 15 ml of 20 mM HEPES buffer (pH 7.4), and 2 ml of saturated ammonium sulfate solution is added before the protein solution is centrifuged to remove insoluble materials.

Step 5: Reversed Phase HPLC on TSK Phenyl-5PW. One-half (~10 ml) of the clear supernatants from Step 4 is applied at a flow rate of 5 ml/min to a HPLC phenyl column (21.5 × 150 cm) previously equilibrated with 20 mM HEPES (pH 7.4) containing 1 M ammonium sulfate. The proteins are eluted at a flow rate of 5 ml/min by a decreasing ammonium sulfate gradient from 1 to 0 M for 40 min. Two peaks of protector activity are seen. Since the protector proteins purified separately from these two fractions exhibit identical physical properties, the two fractions are pooled. The protein solution is concentrated on an Amicon concentrator, washed with 50 mM Tris-HCl buffer (pH 7.6), and finally concentrated to 2 ml. This reversed-phase chromatography step is repeated for the remaining one-half of the post-DEAE pool. The resultant washed and concentrated protector fractions from both runs are combined.

Step 6: Ion-Exchange HPLC on TSK DEAE-5PW. The concentrated samples (4 ml) are applied to a TSK DEAE-5PW column (21.5 × 150 mm) previously equilibrated with 50 mM Tris-HCl buffer (pH 7.6) and eluted at a flow rate of 5 ml/min by successive application of a KCl linear gradient from 0 to 0.1 M for 5 min and a KCl gradient from 0.1 to 0.25 M for 40 min. Fractions of 5 ml are collected, and 20 μl of each fraction is used directly for the protector assay. Peak fractions are concentrated to 0.2 ml.

Step 7: HPLC Gel Filtration. Concentrated fractions from the HPLC-DEAE column are applied to a TSK gel SW3000 (7.5 × 600 mm) previously equilibrated with 20 mM HEPES buffer (pH 7) containing 200 mM

KCl. When elution is continued at a flow rate of 1 ml/min with the same buffer, two peaks of activity emerge. The fractions from each peak are pooled separately, concentrated, divided into aliquots, and stored at $-20°$.

Properties

Purification Step 7, HPLC gel filtration, yields two protector activity peaks, I and II, at retention volumes of 14.8 and 19.5 ml, respectively. The molecular weights of these proteins determined from a plot of the logarithm of molecular weight versus mobility are approximately 500,000 and 90,000, respectively. Sodium dodecyl sulfate–polyacrylamide gel electrophoresis of peak I shows a single band with a molecular weight of 27,000. Peak II yields a major band of 27,000 and an impurity band of higher molecular weight. Although peak II is not completely homogeneous, it is possible to identify the 27,000-MW band as the polypeptide of protector protein in several preparations since the relative amounts of the band corresponds closely to the relative protector activities in the represented fractions across peak II. Although the semiquantitative nature of the protector assay does not allow exact evaluation of protector-specific activity, there is no significant difference in specific protector activity between the large and small size protectors. It appears, therefore, that the small protein contains 3–4 identical subunits, whereas the larger protein consists of 15–20 subunits.

The purification is complicated further by the fact that protector activity is resolved into two peaks by reversed-phase chromatography. Since both of these peaks exhibit similar gel filtration elution profiles, they are combined prior to further purification. This reproducible behavior with respect to reversed-phase column chromatography may also reflect different states of oligomerization of the protector as observed in gel filtration chromatography. Perhaps one form of protector disproportionates to the other during the subsequent purification steps. The purified protein is stable at $-20°$. The protecting capacity of both the partially purified and homogeneous preparations is completely lost on incubation for 2 min at 100°.

In addition to *E. coli* glutamine synthetase, yeast glutamine synthetase and rabbit muscle enolase have been subjected to the $DTT/Fe^{3+}/O_2$ MFO system.[6] All three inactivation reactions are time dependent and are inhibited progressively by the presence of increased amounts of the purified yeast antioxidant protein. For example, incubation of yeast glutamine synthetase for 14 min under the standard inactivation conditions (3 μM Fe^{3+}, 10 mM DTT, 30°) causes 80, 60, 32, and 4% inactivation of glutamine synthetase in the presence of 0, 0.046, 0.09, and 0.18 mg/ml of antioxidant, respectively.

As the concentration of Fe^{3+} is increased, the rate of inactivation increases rapidly, and much more protector is required to provide the same degree of protection. For example, after 14 min of incubation with 5 μM Fe^{3+}, in the presence (0.09 mg/ml) and absence of antioxidant, 25 and 0%, respectively, of the initial yeast glutamine synthetase activity remain; with 3 μM Fe^{3+}, the corresponding values are 68 and 20%. In the presence of 5 μM Fe^{3+}, complete protection (>95%) requires at least 0.45 mg/ml of antioxidant. The rate of mercaptan-dependent enzyme inactivation and the protection from inactivation by the antioxidant protein are independent of DTT or glutamine synthetase concentrations in the ranges of 1–10 mM and 2–100 μg/ml, respectively.[6]

Cu^{2+} can replace Fe^{3+} in the inactivation reaction, and the Cu^{2+}-provoked inactivation can be prevented by the antioxidant protein. When other mercaptans such as 2-mercaptoethanol, 2-mercaptoethylamine, thioglycerol, glycerol dimercaptoacetate, glutathione, dihydrolipoic acid, and cysteine are used instead of DTT, dall except cysteine inactivate glutamine synthetase and all the inactivation reactions are inhibited by the antioxidant protein. However, with the same amounts of antioxidant, the protection efficiency varies depending on the mercaptan used. Cysteine not only causes little or no inactivation of glutamine synthetase by itself, but it also decreases the rate of inactivation when added together with DTT. This is probably due to the fact that Fe^{2+} forms a very tight complex (log K = 6.2) of low solubility; the solubility product of Fe · Cys at pH 5–7 is 2.6 × 10^{-11} at 25°.[16]

Other nonmercaptan systems such as ascorbate, NADH oxidase/ NADH, and xanthine oxidase/hypoxanthine can inactivate enzymes in the presence of Fe^{3+} and O_2. These inactivation reactions can be prevented by catalase or glutathione peroxidase (which removes H_2O_2) or by EDTA (which chelates Fe). However, the protein purified from yeast does not prevent the nonmercaptan-dependent inactivation at all. Furthermore, the yeast protein does not possess catalase, glutathione peroxidase, superoxide dismutase, or iron chelation activities under conditions where this yeast protein inhibits the inactivation.[6] These observations led us to suggest that the 27-kDa protein might be an antioxidant enzyme specific to reactive sulfur species. Nevertheless, neither the role of this 27-kDa protein *in vivo* nor the mechanism of protection *in vitro* is known.

As a step toward clarifying the role of the 27-kDa protein *in vivo*, the effect of O_2, mercaptan, and iron on the level of the 27-kDa protein is

[16] N. Tanaka, I. M. Kolthoff, and W. Stricks, *J. Am. Chem. Soc.* **73**, 1990 (1955).

studied by using ^{125}I-labeled monospecific antibody derived against the 27-kDa protein.[17] Addition of each of the MFO components results in an induction of this protein. Radioimmunoassay results also show that the antioxidant protein is an abundant protein as it constitutes 0.7% of total soluble protein from yeast grown aerobically. Immunoblotting experiments reveal that rat tissues also contain a 27-kDa protein which can be specifically recognized by antibodies derived against the yeast protein. These results suggest that *in vivo* induction in yeast of the 27-kDa protein may represent an adaptive response evolved to protect cells against damage caused by mercaptan-dependent MFO systems, and the antioxidant protein is conserved in mammalian tissues.

[17] I. H. Kim, K. Kim, and S. G. Rhee, *Proc. Natl. Acad. Sci. U.S.A.* **86**, 6018 (1989).

[51] Protein Degradation as an Index of Oxidative Stress

By Robert E. Pacifici and Kelvin J. A. Davies

Introduction

Oxygen radicals and other activated oxygen species are known to participate in numerous physiological and pathological processes. Situations which augment oxidant exposure, or compromise antioxidant capacity, are commonly referred to as oxidative stress. Oxidative stress can result from exogenous sources (i.e., redox-active xenobiotics)[1-3] or from increases in endogenous oxidative metabolism (i.e., mitochondrial electron transport).[4,5] Regardless of its source, oxidative stress has been found to affect the behavior of several different cell types.

With oxidative stress implicated as a component of (or perhaps the cause of) several disease states, investigators have turned their attention to developing methods for more accurate quantification of oxidative stress. Standard cytotoxicity assays are relatively insensitive since they usually measure gross losses of cell integrity (trypan blue exclusion) or

[1] K. J. A. Davies and J. H. Doroshow, *J. Biol. Chem.* **261**, 3060 (1986).
[2] J. H. Doroshow and K. J. A. Davies, *J. Biol. Chem.* **261**, 3068 (1986).
[3] O. Marcillat, Y. Zhang, and K. J. A. Davies, *Biochem. J.* **254**, 677 (1989).
[4] K. J. A. Davies, A. T. Quintanilha, G. A. Brooks, and L. Packer, *Biochem. Biophys. Res. Commun.* **107**, 1198 (1982).
[5] K. J. A. Davies and P. Hochstein, *Biochem. Biophys. Res. Commun.* **107**, 1292 (1982).

viability (colony counts). Another approach relies on assaying end products of oxidative damage.

Oxidative stress has long been known to cause lipid peroxidation. Many investigators have, therefore, developed assays to detect products of lipid peroxidation in order to quantify oxidative stress. Colorimetric assays (thiobarbituric acid for malonyldialdehyde) as well as spectrophotometric assays (e.g., absorbance at 234 nm for conjugated dienes) are commonly used. While these assays are of obvious importance in characterizing membrane lipid oxidation, we believe that they may not be ideal measures of oxidative stress. Lipid peroxidation products typically appear after a lag time which may reflect initiation reactions. Lipids are also oxidized to form many different species of compounds. Hence, quantification of individual products of lipid peroxidation is not likely to be highly sensitive, nor are such measures likely to represent the extent to which the cell is oxidized.

The scheme shown in Fig. 1 illustrates the possible fates of oxidized cellular proteins. In addition to forming cross-linked aggregates, oxidative treatment can also give rise to modified proteins and peptide fragments. Our laboratory[6-31] as well as others[32-44] have shown that oxidatively modi-

[6] K. J. A. Davies, *J. Free Radicals Biol. Med.* **2,** 155 (1986).

[7] K. J. A. Davies and A. L. Goldberg, *J. Biol. Chem.* **262,** 8220 (1987).

[8] K. J. A. Davies and A. L. Goldberg, *J. Biol. Chem.* **262,** 8227 (1987).

[9] K. J. A. Davies, *J. Biol. Chem.* **262,** 9895 (1987).

[10] K. J. A. Davies, M. E. Delsignore, and S. W. Lin, *J. Biol. Chem.* **262,** 9902 (1987).

[11] K. J. A. Davies and M. E. Delsignore, *J. Biol. Chem.* **262,** 9908 (1987).

[12] K. J. A. Davies, S. W. Lin, and R. E. Pacifici, *J. Biol. Chem.* **262,** 9914 (1987).

[13] A. Taylor and K. J. A. Davies, *Free Radical Biol. Med.* **3,** 371 (1987).

[14] K. J. A. Davies and S. W. Lin, *Free Radical Biol. Med.* **5,** 215 (1988).

[15] K. J. A. Davies and S. W. Lin, *Free Radical Biol. Med.* **5,** 225 (1988).

[16] O. Marcillat, Y. Zhang, S. W. Lin, and K. J. A. Davies, *Biochem. J.* **254,** 677 (1988).

[17] K. Murakami, S. W. Lin, K. J. A. Davies, and A. Taylor, *Free Radical Biol. Med.* **8,** 217 (1990).

[18] K. J. A. Davies, in "Cellular and Molecular Aspects of Aging: The Red Cell as a Model" (J. W. Eaton, D. K. Konzen, and J. G. White, eds.), p. 15. Alan R. Liss, New York, 1985.

[19] P. Hochstein, A. Sevanian, and K. J. A. Davies, in "Purine and Pyrimidine Metabolism in Man" (E. J. Sigmeuler, ed.), Vol. 5, p. 325. Plenum, New York, 1986.

[20] K. J. A. Davies, in "Superoxide and Superoxide Dismutase in Chemistry, Biology, and Medicine" (G. Rotillio, ed.), p. 443. Elsevier, Amsterdam, 1986.

[21] K. J. A. Davies, in "Lipofuscin—1987: State of the Art" (I. Zs.-Nagy, ed.), p. 109. Elsevier, Amsterdam, 1988.

[22] K. J. A. Davies, in "Lipid Peroxidation in Biological Systems" (A. Sevanian, ed.), p. 100. American Oil Chemists' Society, Champaign, Illinois, 1988.

[23] K. J. A. Davies, in "Cellular Antioxidant Defense Mechanisms" (C. K. Chow, ed.), Vol. 2, p. 320. CRC Press, Boca Raton, Florida, 1988.

[24] K. J. A. Davies, in "The Role of Oxygen Radicals in Cardiovascular Diseases" (A. L'Abbate and F. Ursini, eds.), p. 143. Kluwer Academic Pub., Dordrecht, 1988.

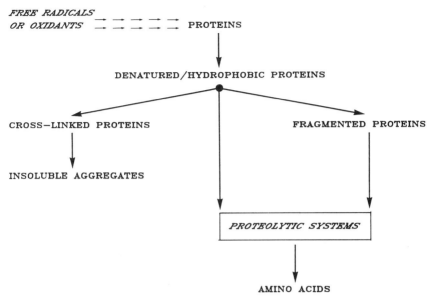

FIG. 1. Proposed scheme of protein damage, fragmentation, and aggregation by free radicals or oxidants and protein degradation by proteolytic systems.

[25] R. E. Pacifici, S. W. Lin, and K. J. A. Davies, *in* "Basic Life Sciences: Oxygen Radicals in Biology and Medicine" (M. G. Simic and K. A. Taylor, J. F. Ward, and C. von Sonntag, eds.), Vol. 49, p. 531. Plenum, New York, 1988.

[26] K. J. A. Davies, *in* "Basic Life Sciences: Oxygen Radicals in Biology and Medicine" (M. G. Simic and K. A. Taylor, J. F. Ward, and C. von Sonntag, eds.), Vol. 49, p. 575. Plenum, New York, 1988.

[27] D. C. Salo, S. W. Lin, R. E. Pacifici, and K. J. A. Davies, *Free Radical Biol. Med.* **5,** 335 (1988).

[28] D. C. Salo, R. E. Pacifici, and K. J. A. Davies, *J. Biol. Chem.* in press (1990).

[29] R. E. Pacifici, D. C. Salo, and K. J. A. Davies, *Free Radical Biol. Med.* **7,** 521 (1989).

[30] R. E. Pacifici and K. J. A. Davies, *FASEB J.* **2,** A1007 (Abstr. 4135) (1988).

[31] K. J. A. Davies, S. W. Lin, and R. E. Pacifici, *in* "International Committee on Proteolysis Newsletter" (J. S. Bond, ed.), p. 3. August, 1988.

[32] S. P. Wolf, A. Garner, and R. T. Dean, *Trends Biochem. Sci.* **11,** 27 (1986).

[33] R. T. Dean and J. K. Pollak, *Biochem. Biophys. Res. Commun.* **126,** 1082 (1985).

[34] R. T. Dean, S. M. Thomas, and A. Garner, *Biochem. J.* **240,** 489 (1986).

[35] S. P. Wolf and R. T. Dean, *Biochem. J.* **234,** 399 (1986).

[36] R. L. Levine, C. N. Oliver, R. M. Fulks, and E. R. Stadtman, *Proc. Natl. Acad. Sci. U.S.A.* **78,** 2120 (1981).

[37] L. Fucci, C. N. Oliver, M. J. Coon, and E. R. Stadtman, *Proc. Natl. Acad. Sci. U.S.A.* **80,** 1521 (1983).

[38] R. L. Levine, *J. Biol. Chem.* **258,** 11823 (1983).

[39] R. L. Levine, *J. Biol. Chem.* **258,** 11828 (1983).

[40] K. Nakamura and E. R. Stadtman, *Proc. Natl. Acad. Sci. U.S.A.* **81,** 2011 (1984).

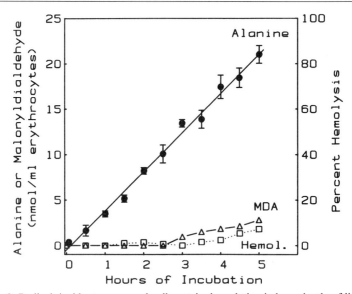

FIG. 2. Radicals/oxidants cause red cell protein degradation independently of lipid peroxidation and hemolysis. Rabbit erythrocytes were incubated in the presence of xanthine and xanthine oxidase (0.2 U/3.2 ml reaction volume). Alanine formation was measured as described in the text, and malonyldialdehyde (MDA) was estimated by the thiobarbituric acid assay as an indicator of lipid peroxidation. Percent hemolysis was determined by percent hemoglobin release. (Reprinted, with permission, from Ref. 23.)

fied proteins are preferred substrates for proteolytic degradation. This correlates well with the observation that intact cells exposed to oxidants exhibit increased rates of protein turnover.[6-8,18-23,36]

Protein degradation is apparent immediately (no lag phase) after red blood cells (RBC) (Fig. 2), muscle cells, or bacteria are treated with an oxidant.[6-8,18-23,36] The response appears to be linear over at least a 3-hr treatment time (Fig. 2). Protein degradation assays provide powerful advantages over other techniques. For example, it is possible to perform amino acid analysis to obtain a more complete profile of products. The assays are inherently more sensitive due to the homogeneity of products formed (i.e., amino acids) and the amplification of signal that fluorescence and radiolabeling techniques provide.

Since the study of protein degradation following oxidative stress is a

[41] A. J. Rivett, *J. Biol. Chem.* **260,** 300 (1985).
[42] E. R. Stadtman and M. E. Wittenberger, *Arch. Biochem. Biophys.* **239,** 379 (1985).
[43] J. E. Roseman and R. L. Levine, *J. Biol. Chem.* **363,** 3203 (1987).
[44] J. Cervera and R. L. Levine, *FASEB J.* **2,** 2591 (1988).

relatively new area, published methods are often difficult to find. We have developed and compiled a variety of methods for measuring protein modification and degradation both *in vitro* and *in vivo*. Each method has its advantages, as well as limitations; it is therefore recommended that the investigator scrutinize each method with his/her question in mind to see which is most suitable. We hope that these methods will prove useful in a variety of systems, both as independent indices of oxidative stress and as components in more complete studies of the cell biology of oxidative stress.

Exposure of Cells and Proteins to Oxidative Stress

Most readers are probably well versed in the large variety of electrochemical, photochemical, enzymatic, and radiolytic, radical/oxidant generating systems available. We would like, however, to briefly outline some of the most common systems, with an emphasis on their relation to proteolysis. Many of the reagents used to form oxygen radicals may also affect proteolytic enzymes; we therefore caution investigators to choose their generating systems carefully.

Exposure of Intact Cells to Radicals/Oxidants

Intact cells can be subjected to oxidative stress by the direct addition of oxidants. This can be as experimentally simple as growing cells at elevated O_2 tensions. We have found hydrogen peroxide (H_2O_2) to be a convenient reagent as it is commercially available in high purity and is easy to quantify. Redox-active compounds can also be added to cells to generate free radicals. Quinones such as menadione and paraquat have been used extensively to generate free radicals; however, one must always ascertain if it is the compound or the radicals/oxidants generated which produce the effect. Enzymatic production of oxygen radicals is commonly achieved using the xanthine (X)–xanthine oxidase (XO) system. While this system is simple and versatile, it must be used with extreme caution when studying proteolysis since commercial preparations of XO are heavily contaminated with proteases. This difficulty can be overcome by performing the proper controls of incubating XO without substrate or by preparing XO from buttermilk in the laboratory.

Exposure of Purified Proteins to Radical/Oxidants

Changes in the structure and function of individual proteins can be determined by exposing purified proteins to radical-generating systems. Radiolysis of water can be manipulated to generate a variety of oxygen radicals.[9–12] Protein solutions in deionized double-distilled water (no buffer) can be subjected to γ-irradiation (usually from a ^{60}Co source), under 100% O_2, to generate O_2^- + $\cdot$OH. The same system under N_2O yields essentially all $\cdot$OH. O_2^- can be produced by irradiating under 100% O_2 in the presence of 10 mM sodium formate. By changing the period of irradiation, different radical to protein molar ratios can be achieved. We[6,27,28] have also exposed proteins to H_2O_2 or to mixtures of H_2O_2 plus iron or copper ($\pm$ chelating agents). Several groups have investigated the effects of H_2O_2 alone[45–47] and in conjunction with peroxidases[48,49] on proteins. Stadtman, Levine, and co-workers have treated bacterial glutamine synthetase (glutamate–ammonia ligase) with an ascorbate–iron system to mimic mixed-function oxidase modifications.[36–44]

Exposure of proteins to $\cdot$OH (or mixtures of $\cdot$OH and O_2^-) represents a relatively indiscriminate damaging system; in other words, most amino acid residues are good targets for modification. In the absence of transition metals, H_2O_2 and O_2^- do not appear to be effective agents in oxidizing proteins. Random protein damage may occur when H_2O_2 or O_2^- are added in concert with exogenous metals (probably via $\cdot$OH, ferryl, or perferryl species). Proteins which contain transition metals (such as superoxide dismutase and hemoglobin) may suffer site-specific damage when H_2O_2 or O_2^- is added.[27,28,47,50] Peroxidases have also been shown to potentiate H_2O_2–protein interactions.[48,49]

Regardless of the source of radical/oxidant, working with isolated proteins provides a simple system with which to model *in vivo* protein oxidation.

Characterization of Oxidative Modification to Proteins

The products formed on oxidation of proteins are clearly a function of several factors, including which proteins are being studied, the dose and type of radical/oxidant to which proteins are exposed, and the environment under which the exposure occurs. There are, however, certain characteristic products which are formed in most situations. A brief list is presented below for readers who may not be familiar with protein oxidation.

[45] D. Tew and P. R. Ortiz de Montellano, *J. Biol. Chem.* **263**, 17880 (1988).
[46] S. E. Mitsos, D. Kim, B. R. Lucchesi, and J. C. Fantone, *Lab Invest.* **59**, 824 (1988).
[47] C. L. Borders, Jr., and I. Fridovich, *Arch. Biochem. Biophys.* **241**, 472 (1985).
[48] F. LaBella, P. Waykole, and G. Queen, *Biochem. Biophys. Res. Commun.* **30**, 333 (1968).
[49] S. Ohtaki, H. Nakagawa, M. Nakamura, and I. Yamazaki, *J. Biol. Chem.* **257**, 13398 (1982).
[50] S. Jewett, *Eur. J. Biochem.* **180**, 569 (1989).

Altered Amino Acids

With low concentrations of radical/oxidant, the first type of changes observed in proteins are changes in amino acid R groups.[6,10] Loss of native amino acids is accompanied by the formation of certain character-istic products (e.g., phenylkynurenine and bityrosine) as well as amino acid interconversion (e.g., hydroxylation of phenylalanine to form tyro-sine). One well-characterized example of amino acid interconversion is the oxidation of cysteine (SH) to cystine (SS).[51] Carbonyl formation may also be an early marker for protein oxidation.[11,38,39] We believe that these early alterations in primary structure give rise to changes in higher order structures.

Changes in Secondary, Tertiary, and Quaternary Structure

As a protein is progressively oxidized, more and more of its amino acids become altered. This change in primary structure alters the overall charge (isoelectric point), folding, and hydrophobicity of the pro-teins.[6,11,12,14,15,23,26,29–31] At higher radical to protein molar ratios, in the absence of O_2, · OH treatment of proteins results in protein aggregation. Formation of bityrosine between two proteins is an example of intermo-lecular covalent cross-linking, and formation of bityrosine within a pro-tein is an example of intramolecular cross-linking.[9–12] Hydrophobic patches on proteins may also combine to give rise to protein aggre-gates.[9–12] Formation of aggregates can be monitored using electrophoretic techniques. Covalent cross-links can be differentiated from hydrophobic interactions by the addition of denaturants [sodium dodecyl sulfate (SDS) urea, 2-mercaptoethanol] to the system. In the presence of 100% O_2, there is dramatic fragmentation of proteins exposed to · OH.[6,9–12,32–35] Many of the polypeptide fragments are smaller than 5000 MW and are, therefore, acid soluble. Investigators are urged to monitor increases in acid-soluble counts, both in the absence and presence of proteases.

Increased Susceptibility to Proteolysis

Virtually all the proteins we have tested (>20) exhibit enhanced rates of proteolytic degradation following oxidative denaturation.[6–31] Although it is not entirely clear by what mechanism proteins become better sub-strates for degradation, we have found an excellent correlation between increases in hydrophobicity and degradation (Fig. 3).[6,11,12,14,15,23,26,29–31,52]

---

[51] M. J. McKay and J. S. Bond, *in* "Intracellular Protein Catabolism," p. 351. Alan R. Liss, New York, 1985.

[52] R. H. Rice and G. E. Means, *J. Biol. Chem.* **246,** 831 (1971).

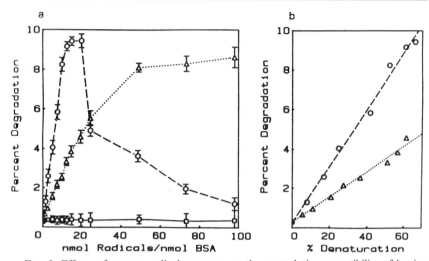

FIG. 3. Effects of oxygen radical exposure on the proteolytic susceptibility of bovine serum albumin (BSA) in extracts of *Escherichia coli*. BSA was radiolabeled by reductive methylation.[52] The [³H]BSA was then exposed to ·OH (O), to $O_2^- + O_2$ (□), or to ·OH + $O_2^- + O_2$ (△) (see Exposure of Purified Proteins to Radicals/Oxidants). Oxygen radical exposures are expressed as all ·OH, all O_2^-, or 50% ·OH + 50% O_2^-. Both treated and untreated samples of BSA were extensively dialyzed and then incubated with cell-free extracts of *E. coli* (60 min at 37°) as described for Bacterial Extracts in the section on Preparation of Cell-Free Extracts. Percent [³H]BSA degradation is shown in (a). (b) shows a least-squares linear regression for the relationship between BSA denaturation/hydrophobicity and proteolytic susceptibility. Denaturation was measured by [³H]BSA solubility in 3.0 M KCl plus 50 mM sodium succinate at pH 4.0, following oxygen radical exposure in the range 1–15 nmol radicals/nmol BSA. Values are means ± S.E. of three independent determinations. BSA becomes a progressively better substrate for degradation by *E. coli* extracts at low oxidant to protein ratios. At higher oxidant to protein ratios, protein cross-linking prevents BSA degradation (see Changes in Secondary, Tertiary, and Quaternary Structure). (Reprinted with permission, from Ref. 14.)

Based on such results, we first proposed that denaturation/hydrophobicity is the key to the increased proteolytic susceptibility of oxidatively modified proteins. Levine has also recently demonstrated a hydrophobic subpopulation of oxidatively modified glutamine synthetase molecules which are preferred substrates for proteolytic digestion by bacterial extracts.[44] Further support for increases in hydrophobicity acting as a generalized trigger for proteolysis has recently been obtained in our laboratory.[27-31] It appears that the enzymes [e.g., macroxyproteinase (MOP), see below] which may be responsible for degrading oxidatively denatured proteins prefer hydrophobic residues at or near the cleavage site.

Increased Hydrophobicity

In our previous studies on protein oxidation [using bovine serum albumin (BSA) as a model protein], we were able to assess increases in hydrophobicity by quantitating losses in solubility when BSA was suspended in 3.0 M KCl at its isoelectric point. Several chromatographic resins are now available which separate proteins based on their overall hydrophobicity. Proteins are typically eluted from these resins under gentle conditions. The proteolytic susceptibility of proteins eluted from such columns can be determined and correlated with their hydrophobicity.

Macroxyproteinase

We have recently isolated a 600-kDa protease complex [macroxyproteinase (MOP)] which appears to be largely responsible for the degradation of oxidatively modified hemoglobin, superoxide dismutase (SOD), and other proteins in red blood cells.[27–31] The enzyme seems to be identical to the 600- to 700-kDa multicatalytic protease complex (for which no physiological function was previously proposed) isolated from a variety of tissues by other investigators.[53,54] We have been able to inhibit the degradation of oxidatively denatured proteins by RBC extracts and purified MOP with three classes of proteolytic inhibitors: serine reagents, sulfhydryl reagents, and transition metal chelators. These results suggest that three active sites in the complex may work in concert to remove aberrant proteins from the cell. This proteolytic complex prefers hydrophobic and basic residues.[27–31,55] Intracellular proteins may resist degradation by MOP until they are oxidatively denatured. Oxidative denaturation exposes the previously buried hydrophobic residues which MOP requires to recognize its substrates. We have suggested the name macroxyproteinase for this proteolytic complex based on its large size (macro-) and its selectivity for oxidatively denatured proteins (-oxy-), adding the generic term -proteinase since the enzyme has such a broad inhibition profile.

Measurement of Protein Degradation in Intact Cells

Following oxidative stress, rapid increases in protein turnover have been observed. Several techniques which rely on metabolic labeling or amino acid detection have been developed to assay for proteolysis in

[53] R. Hough, G. Pratt, M. Rechsteiner, J. S. Bond, and M. Orlowski, *in* "International Committee on Proteolysis Newsletter" (J. S. Bond, ed.), p. 3. January, 1988.

[54] A. J. Rivett, *Arch. Biochem. Biophys.* **268**, 1 (1989).

[55] R. Hough, G. Pratt, and M. Rechsteiner, *J. Biol. Chem.* **262**, 8303 (1987).

intact cells. We present methods for the metabolic labeling of cells and methods to detect the release of amino acids from cells.

Radioactive Tracer Techniques in Intact Cells

Any cell with the capacity for protein synthesis can be metabolically labeled by incubation with an amino acid carrying a radioisotope. The newly synthesized proteins containing the radiolabel then serve as *in situ* substrates for cellular proteolytic enzymes.

Nearly all amino acids are commercially available in radioactive form. Leucine is commonly used since it is a major constituent of proteins and therefore exhibits extensive incorporation. Leucine is available as [³H]leucine as well as [¹⁴C]leucine. Either form can be used independently, or the two can be used together in dual-label experiments to follow the fates of proteins metabolically labeled at different times in the same cell population. Methionine is often used to label cell proteins since it is available in the [³⁵S]methionine form. Investigators should be aware that not all proteins contain methionine [e.g., heat-shock protein (HSP30)], and therefore such proteins would elude detection. One solution to this problem is to label cells with a cocktail of [³⁵S]methionine and [³⁵S]cysteine which is now commercially available from ICN as "Trans label."

Reticulocytes. Red blood cells are easily obtained from animals in high abundance and purity. However, erythrocytes (which make up >95% of circulating red cells) lose their ability to synthesize proteins during maturation from reticulocytes.

Reticulocytes can be obtained from rabbits in which reticulocytosis has been induced by daily bleedings or by repeated phenylhydrazine treatments. Reticulocyte counts should always be made. The phenylhydrazine procedure may not be ideal for studies of free radicals/oxidants because phenylhydrazine itself is a red cell oxidant which may induce unwanted effects.[56] Reticulocytes are collected and diluted approximately 1 : 1 with 0.9% (w/v) saline containing 10 U/ml heparin at 4°. The diluted blood is centrifuged at 500 g for 10 min (4°), and the supernatant, "buffy coat," and upper 10% of the packed cells (including the white blood cells) are carefully removed by aspiration. The remaining packed cells are resuspended in 4 volumes of Krebs–Ringer phosphate buffer (KRP) consisting of 143 mM NaCl, 5.7 mM KCl, 1.4 mM MgCl₂, and 18 mM sodium phosphate (pH 7.4) and centrifuged at 500 g for 10 min (4°); this centrifugal wash is repeated twice.

The packed reticulocytes are then suspended in 3 to 4 volumes of KRP

[56] S. Rapoport and W. Dubiel, *Biomed. Biochim. Acta* **43**, 23 (1984).

(± 3 mM CaCl$_2$) supplemented with 5 mM glucose, 30 μg/ml FeSO$_4$, and amino acids at plasma concentrations.[57] The cell suspension should be preincubated for 15 min, at 37°, in a shaking water bath. At this time, a radiolabeled amino acid is added for incorporation into new proteins, using a pulse–chase technique. The amount of radioactivity added will depend on the isotope employed, typically 5 to 10 μCi/ml cells for ^{3}H, 0.1 to 1 μCi/ml cells for ^{14}C and ^{35}S. For example, the reticulocytes are incubated for 15–30 min with [^{3}H]leucine (or [^{35}S]methionine) and then washed (800 g for 10 min) 3 times in KRP supplemented with 5 mM glucose, 30 μg/ml FeSO$_4$, plasma amino acids, and 10 mM unlabeled ("cold") leucine (or methionine). This pulse–chase technique ensures that unincorporated radiolabeled amino acids are removed. Should the labeled proteins be degraded, the "cold chase" will prevent reincorporation of the label. These labeled cells are incubated with or without oxidant, and protein degradation is measured with high sensitivity by the release of acid-soluble counts as described below.

Following incubation ($\pm$ radical/oxidants) sample tubes are placed in an ice-water bath for 5 min. Cell aliquots (e.g., 50 μl) are then diluted 1 : 20 with 10% (w/v) trichloroacetic acid (TCA) on ice in order to cause cell lysis and precipitate cell membranes and intact proteins. After at least 20 min on ice (convenient stopping point), the tubes are centrifuged at 3000 g for 15 min to sediment membranes and remaining intact proteins. Protein degradation is assessed by measuring acid-soluble counts in the supernatant. Amino acids and small peptides (<5000 MW), formed by protein degradation, are acid soluble. Thus, if newly synthesized proteins are degraded, some of the radiolabeled amino acid will be in the supernatant. Aliquots of the acid supernatants are diluted approximately 10-fold with liquid scintillation cocktail and counted for 5 min.

It is possible to estimate the percentage of newly synthesized protein which has been degraded if the total counts incorporated after the radiolabeling stage are measured. For this procedure, aliquots of labeled (but untreated) cells are precipitated with TCA and centrifuged as above. The pellet is washed twice by centrifugation in acid and once by centrifugation in a 1 : 1 ethanol–ether mixture. The pellet is solubilized by the addition of Solubilizer (Nuclear-Chicago), diluted in scintillation cocktail, and counted as above. Background acid-soluble counts (before the experiment, or from samples on ice) are subtracted from experimental acid-soluble counts. Percent degradation is then estimated (with corrections for volume differences, if necessary) as follows:

% Degradation = acid-soluble counts/acid-precipitable counts

[57] L. E. Mallette, J. H. Exton, and C. R. Park, *J. Biol. Chem.* **244,** 5713 (1969).

Bacteria. Intact bacteria can be studied[14,15,36] by essentially the same labeling procedures as those described above for reticulocytes. Prokaryotes offer several advantages for studies of protein degradation. Bacteria are readily maintained in culture and divide rapidly. This allows rapid labeling of proteins with high specific activities. Bacterial studies also provide the versatility of allowing for selection of mutant strains with desired phenotypes.

Wild-type *Escherichia coli* are grown in suspensions of M9 minimal media plus glucose (2 g/liter). Mutant strains may require specialized growth conditions. When the culture reaches an OD_{595} of 0.2, 0.08 $\mu Ci/ml$ of ^3H-labeled amino acid is added. After 2 to 3 hr of growth, the cells are washed twice by centrifugation (2000–3000 g for 10 min) in the presence of glucose–M9 plus 75 $\mu g/ml$ of unlabeled ("cold") amino acid. The radiolabeled bacterial cells are used to assess differences in protein turnover following oxidant exposure, as described below.

At the end of the experimental exposure, 0.1 ml of 50% (w/v) TCA is added per milliliter of cell suspension. After at least 30 min on ice the sample tubes are centrifuged (3000 g, 15 min). To measure acid-soluble counts (protein degradation), 0.5 ml of supernatant is diluted with 5 to 10 ml of liquid scintillation cocktail and counted for 5 min.

Total incorporation can be measured as follows: 1 ml of untreated radiolabeled cell suspension is precipitated with 0.1 ml of 50% (w/v) TCA. The pellet is washed twice by centrifugation in 5% (w/v) TCA. The pellet is then solubilized by treatment with 0.1 ml of 100% formic acid at 55°, and diluted 10-fold with scintillation cocktail for counting. The equation given above can be used to calculate percent degradation.

Other Cells and Organelles. The procedures outlined above are amenable to a variety of other systems. Cultured cells (primary cultures, strains, and cell lines) can all be metabolically labeled and subjected to oxidative stress.[58] The effects of radical/oxidants on protein degradation in mitochondria can be studied using intact, well-coupled preparations. Rat liver and bovine heart mitochondria have been used for such studies. Mitochondria offer the advantage that only 12–15 polypeptides are synthesized *de novo,* thus simplifying the task of identifying degraded proteins. The polypeptides synthesized by mitochondria undergo rapid degradation *in vitro,* thus increasing the background levels of proteolysis in control samples.[16]

Nonradioactive Methods in Intact Cells

While radioisotopes provide powerful and sensitive techniques for studying protein degradation, it may not always be possible to use labeled

[58] H. Joenje, J. J. P. Gille, A. B. Oostra, and P. van der Valk, *Lab. Invest.* **52,** 420 (1985).

cells. For this reason, many other techniques for detecting proteolysis have been developed.

Alanine Formation. Mammalian cells are not able to synthesize all of the amino acids they require. Red blood cells, for example, are unable to synthesize alanine *de novo* or by metabolic interconversion. Alanine appearance in red cells is, therefore, indicative of protein degradation. Alanine formation is measured in perchloric acid extracts of red cells. Perchloric acid (0.4 M final) is added to lyse the cells and precipitate membranes and intact proteins (e.g., add 1.0 ml of 1.6 M perchlorate to 3.0 ml of red cell suspension). Tubes are left in ice-water for 10 min and then centrifuged at 3000 g for 15 min. Next, 1.0-ml aliquots of the supernatant are adjusted to pH 9.0 by addition of 0.2 ml of 2.0 M KOH and maintained at pH 9.0 by the addition of 0.8 ml of 0.5 M Tris-HCl. After 1 hr on ice, during which time the perchlorate crystallizes out of solution, alanine formation (release from intact proteins) can be measured. Alanine is measured, in an enzyme-linked fluorescence assay, by the reduction of NAD$^+$ to NADH. For this procedure 0.5 ml of the perchlorate–KOH–Tris supernatants are added to 0.5 ml of a buffer containing 0.8 M Tris (pH 9.0), and 40 mM ethylenediaminetetraacetic acid (EDTA). Next, 0.5 ml of 6.6% hydrazine hydrate solution (pH 9.0), 0.1 ml of 20 mg/ml NAD$^+$, and 0.1 ml of 0.2 mg/ml alanine dehydrogenase (~30 U/mg) are added. Each tube is incubated for 1 hr at 37° in a shaking water bath. NADH is detected fluorometrically at 340 nm excitation and 450 nm emission, in comparisons with standards. The assay gives a linear response for alanine contents up to 25 nM, where 1 mol of NADH is equivalent to 1 mol of alanine.

The same type of protocol can be adapted to any cell, using essential amino acids as markers. Investigators should choose to assay amino acids that are relatively resistant to oxidative damage (to prevent false negatives) and that are not formed by oxidation (to prevent false positives). Initial studies should include recovery experiments with known amounts of exogenously added amino acid standards.

Fluorescamine Reactivity. The amino terminus and the ε-amino group of lysine residues are the only primary amines present in intact proteins. As a protein is degraded into its constituent amino acids, 1 mol of primary amine is exposed (the amino group) for each amino acid released. Colorimetric or fluorometric reagents which derivatize primary amines consequently provide sensitive assays to monitor protein degradation. We have used fluorescamine to detect protein degradation in neutralized TCA supernatants of cells with or without radical treatment.[7–12] For this protocol, HEPES is used instead of Tris (which is itself a primary amine). Neutralized supernatant (0.25 ml) is taken to 1.5 ml by the addition of 50 mM HEPES (pH 9.0). Next, 0.5 ml of a 0.3 mg/ml fluorescamine solution in acetone is slowly added while vortexing each tube. The results are evaluated by fluorimetry using an excitation wavelength of 390 nm and an emission wavelength of 475 nm. Any amino acid can be used to construct a concentration curve. Several other reagents (e.g., *o*-phthalaldehyde) are also commonly used for highly sensitive detection of primary amines.

Complete Amino Acid Analysis. With recent advances in HPLC technology, it is now possible to easily conduct complete amino acid analysis with multiple samples. This type of analysis is especially powerful in detecting oxidant-induced protein degradation, where individual amino acid concentrations may not be reliable. Obtaining a complete profile of amino acids gives higher specificity and greater reliability. Amino acids which are not normally found in proteins (i.e., ornithine) can also be monitored to indicate changes in metabolism. Release of posttranslationally modified amino acids (i.e., hydroxyproline) can also be followed using amino acid analysis, and such an approach may be useful for quantifying degradation of specific proteins which contain these unusual amino acids (i.e., collagen).

Immunological and Electrophoretic Methods

There may be instances where investigators are particularly interested in the fate of one or a few proteins *in situ* following oxidative treatment. The methods outlined below should provide the basis for such analyses.

Two-Dimensional Gels/Autoradiography. Two-dimensional (2D) electrophoresis is a powerful method for separating proteins based on charge and size. Proteins separated by 2D electrophoresis can be visualized using standard Coomassie blue and silver staining techniques, or by autoradiography if the cells are radiolabeled. Resolutions on 2D gels are typically sufficiently high so that each spot is representative of one protein type. Large decreases in spot intensity ($>10\%$) can be quantified using scanning densitometry. Disappearance of a protein spot may be due to enzymatic hydrolysis of that polypeptide.

Western Blots and Immunoprecipitation. Recent advances in antibody technology can be adapted to study the *in situ* turnover of individual proteins. Individual proteins can be immunoprecipitated from radiolabeled cell extracts and quantified using liquid scintillation techniques. Alternatively, unlabeled extracts can be separated using SDS–PAGE. The protein of interest can be visualized and quantified using Western blot analysis. Interactions between antibodies and their antigens are generally highly specific. Investigators must be aware that loss of antibody binding may be due to protein degradation or to a loss of antigenicity caused by protein oxidation. Stadtman and co-workers[36] used immunologic techniques to demonstrate a loss of glutamine synthetase in *E. coli* after oxidative stress. These results were later confirmed to be due to proteolysis (rather than loss of antigenicity) by *in vitro* immunoprecipitation of oxidized glutamine synthetase.[59] Polyclonal antibody preparations, which recognize a variety of epitopes, should be used in these studies (rather than monoclonal antibodies) to ensure that results are not caused by the loss of only one critical epitope.

Measurement of Protein Degradation with Cell-Free Systems and Purified Proteins

Cell-free extracts provide versatile systems for studying oxidative protein degradation. Extracts which contain the soluble (nonlysosomal) proteolytic systems which exist in intact cells are easy to prepare and can be stored for future studies. Extracts can be made from a variety of cell types as well as from both control and oxidant-treated cells (to test for the potential induction of proteolytic activities). Radicals or oxidants may be added directly to the extracts for studies of protein degradation, or preexposed (labeled) proteins may be studied.

Labeling Techniques with Purified Proteins

Studies which monitor the degradation of exogenous proteins by cell-free extracts require that the protein of interest be labeled. Using chemical methods, it is possible to tag purified proteins *in vitro* with radioactive or fluorescent labels.

Radiolabels. Purified proteins can be radiolabeled for studies of radicals/oxidants and degradation by reductive methylation procedures.[8–16,27–31] In such methods, a $[^3H]$-, $[^{14}C]$-, or $[^{13}C]$methyl group is bound to free amino groups on the protein. The ε-amino groups of lysine residues are most likely to be labeled, since the only other primary amine is at the amino terminus. High labeling efficiency is not necessary since this is a tracer method for protein degradation. Significant alteration of proteolytic susceptibility (or other properties) may result if too many amino groups are modified.

Proteins which are soluble at pH 9.0 can be labeled by reductive methylation with formaldehyde (as the source of isotope) and sodium borohydride (as the reductant). For proteins which are soluble only below pH 9.0, a closely related (and more gentle) procedure involving formaldehyde and sodium cyanoborohydride can be employed. These labeling procedures have been described in detail by Rice and Means[52] and Jentoft and Dearborn.[60]

---

[59] C. N. Oliver, personal communication (1989).
[60] N. Jentoft and D. G. Dearborn, *J. Biol. Chem.* **254,** 8194 (1979).

Several kits are commercially available for radioiodinating proteins. ^{125}I- or ^{131}I-labeled proteins can be produced quickly and with very high specific activities. When studying radical damage to proteins, it is highly undesirable to have a γ source intimately associated with the protein for long periods of time. Yallow and Berson[61] were the first to notice that radioiodinated serum albumin became a better substrate for degradation in humans as a function of storage time. This type of protocol should be reserved for proteins in very low abundance, since high specific activities can be achieved.

Fluorescent Dyes. Investigators without radioisotope facilities may wish to use fluorescent dyes to label purified proteins. Fluorescent dyes can be easily coupled to the R groups of many amino acids within proteins. While the radiolabeling techniques are limited to lysine and tyrosine, fluorescent probes can be chosen which react with other amino acids.

Preparation of Cell-Free Extracts

Red Blood Cells. Erythrocytes or reticulocytes are harvested and washed as described for Radioactive Tracer Techniques in Intact Cells. Washed packed RBC are hypotonically lysed by suspension in 1.5 volumes of 1 mM DTT. To insure complete disruption, the cells are vigorously stirred for 1 hr (4°). Next, the suspension is centrifuged at 10,000 g for 10 min (4°) to remove most of the membranes and any unbroken cells. Membranes (and reticulocyte organelles) are removed by high-speed centrifugation (20,000 g for erythrocytes, 100,000 g for reticulocytes) for 90 min at 4°.

A cell-free "extract" is prepared from the supernatant (lysate) by extensive dialysis against a suitable buffer. The dialysis removes small molecules such as ATP, ubiquitin, and other ions. Some or all of these may play roles in intracellular protein catabolism (ATP/ubiquitin pathway). We have not found degradation of any of our oxidatively modified substrates to require these cofactors; however, investigators studying new systems should always check both lysates and extracts for the effect of potential cofactors. Extracts can be stored for several months (at −70°) without appreciable loss of activity in buffers which contain 0.5 mM DTT and 10% (v/v) glycerol.

Bacterial Extracts. *Escherichia coli* is grown as described previously. The cells are harvested by centrifugation at 16,000 g for 15 min (4°). The cell pellet is resuspended in a buffer containing 10% sucrose and 50 mM

[61] R. S. Yallow and S. A. Berson, *J. Clin. Invest.* **36**, 44 (1957).

triethanolamine (pH 8.0) at a ratio of 5 ml of buffer per liter of original cell suspension. The cells are frozen in dry ice–ethanol and allowed to thaw at 0–4°. Lysozyme (100 μg/ml cells) treatment (to disrupt the cell wall) for 30 min at 0–4° is followed by addition of 2-mercaptoethanol (10 mM) and KCl (500 mM). The suspension is centrifuged at 40,000 g for 30 min, and the supernatant (lysate) is dialyzed against a suitable buffer in order to produce a cell-free extract.

Cultured Cells. Cultured cells in suspension can be harvested by centrifugation. Attached cells should be collected by scraping with a rubber policeman (avoiding trypsin for obvious reasons). Lysates and extracts can be prepared as described in above for RBC. Most cells will require successive rounds of freezing and thawing in order to ensure complete disruption. Still others may require mechanical disruption.

Other Tissues and Organelles. Similar procedures have been published for obtaining cell-free extracts from a variety of other tissues including bovine heart, bovine lenses, rat liver, and mitochondria.[3,17,23,24,41]

Assaying Proteolysis in Cell-Free Systems

Proteolysis assays can be performed quickly and without complex instrumentation. We have found 3- to 4-ml plastic conical centrifuge tubes to be the most effective reaction vessel. Typically, 3–10 μg of radiolabeled protein substrate is added to a reaction mixture (on ice) containing a suitable buffer (50 mM Tris-HCl), any appropriate cofactors, and the source of proteolytic enzyme (final volume 200 μl). The enzyme concentration should be manipulated to give no more than 5–10% degradation to prevent substrate limitations. Investigators should take care in pipetting cell-free extracts which contain glycerol, as they are very viscous. The samples are then incubated in a shaking water bath at 37° for 15 min to 1 hr. Next, samples are placed on ice, and the reaction is quenched by the addition of 0.8 ml of 12% (w/v) TCA (4°). Intact (undegraded) protein is pelleted by centrifugation for 15 min at 3000 g. A 0.5-ml aliquot of the supernatant is diluted 10-fold with scintillation cocktail and counted for 5 min. Total counts added should be determined from a separate set of tubes containing identical additions, except that water is substituted for the TCA. Background counts should also be determined from a set of tubes incubated without protease. Percent degradation is then calculated using the following formula:

% Degradation =
        (acid-soluble counts − background)/(total counts − background)

Measurement of Protein Degradation with Purified Proteases

Proteolytic enzymes from cell extracts have been isolated using oxidatively modified proteins as substrates.[14,15,27–31,41,43] Proteases isolated by such schemes (e.g., MOP) seem to be responsible for the degradation of oxidatively modified proteins *in situ*. In studies of oxidative protein degradation several commercially available protease preparations have also been used to degrade purified proteins which had been preexposed to oxidants.[12,23,32,35] We found that several proteases were able to degrade the oxidatively modified proteins more rapidly than control (untreated) proteins. Investigators can try common enzymes such as trypsin or chymotrypsin to test for oxidative denaturation of a protein *in vitro*. In assays using purified proteases, the overall protein concentration will be low; to aid precipitation, 50 μl of 3% (w/v) BSA should be added as a carrier.

Conclusions

Protein degradation provides a rapid, versatile, and sensitive method for quantifying oxidative stress. Protein degradation can be assayed in a variety of *in vivo* and *in vitro* systems. The protocols outlined above are readily adaptable to most biological systems. We hope that the methods outlined in this chapter inspire investigators to use protein degradation as an index of oxidative stress.

Acknowledgments

This work was supported by Grants ES 03598 and ES 03785 from the National Institutes of Health to K.J.A.D.

[52] Photodynamic Methods for Oxy Radical-Induced DNA Damage

By J. R. WAGNER, J. E. VAN LIER, C. DECARROZ, M. BERGER, and J. CADET

DNA damage in a cell associated with oxidative stress or exposure to ionizing radiation is in part induced by initial OH radical attack on DNA constituents. These reactions modify the chemical structure of DNA subunits (nucleobases and deoxyribose moieties), and they mark the onset of subsequent biochemical and biological effects observed in OH-generating systems.[1-4] The purpose of this chapter is to define the types of stable products which arise from the reactions of OH radicals with pyrimidine nucleosides of DNA and to provide some insight into the rather complicated mechanism by which they are formed.

Thymidine (dThd) and 2'-deoxycytidine (dCyd) are degraded into a mixture of products by photooxidation using 2-methyl-1,4-naphthoquinone (MQ) as a sensitizer and near-UV light ($\lambda > 320$ nm). The same types of pyrimidine products are formed by this photochemical reaction as by OH radical-induced degradation in aqueous oxygenated solutions. However, the attack of OH radicals on nucleosides takes place at several sites on both the pyrimidine base (OH radical addition either at C-5 or C-6; H-abstraction from the C-5 methyl group of thymidine) and the deoxyribose moiety (H-abstraction from several sites), whereas in the photochemical reaction the primary precursor of all products (pyrimidine radical cations) is generated specifically. Consequently, the product mixtures are cleaner and the products can be obtained in higher yields compared to using OH-generating systems. The chromatography techniques including high-performance liquid chromatography (HPLC) and thin-layer chromatography (TLC) are presented in order to provide simple and unambiguous analysis of these products.

Materials and Methods

Thymidine or 2'-deoxycytidine (10 mM) and 2-methyl-1,4-naphthoquinone (0.5 mM) which is recrystallized once from methanol are mixed in 1 liter of doubly distilled water and irradiated with 2 blak-ray lamps (λ_{max} 365 nm, I_0 10^{16} photons/cm^2/sec, UV Photoproducts Ltd.) for a period of 2 hr. The solutions are constantly purged with O$_2$ and cooled by circulating water ($\sim 4°$) during photolysis. The solutions are dried by rotary evaporation and dissolved in the appropriate solvent used in semipreparative HPLC. Roughly 50 mg in 200 μl is injected on HPLC for each run. The HPLC equipment consisted of an M6000 pump (Waters), with injector (Rheodyne) with R13 refractive index detector (Varian). The semipreparative columns are either reversed-phase (C$_{18}$ ODS 5 μm Ultrasphere, 75 mm i.d. × 250 mm, Beckman) operating with aqueous mobile phases or normal phase (10 μm Partisil, 75 mm i.d. × 300 mm, home

[1] R. Téoule and J. Cadet, in "The Effects of Ionizing Radiation on DNA" (J. Hüttermann, W. Köhnlein, R. Téoule, and A. J. Bertinchamps, eds.), p. 171. Springer-Verlag, Berlin, 1978.

[2] H. Sies (ed.), "Oxidative Stress." Academic Press, New York, 1985.

[3] C. von Sonntag, "The Chemical Basis of Radiation Biology." Taylor & Francis, London, 1987.

[4] G. Teebor, R. J. Boorstein, and J. Cadet, Int. J. Radiat. Biol. **54**, 131 (1988).

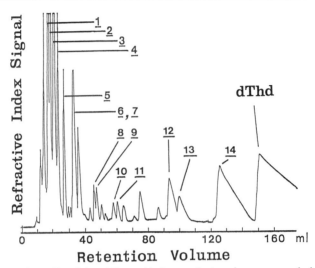

FIG. 1. HPLC profile of thymidine oxidation products using a reversed-phase column with 3% methanol in water as the mobile phase. Photoproducts are numbered in order of elution. **1**, N-dR-Formamide, where dR is 2-deoxy-β-D-*erythro*-pentofuranosyl; **2**, (−)-*trans*-(5S,6S)5,6-dihydroxy-5,6-dihydrothymidine; **3**, (+)-*trans*-(5R,6R)5,6-dihydroxy-5,6-dihydrothymidine; **4**, *trans*-(5S,6S)5-hydroperoxy-6-hydroxy-5,6-dihydrothymidine; **5**, *trans*-(5R,6R)5-hydroperoxy-6-hydroxy-5,6-dihydrothymidine; **6**, (−)-*cis*-(5R,6S)5,6-dihydroxy-5,6-dihydrothymidine; **7**, (+)-*cis*-(5S,6R)5,6-dihydroxy-5,6-dihydrothymidine; **8**, *cis*-(5R,6S)5-hydroperoxy-6-hydroxy-5,6-dihydrothymidine; **9**, *cis*-(5S,6R)5-hydroperoxy-6-hydroxy-5,6-dihydrothymidine; **10**, (5R*)1-dR-5-hydroxy-5-methylhydantoin; **11**, (5S*)1-dR-5-hydroxy-5-methylhydantoin; **12**, 5-hydroxymethyl-2'-deoxyuridine; **13**, 5-hydroperoxymethyl-2'-deoxyuridine; **14**, 5-formyl-2'-deoxyuridine; dThd, thymidine.

packed) with ethyl acetate–2-propanol–water (75 : 16 : 9) as the mobile phase. Two-dimensional TLC is performed on silica gel plates (0.25 mm thick, Machery-Nagel). Peroxides are detected by mixing the eluent with xylenol orange reagent[5] or by spraying the TLC plate with 1% (w/v) N, N-dimethyl-1,4-phenylenediamine in 50% aqueous methanol. Nucleosides are visualized on TLC by spraying with 0.4% (w/v) cysteine in 3 N H_2SO_4 followed by heating at 100° for a couple of minutes.

Isolation and Properties of Thymidine Photoproducts

Two complementary chromatography systems able to resolve the complex mixtures of thymidine oxidation products include reversed-phase HPLC (Fig. 1) and two-dimensional silica gel TLC[6] (Fig. 2). The

[5] B. L. Gupta, *Microchem. J.* **18**, 363 (1973).
[6] J. Cadet and R. Téoule, *J. Chromatogr.* **76**, 407 (1973).

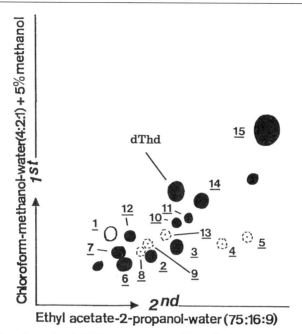

FIG. 2. Two-dimensional TLC separation on silica gel plates of thymidine oxidation products. Numbers correspond to compounds identified in the legend to Fig. 1. Products were localized by autoradiography with $^{14}CH_3$-labeled product mixtures (dark spots) or by spraying with cysteine reagent (all spots except pyruvamide, **15**). The approximate position of the hydroperoxides are given as dashed circles but are not clearly observed from analysis of total mixtures.

absolute identity of the numbered products (**1–14**) is confirmed by their characteristic ^{1}H-NMR spectra which are known from compounds obtained by synthesis or by exposure of thymidine in aqueous oxygenated solutions to ionizing radiation.[7,8]

The formamide-type compound **1** is a very polar photoproduct because it lacks the pyrimidine ring and elutes immediately on reversed-phase HPLC, yet **1** is well resolved on TLC and is visualized by its reaction with the cysteine spray reagent. The 5,6-saturated products of thymidine including the 5,6-dihydroxy-5,6-dihydrothymidine isomers (referred to as thymidine 5,6-glycols; **2, 3, 6, 7**) and the 5-hydroperoxy-6-hydroxy-5,6-dihydrothymidine isomers (**4, 5, 8, 9**) elute next on reversed-phase HPLC. The trans diastereoisomers elute before the corresponding cis analogs of each series, and, of these, *6S* elute before *6R* isomers, with

[7] J. Cadet and R. Téoule, *Tetrahedron Lett.*, 3325 (1972).
[8] J. Cadet and R. Téoule, *Bull. Soc. Chim. Fr.* 885 (1975).

the exception of cis-5,6-glycols (6 and 7) which are inseparable on reversed-phase chromatography. An important property of thymidine 5,6-glycols is their specific isomerization involving the N-1–C-6 bond (ring–chain tautomerism)[9]; as such, the less stable trans isomers readily convert to the cis forms (2 → 7 and 3 → 6).

The hydroperoxides are unambiguously located on HPLC or on TLC by their specific reaction with peroxide-sensitive reagents. Postcolumn reaction HPLC using Fe^{2+}–xylenol orange reagent has been shown to be an effective peroxide detection system for the analysis of thymidine hydroperoxides.[10] Hydroperoxides may also be separated on TLC although there is some decomposition of the hydroperoxide during analysis. The absolute configurations of the various diastereoisomers of thymidine 5,6-glycols and hydroperoxides have been established by chemical methods.[11] This recently received support from X-ray crystallography of cis-(5R,6S)-thymidine 5,6-glycol.[12] The hydantoin-type photoproducts 10 and 11 elute on HPLC or migrate on TLC as equally important compounds in intermediate zones. The remaining photoproducts including 5,6-unsaturated compounds (12–14) behave similarly on HPLC with thymidine as a result of the lipophilic 5,6 pyrimidine double bond in their structure. They are easily distinguished from the rest of the products by their characteristic UV absorption and by their ^{1}H-NMR spectra.

Mechanism of Sensitized Photooxidation of Thymidine

Thymidine undergoes sensitized photooxidation with an high efficiency [Φ(-dThd) ~ 0.6] and up to nearly complete substrate conversions (~90%). The photoproducts are distributed into two major groups with 1–11 representing about 70% and 12–14 the remaining 30% of the total decomposition (the proportions of each product can be roughly estimated from Figs. 1 and 2). The hydroperoxide yields are variable since these compounds are unstable.

The proposed mechanism of thymidine photooxidation is shown in Fig. 3. The initial photochemical step involves electron transfer from the pyrimidine to the triplet excited state of MQ, giving in this case the thymidine radical cation. All of the photoproducts are subsequently derived from hydration or deprotonation of this species.[13] Hydration leads

[9] J. Cadet, J. Ulrich, and R. Téoule, Tetrahedron 31, 2057 (1975).
[10] J. R. Wagner, J. E. van Lier, C. Decarroz, and J. Cadet, Bioelectrochem. Bioenerg. 18, 155 (1987).
[11] J. Cadet, R. Ducolomb, and R. Téoule, Tetrahedron 33, 1603 (1977).
[12] F. E. Hruska, R. Sebastian, A. Grand, L. Voituriez, and J. Cadet, Can. J. Chem. 65, 2618 (1987).

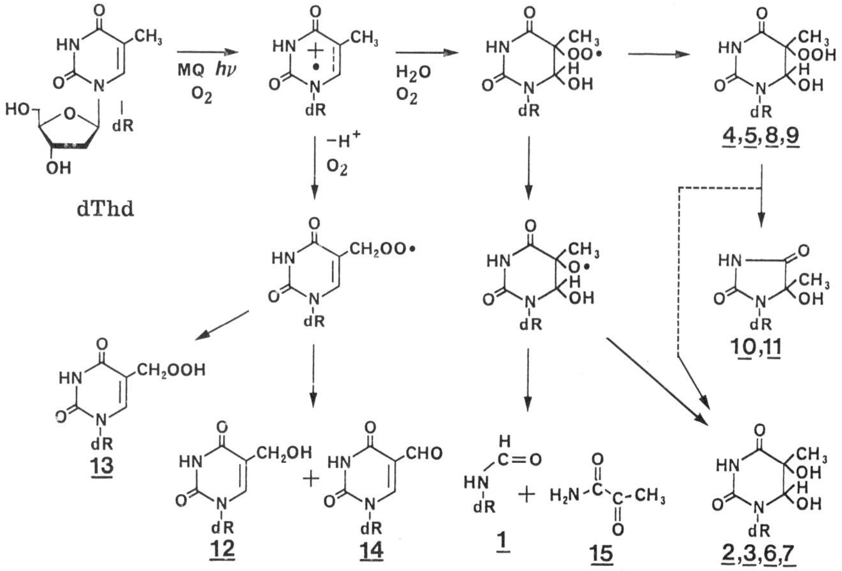

FIG. 3. Proposed mechanism for the MQ-sensitized photooxidation of thymidine.

to β-hydroxy-substituted carbon-centered radicals (analogous to OH radical adducts), and then oxygen adds to the free radical site to give a peroxyl radical. These species are relatively unreactive in solution but undergo radical–radical reactions, thereby becoming reduced to give hydroperoxides (possibly by superoxide radicals produced during photolysis), or they decay bimolecularly to give alkoxyl radicals. The presence of these particular hydroperoxide diastereoisomers (**4, 5, 8, 9**) and not the other four isomeric 5-hydroxy-6-hydroperoxides strongly suggests that hydration occurs selectively at C-6 of the thymidine radical cation (this compares to the presence of all eight isomers in OH-induced oxidation).

The chief source of 5,6-glycols (**2, 3, 6, 7**) is most likely the reduction of alkoxyl radicals during photolysis, but it should be noted that some 5,6-glycol may arise from decomposition of the hydroperoxides. The formamide derivative (**1**) and pyruvamide (**15**) are likely formed by initial β-cleavage of the above alkoxyl radicals, leading to fragmentation of the pyrimidine ring.[13] Last, the hydantoin derivatives **10** and **11** are either formed by the above free radical pathway or by decomposition of the hydroperoxides. The second pathway of the thymidine radical cation is

[13] C. Decarroz, J. R. Wagner, J. E. van Lier, C. M. Krishna, P. Riesz, and J. Cadet, *Int. J. Radiat. Biol. Relat. Stud. Phys. Chem. Med.* **50**, 191 (1986).

deprotonation from the methyl group, which accounts for the formation of **12–14** by similar peroxyl radical chemistry. The high yield of **14** underlines the tendency of these primary peroxyl radicals to decay by a Russel mechanism which would involve the concerted loss of H_2O_2 from transient tetroxide intermediates.

Photoproducts of 2'-Deoxycytidine

The mixture of 2'-deoxycytidine photoproducts is not clearly resolved on reversed-phase HPLC because of their much greater polarity compared to the thymidine products. Therefore, it is necessary to employ normal phase HPLC to initially separate the products and then reversed-

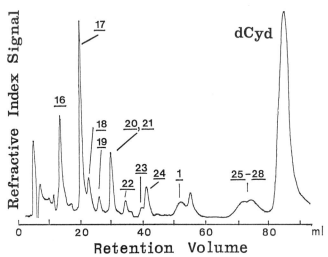

FIG. 4. HPLC profile of 2'-deoxycytidine oxidation products using the normal phase column with ethyl acetate–2-propanol–water (75 : 16 : 9) as the mobile phase. Mixtures after normal phase separation were repurified by reversed-phase HPLC using the same system as for thymidine products except that water was used as the mobile phase. The retentions of these products on reversed-phase HPLC are given in parentheses after the compound and are expressed as the k' value, which is equal to $(V_r - V_0)/V_0$ where V_r is the retention volume and V_0 the void volume of the column (6 ml). **16**, 2-Deoxy-1,4-ribonolactone; **17**, 2'-deoxyuridine; **18**, (5S*)1-dR-5-hydroxyhydantoin, where dR is 2-deoxy-β-D-*erythro*-pentofuranosyl; **19**, (5R*)1-dR-5-hydroxyhydantoin; **20**, (+)-*trans*-(5R,6R)5,6-dihydroxy-5,6-dihydro-2'-deoxyuridine (2.5); **21**, *trans*-(4R*,5R*)1-carbamoyl-3-dR-4,5-dihydroxyimidazolidin-2-one (4.7); **22**, *trans*-(4S*,5S*)1-carbamoyl-3-dR-4,5-dihydroxyimidazolidin-2-one; **23**, 1-dR-biuret; **24**, (−)-*trans*-(5S,6S)5,6-dihydroxy-5,6-dihydro-2'-deoxyuridine; **1**, N-dR-formamide; **25**, (+)-*cis*-(5S,6R)5,6-dihydroxy-5,6-dihydro-2'-deoxyuridine (2.7); **26**, (−)-*cis*-(5R,6S)5,6-dihydroxy-5,6-dihydro-2'-deoxyuridine (2.7); **27**, 5-hydroxy-2'-deoxycytidine (1.2); **28**, cytosine (1.8); dCyd, 2'-deoxycytidine.

phase HPLC to further purify some of the normal phase fractions down to single compounds. The normal phase separations result in a fairly good resolution of most of the photoproducts (Figs. 4 and 5).

The characterization of OH-induced oxidation products of 2′-deoxycytidine has not previously been done owing in part to the complexity of the product mixture; thus, the use of MQ photosensitization as a means to generate them has facilitated the identification of several modified 2′-deoxycytidine products. The identity of **16**, **17**, and **28** is confirmed by their ¹H NMR with standard compounds. Each of the 2′-deoxyuridine 5,6-glycols (**20** and **24–26**) and 5-hydroxy-2′-deoxycytidine (**27**) are assigned by comparison with compounds obtained by chemical synthesis. The *trans*-5,6-glycols convert rapidly to the cis isomers by ring–chain tautomerism as for thymidine 5,6-glycols but at a faster rate. The remaining photoproducts are identified by extensive analysis including several NMR and MS techniques.

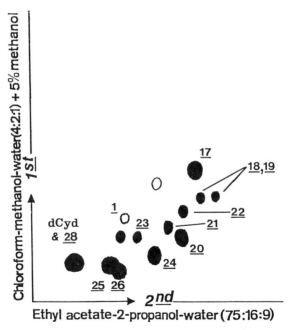

FIG. 5. Two-dimensional TLC separation on silica gel plates of 2′-deoxycytidine oxidation products. Numbers correspond to compounds identified in the legend to Fig. 4. Products were localized by autoradiography with 2-¹⁴C-labeled product mixtures (dark spots) or by spraying with cysteine reagent (all spots).

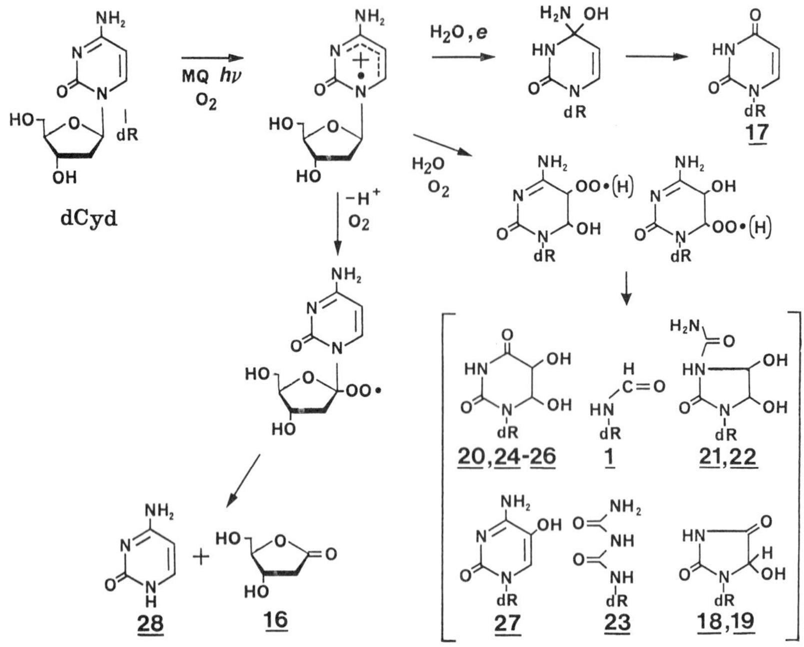

FIG. 6. Proposed mechanism for MQ-sensitized photooxidation of 2′-deoxycytidine.

Mechanism of Sensitized Photooxidation of 2′-Deoxycytidine

Sensitized photooxidation of 2′-deoxycytidine is less efficient than that of thymidine [Φ(-dCyd) ~ 0.4], and substrate decomposition reaches a plateau at about 50%. The different yields of the photoproducts are roughly estimated from Figs. 4 and 5. The main pathways are formation of 2′-deoxyuridine (20%), 2′-deoxyuridine 5,6-glycols, and other rearrangement products which are present in OH-induced reactions (60%), and a minor pathway gives cytosine and 1,4-ribonolactone (5%). Again, all photoproducts can be explained by a common radical cation precursor (Fig. 6). The major pathway is hydration of the radical cation. Although hydroperoxides could not themselves be detected, electron spin resonance and particularly ^{18}O-labeling experiments strongly suggest that peroxyl radicals or hydroperoxides are involved in the formation of stable products.[14] From this work, C-5 peroxyl intermediates appear to predominate over possible C-6 peroxyl radicals, which are key intermediates in the OH-induced decomposition of 2′-deoxycytidine. The 2′-deoxyuridine 5,6-gly-

---

[14] C. M. Krishna, C. Decarroz, J. R. Wagner, J. Cadet, and P. Riesz, *Photochem. Photobiol.* **46,** 175 (1987).

cols (**20, 24–26**) likely arise from initial 2′-deoxycytidine 5,6-glycols which are known to readily undergo deamination in aqueous solution. Thus, similar processes in the formation of 5,6-glycols can be implicated as in the thymidine case. Likewise, this applies to the formamide derivative (**1**).

The photoproducts that do not have precedent in the thymidine reactions are **21–23**. The imadazolin-2-one and biuret derivatives likely arise from intramolecular attack of the peroxyl radical or hydroperoxide group with the N-3–C-4 double bond, leading to C-4–C-5 bond cleavage as was suggested previously for the OH-induced oxidation of cytosine.[15] The other pathways of the radical cation appear to involve hydration at C-4 which, followed by reduction and deamination, would give 2′-deoxyuridine (it should be noted that this nucleoside itself is not subsequently photooxidized during photolysis). The deprotonation reaction in the case of 2′-deoxycytidine likely occurs at the anomeric position since this would lead to 2-deoxy-1,4-ribonolactone (**16**) and cytosine (**27**), which are formed in stoichiometric amounts during the photolysis.[16]

[15] M. Polverelli and R. Téoule, *C. R. Acad. Sci.* **277C,** 747 (1973).
[16] C. Decarroz, J. R. Wagner, and J. Cadet, *Free Radical Res. Commun.* **2,** 295 (1987).

[53] Radical-Mediated DNA Damage in Presence of Oxygen

By CLEMENS VON SONNTAG and HEINZ-PETER SCHUCHMANN

Introduction

Radical-mediated DNA damage is caused by ionizing radiation and transition metal ion-induced reactions involving hydrogen peroxide and superoxide radicals, by drugs, such as bleomycin or neocarzinostatin (often termed radiomimetic drugs), by the action of light in the presence of photosensitizers, such as menadione, and also by activated leukocyte-derived oxidative species. Traditionally, there have been two ways of viewing DNA damage, namely, structurally and chemically. We emphasize the latter; it is also the more encompassing as any structural change (e.g., strand breakage) is preceded or accompanied by chemical change. The products and the mechanisms leading to these products have been discussed recently in some detail.[1] Here, methods to detect such damage are described.

[1] C. von Sonntag, "The Chemical Basis of Radiation Biology." Taylor & Francis, London, 1987.

Much of our knowledge regarding radical-mediated DNA damage has been obtained using low molecular weight model systems including nucleobases, the sugars deoxyribose or ribose, ribose 5-phosphate, nucleosides, and nucleotides (see Ref. 1). With these systems, products can be separated by chromatographic techniques such as gas chromatography (GC) or high-performance liquid chromatography (HPLC) and identified by mass spectrometry and nuclear magnetic resonance. Regarding DNA itself, only a minor part of the products are released from the damaged strand without further treatment, and methods have been developed to hydrolyze the damaged DNA to the level of the subunits or to detach damaged bases from the strand with specific glycolytic enzymes (see Ref. 2) and determine the products by the techniques indicated. As a first step in *in vivo* experiments, the DNA must be isolated after cell lysis.

Apart from chemical identification, special assays are used that are directed at the functional groups of the damaged sites. Antibodies to a variety of lesions are raised (see Ref. 1); however, immunoassay kits are not yet commercially available, so they are not dealt with here. It might be mentioned that the original immunoassays were not superior to the conventional techniques. More recent developments in this field are very encouraging, and in the near future this approach may well be superior to the chemical assays reported here.

The subject of this chapter is the radical-induced DNA damage formed *in the presence of oxygen*. Although a number of DNA products are exclusively formed under this condition (e.g., hydroperoxides), there are a large number of products which are formed both in the presence and the absence of oxygen. Thus, their formation is not necessarily an indication of the involvement of oxygen. In *in vivo* studies of DNA damage, nucleobase products are released by the DNA into the cellular environment. These products might be formed in some kind of degradation reaction of the bases present in the cellular pool, or they might be metabolically generated in parallel to base synthesis in the cellular pool. Both constituents making up the DNA (the nucleobases and the sugar phosphate moiety) are subject to oxidative damage. We first show how the damaged constituents are made accessible to the various assays (by hydrolysis of the DNA) and then introduce procedures that have been used to analyze various base- and sugar-derived oxidation products.

Hydrolysis of Damaged DNA

In those cases where the damaged constituent is neither promptly released nor easily liberated by other means, complete digestion may be necessary. This may be done by acid hydrolysis or enzymatically.

[2] G. W. Teebor, R. J. Boorstein, and J. Cadet, *Int. J. Radiat. Biol.* **54,** 131 (1988).

Acid Hydrolysis

Under the severe conditions of acid hydrolysis some products do not survive at all, others only partially survive. As long as the conditions are reproducible, linear yield–dose relationships can be obtained, but this does not necessarily constitute proof that the damaged sites are excised quantitatively.

For acid hydrolysis, first treat the DNA with 95% formic acid at 90° for 16 hr.[3] Alternatively, treat the DNA with 72% formic acid at 140° for 30 min,[4] or, treat the DNA with 6 M HCl at 120° for 10 min to 4 hr.[5]

Enzymatic Hydrolysis

The search for repair enzymes that cope *in vivo* with DNA damage has revealed the existence of a number of more or less specific glycosylases which are capable of excising damaged nucleobases (see Refs. 1 and 2). (At present, such enzymes are not commercially available; for this reason, experimental procedures involving their use are not referred to here.) Sometimes, a crude bacterial enzyme extract may serve adequately. Under the milder conditions of enzymatic hydrolysis, products survive that would have been destroyed in the acid digest.

It has been noted that an enzymatic cocktail of phosphodiesterases and phosphomonoesterases that fully degrades undamaged DNA to the nucleosides leaves untouched some damaged sites which remain bound in the form of di- and oligonucleotides.[6] This raises the question of whether enzymes that can excise a given damage could fail to do so if there are two damaged sites close to each other or if their matrix is no longer double-stranded. Nevertheless, this method is often used. It might be possible to complement the method by first treating the oxidized or otherwise damaged DNA with a mixture of damage-specific endonucleolytic enzymes, or a bacterial extract, *prior* to the application of the phosphodiesterase/monoesterase cocktail.

Degradation to Deoxyribosides with Well-Defined Enzymes. For enzymatic hydrolysis,[7] treat the sample with DNase I, snake and spleen phosphodiesterases, and alkaline phosphatase overnight. Add 5 volumes of cold acetone, centrifuge, and concentrate the supernatant (e.g., for analysis by HPLC).

[3] R. Teoule, A. Bonicel, C. Bert, and B. Fouque, *J. Am. Chem. Soc.* **100**, 6749 (1974).

[4] M. Dizdaroglu and D. S. Bergtold, *Anal. Biochem.* **156**, 182 (1986).

[5] K. A. Schellenberg, J. Shaeffer, R. K. Nichols, and D. Gates, *Nucleic Acids Res.* **9**, 3863 (1981).

[6] M. Dizdaroglu, W. Hermes, D. Schulte-Frohlinde, and C. von Sonntag, *Int. J. Radiat. Biol. Relat. Stud. Phys. Chem. Med.* **33**, 563 (1978).

[7] G. Teebor, A. Cummings, K. Frenkel, A. Shaw, L. Voituriez, and J. Cadet, *Free Radical Res. Commun.* **2**, 303 (1987).

In an alternative procedure,[8] treat 25 μg of DNA with 5 μg nuclease P1 in 50 μl of a solution of 25 μM sodium acetate, 5 mM MgCl$_2$, 0.1 mM ZnCl$_2$ (pH 5.5). Incubate at 50°C for 2 hr. Raise the pH to 9.0 with NH$_4$OH and treat with alkaline phosphatase at 37° for 20 hr. Pass the hydrolyzate through a Sep-Pak cartridge for subsequent HPLC analysis.

Enzyme Extract from Escherichia coli.[9] Grind freshly grown log-phase *E. coli* B with alumina. Extract with 7 mM magnesium acetate, 70 mM glycine buffer (pH 9.2). Centrifuge at 12,000 g.

Enzyme Extract from Micrococcus luteus.[10] Wash commercially available cell powder (6 g) of ATTC Strain 4698 via suspension in 125 ml of 10 mM Tris-HCl buffer (pH 8). Centrifuge at 10,000 g for 5 min at 5°. Resuspend the pellet in 125 ml of 10 mM Tris-HCl buffer (pH 8) containing 0.2 M sucrose and 25 g hen egg-white lysozyme and incubate for 45 min at 30°. Add an equal amount of distilled water at 0°. Stir vigorously. Cool to 0°, then sonicate for 4 min at ≤15°. Centrifuge at 20,000 g for 45 min at 5°. Slowly add (1 hr) 108 g ammonium sulfate to 250 ml of supernatant at room temperature. Centrifuge at 20,000 g for 45 min at 5°. Dissolve the wet protein paste in 5 ml of 5 mM phosphate buffer (pH 7.5).

Note: This solution can be diluted to portions containing 50 mg protein/ml and stored at −20° for at least 1 year. The preparation acts selectively on certain modified sites.

Analysis of Base Damage

Gas Chromatographic Determination of Damaged Nucleobases[4]

Damaged nucleobases excised by one method or another can be determined by GC after trimethylsilylation. Detection by single-ion monitoring is much more sensitive than conventional detection by flame ionization. For determination, add 50 μl N,O-bis(trimethylsilyl)trifluoroacetamide (BSTFA) and 50 μl acetonitrile to vacuum-dried DNA hydrolyzates and heat in polytetrafluoroethylene (PTFE)-capped vials at 130° for 15 min. Perform GC on fused silica capillary columns (e.g., cross-linked SE-54, 25 m, 0.2 mm i.d., film thickness 0.11 μm, 150–250°, 10°/min).

[8] C. J. Chetsanga and C. Grigorian, *Int. J. Radiat. Biol. Relat. Stud. Phys. Chem. Med.* **44,** 321 (1983).
[9] Z. Trgovcevic and Z. Kucan, *Int. J. Radiat. Biol. Relat. Stud. Phys. Chem. Med.* **12,** 193 (1967).
[10] M. C. Paterson, B. P. Smith, and P. J. Smith, *in* "DNA Repair" (E. C. Friedberg and P. C. Hanawalt, eds.), Vol. 1A, p. 99. Dekker, New York, 1981.

Assays for Specific Base Damage

Although many altered bases can be detected by GC, there are other techniques available, some of which do not require the hydrolysis of DNA because the damaged constituent or a relevant part of it is released under the conditions of the assay. In some assays of thymine glycol the related hydroperoxides are transformed to the glycols.

Determination of Thymine Glycol

cis-Thymine glycol 2'-deoxyribosides.[7] Perform the enzymatic degradation of DNA as outlined above. Precipitate the digest by adding 5 volumes of cold acetone, centrifuge, reduce the volume of supernatant. Analyze the supernatant by HPLC (Sephadex LH-30, water or 50 m*M* sodium borate–boric acid buffer, pH 8.6, as the eluent, or, alternatively, reversed-phase column, ODS-C18, with water as the eluent).

Acetylate the isolated *cis*-thymine glycol deoxyribosides in 500 μl dry pyridine, 600 μl acetic anhydride at room temperature overnight (16 hr). Add 5 ml water and reduce the volume by vacuum evaporation. Perform HPLC (200 μl) on ODS column 25 $\times$ 1.4 cm (with acetonitrile–water, 30 : 70, as the eluent at 2 ml/min) and collect four fractions representing the tri- and tetraacetyl-($\pm$)-*cis*-thymine glycol 2'-deoxyribosides.

Note: This method uses radioactively labeled DNA. Although the retention time of *cis*-thymine glycol is generally consistent, it is best to add 30–100 μg of marker, nonradioactive *cis*-thymine glycol[11] and monitor absorbance at 220 nm. This compound is stable in aqueous solution for months or may be stored dry indefinitely at 0°.

2,3-Dihydroxy-2-methylpropanoic acid.[5] Treat the DNA with 0.1 *M* NaOH at room temperature for 15 min. React with 5 m*M* tritiated sodium borohydride for 5 min. Hydrolyze with 6 *M* HCl at 120° for 10 min. Analyze labeled 2,3-dihydroxy-2-methylpropanoic acid by TLC on cellulose plates (2-propanol–concentrated NH$_4$OH–water, 7 : 1 : 2, as eluent; R_f 0.42).

Thymine.[12] Add 0.1 ml 55% HI (without preservative) to lyophilized (thymine/thymidine-free) fractions containing thymine glycol and thymidine glycol, stopper, and heat at 100° for 35 min. Determine thymine by HPLC (C$_{18}$ column, 25 $\times$ 0.45 cm; 0.1 *M* ammonium acetate, pH 6, as eluent at 2 ml/min, elution time 7.8–8.0 min]. Optical detection is at 265 nm.

[11] K. Frenkel, M. S. Goldstein, N. Duker, and G. W. Teebor, *Biochemistry* **20,** 750 (1981).
[12] R. Cathcart, E. Schwiers, R. L. Saul, and B. N. Ames, *Proc. Natl. Acad. Sci. U.S.A.* **81,** 5633 (1984).

Note: This method has been developed to determine thymine glycol in human and rat urine. The yield of the conversion step from thymine glycol to thymine in this assay is 25–35%.

Acetol.[13] Treat the DNA in 1 ml of 0.2 M KOH at room temperature for 1.5 hr. Add 0.33 ml of 2 M HCl; treat at 70° for 15 min. Centrifuge and neutralize the supernatant with KOH. In DNA labeled with ^{3}H at the methyl group of thymine, the resulting acetol has been determined by ion-exchange chromatography [column diameter 1 cm; bottom 6-cm layer of Bio-Rad AG50W-X8 (H^+), layer of sand, middle 6-cm layer of Bio-Rad AG1-X10 (OH^-), layer of sand, top 3-cm layer of DEAE-Sephadex A-25]. Acetol yield is 20% with respect to thymine glycol in DNA.

Note: When nonradioactive DNA is assayed it might be possible to determine the acetol by GC after trimethylsilylation. Experimental procedures for this are not yet established.

Acetol by fluorimetry[14] (see Refs. 15 and 16). Prepare 0.1 M EDTA solution in 2 M NaOH. Dissolve 3 mg of *o*-aminobenzaldehyde in 5 ml distilled water, stir, and filter off any insoluble residue. Bring the thymine glycol-containing sample to pH 13. Leave at room temperature for 90 min. To 5 ml of the pH 13 sample solution, add 70 μl of the EDTA solution and 30 μl of the *o*-aminobenzaldehyde solution; incubate at 65° for 20 min. Make the volume up to 10 ml with 0.1 M phosphate buffer (pH 6.9). Analyze by HPLC [column 0.4 × 24 cm, Nucleosil 5 C_{18}; mixture of 45 vol% of a solution (pH 7) of 4.54 g KH_2PO_4 and 5.84 g Na_2HPO_4 in 1000 ml water, 55 vol% methanol as eluent]. Detection of the condensation product 3-hydroxyquinaldine is by fluorescence spectrophotometry; excite at 375 nm, record emission at 444 nm. Keep the column thermostatted at 0° for adequate separation from unreacted *o*-aminobenzaldehyde.

Note: Run a blank in the same way except use 80 mM ammonium hydroxide instead of *o*-aminobenzaldehyde. It is necessary to calibrate the procedure with authentic acetol as well as thymidine glycol or thymine glycol in order to establish the yield of recovery. We have found that conversion of acetol is far from complete (only a few percent), even in the presence of excess *o*-aminobenzaldehyde.

Determination of 5-Hydroxymethyl-2'-deoxyuridine.[17] Perform the enzymatic degradation of DNA as described above. Precipitate the hydroly-

---

[13] P. A. Cerutti, *in* "DNA Repair" (E. C. Friedberg and P. C. Hanawalt, eds.), Vol. 1A, p. 57. Dekker, New York, 1981.

[14] Procedure used in these laboratories.

[15] D. Roberts and M. Friedkin, *J. Biol. Chem.* **233**, 483 (1958).

[16] K. Pfeilsticker and J. Lucas, *Angew. Chem.* **99**, 341 (1987).

[17] K. Frenkel, A. Cummings, J. Solomon, J. Cadet, J. J. Steinberg, and G. W. Teebor, *Biochemistry* **24**, 4527 (1985).

zate with acetone, centrifuge, and evaporate supernatant to dryness. Dissolve the residue in water and filter through a Millipore 0.22-μm filter; add the appropriate marker. Analyze the sample by HPLC (5 μm Ultrasphere-ODS column, 1 × 25 cm, water as eluent at a flow rate of 2 ml/min).

It is possible to determine this product as its acetylated derivative. For this, combine fractions which coelute with authentic marker 5-hydroxymethyl-2'-deoxyuridine, evaporate to dryness, and dissolve in dry pyridine. React with acetic anhydride at room temperature for 15 min. Add water to decompose excess acetic anhydride and evaporate to dryness *in vacuo*. Dissolve in 25% acetonitrile–water and filter through Millipore filter. Chromatograph on an ODS column with 25% acetonitrile–water. The yield of the acetylation procedure is between 70 and 80%.

Determination of 8-Hydroxy-2'-deoxyguanosine.[18] Among the DNA products, 8-hydroxy-2'-deoxyguanosine is readily oxidized electrochemically and hence can be determined selectively by HPLC with electrochemical detection, which in this case is more sensitive than optical detection. The column is Ultrasphere ODS (5 μm, 15 × 0.46 cm); the eluent is 12.5 mM citric acid, 25 mM sodium acetate, 30 mM NaOH, 10 mM acetic acid (final pH 5.1)–methanol (100:15); the flow rate is 0.5 ml/min. Detection is with a glassy carbon electrode[19] at approximately 500 mV.

Detection of 4,6-Diamino-5-N-formamidopyrimidine[8]

Remove nucleases and any unhydrolyzed oligonucleotides by filtering the DNA digest through Sep-Pak. Perform HPLC analysis at 22° on a C_{18} μBondapak 5 μm column (0.4 × 25 cm; guard column, LC-18) with 3% methanol in 10 mM $NH_4H_2PO_4$ (pH 5.1) as the eluent.

Detection of 7,8-Dihydro-8-oxoadenine[20]

Evaporate the formic acid–DNA hydrolyzate to dryness and take up in methanol. Centrifuge, filter, and reduce volume by evaporation. Perform HPLC on a Nucleosil C_8 (10 μm) column (0.4 × 25 cm) with water (pH 6.8) as eluent at a flow rate 0.7 ml/hr. Detection is at 254 nm.

[18] R. A. Floyd, J. J. Watson, P. K. Wong, D. H. Altmiller, and R. C. Rickard, *Free Radical Res. Commun.* **1**, 163 (1986).
[19] R. A. Floyd and P. K. Wong, *in* "DNA Repair" (E. C. Friedberg and P. C. Hanawalt, eds.), Vol. 3, p. 419. Dekker, New York, 1988.
[20] A. Bonicel, N. Mariaggi, E. Hughes, and R. Teoule, *Radiat. Res.* **83**, 19 (1980).

Analysis of Sugar Damage

Determination of Apurinic/Apyrimidinic Sites[21]

Apurinic/apyrimidinic sites are formed when damaged bases are released hydrolytically as such or by enzymatic action. In the methoxyamine test the reagent binds to the carbonyl function at C-1'. Carbonyl functions are also formed upon damage to the sugar moiety,[1] and these modifications are dosed by this assay as well.

Incubate the isolated DNA (100–200 μg nucleotides/ml) with 5 mM [^{14}C]methoxyamine at pH 7.2 (10 mM phosphate buffer or 0.1 M borate buffer) for 30 min at 37°. Spot 10–50 μl of the reaction mixture on glass fiber disks, which are immediately placed in ice-cold 1 M HCl. Wash the disks with 10 ml of 1 M HCl 5 times. Rinse with ethanol 3 times (total 10 ml) and dry the disks under an infrared lamp. Count the radioactivity in vials containing the disks and 4 ml of toluene supplemented with 16 mg of Omnifluor. The optimal pH range for the methoximation reaction is 7 to 8.

2-Thiobarbituric Acid Assay[14]

Especially in the presence of oxygen, the sugar fragments consist of compounds that carry one or several carbonyl functions. Some of them, for example 2'-deoxyguanosine-5'-aldehyde,[22] have been mistaken for malondialdehyde since they give a very similar 2-thiobarbituric acid (TBA) reaction. TBA-reactive compounds are also formed in the peroxidation of lipids. Nevertheless, the assay may be useful in many instances.

Mix one aliquot of the DNA solution with one aliquot of a solution containing 0.6% 2-thiobarbituric acid. Heat at 90° for 20 min. Measure the red color at 537 nm. *Note:* The extinction coefficient of the red coloration generated by malondialdehyde is 1.5×10^5 dm^3 mol^{-1} cm^{-1}. In the case of the altered DNA sugars, the extinction coefficients may not have the same value.

Reactivity toward TBA is not only shown by certain base-free sugar fragments. For instance, base propenals, which are typical products when DNA is treated with bleomycin may also be formed by ionizing radiation in the presence of oxygen (see Ref. 1), are labile to hydrolysis and thus assayable by the TBA test.

[21] M. Liuzzi and M. Talpaert-Borlé, *in* "DNA Repair" (E. C. Friedberg and P. C. Hanawalt, eds.), Vol. 3, p. 443. Dekker, New York, 1988.
[22] D. Langfinger and C. von Sonntag, Z. *Naturforsch, C: Biosci.* **40,** 446 (1985).

Determination of Various Altered Sugars by GC

A number of altered sugars are released from (others remain bound to) the DNA after free-radical attack.[1] It is advantageous to reduce these products to the corresponding polyhydric alcohols which are readily trimethylsilylated and analyzed by GC.

Free Altered Sugars[23] (see Ref. 24). Reduce the aqueous sample with a few milligrams of NaB^2H_4 per 20 mg altered DNA at room temperature for at least 2 hr. Destroy excess $NaBD_4$ with formic acid. Chase the boric acid by repeated (≥ 5 times) evaporation with methanol. Trimethylsilylate the dry residue with 100 μl BSTFA–trimethylchlorosilane (100 : 3) in 1 ml dry pyridine at room temperature overnight. Perform GC on a 70 m glass capillary, 0.25 mm i.d., liquid phase, e.g., OV-101, at 70–250°, increasing the temperature 10°/min.

Note: For identification, GC–MS is suitable because the fragmentation pattern of this class of compounds is fairly well known.[25] The incorporation of deuterium marks the location of carbonyl functions before reduction.

Bound Altered Sugars.[23] Reduce the aqueous sample with a few milligrams of NaB^2H_4 per 20 mg altered DNA at room temperature for at least 2 hr. Evaporate to dryness and incubate with 10 ml saturated barium hydroxide solution (pH 13) at 85° for 2 days with exclusion of oxygen. Neutralize to pH 8 with sulfuric acid. Incubate with alkaline phosphatase (0.8 U/ml) at 37° for at least 12 hr. Neutralize with formic acid and centrifuge. Further treatment is as above from the chase onward.

Note: Bound 2-deoxyribonic acid is one of the alterations at the sugar moiety. It is more readily determined by omitting the reduction step.[24]

Determination of Hydroperoxide Functions

Reduction by Iodide

Hydroperoxides in DNA.[26] Prepare a sodium thiosulfate stock solution in quintuply distilled water. Standardize against potassium thiosulfate. Prepare a 3% potassium iodide solution and a 5% Thyodene (soluble starch) solution in quintuply distilled water. To 10 ml of hydroperoxide-

[23] M. Isildar, M. N. Schuchmann, D. Schulte-Frohlinde, and C. von Sonntag, *Int. J. Radiat. Biol. Relat. Stud. Phys. Chem. Med.* **40,** 347 (1981).

[24] M. Dizdaroglu, D. Schulte-Frohlinde, and C. von Sonntag, *Int. J. Radiat. Biol. Relat. Stud. Phys. Chem. Med.* **32,** 481 (1977).

[25] M. Dizdaroglu, D. Henneberg, and C. von Sonntag, *Org. Mass Spectrom.* **8,** 335 (1974).

[26] H. B. Michaels and J. W. Hunt, *Anal. Biochem.* **87,** 135 (1974).

containing sample, add 1 ml concentrated H_2SO_4 and 2 ml KI solution. After 15 min, titrate with thiosulfate solution (1×10^{-4} M, diluted from stock solution). Just before the end point, add 2 ml starch solution. Solutions must be air-free to avoid high readings.

Hydroperoxides in DNA Model Compound Poly(U).[27] Determine the sum of organic hydroperoxides and hydrogen peroxide using Allen's reagent (KI/molybdate).[28] To another sample (pH near 7), add catalase (100–200 U/10 ml sample); any H_2O_2 is destroyed after 5 min. Determine organic hydroperoxides as before.

Note: Iodine reacts rapidly with some pyrimidine radiolysis products, e.g., barbituric acid; therefore, the values obtained by this method may underestimate the true yield. The catalase preparation may not be completely inert toward some organic hydroperoxides, and excessive contact times should be avoided.

Coupled Oxidation of NADPH[29]

Glutathione peroxidase catalyzes the reduction of hydroperoxides with glutathione (GSH) as the reductant. Prepare stock solutions of reduced glutathione, glutathione reductase (type III, yeast), NADPH (tetrasodium salt, type X), and glutathione peroxidase (bovine erythrocyte) on the day of use. Mix the following in a 1-cm-path length optical cell: 1 ml of 0.1 M phosphate buffer (pH 7.75), 0.1 ml of 10 mM EDTA, 0.2 ml of 52 mM sodium azide, 0.2 ml of 1 μM glutathione peroxidase, 0.2 ml of 38 mM reduced glutathione, 0.1 ml of 1.4 mM NADPH, and 0.2 ml glutathione reductase (5 U/ml). Flush the solution with N_2 for 5 min. Stopper the cell and record the NADPH absorption at 340 nm at 25°. Add 0.5 ml of peroxide solution (saturated with N_2) and monitor the extent of NADPH oxidation at 25°. Determine the hydroperoxide-independent oxidation of NADPH by adding 0.5 ml of water instead of hydroperoxide sample.

Note: The molar extinction coefficient of NADPH at 340 nm is 6200 M^{-1} cm^{-1}. Oxidation of NADPH is complete within 2–5 min at 25° with most hydroperoxides. The azide inhibits the activity of traces of catalase present in the commercial glutathione peroxidase preparation.

[27] D. J. Deeble and C. von Sonntag, *Int. J. Radiat. Biol. Relat. Stud. Phys. Chem. Med.* **49,** 927 (1986).
[28] A. O. Allen, C. J. Hochanadel, J. A. Ghormley, and T. W. Davis, *J. Phys. Chem.* **56,** 575 (1952).
[29] J. E. Frew, P. Jones, and G. Scholes, *Anal. Chim. Acta* **155,** 139 (1983).

[54] *In Vivo* Oxidative DNA Damage: Measurement of 8-Hydroxy-2′-deoxyguanosine in DNA and Urine by High-Performance Liquid Chromatography with Electrochemical Detection

By MARK K. SHIGENAGA, JEEN-WOO PARK, KENNETH C. CUNDY, CARLOS J. GIMENO, and BRUCE N. AMES

Introduction

Reactive oxygen species, which are generated as by-products of cellular metabolism[1] and ionizing radiation,[2] produce irreversible modifications to DNA. The damage caused by these reactive free radical species has been proposed to contribute to aging, cancer, and other age-related degenerative processes. In order to study the relationship between endogenous oxidative DNA damage and aging it is desirable to acquire techniques that can detect such damage. Analytical approaches such as GC–MS and immunochemical techniques have been shown to be effective methods for detecting a number of specific lesions caused by oxidative damage, including the modified bases 8-hydroxyguanine,[3] 8-hydroxyadenine,[4] and thymine glycol.[5] In general, these methods are useful for detecting the effects of high levels of damage caused by irradiation but are less useful for measuring endogenous levels of oxidative damage. In an effort to quantitate endogenous levels of oxidative DNA damage, a recent method employing HPLC combined with UV detection has been used to measure the urinary levels of the oxidatively modified deoxynucleoside, thymidine glycol.[6] However, this method also has certain limitations, principally the time required to perform the assay and the quantity of sample required to obtain sufficient sensitivity.

The greatest obstacle associated with the routine quantitative analysis of oxidized bases and deoxynucleosides, such as those mentioned above, is the relatively poor sensitivity obtained by these techniques. Recent studies, however, have shown that a number of purine bases can be

[1] B. N. Ames, *Science* **221**, 1256 (1983).

[2] R. Teoule, *Int. J. Radiat. Biol. Relat. Stud. Phys. Chem. Med.* **51**, 573 (1987).

[3] M. Dizdaroglu, *Biochemistry* **24**, 4476 (1985).

[4] G. C. West, I. W.-L. West, and J. W. Ward, *Int. J. Radiat. Biol. Relat. Stud. Phys. Chem. Med.* **42**, 481 (1982).

[5] S. A. Leadon and P. C. Hanawalt, *Mutat. Res.* **112**, 191 (1983).

[6] R. Cathcart, E. Schwiers, R. L. Saul, and B. N. Ames, *Proc. Natl. Acad. Sci. U.S.A.* **81**, 5633 (1984).

oxidized and detected with excellent sensitivity by electrochemical means.[7] Application of this detection technique following separation of enzymatic hydrolyzates of DNA by HPLC has been shown by Floyd *et al.*[8] to be an effective approach for the detection of 8-hydroxy-2'-deoxyguanosine (oh^{8}dG). The sensitivity of this method has been estimated to be 2 to 3 orders of magnitude greater than that obtained by optical methods.[8] oh^{8}dG is one of about 20 known primary oxidative DNA damage products produced by reactive oxygen species[2] and has been found by this detection technique to be present in nuclear and mitochondrial DNA and in urine.[9-11] Thus, HPLC with electrochemical detection (HPLC-EC) provides a new and powerful tool for the detection and quantitative analysis of the oxidatively modified deoxynucleoside, oh^{8}dG.

Analysis of 8-Hydroxy-2'-deoxyguanosine in Enzyme Hydrolyzates
of DNA

Isolation and Preparation of DNA Samples for HPLC-EC

DNA from different species and their tissues are isolated by the procedure described by Gupta.[12] Since peroxides may be present in some phenol preparations, it is necessary to use the highest purity distilled phenol (i.e., Applied Biosystems, Foster City, CA, or International Biotechnologies, Inc., New Haven, CT) available for DNA extractions in order to avoid artifactual oxidation of DNA during the workup procedure. Following the isolation of DNA from tissues, samples (4–5 A_{260} units/200 μl), prepared in 20 mM sodium acetate buffer (pH 4.8), are digested to nucleotides with 20 μg of nuclease P1 (Sigma, St. Louis, MO) at 37° for 30 min, and then treated with 1.3 units of *Escherichia coli* alkaline phosphatase (Sigma) in 0.1 M Tris-HCl buffer (pH 7.5) at 37° for 1 hr to liberate the corresponding nucleosides from phosphate residues.

HPLC-EC of Enzyme Hydrolyzates of DNA

For the separation of oh^{8}dG from other components present in the hydrolyzed DNA samples, a Waters Associates (Milford, MA) Model 510

[7] R. A. Kenley, S. E. Jackson, J. C. Martin, and G. C. Visor, *J. Pharm. Sci.* **74,** 1082 (1985).
[8] R. A. Floyd, J. J. Watson, and P. K. Wong, *Free Radical Res. Commun.* **1,** 163 (1986).
[9] H. Kasai and S. Nishimura, *Nucleic Acids Res.* **12,** 2137 (1984).
[10] C. Richter, J.-W. Park, and B. N. Ames, *Proc. Natl. Acad. Sci. U.S.A.* **85,** 6465 (1988).
[11] M. K. Shigenaga, C. J. Gimeno, and B. N. Ames, *Proc. Natl. Acad. Sci. U.S.A.* **86,** 9697 (1989).
[12] R. C. Gupta, *Proc. Natl. Acad. Sci. U.S.A.* **81,** 6943 (1984).

solvent delivery system equipped with a Rheodyne (Cotati, CA) 7125 injector and a 3 μm Supelcosil (Supelco, Bellefonte, PA) LC-18-DB column (15 cm × 4.6 mm) is used. The mobile phase consists of filtered and degassed 50 mM KH$_2$PO$_4$ buffer (pH 5.5)–methanol (90:10, v/v). The separations are performed at a flow rate of 0.8 ml/min with a back pressure of approximately 2000 psi. Samples are analyzed by separate UV and electrochemical detection systems linked in series via low dead-volume polytetrafluoroethylene (PTFE) tubing. Optical detection utilizes a Kratos (Westwood, NJ) Model 773 UV detector; electrochemical detection is accomplished by a Bioanalytical Systems (West Lafayette, IN) LC-4B amperometric detector with a glassy-carbon working electrode and an Ag/AgCl reference electrode. Since the voltage at which oh^{8}dG is oxidized maximally appears to depend largely on the individual working and reference electrodes, a hydrodynamic voltammogram of the modified deoxynucleoside is constructed for each amperometric detector system. For the detector used in these analyses, an oxidation potential of +0.6 V is judged to provide the best balance between the current generated from the electrochemical oxidation of oh^{8}dG and the noise associated with the baseline current. Coulometric EC detection may also be used for the quantitative analysis of oh^{8}dG from DNA hydrolyzates. An ESA (Bedford, MA) Model 5100 Coulochem detector equipped with a 5011 analytical cell may be used for this purpose. The potentials set for the dual coulometric detector are 0.12 and 0.30–0.40 V for electrodes 1 and 2, respectively.

As shown in Fig. 1A, analysis of 40 μg of an enzyme hydrolyzate of rat liver DNA following HPLC separation produces four prominent peaks absorbing at 260 nm which correspond to the normal deoxynucleosides deoxycytidine (retention time 3.8 min), deoxyguanosine (6.6), thymidine (8.2), and deoxyadenosine (13.9). Minor peaks corresponding to enzyme hydrolyzates of RNA, the principal contaminant in these DNA samples, are also observed and include cytidine (3.3), uridine (3.8), guanosine (5.7), and adenosine (12.1). The minor deoxynucleoside, 5-methyldeoxycytidine, elutes with the same retention time as guanosine (5.7 min). At the sensitivity used for the UV detection of the normal nucleosides (0.5 absorbance units full scale at 260 nm), a signal for oh^{8}dG (9.0), which is detected by electrochemical detection (see below), is not observed. The chromatogram (Fig. 1B) generated from electrochemical oxidation of the DNA hydrolyzate separated by HPLC is qualitatively much different compared to that obtained by optical detection (Fig. 1A). The peaks corresponding to normal deoxy- and ribonucleosides which are observed in the UV chromatogram are either not observed or are represented by a slight transient loss in current. Using this amperometric detection system, oh^{8}dG produces a detectable signal current at 9.0 min. Also shown in Fig.

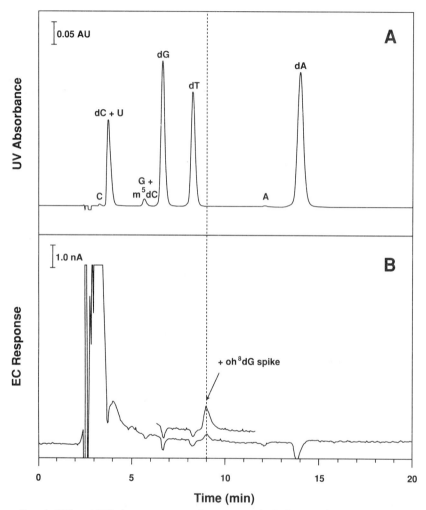

FIG. 1. UV and EC chromatograms of an enzyme hydrolyzate of DNA following isocratic reversed-phase HPLC. An enzyme hydrolyzate of rat liver DNA (40 μg) is separated by reversed-phase HPLC and analyzed by two detection systems linked in series. (A) HPLC chromatogram with UV detection at 260 nm. C, Cytidine; dC, deoxycytidine; U, uridine; G, guanosine; m⁵dC, 5-methyldeoxycytidine; dG, deoxyguanosine; dT, thymidine; A, adenosine; dA, deoxyadenosine. (B) HPLC chromatogram with EC detection at +0.6 V. Also shown is a partial EC chromatogram of the same sample spiked with 4 pmol of authentic oh⁸dG. The elution time of oh⁸dG is designated by the vertical dashed line. Other details are described in the text.

1B is a partial HPLC-EC chromatogram of the same DNA hydrolyzate sample spiked with 4 pmol of authentic oh[8]dG.

Depending on the source of DNA, quantitation of 5–50 fmol of oh[8]dG/μg can be achieved readily on 40–100 μg samples of DNA following enzyme hydrolysis. Levels of oh[8]dG in various DNA samples measured by EC detection are expressed as values relative to the amount of deoxy-guanosine detected by UV absorbance at 260 nm. Alternatively, the amount of oh[8]dG may be normalized to the amount of DNA analyzed. The amount of DNA can be determined by comparing the peak area of deoxyguanosine obtained from the sample DNA hydrolyzates to a cali-bration curve for deoxyguanosine obtained from the analysis of known amounts of calf thymus DNA. The amount of oh[8]dG may therefore be expressed in terms of the molar ratio of oh[8]dG/deoxyguanosine or femto-moles oh[8]dG per microgram DNA. The value obtained from this analysis is approximately 2.5 residues of oh[8]dG/10^5 residues of deoxyguanosine.[10] The normal range found in various tissues is 6–40 fmol oh[8]dG/μg of DNA.[8–10,13,14] The combination of HPLC with UV and amperometric elec-trochemical detection has been applied successfully to the identification and quantitation of oh[8]dG in DNA isolated from various rat tissues, bacte-ria, and human blood.[8–10,13,14]

Analysis of oh[8]dG in Urine

Synthesis of Standards

The synthesis of oh[8]dG follows the procedure described by Kasai and Nishimura.[9] High specific activity [^{3}H]oh[8]dG is synthesized using a modi-fication of the protocol and serves two purposes in the assay. It allows one to calculate the recovery of oh[8]dG following solid-phase extraction (SPE) of urine samples and minimizes the contribution of the radiolabeled standard to the electrochemical signal obtained from urinary oh[8]dG. Syn-thesis of [^{3}H]oh[8]dG is accomplished readily by chemical oxidation of [^{3}H]deoxyguanosine. Deoxyguanosine labeled at the nonexchangeable C-1' and C-2' positions may be purchased commercially (Amersham, Arlington Heights, IL) in the form of the deoxynucleoside or the corre-sponding deoxynucleoside triphosphate.

Synthesis of radiolabeled oh[8]dG is as follows. [1',2'-^{3}H]dGTP is di-gested enzymatically to [1',2'-^{3}H]deoxyguanosine. The procedure for the

---

[13] H. Kasai, P. F. Crain, Y. Kuchino, S. Nishimura, A. Ootsuyama, and H. Tanooka, *Carcinogenesis* **7**, 1849 (1986).

[14] C. G. Fraga, M. K. Shigenaga, J.-W. Park, P. Degan, and B. N. Ames, *Proc. Natl. Acad. Sci. U.S.A.* (in press).

enzymatic digestion employs the addition of 900 μl of 0.1 M Tris buffer (pH 7.5) and 20 μl of 100 U/ml bacterial alkaline phosphatase to a 100 μl sample of [1',2'-^{3}H]dGTP (250 μCi, 28 Ci/mmol, 8.9 nmol) which has been concentrated to remove ethanol. Following incubation of this mixture at 37° for 3.5 hr the digest is passed through a 0.2-μm nylon fiber and purified by HPLC using a 5-μm Supelcosil LC-18 (25 cm × 4.6 mm) column with a 5-μm Supelcosil LC-18 precolumn cartridge assembly. oh^{8}dG is eluted with a mobile phase consisting of 10% methanol in water at a flow rate of 1.0 ml/min. With commercially available [1',2'-^{3}H]deoxyguanosine, the steps prior to HPLC purification are not required and instead the following procedure is used. [1',2'-^{3}H]Deoxyguanosine (250 μCi, 49 Ci/mmol, 5.1 nmol) is transferred to a 1-ml Reactivial (Pierce, Rockford, IL) and concentrated under a stream of argon to remove ethanol then purified by HPLC as described above.

The fractions corresponding to [1',2'-^{3}H]deoxyguanosine (retention time 10 min) are pooled and concentrated to dryness prior to chemical oxidation. Hydroxylation of the deoxynucleoside at the C-8 position is achieved by adding to 5.1 nmol [1',2'-^{3}H]deoxyguanosine in 30 μl of water the following in order: 71 μl of freshly prepared 1.7 M ascorbic acid, 2 μl of 0.2 M CuSO$_4$, and 145 μl of a 31% solution of H$_2$O$_2$. The extremely vigorous reaction is essentially complete in 10 sec. The reaction, which yields approximately 15% of the desired standard, must be purified immediately by HPLC or decomposition of [1',2'-^{3}H]oh^{8}dG will result. [1',2'-^{3}H]oh^{8}dG is purified using a 5-μm Supelcosil LC-18-S column (25 cm × 4.6 mm) and eluted with 5% methanol in water at a flow rate of 1.7 ml/min. Fractions are collected every 30 sec. [1',2'-^{3}H]oh^{8}dG elutes at approximately 16.5 min (Fractions 32–33). [1',2' − ^{3}H]oh^{8}dG is stored in the HPLC solvent (5% methanol in water) at 4° and is used directly. Under these storage conditions, the degradation rate of [1',2'-^{3}H]oh^{8}dG is typically less than 1% per month. HPLC combined with a Flo-1β (Radiomatic, Tampa, FL) radiochemical detector is used to monitor the purity of the standard as needed.

Sample Preparation: Solid-Phase Extraction of Urine

An efficient cleanup of urinary oh^{8}dG is achieved using a combination of reversed and normal solid-phase extraction. We chose as the initial solid phase a non-end-capped Bond Elut (Analytichem, Harbor City, CA) C$_{18}$-OH SPE column since these are found to retain oh^{8}dG much better than conventional C$_{18}$ SPE columns. The improved retention of oh^{8}dG to this solid phase may be explained, in part, by hydrogen bonding interactions between the polar hydroxyl/amino groups of oh^{8}dG and the non-end-capped silanol residues.

Urine samples are stored at $-20°$ prior to workup and analysis. To a volume of 1–5 ml urine, an equal volume of 1 M NaCl and approximately 20,000 cpm of $[1',2'-{}^{3}H]oh^{8}dG$ is added. The mixture is vortexed and applied to a Bond Elut 500-mg C_{18}-OH SPE column which has been preconditioned with 10 ml each of methanol, water, and 50 mM KH$_{2}$PO$_{4}$ buffer, pH 7.5 (buffer A). The SPE columns are arranged on a vacuum manifold, and elution of solvents and eluants containing the analyte is effected by applying a slight vacuum. The eluant from all washes described subsequently are discarded. The column is washed in succession with 4 ml of buffer and depending on the sample, between 2 and 5 ml of buffer B (5% methanol in buffer A, v/v). To elute $oh^{8}dG$, 3 ml of buffer C (15% methanol in buffer A, pH 5.5, v/v) is applied to the column. This eluant is transferred to a second C_{18}-OH column previously preconditioned as described above. After the eluant is applied, an additional 1 ml of water is added to remove most of the remaining buffer salts. The column is dried thoroughly under vacuum for 10–15 min, then $oh^{8}dG$ is eluted from this column with 1 ml methanol.

The eluant containing $oh^{8}dG$ is placed in a water bath at a temperature of 35–40° and concentrated to dryness under a gentle stream of argon. The concentrated sample is resuspended in 100 μl methanol and applied to a Bond Elut 500-mg silica SPE column previously preconditoned with 3 ml of dichloromethane. This column is washed in succession with 6 ml of acetonitrile, 2 ml of 5% methanol in acetonitrile (v/v), 2 ml of 10% methanol in acetonitrile (v/v), and 1 ml of 20% methanol in acetonitrile (v/v). $oh^{8}dG$ is then eluted from this column with a further 3-ml aliquot of 20% methanol in acetonitrile. The eluant is concentrated to dryness, and the white residue is resuspended in 500 μl of 50 mM KH$_{2}$PO$_{4}$ buffer, pH 5.5. From the sample of 500 μl, a 25- to 100-μl aliquot is analyzed by HPLC-EC. The amount of radiolabeled tracer present in the remaining sample is quantitated by scintillation counting to determine the volume of urine analyzed. The overall recovery of radioactivity following solid-phase extraction of urine is approximately 35%.

HPLC-EC Analysis of Urinary $oh^{8}dG$

Waters Associates Models 510 and 6000A solvent delivery systems equipped with a Waters WISP autoinjector are used. Separation of $oh^{8}dG$ by HPLC employs a 5-μm Supelcosil LC-18-S precolumn cartridge assembly (2 cm $\times$ 4.6 mm) linked to two 5-μm Supelcosil LC-18-S analytical columns (25 cm $\times$ 4.6 mm) attached in series and eluted at a flow rate of 1 ml/min. The solvents used for gradient elution are as follows: solvent A, 50 mM KH$_{2}$PO$_{4}$ buffer, pH 5.5; solvent B, acetonitrile–methanol (7:3, v/v)–solvent A, 1:1. The conditions for gradient elution are as follows: 0–

80 min, 0–8% solvent B; 80–85 min, 8% solvent B isocratic; 85–90 min, 8–50% solvent B; 90–100 min, 50% solvent B isocratic; 100–105 min, 50–0% solvent B. In all cases, a linear ramp is used when solvent compositions are changed. Before proceeding with the analysis of each subsequent sample, a 45-min equilibration under the initial conditions is programmed. Under these conditions the backpressure is approximately 3000–3500 psi.

For electrochemical detection of SPE-processed urine samples, an ESA Model 5100 Coulochem detector equipped with a 5011 high sensitivity analytical cell is used. The 5011 high sensitivity analytical cell is equipped with coulometric (electrode 1) and amperometric (electrode 2) electrodes linked in series. It is therefore possible to screen or minimize the electrochemical signals generated from interfering substances by oxidizing these substances extensively, if not completely, by the coulometric electrode. For the purpose of this assay, the potential set for the coulometric electrode ranges from 0.10 to 0.15 V depending on the individual analytical cell used. Substances, including the desired analyte oh^{8}dG, that

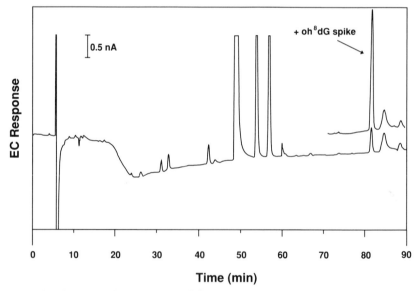

FIG. 2. Electrochemical detection of SPE-processed human urine following gradient reversed-phase HPLC; showing the electrochemical response for a sample of human urine, which is analyzed using a screening potential of 0.1 V (electrode 1, coulometric) and an electrode potential of 0.2 V (electrode 2, amperometric). Also shown is a partial EC chromatogram of the same sample spiked with 2.5 pmol of authentic oh^{8}dG. Approximately 650 fmol of oh^{8}dG is detected from the equivalent of 59 μl of normal human urine. The contribution of the radiolabeled standard (7.2 fmol) is negligible, accounting only for approximately 1% of the total signal obtained from this sample.

are not oxidized at these lower potentials are oxidized by the ampero-metric electrode, which is set to potentials ranging from 0.20 to 0.40 V, again depending on the individual analytical cell used. A preinjector guard cell set at a potential of 0.40 V is also employed. A Knauer dynamic mobile phase mixer together with two pulse dampeners (SSI, State College, PA) attached in series minimizes the fluctuations in baseline signal current and thus enhances the overall detection sensitivity. Data are digitized by a Nelson 760 (Cupertino, CA) analytical interface and processed by Nelson Analytical series 4400 data acquisition software on a Hewlett-Packard 9816 computer.

Figure 2 shows a typical chromatogram produced from a SPE-processed sample of normal human urine. Under the conditions described, oh^{8}dG elutes at 81.8 min as a well-defined peak. The inset shows a partial chromatogram of the same sample which has been spiked with 2.5 pmol of authentic oh^{8}dG. The detection is normally performed at 200 nanoamperes (nA) full scale (gain = 500) and reprocessed according to the signal obtained for oh^{8}dG. In Fig. 2, reprocessed data is shown at 5 nA full scale. Detection of approximately 250–1500 fmol of oh^{8}dG is achieved readily by analyzing 25- to 100-μl aliquots of SPE-processed human urine. The value obtained is normalized to the volume of urine excreted in a 24-hr period and the body weight of the individual. Results obtained by this method indicate that humans excrete approximately 130–300 pmol of oh^{8}dG/kg of body weight/24 hr.

Summary

HPLC with electrochemical detection is a highly sensitive and selective method for detecting the oxidatively modified DNA residue oh^{8}dG. By this method, the detection of oh^{8}dG from DNA and urine offers a powerful approach for assessing *in vivo* oxidative damage. Application of this technique to the detection of oh^{8}dG from DNA permits the quantitation of the steady-state levels of this oxidatively modified deoxynucleoside and overcomes the detection problems associated with the extremely low levels present in DNA. In addition, the selectivity gained by this detection method eliminates the problem of separating the signal for oh^{8}dG from normal deoxynucleosides. The quantitation of oh^{8}dG in urine complements the measurement of oh^{8}dG in DNA by estimating cumulative oxidative DNA damage in the body. In addition, the urinary assay provides a noninvasive means of measuring this type of damage in laboratory animals and human populations. Thus, an individual animal or human subject may be monitored over time, possibly under various prooxidant conditions, using oh^{8}dG as a sensitive marker for oxidative DNA damage. This analytical approach may allow one to estimate the exposure of an individual to prooxidant conditions associated with lifestyle, genetic predisposition, degenerative diseases, and environmental toxins.

Acknowledgments

This work was supported by National Institutes of Health Grant CA39910 to B.N.A. and by National Institute of Environmental Health Sciences Center Grant ES01896. J.-W.P. was supported by the Health Effects Component of the University of California Toxic Substances Research and Teaching Program.

[55] Selected-Ion Mass Spectrometry: Assays of Oxidative DNA Damage

By MIRAL DIZDAROGLU and EWA GAJEWSKI

Introduction

Oxidative damage to DNA *in vivo* caused by free radicals appears to play an important role in a number of human diseases.[1] Metal ion-dependent formation of highly reactive hydroxyl (OH) radicals is expected in systems which generate superoxide radicals and hydrogen peroxide.[2] The interaction of ionizing radiation with cellular water can also form OH radicals and other free radicals in living cells.[3] Reactions of OH radicals with DNA in aqueous solution are known to produce a large number of sugar and base products.[3] Hydroxyl radicals also cause formation of DNA–protein cross-links in nucleoprotein.[4-7] Lesions produced in DNA in living cells are subject to cellular repair processes and, unless repaired, may have detrimental biological consequences.[8] Chemical characteriza-

---

[1] B. Halliwell and J. M. C. Gutteridge, *Mol. Aspects Med.* **8,** 89 (1985).

[2] I. Fridovich, *Arch. Biochem. Biophys.* **247,** 1 (1986).

[3] C. von Sonntag, "The Chemical Basis of Radiation Biology." Taylor & Francis, London, 1987.

[4] L. K. Mee and S. J. Adelstein, *Proc. Natl. Acad. Sci. U.S.A.* **78,** 2194 (1981).

[5] N. L. Oleinick, S. Chiu, L. R. Friedman, L. Xue, and N. Ramakrishnan, *in* "Mechanisms of DNA Damage and Repair: Implications for Carcinogenesis and Risk Assessment" (M. G. Simic, L. Grossman, and A. C. Upton, eds.), p. 181. Plenum, New York, 1986.

[6] E. Gajewski, A. F. Fuciarelli, and M. Dizdaroglu, *Int. J. Radiat. Biol.* **54,** 445 (1988).

[7] M. Dizdaroglu, E. Gajewski, P. Reddy, and S. A. Margolis, *Biochemistry* **28,** 3625 (1989).

[8] E. C. Friedberg, "DNA Repair." Freeman, New York, 1985.

tion and quantitative measurement of DNA lesions are necessary for an understanding of their biological consequences and cellular repairability. This chapter describes the chemical characterization and quantitative measurement of OH radical-induced DNA damage by the technique of gas chromatography–mass spectrometry (GC–MS).

Materials and Methods

Reagents and some reference compounds used in the methodology described here are available commercially as listed below.[9]

Reagents. Acetonitrile, bis(trimethylsilyl)trifluoroacetamide (BSTFA) (containing 1% trimethylchlorosilane), and 6 *M* HCl are purchased from Pierce Chemical Company (Rockford, IL); formic acid (88%) from Mallinckrodt (Paris, KY); deoxyribonuclease I, snake venom exonuclease, spleen exonuclease, and alkaline phosphatase from Boehringer-Mannheim (Indianapolis, IN).

Reference Compounds. Adenine, cytosine, guanine, thymine, 6-azathymine, 8-azaadenine, 2'-deoxyadenosine, 2'-deoxyguanosine, 8-bromoadenine, isobarbituric acid (5-hydroxyuracil), 5,6-dihydrothymine, 5-hydroxymethyluracil, 4,6-diamino-5-formamidopyrimidine, and 2,5,6-triamino-4-hydroxypyrimidine are purchased from Sigma Chemical Company (St. Louis, MO), and 2-amino-6,8-dihydroxypurine (8-hydroxyguanine) from Chemical Dynamics Corporation (South Plainfield, NJ). 6-Amino-8-hydroxypurine (8-hydroxyadenine) is synthesized from 8-bromoadenine by treatment with formic acid and is purified by recrystallization from water.[10] 2,6-Diamino-4-hydroxy-5-formamidopyrimidine is synthesized by treatment of 2,5,6-triamino-4-hydroxypyrimidine with formic acid and is recrystallized from water.[11] 5,6-Dihydroxy-5,6-dihydrothymine (thymine glycol) is synthesized by treatment of thymine with osmium tetroxide.[12]

Hydrolysis

DNA and nucleoprotein must be hydrolyzed so that the GC–MS technique can be applied to characterization and quantitative measurement of

---

[9] Certain commercial equipment or materials are identified in this paper in order to specify adequately the experimental procedure. Such identification does not imply recommendation or endorsement by the National Institute of Standards and Technology, nor does it imply that the materials or equipment identified are necessarily the best available for the purpose.

[10] M. Dizdaroglu and D. S. Bergtold, *Anal. Biochem.* **156,** 182 (1986).

[11] L. F. Cavalieri and A. Bendich, *J. Am. Chem. Soc.* **72,** 2587 (1950).

[12] M. Dizdaroglu, E. Holwitt, M. P. Hagan, and W. F. Blakely, *Biochem. J.* **235,** 531 (1986).

base products in DNA and of cross-linked components of DNA and proteins in nucleoprotein. Acid hydrolysis cleaves the glycosidic bonds between bases and sugar moieties in DNA and thus frees intact and modified bases. Enzymatic hydrolysis is used to hydrolyze DNA to nucleosides. In the case of DNA–protein cross-links, the simplest approach to hydrolysis of nucleoprotein appears to be the standard method of protein hydrolysis, i.e., hydrolysis with 6 M HCl, which cleaves peptide bonds in proteins as well as glycosidic bonds in DNA to free base–amino acid cross-links.[6,7] DNA and nucleoprotein samples should be extensively dialyzed against water prior to hydrolysis and subsequently lyophilized.

Acid Hydrolysis. Of the various acids, formic acid appears to be the most suitable for hydrolysis of DNA.[13] The following conditions are used for analysis of products of all four DNA bases in a single GC–MS run. An aliquot of DNA (usually 0.1–1 mg) is treated with 1 ml of formic acid (88%) in evacuated and sealed tubes at 150° for 30 min, then lyophilized.[13] This procedure does not alter the base products of DNA. Exceptions are the following compounds: 5,6-dihydroxy-5,6-dihydrocytosine (cytosine glycol) yields a mixture of 5-hydroxycytosine and 5-hydroxyuracil, the former by dehydration and the latter by deamination and dehydration.[12,13] 5-Hydroxy-5,6-dihydrocytosine and 5,6-dihydroxycytosine yield 5-hydroxy-5,6-dihydrouracil and 5,6-dihydroxyuracil, respectively, upon deamination.[13] Nucleoprotein (~2 mg) is hydrolyzed with 1 ml of 6 M HCl in evacuated and sealed tubes at 120° for 18 hr. The sample is then lyophilized.

Enzymatic Hydrolysis. Enzymatic hydrolysis of DNA is used specifically for identification and quantitative measurement of 8,5′-cyclopurine-2′-deoxynucleoside moieties in DNA by GC–MS.[14,15] Although the use of GC–MS for identification of various other modified DNA bases as nucleosides following enzymatic hydrolysis of DNA has been reported, quantitative measurement by GC–MS has not been demonstrated.[16] The following conditions are used to hydrolyze DNA to nucleosides. An aliquot of DNA (0.5–1 mg) is incubated in 0.5 ml of 10 mM Tris-HCl buffer (pH 8.5) containing 2 mM MgCl$_2$) with deoxyribonuclease I (100 units), spleen exonuclease (0.01 unit), snake venom exonuclease (0.5 units), and alkaline phosphatase (10 units) at 37° for 24 hr. After hydrolysis, the sample is lyophilized. Enzymatic hydrolysis of nucleoprotein to release DNA base–amino acid cross-links for GC–MS analysis has not been reported.

[13] M. Dizdaroglu, *Anal. Biochem.* **144,** 593 (1985).

[14] M. Dizdaroglu, *Biochem. J.* **238,** 247 (1986).

[15] M.-L. Dirksen, W. F. Blakely, E. Holwitt, and M. Dizdaroglu, *Int. J. Radiat. Biol.* **54,** 195 (1988).

[16] M. Dizdaroglu, *J. Chromatogr.* **367,** 357 (1986).

Derivatization

Bases, nucleosides, and DNA base–amino acid cross-links are not sufficiently volatile for GC–MS analysis and thus must be converted to volatile derivatives. Trimethylsilylation is used to convert bases or nucleosides to their volatile trimethylsilyl (Me₃Si) derivatives. The use of *tert*-butyldimethylsilylation has been also reported, but so far only for qualitative analysis of bases.[13,17,18]

Trimethylsilylation. Lyophilized hydrolyzates of DNA or nucleoprotein are trimethylsilylated with 0.25 ml of a mixture of BSTFA and acetonitrile (1.5:1, by volume) at 140° for 30 min in polytetrafluoroethylene-capped vials (sealed under nitrogen). The amount of the reagents can be modified according to the amount of DNA or nucleoprotein used. After cooling, the mixture is directly injected into the injection port of the gas chromatograph without further treatment.

Apparatus

Any gas chromatograph equipped with a capillary inlet system and interfaced to a mass spectrometer can be used. A Mass Selective Detector interfaced to a gas chromatograph and a computer work station (all from Hewlett-Packard) has been used throughout the studies reported here. The present methodology applies fused silica capillary columns, which are commercially available (e.g., Hewlett-Packard; J. W. Scientific, Inc.). These types of columns provide high inertness, separation efficiency, and sensitivity. The use of columns coated with a cross-linked 5% phenyl methylsilicone gum phase is recommended. Column length and internal diameter vary depending on the type of analysis. Base products in DNA and DNA base–amino acid cross-links are analyzed on a 12-m column with an internal diameter of 0.20 or 0.32 mm (film thickness, e.g., 0.11–0.33 μm). Analysis of 8,5′-cyclopurine-2′-deoxynucleosides is performed on a 25-m column with an internal diameter of 0.32 mm (film thickness, e.g., 0.17 μm). Helium (ultrahigh purity) is used as the carrier gas at an inlet pressure of 20–40 kPa. Generally, the split mode of injection is used, and the split ratio (i.e., ratio of carrier gas flow through the splitter vent to flow through the column) is adjusted according to the concentration of the analyte(s) in a given mixture. The injection port of the gas chromatograph, the GC–MS interface, and the ion source of the mass spectrometer are kept at 250°. The glass liner in the injection port is filled with silanized glass wool. Mass spectra are recorded in the electron-impact mode at 70 eV.

[17] M. Dizdaroglu, *J. Chromatogr.* **295**, 103 (1984).
[18] M. Dizdaroglu, *BioTechniques* **4**, 536 (1986).

GC–MS of Oxidative DNA Damage

Characterization of Base Products in DNA. Figure 1 illustrates a typical GC separation of a formic acid hydrolyzate of DNA after trimethylsilylation.[13] Prior to GC analysis, DNA was exposed to ionizing radiation in aqueous solution. Peak identification is given in the legend to Fig. 1. As can be seen from this chromatogram, the base products of DNA are well separated from one another and from the four intact DNA bases. The Me_3Si derivatives of these products possess suitable GC properties as indicated by sharp and symmetrical peaks in Fig. 1. Moreover, these compounds provide characteristic mass spectra with an intense molecular ion (M^+) and an intense $[M - CH_3]^+$ ion. The latter ion is typical of Me_3Si

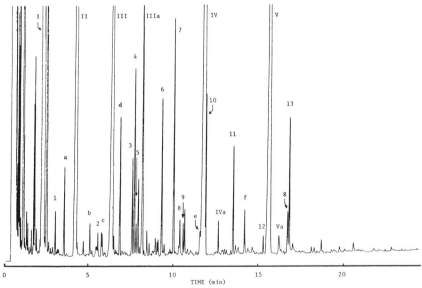

FIG. 1. Gas chromatogram of a formic acid hydrolyzate of DNA after trimethylsilylation. Prior to analysis, DNA was exposed to ionizing radiation in aqueous solution [dose, 330 Gray (joule/kg)]. Column, fused silica capillary (12 m; 0.2 mm i.d.) coated with cross-linked 5% phenyl methylsilicone gum phase (film thickness, 0.11 μm). Temperature program, 100 to 250° at 7°/min after 3 min at 100°. Peaks: **I**, phosphoric acid; **1**, uracil; **II**, thymine; **2**, 5,6-dihydrothymine; **III**, cytosine; **d**, 5-methylcytosine; **3**, 5-hydroxy-5,6-dihydrothymine; **4**, 5-hydroxyuracil; **5**, 5-hydroxy-5,6-dihydrouracil; **IIIa**, cytosine; **6**, 5-hydroxycytosine; **7** and **8**, *cis*- and *trans*-thymine glycol; **9**, 5,6-dihydroxyuracil; **IV**, adenine; **10**, 4,6-diamino-5-formamidopyrimidine; **IVa**, adenine; **11**, 8-hydroxyadenine; **12**, 2,6-diamino-4-hydroxy-5-formamidopyrimidine; **V**, guanine; **Va**, guanine; **13**, 8-hydroxyguanine. Compounds represented by peaks **a** to **g** were also present in control samples. (From Ref. 13, with permission.)

derivatives, and is formed from M^+ by loss of a methyl radical. Mass spectra of the base products of DNA listed in Fig. 1 have been published elsewhere.[13,17]

The characterization of components of a complex mixture at low concentrations (picomole to femtomole) is performed by GC–MS using the technique of selected-ion monitoring (SIM).[19] In this mode, a mass spectrometer is set to monitor a number of ions of the analyte during a time interval during which the analyte elutes from the GC column. The analyte is identified when the signals of the monitored ions with typical abundances all line up at its retention time. For this purpose, the knowledge of the mass spectrum and the retention time of the analyte is required. The use of this technique for characterization at low concentrations of the base products of DNA shown in Fig. 1 has been recently demonstrated.[10,18]

Figure 2 illustrates the use of GC–MS/SIM for detection of base products of DNA in a formic acid hydrolyzate of nucleoprotein (commercial nucleohistone, Sigma) after trimethylsilylation. Prior to analysis, nucleoprotein was exposed to ionizing radiation in buffered aqueous solution. The purpose of the use of nucleohistone here rather than DNA alone was to demonstrate that base products of DNA can be analyzed in the presence of histone proteins by GC–MS/SIM without difficulty. Nucleohistone certainly presents a biologically more relevant model system than DNA alone for *in vitro* studies of free radical damage to DNA. Furthermore, if DNA of living cells exposed to free radical-producing agents (e.g., ionizing radiation) were to be analyzed by GC–MS/SIM, it would be desirable to analyze chromatin rather than DNA alone. This is because DNA, which is covalently cross-linked to proteins in exposed cells, may not be extracted efficiently,[20] and may also contain a significant portion of damaged bases.

In Fig. 2, arrows indicate the time intervals where ions typical of compounds eluting during those intervals are recorded. The insert illustrates a replot of the interval between 4.35 and 4.7 min to show the separation of ion profiles more clearly. As can be seen from Fig. 2, a large number of base products and the internal standards are monitored in a single run within a reasonably short time. The amount of DNA in nucleohistone injected onto the GC column was approximately 0.7 μg. Although Fig. 2 illustrates only one ion for each compound, a number of ions typical of a compound are monitored in the same interval for an

[19] J. T. Watson, "Introduction to Mass Spectrometry." Raven, New York, 1985.
[20] O. Yamamoto, *in* "Aging, Carcinogenesis, and Radiation Biology" (K. C. Smith, ed.), p. 165. Plenum, New York, 1976.

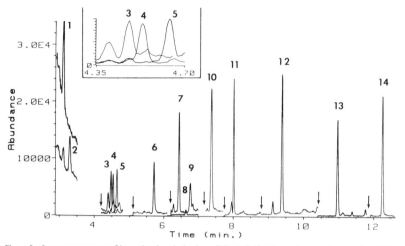

Fig. 2. Ion-current profiles obtained during GC–MS/SIM analysis of a nucleohistone sample, which was hydrolyzed by formic acid and trimethylsilylated. Prior to analysis, nucleohistone was exposed to ionizing radiation in aqueous solution [0.2 mg/ml nucleohistone in 10 mM phosphate buffer (pH 7.2), N_2O-saturated prior to irradiation; dose, 50 Gray; dose rate, 150 Gray/min]. Amount of nucleohistone injected onto the column, approximately 1.8 μg (containing 0.7 μg of DNA). Column as in Fig. 1 except for film thickness, 0.33 μm. Temperature program, 150 to 250° at 8°/min after 2 min at 150°. Peaks (ion monitored): **1,** 6-azathymine (m/z 256) [internal standard]; **2,** 5,6-dihydrothymine (m/z 257); **3,** 5-hydroxyuracil (m/z 329); **4,** 5-hydroxy-5,6-dihydrothymine (m/z 345); **5,** 5-hydroxy-5,6-dihydrouracil (m/z 331); **6,** 5-hydroxycytosine (m/z 343); **7,** cis-thymine glycol (m/z 259); **8,** 5,6-dihydroxyuracil (m/z 417); **9,** trans-thymine glycol (m/z 259); **10,** 8-azaadenine (m/z 265) [internal standard]; **11,** 4,6-diamino-5-formamidopyrimidine (m/z 354); **12,** 8-hydroxyadenine (m/z 352); **13,** 2,6-diamino-4-hydroxy-5-formamidopyrimidine (m/z 442); **14,** 8-hydroxyguanine (m/z 440).

unequivocal identification. Subsequently, a partial mass spectrum is obtained on the basis of the monitored ions and their relative abundances, and the spectrum compared with that of authentic compounds. For this purpose, the mass spectrum of the desired authentic compound should be recorded under the same tuning conditions of the mass spectrometer as are used to monitor actual samples. Differences in the relative abundances of ions in a mass spectrum may occur depending on the tuning conditions of the mass spectrometer.

An example of this technique is illustrated in Fig. 3. Shown are ion-current profiles of six ions of the Me_3Si derivative of 2,6-diamino-4-hydroxy-5-formamidopyrimidine [2,6-diamino-4-hydroxy-5-formamidopyrimidine($Me_3Si)_4$] from GC–MS/SIM analyses of nucleohistone samples, which were hydrolyzed by formic acid and trimethylsilylated.

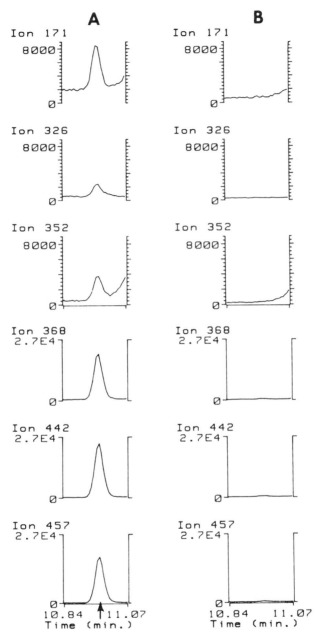

FIG. 3. Ion-current profiles of some ions of 2,6-diamino-4-hydroxy-5-formamidopyrimi-dine(Me₃Si)₄ obtained during GC–MS/SIM analysis of nucleohistone samples, which were hydrolyzed by formic acid and trimethylsilylated. (A) Nucleohistone exposed to ionizing radiation in buffered aqueous solution (dose, 50 Gray); (B) control. Column details as in Fig. 2. It should be noted that the full scales of the m/z 171, 326, and 352 ions are 3 times those of the m/z 368, 442, and 457 ions.

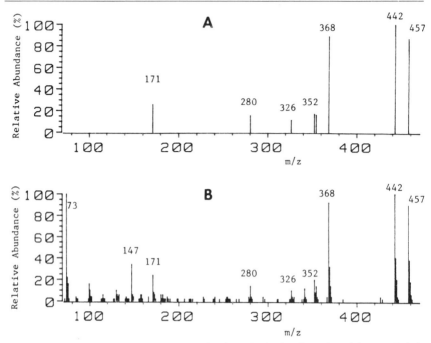

FIG. 4. (A) Partial mass spectrum obtained on the basis of monitored ions and their relative abundances at the elution position of 2,6-diamino-4-hydroxy-5-formamidopyrimidine(Me₃Si)₄ in Fig. 3A. (B) Mass spectrum of 2,6-diamino-4-hydroxy-5-formamidopyrimidine(Me₃Si)₄.

Eight ions were monitored in the same interval; however, profiles of six ions are illustrated in Fig. 3 for practical reasons. The readily discernible peaks of monitored ions are seen in Fig. 3A at the expected retention time (indicated with an arrow) of 2,6-diamino-4-hydroxy-5-formamidopyrimidine(Me₃Si)₄. A partial spectrum obtained from the signals of eight ions and their relative abundances in Fig. 3A is illustrated in Fig. 4A. For identification, this spectrum is compared with the mass spectrum of authentic 2,6-diamino-4-hydroxy-5-formamidopyrimidine(Me₃Si)₄ shown in Fig. 4B. It is clear that these two spectra are directly correlatable to each other.

Analysis of 8,5'-Cyclopurine-2'-deoxynucleosides in DNA. Hydroxyl radical-induced formation of 8,5'-cyclopurine-2'-deoxynucleosides in DNA *in vitro* and *in vivo* has been demonstrated recently.[14,15,21] Moreover, the GC–MS/SIM technique enabled the study of the effect of DNA

[21] M. Dizdaroglu, M.-L. Dirksen, H. Jiang, and J. H. Robbins, *Biochem. J.* **241,** 929 (1987).

conformation on the yields of these compounds and the ratios of their
(5'R) and (5'S) diastereomers.[15] Figure 5 illustrates SIM plots obtained
during GC–MS/SIM analysis of trimethylsilylated enzymatic hydrolyzates of DNA exposed to ionizing radiation in aqueous solution. Shown
are the ion-current profiles of m/z 465 and 553, which are the molecular
ions of 8,5'-cyclo-2'-deoxyadenosine(Me$_3$Si)$_3$ and 8,5'-cyclo-2'-deoxy-
guanosine(Me$_3$Si)$_4$, respectively. These plots clearly show that the (5'R)

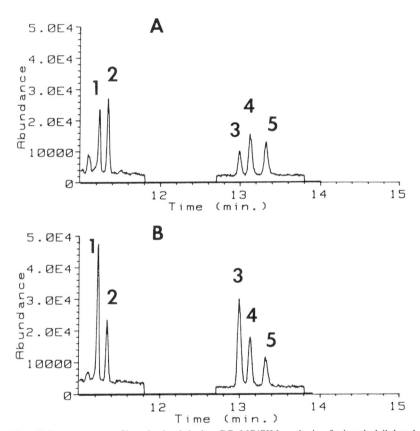

FIG. 5. Ion-current profiles obtained during GC–MS/SIM analysis of trimethylsilylated
enzymatic hydrolysates of DNA exposed to ionizing radiation in buffered aqueous solution
(dose, 40 Gray). (A) Double-stranded DNA; (B) single-stranded DNA. Amount of DNA
injected onto the column, approximately 1 μg. Column, fused silica capillary (25 m; 0.32 mm
i.d.), coating as in Fig. 1 (film thickness, 0.17 μm). Peaks (ions monitored): 1 and 2, (5'R) and
(5'S) diastereomers of 8,5'-cyclo-2'-deoxyadenosine (m/z 465), respectively; 3 and 5, (5'R)
and (5'S) diastereomers of 8,5'-cyclo-2'-deoxyguanosine (m/z 553), respectively; 4, (5'S)-
8,5'-cycloadenosine (m/z 553) (added to the sample as an internal standard). (From Ref. 15,
with permission.)

and (5'S) diastereomers of each compound are well separated from each other. Again, a number of ions from the known mass spectra of these compounds[14,15] are monitored to obtain a partial mass spectrum for comparison with the spectrum of the authentic material for an unequivocal identification, as explained above.

Identification of DNA–Protein Cross-Links in Nucleohistone. A number of OH radical-induced DNA–protein cross-links in nucleohistone *in vitro* have been elucidated recently.[6,7,22] Initially, OH radical-induced cross-linking of a DNA base (thymine, Thy) to a number of amino acids has been investigated in model systems, i.e., an aqueous mixture of Thy and an amino acid, since there was no authentic material available. Using this approach, gas chromatographic and mass spectral properties of possible thymine–amino acid cross-links have been obtained. Subsequently, the GC–MS/SIM technique has been used to identify OH radical-induced DNA–protein cross-links in acidic hydrolyzates of nucleohistone. Figure 6 illustrates representative SIM plots obtained during GC–MS/SIM analyses of nucleohistone samples, which were hydrolyzed by HCl and trimethylsilylated. Ions monitored here are characteristic ions of Me_3Si derivatives of Thy–Gly (*m/z* 370), Thy–Ala (*m/z* 384), Thy–Val (*m/z* 412), Thy–Leu (*m/z* 426), Thy–Ile (*m/z* 426), and Thy–Thr (*m/z* 385) cross-links.[6] Arrows indicate the positions of time intervals for each ion. The profile of the *m/z* 385 ion is illustrated here separately, because its time interval coincides with that of *m/z* 426 ion. In some cases, the same ion is represented by several peaks, which correspond to several isomers of the same compound (e.g., peaks **6, 8,** and **9** correspond to isomers of Thy–Leu cross-links). Compounds represented by peaks **A** to **G** were also present in control samples as illustrated in Fig. 6B.

Figure 6 illustrates only one ion for each compound. For identification purposes, a number of ions from the known mass spectrum of a cross-link are monitored in the same time interval. An example is shown in Fig. 7. The partial mass spectrum (Fig. 7A) obtained on the basis of monitored ions and their relative abundances at the elution position of peak 8 in Fig. 6 correlates directly to the mass spectrum of the Me_3Si derivative of the Thy–Leu cross-link (Fig. 7B).

Quantitative Measurements. The GC–MS/SIM technique is also well suited for quantitative measurement of components of a complex mixture.[19] For this purpose, an internal standard is used. Ideally, a stable isotope-labeled analog of an analyte serves as the internal standard. However, such analogs of the compounds, of which analysis is described here, are not available. Hence, structurally similar compounds must be used as internal standards. Prior to a quantitative measurement, a calibration plot

[22] S. A. Margolis, B. Coxon, E. Gajewski, and M. Dizdaroglu, *Biochemistry* **27,** 6353 (1988).

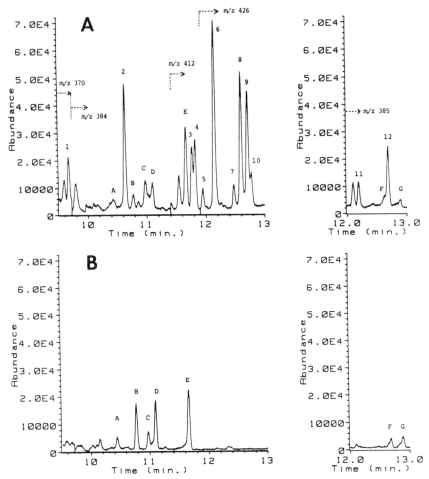

FIG. 6. Ion-current profiles obtained during GC–MS/SIM analysis of nucleohistone samples, which were hydrolyzed by HCl and trimethylsilylated. (A) Samples exposed to ionizing radiation in buffered aqueous solution (dose, 436 Gray); (B) control. Column, fused silica capillary (12 m, 0.32 mm i.d.), coating as in Fig. 1 (film thickness, 0.17 μm). Temperature program, 120 to 250° at 8°/min after 2 min at 120°. Peaks (ion monitored): **1**, Thy–Gly (m/z 370); **2**, Thy–Ala (m/z 384); **3** and **4**, Thy–Val (m/z 412); **6, 8,** and **9**, Thy–Leu (m/z 426); **5, 7,** and **10**, Thy–Ile (m/z 426); **11** and **12**, Thy–Thr (m/z 385). Origin of other peaks is not known. (From Ref. 6, with permission.)

is obtained for the response of the mass spectrometer to standard quantities of both the analyte and internal standard by monitoring an intense and typical ion of each compound during GC–MS/SIM analysis. The ratio of peak areas of an analyte ion and an ion of the internal standard (A/A_{st}) is measured and plotted as a function of the ratio of the molar amounts of

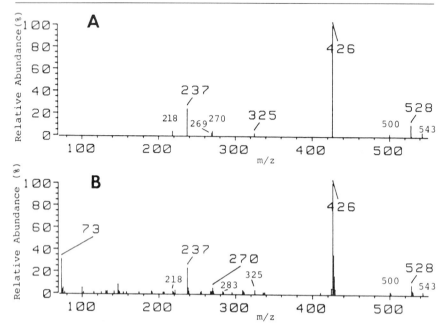

FIG. 7. (A) Partial mass spectrum obtained on the basis of monitored ions and their relative abundances at the elution position of peak **8** in Fig. 6A. (B) Mass spectrum of the Me₃Si derivative of the Thy–Leu cross-link. (From Ref. 6, with permission.)

the analyte and the internal standard (m/m_{st}). The slope of such a plot is the relative molar response factor [RMRF = $(A/A_{st})(m_{st}/m)$]. RMRFs generally vary depending on experimental and instrumental conditions and should be determined in each laboratory. Tuning conditions of a mass spectrometer may also affect the RMRF values. It is therefore recommended that the RMRFs be measured after each tuning of the mass spectrometer. A known amount of an internal standard must be added prior to hydrolysis to DNA or nucelohistone samples with known amounts. After GC–MS/SIM analysis, peak areas of corresponding ions are integrated, and the amount of an analyte in a sample is calculated using its RMRF and the known amount of the internal standard.

Quantitative measurement of DNA base products, 8,5'-cyclopurine-2'-deoxynucleosides, and DNA–protein cross-links by GC–MS/SIM has been demonstrated recently.[6,7,15,23] Examples of some quantitative measurements are illustrated in Figs. 8–10. These data clearly demonstrate the excellent reproducibility achieved by the use of GC–MS/SIM for quantitative measurement of the corresponding compounds.

[23] A. F. Fuciarelli, B. J. Wegher, E. Gajewski, M. Dizdaroglu, and W. F. Blakely, *Radiat. Res.* **119**, 219 (1989).

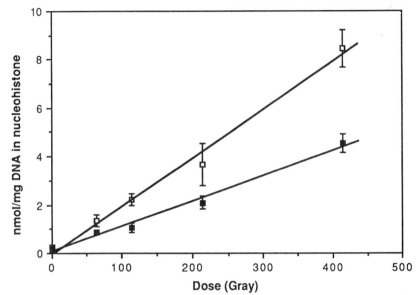

FIG. 8. Dose–yield plots of some DNA base products in nucleohistone: □, 2,6-diamino-4-hydroxy-5-formamidopyrimidine; ■, 8-hydroxyadenine. Each point represents the mean (± standard deviation) from three independent experiments.

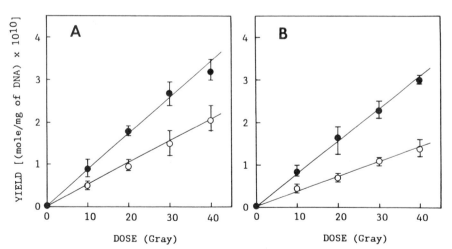

FIG. 9. Dose–yield plots of 8,5'-cyclo-2'-deoxynucleosides in DNA: ●, in single-stranded DNA; ○, in double-stranded DNA. (A) 8,5'-Cyclo-2'-deoxyadenosine; (B) 8,5'-cyclo-2'-deoxyguanosine. Each point represents the mean (± standard deviation) from three independent experiments. (From Ref. 15, with permission.)

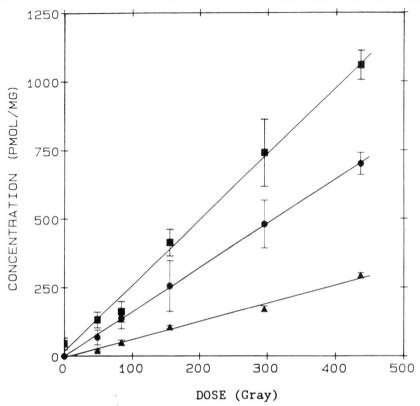

FIG. 10. Dose–yield plots of some Thy–amino acid cross-links in nucleohistone: ■, Thy–Thr; ●, Thy–Leu; ▲, Thy–Val. Each point represents the mean (± standard deviation of the mean) from three independent experiments. (From Ref. 6, with permission.)

Conclusions

The GC–MS/SIM technique is well suited for assessment of oxidative damage to DNA caused by free radicals. Base products and 8,5'-cyclo-purine-2'-deoxynucleosides in DNA as well as DNA–protein cross-links in nucleohistone are unequivocally detected and quantitatively measured. Furthermore, the high selectivity of this technique permits the detection of base products of DNA even in a DNA–protein complex in the presence of histone proteins. With its selectivity, sensitivity, and speed, GC–MS/SIM technique should be the technique of choice for assaying oxidative damage to DNA.

[56] Footprinting Protein–DNA Complexes With γ-Rays

By Jeffrey J. Hayes, Laurance Kam, and Thomas D. Tullius

Introduction

Proteins that bind to specific sequences of DNA are important components of gene regulatory systems. Protein–DNA complexes are commonly studied by *in vitro* techniques such as DNase I footprinting[1] and methylation interference.[2] Another innovative technique uses UV light to create photochemical lesions in DNA, which are susceptible to later chemical cleavage.[3] Bound proteins affect the yield of these photoproducts, giving rise to "photofootprints." Each of these techniques, however, suffers from the inherent DNA base or sequence specificity of the probe used.

A reagent system consisting of $[Fe(II)EDTA]^{2-}$, hydrogen peroxide, and sodium ascorbate has been shown to cleave DNA with little sequence specificity, thus allowing investigation of protein contact with every DNA sequence position.[4,5] The cleavage reagent in this system is likely the hydroxyl radical, generated by reduction of hydrogen peroxide by iron(II). A key advantage of the hydroxyl radical is that its small size (with respect to the protein–DNA system of interest) results in very high-resolution footprints. This footprinting method is not applicable in all solution conditions, though, and it requires the addition of several reagents to a DNA–protein sample that may interfere with the function of the DNA-binding protein, or otherwise be inconvenient.[5]

We describe here an analog of the recently introduced method of hydroxyl radical footprinting.[4,5] We have duplicated the high-resolution footprints of this technique, while substituting γ-radiation for the chemical reagents used heretofore. γ-Radiation has previously been shown to mediate DNA cleavage with no apparent base or sequence specificity.[6] The primary cutting reagent, as in the analogous chemical reagent system, is thought to be the hydroxyl radical.[7] This new method requires no addition of reagents to the protein–DNA binding solution.

[1] D. J. Galas and A. Schmitz, *Nucleic Acids Res.* **5**, 3157 (1978).
[2] U. Siebenlist and W. Gilbert, *Proc. Natl. Acad. Sci. U.S.A.* **77**, 122 (1980).
[3] M. M. Becker and J. C. Wang, *Nature (London)* **309**, 682 (1984).
[4] T. D. Tullius and B. A. Dombroski, *Proc. Natl. Acad. Sci. U.S.A.* **83**, 5469 (1986).
[5] T. D. Tullius, B. A. Dombroski, M. E. A. Churchill, and L. Kam, this series, Vol. 155, p. 537.
[6] W. D. Henner, S. M. Grunberg, and W. A. Haseltine, *J. Biol. Chem.* **257**, 11750 (1982).
[7] T. D. Tullius and B. A. Dombroski, *Science* **230**, 679 (1985).

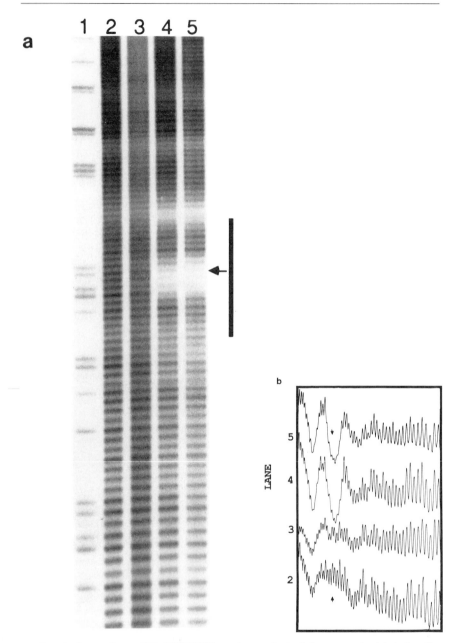

FIG. 1. Comparison of [Fe(II)EDTA]$^{2-}$ and γ-ray footprinting reagents. (a) Autoradiograph of denaturing polyacrylamide electrophoresis gel on which was separated the DNA products from the [Fe(II)EDTA]$^{2-}$ and γ-ray footprinting reactions: lane 1, Maxam–Gilbert

Materials and Methods

Footprinting of the bacteriophage λ repressor/operator complex with the $[Fe(II)EDTA]^{2-}$ reagent system has been described in detail.[4,5] γ-Ray footprinting is carried out in the same manner and under the same conditions, except that the volume of the DNA–protein binding solution is larger. This increases the effective target volume for the radiation and thus increases the rate of production of hydroxyl radicals.

Typically, 20 fmol of ^{32}P-end-labeled DNA is dissolved in 180 μl of repressor binding buffer [10 mM Bis-Tris-HCl (pH 7.0), 50 mM KCl, 1 mM CaCl$_2$]. λ repressor (20 μl, 12.5 pmol) (a generous gift from M. Brenowitz and G. Ackers), diluted in the same buffer, is added, and the mixture is incubated at room temperature for 15 min. The sample is irradiated at a dose rate of approximately 5000 rads/min for 7 min at room temperature in a Shepherd ^{137}Cs irradiator. Control DNA samples (without added protein) are irradiated for 3 min because DNA damage accumulates more quickly in these samples. The total dose is empirically determined, and it varies as a function of sample volume, temperature, and protein and glycerol content. The irradiated samples are analyzed by denaturing gel electrophoresis, autoradiography, and densitometry, as described.[5]

Comments

We first obtained cleavage patterns for a 692-base pair restriction fragment containing the O_R1 operator site, using both γ-radiation and the $[Fe(II)EDTA]^{2-}$ reagent system. Conventional strategy was followed in that the fragment was uniquely end-labeled prior to the cleavage experiment.[4,5] The patterns of cleavage obtained by the two methods are shown in Fig. 1a, lanes 2 and 3. As has been discussed before,[7] the cutting in the $[Fe(II)EDTA]^{2-}$-treated sample occurs to nearly an equal extent at each base position, resulting in a cleavage pattern which is not directly a function of DNA sequence. The cutting pattern of the γ-ray-treated sample also is evenly distributed over all of the nucleotide positions and is nearly identical to that observed with the $[Fe(II)EDTA]^{2-}$ method.

G-specific reaction; lanes 2 and 3, $[Fe(II)EDTA]^{2-}$ and γ-ray cleavage, respectively, of naked DNA control; lanes 4 and 5, $[Fe(II)EDTA]^{2-}$ and γ-ray footprinting of λ repressor/operator complex, respectively. The arrow points to the central base in the operator sequence O_R1, and the bar spans the 17 bases of the operator site. (b) Densitometer scans of the gel shown in Fig. 1a. Scans are numbered corresponding to the lanes in Fig. 1a. The arrow points to the central position (the dyad) in the 17 base-pair-long operator sequence.

A closer inspection of the densitometer scans (Fig. 1b) reveals that even the subtle differences in cleavage frequency at each position found in the [Fe(II)EDTA]$^{2-}$-treated sample are exactly reproduced in the γ-ray-treated sample. These small fluctuations are thought to be due to sequence-dependent conformational variation in the DNA helix,[7] most likely the width of the minor groove.[8] This result is somewhat surprising since the point of origin of the reactive species produced by ionizing radiation is not constrained to be some distance away from the DNA helix as in the [Fe(II)EDTA]$^{2-}$ method.[9] It therefore appears that the majority of hydroxyl radical flux resulting in DNA backbone cleavage in both cases comes from the bulk solvent, and that sensitivity to structural irregularities in the helix is not lost by the use of γ-radiation as a source of hydroxyl radicals. This result suggests that water molecules bound in the grooves of the DNA helix (which could be homolytically cleaved by the high-energy radiation and subsequently react with the DNA backbone) contribute very little to the cutting pattern in the γ-ray-treated sample.

The advantages of the hydroxyl radical as a footprinting reagent have been previously discussed.[10] The [Fe(II)EDTA]$^{2-}$ system has been used to demonstrate clearly that the λ repressor binds to one side of an O_R1-containing DNA molecule.[4] Thus, this system is well suited to test the use of γ-radiation as a footprinting reagent. The amount of radiation used was empirically determined, since the extent of DNA damage by radiation is influenced by salt conditions[11] and the presence of radioprotectants (such as proteins). Because of the nature of the experiment, the goal is to create no more than one strand break per molecule.

The pattern of DNA fragments which results from hydroxyl radical cleavage of the repressor–DNA complex (Fig. 1a,b, lanes 4 and 5) is clearly different from the pattern of the controls (no repressor added, Fig. 1a,b, lanes 2 and 3) and is qualitatively similar to that observed in the earlier study.[4] Both the [Fe(II)EDTA]$^{2-}$ system and γ-radiation give very similar footprinting patterns. The presence of nucleotides that are readily cleaved between sites of protection leads directly to the conclusion that repressor binds to one side of the DNA molecule, as shown by X-ray crystallography of a cocrystal of the repressor–DNA complex.[12]

---

[8] T. D. Tullius and A. M. Burkhoff, in "Structure and Expression, Volume 3: DNA Bending and Curvature" (W. K. Olson, M. H. Sarma, R. H. Sarma, and M. Sundaralingam, eds.), p.77. Adenine Press, Guilderland, New York, 1988.

[9] T. D. Tullius, Trends Biochem. Sci. 12, 297 (1987).

[10] T. D. Tullius, Nature (London) 332, 663 (1988).

[11] J. Ward and I. Kuo, Int. J. Radiat. Biol. Relat. Stud. Phys. Chem. Med. 18, 381 (1970).

[12] S. R. Jordan and C. O. Pabo, Science 242, 893 (1988).

An identical result is obtained whether the O_R1 site is contained in a small linear DNA fragment or in a form I plasmid during the footprinting experiment (data not shown).

Perspectives

We have shown here that γ-radiation is a suitable alternative to chemical systems for generating hydroxyl radicals for footprinting purposes. The technique can be used to carry out footprinting in protein–DNA-binding solutions that are not conducive to the use of the $[Fe(II)EDTA]^{2-}$ reagent system, such as those that contain moderate amounts of glycerol,[5] or those that cannot tolerate the addition of one of the reagent components, such as hydrogen peroxide.[5,13]

Morever, the ability of γ-radiation to penetrate matter suggests that extention of the technique to *in vivo* systems might be possible. Such experiments could be performed on a model system constructed on a plasmid or could be applied to interactions of proteins with chromosomal DNA when used in combination with an appropriate detection scheme.[14] An *in vivo* application would offer the opportunity of determining high-resolution footprints of proteins bound to DNA inside cells. On the debit side, the delivery of large radiation doses to an organism causes a great deal of collateral damage, which over an extended period of time may not be acceptable. This problem may be rectified by the use of particle accelerators, such as those in use for pulse radiolysis experiments, to generate the required radiation dose in the shortest time period possible.

Acknowledgments

This research was supported by grants from the American Cancer Society (Institutional Research Grant IN-11Z) and the National Cancer Institute of the National Institutes of Health (CA 37444). T.D.T. is a fellow of the Alfred P. Sloan Foundation, a Camille and Henry Dreyfus Teacher–Scholar, and the recipient of a Research Career Development Award from the National Cancer Institute (CA 01208).

[13] K. E. Vrana, M. E. A. Churchill, T. D. Tullius, and D. D. Brown, *Mol. Cell. Biol.* **8**, 1684 (1988).
[14] G. M. Church and W. Gilbert, *Proc. Natl. Acad. Sci. U.S.A.* **81**, 1991 (1984).

[57] Fluorometric Analysis of DNA Unwinding to Study Strand Breaks and Repair in Mammalian Cells

By H. Chaim Birnboim

Introduction

The early work of Ahnström and Erixon and of Kohn and colleagues[1,2] demonstrated that the rate of unwinding in alkali of the two strands of DNA can be used as a sensitive measure of the number of breaks induced in the DNA backbone. While separation of the two strands of a typical purified viral DNA occurs in seconds, the rate of strand separation of DNA from a mammalian cell is measured in hours. Introduction of breaks into the DNA backbone by various treatments increases the rate of unwinding. Although theories have been developed to explain the kinetics of the unwinding process, it is common practice to gauge the absolute number of breaks induced by any given treatment by reference to the number of breaks induced by ionizing radiation.

The FADU (fluorometric analysis of DNA unwinding) technique was initially developed to allow study of DNA strand breaks in cells such as leukocytes in which DNA cannot be readily radiolabeled.[3] The principle of the method is that, under certain conditions, double-stranded DNA in crude cellular extracts can be measured using a fluorescent dye, ethidium bromide, with relatively little interference from single-stranded DNA and RNA. Different suspensions of cells are exposed either to no treatment or to treatments such as ionizing radiation, hydrogen peroxide, or other agents which cause strand breaks. An extract of each cell suspension is exposed to alkaline denaturing conditions for a fixed period of time; the pH is then lowered to stop further unwinding, and the amount of residual double-stranded DNA is determined using the fluorescence of ethidium bromide.

Materials

Stock Solutions for FADU Procedure

10 M urea: for 1.0 liter, slowly add 600.6 g urea to 400 ml of water and stir under low heat (<40°) until dissolved; make to volume (store at

[1] G. Ahnström and K. Erixon, *Int. J. Radiat. Biol. Relat. Stud. Phys. Chem. Med.* **23**, 285 (1973).
[2] K. W. Kohn and R. A. G. Ewig, *Cancer Res.* **33**, 1849 (1973).
[3] H. C. Birnboim and J. J. Jevcak, *Cancer Res.* **41**, 1889 (1981).

4° in 49.5- or 99.0-ml aliquots; urea will crystallize out under these conditions, and an increase in conductivity with time indicates generation of ammonium cyanate whose formation is slowed under these storage conditions)

0.5 M CDTA (diaminocyclohexanetetraacetic acid), a chelating agent similar to EDTA, available from Sigma (St. Louis, MO) or Aldrich (Milwaukee, WI)

10% (w/v) SDS (sodium dodecyl sulfate)

1.0 M glucose (store at 4° over a few drops of chloroform)

0.5 M *myo-inositol* (stored at 4° over a few drops of chloroform)

1 mg/ml ethidium bromide (store at 4° in the dark)

All solutions (except ethidium bromide) are filtered through a 0.45-μm membrane after preparation.

Working Solutions for FADU (Prepared Daily from Stock Solutions)

Solution I (a buffered isotonic solution): 0.25 M *myo*-inositol–10 mM sodium phosphate–1 mM MgCl$_2$; adjust to pH 7.2 with 1 N NaOH

Solution II (for lysis of cells and disruption of chromatin): 9 M urea (from 10 M urea stock)–10 mM NH$_4$OH–2.5 mM CDTA–0.1% SDS

Solution IIIa: 0.55 volume solution II, 0.2 volume of 1 N NaOH, 0.25 volume water

Solution IIIb: 0.45 volume solution II, 0.2 volume of 1 N NaOH, 0.35 volume water

Solutions IIIa and IIIb contain NaOH (to denature the DNA at pH 12.6) but differ in density to aid in diffusion of NaOH into the cell lysate.

Solution IV (to drop the pH to 11.0): 1 M glucose–14 mM 2-mercaptoethanol

Solution V: 6.67 μg/ml ethidium bromide–13.3 mM NaOH

Other Solutions

Ammonium chloride (for red cell lysis): 0.87% NH$_4$Cl–10 mM Tris-HCl (pH 7.2)–10 mM NaHCO$_3$

BSS: 137 mM NaCl, 5 mM KCl, 0.8 mM MgSO$_4$, 10 mM HEPES, 5 mM glucose (pH 7.4)

PMA (phorbol 12-myristate 13-acetate) stock solutions: 1 and 10 mM PMA in DMSO (dimethyl sulfoxide, HPLC or spectrophotometric grade) (store in aliquots in DMSO-rinsed glass vials with Teflon-lined caps at −20°; stock solutions of PMA from Sigma have been stable for several years under these conditions)

Isolation of Leukocytes and Stimulation of Respiratory Burst

Human leukocytes are isolated from freshly drawn blood using 10-ml Vacutainer tubes (Becton-Dickinson) containing EDTA as an anticoagulant. The blood samples need to be carefully mixed to prevent formation of microclots. A granulocyte-rich cell suspension is prepared by mixing 1 volume of blood with 3 volumes of ammonium chloride solution. The blood is incubated at 15–20° until it turns dark in color, indicating lysis of the red cells. The lysate is centrifuged in polypropylene tubes (50 ml) or bottles (200 ml) at 2° for 10–15 min (depending on the volume being processed) at 600 g. The white cell pellet is washed with BSS, and residual red cells are removed by hypotonic lysis (2 volumes water, 0°, 60 sec). Isotonicity is restored with 0.2 volume of 1.7 M NaCl solution. Typically, the yield is about 1.5 × 10^8 cells from 50 ml of blood. Cells are centrifuged, suspended in cold BSS at a concentration of 1.0 × 10^6/ml, and 10 ml of cell suspension is distributed to a series of 15-ml polypropylene or siliconized glass centrifuge tubes. Polystyrene tubes are unsatisfactory.

More recently, we have found that an essentially pure preparation of granulocytes can be recovered in high yield by carrying out the above ammonium chloride–hypotonic lysis technique on blood cells from which mononuclear cells have been removed by conventional centrifugal banding techniques. Fifteen milliliters of blood is diluted with 7.5 ml of BSS and layered over 12.5 ml of a Metrizoate–Ficoll solution, density 1.077, in a 50-ml polypropylene or polyethylene tube. After centrifugation at 450 g for 15 min at room temperature, the top layers are removed by aspiration, leaving the lower layer containing red blood cells and granulocytes. The wall of the tube is washed with 5 ml of BSS which is then removed by aspiration. The pellet volume is made up to 12 ml with BSS, and 38 ml of ammonium chloride solution is added. The suspension is incubated at 30° for about 10 min until red cell lysis occurs. Further processing is as described above. A typical yield is 1.2–1.4 × 10^8 granulocytes from 60 ml of blood. Microscopic examination of stained cytocentrifuged samples shows no more than 2% contamination with mononuclear cells.

For experiments involving PMA, the PMA is freshly diluted 1:100 from stock to 10 μM in DMSO and then added to the cold cell suspension at 10 nM. In order to study O_2^--specific breaks, catalase is added to 20 μg/ml to eliminate H_2O_2. Cells start to produce O_2^- on warming to 37°; incubation is continued for 40 min with mixing by inversion at 10-min intervals. Control tubes without PMA are included for reference in each experiment.

For experiments involving glucose oxidase to produce a flux of H_2O_2, the enzyme is added at 50–100 mU/ml to a cold cell suspension; cell treatment is begun by warming cells to 37° as above. If repair of breaks is to be studied, catalase is added to some tubes after a period of time to eliminate further H_2O_2 produced by the glucose oxidase.

For experiments involving ionizing radiation, the cells are held in an ice-water bath during exposure to ^{60}Co γ-rays to prevent/minimize repair processes. Rejoining of strand breaks induced by radiation can be followed by incubating the cells at 37° for varying lengths of time.

Following the treatments described above, incubations are terminated by chilling the suspensions and pelleting cells at 2°.

Standard FADU Procedure

Following centrifugation, each cell pellet is suspended in 3 ml of cold solution I using a 5-ml plastic pipet tip. The suspension is quickly transferred to a 5-ml Eppendorf "combitip" or other similar repeating pipettor. Cells are distributed rapidly and uniformly in 0.2-ml aliquots to a series of 12 glass culture tubes (16 × 100 mm) at ice temperature. The tubes are held in a metal block (aluminum or copper) containing a series of wells which fit the tubes closely. The holder allows tubes to be chilled and warmed as necessary without wetting them; this is useful as the fluorescence of the tube contents will be read in the original tube. Four of the 12 tubes are designated as T (total), 4 as P (partial), and the remaining 4 as B (blanks). Sets of 4 are used as quadruplicates to improve precision. Using a repeating pipettor, 0.2 ml of solution II is added to all tubes briskly enough to mix the contents as the solution is injected.

To the P tubes are *gently* added 0.1 ml of solution IIIa and 0.1 ml of solution IIIb. At this point, the lysate is sensitive to shear, and handling may artifactually introduce breaks into the DNA. The tubes are incubated on ice for 30 min to allow alkali to diffuse from below and above into the cell lysate; some DNA unwinding occurs. The tubes are incubated for 60 min at 15° to allow unwinding to proceed, then chilled for 5 min. The alkali is "neutralized" (brought from pH 12.6 to 11.0) by rapidly injecting 0.4 ml of solution IV and mixing immediately on a vortex mixer. The pH must be dropped quickly to below the denaturing pH (>11.8). The lysates are sonicated with a probe at low power for 1–2 sec to render them homogeneous. Then 1.5 ml of solution V is added and samples are mixed on a vortexer or vortexing platform.

The T tubes are brought to pH 11 without exposure to pH 12.6. This is done by combining solutions IIIa, IIIb, and IV (in the same ratio as used for the P tubes) and adding 0.6 ml to each T tube. Thereafter, they are treated similarly to the P tubes.

The B tubes receive solutions IIIa and IIIb (combined) as do the P tubes, except that these solutions can be added quickly. The lysates are sonicated for a few seconds to introduce a large number of strand breaks, so that there will be complete unwinding of the DNA. After about 10 min, solution IV is added and the lysates are again sonicated. Thereafter, they are treated similarly to the P tubes.

The tubes in the metal holder are transferred to a styrofoam box containing tap water at 2° above room temperature. The fluorescence of each tube is read directly (without transfer to a cuvette) in a spectro-fluorometer (λ_{ex} 520 nm, λ_{em} 590 nm) at room temperature using a cell holder that will accommodate a 16-mm-diameter tube. We currently use a Perkin-Elmer LS-5 with a cell holder fabricated from an aluminum block. With this procedure, the fluorescence of a large number of samples can be read while controlling the temperature at ±2°; larger fluctuations in tem-perature affect the fluorescence readings significantly. For convenience, the spectrofluorometer is set to 0 with a "no-cell blank" (i.e., containing all reagents but no cells). One of the T tubes is set to 150 (arbitrary) units. At the completion of the readings, all tubes are read a second time in the same order. We use an IBM-compatible computer and a program devel-oped locally to record fluorescence readings automatically.

Modifications of Standard FADU Procedure

Lowering Sensitivity. In studies of treatments producing a large num-ber of DNA strand breaks, it is possible to decrease the sensitivity about 3-fold and extend the linear range by omitting the 60-min incubation step at 15°.[4]

Detection of Cross-Linking Agents. As in the Kohn technique,[5] cross-linking agents such as nitrogen mustard (NH2) can be detected by their ability to slow down the rate of DNA unwinding both in control cells and in cells exposed (after NH2 treatment) to ionizing radiation at 0°.

Removal of Protein Cross-Links Which Interfere with DNA Unwind-ing. Proteinase K can be used to remove protein cross-links by adding it to solution II (at 0.8 mg/ml) immediately before use. The lysate is incu-bated at 37° for 40 min instead of at 0° for 10 min.

Calculations and Presentation of Results

The effect of any given treatment is determined by subtracting the number of breaks in untreated cells from the number in treated cells. Both

[4] D. Thierry, O. Rigaud, I. Duranton, E. Moustacchi, and H. Magdelenat, *Radiat. Res.* **102**, 347 (1985).
[5] K. W. Kohn, *in* "Methods in Cancer Research" (V. T. Devita, Jr., and H. Busch, eds.), Vol. 16, p. 291. Academic Press, New York, 1979.

sets of readings of the quadruplicates are averaged to give a single T, P, and B value for each cell sample. Percent double-stranded DNA (D or D_c for treated or control samples, respectively) is given by $100(P - B)/(T - B)$. The quantity of DNA strand breaks induced by a given treatment (Q_d) is $100[\log(D_c) - \log(D)]$. An essentially linear dose–response curve for ^{60}Co γ-radiation is given by this equation in the range 0–4 Gy.[4,6] One Q_d unit is estimated to correspond to about 100 strand breaks per cell.[6]

Concluding Remarks

The FADU technique has proved to be a highly useful method for detecting and quantifying DNA strand breaks induced by a wide variety of treatments in many cell types including granulocytes, lymphocytes, bone marrow cells, leukemia cells, endothelial cells, and fibroblasts.[3,6–13] When semiautomated by using a computer to record, calculate, and plot data, 16 or more samples can be analyzed on a daily basis by a single experienced technician.

Acknowledgments

This work has been supported by a grant from the Medical Research Council of Canada.

[6] R. S. McWilliams, W. G. Cross, J. G. Kaplan, and H. C. Birnboim, *Radiat. Res.* **94**, 499 (1983).

[7] H. C. Birnboim, *Science* **215**, 1247 (1982).

[8] H. C. Birnboim, *Carcinogenesis* **7**, 1511 (1986).

[9] W. L. Greer and J. G. Kaplan, *Exp. Cell Res.* **166**, 399 (1986).

[10] G. E. Francis, A. D. Ho, D. A. Gray, J. J. Berney, M. A. Wing, J. J. Yaxley, D. D. F. Ma, and A. V. Hoffbrand, *Leukemia Res.* **8**, 407 (1984).

[11] H. G. Prentice, G. Robbins, D. D. F. Ma, and A. D. Ho, *Cancer Treat. Rev.* **10**, 57 (1983).

[12] M. Lorenzi, D. F. Montisano, S. Toledo, and A. Barrieux, *J. Clin. Invest.* **77**, 322 (1986).

[13] R. D. Snyder, *Mutat. Res.* **193**, 237 (1988).

[58] Clastogenic Factors: Detection and Assay

By Ingrid Emerit

Introduction

Chromosome breakage factors, or clastogenic factors (CF), were first described by radiobiologists, who reported a chromosome-damaging effect of plasma from irradiated persons (for review, see Ref. 1). Renewed

[1] G. B. Faguet, S. M. Reichard, and D. A. Welter, *Cancer Genet. Cytogenet.* **12**, 73 (1984).

interest in CF paralleled the recognition of clastogenic activity in the plasma of patients with so-called spontaneous chromosomal instability. These observations include not only the congenital breakage syndromes ataxia telangiectasia[2] and Bloom's syndrome,[3] but also chronic inflammatory diseases such as rheumatoid arthritis, systemic lupus erythematosus, progressive systemic sclerosis, Crohn's disease, and multiple sclerosis (for review, see Refs. 4 and 5). Recent work in our laboratory has shown that CF may also be isolated from the blood of patients subjected to ischemia/reperfusion.[6]

In all of these conditions, chromosome damage is prevented by superoxide dismutase (SOD), indicating a role for the superoxide anion radical in the breakage phenomenon. The possibility of producing CF *in vitro* by exposure of cells to a xanthine oxidase reaction[7] or to a respiratory burst[8] and of preventing CF induction in the culture system by addition of SOD is another consideration.

Clastogenic factors are composed of low molecular weight substances, filterable through Diaflo ultrafiltration filters with a cutoff at 10,000 daltons. The presence of TBA-reactive material and of substances with conjugated diene structures in CF preparations[7,8] suggests a relationship between CF formation and lipid peroxidation of membranes. However, in addition to the lipophilic components, other clastogenic material was detected by fractionation of CF preparations using HPLC. The exact chemical nature of this material is presently being studied.

Preparation of Clastogenic Factors

Clastogenic factors can be isolated from serum (plasma) of patients or from cell culture media (Fig. 1). Synovial or spinal fluid may also contain CF (e.g., rheumatoid arthritis, multiple sclerosis). After centrifugation at 3000 rpm for 10 min, the samples are filtered through an Amicon YM10 (Danvers, MA) filter with a cutoff at 10,000 daltons. As indicated above, all CF studied to date pass through this pore size. If the retentate is restored to the initial volume, it is not clastogenic. In previous studies, we

[2] M. Shaham, Y. Becker, and M. M. Cohen, *Cytogenet. Cell Genet.* **27**, 155 (1980).

[3] I. Emerit and P. Cerutti, *Proc. Natl. Acad. Sci. U.S.A.* **78**, 1868 (1981).

[4] I. Emerit, *Prog. Mutat. Res.* **4**, 61 (1982).

[5] I. Emerit, in "Free Radicals, Aging and Degenerative Diseases" (J. E. Johnson, ed.), p. 307. Alan R. Liss, New York, 1986.

[6] I. Emerit, J. N. Fabiani, O. Ponzio, A. Murday, F. Lunel, and A. Carpentier, *Ann. Thorac. Surg.* **46**, 619 (1988).

[7] I. Emerit, S. H. Khan, and P. Cerutti, *J. Free Radicals Biol. Med.* **1**, 51 (1985).

[8] S. H. Khan and I. Emerit, *J. Free Radicals Biol. Med.* **1**, 443 (1985).

CF Preparation

Obtain starting material:
  Plasma or cerebrospinal or synovial fluid
  Supernatants of blood or fibroblast cultures from patients, or
  Supernatants of cultures exposed to irradiation, xanthine–xanthine oxidase reaction,
    TPA, or other CF-inducing chemicals
  Ultrafiltrate through Diaflo YM10 filter
  Concentrate through Diaflo YM2 filter, or
  Lyophilize

Activity Assay

Add 100–500 μl ultrafiltrate (or concentrate) to blood culture (0.5 ml blood, 5 ml TCM 199,
  1.5 ml FCS, PHA) from healthy donor (duplicate)
Arrest cultures after 68–72 hr with colchicine (last 2 hr of cultivation)
Centrifuge at 3000 rpm for 10 min, resuspend in hypotonic KCl (75 mM) for 10 min;
  centrifuge
Add ethanol–glacial acetic acid (3 : 1)
Fix for at least 60 min; centrifuge; resuspend in 3 drops of fresh fixative
Spread on cold, wet slides; after air drying, stain with Giemsa
Examine 50–100 mitoses on photographs or under the microscope for presence of breaks,
  fragments, and exchange figures (or count SCE)
Calculate results as percentage of mitoses with aberrations (and total number of aberra-
  tions/100 mitoses)
Compare to aberration rate of control cultures

FIG. 1. Scheme for preparation of CF and assay for clastogenic activity.

used a second filtration through Amicon YM2 to concentrate the CF. However, retention is not complete, so we now use a lyophilization technique. Clastogenic factors are not inactivated by lyophilization if it is done rapidly using small aliquots.

Blood samples should be centrifuged rapidly and the serum frozen if the ultrafiltration cannot be done the same day. Repeated freezing and thawing of the samples should be avoided. The same caution is necessary for CF isolated from conditioned cell culture media. In addition, the following factors should be kept in mind:

1. Because CF production is related to oxy radical generation, the culture medium and the serum used for supplementation should contain only low levels of antioxidants. In a previous study concerning the influence of culture medium composition on the incidence of chromosomal breakage (Table I), TCM 199 yielded the highest aberration rates.[9] Considerable variation in SOD content exists between batches of fetal calf serum.[10]

[9] M. Keck and I. Emerit, *Hum. Genet.* **50,** 277 (1979).
[10] A. Baret and I. Emerit, *Mutat. Res.* **121,** 293 (1983).

TABLE I
INFLUENCE OF CULTURE MEDIUM COMPOSITION ON
CHROMOSOMAL BREAKAGE AND SISTER CHROMATID
EXCHANGE RATES

Medium	Chromosomal breakage			SCE rate	
	PSS	CD	Controls	TPA	Controls
TCM 199	23.7	25.8	7.2	8.7	4.9
MEM	9.5	14.2	4.2	6.2	4.6
RPMI 1629	6.9	7.9	2.3	5.9	4.5

[a] DNA damage was induced in cells of same donor by the same CF preparation. The figures represent percentages of mitoses with aberrations in cells exposed to CF from patients with progressive systemic sclerosis (PSS) or Crohn's disease (CD) and number of SCE per mitosis after exposure of cells to CF from TPA-treated cultures (100 ng/ml). Laboratory standard for normal persons: breakage studies, 4–6 aberrations per 100 mitoses studied; SCE studies, 5–6 SCE per mitosis.

2. When lymphocytes are used for CF production, they must be stimulated with phytohemagglutinin (1 μg/ml medium). The percentage of monocytes in the mononuclear cell population influences the results. This has been shown not only for CF production by cultures exposed to the tumor promoter phorbol 12-myristate B-acetate (TPA),[11] but also for CF from blood cultures of patients with rheumatoid arthritis.[12] Clastogenic potency reaches a maximum after 48 hr with no further increase thereafter. Formation of CF in hypoxanthine plus xanthine oxidase-treated cultures and in TPA-treated cultures requires about 15 hr to reach detectable levels.

3. When fibroblasts from patients with Bloom's syndrome are cultivated or Chinese hamster lung fibroblasts (V79) are exposed to TPA, CF concentrations reach a maximum at the end of the exponential growth phase.

4. Good culture growth is a prerequisite for CF formation. Cytotoxic effects therefore have to be avoided. For this reason smaller doses of CF-inducing chemicals may result in more clastogenic activity of the culture supernatant than higher doses with cytotoxic effects. For TPA, the optimal dose varies between 10 and 100 ng/ml[13]; for hypoxanthine plus xan-

[11] I. Emerit and P. Cerutti, *Carcinogenesis* **4**, 1313 (1983).
[12] I. Emerit, A. Levy, and J. P. Camus, *J. Free Radicals Biol. Med.* **6**, 245 (1989).
[13] I. Emerit and P. Cerutti, *Proc. Natl. Acad. Sci. U.S.A.* **79**, 7509 (1982).

thine oxidase-treated cultures the optimal quantities are 7 and 10 μg/ml, respectively. Relatively low doses of irradiation (400–1200 rads) are more efficient for CF production *in vitro* than higher doses.[4]

5. Since the quantity of CF produced is variable from one culture to the other, it is preferable to pool the supernatants from several cultures.

6. The quantities of medium and serum used for the cultures are dependent on the cell type. Lymphocytes (2.5×10^6) or whole blood samples (0.5 ml) are cultivated in 5 ml TCM 199 supplemented with 1 ml fetal calf serum. The quantities for other cell cultures should be those giving optimal growth conditions.

7. It is of course crucial to determine that CF-inducing chemicals are not present in CF preparations. This is easy for substances with a molecular weight of more than 10,000 such as xanthine oxidase, but not for low molecular weight chemicals which pass through the Amicon YM10 membrane along with CF. For TPA, we could confirm, by use of a radioactive compound, that this phorbol ester is retained by the YM10 membrane.[13]

8. No CF is produced in medium or serum exposed to irradiation, TPA, or xanthine oxidase in the absence of cells.

Assay for Detection of Clastogenic Activity

Cytogenetic techniques detecting an increased frequency of chromosomal breaks and rearrangements or an increase in sister chromatid exchanges (SCE) are widely used for the demonstration of genotoxic effects, even though the exact mechanisms of break and SCE induction are not fully understood. Microscopically visible chromosome damage is considered as an indication of lesions occurring at the molecular level in DNA.

For most studies of chromosome damage or SCE induction, blood culture procedures are convenient (Fig. 1). As described above for *in vitro* induction of CF, whole blood (0.5 ml) or isolated lymphocytes (2.5×10^6) from a healthy blood donor are incubated in TCM 199 (5 ml) supplemented with fetal calf serum (FCS) (1.5 ml). Cell division is promoted by the addition of phytohemagglutinin (PHA, Wellcome, Datford, UK; 1 μg/ml). All cultures are set up in duplicate in Nunclon Delta tissue culture tubes (110 × 16 mm, ~10 ml; No. 156758, Nunc, Roskilde, Denmark), which have one flat side for cultivation in the horizontal position.

If the clastogenic potential of a patient's serum is to be studied, one may use it for supplementation of the medium. The control cultures are supplemented with serum of the blood donor. However, it is preferable to cultivate CF- and control-treated cells in the same serum and to add ultrafiltrates to the FCS-supplemented culture system. For detection of CF after reperfusion only, whole serum may be used, collected from the same patient before ischemia and after reperfusion.

In general, aliquots of 0.1–0.5 ml ultrafiltrate are added to a total volume of 6.5 ml per tube. The percentage of ultrafiltrate should not exceed 10% of the total volume. Controls are treated with the same quantity of control ultrafiltrate derived from normal serum or sham-treated culture supernatants. The quantity of CF necessary to produce a significant increase in chromosome aberrations without disturbing culture growth has to be determined because the clastogenic potential of CF preparations is variable. A low mitotic index in treated cultures compared to control cultures may be due to the cytotoxic effect of high CF concentrations. In this case, the assay should be repeated with a smaller quantity. Cytogenetic analysis should not be done on poor cultures.

The ultrafiltrates may be added to the growth medium at any time of the cultivation period. The highest aberration rates are obtained when the ultrafiltrates are added at the beginning. The duration of culture is 68–72 hr at 37° and may be reduced to 48 hr, if only the first cell divisions are desired for study. At 48 hr, however, one will dispose of less mitoses. Two hours before harvesting, colchicine (1–2 μg/ml) is added to the cultures in order to arrest mitoses in metaphase. The tubes are then centrifuged at 1000 rpm for 10 min. The supernatant is discarded, and the cell pellet is suspended in hypotonic shock solution (75 mM KCl) for 10 min at 37° or room temperature. After centrifugation, the cells are fixed in ethanol–glacial acetic acid (3:1).

The exact quantities of shock solution and fixative are not important. In general, the tubes are filled. However, to obtain good quality preparations the fixation step should be repeated 2–3 times, and the first drops of fixative should fall into a small quantity of residual KCl. Otherwise, the cells stick together and spreading is not good. A minimal fixation time of 60 min is necessary. In general, we leave the cells overnight at 4° after the second addition of fixative.

The next day, cells are centrifuged and resuspended in a small quantity of fixative, sufficient to prepare three slides, each receiving 2 drops of cell suspension. Iced wet slides, carefully cleaned before use, allow good spreading of the metaphase plates. The drops should fall on the slide from a distance of about 10 cm. After drying at room temperature (accelerated, if necessary, by a stream of warm air), the metaphase plates are stained with Giemsa.

When SCE are studied instead of chromosomal aberrations, bromodeoxyuridine (BrdU) has to be added to the culture medium at a final concentration of 5 μg/ml at hour 0 of the cultivation period. After two cell

TABLE II
COMPARISON OF BREAKAGE AND SISTER
CHROMATID EXCHANGE RATES WITH
CLASTOGENIC FACTORS FROM PATIENTS WITH
BLOOM'S SYNDROME[a]

Breakage	SCE
Supernatants of BS fibroblast cultures	
22.0	10.2
18.0	8.4
14.0	7.7
BS patient plasma	
26.0	10.5

[a] The figures represent percentages of cells with aberrations for the breakage study and number of SCE per mitosis for the SCE study. Laboratory standards: 4–6% mitoses with aberrations and 5–6 SCE per mitosis.

cycles (cultivation time at least 68 hr), slides are prepared as indicated above. Differential staining of the sister chromatids is accomplished, after brief exposure to UV light, with Hoechst 33258 and Giemsa stain according to Perry and Wolff.[14]

When chromosomal aberrations are 3 to 5 times more frequent in CF-exposed cultures than in controls, SCE rates, though consistently increased, are generally not even doubled (Table II). In studying genotoxic effects, SCE studies are easier, since they can be done without cytogenetic training by the technician. However, because the CF-induced chromosome damage is mainly of the chromatid type, as already noted in the first reports,[1] chromosome type aberrations are rare, and the observer may score only open breaks and fragments, which cannot be overlooked (Figs. 2 and 3). A minimum of 50 mitoses should be studied for each experiment on coded slides. Even if many mitoses are visible, they should be examined on all the slides; in general, use 2 × 3 slides for duplicate cultures.

Addition of SOD (10 μg, equivalent to 30 cytochrome c units/ml) prevents the induction of chromosome damage and SCE by CF. In addition to the predominance of chromatid-type damage, the protective effect of SOD is an indication for the indirect action of CF. Damage can be induced only in living cells, not in isolated DNA.

[14] P. Perry and S. Wolff, *Chromosoma* **48,** 341 (1974).

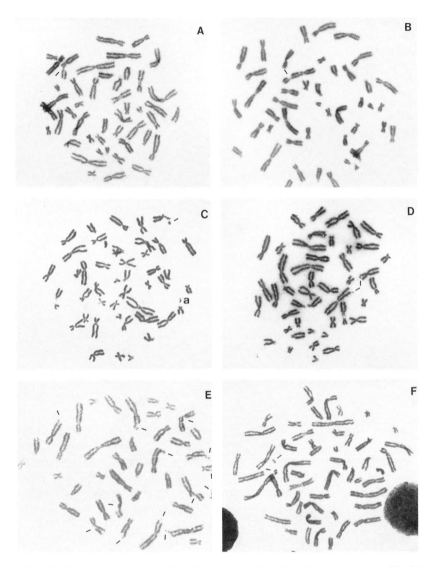

FIG. 2. Examples of chromatid breaks in mitoses from lymphocyte cultures of healthy donors exposed to CF preparations of various origin. Metaphase plates A, B, and C show one chromatid break each. In C, the break resulted in extrusion of a small round fragment. Telomeric extrusions may be as tiny as those shown in D and should not be confused with small particles of nonchromosomal origin (marked a for artifact in C). Metaphase plate E exhibits numerous breaks, which is a relative rare finding (~0.4% of aberrations observed). In F, two acentric fragments derived from isochromatid breaks are shown.

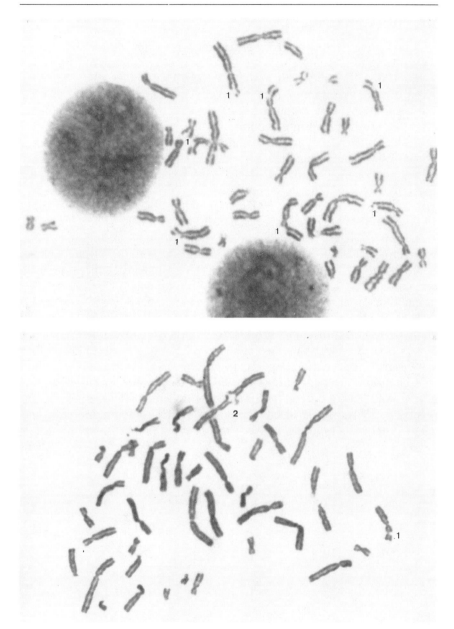

FIG. 3. Increased chromosome damage with the presence of exchange figures is one of the features of Bloom's syndrome. Similar damage is induced in lymphocytes of healthy donors by CF isolated from serum or supernatants of fibroblast cultures from patients with Bloom's syndrome. Chromatid breaks; 2, quadriradial exchange figure.

Significance of Clastogenic Factors

Clastogenic factors are biomarkers for a prooxidant state and may be responsible for oxidative damage at distance from the site of free radical generation. These diffusible circulating factors are released by cells after exposure to oxygen free radicals, which may be generated by irradiation, the respiratory burst of inflammatory cells, or other chemical reactions. Because the mechanism of action also appears to be mediated by free radicals, an autocatalytic self-sustaining process may explain the persistence of the clastogenic effect. As already suggested,[1] the high incidence of malignancy after irradiation might relate to the repetitive DNA-damaging effect of CF. The carcinogenicity of radiation-induced CF can be confirmed experimentally.[15] Elucidation of the exact nature of CF may be of considerable importance for radiation protection.

CF may also be responsible for certain pathologic manifestations in connective tissue diseases, for instance, DNA antibody formation and photosensitivity in lupus erythematosus.[16] The diagnostic and prognostic interest in CF is documented in patients with Raynaud's disease, a rather frequent disorder of various origins. Raynaud's phenomenon is observed in 80% of patients with progressive systemic sclerosis (PSS) and may precede the onset of cutaneous sclerosis and visceral involvement by a long period. Six of 21 patients studied had CF, and all 6 developed other symptoms of PSS during the 5 years of follow up.[17]

Clastogenic factors can also be used as a test system for the efficacy of drugs, as indicated by our recent study of patients undergoing ischemia/reperfusion.[6] The study of CF formation and CF action is therefore not only of theoretical interest but also has practical implications.

[15] J. Souto, *Nature (London)* **195,** 1317 (1962).
[16] I. Emerit and A. M. Michelson, *Proc. Natl. Acad. Sci. U.S.A.* **78,** 2537 (1981).
[17] I. Emerit, *Z. Rheumatol.* **39,** 84 (1980).

Section IV

Leukocytes and Macrophages

[59] Kinetic Microplate Assay for Superoxide Production by Neutrophils and Other Phagocytic Cells

By LAURA A. MAYO and JOHN T. CURNUTTE

Introduction

Phagocytic cells recognize and respond to a remarkable variety of stimuli which include opsonized microorganisms, chemotactic peptides, immune complexes, protein kinase C activators, membrane perturbants, and unsaturated fatty acids. One of the most dramatic responses of phagocytic cells to these types of stimuli is the respiratory burst, in which the rate of oxygen consumption abruptly increases by a factor of 10 to over 100.[1] This oxygen participates in a nonmitochondrial reaction in which it is reduced by one electron to O$_2^-$ by a plasma membrane-bound NADPH oxidase according to reaction (1).[1] Superoxide then serves as the pre-

$$NADPH + 2 O_2 \rightarrow NADP^+ + H^+ + 2 O_2^- \tag{1}$$

cursor for a series of lethal oxidants such as H$_2$O$_2$, HO$\cdot$, and HOCl. Because superoxide is the initial product formed by NADPH oxidase and a substantial portion of it is released extracellularly, its measurement serves as a sensitive and convenient way of monitoring respiratory burst activity in phagocytic cells.

The most widely used assay for monitoring O$_2^-$ is the cytochrome c assay in which O$_2^-$ reduces ferric cytochrome c to its ferrous form, a change which can be sensitively monitored at 550 nm[2-4]:

$$\text{Oxidized cytochrome } c \text{ (Fe}^{3+}) + O_2^- \rightarrow \text{reduced cytochrome } c \text{ (Fe}^{2+}) + O_2 \tag{2}$$

In complex cellular reaction mixtures, the extent to which O$_2^-$ is responsible for the cytochrome c reduction can be determined by measuring the amount of this reduction that is sensitive to superoxide dismutase (SOD).[2-4] While the activity of the respiratory burst can be monitored by a variety of assays for either O$_2^-$ or other products of the burst, none offers the combined simplicity, specificity, and sensitivity of the cytochrome c assay when performed in the absence and presence of SOD.

[1] J. T. Curnutte and B. M. Babior, *Adv. Hum. Genet.* **16**, 229 (1987).
[2] J. M. McCord and I. Fridovich, *J. Biol. Chem.* **244**, 6049 (1969).
[3] B. M. Babior, R. S. Kipnes, and J. T. Curnutte, *J. Clin. Invest.* **52**, 741 (1973).
[4] J. T. Curnutte and B. M. Babior, *J. Clin. Invest.* **53**, 1662 (1974).

A variety of assay procedures for measuring O_2^- production by phagocytic cells have been described in previous volumes of this series.[5-10] Rather than reviewing or describing modifications of these assays, this chapter focuses on the adaptation of the cytochrome c assay to recently available kinetic microplate readers.

Instrumentation

Superoxide production by phagocytic cells is usually monitored by either continuous or end-point assays in which cytochrome c reduction is measured with single- or dual-beam spectrophotometers. The continuous assay is the preferred method since the generation of O_2^- by phagocytes is not linear with respect to time. Depending on the stimulus, there may be a lag of between 5 sec and 10 min before the rate of production of O_2^- reaches its maximal velocity. Arachidonic acid and the chemotactic peptide N-formyl-L-methionyl-L-leucyl-L-phenylalanine (FMLP), for example, have lag times of only several seconds, whereas fluoride anion does not cause O_2^- production for nearly 10 min. The duration of O_2^- production also varies considerably depending on the stimulus and is yet another reason why the continuous assay is preferable.

In principle, either the continuous or end-point assay should be readily adaptable to 96-well microplates by simply decreasing proportionately the volumes of each of the reaction constituents and then measuring the absorbance change in a 96-well microplate photometer. An end-point assay for measuring O_2^- was developed along these lines and was described in this series by Pick in 1986.[10] Since that time, the development of microplate photometers capable of reading all 96 wells every 5 to 12 sec has made it possible to follow the kinetics of a wide variety of colorimetric assays in microplates.

In developing a kinetic microplate assay for phagocyte O_2^- production, a number of technical problems had to be addressed, some of which were unique to the kinetic method. First, the temperature of the reaction wells has to be tightly controlled in order to optimize O_2^- production and minimize well-to-well variability arising from inhomogeneities in temperature. Therefore, the microplate reader has to fit comfortably in a 37° incubator or have a thermostatted sample compartment. To further mini-

[5] M. A. Trush, M. E. Wilson, and K. Van Dyke, this series, Vol 57, p. 462.

[6] M. Markert, P. C. Andrews, and B. M. Babior, this series, Vol. 105, p. 358.

[7] R. B. Johnston, Jr., this series, Vol. 105, p. 365.

[8] P. J. O'Brien, this series, Vol. 105, p. 370.

[9] D. P. Clifford and J. E. Repine, this series, Vol. 105, p. 393.

[10] E. Pick, this series, Vol. 132, p. 407.

mize temperature variability, it is preferable that the photometer light source be located within the instrument in such a position that it does not cause additional heating of the sample chamber. Second, turbidity caused by the presence of intact cells in each reaction mixture decreases the light signal that reaches the detector. Therefore, the light source must have sufficient intensity to penetrate reaction mixtures containing relatively high concentrations of cells. A third problem, also related to the presence of intact cells in each reaction, is that substantial light scatter can occur and result in "cross-talk" between wells. Fourth, in all but the briefest incubations, the reaction wells must be agitated regularly to keep the cells in a homogeneous suspension and thereby optimize the reduction of cytochrome c by O_2^-. Fifth, the light path in each reaction mixture has to be uniform and not vary during the course of the incubation. This is best accomplished using microplates made of hydrophilic plastic in which the meniscus is controlled by gravity and is therefore uniform in each well. Sixth, the absorption peak at 550 nm of the reduced minus oxidized cytochrome c spectrum is relatively narrow, necessitating the use of a narrow bandwidth optical filter.

In light of the above considerations, the microplate reader chosen for the development of the kinetic O_2^- assay was the Vmax kinetic reader manufactured by Molecular Devices (Menlo Park, CA). Other commercially available kinetic microplate readers were not tested in this study. The Vmax was used in a custom-built plexiglass 37° incubator equipped with a front access door. The instrument performed well under these conditions. (The manufacturer of the Vmax can be contacted for information regarding a recently available thermostatted microplate reader.) The reading chamber in the Vmax has a temperature variation of less than ±1.0% under these conditions, owing in part to the location of the low-wattage light source away from the sample chamber. Individual wells are sequentially illuminated by an array of fiber optic channels. Besides contributing to the temperature stability of the samples, this fiber optic arrangement allows 100% of the available light to be delivered to each well independently, thereby improving the accuracy and linearity of the absorbance measurements, particularly with turbid samples. A focal lens array serves to focus the light transmitted through each well onto its own detector. The microplate carriage is capable of agitation both before and between readings, still allowing measurements to be performed every 8 sec. Uniformity of each light path was achieved using constant reaction volumes in hydrophilic plates (Nunclon, Vangard, Inc., Neptune, NJ).

The Vmax microplate reader was equipped with separate 550-nm optical filters with bandwidths of either 10 or 1 nm. Figure 1 shows the absorbance difference between dithionite-reduced and oxidized cyto-

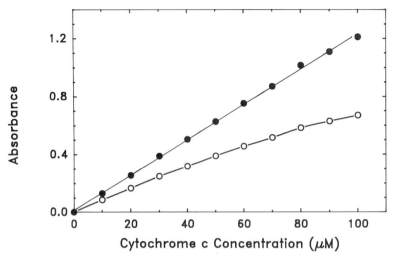

FIG. 1. Comparison of 550-nm filter performance. Each well contained 250 µl of oxidized cytochrome c dissolved in PBSG at the concentrations shown. Two different 550-nm optical fibers were used to measure the absorbance of each well: one had a 1-nm bandwidth and the other a 10-nm bandwidth. The cytochrome c solutions were reduced with a few grains of dithionite, the absorbance determined again, and the difference between the reduced and oxidized values calculated. The filled circles show the results obtained using the 1-nm filter, whereas the open circles show those obtained with the 10-nm filter.

chrome c at various cytochrome concentrations using the two filters. The relationship between absorbance and cytochrome c concentration was linear over the entire range tested using the 1-nm bandwidth filter (filled circles) up to 1.2 absorbance units. In contrast, the 10-nm bandwidth filter showed the expected decrease in sensitivity as well as a loss of linearity at the higher cytochrome c concentrations. Therefore, the 1-nm filter is used routinely for the O_2^- assay. The calculated extinction coefficient for the 1-nm bandwidth filter is 20.5 ± 0.1 S.D. µmol/cm^2 ($n = 3$) based on a light path of 0.6 cm in wells containing 250 µl of cytochrome c solution. This value agrees closely with the value of 21.7 µmol/cm^2 obtained on a Uvikon double-beam spectrophotometer (Kontron Analytical, Zurich, Switzerland).

Isolation of Human Neutrophils

Human neutrophils of over 98% purity are prepared as previously described in this series,[6] using acid–citrate–dextrose as the anticoagulant, dextran to sediment erythrocytes, and Ficoll–Hypaque density gradient centrifugation to separate mononuclear cells from neutrophils. Neutrophils are suspended in phosphate-buffered saline (PBS) [138 mM NaCl, 2.7 mM KCl, 8.1 mM Na$_2$HPO$_4$, and 1.47 mM KH$_2$PO$_4$ (final pH 7.37–7.42)] at a concentration of 5 × 10^6 cells/ml and kept at 4° for up to 6 hr prior to use. For experiments requiring higher cell concentrations in each reaction, neutrophils are suspended at appropriately increased concentrations before use.

Monocytes or macrophages can be substituted for neutrophils in the microplate O_2^- assay described below. If desired, the mononuclear phagocytes can be cultured in the same 96-well microplates eventually used for the O_2^- assay. Under these conditions, a series of wells should then be reserved for measurement of cellular protein, the average of which can serve as the denominator for the rate of O_2^- production.

Kinetic Microplate Assay

Reagents. The physiologic salt solution used in the O_2^- assay is PBS supplemented with 0.9 mM CaCl$_2$, 0.5 mM MgCl$_2$, and 7.5 mM glucose (PBSG). Ferricytochrome c from horse heart mitochondria (Type VI, Sigma Chemical Company, St. Louis, MO) is dissolved in PBSG to make a 1.5 mM stock solution. This solution is prepared fresh every 3 days and is stored at 4°. Superoxide dismutase (SOD) derived from bovine erythrocytes and containing 3,000 units/mg protein is obtained from Sigma. A stock solution containing 3 mg SOD/ml water is prepared weekly and stored at 4°. Alternatively, the stock solution can be stored in aliquots at −20° for at least several months.

Phorbol 12-myristate 13-acetate (PMA) is obtained from Sigma and stored as a stock solution in dimethyl sulfoxide (DMSO) at a concentration of 2 mg/ml. This stock solution can be stored at −20° for at least 1 year. A given aliquot of PMA can be thawed and refrozen multiple times without any measurable loss of activity. For use in the O_2^- assay, a working stock solution is prepared by diluting the frozen stock with DMSO to a concentration of 80 μg/ml. This solution is in turn diluted with water to 8 μg/ml. The working stock solution is prepared fresh each day. Since PMA is a cocarcinogen, it should be handled with great care and disposed of in its own waste container.

The chemotactic peptide FMLP is obtained from Sigma and stored at −70° in aliquots of a 4 mM stock solution in DMSO for up to 1 year. Each aliquot is thawed only once and diluted to 0.4 mM with PBSG each day.

Procedure

Prior to the start of the assay, the microplate reader, a 96-well microplate, PBSG, and the cytochrome c stock solution are all equilibrated to

37°. Immediately prior to the start of the assay, the following reagents are added to each assay well (final concentrations are indicated in parentheses): 176 μl PBSG, 12.5 μl cytochrome c stock (75 μM), 5 μl of either SOD stock solution (60 μg/ml) or 5 μl water, and 50 μl cell suspension. The contents of each well are mixed thoroughly by repeated pipetting with a micropipettor and then placed in the 37° incubator. The reaction mixtures are allowed to reequilibrate at 37° for 3 min. The unstimulated rate of O_2^- production is then determined by placing the 96-well plate in the Vmax reading chamber and following the reaction for 3 min with the instrument set to agitate the plate before the first measurement and then briefly between measurements, which are taken at 8-sec intervals. The absorbance measurement for each of the wells is stored in the kinetic software program for the Vmax. The neutrophils are then activated with PMA or FLMP by adding 6.5 μl of either stimulus to each well. The microplate is quickly placed in the reading chamber of the Vmax, agitated, and then read at 550 nm in the kinetic mode for at least 5 min with intermittent agitation.

If a multichannel micropipettor is used to deliver the stimulus to each well, 16 pairs of reactions (with and without SOD) can be conveniently performed. For the assays described below, an 8-channel electronic pipet is used (Rainin Instrument Company, Woburn, MA, Model EDP-M8). The pipet tip of each channel is loaded with 26 μl of the stimulus with 6.5-μl aliquots dispensed sequentially into each well (the concentrations of stock solutions of each of the stimuli are adjusted slightly since the pipettor could only deliver solutions in increments of 0.5 μl). It is important that the stimulus be added to the wells not containing SOD first so as to avoid contamination of these wells with trace amounts of SOD that might be carried along on the pipet tips if the SOD wells were done first.

Calculations

To facilitate the processing and calculation of the rates of O_2^- production, the kinetic software designed for the Vmax, SOFTmax Version 2.0 (Molecular Devices), is used. This software is capable of determining the maximal rate of absorbance change by calculating the first derivative of the absorbance time course in each well. This feature is particularly useful for stimuli such as PMA which exhibits an approximately 45-sec lag before O_2^- is generated at a maximal rate. This feature is also convenient for the FMLP experiments since the rate of O_2^- production with this stimulus decreases rapidly after 2 min. The extinction coefficient for the 1-nm bandwidth filter (see above) is then used to calculate the concentration of cytochrome c reduced in each reaction mixture. The difference in cytochrome c reduction in the presence and absence of SOD is used as a measure of O_2^- production since one cytochrome c is reduced for every O_2^- detected [Eq. (2)].

Results

Figure 2 shows that the kinetic microplate O_2^- assay exhibits excellent linearity with respect to neutrophil concentration up to the highest concentration tested, 3.5×10^6 cells/ml. For most respiratory burst studies, this concentration of neutrophils is generally sufficient. The inset (Fig. 2) shows the effect of increasing concentrations of cytochrome c in the assay mixture on the measured rate of O_2^- generation. Optimal detection of O_2^- is seen at approximately 25 μM cytochrome c, a value well below the 75 μM concentration used routinely in the assay.

Figure 3 demonstrates the sensitivity and reproducibility of the kinetic microplate assay when various concentrations of FMLP are used as the

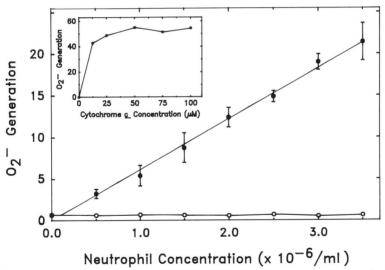

FIG. 2. Effects of neutrophil and cytochrome c concentrations on O_2^- generation. Neutrophils were incubated either in the absence (open circles) or presence (filled circles) of PMA, and O_2^- generation was measured by the kinetic microplate assay described in the text. The units of O_2^- generation are nanomoles per minute. The neutrophil concentration data represent the mean ± S.E. of three experiments, each performed with a different preparation of neutrophils. The inset shows the effect of increasing concentrations of cytochrome c in the assay mixture on the measured rate of O_2^- generation, here expressed as nanomoles per minute per 10^7 neutrophils. The data shown are from a representative experiment performed in triplicate using 10^6 neutrophils/ml.

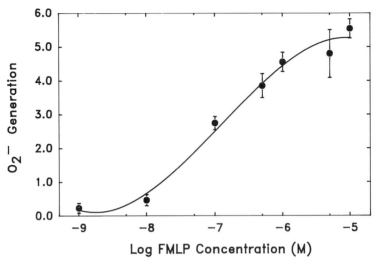

FIG. 3. Effect of FMLP concentration on O_2^- generation. Neutrophils (2×10^6/ml) were stimulated with various concentrations of FMLP as shown, and the rate of O_2^- generation was determined using the kinetic microplate assay described in the text. The units of O_2^- generation are nanomoles per minute per 10^7 neutrophils. The data points represent the means ± S.E. of three experiments, each performed with a different preparation of neutrophils.

TABLE I

REPRODUCIBILITY OF KINETIC MEASUREMENTS OF
CYTOCHROME *c* REDUCTION
BY HUMAN NEUTROPHILS[a]

Reaction conditions	Cytochrome *c* reduction (nmol/min/10^7 cells)		
	Exp. 1	Exp. 2	Exp. 3
Unstimulated			
No SOD	1.3 ± 1.7	0.6 ± 0.6	0.1 ± 0.1
Plus SOD	0	0	0
n	16	16	8
PMA stimulated			
No SOD	89.3 ± 5.0	55.9 ± 5.5	100.4 ± 4.3
Plus SOD	0.3 ± 0.2	0.3 ± 0.7	0.1 ± 0.1
n	16	16	8

[a] Cytochrome *c* reduction was measured by the kinetic microplate method described in the text using a different preparation of neutrophils in each of three experiments. Within each experiment the number of reaction wells for each set of conditions is indicated by *n*.

stimulus. These results are noteworthy in that FMLP is a relatively weak and short-lived stimulus of the respiratory burst. The assay appears to be sufficiently sensitive to consistently detect the low rates of FMLP-induced O_2^- production.

The well-to-well reproducibility of the kinetic microplate assay is demonstrated in Table I. In each of the three experiments, unstimulated neutrophils generated negligible rates of O_2^- with relatively small standard deviations for each set of reactions. The assay also exhibited excellent reproducibility using PMA-stimulated neutrophils. Moreover, these data show that the bulk of cytochrome c reduction that occurs in the assay is superoxide mediated since SOD inhibited over 99% of this reduction in each of the experiments.

Finally, the data in Table I show that the absolute rates of O_2^- generation measured in the kinetic microplate assay are in excellent agreement with those obtained with single reactions on a dual-beam spectrophotometer. Using a cytochrome c assay comparable to the microplate assay described above, PMA stimulated human neutrophils produce O_2^- at a rate of 117 ± 30 nmol O_2^-/min/10^7 cells.[11]

Acknowledgments

This work was supported by U.S. Public Health Service Grants AI-24838 and RR-00833. Dr. Curnutte is an Established Investigator of the American Heart Association with funds contributed in part by the California affiliate of the AHA.

[11] J. T. Curnutte, J. A. Badwey, J. M. Robinson, M. J. Karnovsky, and M. L. Karnovsky, *J. Biol. Chem.* **259**, 11851 (1984).

[60] Measurement of Oxidized Proteins in Systems Involving Activated Neutrophils or HL-60 Cells

By CYNTHIA N. OLIVER

Introduction

Considerable evidence has been presented that enzymatic and nonenzymatic metal-catalyzed oxidation (MCO) reactions inactivate enzymes and oxidize proteins *in vitro*[1–3] and *in vivo*.[4–6] Because NAD(P)H oxidase

[1] R. L. Levine, C. N. Oliver, R. M. Fulks, and E. R. Stadtman, *Proc. Natl. Acad. Sci. U.S.A.* **78**, 2120 (1981).

readily catalyzes these oxidation reactions *in vitro* and because the neutrophil NADPH oxidase is activated when neutrophils are stimulated to undergo a respiratory burst,[7] we have examined the capacity of activated neutrophils to catalyze similar oxidation reactions. Neutrophil-mediated protein oxidation is likely to be important in a variety of physiological conditions such as microbicidal activity of neutrophils, rheumatoid arthritis, ischemia/reperfusion injury, emphysema, and arthrogenesis. Because the levels of oxidized proteins are low in normal cells, the occurrence of oxidatively modified proteins in various tissues is an indicator of cell damage. Thus, incubation of activated neutrophils with potential target cells or proteins may provide useful *in vitro* model systems to examine neutrophil-mediated tissue damage and inhibitors of damage, susceptibility of proteins to oxidation leading to loss of biological activity, or the susceptibility of cells to damage leading to loss of viability.

Preparation of Cells

Studies have been carried out with both neutrophils and HL-60 cells. HL-60 cells are a continuous cell line derived from a patient with promyelocytic leukemia.[8] These cells can be differentiated in culture to form neutrophil-like cells which are capable of phorbol 12-myristate 13-acetate (PMA)-stimulated respiratory burst activity. The advantage of these cells is that they can be grown in large batches and harvested easily by centrifugation, but these properties are countered by lower respiratory burst activity compared to freshly isolated neutrophils. The growth and differentiation of these cells has been described in detail, and these cells can be adapted to serum-free medium.[9]

For our studies, HL-60 cells are grown in a 1:1 mixture of Dulbecco's

[2] L. Fucci, C. N. Oliver, M. J. Coon, and E. R. Stadtman, *Proc. Natl. Acad. Sci. U.S.A.* **80,** 1521 (1983).

[3] R. L. Levine, *J. Biol. Chem.* **258,** 11828 (1983).

[4] C. N. Oliver, *Arch. Biochem. Biophys.* **253,** 62 (1987).

[5] C. N. Oliver, B.-W. Ahn, E. J. Moerman, S. Goldstein, and E. R. Stadtman, *J. Biol. Chem.* **262,** 5488 (1987).

[6] P. E. Starke, C. N. Oliver, and E. R. Stadtman, *FASEB J.* **1,** 36 (1987).

[7] S. J. Klebanoff and R. A. Clark, eds., *in* "The Neutrophil: Function and Clinical Disorders," p. 283. North-Holland, New York, 1978.

[8] S. J. Collins, F. W. Ruscetti, R. E. Gallagher, and R. C. Gallo, *J. Exp. Med.* **149,** 969 (1979).

[9] T. R. Breitman, B. R. Keene, and H. Hemmi, *in* "Methods of Serum-Free Culture of Neuronal and Lymphoid Cells," (D. W. Barnes, D. A. Girbasku, and G. H. Sato, eds.), p. 215. Alan R. Liss, New York, 1984.

modified Eagle's medium (GIBCO, Grand Island, NY) and Ham's F12 nutrient mixture (GIBCO), 20 mM HEPES buffer (M. A. Bioproducts, Walkersville, MD), 2 mM L-glutamine (M. A. Bioproducts), and 10% heat-inactivated fetal bovine serum (Hyclone, Sterile Systems, Inc., Logan, UT). At the time these studies were initiated, we were able to obtain sera which contained low retinoid concentrations and high insulin and transferrin levels, and under these conditions excellent growth is obtained with 2% heat-inactivated fetal bovine serum. Under serum-free conditions the medium is supplemented with albumin (100–300 μg/ml), insulin (5 μg/ml), and transferrin (5 μg/ml).[9] Cultures are seeded at 2–5 × 10^5 cells/ml and grown in suspension in 5% CO_2/air for 4–6 days.

Although cells can be differentiated with a variety of agents,[9] 1% dimethyl sulfoxide (Sigma Chemical Co., St Louis, MO) is used here. Cultures are 70% differentiated in 4 days, as judged by nitroblue tetrazolium reduction,[8] with 90% viability, as determined by trypan blue exclusion.[10] Control, untreated cultures exhibit 5% spontaneous differentiation and 95% viability. Cells are harvested by centrifugation, washed thoroughly, and resuspended in isotonic buffer such as phosphate-buffered saline (pH 7.0–7.4) without calcium or magnesium or Hanks' balanced salt solution (pH 7.0–7.4) without calcium or magnesium. HL-60 cells can be obtained from the American Type Culture Collection (Rockville, MD).

Normal human neutrophils are obtained from either heparinized whole blood or leukophoresis preparations from normal donors provided by the blood bank of the National Institutes of Health. Neutrophils are isolated using Ficoll–Paque (Pharmacia, Piscataway, NJ) discontinuous density gradients followed by dextran (Sigma) sedimentation according to standard procedures.[11,12] If hypotonic lysis of residual red blood cells is required, the isolated neutrophils should be washed thoroughly to remove hemoglobin. The supernatant fluid should be colorless, and the presence of residual hemoglobin can be assessed by visible absorbance in the range of 400–430 nm. However, a new method adapted from the Mono-Poly Resolving Media (MPRM) method (Flow Labs) and recently reported by McFaul[13] is likely to provide excellent separation of neutrophils from larger working volumes of whole blood without hypotonic shock. Again, cells should be resuspended in an isotonic buffer at pH 7.0–7.4, without calcium or magnesium, until use.

[10] M. K. Patterson, Jr., this series, Vol. 58, p. 141.
[11] A. Böyum, *Scand. J. Clin. Lab. Invest.* **98** (Suppl. 29), 77 (1968).
[12] J. A. Metcalf, J. I. Gallin, W. M. Nauseef, and R. K. Root, "Laboratory Manual of Neutrophil Function." Raven, New York, 1986.
[13] S. J. McFaul, *Cell Biol.* **107**, 573a (1988).

Measurement of Oxidized Proteins

The measurement of oxidized proteins in systems involving activated neutrophils is not fundamentally different from the determination of oxidized proteins in crude extracts.[14] In general, similar experiments are carried out with either freshly isolated neutrophils or HL-60 cells, but HL-60 cells are used at densities 2–3 times greater than that of neutrophils because of lower respiratory burst activity. Assays must be adapted for each *in vitro* system depending on the cells or proteins used and the information required. For example, when oxidized proteins and enzymatic activity are determined in activated neutrophils alone, 10^8 to 10^9 cells/ml are suspended in an isotonic buffer with 1 mM CaCl$_2$, 1 mM MgCl$_2$, and 1 mg/ml glucose and stimulated with 100 ng of PMA (Sigma). For these experiments, cells are isolated from leukophoresis preparations enriched for neutrophils, and although high yields of cells are obtained by this method, the levels of spontaneous activation are higher than cells obtained from whole blood without prior leukophoresis. Following an incubation of 30 min to 1 hr, neutrophils are collected by centrifugation[12] and disrupted by sonication (2 bursts of 30 sec each) using a Heat Systems Sonifier Cell Model 185 equipped with a microtip. Insoluble cellular debris is removed by centrifugation, and oxidized protein is determined by 2,4-dinitrophenylhydrazine reactivity using the filter paper method.[14]

Purified proteins and enzymes are incubated with activated neutrophils of HL-60 cells in order to determine the possible susceptibility to neutrophil-mediated oxidative modification *in vivo*. In these experiments, both loss of enzymatic activity and generation of protein carbonyl derivatives are examined. The choice of protein concentrations to be used in such experiments is dependent on the amount protein required for subsequent analytical procedures such as assay of enzymatic activity, electrophoresis and blotting, peptide mapping, or acid hydrolysis and amino acid analysis. Conditions for the assay of protein carbonyl derivatives are subject to the same considerations outlined by Levine *et al.*[14]

For experiments using unopsonized bacterial cells in which protein oxidation is assessed in intact bacterial cells, 10^8 *Escherichia coli* cells are incubated with 2×10^6 neutrophils and activated with 10–100 ng PMA. Following separation of neutrophils and *E. coli* by differential centrifugation, the enzymatic activity of glutamine synthetase (GS) could be detected directly in bacterial cells using CTAB (cetyltrimethylammonium

[14] R. L. Levine, D. Garland, C. N. Oliver, A. Amici, I. Climent, A. G. Lenz, B.-W. Ahn, S. Shaltiel, and E. R. Stadtman, this volume [49].

bromide) permeabilization treatment without disruption of the *E. coli*.[15] In the case of GS, the availability of a overproducing strain facilitates these studies.[4] Because relatively small numbers of bacterial cells are used in these experiments, the best method for measurement of oxidized proteins is labeling with tritiated sodium borohydride[14] followed by acid hydrolysis and Dowex chromatography. Similar studies are designed for the determination of oxidized proteins in other target cells such as mammalian cells and, in particular, adherent cells which could be treated in culture dishes.

[15] R. Backman, Y.-M. Chen, and B. Magasanik, *Proc. Natl. Acad. Sci. U.S.A.* **78**, 3743 (1981).

[61] Xenobiotic Activation by Stimulated Human Polymorphonuclear Leukocytes and Myeloperoxidase

By DAVID A. EASTMOND and MARTYN T. SMITH

Introduction

Numerous pharmacological and environmental chemicals (xenobiotics) are enzymatically converted in the body to reactive intermediates which may be toxic and/or carcinogenic. Although a significant proportion of this metabolic activation occurs via the cytochrome *P*-450 monooxygenases, numerous studies indicate that other enzymes and cellular components are capable of bioactivation. For example, prostaglandin synthase has been implicated in the nephrotoxicity of phenacetin[1] and acetaminophen[2] and in the induction of bladder cancer by nitrofurans[3] and aromatic amines.[4] Hydroperoxide-dependent bioactivation may also be involved in the induction of skin cancer by polycyclic aromatic hydrocarbons.[5]

Recent studies have demonstrated that the metabolic activation of xenobiotics can occur during the oxidative burst of polymorphonuclear

[1] B. Andersson, M. Nordenskjold, A. Rahimtula, and P. Moldeus, *Mol. Pharmacol.* **22**, 479 (1982).

[2] P. Moldeus, B. Andersson, A. Rahimtula, and M. Berggren, *Biochem. Pharmacol.* **31**, 1363 (1982).

[3] S. M. Cohen, T. V. Zenser, G. Murasaki, S. Fukushima, M. B. Mattammal, N. S. Rapp, and B. B. Davis, *Cancer Res.* **41**, 3355 (1981).

[4] R. W. Wise, T. V. Zenser, F. F. Kadlubar, and B. B. Davis, *Cancer Res.* **44**, 1893 (1984).

[5] L. J. Marnett, *Carcinogenesis* **8**, 1365 (1987).

METHODS IN ENZYMOLOGY, VOL. 186

leukocytes (PMNs).[6-14] The oxidative burst of PMNs can be triggered by various chemicals and physical agents and is part of the inflammatory response.[15] Numerous lysosomal and peroxidative enzymes, as well as large quantities of active oxygen species, are released during this process.[16] Chemical bioactivation occurs primarily through peroxidase-mediated oxidations although some direct oxygen radical effects have been reported.[6,7] This bioactivation mechanism may be involved in the myelotoxicity induced by benzene,[7,8] the agranulocytosis caused by procainamide and other drugs,[9] and the genotoxicity and carcinogenicity of polycyclic aromatic hydrocarbons,[10] diethylstilbestrol,[11] and azo dyes[12] as well as the antiinflammatory action of numerous drugs.[13,14]

Conversion of xenobiotics to reactive and genotoxic species can be studied through the use of various *in vitro* cellular and enzymatic systems. These methods generally rely on the conversion of the chemical of interest to a reactive species and the subsequent quantitation of protein or DNA binding, mutagenicity, or identification of the reaction products. In this chapter, we focus primarily on the techniques that have been used in our research to follow this process of bioactivation. Detailed protocols and specific information on various aspects of neutrophil isolation and function can be obtained from Ref. 17.

Studies with Human Polymorphonuclear Leukocytes

Isolation Procedure. Venous blood is obtained from healthy volunteers using acid–citrate–dextrose (ACD)-containing Vacutainers. PMNs

[6] M. A. Trush, *Toxicol. Lett.* **20,** 297 (1984).

[7] D. A. Eastmond, R. C. French, D. Ross, and M. T. Smith, *Chem.–Biol. Interact.* **63,** 47 (1987).

[8] D. A. Eastmond, M. T. Smith, and R. D. Irons, *Toxicol. Appl. Pharmacol.* **91,** 85 (1987).

[9] J. Uetrecht, N. Zahid, and R. Rubin, *Chem. Res. Toxicol.* **1,** 74 (1988).

[10] M. A. Trush, J. L. Seed, and T. W. Kensler, *Proc. Natl. Acad. Sci. U.S.A.* **82,** 5194 (1985).

[11] D. A. Eastmond, R. C. French, D. Ross, and M. T. Smith, *Cancer Lett.* **35,** 79 (1987).

[12] K. Takanaka, P. J. O'Brien, Y. Tsuruta, and A. D. Rahimtula, *Cancer Lett.* **15,** 311 (1982).

[13] S. Ichihara, H. Tomisawa, H. Fukazawa, and M. Tateishi, *Biochem. Pharmacol.* **34,** 1337 (1985).

[14] M. Wasil, B. Haliwell, C. P. Moorhouse, D. C. S. Hutchinson, and H. Baum, *Biochem. Pharmacol.* **36,** 3847 (1987).

[15] S. T. Test and S. J. Weiss, *Adv. Free Radical Biol. Med.* **2,** 91 (1986).

[16] M. J. Karnovsky and J. M. Robinson, *in* "Histochemistry: The Widening Horizons" (P. J. Stoward and J. M. Polak, eds.), p. 46. Wiley, New York, 1981.

[17] J. A. Metcalf, J. I. Gallin, W. M. Nauseef, and R. K. Root, "Laboratory Manual of Neutrophil Function." Raven, New York, 1986.

are isolated by dextran sedimentation following standard protocols.[18] Cell counts are performed using a hemocytometer or a cell counter, and viability is determined by trypan blue exclusion.[18] Following isolation, the capacity of each batch of PMNs to undergo the oxidative burst should be determined by measuring the production of O_2^- from the cells when stimulated by phorbol 12-myristate 13-acetate (PMA) or another stimulating agent. Typical values for neutrophils isolated by this protocol are 1–3 nmol O_2^-/min/10^6 cells.

Standard Incubations. PMNs (10^7 cells/ml) and the chemical (radiolabeled for binding studies) at nontoxic concentrations are incubated in Dulbecco's phosphate-buffered saline (PBS; pH 7.1; 1 ml final volume) at 37° in a shaking water bath. PMA (0.1–1 μg/ml; prepared as described by Markert *et al.*[18]) is added to initiate the oxidative burst. Under these conditions, the enhanced effect of the stimulated PMNs on the end point of interest is determined and compared with the appropriate controls. Appropriate controls may include incubations without PMA or without the chemical, incubations performed under anaerobic conditions or with heat-killed cells, or incubations containing superoxide dismutase (SOD, 10 μg/ml; from frozen stock 1 mg/ml in phosphate buffer; 3000 U/mg), catalase (650 U/ml), sodium azide (10 m*M*), or combinations of the above treatments. The direct measurement of chemical disappearance and metabolite formation[10,19–21] is preferred. Since this is often impossible for reactive metabolites, other indirect measurements of xenobiotic activation have been employed, such as macromolecular binding (Table I). A useful technique for measuring xenobiotic binding to cellular macromolecules (primarily proteins) is described below.

PMN-Mediated Macromolecular Binding Assay. The conversion of the xenobiotic to reactive species can be determined by measuring the covalent binding of the activated radiolabeled compound to cellular proteins or other macromolecules. The use of SOD, catalase, and azide treatments can give valuable insights into the mechanism by which metabolic activation is occurring. The standard incubation employs a radiolabeled compound (~0.01–0.1 μCi), and the reaction is terminated at 20 min by the addition of trichloroacetic acid (TCA; 50 μl of a 100% solution; 5% final concentration). The samples are maintained on ice and centrifuged (550 *g* for 5 min) prior to the determination of binding.

---

[18] M. Markert, P. C. Andrews, and B. M. Babior, this series, Vol. 105, p. 358.

[19] M. S. Alexander, R. M. Husney, and A. L. Sagone, Jr., *Biochem. Pharmacol.* **35**, 3649 (1986).

[20] B. Kalyanaraman and P. G. Sohnle, *J. Clin. Invest.* **75**, 1618 (1985).

[21] J. Uetrecht, N. Zahir, N. H. Shear, and W. D. Biggar, *J. Pharmacol. Exp. Therapeut.* **245**, 274 (1988).

TABLE I
Indirect Measurement of Metabolism and Bioactivation by
Polymorphonuclear Leukocytes

End point	Chemical	Refs.
Macromolecular binding	Phenols, estrogens, phenytoin	a–e
DNA/RNA binding	Polycyclic aromatic hydrocarbons (PAH), N-methylaminoazobenzene, phenol, aromatic amines	f–k
Cytochrome c reduction	Phenols, estrogens	a, b
Chemiluminescence	PAH, phenols, estrogens, imipramine	a, b, f, l
DNA breakage	Bleomycin A₂	m
Mutagenicity in Salmonella typhimurium	PAH	f
Sister chromatid exchange in V-79 cells	PAH	n

[a] D. A. Eastmond R. C. French, D. Ross, and M. T. Smith, *Chem.–Biol. Interact.* **63**, 47 (1987).

[b] D. A. Eastmond, R. C. French, D. Ross, and M. T. Smith, *Cancer Lett.* **35**, 79 (1987).

[c] D. A. Eastmond, M. T. Smith, and R. D. Irons, *Toxicol. Appl. Pharmacol.* **91**, 85 (1987).

[d] S. J. Klebanoff, *J. Exp. Med.* **145**, 983 (1977).

[e] J. Uetrecht and N. Zahid, *Chem. Res. Toxicol.* **1**, 148 (1988).

[f] M. A. Trush, J. L. Seed, and T. W. Kensler, *Proc. Natl. Acad. Sci. U.S.A.* **82**, 5194 (1985).

[g] K. Takanaka, P. J. O'Brien, Y. Tsuruta, and A. D. Rahimtula, *Cancer Lett.* **15**, 311 (1982).

[h] P. J. O'Brien, *in* "Microsomes and Drug Oxidations" (A. R. Boobis, J. Caldwell, F. De Matteis, and C. R. Elcombe, eds.), p. 284. Taylor & Francis, London and Philadelphia, 1985.

[i] Y. Tsuruta, V. V. Subrahmanyam, W. Marshall, and P. J. O'Brien, *Chem.–Biol. Interact.* **53**, 25 (1985).

[j] M. O. Corbett and B. R. Corbett, *Chem. Res. Toxicol.* **1**, 356 (1988).

[k] M. O. Corbett, B. R. Corbett, M. Hannothiaux, and S. J. Quintana, *Chem. Res. Toxicol.* **2**, 260 (1989).

[l] M. A. Trush, M. J. Reasor, M. E. Wilson, and K. Van Dykes, *Biochem. Pharmacol.* **33**, 1401 (1984).

[m] M. A. Trush, *Toxicol. Lett.* **20**, 297 (1984).

[n] M. A. Trush, T. W. Kensler, and J. L. Seed, *Adv. Exp. Med. Biol.* **197**, 311 (1986).

Macromolecular binding is determined by liquid scintillation counting using the following modification of the method of Jollow *et al.*[22] The supernatant is discarded, and the pellet is resuspended in 4 ml of 5% TCA by vigorous mixing on a vortex mixer. Occasionally, the use of a wooden applicator stick is necessary to break clumps. The suspension is recentri-

[22] D. J. Jollow, J. R. Mitchell, W. Z. Potter, D. C. Davis, J. R. Gillette, and B. B. Brodie, *J. Pharmacol. Exp. Therapeut.* **187**, 195 (1973).

fuged to form a pellet, and the supernatant is discarded. This procedure is repeated 2 additional times. Similar washes are performed 2–3 times using methanol–water (80:20) and 2–3 times using ethanol–ether (1:1; chilled to −20°). By this time, the radioactivity in the supernatant should be similar to background levels. The pellet is then solubilized by incubation at 55° for at least 1 hr in a solution containing 1 ml of 0.5% sodium lauryl sulfate in Tris buffer (50 mM; pH 7.4) and 1 ml of 2 N NaOH. The pH is then adjusted to 7.0–7.5 by the addition of 12 N HCl, using 1- to 3-μl volumes as the pH approaches the neutral range. Aliquots are removed for liquid scintillation counting (1 ml) and for protein determinations (0.2 ml) as described by Lowry et al.[23] Sodium lauryl sulfate can interfere with the Lowry assay so appropriate controls should contain this compound.

Additional Considerations

With many of these end points, it is prudent to monitor cell viability or cellular respiration (O_2 uptake) to ensure that the observed effect is due to metabolism rather than toxicity. In addition, owing to the indirect nature of many of the end points of interest, it is wise to use additional lines of evidence to support the results. The use of metabolic inhibitors should be interpreted with caution. For example, sodium azide, a "specific" inhibitor of myeloperoxidase, may affect other cellular systems and react directly with bioactive metabolites.[24] The use of isolated enzyme systems such as myeloperoxidase (MPO), horseradish peroxidase, xanthine oxidase, and NADPH oxidase or chemicals such as HOCl, H_2O_2, and taurine chloramine can be used to more thoroughly investigate and confirm proposed mechanisms of xenobiotic activation by PMNs (see Refs. 7, 10, and 11 for examples).

Studies with Myeloperoxidase

Initial studies of metabolic activation by neutrophils can be followed by additional experiments to identify the mechanisms of bioactivation and identify reactive metabolites.[7,8,11,25–27] A typical sequence of experiments

[23] O. H. Lowry, N. J. Rosebrough, A. L. Farr, and R. J. Randall, *J. Biol. Chem.* **193**, 265 (1951).

[24] R. V. Bhat, V. V. Subrahmanyam, A. Sadler, and D. Ross, *Toxicol. Appl. Pharmacol.* **94**, 297 (1988).

[25] D. A. Eastmond, M. T. Smith, L. O. Ruzo, and D. Ross, *Mol. Pharmacol.* **30**, 674 (1986).

[26] P. J. O'Brien, *in* "Microsomes and Drug Oxidations" (A. R. Boobis, J. Caldwell, F. De Matteis, and C. R. Elcombe, eds.), p. 284. Taylor & Francis, London and Philadelphia, 1985.

[27] J. Uetrecht and N. Zahid, *Chem. Res. Toxicol.* **1**, 148 (1988).

to complement the experiments describing xenobiotic activation and protein binding by neutrophils (see Refs. 7, 8, and 25) is as follows.

Incubation Conditions and Analytic Approaches. Protein binding studies are performed by incubating purified MPO (EC 1.11.1.7; Calbiochem, San Diego, CA) or a PMN lysate (1.5 U/ml)[25,28] with the chemical of interest (0.5 mM), an excess of H_2O_2 (1 mM), and boiled rat liver protein (0.9 mg). These conditions result in a rapid and complete oxidization of most chemical substrates within 10 min. The reaction is stopped by the addition of 5% TCA (final concentration) and placed on ice. The removal of the chemical from the incubation can be monitored by high-performance liquid chromatography (HPLC), and the recovery of radioactive equivalents bound to protein can be determined as described above. The rat liver protein is prepared from the 9000 g supernatant by previously described standard methods[29] and boiled for 30 min. It is important that the blood be removed from the liver by perfusion prior to microsome preparation because oxyhemoglobin is capable of catalyzing peroxidase reactions.[30] The use of bovine serum albumin is not recommended as it is extensively removed during the washing steps.

By omitting the boiled rat liver protein and by stopping the reaction with catalase, the formation of metabolites can be monitored at various time points with subsequent analysis by HPLC or gas chromatography–mass spectrometry. The identity of protein-binding species can be determined by the addition of glutathione (5 mM) or other small nucleophiles and the subsequent identification of the chemical–nucleophile conjugate. Glutathione should be added at the completion rather than the beginning of the incubation because gluthatione added at the beginning of peroxidase reactions can act as an antioxidant rather than a nucleophile. The chemical–glutathione conjugates can be isolated by HPLC and identified by the use of fast atom bombardment mass spectrometry and nuclear magnetic resonance spectroscopy. In addition, the identity of chemical–glutathione conjugates can also be confirmed by performing the above reaction with the ^{14}C-labeled chemical and [3H] glutathione and recovering both radioisotopes in the isolated conjugate (see Ref. 25 for example).

Concluding Comments

The use of stimulated polymorphonuclear leukocytes combined with *in vitro* studies with isolated enzymes and chemicals can provide valuable

[28] S. J. Klebanoff, A. M. Waltersdorf, and H. Rosen, this series, Vol. 105, p. 399.

[29] L. Ernster, P. Siekevitz, and G. E. Palade, *J. Cell Biol.* **15,** 541 (1962).

[30] T. Sawahata, D. E. Rickert, and W. F. Greenlee, *in* "Toxicology of the Blood and Bone Marrow" (R. D. Irons, ed.) p. 141. Raven, New York, 1985.

insights into the mechanisms of xenobiotic activation of numerous pharmaceutical and environmental agents. Furthermore, this model of bioactivation may be particularly useful in understanding the observed interaction between inflammation and cancer[10,31] as well as the mechanisms of toxicity and leukemogenesis in leukocyte-rich organs such as the bone marrow and blood.[8,32] Since 90% of the immature granulocytes of the body reside in the bone marrow and contain high levels of myeloperoxidase, this mechanism of xenobiotic activation may be particularly important in this organ.

Acknowledgments

This work was supported by National Institutes of Health Grants P42 ES04705 and P30 ES01896 and the National Foundation for Cancer Research. David A. Eastmond was supported by an appointment to the Alexander Hollaender Distinguished Postdoctoral Program administered by the U.S. Department of Energy and Oak Ridge Associated Universities. Work was performed in part under the auspices of the U.S. Department of Energy by the Lawrence Livermore National Laboratory under Contract W-7405-ENG-48. We thank Dr. David Ross for valuable contributions.

[31] T. W. Kensler, P. A. Egner, K. G. Moore, B. G. Taffe, L. E. Twerdok, and M. A. Trush, *Toxicol. Appl. Pharmacol.* **90**, 337 (1987).
[32] L. E. Twerdok and M. A. Trush, *Chem.–Biol. Interact.* **65**, 261 (1988).

[62] Determination of Superoxide Radical and Singlet Oxygen Based on Chemiluminescence of Luciferin Analogs

By Minoru Nakano

Principle

2-Methyl-6-phenyl-3,7-dihydroimidazo[1,2-a]pyrazin-3-one or 2-methyl-6-(p-methoxyphenyl)-3,7-dihydroimidazo[1,2-a]pyrazin-3-one reacts with O_2^- or $^1O_2$ to emit light, probably via the dioxetanone analog. Superoxide dismutase (SOD, a scavenger of O_2^-) or NaN_3 (a quencher of $^1O_2$) can be used for differentiation between O_2^- and $^1O_2$-dependent luminescence. The maximum light intensity or integrated light intensity is detected for the assay of $^1O_2$ or O_2^- generation in biological systems.

Reagents

2-Methyl-6-phenyl-3,7-dihydroimidazo[1,2-*a*]pyrazin-3-one (CLA)[1] or 2-methyl-6-(*p*-methoxyphenyl)-3,7-dihydroimidazo[1,2-*a*]pyrazin-3-one (MCLA),[2] 56–60 μg/ml in doubly distilled water (the solution is stored in 1.0-ml aliquots at −80° until needed; CLA or MCLA concentrations are based on $\varepsilon_{410\,nm} = 8900\ M^{-1}\ cm^{-1}$ and $\varepsilon_{430\,nm} = 9600\ M^{-1}\ cm^{-1}$, respectively)

Xanthine oxidase (XO) (grade III buttermilk) (Sigma, St. Louis, MO); activity can be determined by the method described by Roussos[3]

Peroxidases, lactoperoxidase (LPO), myeloperoxidase (MPO), chloroperoxidase (CPO), and horseradish peroxidase (HRP); concentrations are based on measured coefficients of $1.14 \times 10^5\ M^{-1}\ cm^{-1}$ at 412 nm for LPO,[4] $9.2 \times 10^4\ M^{-1}\ cm^{-1}$ at 430 nm for MPO,[5] $7.53 \times 10^4\ M^{-1}\ cm^{-1}$ at 403 nm for CPO,[6] and $1.02 \times 10^5\ M^{-1}\ cm^{-1}$ at 403 nm for HRP[7]

Determination of O_2^-

Estimation of Ability of Human Granulocytes or Monocytes to Generate O_2^-

Preparation of Opsonized Zymosan. Zymosan (Sigma), 4.0 mg/ml, is suspended in 20 m*M* veronal buffer at pH 7.4, boiled for 100 min, and cooled to room temperature. Zymosan is collected after centrifugation for 10 min at 1900 *g* and then added to pooled human blood serum at a concentration of 10 mg/ml and incubated at 37° for 30 min. After incubation the zymosan is collected by centrifugation and washed twice with Hanks' balanced salt solution (HBSS) or a modified Hanks' balanced salt solution (mHBSS). The resulting opsonized zymosan (OZ) is suspended to a concentration of 20 mg/ml in HBSS or mHBSS.[2]

Preparation of HBSS or mHBSS. HBSS, 0.98% (w/v) (Nissui Pharmaceutical, Tokyo) is dissolved in double-distilled water, and the pH value is adjusted to 7.4 with sodium bicarbonate. mHBSS is prepared by dissolv-

---

[1] S. Inoue, S. Sugiura, M. Kakoi, and T. Goto, *Tetrahedron Lett.,* 1609 (1969).
[2] A. Nishida, H. Kimura, M. Nakano, and T. Goto, *Clin. Chim. Acta* (in press); M. Nakano, this volume [20].
[3] G. G. Roussos, this series, Vol. 12, p. 5.
[4] L. P. Hager, this series, Vol 17A, p. 648.
[5] J. Schultz and H. W. Shmukler, *Biochemistry* **3,** 1234 (1964).
[6] D. R. Morris and L. P. Hager, *J. Biol. Chem.* **241,** 1763 (1966).
[7] G. R. Schonbaum and S. Lo, *J. Biol. Chem.* **247,** 3353 (1972).

ing HBSS in water as described above and adjusting the pH to 7.4 with 280 mOsm disodium phosphate, instead of sodium bicarbonate. mHBSS is better than HBSS for keeping pH values constant during long incubation experiments.

Preparation of Granulocytes and Monocytes. Blood is drawn into a heparinized syringe (20 units/ml) from healthy human volunteers and patients. Leukocytes are isolated by sedimentation in the presence of dextran followed by brief hypotonic lysis of contaminating erythrocytes.[8] The resulting leukocytes, which contain 65–91% granulocytes, are suspended (1×10^7 cells/ml) in HBSS and kept at 0° for no longer than 3 hr prior to use. Human peripheral blood monocytes are prepared according to Kumagai *et al.*[9] and suspended in HBSS.

Preparation of Standard Curve. For the measurement of XO-induced luminescence, the incubation mixture contains different concentrations of XO, 43 μM hypoxanthine, 0.5 μM MCLA (or CLA), and mHBSS to a total volume of 2 ml. After a 3-min preincubation of the mixture without XO and MCLA (or CLA), the reaction is initiated by the simultaneous addition of XO and MCLA (or CLA). During the luminescence measurement, the incubation mixture is agitated by rotation at 37° in the luminescence reader. CLA- or MCLA-dependent luminescence in the XO–hypoxanthine system (corrected for control without the enzyme), in terms of maximum light intensity, is a linear function of XO concentration.[10] Thus, the following equation can be obtained: Maximum light intensity corrected for control = factor times XO units (with our chemiluminescence detector). MCLA-dependent luminescence is 4.6 times brighter than CLA-dependent luminescence. The factor will alter depending on the chemiluminescence detector used.

Assay of O_2^- Generation in Activated Granulocytes or Monocytes. CLA- or MCLA-dependent luminescence is measured with a luminescence reader (Aloka, BLR-101). Reaction mixtures typically contain 0–4×10^5 granulocytes or monocytes, 1.6 mg OZ, 0.5 μM MCLA or CLA, and HBSS or mHBSS for a total volume of 2.0 ml. In some cases, 0.5 mM NaN$_3$, 0.5 μM SOD, or other substances are added. The volume of each additive (MCLA, NaN$_3$, SOD, or OZ) is 0.1–0.2 ml. All contents, except for OZ and MCLA (or CLA), are preincubated for 3 min, and the reaction is initiated by the simultaneous addition of OZ and MCLA (or CLA). The light emission by CLA or MCLA in the presence of activated granulo-

[8] H. Rosen and S. J. Klebanoff, *J. Clin. Invest.* **58,** 50 (1976).

[9] K. Kumagai, K. Itoh, S. Hinuma, and H. Tada, *J. Immunol. Methods* **29,** 17 (1979).

[10] K. Sugioka, M. Nakano, S. Kurashige, Y. Akuzawa, and T. Goto, *FEBS Lett.* **197,** 27 (1986).

cytes or monocytes is inhibited by SOD but not by catalase (20 μg/ml) or 2 mM benzoate. Azide at 0.5 mM does not inhibit the light emission significantly. These results indicate that O_2^-, rather than H_2O_2, $\cdot$OH, $^1O_2$, or HOCl, is the agent responsible for eliciting the chemiluminescence of CLA or MCLA.[2,11]

A suitable standard reaction mixture contains 7×10^4 granulocytes or monocytes suspended in mHBSS containing 0.5 μM MCLA (or CLA) and 1.6 mg OZ (100 μl of serum/mg of zymosan) for a total volume of 2.0 ml. According to the above equation, the maximum rate of O_2^- production in activated granulocytes or monocytes, corresponding to maximum light intensity, can be expressed as XO units. Twenty normal humans, ages 23–25 years, donated granulocytes which were activated and assayed. Their ability to generate O_2^- was determined as 217.1 $\pm$ 61.1 (mean $\pm$ S.D.) XO units for the standard MCLA system and 231.5 $\pm$ 69.7 XO units for the standard CLA system.[2]

Caution. MPO (a host enzyme in granulocytes) and hydrogen peroxide can similarly evoke CLA or MCLA luminescence. Treatment of 2.5×10^6 granulocytes/ml with 1 mg OZ results in the release of 1.83 nM MPO (6% of the total MPO) during a 15-min incubation.[11] This amount of MPO could account for only a small fraction (~5%) of the light emission by CLA or MCLA in the presence of OZ-stimulated granulocytes. Thus, it is not necessary to consider the MPO-dependent luminescence, if 4×10^5 cells/2.0 ml are used for the luminescence assay. Of course, granulocytes, which are highly contaminated with colored substance such as hemoglobin, cannot be used for this assay, because of absorbance of the light.

Estimation of O_2^- Generation in Subcellular Fractions

Preparation of Subcellular Fractions. Mitochondrial fractions are prepared from beef heart,[12] submitochondrial fractions are prepared by an established method,[13] and microsomal fractions are obtained from rat liver.[14]

Procedure. The submitochondrial system contains 0.3 mM NADH or 7 mM sodium succinate, 12.5–100 μg protein (submitochondrial fraction), 4 μM MCLA, 0.25 M sucrose, and 20 mM Tris-HCl buffer (pH 7.4) for a total volume of 1.0 ml. The volume of each additive (NADH, succinate, or MCLA) is 10 μl. The reaction is initiated by the addition of NADH or succinate. The incubation is carried out at 37° in the chemiluminescence

[11] M. Nakano, K. Sugioka, Y. Ushijima, and T. Goto, *Anal. Biochem.* **159,** 363 (1986).
[12] P. V. Blair, this series, Vol. 10, p. 78.
[13] H. Low and I. Vallin, *Biochim. Biophys. Acta* **63,** 361 (1963).
[14] H. E. May and P. B. McCay, *J. Biol. Chem.* **243,** 2288 (1968).

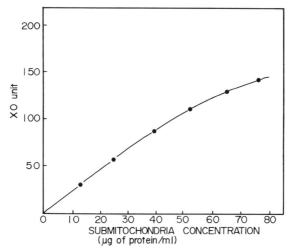

Fig. 1. Relationship between O_2^- generation (XO units) and submitochondrial concentration in the NADH-containing system.

reader (Aloka, BLR-102). The 10-min integrated light intensity corrected for control without substrate is calculated[15] and expressed as XO units (Fig. 1). The XO units can be obtained using the system containing 43 μM hypoxanthine, XO (50–400 units), 4 μM MCLA, 20 mM Tris-HCl buffer (pH 7.4), and 0.2 M sucrose for a total volume of 1 ml. The chemiluminescence is measured, and 10-min integrated light intensity obtained with 100 units of XO is defined as 1 unit.

The microsomal system contains 1.0 mM NADPH, 10 μM MCLA, microsomes (20–100 μg of protein), and 0.1 M Tris-HCl buffer (pH 7.4) for a total volume of 1.0 ml. The volume of NADPH, succinate, or MCLA is 30 μl. The reaction is initiated by the addition of NADPH after a 3-min preincubation. The incubation is carried out at 37°. The generation of O_2^- in the microsomal systems can be expressed as xanthine oxidase units, identical to the mitochondrial systems. Under the experimental conditions, O_2^- generation in the microsomal system containing NADPH as substrate is approximately 3.5 times higher than that in the mitochondrial system containing NADH as substrate (Fig. 1).

Caution. Submitochondrial and microsomal fractions contain colored materials which absorb the chemiluminescence. Thus, integrated light intensity increases linearly with increasing concentrations of microsomes

15 K. Takayama, M. Nakano, K. Zinner, C. C. C. Vidigal, N. Duran, Y. Shimizu, and G. Cilento, *Arch. Biochem. Biophys.* **176,** 663 (1976).

or submitochondria, but only in the range between 20 and 100 μg of protein in the reaction mixture. Since microsomal and submitochondrial fractions contain cytochrome c-reducing systems, a conventional method for the detection of O_2^- (cytochrome c method[16]) cannot be used. The sensitivity of the chemiluminescence method is about 10 times higher than that of the adrenochrome method,[16] which is widely used for O_2^- generation in mitochondrial systems.[17]

Determination of $^1O_2$

Determination of Generation of $^1O_2$ in Peroxidase–H_2O_2–Halide Systems

Preparation of Standard Curve for $^1O_2$ Generation in NaOCl–H_2O_2 System. The reaction mixture contains 20 mM H_2O_2, NaOCl (10–30 μM), 10 μM MCLA, and 0.1 M acetate buffer (pH or pD 4.5) for a total volume of 1 ml. The reaction is initiated by the rapid injection of 20 μl of NaOCl 20 sec after the addition of 20 μl of MCLA. For experiments in 93.5% D_2O, apparent pH, measured with a glass electrode, is adjusted to 4.8 to give a pD of 4.5.[18] Chemiluminescence is measured in a chemiluminescence reader at 25°. Integrated intensity is calculated[15] and plotted against NaOCl concentration. Linearity could be obtained up to 30 μM NaOCl. The integrated light intensity in D_2O is about 4 times that obtained in H_2O.

Procedure. The reaction mixture contains 3 nM peroxidase, 20 mM H_2O_2, 20 mM KBr, 10 μM MCLA, and 0.1 M acetate buffer (pH 4.5 or pD 4.5) for a total volume of 1.0 ml. The reaction is initiated by rapid injection of 20 μl of peroxidase using a microsyringe 20 sec after the addition of 20 μl MCLA. With peroxidases in the range of 1 to 10 nM, MCLA-dependent luminescence, similar to the emission in the NaOCl–H_2O_2 system, can be observed in the peroxidase–H_2O_2–KBr system and increases linearly with increasing enzyme concentration in both H_2O and D_2O. Chemiluminescence measurements and incubation conditions are essentially the same as those for the preparation of the standard curve. Under these experimental conditions, the replacement of H_2O with D_2O would enhance the integrated chemiluminescence, i.e., a 4.7 (± 0.6)-fold increase for the LPO system, a 4.3 (± 0.3)-fold increase for MPO system, and a 2.8 (± 0.5)-fold increase for CPO. The HRP system does not emit, because no HOBr is produced. On the basis of the generation of 1 mol of $^1O_2$ from 1

[16] J. M. McCord and I. Fridovich, *J. Biol. Chem.* **257**, 2713 (1969).
[17] J. F. Turrens and A. Boveris, *Biochem. J.* **191**, 421 (1980).
[18] P. Salomaa, L. L. Schaleger, and F. A. Long, *J. Am. Chem. Soc.* **86**, 1 (1964).

mol of NaOCl in the presence of excess of H_2O_2 in the H_2O_2–NaOCl systems in D_2O and the participation of $^1O_2$ generated from the HOBr + H_2O_2 reaction in the MCLA-dependent luminescence, the $^1O_2$ concentration generated in peroxidase–H_2O_2–KBr systems in D_2O can be calculated as 4.6 ± 0.9 μM for the LPO system, 12.4 ± 0.6 μM for the MPO system, and 14.5 ± 0.6 μM for the CPO system using the standard curve.

Caution. Judging from the increase in integrated light intensity in D_2O by a factor of more than 4, MCLA-dependent luminescence in the LPO or MPO system containing KBr is considered to be induced only by $^1O_2$ generated in the system, according to the reaction:

$$H_2O_2 + HOBr \rightarrow {}^1O_2 + HBr + H_2O$$

Most of the emission from MCLA in the CPO–H_2O_2–KBr system would be involved in $^1O_2$ generation, but also partially in another oxidant. MCLA is also oxidized by HOCl, and probably by HOBr, too, without light emission [reaction (1)]. However the HOCl + H_2O_2 reaction to produce $^1O_2$ [reaction (2)] is much faster than reaction (1) and probably also faster than the $^1O_2$ + MCLA reaction [reaction (3)]. Since peroxidase–

$$MCLA + HOCl \rightarrow products \tag{1}$$
$$HOCl + H_2O_2 \rightarrow {}^1O_2 + HCl + H_2O \tag{2}$$
$${}^1O_2 + MCLA \rightarrow \text{dioxetanone analog} \rightarrow h\nu \tag{3}$$

H_2O_2–halide systems do not produce O_2^-, it is not necessary to consider O_2^--dependent luminescence in these systems.

Section V

Organ, Tissue, and Cell Damage and Medical Applications

[63] Visible-Range Low-Level Chemiluminescence in Biological Systems

By Michael E. Murphy and Helmut Sies

Introduction

Electronically excited oxygen species are of interest in a variety of cellular processes in normal and pathological conditions. The characterization of these short-lived intermediate compounds and identification of their sources remain important research problems. A variety of molecules in biological systems can become excited to higher energy states, with molecular oxygen and carbonyl (aldehyde and ketone) groups as examples. This excitation can be via light absorption, termed photoexcitation, or via chemiexcitation, which can result from the interaction with free radicals or other activated species.[1,2,2a] Two photoemissive processes, fluorescence and phosphorescence, can follow excitation and are collectively termed low-level chemiluminescence or bioluminescence.

Singlet oxygen and triplet carbonyls are likely to account for most of the chemiluminescence in biological systems,[1] but it should be mentioned that the formation and emission from excited species probably represent only minor side reactions of the biological pathways that act as their source. In contrast to the view that these side reactions are solely accidental, an interesting possibility is that some excited triplet species may transfer their energy to molecules involved in other processes, thus promoting what has been termed "photochemistry without light,"[1,3,4] but this topic is not covered here.

Highly sensitive photodetectors can measure the low-level (or ultra-weak) chemiluminescence accompanying these processes even in complex biological systems. The monitoring of this low-level chemiluminescence has been used together with other methods of assessing oxidative processes within cells, but it has the advantages of being noninvasive and providing continuous monitoring. In addition to the procedures for the direct measurement of ongoing chemiluminescence, specific enhancers and quenchers of activated compounds can aid in the identification of the

[1] G. Cilento, *in* "Chemical and Biological Generation of Excited States" (W. Adam and G. Cilento, eds.), p. 221. Academic Press, New York, 1982.

[2] J. Slawínski, *Experientia* **44**, 559 (1988).

[2a] A. A. Krasnovsky, *Photochem. Photobiol.* **29**, 29 (1979).

[3] G. Cilento, *Pure Appl. Chem.* **56**, 1179 (1984).

[4] G. Cilento and W. Adam, *Photochem. Photobiol.* **48**, 361 (1988).

luminescent species in some biological systems.[1,5] Methodology and equipment for such analyses have been described before in this series[6] and by others[7,8] and are summarized and extended here.

Chemistry and Sources of Chemiluminescent Species

Singlet Oxygen. Singlet oxygen can be formed as a by-product of reactions found in biological systems.[9,10] These include (1) the Russell mechanism of peroxyl radical recombination reactions [reaction (1)], in which singlet oxygen can be formed directly [reaction (2a)] or by a reaction between triplet carbonyl and triplet oxygen [reaction (2b)] to yield singlet oxygen and ground state carbonyl [reaction (2a)][11,12]; (2) the interactions of activated oxygen intermediates (superoxide, hydrogen peroxide, hydroxyl radical); or (3) reactions via enzymes that act to dismutate peroxides, such as myelo- and lactoperoxidase and prostaglandin hydroperoxidase.[13-16]

$$2 \quad CHOO \cdot \rightarrow [\overset{\lceil----H}{=CHOOOOC=}] \tag{1}$$
$$\rightarrow \,\supset CHOH + {}^1O_2 + \supset C{=}O \tag{2a}$$
$$\rightarrow \,\supset CHOH + {}^3O_2 + \supset C{=}O^3 \tag{2b}$$

The dimol reaction of singlet oxygen ($^1\Delta_g$) reverting to ground state ($^3\Sigma_g$) generates light of 634 and 703 nm, while the monomal reaction generates a characteristic 1268-nm infrared photoemission. Less common transitions to or from other vibrational levels can generate light of other wavelengths.[2]

Triplet Carbonyls. The C=O groups of ketones and aldehydes are polar and have relatively low energy levels, so that the excited triplet state is easily reached through some biologically relevant pathways.[2,4] At

[5] W. Adam and G. Cilento, *Angew. Chem. Int. Ed. Engl.* **22**, 529 (1983).
[6] E. Cadenas and H. Sies, this series, Vol. 105, p. 221.
[7] H. Inaba, *Experientia* **44**, 550 (1988).
[8] A. Boveris, E. Cadenas, and B. Chance, *Fed. Proc., Fed. Am. Soc. Exp. Biol.* **40**, 195 (1981).
[9] H. Wefers, *Bioelectrochem. Bioenerg.* **18**, 91 (1987).
[10] E. Cadenas, *Annu. Rev. Biochem.* **58**, 79 (1989).
[11] G. S. Russell, *J. Am. Chem. Soc.* **79**, 3871 (1957).
[12] J. K. Howard and K. U. Ingold, *J. Am. Chem. Soc.* **90**, 1058 (1968).
[13] E. Cadenas, H. Sies, W. Nastainczyk, and V. Ullrich, *Hoppe-Seyler's Z. Physiol. Chem.* **364**, 519 (1983).
[14] A. U. Khan, P. Gebauer, and L. P. Hager, *Proc. Natl. Acad. Sci. U.S.A.* **80**, 5195 (1983).
[15] A. U. Khan, *Biochem. Biophys. Res. Commun.* **122**, 668 (1984).
[16] N. I. Krinsky, in "Singlet Oxygen" (H. H. Wasserman and W. A. Murray, eds.), p. 597. Academic Press, New York, 1979.

least four routes to triplet carbonyls also occur during lipid peroxidation.[10] These include (1) the addition of singlet oxygen to an olefin yielding a dioxetane [reaction (3)] which breaks down to carbonyls [reaction (4)], (2) a Hock cleavage of lipid hydroperoxides proceeding through a hypothetical dioxetane intermediate, (3) the disproportionation of two peroxyl radicals through a Russell mechanism [reaction (2b)], or (4) a similar disproportionation of alkoxyl radicals. Triplet carbonyls exhibit a heterogeneous emission spectrum with a maximum in biological systems reported to be around 500–560 nm,[7] whereas there was a broad emission of 375–455 nm in model systems.[3,17]

$$>\!C\!=\!C\!<\; +\; {}^1O_2 \;\rightarrow\; \begin{matrix} >\!C\!-\!C\!< \\ |\quad| \\ O\!-\!O \end{matrix} \tag{3}$$

$$\rightarrow\; >\!C\!=\!O\; +\; >\!C\!=\!O^3 \tag{4}$$

Procedures

Standard Detection Techniques

The high-sensitivity single-photon counting systems used in the detection of low-level emission from biological samples have been reviewed[6,7,18] and are summarized below. The basic system contains a photomultiplier placed in a light-tight chamber to detect photons coming from an organ, suspension of cells, subcellular fraction, or enzyme solution.

Microsomes. Microsomes prepared from liver or other organs by standard techniques are a common model system for the study of chemiluminescent processes. A basic reaction suspension includes microsomes (0.5 mg protein/ml), ADP-chelated ferric iron (4 mM ADP, 10 μM FeCl$_3$, mixed as 65× stock before addition to the suspension), an NADPH-generating system (0.4 mM NADP$^+$, 10 mM glucose 6-phosphate, 5 U/ml glucose-6-phosphate dehydrogenase) suspended in a final volume of 6.5 ml of 0.1 M potassium phosphate buffer (pH 7.4) equilibrated to 37° and pregassed with 100% oxygen. All reactants are kept separate until mixed in a sample cuvette (35 × 56 × 5 mm) placed with the broad surface facing the detector, and oxygen is bubbled continuously through the mixture (Fig. 1). Reactions leading to chemiluminescence are temperature dependent, and so a sample cell is maintained at 37° by connection to a circulating water bath. The reaction is started by addition of the NADP$^+$ through a

---

[17] T. Schulte-Herbrüggen and H. Sies, *Photochem. Photobiol.* **49**, 705 (1989).
[18] W. R. Seitz and M. P. Neary, *in* "Methods of Biochemical Analysis, Vol. 23" (D. Glick, ed.), p. 161. Wiley, New York, 1976.

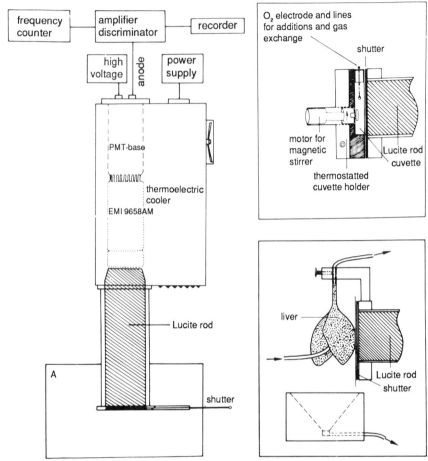

FIG. 1. Single-photon counting apparatus for the measurement of low-level chemiluminescence. Exposed organs of test animals can be placed near the Lucite rod within a light-tight chamber (A), or emission from a perfused organ or from a cuvette containing cells or microsomes can be recorded. PMT, Photomultiplier tube. [Modified from E. Cadenas and H. Sies, this series, Vol. 105, p. 221, and A. Boveris, E. Cadenas, and B. Chance, *Fed. Proc., Fed. Am. Soc. Exp. Biol.* **40,** 195 (1981).]

polyethylene tube running to the cuvette from a syringe outside the apparatus.

Microsomes from normal rats show a lag phase (or induction period) of 10–30 min before a rapid acceleration of chemiluminescence begins, which peaks after a further 10–15 min. The time of onset and acceleration of chemiluminescence is quite sensitive to the buffer composition (e.g., Tris-based buffers shorten the lag phase) and to temperature, oxygen content, and iron content (decreases in the latter three all delay chemiluminescence). The NADPH-generating system can be replaced by a final concentration of ascorbate (0.4 mM), which can be added last in place of the NADP$^+$, or the reaction can be started without ADP/iron by 25 mM AAPH [2,2'-azobis(2-amidinopropane) dihydrochloride].[19] The lag phase before chemiluminescence is longer using ascorbate, and shorter using AAPH, as compared to the NADPH system. We have presented detailed comparisons between prooxidants elsewhere.[19]

Perfused or Exposed Organs. Alternatively, intact organs of test animals can be placed closely to the collector of the photomultiplier (Fig. 1).[6] In this case it is important to shield any surrounding bone or skin which may contribute to the chemiluminescence signal.[20] Inlets to the chamber for circulating water, perfusion media, liquid or gas infusions, and wires for other detectors, such as an oxygen electrode, can be easily allowed using coiled back rubber hoses.

Detection Equipment. The sensitivity of the detection is directly correlated to the surface area of the photomultiplier, but the sensitivity also depends on a low noise background. Low noise can be attained by shielding the entire detection system from stray light and electrical interference, and by cooling the photomultiplier to $-25°$ by a thermoelectric cooler (e.g., a FACT 50 MKIII from EMI-Gencom, Plainview, NY). This latter requirement necessitates an insulating space between the sample and detector, usually a cylindrical Lucite rod or semiellipsoid light reflector, which also channels the light. A shutter is included between the sample and channeling rod so that manipulation within the chamber can be made without shutting down the photodetector and necessitating repeated warmup periods.

The most sensitive photomultipliers use single-photoelectron counting. The red-sensitive photomultiplier Model 9658A from EMI-Gencom,[6] the HTV R 878, R 550, and R 375 photomultipliers manufactured by Hamamatsu Photonics (Hamamatsu City, Japan),[7] or the Hamamatsu HTV R374 and RCA 4832 or 8852[21] are suitable. In addition, complete detection systems including a range of peripheral equipment have been developed by Tohoku Electronic Industrial Co. (Sendai, Japan). The photomultiplier is connected to an amplifier discriminator (EG & G Princeton Applied Research, Princeton, NJ). The output of the discriminator may be quantified by a frequency counter and/or connected to a chart recorder.

[19] A. Scheschonka, M. E. Murphy, and H. Sies, *Chem.–Biol. Interact.* in press (1990).
[20] J. F. Turrens, C. Giulivi, C. R. Pinus, C. Lavagno, and A. Boveris, *Free Radical Biol. Med.* **5,** 319 (1988).
[21] C. F. Deneke and N. I. Krinsky, *Photochem. Photobiol.* **25,** 299 (1977).

Additional Techniques

Spectral Analysis. The production of low-level chemiluminescence alone is not direct evidence for singlet oxygen or a dimol emission, since other species may emit light in the visible region in addition to oxygen. The monomol emission of oxygen in the 1268-nm region is more specific but requires different types of detectors, such as more sensitive germanium diodes, before it can be feasibly applied to biological systems.[14,22,23]

The emission spectrum of low-level chemiluminescence can be useful in determining the nature of the reacting species, particularly when used in combination with quenchers or potentiators of specific emissive species. The greatest limitation to this analysis remains the low level of light.[18] This makes the use of interference filters difficult because of their low maximal transmission (45–55%) and their narrow bandwidth (12–22 nm). The shift in transmission wavelength that accompanies the multidirectional emission pattern of biological samples and wide-acceptance angle of photomultipliers also complicates the use of interference filters. The use of prisms is also excluded by the need for using a large surface area to collect light from a sample.

Colored glass filters with sharp wavelength cutoffs are also convenient and allow a higher sensitivity as compared to spectral analyses which provide higher resolution.[7] Each cutoff filter is designated by its 50% transmission wavelength, but filters typically show a range of 30–40 nm between 1 and 99% transmission. Sets of filters are produced with 15–20 nm between successive filters (Wratten gelatin filters, Eastman-Kodak, Rochester, NY, or glass colored filters, Jenaer Glaswerke Schott, Mainz, FRG or Toshiba Electric Co., Minato ku, Japan). These physical limitations of cutoff filters make a high resolution of emission spectra impossible, but they are often satisfactory for biological analyses.

The basic analysis of a spectrum involves measuring the sequential change in light intensity as a series of filters are brought between the sample and the photomultiplier. For convenience this is often accomplished using a remote-controlled filter insertion device or rotating wheel which brings successive filters into the light path. The change in intensity between two filters represents the emission of light with an average wavelength halfway between the 50% cutoff wavelengths of the filters but includes light with higher and lower wavelengths as determined by the transmission characteristics of the filters. Other factors to consider when preparing emission spectra include correction for the photomultiplier re-

[22] E. Lengfelder, E. Cadenas, and H. Sies, *FEBS Lett.* **164,** 366 (1983).
[23] P. Di Mascio and H. Sies, *J. Am. Chem. Soc.* **111,** 2909 (1989).

sponse at different wavelengths and the changing chemiluminescence intensity during time with biological samples.[7]

Emission Enhancers. Chemiluminescence enhancers exhibit a variety of characteristics and reaction specificities (Table I). The use of one or more of these agents may provide information regarding the nature of the emissive species. The emission from singlet oxygen can be enhanced by D_2O, which can prolong the half-life before its nonemissive quenching by the solvent and increases its monomol and dimol emission.[23,24] DABCO (1,4-diazabicyclo[2.2.2]octane) also enhances dimol emission of singlet oxygen, but there is a decrease in monomol emission.[22,23] As with the effect of D_2O, DABCO does not affect the spectral characteristics of the dimol reaction, but the mechanism of DABCO enhancement remains unclear. Unfortunately, the maximum effect of DABCO in a model system was around 100 mM. Although effects were noted with as little as 5 mM, this is near concentrations that interfered with the activity of horseradish peroxidase.[25]

The breakdown of dioxetane or dioxetanone (4-membered ring) and other reaction pathways may yield excited carbonyl species.[3,5] Singlet carbonyls or lower energy triplet carbonyls may be the predominant product when the substituent groups contain extensive conjugation, and higher energy triplet groups when there is no conjugation.[3,5] The carbonyls can promote chlorophyll *a*-sensitized emission without affecting other parameters of lipid peroxidation (such as malondialdehyde formation and oxygen uptake).[26] The peak of its fluorescence was around 700 nm but was difficult to detail further in microsomal systems, and quantitative data are not easily reached in biological systems using this method.[26]

The higher energy triplet carbonyls can promote both chlorophyll *a*- and DBA- or DBAS-sensitized emission (Table I). The related compound anthracene 2-sulfonate, lacking the necessary heavy-atom addition, does not sensitize the emission of triplet carbonyls and can act as an experimental control.[3,5] The possible formation of singlet carbonyls can be indicated experimentally by DPA-enhanced photoemission.[3,5,27]

Luminol and lucigenin are used to produce chemiluminescent signals from activated oxygen species, including some which would not otherwise chemiluminesce. Luminol emission is stimulated by oxidants such as

[24] D. R. Kearns, *Chem. Rev.* **71**, 395 (1971).
[25] H. Wefers, E. Riechmann, and H. Sies, *J. Free Radicals Biol. Med.* **1**, 311 (1985).
[26] E. Cadenas, H. Sies, A. Campa, and G. Cilento, *Photochem. Photobiol.* **40**, 661 (1984).
[27] E. Rivas, D. Mira, G. Cilento, and A. Boveris, in "Superoxide and Superoxide Dismutase in Chemistry, Biology and Medicine" (G. Rotilio, ed.), p. 108. Elsevier, Amsterdam, 1986.

TABLE I

ENHANCERS OF LOW-LEVEL CHEMILUMINESCENCE

Enhancer	Excited species	Concentration[a] (μM)	Emission wavelength (nm)	Refs.
DBA (9,10-dibromoanthracene)	Triplet C=O	5	415,435	b–d
DBAS (9,10-dibromoanthracene-2-sulfonate)	Triplet C=O	50	440	b–g
Eosin Y (tetrabromofluorescein)	Triplet C=O	58	550	c, g, h
Rose Bengal (tetrachlorotetra-iodofluorescein)	Triplet C=O	38	600	c, g, h
Chlorophyll a	Conjugated triplet C=O	10	682	d–f, i
Diphenylanthracene	Singlet C=O	25	Not reported	c, d
D_2O (heavy water)	$^1O_2$	Replaces H_2O	Unchanged	j
DABCO (1,4-diazabicyclo[2.2.2] octane)	$^1O_2$	60×10^3	Unchanged	k, l
Europium tetracycline	Lipid peroxides	100	620	m
Luminol (5-amino-2,3-dihydro-1,4-phthalazinedione)	Nonspecific	50	427	f, m–q
Lucigenin (N',N-dimethyl-9,9'-biacridinium dinitrate)	Superoxide	400	470	o–q

[a] Effective concentration of enhancer varied in different systems, and a typical value is given. Values listed are not always the minimal effective concentration nor that proved to provide the maximal effect.

[b] G. Cilento, Pure Appl. Chem. **56**, 1179 (1984).

[c] W. Adam and G. Cilento, Angew. Chem. Int. Ed. Engl. **22**, 529 (1983).

[d] E. Rivas, D. Mira, G. Cilento, and A. Boveris, in "Superoxide and Superoxide Dismutase in Chemistry, Biology and Medicine" (G. Rotilio, ed.) p. 108. Elsevier, Amsterdam, 1986.

[e] E. Cadenas, Annu. Rev. Biochem. **58**, 79 (1989).

[f] T. Schulte-Herbrüggen and H. Sies, Photochem. Photobiol. **49**, 705 (1989).

[g] M. Haun, N. Duran, O. Augusto, and G. Cilento, Arch. Biochem. Biophys. **200**, 245 (1980).

[h] N. Duran and G. Cilento, Photochem. Photobiol. **32**, 113 (1980).

[i] E. Cadenas, H. Sies, A. Campa, and G. Cilento, Photochem. Photobiol. **40**, 661 (1984).

[j] D. R. Kearns, Chem. Rev. **71**, 395 (1971).

[k] E. Lengfelder, E. Cadenas, and H. Sies, FEBS Lett. **164**, 366 (1983).

[l] P. Di Mascio and H. Sies, J. Am. Chem. Soc. **111**, 2909 (1989).

[m] V. S. Sharov, V. A. Kazamanov, and Y. A. Vladimirov, Free Radical Biol. Med. **7**, 237 (1989).

[n] J. Slawinski, Experientia **44**, 559 (1988).

[o] W. R. Seitz and M. P. Neary, in "Methods of Biochemical Analysis, Vol. 23" (D. Glick, ed.), p. 161. Wiley, New York, 1976.

[p] A. Weimann, A. G. Hildebrandt, and R. Kahl, Biochem. Biophys. Res. Commun. **125**, 1033 (1984).

[q] R. C. Allen, in "Chemical and Biological Generation of Excited States" (W. Adam and G. Cilento, eds.), p. 309. Academic Press, New York, 1982.

hydrogen peroxide and hydroxyl radicals, and it has been used in a postcolumn reaction to quantify lipid hydroperoxides separated by HPLC.[18,28,28a] Lucigenin emission occurs after reductive oxygenation by compounds including superoxide anions.[18,28] Interestingly, europium tetracycline had a complementary reactivity, preferentially enhancing emission from intermediates of lipid peroxidation rather than activated oxygen species.[29]

In another study, the fluorescent dyes rhodamine 123, eosin Y, and NADPH were used as emission enhancers in a mitochondrial system.[30] The use of the latter compound must be viewed cautiously since it will likely produce a range of effects in biological systems. We note, for example, that the total chemiluminescence arising from microsomal peroxidation is greater in systems initiated with ascorbate/Fe/ADP as compared to NADPH/Fe/ADP. The use of xanthine derivatives, such as eosin Y and Rose Bengal, must also be used cautiously, since they can also produce singlet oxygen via photoexcitation.[31]

Quenchers. Compounds which diminish chemiluminescence may also be used in experiments, but they often demonstrate multiple interactions in complex systems, such as free radical-scavenging properties which may interfere with the generating processes.[13] Singlet oxygen can be efficiently quenched by compounds, e.g., 5–20 mM azide,[13,25] 2–10 μM cyanidanol [(+)-catechin],[25,32–34] 100 μM hydroquinone,[13] 300 μM tocopherols,[7,33] 5 μM β-carotene,[7,35] and related carotenoid derivatives such as lycopene.[35] Azide quenches singlet oxygen but can have more pronounced chemiluminescence-enhancing effects on complex systems.[30] It is worth noting that oxygen can quench triplet carbonyl phosphorescence too quickly for efficient direct measurement, but in cellular environments lacking oxygen, such as in the protected reaction centers of enzymes, the emission can proceed.[3]

[28] R. C. Allen, *in* "Chemical and Biological Generation of Excited States" (W. Adam and G. Cilento, eds.), p. 309. Academic Press, New York, 1982.

[28a] T. Miyazawa, K. Yasuda, and K. Fujimoto, *Anal. Lett.* **20**, 915 (1987).

[29] V. S. Sharov, V. A. Kazamanov, and Y. A. Vladimirov, *Free Radicals Biol. Med.* **7**, 237 (1989).

[30] R. H. Steele, J. Sabik, R. R. Benerito, and S. W. O'Dea, *Arch. Biochem. Biophys.* **267**, 125 (1988).

[31] E. Gandin, Y. Lion, and A. van de Vorst, *Photochem. Photobiol.* **37**, 271 (1983).

[32] C. G. Fraga, V. S. Martino, G. E. Ferraro, J. D. Coussio, and A. Boveris, *Biochem. Pharmacol.* **36**, 717 (1987).

[33] E. Cadenas, A. Müller, R. Brigelius, H. Esterbauer, and H. Sies, *Biochem. J.* **214**, 479 (1983).

[34] G. M. Bartoli, A. Müller, E. Cadenas, and H. Sies, *FEBS Lett.* **164**, 371 (1983).

[35] P. Di Mascio, P. S. Kaiser, and H. Sies, *Arch. Biochem. Biophys.* **274**, 532 (1989).

Two-Dimensional Imaging. Some attempts have been made to channel luminescent light onto a double-microchannel plate in order to produce a time-resolved two-dimensional image of chemiluminescent sample. This technique may have some application to analysis of oxidative processes in perfused or intact organs, but so far its sensitivity limits its use to chemiluminescence stimulated by lucigenin or other chemiluminescence-generating reactions.[36]

Chemical Verification of Singlet Oxygen. A putative identification of singlet oxygen by chemical procedures has been explored with varying success, e.g., by measuring the formation of the 5α-endoperoxide product of cholesterol, which is relatively characteristic of attack by this species.[37] Singlet oxygen can also be trapped by 2,2,6,6-tetramethylpiperidine, forming a stable nitroxide radical that can be monitored by electron spin resonance (ESR) in *in vitro* systems, although its use in biological systems may be limited.[38] Other singlet oxygen traps include 1,3-diphenylisobenzofuran, naphthalene, and anthracene. We have employed 9,10-bis(2-ethylene)anthracene disulfate.[23,39]

Applications to Biologically Relevant Systems

Analysis of Light-Generating Reactions

A number of cellular reactions generate chemiluminescent products (Table II). For example, hydroperoxidase enzymes in mammalian cells may generate singlet oxygen as a by-product.[10] The eosinophil peroxidase, under physiological conditions and substrate concentrations, appeared to produce 20% of the theoretical yield of singlet oxygen, which was higher than the production by myeloperoxidase and lactoperoxidase.[40] Reduced glutathione (GSH) can react with horseradish peroxidase in the absence of exogenous H_2O_2 to generate chemiluminescence. This is likely to be due to singlet oxygen formation via thiyl radicals and their further chain reactions with other thiols and oxygen.[25] The related enzyme soybean lipoxygenase also shows a dimol emission in the presence of glutathione and fatty acids, but in the absence of glutathione an emission from excited carbonyls is evident.[17] However, when H_2O_2 replaces the fatty acids in the presence of glutathione, an emission with a

[36] F. E. Maly, A. Urwyler, H. P. Rolli, C. A. Dahinden, and A. L. de Weck, *Anal. Biochem.* **168,** 462 (1988).

[37] L. L. Smith and J. P. Stroud, *Photochem. Photobiol.* **28,** 479 (1978).

[38] Y. Lion, M. Delmelle, and A. van de Vorst, *Nature (London)* **263,** 442 (1976).

[39] B. A. Lindig, M. A. J. Rodgers, and A. P. Schaap, *J. Am. Chem. Soc.* **102,** 5590 (1980).

[40] J. R. Kanofsky, H. Hoogland, R. Wever, and S. J. Weiss, *J. Biol. Chem.* **263,** 9692 (1988).

TABLE II
Low-Level Chemiluminescence from Biological Systems

Sample type	Initiator of chemiluminescence	Emission characteristics	Ref.
'hole plant	None	Dimol emission indicated by spectral peaks at 382, 480, 580, and 640 nm	a
situ rat liver	Dietary deficiencies	Vitamin E or Se deficiency increased spontaneous emission	b
	Triiodothyronine (T_3), 0.1 mg/kg/day for 7 days	Treatment led to decreased endogenous antioxidants and higher spontaneous emission	c
situ mouse liver	CCl_4, 5 ml/kg	Decreased by polyphenolic quenchers which did not affect spontaneous emission	d
erfused rat liver	25 μM menadione	Resulted from redox cycling, emission not affected by glutathione conjugation of quinone	e
situ rat lung	Paraquat, 30 mg/kg	Polymorphonuclear white blood cells responsible for emissive species	f
situ rat erythrocytes	15-W UV bulb radiation, 4 days, 40 cm	Spontaneous emission also increases with age	g
	2 mM tert-butyl hydroperoxide	Inhibited by antioxidants, but dimol spectrum not evident	h
at liver hepatoyctes	54 μM Fe^{2+}/2 mM ADP or 0.5 mM tert-butyl hydroperoxide	Correlated to low cellular glutathione, but not to malondialdehyde formation	i
	2 mM 4-hydroxynonenal	Correlated to glutathione and vitamin E content, but not to malondialdehyde and alkanes	j
uman eosinophils	100 μM Br^-, 10 μg/ml phorbol myristate acetate	Dependent on H_2O_2 and peroxidase, monomal spectra was evident	k
at liver homogenate	Autoxidized linseed oil	Inhibited by vitamin E, BHT, and β-carotene, enhanced by DABCO, unaffected by superoxide dismutase, catalase, and mannitol	g
at brain homogenates	None	Inhibited only partially by antioxidants	l
at heart mitochondria	1 mM CN^-, 5 μM Cu^{2+}, 23.7 mM acetaldehyde	Dependent on O_2 and sulfhydryl groups, evidence of dimol spectrum, complex kinetics	m
at liver microsomes	Chronic ethanol	Inhibited by glutathione but not by enzymatic antioxidants, increased O_2 sensitivity	n
	1.5 mM iodosobenzene and other oxene donors	Inhibited by NADPH or NADH, partial dimol spectrum, not correlated to malondialdehyde	o
	ADP/Fe^{3+}, ascorbate	Maximal emission not affected by glutathione or other thiol compounds	p
	ADP/Fe^{3+}, NADPH	Maximal emission diminished by dihydrolipoate	q

(continued)

TABLE II (*continued*)

Sample type	Initiator of chemiluminescence	Emission characteristics	R
Lacto-, myelo-, and chloroperoxidases	Substrates, halide ions, and H_2O_2	Evidence of dimol spectra in some cases	r
Prostaglandin endoperoxide synthase and hematin	54 μM arachidonic acid and O_2, or 17 μM prostaglandin G_2	Inhibited by azide and β-carotene, enhanced by DABCO, evidence of dimol but not monomol spectrum	u
Horseradish peroxidase and hemin	5 mM reduced glutathione	Enhanced by O_2, D_2O, and hemin, inhibited by ascorbic acid and enzymatic antioxidants, emission mostly beyond 610 nm	w
Soybean lipoxygenase	1 mM fatty acid and O_2, or 1 mM glutathione	Enhanced by DBAS or chlorophyll a, inhibited x by quenchers or superoxide dismutase, dimol spectrum in glutathione-stimulated emission	x

[a] J. Slawinski, E. Grabikowski, and L. Ciesla, *J. Lumin.* **24/25,** 791 (1981).

[b] C. G. Fraga, R. F. Arias, S. F. Llesuy, O. R. Koch, and A. Boveris, *Biochem. J.* **242,** 383 (1987)

[c] V. Fernandez, S. Llesuy, L. Solari, K. Kipreos, L. A. Videla, and A. Boveris, *Free Radical R* *Commun.* **5,** 77 (1988).

[d] C. G. Fraga, V. S. Martino, G. E. Ferraro, J. D. Coussio, and A. Boveris, *Biochem. Pharmacol.* 717 (1987).

[e] H. Wefers and H. Sies, *Arch. Biochem. Biophys.* **224,** 568 (1983).

[f] J. F. Turrens, C. Giulivi, C. R. Pinus, C. Lavagno, and A. Boveris, *Free Radical Biol. Med.* **5,** ? (1988).

[g] H. Inaba, *Experientia* **44,** 550 (1988).

[h] L. A. Videla, M. I. Villena, G. Donoso, J. de le Fuente, and E. Lissi, *Biochem. Int.* **8,** 821 (1984)

[i] F. J. Romero and E. Cadenas, *Pharmacol. Ther.* **33,** 179 (1987).

[j] E. Cadenas, A. Müller, R. Brigelius, H. Esterbauer, and H. Sies, *Biochem. J.* **214,** 479 (1983).

[k] J. R. Kanofsky, H. Hoogland, R. Wever, and S. J. Weiss, *J. Biol. Chem.* **263,** 9692 (1988).

[l] E. A. Lissi, T. Cáceres, and L. A. Videla, *Free Radical Biol. Med.* **4,** 93 (1988).

[m] R. H. Steele, J. Sabik, R. R. Benerito, and S. W. O'Dea, *Arch Biochem. Biophys.* **267,** 125 (1988)

[n] S. Puntarulo and A. I. Cederbaum, *Arch. Biochem. Biophys.* **266,** 435 (1988).

[o] E. Cadenas, H. Sies, H. Graf., and V. Ullrich, *Eur. J. Biochem.* **130,** 117 (1983).

[p] G. M. Bartoli, A. Müller, E. Cadenas, and H. Sies, *FEBS Lett.* **164,** 371 (1983).

[q] H. Scholich, M. E. Murphy, and H. Sies, *Biochim. Biophys. Acta* **1001,** 256 (1989).

[r] E. Cadenas, *Annu. Rev. Biochem.* **58,** 79 (1989).

[s] A. U. Khan, P. Gebauer, and L. P. Hager, *Proc. Natl. Acad. Sci. U.S.A.* **80,** 5195 (1983).

[t] A. U. Khan, *Biochem. Biophys. Res. Commun.* **122,** 668 (1984).

[u] E. Cadenas, H. Sies, W. Nastainczyk, and V. Ullrich, *Hoppe-Seyler's Z. Physiol. Chem.* **364,** ? (1983).

[v] J. R. Kanofsky, *Photochem. Photobiol.* **47,** 605 (1988).

[w] H. Wefers, E. Riechmann, and H. Sies, *J. Free Radicals Biol. Med.* **1,** 311 (1985).

[x] T. Schulte-Herbrüggen and H. Sies, *Photochem. Photobiol.* **49,** 705 (1989).

maximum of 490–520 nm occurs, which may result from the semiquinone reactions of the enzyme's pyrroquinoline quinone cofactor.[41]

The prostaglandin endoperoxidase-mediated pathway may proceed through a $[Fe=O]^{3+}$ intermediate and a similar intermediate in connection with the hydroperoxide activity of cytochrome P-450 or myoglobin.[10] The rise in spontaneous chemiluminescence in liver microsomes from rats administered ethanol chronically seems to depend partly on this pathway, since the induction of cytochrome P-450 correlated to the higher emission and CO inhibited the chemiluminescence.[42] The reaction of some oxene donors with rat liver microsomes or isolated cytochrome P-450 also yielded chemiluminescence by pathways not associated with lipid peroxidation. It appeared to first involve a heterolytic cleavage of the oxene to form an active oxygen complex which reacted with a second oxene donor to produce singlet oxygen as identified by the emission at 634 and 703 nm. Reduction of the cytochrome P-450 with NADH or NADPH resulted in less chemiluminescence by the proposed mechanism.[43]

In addition to endogenous pathways, exogenous compounds can elicit chemiluminescence, with menadione and paraquat as typical examples. The emission spectrum from menadione-perfused rat liver showed nearly all light at wavelengths beyond 600 nm and is thought to result from redox cycling of the quinone. Although the emission peaked about 700 nm, the matching 634-nm emission expected of a dimol reaction was missing. A small peak at 480 nm may have been from a singlet–triplet dimol emission of oxygen.[44] In contrast, the spontaneous chemiluminescence of in situ lungs from rats administered paraquat did not appear to result from redox cycling of the drug, since the higher chemiluminescence was only associated with the infiltration of polymorphonuclear blood cells that followed many hours later.[20]

A standard criterion for the dimol reaction of singlet oxygen are peaks of emission centered around 634 and 703 nm with a concurrent trough of emission intensity at 668 nm.[6,21] The apparent lack of monomal emission in the prostaglandin endoperoxide synthase reaction[45] where our laboratory had reported the characteristic dimol emission[13] may call into question whether this criterion is sufficient. The visible-range spectrum from the preparation lacking the monomol emission did show some dimol emission bands, such as the 634-nm band and the alternate dimol emission

[41] R. A. van der Meer and J. A. Duine, FEBS Lett. **235**, 194 (1988).
[42] S. Puntarulo and A. I. Cederbaum, Arch Biochem. Biophys. **266**, 435 (1988).
[43] E. Cadenas, H. Sies, H. Graf, and V. Ullrich, Eur. J. Biochem. **130**, 117 (1983).
[44] H. Wefers and H. Sies, Arch. Biochem. Biophys. **224**, 568 (1983).
[45] J. R. Kanofsky, Photochem. Photobiol. **47**, 605 (1988).

bands of 480, 520, and 580 nm, even though the 703-nm band was not distinct.[45] It was noted that the dimol emission was not distinct in the light emission when microsomes from a vesicular gland replaced the purified the endoperoxidase unless DABCO was present.[13] The model of autoxidized linseed oil also produced a spectrum with several peaks matching those of dimol emissions, although the 703-nm peak was not characterized; it appeared the alternate dimol emission bands of 480, 520, and 580 nm associated with other transitional states predominated.[7] This suggests that complex biological systems could show emission profiles that differ from simple chemical systems.

Effects of Antioxidants

Because of its importance in other antioxidant pathways, the effect of glutathione on light emission has been studied in several models. 4-Hydroxynonenal-induced chemiluminescence was increased in both control and glutathione-depleted rat hepatocytes.[33] In the same system, low doses of *tert*-butyl hydroperoxide caused a rapid oxidation of glutathione that was paralleled by increased chemiluminescence, whereas both chemiluminescence and glutathione returned to near-initial levels when the hydroperoxide was consumed.[46] In contrast, glutathione depletion decreased the light emission from menadione-treated perfused livers and hepatocytes, suggesting that glutathione binds to the menadione to form the thioether but does not affect its redox cycling and oxygen radical formation, whereas the light-generating reactions of *tert*-butyl hydroperoxide can be blocked by glutathione.[44] Glutathione inhibited the chemiluminescence from the prostaglandin endoperoxide synthase reaction,[13] while lower levels of glutathione delayed iron/ADP/NADPH-induced lipid peroxidation in rat liver microsomes but did not diminish the final level of chemiluminescence.[34] The reactions of glutathione with superoxide or thiyl radicals leads to reactions generating singlet oxygen and chemiluminescence.[25]

Other cellular antioxidants such as vitamin E, carotenoids, ascorbic acid, and enzymatic antioxidants can also affect chemiluminescence. For example, the spontaneous chemiluminescence from livers of rats fed diets low in selenium (leading to lower glutathione peroxidase activities) or vitamin E was higher than that of controls.[47] Vitamin E pretreatment

[46] F. J. Romero and E. Cadenas, *Pharmacol. Ther.* **33,** 179 (1987).
[47] C. G. Fraga, R. F. Arias, S. F. Llesuy, O. R. Koch, and A. Boveris, *Biochem. J.* **242,** 383 (1987).

decreased the chemiluminescence in rat hepatocytes treated with 4-hydroxynonenal.[33]

Exogenous antioxidants can also decrease the light emission from biological systems. The CCl_4-induced chemiluminescence of *in situ* mouse liver and the *tert*-butyl hydroperoxide-generated chemiluminescence in liver homogenates was decreased by eriodyctiol and cyanidanol.[32] Cyanidanol also diminished chemiluminescence without delaying or diminishing other parameters of microsomal lipid peroxidation.[33,34] In contrast, the sulfhydryl-containing agents diethyl dithiocarbamate, penicillamine, and dithioerythritol all delayed peroxidation without inhibiting chemiluminescence.[34] Of particular interest is the report that during the spontaneous lipid peroxidation and chemiluminescence of rat brain homogenates about 50% of the emission was stopped within about 1 min by the addition of a variety of antioxidants, while the remaining emission continued unaffected. The ratio of these two components of light emission did not change with time during incubation, suggesting a precursor–product relationship.[48]

Relationship to Other Parameters of Lipid Peroxidation

The processes leading to chemiluminescence may relate to other parameters of lipid peroxidation, such as oxygen uptake and the formation of malondialdehyde (or thiobarbituric acid-reactive substances), alkanes, or conjugated dienes, but simple correlations do not seem to apply. The onset of chemiluminescence during microsomal peroxidation occurred following a lag phase that was similar to those before accelerated oxygen uptake and malondialdehyde formation, but chemiluminescence continued after the latter two parameters had reached end points.[49,50] The chemiluminescence from tissue homogenates exposed to radiation also correlated with the production of malondialdehyde.[7] In contrast, in hepatocytes treated with *tert*-butyl hydroperoxide malondialdehyde increased steadily to maximal levels after the hydroperoxide was consumed, whereas chemiluminescence was high only while the hydroperoxide was present.[46] The changes in chemiluminescence in *in situ* lung after paraquat treatment, which occurred only after polymorphonuclear infiltration, were not paralleled by malondialdehyde accumulation.[20] Similarly, the singlet oxygen produced by the reaction of cytochrome *P*-450 with an

[48] E. A. Lissi, T. Cáceres, and L. A. Videla, *Free Radical Biol. Med.* **4,** 93 (1988).
[49] T. Noll, H. De Groot, and H. Sies, *Arch. Biochem. Biophys.* **252,** 284 (1987).
[50] H. Scholich, M. E. Murphy, and H. Sies, *Biochim. Biophys. Acta* **1001,** 256 (1989).

oxene donor (iodosobenzene) was accompanied by chemiluminescence but no malondialdehyde formation, while other oxene donors (such as *tert*-butyl hydroperoxide) produce much less intense chemiluminescence but more malondialdehyde.[43] In rat hepatocytes treated with 4-hydroxynonenal, increased chemiluminescence correlated to alkane formation, but not to malondialdehyde or conjugated diene formation.[33]

Chemiluminescence in Nitric Oxide Measurement

Chemiluminescence has been applied recently to measure nitric oxide released from biological tissues and cultured cells.[51,52] In these experiments, perfusate or medium is injected into refluxing acetic acid, and the nitric oxide vapor is carried via an inert gas to a chemiluminescence chamber where it is mixed with ozone.[53] The ozone and nitric oxide react to form nitrogen dioxide in an excited state, with an emission in the near-infrared region upon relaxation to the ground state.[54]

Concluding Remarks

The monitoring of chemiluminescence is a useful tool for research concerning excited species in biological systems. The available equipment is sufficiently sensitive to measure the ultraweak chemiluminescence emitted by normal cellular processes. The sources of the emission are varied but may be further characterized by analysis of the spectrum or by using a number of enhancers and quenchers, some of which interact with only specific excited species. Care must be taken using these compounds, though, since they may sometimes affect the very processes under study. Likewise, consideration must also be given in complex systems to the effects of endogenous compounds, such as tocopherols, carotenoids, and glutathione, which can affect chemiluminescence. An important finding has been that the emission of light does not always correlate with lipid peroxidation, so this measurement cannot be used as a sole criterion for such processes. In this regard, however, the simultaneous comparison of chemiluminescence with other parameters of lipid peroxidation can lead to a better understanding of the pathways involved in these processes.

[51] R. M. J. Palmer, A. G. Ferrige, and S. Moncada, *Nature (London)* **327,** 524 (1987).
[52] M. A. Marletta, P. S. Yoon, R. Iyengar, C. D. Leaf, and J. S. Wishnok, *Biochemistry* **27,** 8706 (1988).
[53] M. J. Downes, M. W. Edwards, T. S. Elsey, and C. L. Walters, *Analyst* **101,** 742 (1976).
[54] D. H. Fine, F. Rufeh, and B. Gunther, *Anal. Lett.* **6,** 731 (1973).

[64] Spin Trapping Biologically Generated Free R Correlating Formation with Cellular Injury

By GERALD M. ROSEN and HOWARD J. HALPERN

Introduction

One of the most severe limitations confronting the field of free radical biology is our inability to correlate the formation of specific free radicals with cellular injury. Despite the importance of these data, surprisingly little has been published on this topic. In this chapter we discuss approaches we have used to overcome this problem.

Several years ago, we reported the spin trapping of superoxide during the metabolism of xenobiotics by rat enterocytes.[1] On incubation of menadione (0.1 mM) with the spin trap 5,5-dimethyl-1-pyrroline N-oxide (DMPO, 0.1 M) we recorded an electron spin resonance (ESR) spectrum characteristic of 2,2-dimethyl-5-hydroxy-1-pyrrolidinyoxyl (DMPO-OH), which was not inhibited by the addition of either SOD or catalase. Although it would appear that, intracellularly, hydroxyl radical was spin trapped during the metabolism of menadione, additional experiments determined that the ESR spectrum corresponding to DMPO-OH resulted from the initial spin trapping of superoxide, giving 2,2-dimethyl-5-hydroperoxy-1-pyrrolidinyoxyl (DMPO-OOH), followed by rapid bioreduction, catalyzed by the enzyme glutathione peroxidase, to DMPO-OH.[1-3]

Given that superoxide is generated and spin trapped, as the result of enterocyte metabolism of menadione, does it follow that cellular injury will result? If so, is this toxicity related to production of this free radical? What criteria can be employed as a measure of cellular toxicity? A literature search on this topic discloses numerous indices of cell injury. For our studies, we have chosen several indicators of membrane injury, including trypan blue exclusion, leakage of lactate dehydrogenase (LDH), and release of ^{51}Cr. We have also used more subtle measures of cell toxicity, such as inhibition of glucose metabolism by measuring $^{14}CO_2$ release, diminition in [^{14}C] leucine incorporation into proteins, and alterations in glutathione (reduced and oxidized) and NADPH levels.

[1] C. M. Mansbach II, G. M. Rosen, C. A. Rahn, and K. E. Strauss, *Biochim. Biophys. Acta* **888**, 1 (1986).
[2] G. M. Rosen and B. A. Freeman, *Proc. Natl. Acad. Sci. U.S.A.* **81**, 7269 (1984).
[3] G. M. Rosen and M. J. Turner III, *J. Med. Chem.* **31**, 428 (1988).

TABLE I
Metabolism of ^{14}C Glucose to $^{14}CO_2$ in Rat
Enterocytes: Effect of Menadione[a]

O$_2$ content of incubation flask	$^{14}CO_2$ (cpm/hr)	
	Control	Menadione (50 μM)
20%	1895 ± 130	1034 ± 109[b]
10%	1846 ± 170	1310 ± 70[c]
Saturating	2075 ± 192	830 ± 81[b]

[a] Adapted from Ref. 1.
[b] $p < .01$ as compared to controls.
[c] $p < .05$ as compared to controls.

One of the reasons we have chosen to study menadione-mediated cell injury is that formation of menadione semiquinone via the one-electron reduction by cellular reductases is oxygen independent, whereas the generation of superoxide during the metabolism of this quinone is oxygen dependent.[4] This allows us to separate effects associated with menadione reduction from those due to superoxide generation. As shown in Table I, production of CO_2 by rat enterocytes is not affected by oxygen concentration in the cell culture medium. However, in the presence of menadione (50 μM, final concentration) equilibrated with 20% oxygen, CO_2 production was inhibited by 45%. With 10% oxygen, inhibition of CO_2 was only 29%. With saturating levels of oxygen, menadione-mediated inhibition of CO_2 was 60%.

Using the same experimental model with DMPO (0.1 M) added to the reaction mixture in place of [^{14}C]glucose, we measured the rate of intracellular spin trapping of superoxide as a function of oxygen concentration. Fig. 1 shows a plot of the peak height at the downfield line of the DMPO-OH actually observed in these experiments. DMPO-OH is seen rather than DMPO-OOH as a result of the rapid reduction of DMPO-OOH by cellular peroxidases, including glutathione peroxidase. As shown in Fig. 1, peak height of DMPO-OH increases rapidly over a 5 to 7 min period at a rate which is independent of oxygen concentration. Thereafter, a decrease in signal intensity is observed, the slope of which is dependent on the concentration of oxygen in the reaction mixture. The biphasic curve is the result of the following: the increase in signal intensity comes

[4] H. Thor, M. T. Smith, P. Hartzell, G. Bellomo, S. A. Jewel, and S. Orrenius, *J. Biol. Chem.* **257**, 12419 (1982).

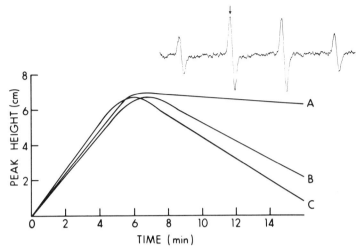

FIG. 1. The rate of spin trapping of superoxide was obtained when cells were incubated with DMPO in the presence of menadione using medium with various oxygen tensions: (A) saturating concentration of O_2, (B) 20% O_2, (C) 10% O_2. The magnetic field was set at the top of the second peak (note arrow in inset), and peak height was monitored as a function of time.

from the intracellular spin trapping of superoxide, as observed as DMPO-OH. However, there are a number of enzymes which can reduce the nitroxide functional group of DMPO-OH to give the corresponding hydroxylamine (for a review of nitroxide bioreduction, see Rauckman et al.[5]). After 5–7 min, the concentration of DMPO-OH is sufficiently great so that nitroxide reductase activity dominates. Thus, the decrease in peak height is dependent on the rate of nitroxide reduction minus the rate of nitroxide formation (superoxide spin trapping).

It is of interest to note that we have previously demonstrated that nitroxide bioreduction is oxygen independent.[6] Therefore, the more rapidly superoxide is formed and spin trapped, the slower the rate of fall of the curve in the second phase. In the case of 10% oxygen, the rate of bioreduction is greater than either 20% or saturating levels of oxygen, suggesting that the rate of DMPO-OH formation has markedly decreased. For saturating concentrations of oxygen, the rate of bioreduction is not that much greater then the rate of spin trapping. For 20% oxygen concentration, the rate of bioreduction lies between the two extremes.

[5] E. J. Rauckman, G. M. Rosen, and L. K. Griffeth, in "Spin Labeling in Pharmacology" (J. L. Holtzman, ed.), p. 175. Academic Press, New York, 1984.
[6] G. M. Rosen and E. J. Rauckman, Biochem. Pharmacol. 26, 675 (1977).

It is conceivable that at the high concentration of DMPO (0.1 M) used in our spin-trapping studies DMPO may have inhibited cytochrome P-450 such that the data presented above are not due to changes in oxygen tension but are the result of DMPO antagonism. Therefore, we examined the effect of DMPO, at 0.1 M, on the metabolism of menadione using the well-accepted model for drug metabolism, hepatic microsomes and NADPH. Under our experimental conditions, we observed no inhibition in the biotransformation of menadione.

Despite the obvious importance of these types of experiments, there is a dearth of literature[7] on how spin traps may alter (either inhibit or potentiate) the function of the biological system under study. For example, 2-methyl-2-nitrosopropane (MNP) has been reported to be cytotoxic to cultured fetal mouse hepatocytes even at concentrations (5 mM) below which measurable levels of spin-trapped adducts can be detected.[8] Likewise, the toxicity of N-tert-butyl-α-phenylnitrone (PBN) is legendary as this nitrone has been found to inhibit a number of enzymes, including cytochrome P-450.[9] From our experience, DMPO and congeners appear to exhibit limited pharmacological activity. Recently, we have discovered that DMPO (0.1 M) inhibits by over 90% neutrophil superoxide generation induced by the chemotatic peptide N-formylmethionylleucylphenylalanine (FMLP; B. Britigan, G. M. Rosen, M. S. Cohen, and S. Pou, unpublished results). In sum, data presented point to the fact that it is possible to correlate the intracellular generation of superoxide with cellular injury, by varying the oxygen tension on the cell culture medium.

While whole organ preparations provide a better model of the *in vivo* situation than isolated cells, the present generation of commercial ESR spectrometers limit detection of free radicals to small volumes of dissipative tissues.[10–16] Recently, one of us (H.J.H.) has developed a low-fre-

---

[7] S. Pou, D. J. Hassett, B. E. Britigan, M. S. Cohen, and G. M. Rosen, *Anal. Biochem.* **177**, 1 (1989).

[8] D. D. Morgan, C. L. Mendenhall, A. M. Bobst, and S. D. Rouster, *Photochem. Photobiol.* **42**, 93 (1985).

[9] O. Augusto, H. S. Beilan, and P. R. Oritz de Montellano, *J. Biol. Chem.* **257**, 11288 (1982).

[10] M. J. Barber, G. M. Rosen, L. M. Siegel, and E. J. Rauckman, *J. Bacteriol.* **153**, 1282 (1983).

[11] B. E. Britigan, G. M. Rosen, Y. Chai, and M. S. Cohen, *J. Biol. Chem.* **261**, 4426 (1986).

[12] J. K. Horton, R. Brigelius, R. P. Mason, and J. R. Bend, *Mol. Pharmacol.* **29**, 484 (1986).

[13] I. E. Blasig, B. Ebert, and H. Love, *Stud. Biophys.* **116**, 35 (1986).

[14] C. M. Arroyo, J. H. Kramer, B. F. Dickens, and W. B. Weglicki, *FEBS Lett.* **221**, 101 (1987).

[15] J. L. Zweier, J. T. Flaherty, and M. L. Weisfeldt, *Proc. Natl. Acad. Sci. U.S.A.* **84**, 1404 (1987).

[16] J. L. Zweier, *J. Biol. Chem.* **263**, 1353 (1988).

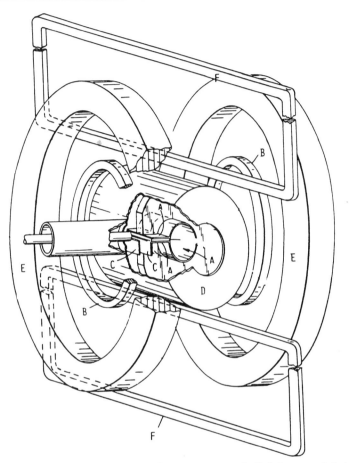

FIG. 2. Design of the low-frequency ESR spectrometer. A, Stripline sample holder; B, modulation coils; C, capacitive coupling; D, radiofrequency shield; E, main magnetic; F, horizontal plane defining coils.

quency ESR spectrometer[17] which will allow the study of free radical processes in isolated organs. The spectrometer layout is shown in Fig. 2. Whereas most ESR spectrometers are microwave devices, this operates at radiofrequency, 250 MHz. At these frequencies the skin depth of the exciting radiation is greater than 6 cm. This reduces the artifacts associated with superficial absorption of electromagnetic radiation and phase shifts. The design and performance criteria for the spectrometer include standard and nonstandard electronic feedback stabilization of frequency and match, open design, oversized modulation coils, and highly uniform

[17] H. J. Halpern, U.S. Patent 4,714,886 (1985).

magnetic fields of very low magnitude, with high spectral resolution. It is capable of imaging the source of its spectral signal, although this has not been implemented in present studies.

Before *in situ* spin-trapping investigations can be undertaken, there are a number of issues that must be addressed. First, in isolated cells the spin trap rapidly distributes between two compartments, the intra- and extracellular spaces. Second, rarely are spin-trapping experiments conducted on a multicellular system, where metabolism may be vastly different among the various cell types. Third, in contrast to cells, for which spin trap equilibration between extra- and intracellular environments is rapid and uniform, the intravascularly administered spin trap in an intact heart, for example, must cross the vascular endothelium to reach the myocyte or whatever location represents the site of free radical formation. Finally, dynamic factors such as rate of perfusion must be considered. We have recently completed an extensive series of studies, examining these issues.[18] Herein, we briefly touch on several key elements of these studies.

Because nitrones are not free radicals, they are not observable by ESR spectroscopy. Therefore, in order to study the pharmacodynamics and metabolism of spin traps in isolated organs situated in the low-frequency ESR spectrometer, we synthesized a family of nitroxides whose structure and lipophilicity approximates DMPO and its spin trapped adduct, DMPO-OH.[19] We examined the biological affects of these spin labels on isolated perfused rat hearts placed in a modified Langendorf apparatus to determine, with otherwise similar stable compounds, the effects and correctable artifacts of *in situ* measurements. At 1 mM, temporary bradycardia and decreased contractile force were observed. Within 1 min after administration of the nitroxides, the heart completely recovered.

Uptake of nitroxides into the beating heart is primarily dependent on the perfusion flow rate. This finding has significant implications for studies where variable flow rates are observed under various experimental conditions. For instances, short periods of ischemia result in reactive hyperemia and increased coronary flow rates upon reperfusion. Longer periods of ischemia are associated with increased tissue pressure and vascular endothelial swelling, leading to reduced coronary flow. We have found that differences in perfusate flow rates may actually result in artifactual differences in spin trap adduct formation and metabolism. From our studies, we have discovered that early sampling may be required so as not to underestimate spin trap adduct formation when higher perfusion rates are used. Decreased perfusion flow may result in longer contact time

[18] G. M. Rosen, H. J. Halpern, L. A. Brunsting, D. P. Spencer, K. E. Strauss, M. J. Bowman, and A. S. Wechsler. *Proc. Natl. Acad. Sci. U.S.A.* **85,** 7772 (1988).
[19] M. J. Turner III and G. M. Rosen, *J. Med. Chem.* **29,** 2439 (1986).

between the spin trap and the heart tissue, resulting in altered levels of spin trap adduct formation without a corresponding change in the real rate of free radical generation. These factors are of major importance when one considers that the spin trap DMPO readily oxidizes to DMPO-OH in oxygenated solutions, providing an ever present, slowly increasing background level of spin trap adduct.

One of our initial studies was to compare pharmacokinetic curves for the distribution and metabolism of nitroxides in our recirculating heart model by sampling either from the perfusion line, using the commercial ESR spectrometer, or *in situ,* using the low-frequency ESR spectrometer.[18] Using 2,2,3,3,5,5-hexamethylpyrrolidinyloxy (HMPO), whose partition coefficient is similar to DMPO-OH, and the standard ESR spectrometer data (Fig. 3A) are consistent with a two-compartment open-system model that has an initial distribution half-life of 2.4 min and a metabolic half-life (e.g., one-electron reduction to the corresponding hydroxylamine) of 22 min. When identical studies were conducted measuring nitroxide concentration *in situ,* the early distribution phase was not observed (Fig. 3B). This was the result of the time (~3 min) required to retune the low-frequency ESR spectrometer after placement of the perfused heart in the spectrometer, along with scatter in the data. However, the metabolic half-life for HMPO was calculated to be 25 min, in excellent agreement with the value obtained using the commercial spectrometer.

Similar studies were conducted using the nitroxide 3-carboxy-2,2,5,5-tetramethyl-1-pyrrolidinyloxy (CTMPO), whose partition coefficient was roughly that of DMPO.[18] These data indicate that the pharmacokinetics of the two nitroxides are quite disparate, based largely on differences in partition coefficients and experimental conditions (flow or ischemia). Care is, therefore, necessary in interpreting the time dependence of the ESR signals from even these stable nitroxides, *in situ,* given differences between their pharmacokinetic properties. Clearly, equal care should be employed when spin traps like DMPO are used.

Another serious problem that must be overcome is the decreased sensitivity one observes, going from a commercial ESR spectrometer at 9.5 GHz to our low-field ESR spectrometer at 250 MHz. This is especially troublesome when one considers that under the best of conditions, concentrations of spin-trapped adducts are rarely more than 1 μM. One approach is to increase the sensitivity of our spin-trapping procedure. Based on the pioneering work of Beth *et al.*[20] and Janzen *et al.,*[21] a significant

[20] A. H. Beth, S. D. Venkataramu, K. Balasubramanian, L. K. Dalton, B. H. Robinson, D. E. Pearson, C. R. Park, and J. H. Park, *Proc. Natl. Acad. Sci. U.S.A.* **78,** 967 (1981).
[21] E. G. Janzen, U. M. Oehler, D. L. Haire, and Y. Kotake, *J. Am. Chem. Soc.* **108,** 6858 (1986).

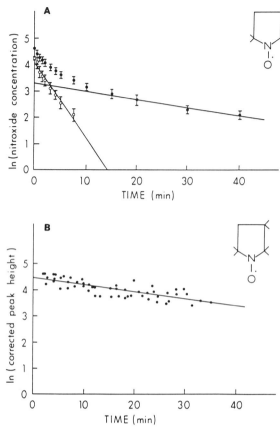

FIG. 3. Semilogarithmic plot of HMPO concentration versus time after administration of the nitroxide to isolated rat hearts placed in a recirculating retrograde modified Langendorf apparatus. Samples were removed at defined times, added to a quartz flat cell, and assayed in a commercial ESR spectrometer. Each point represents the nitroxide concentration for six hearts. The distribution half-life (open circles) was 2.4 min, and the metabolic half-life (filled circles) was 22 min. (B) Semilogarithmic plot of HMPO concentration versus time after administration of the nitroxide to isolated rat hearts placed in a recirculating retrograde modified Langedorf apparatus housed in the low-frequency ESR spectrometer. Each datum point (four experiments) is plotted because of difficulties in obtaining ESR signals at the same time for each heart. The metabolic half-life was calculated to 25 min.

increase in sensitivity may be obtained by preparing spin traps containing deuterium and ^{15}N in place of hydrogen and [14]N. We have recently synthesized a family of pyrroline-1-oxides containing these isotopes.[22] As shown in Fig. 4, incorporation of deuterium in positions 2 and 3, indepen-

[22] S. Pou, G. M. Rosen, X. Wu, and J. F. W. Keana, *J. Org. Chem.* submitted (1990).

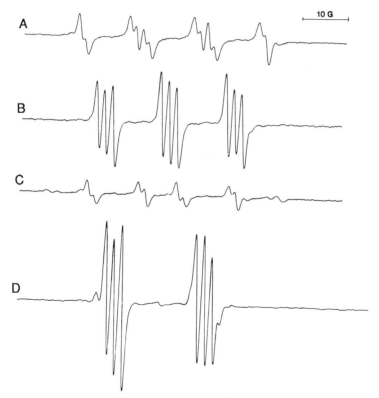

3; R = CD₃, N = ¹⁵N   2; R = CD₃, N = ¹⁵N
DMPO; R = CH₃, N = ¹⁴N   1; R = CH₃, N = ¹⁴N

FIG. 4. Structures of deuterium- and ^{15}N-containing DMPO congeners.

FIG. 5. ESR spectra obtained from the spin trapping of superoxide, generated by the oxidation of xanthine by xanthine oxidase, in the presence of (A) DMPO, (B) **1**, (C) **3**, and (D) **2**. Microwave power was 20 mW; modulation frequency was 100 kHz, with an amplitude of 0.63 G; sweep time was 12.5 G/min; and the receiver gain was 2.3×10^3, with a response of 1 sec.

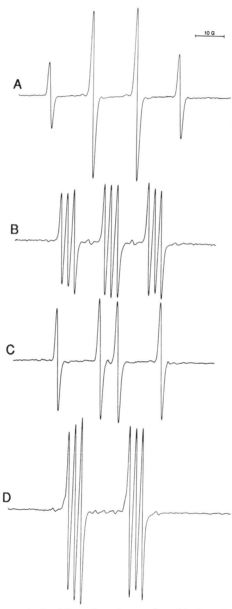

FIG. 6. ESR spectra obtained from the spin trapping of hydroxyl radical, generated by the reaction of ferrous sulfate with H_2O_2 and superoxide, produced by the oxidation of xanthine by xanthine oxidase, in the presence of (A) DMPO, (B) **1**, (C) **3**, and (D) **2**. Instrumentation settings were as in Fig. 5.

dent of ^{15}N, demonstrates the utility of this approach. The spin trapping of superoxide, generated from a xanthine/xanthine oxidase system, with nitrone **1** led to a doubling in sensitivity compared to DMPO (Fig. 5). A further increase was achieved by using ^{15}N analog **2** in our model superoxide-generating system. However, when the ^{15}N-DMPO analog **3** was compared to DMPO, an increase was not observed. This result is surprising, considering the decrease in the hyperfine splitting from 3 lines for ^{14}N to 2 lines for ^{15}N. At present, we have no explanation for these findings.

ESR spectra derived from the spin trapping of hydroxyl radical are markedly enhanced when compared to studies using DMPO (Fig. 6). In our model superoxide-generating system, where high steady-state levels of this free radical are easily achieved, increased sensitivity is clearly unnecessary. However, for cell suspensions or *in situ* studies, where rates of free radical generation are very small, a significant increase in sensivity may make it possible to unequivocally identify these reactive species, which might be impossible in the absence of deuterium- and ^{15}N-containing DMPO.

In this review, our goal has been to make the reader cognizant of the fact that spin traps are not benign chemicals, but, in some cases, they may elicit significant biological activity. It is very difficult to correlate free radical formation with cellular injury. This is an especially troublesome issue when the spin trap used to identify the free radical, generated as a consequence of some enzymatic process, is an active participant in the process under investigation. Finally, we have taken the opportunity to mention current areas of active research, which we believe have enormous potential for studying free radical processes in isolated organ preparations and, eventually, *in vivo*.

Acknowledgments

This research was supported in part by grants from the National Institutes of Health (HL-33550), the National Science Foundation (DCB-8616115), The American Cancer Society (BC-453 and PDT-262), and the Council for Tobacco Research U.S.A.

[65] Oxygen Radicals and Drugs: *In Vitro* Measurements

By D. M. ZIEGLER and JAMES P. KEHRER

Drugs that can generate oxygen radicals include many compounds that can transfer one or two electrons to molecular oxygen via enzyme-catalyzed reactions. The most toxic xenobiotics in this category are compounds bearing quinone, dipyridyl, nitro, or azo groups.[1,2] Xenobiotics with these functional groups are particularly susceptible to enzymatic one-electron reduction [Eq. (1)] to form intermediate radical anions that are rapidly reoxidized by O_2 [Eq. (2)].

$$2 \text{ (Xeno)} + \text{NAD(P)H} \rightarrow 2 \text{ (Xeno)}^{\overline{\cdot}} + \text{NAD(P)}^+ + \text{H}^+ \qquad (1)$$
$$\text{(Xeno)}^{\overline{\cdot}} + O_2 \rightarrow \text{(Xeno)} + O_2^{\overline{\cdot}} \qquad (2)$$

Xenobiotic-mediated one-electron transfer from cellular metabolites to molecular oxygen can generate damaging quantities of reactive oxygen species in tissues containing reductases capable of sustaining the reaction. However, two-electron reduction pathways also exist in tissues and may preempt or terminate this process by producing stable, nontoxic xenobiotic metabolites.

The superoxide anion, $O_2^{\overline{\cdot}}$, is usually the precursor (and perhaps an essential intermediate) in the metabolic generation of reactive oxygen species. Velocities of the drug-dependent production of reactive oxygen species *in vitro* are, therefore, usually based on rates of $O_2^{\overline{\cdot}}$ generation. Direct measurements of $O_2^{\overline{\cdot}}$ in crude tissue preparations are complicated by endogenous superoxide dismutase (SOD), and most procedures simply determine drug-dependent generation of its dismutation product, H_2O_2.

Many different methods for measuring H_2O_2 are available,[3] but virtually all procedures couple H_2O_2 formed to the peroxidation of a compound that can be measured spectrophotometrically or fluorimetrically. Fluorimetric methods are usually more sensitive and may permit analyses in single cells. For example, 2,7-dichlorofluorescin diacetate is trapped following deacetylation in living cells and can be oxidized to a fluorescent compound.[4] Continuous monitoring of reaction rates is also possible with fluorescent probes, and rates as low as 1–2 μM/min can be measured by following the horseradish peroxidase-catalyzed oxidation, by H_2O_2, of *p*-hydroxyphenyl acetate to the stable fluorescent 2,2'-dihydroxybiphenyl 5,5'-diacetate with the procedure described by Hyslop and Sklar.[5] However, overlapping absorption of many drugs of interest limits the use of this, as well as other, fluorimetric methods.

The peroxidation of methanol to formaldehyde is one of the more convenient methods for measuring drug-dependent generation of H_2O_2.

[1] K. F. McLane, J. Fisher, and K. Ramakrishan, *Drug Metab. Rev.* **14,** 741-799 (1983).

[2] D. M. Ziegler, *in* "Oxygen Radicals in Biology and Medicine" (M. G. Simic, K. A. Taylor, J. F. Ward, and C. von Sonntag, eds.), p. 729. Plenum, New York, 1989.

[3] A. G. Hilderbrandt, I. Roots, M. Tjoe, and G. Heinemeyer, this series, Vol. 52, p. 342.

[4] J. A. Acott, C. J. Homcy, B. A. Khaw, and C. A. Rabito, *Free Radical Biol. Med.* **4,** 79 (1988).

[5] P. A. Hyslop and L. A. Sklar, *Anal. Biochem.* **141,** 280 (1984).

Modifications that permit its application to relatively crude tissue preparations are described below.

Principle. SOD catalyzes the disproportionation of O_2^- to H_2O_2 [Eq. (3)] which is coupled to the oxidation of methanol by catalase. The latter reaction [Eq. (4)] is essentially stoichiometric with H_2O_2 at low concentrations of peroxide.[6]

$$2\ O_2^- + 2\ H^+ \rightarrow H_2O_2 + O_2 \tag{3}$$
$$CH_3OH + H_2O_2 \rightarrow CH_2O + 2\ H_2O \tag{4}$$

The formaldehyde formed is converted stoichiometrically to 3,5-diacetyl-1,4-dihydrolutidine by the Hantzsch reaction with the reagent described by Nash.[7] The reagent is specific for formaldehyde, and the concentration of the diacetyldihydrolutidine is estimated by its absorption at 412 nm (the millimolar absorptivity in the Nash reagent is 7.7/cm).

Reagents

1.0 M Potassium phosphate, pH 7.5
50 mM NADP$^+$
50 mM NAD$^+$
0.1 M Glucose 6-phosphate
0.1 M n-Octylamine hydrochloride
3.0 M Trichloroacetic acid (TCA)
3.6 M Potassium (or sodium) hydroxide
Methanol (reagent grade)
50 units/ml *Leuconostoc mesenteroides* glucose-6-phosphate dehydrogenase (ZF)[8]
10,000 units/ml Catalase (bovine liver)
500 units/ml Superoxide dismutase (bovine erythrocytes)
Nash reagent: Prepared daily by mixing 0.1 ml 2,4-pentanedione (acetylacetone) with 100 ml of 1.9 M ammonium acetate containing 0.15 ml glacial acetic acid

Tissue Preparation. Organs, removed and chilled as quickly as possible, are rinsed to remove excess blood and then homogenized in 10 volumes of 0.25 M sucrose containing 5 mM phosphate, pH 7.5. The homogenate is centrifuged at 9,000 g for 10 min to sediment nuclei, mitochondria, and any unbroken cells. The 9,000 g supernatant fraction, which contains most of the cellular xenobiotic reductases,[1] is passed

---

[6] P. Nicholls and G. R. Schonbaum, *in* "The Enzymes" (P. D. Boyer, H. Lardy, and K. Myrback, eds.), p. 147. Academic Press, New York, 1963.

[7] T. Nash, *Biochem. J.* **55**, 416 (1953).

[8] One unit is defined as the concentration of enzyme required to catalyze the conversion of 1 μmol substrate per minute in the reaction media used in the metabolic assays.

through a Sephadex G-10 column, or dialyzed, to remove glutathione. The preparations must be essentially free from this tripeptide to prevent loss of H_2O_2 and formaldehyde owing to reactions catalyzed by cytosolic glutathione peroxidase and formaldehyde dehydrogenase.

Procedure. The metabolic reactions are carried out in 10-ml Erlenmeyer flasks in a shaker bath at 37°. Up to six flasks with different drugs or various controls minus drug substrate can be processed at a time, but one flask containing all components except the drug substrate must be included.

Automatic pipets are used for virtually all transfers, and the volumes specified are adapted to equipment routinely used in our laboratory, but they may be changed provided the final concentrations of the reagents are as specified. The following volumes of the stock solutions are added to the flasks in the order listed: water (to a final reaction volume of 2.5 ml), 0.25 ml phosphate buffer, 75 μl, NADP$^+$ and/or NAD$^+$, 0.25 ml glucose 6-phosphate, 50 μl each ZF, catalase, and SOD, and 0.15 ml methanol. (If the drug substrate is added in methanol the latter volume is adjusted accordingly.)

After a 4 to 6-min temperature equilibration period, 0.1–0.5 ml of the 9,000 g supernatant fraction is added (protein concentration 0.4–1.6 mg/ml), and about 1 min later the reaction is started by adding the drug substrate. Aliquots of 0.5 ml withdrawn at regular intervals, usually at 0, 3, 6, and 9 min, are transferred to centrifuge tubes containing 60 μl of 3 M TCA. After all the aliquots are collected, the tubes are centrifuged (2,000 g for 10 min at room temperature), and 0.45 ml of the clear supernatant fraction is transferred to small test tubes containing 50 μl of 3.6 M KOH.[9] One milliliter of the Nash reagent is added, and after incubation in a water bath (60 $\pm$ 1°) for exactly 10 min, the tubes are chilled on ice to room temperature (2–3 min) and absorbance of the contents measured against water at 412 nm in a cuvette with a 1-cm light path.

If the preceding protocol is followed, the concentration of formaldehyde in aliquots of the reaction medium (in nmol/ml) is equal to OD × 485. Since the velocity of formaldehyde formation is virtually identical with that of H_2O_2 generation, the rate of drug-dependent H_2O_2 production is calculated from the difference in formaldehyde in the presence and

---

[9] The solutions at this point must be optically clear and essentially free from compounds absorbing at 412 nm. Many drugs, especially ones bearing azo or nitro groups, do absorb and must be removed before proceeding. Virtually all drugs of interest can be removed by extracting the aliquots 3 times with about 2 volumes of ether saturated with water. After the last extraction, the tubes are warmed to 60° for a few minutes to remove residual ether before adding the Nash reagent.

absence of drug. Drug-independent rates are usually quite low since octylamine blocks turnover of cytochrome P-450 which is the major source of endogenous H_2O_2 produced by 9,000 g fractions supplemented with NADPH. This inhibitor also blocks formaldehyde formation by oxidative demethylation of drugs containing N- or O-methyl groups. However, for accurate determinations of drug-dependent H_2O_2 formation, controls containing all components except methanol and catalase should be carried through the procedure to correct for any H_2O_2-independent formaldehyde formation.

Applications and Limitations. The preceding method has been used to measure rates of H_2O_2 formation induced by a variety of different xenobiotics in reactions catalyzed by 9,000 g fractions of liver, lung, and kidney homogenates. As shown by the results obtained with rat liver preparations (Table I), the method readily measures rates of drug-dependent H_2O_2 generation. As little as 4 μM H_2O_2/mix can be measured quite accurately. Formaldehyde added to the reaction medium containing reduced pyridine nucleotides and glutathione-free 9,000 g supernate (or its subfractions) is recovered quantitatively as long as enough ZF is added to keep the pyridine nucleotides fully reduced, thereby effectively inhibiting alcohol and aldehyde dehydrogenases. However, mitochondria do interfere with recovery of formaldehyde, and attempts to circumvent this problem have not been successful. Addition of inhibitors selective for the mitochondrial electron transport system decreases recovery, suggesting that reduction of formaldehyde, rather than its oxidation by mitochondrial aldehyde dehydrogenases, may be responsible.

The addition of n-octylamine or some other inhibitor (exclusive of the N-alkylimadazole compounds, which oxidize pyridine nucleotides) selective for cytochrome P-450 is essential. The endogenous rate of O_2^- or

TABLE I
XENOBIOTIC-DEPENDENT H_2O_2 FORMATION CATALYZED BY RAT LIVER PREPARATIONS

Preparation	H_2O_2 (nmols/min/mg protein)[a]			
	Menadione	Paraquat	Methyl red	p-Amino azobenzene
9000 g supernate	48	9.1	2.0	1.7
Microsomes	60	42	—	—
Cytosol	36	0.4	—	—

[a] Rates were measured as described in text at 1 mM substrate. Menadione and p-aminoazobenzene were dissolved and added in methanol.

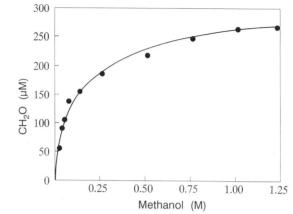

Fig. 1. Paraquat-dependent peroxidation of methanol as a function of methanol concentration. The reactions were carried out as described in the text with rat liver 9000 g supernatant fractions. The ordinate lists the concentration of paraquat-dependent formaldehyde formed in 3 min.

H_2O_2 generation by P-450-dependent monooxygenases can exceed drug-dependent rates, and unless turnover of these hemoproteins is blocked, *in vitro* measurements of drug-dependent generation of reactive oxygen species with crude preparations by any method is virtually impossible. These inhibitors will, however, prevent detection of O_2^- generated by drugs reduced solely by ctyochrome P-450, although, except for a few nitro compounds, the vast majority of drugs are reduced by P-450-independent pathways.[1]

The addition of alcohols other than methanol will also interfere with this procedure since catalase can catalyze peroxidation of a variety of different alcohols.[6] The concentration of methanol required to saturate catalase is also quite high (Fig. 1), but methanol up to 1.75 M does not produce detectable inhibition of drug reductases in crude preparations.

In summary, the method described here provides a simple, convenient means for detecting and quantitating drug-dependent generation of reactive oxygen species catalyzed by relatively crude tissue preparations. Reactive oxygen generation is only one component of multiple metabolic reactions responsible for toxicity *in vivo*, and the rate of O_2^- and/or H_2O_2 generation does not necessarily correlate with toxicity of the xenobiotic. For example, menadione, which produces large amounts of H_2O_2 *in vitro* (Table I) and is quite toxic to cultured cells, is dramatically less toxic *in vivo*, possibly because of other reactions that limit its capacity to generate reactive oxygen species in the intact animal.

[66] Modulation of Cellular Glutathione and Protein Thiol Status during Quinone Metabolism

By Giorgio Bellomo, Hjördis Thor, and Sten Orrenius

Introduction

In mammalian cells the metabolism of various quinones is associated with the consumption of intracellular glutathione (GSH) and the modification of protein thiols.[1–3] Quinone-induced thiol consumption may result from either alkylation or oxidation. Thus, many quinones readily react with GSH to form GSH–quinone conjugates and with protein thiols to form alkylated proteins.[2,4,5] In addition, quinones can serve as substrates for various flavoprotein reductases which catalyze their one-electron reduction to the corresponding semiquinone free radical which, in the presence of molecular oxygen, is reoxidized to the parent quinone yielding a superoxide anion radical and thus entering a redox cycle.[1,6] Further metabolism of O_2^- generates hydrogen peroxide and other active oxygen species which can directly or enzymatically oxidize thiols.

In mammalian cells most of the GSH consumption induced by the metabolism of redox-active quinones is due to the oxidation of GSH to glutathione disulfide (GSSG) linked to H_2O_2 metabolism by glutathione peroxidase.[7] GSSG is subsequently reduced back to GSH by glutathione reductase at the expense of NADPH. However, at cytotoxic quinone concentrations the capacity of the cell to regenerate NADPH becomes exhausted, GSSG accumulates, and a condition of oxidative stress is created.[8]

[1] H. Thor, M. T. Smith, P. Hartzell, G. Bellomo, S. A. Jewell, and S. Orrenius, *J. Biol. Chem.* **257**, 12419 (1982).

[2] D. DiMonte, D. Ross, G. Bellomo, L. Eklöw, and S. Orrenius, *Arch. Biochem. Biophys.* **235**, 334 (1984).

[3] D. DiMonte, G. Bellomo, H. Thor, P. Nicotera, and S. Orrenius, *Arch. Biochem. Biophys.* **235**, 343 (1984).

[4] D. Ross, H. Thor, P. Moldéus, and S. Orrenius, *Chem.–Biol. Interact.* **55**, 177 (1985).

[5] A. J. Streeter, D. C. Dahlin, S. D. Nelson, and T. A. Baille, *Chem.–Biol. Interact.* **48**, 348 (1984).

[6] G. Powis and P. L. Appel, *Biochem. Pharmacol.* **29**, 2567 (1980).

[7] D. P. Jones, L. Eklöw, H. Thor, and S. Orrenius, *Arch. Biochem. Biophys.* **210**, 505 (1985).

[8] L. Eklöw, P. Moldéus, and S. Orrenius, *Eur. J. Biochem.* **138**, 459 (1984).

In this chapter we discuss the relative contribution of alkylation versus oxidation in the modulation of the cellular thiol status during quinone metabolism. For this purpose isolated rat hepatocytes are incubated with either *p*-benzoquinone (an alkylating quinone),[9] 2,3-dimethoxy-1,4-naphthoquinone (a redox-cycling quinone),[10] or menadione (2-methyl-1,4-naphthoquinone; a mixed alkylating/redox-cycling quinone),[4] and the effects on intracellular thiol homeostasis are compared.

Experimental Procedures

Hepatocyte Isolation and Incubation. Male Sprague-Dawley rats (180–250 g), allowed free access to water and food, are used in all experiments. Hepatocytes are isolated by collagenase perfusion of the liver as described by Moldéus *et al.*[11] After isolation, the cell suspension (~30 ml) is centrifuged through a layer (20 ml) of isotonic Percoll in Hanks buffer (final density 1.06) in order to eliminate damaged cells, free organelles, and enzymes released during the isolation procedure. The cell pellet is suspended in a modified Krebs–Henseleit buffer supplemented with 20 mM HEPES (pH 7.4) at a final density of 10^6 cells/ml. All incubations are performed in rotating, round-bottomed flasks at 37° under an atmosphere of 95% O_2/5% CO_2. Cell viability is assessed by trypan blue exclusion and in our studies was always greater than 95%. Quinones are dissolved in dimethyl sulfoxide (DMSO) and added directly to the cell suspension. The final DMSO concentration never exceeds 0.5%.

To inhibit glutathione reductase, hepatocytes are incubated for 30 min with 1,3-bis(2-chloroethyl)-1-nitrosourea (BCNU, 100 μM) in Krebs–Henseleit buffer supplemented with 20 mM HEPES (pH 7.4) and amino acids at the concentrations indicated in Ref. 12, except for serine, glutamine, cysteine, and methionine, which are present at 0.2 mM final concentration.[8] At the end of the incubation, the cells are washed and resuspended in the same medium, but lacking BCNU. The cells are then incubated for 2 hr to allow GSH resynthesis. At the end of this incubation, the cells are harvested, washed once, and finally incubated in a fresh Krebs–Henseleit medium supplemented with 20 mM HEPES (pH 7.4). This treatment usually results in more than 80% inhibition of glutathione

[9] L. Rossi, G. A. Moore, S. Orrenius, and P. J. O'Brien, *Arch. Biochem. Biophys.* **251,** 25 (1986).
[10] T. W. Gant, D. N. R. Rao, R. P. Mason, and G. M. Cohen, *Chem.–Biol. Interact.* **65,** 157 (1988).
[11] P. Moldéus, J. Högberg, and S. Orrenius, this series, Vol. 52, p. 60.
[12] C. Waymouth and R. K. Jakson, *J. Natl. Cancer Inst.* **22,** 1003 (1959).

reductase whereas the intracellular GSH level is comparable to that found in control (non-BCNU-treated) cells.[8]

To deplete intracellular GSH, hepatocytes are incubated for 30 min with 0.5 mM diethyl maleate (DEM) in Krebs–Henseleit buffer supplemented with 20 mM HEPES (pH 7.4). The cells are then pelleted, washed twice, and resuspended in the same medium. As a result of this treatment, the intracellular GSH concentration decreases to 15–20% of the control level without appreciable changes in the amount of protein thiols.

Subcellular Fractionation of Hepatocytes. For isolation of subcellular fractions, 50 ml of a cell suspension is centrifuged at low speed, the pellet is resuspended in 5 volumes of 0.25 M sucrose, containing 10 mM HEPES (pH 7.4), and then the cells are homogenized with 20 passes in a tightly fitted Potter–Elvehjem homogenizer. The mitochondrial fraction is recovered as the 9,000 g pellet after sedimenting unbroken cells and nuclei at 1,500 g. The microsomal fraction is recovered as the 105,000 g pellet and the cytosolic fraction as the 105,000 g supernatant.

Preparation of Hepatocyte Cytoskeleton. For isolation of the cytoskeletal fraction, 10 ml of a cell suspension is centrifuged at low speed, and the pellet is resuspended in 5 ml of a medium containing 40 mM KCl, 5 mM EGTA, 5 mM MgCl$_2$, 50 mM Tris-HCl (pH 7.5), and 1% Triton X-100 (extraction medium) by incubation for 30 min on ice. The suspension is then centrifuged at 4,000 g for 20 min, and the pellet is washed twice in the same medium. Polyacrylamide gel electrophoresis and immunoblotting analysis of the polypeptides isolated following this procedure reveal the presence of several cytoskeletal proteins, including actin and several cytokeratins.[13]

Measurement of GSH, GSSG, and Glutathione–Protein Mixed Disulfides. Intracellular and extracellular GSH and GSSG are measured essentially as described by Reed *et al.*[14] Briefly, 1 ml of the cell suspension is taken, centrifuged at low speed, and the supernatant transferred to another tube. Both the cell pellet and the supernatant are deproteinated by adding 1 ml of 6.5% trichloroacetic acid and 50 μl of 60% perchloric acid (PCA), respectively. After centrifugation, 0.5 ml of the acid-soluble fraction is neutralized with excess NaHCO$_3$ and subsequently incubated in the dark for 60 min with 80 mM iodoacetic acid. One-half milliliter of a 1.5% dinitrofluorobenzene solution in ethanol is added, after which the sample is vigorously mixed and incubated in the dark for an additional 6

[13] F. Mirabelli, A. Salis, V. Marinoni, G. Finardi, G. Bellomo, H. Thor, and S. Orrenius, *Arch. Biochem. Biophys.* **264,** 261 (1988).

[14] D. J. Reed, P. W. Babson, A. Beatty, E. Brodie, and D. W. Potter, *Anal. Biochem.* **106,** 55 (1980).

hr. Separation of GSH and GSSG is performed by HPLC using a NH_2 column with spectrophotometric detection at 350 nm and a linear elution gradient of methanol–ammonium acetate–acetic acid. Identification and quantitation of GSH and GSSG are achieved by parallel running of standard mixtures.

For quantitation of glutathione–protein mixed disulfides, either 10 ml of the cell suspension or 3–5 mg of the subcellular or cytoskeletal fractions is centrifuged at high speed. The pellet is treated with 5% PCA to precipitate protein, and the sample is allowed to stand on ice for 30 min. The pellet is washed twice in 5% PCA and resuspended in 1.5 ml of 0.1 M Tris-HCl (pH 8). $NaBH_4$ (50 mg) is added together with 2 drops of *n*-octanol to prevent foam. The protein suspension is homogenized by treatment for 30 sec with a Polython, set at 6, and then incubated at 40° for 30 min. After cooling on ice, 1 ml of 50% HPO_3 is added to remove $NaBH_4$ and to precipitate proteins, which are pelleted by centrifugation at 105,000 *g* for 15 min at 4°. The treatment with $NaBH_4$ causes the release of protein-bound glutathione into the medium. Subsequent quantitation of the released GSH is performed as described above.

Measurement of Protein Sulfhydryl Groups. Protein thiols are measured essentially as described by DiMonte *et al.*[3] Briefly, 1 ml of the cell suspension or 1–2 mg of cytoskeletal protein is precipitated with 5% PCA and washed twice. The protein pellet is then solubilized in 2.5 ml of 0.5 M Tris-HCl (pH 8.8) containing 5 mM EDTA and 1% sodium dodecyl sulfate (SDS) and divided in 2 aliquots of 1 ml each. One of the aliquots is treated with 25 mM *N*-ethylmaleimide (NEM) for 10 min and used as blank. Dithiodinitrobenzoic acid (DTNB, 250 μM final concentration) is then added, and the color developed is read at 412 nm against the corresponding blank. Quantitation is performed either by using a molar extinction coefficient of 13,600 or by comparing the obtained values with a standard curve made with known amounts of GSH.

Measurement of Protein Concentration. Protein is measured according to Lowry *et al.*[15]

Quinone-Induced Depletion of GSH and Formation of GSSG and Glutathione–Protein Mixed Disulfides

Incubation of isolated rat hepatocytes with increasing concentrations of menadione, *p*-benzoquinone, or 2,3-dimethoxy-1,4-naphthoquinone resulted in a dose-dependent decrease in intracellular GSH level (Fig. 1).

[15] O. H. Lowry, N. J. Rosebrough, A. L. Farr, and R. J. Randall, *J. Biol. Chem.* **193**, 265 (1951).

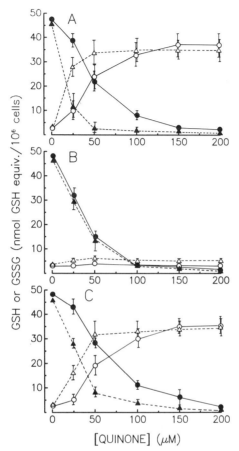

FIG. 1. Quinone-induced GSH depletion and GSSG formation in hepatocytes. Control (●, ○) and BCNU-treated (▲, △) hepatocytes were incubated with the indicated concentrations of menadione (A), *p*-benzoquinone (B), or 2,3-dimethoxy-1,4-naphthoquinone (C). After 30 min, samples were taken and processed as described in the text for the quantitation of total (intracellular + extracellular) GSH (●, ▲) and GSSG (○, △). The results represent the mean ± S.D. of four experiments.

With menadione and 2,3-dimethoxy-1,4-naphthoquinone, the GSH decrease was associated with the concomitant accumulation of GSSG. Previous studies have shown that the intracellular accumulation of GSSG is followed by its active excretion through an ATP-dependent transport system.[3,16] Pretreatment of hepatocytes with BCNU potentiated both GSH

---

[16] P. Nicotera, M. Moore, G. Bellomo, F. Mirabelli, and S. Orrenius, *J. Biol. Chem.* **260,** 1999 (1985).

depletion and GSSG formation, particularly with the lower concentrations of the redox-active quinones. With the alkylating quinone, *p*-benzoquinone, GSH loss was not associated with GSSG formation, and BCNU pretreatment failed to accelerate GSH depletion and to stimulate GSSG formation.

Concomitant with GSH oxidation during metabolism of menadione and 2,3-dimethoxy-1,4-naphthoquinone in hepatocytes, glutathione–protein mixed disulfides were formed in a dose-dependent manner (Fig. 2).

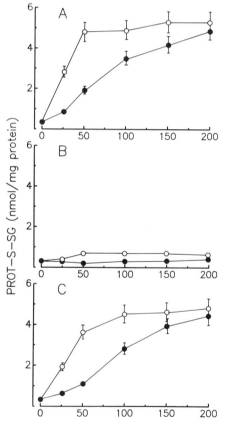

FIG. 2. Quinone-induced formation of glutathione–protein mixed disulfides in hepatocytes. Control (●) and BCNU-treated (○) hepatocytes were incubated with the indicated concentrations of menadione (A), *p*-benzoquinone (B), or 2,3-dimethoxy-1,4-naphthoquinone (C). After 30 min, samples were taken and processed as described in the text for the quantitation of glutathione–protein mixed disulfides. The results represent the mean ± S.D. of three experiments.

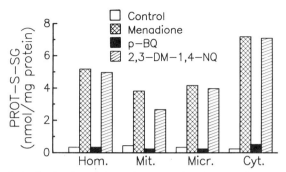

FIG. 3. Subcellular distribution of glutathione–protein mixed disulfides formed during the metabolism of quinones in hepatocytes. Hepatocytes were incubated without or with 200 μM menadione, p-benzoquinone, or 2,3-dimethoxy-1,4-napthoquinone. After 30 min, the cells were collected and processed for subcellular fractionation and quantitation of glutathione–protein mixed disulfides as described in the text. Shown is one experiment typical of three.

Most of the glutathione–protein mixed disulfides formed during the metabolism of menadione or 2,3-dimethoxy-1,4-naphthoquinone in hepatocytes were recovered in the cytosolic fraction, with lesser amounts in microsomes and mitochondria (Fig. 3). Similar to GSSG production, mixed disulfide formation was potentiated by pretreatment of the cells with BCNU. This was probably due either to the increased GSSG level observed under the same conditions or to impaired removal of the mixed disulfides by BCNU-inhibited glutathione reductase. Previous investigations have suggested a possible participation of glutathione reductase in removing glutathione–protein mixed disulfides.[17] In contrast, mixed disulfides were not formed during the metabolism of p-benzoquinone in either control or BCNU-treated hepatocytes.

A pool of protein sulfhydryl groups which appears to be critically related to the morphological intactness of the hepatocyte is associated with the cytoskeleton.[13,18] The modification of this pool of cytoskeletal thiols results in the appearance of surface protrusions (blebs) and leads to the disruption of plasma membrane integrity and to cytotoxicity.[1] During the metabolism of either menadione or 2,3-dimethoxy-1,4-naphthoquinone in hepatocytes, a dose-dependent formation of mixed disulfides

[17] G. Bellomo, F. Mirabelli, D. DiMonte, P. Richemli, H. Thor, C. Orrenius, and S. Orrenius, *Biochem. Pharmacol.* **36,** 1313 (1987).
[18] F. Mirabelli, A. Salis, M. Perotti, F. Taddei, G. Bellomo, and S. Orrenius, *Biochem. Pharmacol.* **37,** 3423 (1988).

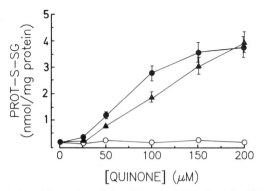

FIG. 4. Quinone-induced formation of glutathione–protein mixed disulfides in hepatocyte cytoskeleton. Hepatocytes were incubated with the indicated concentrations of menadione (●), p-benzoquinone (○), or 2,3-dimethoxy-1,4-naphthoquinone (▲). After 30 min, the cells were collected and processed for isolation of the cytoskeletal fraction and quantitation of glutathione–protein mixed disulfides as described in the text. The results represent the mean ± S.D. of three experiments.

between glutathione and cytoskeletal protein thiols occurred (Fig. 4). No increase in mixed disulfide formation was detected during the metabolism of p-benzoquinone, in accordance with the lack of GSSG formation under the same experimental conditions.

TABLE I

INTRACELLULAR GSH PROTECTING AGAINST QUINONE-INDUCED DEPLETION OF TOTAL CELLULAR AND CYTOSKELETAL PROTEIN SULFHYDRYL GROUPS[a]

	Protein-SH (nmol/mg protein)			
	Control hepatocytes		DEM-treated hepatocytes	
Treatment	Total	Cytoskeletal	Total	Cytoskeletal
Control, no addition	94	48	89	47
Menadione, 100 μM	85	24	52	8
p-Benzoquinone, 100 μM	82	26	42	11
2,3-Dimethyoxy-1,4-naphtho-quinone, 200 μM	88	27	66	14

[a] Hepatocytes were preincubated without or with 0.5 mM diethyl maleate (DEM) for 30 min. After washing and resuspending the cells in fresh medium, the different quinones were added at the concentrations indicated. After incubation for 1 (menadione and p-benzoquinone) or 2 hr (2,3-dimethoxy-1,4-naphthoquinone), samples were taken and processed as described in the text for quantitation of total and cytoskeletal protein sulfhydryl groups.

Role of GSH in Quinone-Induced Depletion of Total Cellular and
Cytoskeletal Protein Thiols

Intracellular GSH plays an important role in the defense against the
toxicity of many xenobiotics including both alkylating and redox-active
quinones.[1,3] In hepatocytes there is a clear relationship between quinone-
induced depletion of protein thiols and cytotoxicity. To investigate the
effect of the intracellular glutathione level on the depletion of protein
thiols triggered by quinone metabolism, GSH-depleted hepatocytes were
incubated with menadione, p-benzoquinone, or 2,3-dimethoxy-1,4-
naphthoquinone, and both total and cytoskeletal protein thiols were assayed. As illustrated in Table I, GSH-depleted cells showed an enhanced
susceptibility to quinone-induced depletion of both total and cytoskeletal
protein thiols, in agreement with the enhanced toxicity and accelerated
appearance of surface abnormalities observed under the same conditions.

Conclusions

The metabolism of cytotoxic levels of quinones in isolated hepatocytes is associated with the depletion of intracellular glutathione. This
results from either alkylation or oxidation, depending on the quinone
employed. During the metabolism of redox-active quinones, GSSG is
formed and glutathione–protein mixed disulfides are detected in several
subcellular fractions, including the cytoskeleton. Moreover, when glutathione reductase is inhibited, the accumulation of GSSG and mixed
disulfides caused by low concentrations of the redox-active quinones is
markedly potentiated. Thus, intracellular glutathione plays a critical role
in the defense against quinone-associated cytotoxicity, particularly by
preventing the quinone-induced modification of protein thiols, an effect
that appears to be closely linked to the occurrence of irreversible cell
damage.

[67] Production of Oxygen Radicals by Photosensitization

By Joseph P. Martin, Jr. and Paula Burch

Introduction

The cytotoxicity of illuminated photosensitizing agents in the presence
of oxygen has long been recognized. It is commonly referred to as the

photodynamic effect.[1-4] A variety of compounds (e.g., synthetic dyes, antipsychotic drugs, antibiotics, and plant secondary compounds) exhibit this effect, and all groups of organisms are susceptible to photodamage. The types of cellular damage that occur include lipid peroxidation and membrane lysis, DNA base modification and DNA strand breakage, mutagenesis, and protein inactivation.[1,3-8]

Dye-mediated photooxidations proceed by two different pathways.[1,2] In the first pathway (type I), excited triplet state dyes react directly with an oxidizable substrate. Electron or hydrogen atom transfer generates a semioxidized substrate radical and a semireduced dye radical. Subsequent reactions of both species with oxygen yield oxidized and modified substrates, regenerated ground state dye, oxygen radicals, and hydrogen peroxide (H_2O_2). When photooxidations occur within cells, damage may be caused during the primary oxidation events[3,4] or by further reactions of oxygen radicals and H_2O_2 with cell components.[8-10] In the second pathway (type II), excited state dye triplets react directly with molecular oxygen, yielding ground state dye and singlet oxygen ($^1O_2$). Photodamage is caused by reactions of singlet oxygen with amino acids, nucleotides, and lipids.[1-6]

Under conditions of high oxygen concentration, characteristic of organic solvents, and in the absence of strong reducing agents, the second pathway is favored by most photosensitizing compounds.[1,2] However, in aqueous buffer solutions, where the concentration of oxygen is low (0.2–0.25 mM), and in the presence of high reductant concentrations, the type I pathway will be favored. Suitable reductants *in vitro* include allylthiourea, ascorbate, aromatic amines, reduced glutathione, tetramethylethylenediamine and reduced pyridine nucleotides.[9-16] Within aerobic cells

[1] C. S. Foote, *in* "Free Radicals in Biology" (W. Pryor, ed.) Vol. 2, p. 86. Academic Press, New York.

[2] C. S. Foote, F. C. Shook, and R. B. Abakerli, this series Vol. 105, p. 36.

[3] J. D. Spikes and R. Livingston, *Adv. Radiat. Biol.* **3**, 29 (1969).

[4] J. D. Spikes and R. Straight, *Annu. Rev. Phys. Chem.* **18**, 409 (1967).

[5] N. Houba-Herin, C. M. Calberg-Bacq, J. Piette, and A. Van de Vorst, *Photochem. Photobiol.* **36**, 297 (1982).

[6] J. Piette, M. Lopez, C. M. Calberg-Bacq, and A. Van de Vorst, *Int. J. Radiat. Biol. Relat. Stud. Phys. Chem. Med.* **40**, 427 (1981).

[7] C. Wallis and H. L. Melnick, *Photochem. Photobiol.* **4**, 159 (1965).

[8] J. P. Martin, K. Colina, and N. Logsdon, *J. Bacteriol.* **169**, 2516 (1987).

[9] J. P. Martin and N. Logsdon, *Arch. Biochem. Biophys.* **256**, 39 (1987).

[10] J. P. Martin and N. Logsdon, *J. Biol. Chem.* **262**, 7213 (1987).

[11] G. R. Beuttner, T. P. Doherty, and T. D. Bannister, *Radiat. Environ. Biophys.* **23**, 235 (1984).

[12] C. Beauchamp and I. Fridovich, *Anal. Biochem.* **44**, 276 (1972).

the type I pathway may be favored because of the low concentration of free oxygen and the high intracellular levels of reductants such as reduced glutathione (1–6 mM)[17,18] and NAD(P)H (1 mM).[19]

In this chapter we present methods of generating superoxide, (O_2^-), hydrogen peroxide, and the hydroxyl radical (OH ·) through photooxidations mediated by synthetic dyes both *in vitro* and *in vivo*, within the cytoplasm of *Escherichia coli*.

In Vitro Photooxidation

Procedures

Dye-Mediated Generation of O_2^-. All reactions are carried out in 3 ml of 50 mM potassium phosphate, 0.1 mM EDTA (pH 7.4–7.8). In the assay designed for O_2^- detection the solution also contains 10 μM oxidized cytochrome c (Sigma type III). The reductant NADH (nicotinamide adenine dinucleotide, reduced form) is added as a 20 mM stock solution made in the phosphate buffer at pH 7.0 and is diluted into the reaction mixture to a final concentration between 0.2 and 2.0 mM. Dye stock solutions of 1 mM are also prepared in the potassium phosphate buffer using published extinction coefficients.[5] Reactions are carried out in fused silica cuvettes. NADH depletion is monitored spectrophotometrically at 340 nm, and cytochrome c reduction is followed at 550 nm. Superoxide generation is estimated by the inhibition of cytochrome c reduction when Cu,Zn-SOD (copper, zinc-superoxide dismutase) is added to the reaction mixture. Cu,Zn-SOD is prepared as a 1 mg/ml stock solution in the potassium phosphate buffer. This corresponds to 3600 McCord and Fridovich units per ml.[20] SOD is added to the reaction to a final concentration of 20 μg/ml.

Light incubation is done by setting cuvettes midway between two 20-W fluorescent lamps (GE F20.T12-p1) positioned 5 cm apart in a foil-lined box. Visible light intensity is measured with a Licor LI-185B radiometer/photometer using an LI-190SB cosine corrected quantum sensor.

[13] H. M. Hassan and I. Fridovich, *Arch. Biochem. Biophys.* **196,** 385 (1979).
[14] I. Kraljic and L. Lindqvist, *Photochem. Photobiol.* **20,** 351 (1974).
[15] G. Oster, J. S. Bellin, R. W. Kimball, and M. S. Schrader, *J. Am. Chem. Soc.* **81,** 5095 (1959).
[16] A. H. Adelman and G. Oster, *J. Am. Chem. Soc.* **78,** 3977 (1956).
[17] R. C. Fahey, W. C. Brown, W. B. Adams, and M. B. Worsham, *J. Bacteriol.* **133,** 1126 (1978).
[18] J. T. Greenberg and B. Demple, *J. Bacteriol.* **168,** 1026 (1986).
[19] K. B. Anderson and K. Von Meyenberg, *J. Biol. Chem.* **252,** 4151 (1977).
[20] J. M. McCord and I. Fridovich, *J. Biol. Chem.* **244,** 6049 (1969).

Reactions are started by addition of dye to the desired concentration followed by illumination. Cuvettes are removed from the light at 1-min intervals and examined spectrophotometrically at either 340 nm or 550 nm in a UV–visible spectrophotometer.

Hydroxyl Radical Generation. The hydroxyl radical is generated in the system described above when the reaction mixture is supplemented with 50 μM ferric chloride, added as a 5 mM acidic stock solution. Also, the buffer concentration is increased to 0.15 M potassium phosphate (pH 7.6) and is supplemented with 4 mM sodium salicylate. Reactions are started as described above and are incubated for 1 hr, after which the degree of salicylate hydroxylation is determined by the method of Halliwell.[21] The hydroxylated diphenolic products are extracted from the reaction mixture with chilled diethyl ether, and the concentration of diphenol is estimated colorimetrically at 510 nm. 2,3-Dihydroxybenzoate is used as a calibration standard and is similarly extracted from mock reaction mixtures that do not contain NADH or salicylate.

Results

Dye-Mediated Cytochrome c Reduction. Figure 1 shows the photosensitized reduction of cytochrome c by the naphthalimide dye lucifer yellow CH. This dye is commonly used to stain and ablate subcellular structures in laser microsurgery.[22,23] Cytochrome c reduction rates are proportional to the dye concentration and are negligible in the absence of dye, reductant, or light. Figure 1B illustrates the inhibitory effect of Cu,Zn-SOD. Cytochrome c reduction rates are diminished as the SOD concentration in the reaction mixture is increased, but equivalent concentrations of bovine serum albumin (BSA) are without effect on the reaction rate. NADH oxidation parallels cytochrome c reduction (Fig. 2A). When various dye classes are compared it is clear that thiazine dyes (i.e., azure C, methylene blue, toluidine blue O) are the most reactive sensitizers under the specified illumination conditions.

Salicylate Hydroxylation by Dye-Mediated Hydroxyl Radical Generation. Figure 2B illustrates that dye-sensitized photooxidation of NADH yields the hydroxyl radical. The relative abilities of the dyes to sensitize the hydroxylation of salicylate correspond to their relative abilities to oxidize NADH under the illumination conditions employed. Hydroxylation rates also depend on the intensity of illumination and on the dye concentration. Although many acridines (e.g., proflavin, acridine orange) are potent photosensitizing agents, the substituted acridine dye quina-

[21] B. Halliwell, *FEBS Lett.* **92**, 321 (1978).
[22] W. W. Stewart, *Cell* **14**, 741 (1978).
[23] J. P. Miller and A. I. Selverston, *Science* **206**, 702 (1979).

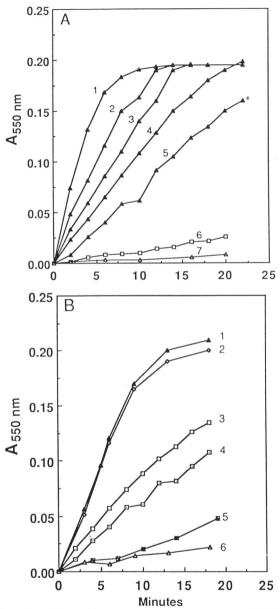

FIG. 1. Lucifer yellow CH-sensitized cytochrome c reduction by NADH and its inhibition by superoxide dismutase (SOD). (A) Cytochrome c reduction was monitored as a function of lucifer yellow CH concentration. Assays contained 3 ml of 50 mM potassium phosphate, 0.1 mM EDTA (pH 7.8 at 25°). Line 1, 47 μM; line 2, 27 μM; line 3, 13 μM; line 4, 10 μM; line 5, 6.6 μM lucifer yellow CH; line 6, NADH, no lucifer yellow CH; line 7, 47 μM lucifer yellow CH, no NADH. (B) Lucifer yellow CH, 27 μM, was incubated with BSA or increasing concentrations of SOD. Line 1, 27 μM lucifer yellow CH, light; line 2, plus 30 μg BSA; line 3, plus 2 units SOD; line 4, plus 4 units SOD; line 5, plus 24 units SOD; line 6, plus 24 units SOD, dark incubation. [From J. P. Martin and N. Logsdon, *Photochem. Photobiol.* **46,** 45 (1987).]

crine does not sensitize either NADH oxidation or salicylate hydroxylation. Hydroxyl radical production by all dyes is completely inhibited by the addition of catalase, by the addition of the hydroxyl radical scavenger thiourea, or by the substitution of desferrioxamine for EDTA in the reaction mixture.

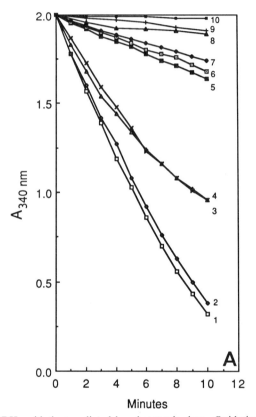

Fig. 2. (A) NADH oxidation mediated by phototoxic dyes. Oxidation of NADH was observed as a decrease in absorbance at 340 nm. Solutions contained 0.32 mM NADH in 3 ml of 50 mM potassium phosphate, 0.1 mM EDTA (pH 7.8 at 25°), and dyes at the indicated concentrations. These were illuminated with visible light, 4.6 mW/cm^2. Line 1, 7 μM azure C; line 2, 7 μM toluidine blue O; line 3, 5 μM methylene blue; line 4, 5 μM rose bengal; line 5, 5 μM acriflavin; line 6, 10 μM neutral red; line 7, 5 μM proflavin; line 8, 10 μM erythrosine; line 9, 10 μM pyronin; line 10, no dye addition. (B) Hydroxyl radical generation by dye-mediated NADH photooxidation. Solutions contained 0.15 M potassium phosphate (pH 7.6), 1.0 mM NADH, 4.0 mM sodium salicylate, 50 μM Fe–EDTA, and dyes (azure C, □; proflavin, ■; neutral red, ×; and quinacrine, ◇) at the indicated concentrations. Hydroxyl radical production was estimated as the hydroxylation of salicylate. [From J. P. Martin and P. Burch, *in* "Oxy-Radicals in Molecular Biology and Pathology" (N. Cerutti, J. McCord, and I. Fridovich, eds.), p. 394. Alan R. Liss, New York, 1988.]

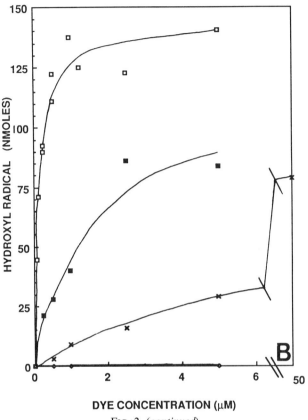

FIG. 2. (*continued*)

In Vivo Photooxidation

Procedures

Intracellular Generation of Oxygen Radicals and SOD Induction Mediated by Synthetic Dyes in Escherichia coli. Escherichia coli B (ATCC 23226) may be obtained from the American Type Culture Collection. Cultures are grown in M9 salts plus 0.4% glucose (Na_2HPO_4, 3 g; NaCl, 0.5 g; NH_4Cl, 1 g; $MgSO_4$, 495 mg; $CaCl_2$, 10.1 mg; glucose, 4.0 g per liter) or in nutrient broth (NB). Media are titrated to pH 7.2. Bacteria are grown in 50-ml cultures in 250-ml Nephelo flasks at 37° with vigorous shaking (200 rpm) in New Brunswick G-76 shakers. Growth is followed using a Klett–Summerson colorimeter.

Preinduction of *E. coli* manganese superoxide dismutase (Mn-SOD) and catalase to levels 10- to 20-fold greater than the basal levels is achieved by the addition of 100 μM manganese and 10 μM paraquat to cultures 1 hr after inoculation into M9 salts plus glucose.[24] The inoculum is 2%, and the cells are from a stationary phase overnight culture grown in M9. Growth in the presence of paraquat and manganese is allowed to continue for 3–5 hr prior to harvest.

SOD induction experiments are carried out by inoculating 50 ml of NB containing dye at the desired concentration to 2% with an overnight culture of *E. coli* B grown in NB. Following inoculation, growth is continued for 6 hr at 37°, 200 rpm in a gyrorotary shaker. Cells are grown either in complete darkness or under a 150-W GE reflector flood lamp enclosed within an aluminum foil tent. The light intensity at the surface of the broth should be 1.9–2.2 mW/cm^2.

Bacterial cells are prepared for superoxide dismutase and catalase assays by centrifugation at 7000 g at 4° for 20 min followed by washing 2 times with M9 salts (pH 7.4). The cells are resuspended in 2 ml of 50 mM potassium phosphate, 0.1 mM EDTA (pH 7.4) and are lysed by sonication in a Heat Systems-Ultrasonics Model 370 sonicator. Sonication is done in a cup horn at 100 W at 4° in six 1-min bursts. Lysates are centrifuged at 15,000 g for 20 min at 4°, and the supernatants are dialyzed for 48 hr against three 400-volume changes of 10 mM potassium phosphate, 0.1 mM EDTA (pH 7.4). Superoxide dismutase is assayed by the xanthine oxidase–cytochrome c method.[20] Xanthine oxidase is purified from unpasteurized cream.[25] Catalase is estimated according to Beers and Sizer.[26] Protein is determined by the Bradford assay[27] using bovine γ-globulin as a standard.

Estimation of Dye-Mediated Lethality in Escherichia coli B. In order to determine lethality, fresh cultures are started with a 2% inoculum of cells grown overnight in M9 salts plus glucose. Cultures are harvested from the mid-log phase of growth by centrifugation at 7000 g for 20 min at 4°. After 2 washes with M9 salts (pH 7.4), cells are resuspended in 3 ml of M9 plus glucose, 80 μg/ml chloramphenicol (pH 7.4) in presterilized cuvettes. Final cell density is 0.5–1 $\times$ 10^8 cells/ml. Hydroxyl radical scavengers are added to the incubation medium prior to cell addition. Filter-sterilized superoxide dismutase, catalase, bovine serum albumin, and

[24] S. Y. R. Pugh and I. Fridovich, *J. Bacteriol.* **162,** 196 (1985).
[25] W. O. Waud, F. O. Brady, R. D. Wiley, and K. V. Rajagopalan, *Arch. Biochem. Biophys.* **169,** 695 (1975).
[26] R. F. Beers and I. W. Sizer, *J. Biol. Chem.* **195,** 133 (1952).
[27] M. Bradford, *Anal. Biochem.* **72,** 248 (1976).

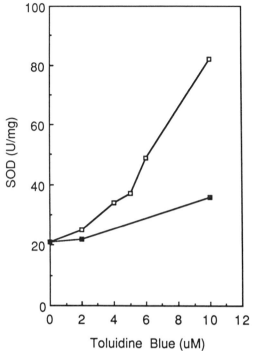

FIG. 3. Induction of SOD in *E. coli* B mediated by toluidine blue O plus light. Cells were grown in NB with toluidine blue O at the indicated concentrations. After 6 hr of growth, cells were harvested and lysed, and the extracts were assayed for SOD as described in the text. □, Cells grown under illumination; ■, cells grown in darkness. [From J. P. Martin and N. Logsdon, *Arch. Biochem. Biophys.* **256,** 39 (1978).]

dyes are added after the addition of cells. Anaerobic incubations are carried out in anaerobic quartz cuvettes.[28] The incubation solutions are scrubbed for 25 min with ultrapure nitrogen prior to sealing the cuvettes and tipping in the photosensitizer from a side arm. All treatment cuvettes are periodically inverted during incubation to assure adequate oxygenation and uniform cell suspensions. Treated cells are diluted into sterile M9 salts (pH 7.4) and incubated for 20 min at 23° to allow diffusion of the dyes from cells prior to plating. Surviving cells are plated on Luria broth (LB) medium solidified with 1.8% Bacto-agar at three dilutions and in triplicate. Plates are incubated at 37° in the dark for 16–24 hr and then counted.

[28] E. K. Hodgson, J. M. McCord, and I. Fridovich, *Anal. Biochem.* **5,** 470 (1973).

Results

Dye-Mediated SOD Induction in Escherichia coli B. Dye-mediated phototoxicity by oxygen radicals involves reduced dye intermediates. Dye reduction by intracellular substrates will be observed *in vivo* only if the dyes penetrate the bacterial cell membranes. Low molecular weight cationic dyes such as the thiazines and acridines should easily pass

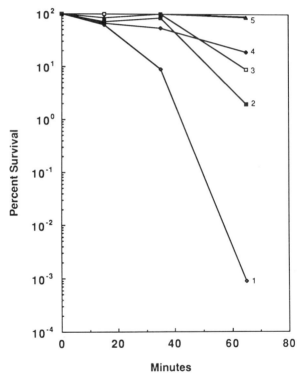

FIG. 4. Protection against azure C phototoxicity conferred by induction of Mn-SOD and catalase, and by addition of oxygen radical scavengers. Incubation conditions were as described in the text. Mn-SOD and catalase levels were elevated in *E. coli* B by addition of 10 μM paraquat and 100 μM Mn to the growth medium. Cell homogenates were assayed for SOD and catalase activities. Line 1, 2 μM azure C, 10.5 U/mg SOD, 0.5 U/mg catalase; line 2, 2 μM azure C, 161 U/mg SOD, 2.2 U/mg catalase; line 3, 2 μM azure C, 161 U/mg SOD, 2.2 U/mg catalase, 300 units exogenous SOD, 3000 units exogenous catalase; line 4, 2 μM azure C, 10.5 U/mg SOD, 0.5 U/mg catalase, 500 mM dimethyl sulfoxide, 300 units exogenous SOD, 3000 units exogenous catalase; line 5, 2 μM azure C, 161 U/mg SOD, 2.2 U/mg catalase, 500 mM dimethyl sulfoxide, 300 units exogenous SOD, 3000 units exogenous catalase. [From J. P. Martin and N. Logsdon, *J. Biol. Chem.* **262,** 7213 (1987).]

through the porin channels of the *E. coli* outer membrane[29] and are readily taken up by the cells at neutral pH.[30] On the other hand, the passage of anionic compounds through porin channels is retarded,[31] and significant accumulation of anionic xanthene dyes by *E. coli* occurs only under acidic conditions.[30]

Escherichia coli is able to induce SOD and overcome the oxidative stress presented by redox-active compounds.[32,33] Analysis of SOD levels in the cultures grown at increasing concentrations of the photosensitizing dye toluidine blue O (Fig. 3) reveals that there is a greater than 4-fold induction of superoxide dismutase at the highest concentration of this thiazine dye. Growth in the dark at high dye concentrations produces a more modest induction.

Dye-Mediated Cell Lethality. Treatment of *E. coli* B with the thiazine dye azure C is lethal. Lethality is dependent on light and oxygen. Bacteria in dark incubations are unaffected by azure C, and anaerobic incubations are only mildly toxic (data not shown). The lethality observed in glucose minimal medium containing chloramphenicol is probably enhanced by the inability of *E. coli* B to induce superoxide dismutase under these conditions. The toxicity of azure C is relieved by the hydroxyl radical scavenger dimethylsulfoxide and the H_2O_2 scavenger catalase (Fig. 4). Furthermore, high intracellular levels of superoxide dismutase and catalase also relieve azure C toxicity. Almost complete protection is obtained by addition of dimethyl sulfoxide and exogenous Cu,Zn-SOD and catalase to *E. coli* B containing elevated intracellular levels of superoxide dismutase and catalase (Fig. 4).

Acknowledgments

This work was supported by Public Health Service Research Grants AI-19695 and GM-07833, Grant C-900 from the Robert A. Welch Foundation, and a grant from the American Heart Association Texas Affiliate.

[29] H. Nikaido and E. Y. Rosenberg, *J. Bacteriol.* **153**, 241 (1983).
[30] J. S. Bellin, L. Lutwick, and B. Jonas, *Arch. Biochem. Biophys.* **132**, 157 (1969).
[31] H. Nikaido, E. Y. Rosenberg, and J. Foulds, *J. Bacteriol.* **153**, 232 (1983).
[32] H. M. Hassan and I. Fridovich, *J. Bacteriol.* **141**, 156 (1980).
[33] H. M. Hassan and I. Fridovich, *J. Biol. Chem.* **252**, 7667 (1977).

[68] Elevated Mutagenesis in Bacterial Mutants Lacking Superoxide Dismutase

By DANIELE TOUATI and SPENCER B. FARR

Introduction

Reactive oxygen species (superoxide radicals, hydrogen peroxide, hydroxyl radicals) are responsible for damage to DNA,[1–3] causing strand breaks and base alterations that can lead to cell death or mutations.[4,5] Although oxygen and a wide variety of oxidants have now been shown to be mutagenic,[6–8] it is often difficult to prove the role of an oxygen derivative in promoting or exacerbating the mutagenic activity of a substance. In particular, a role for superoxide radicals in mutagenesis has been questioned, since intracellular superoxide radical generators, such as paraquat, have not been detected as mutagens by the usual mutagenicity tests, which otherwise have detected numerous oxidants as mutagens.[9] Different assays, however, although not completely convincing, have suggested a role for superoxide radicals in mutagenesis.[10,11]

Difficulties in evaluating the mutagenic potency of reactive oxygen species arise at least from two factors: (1) Rarely is consideration given to the nature of the oxidative DNA lesion(s) (the premutagenic lesions) that can lead to mutations, nor to the repair mechanism that will specifically act on lesions to convert them to mutations. This might lead to nondetection of the oxidative mutagenic events, when certain specific mutagenesis tests are used. (2) The generation of superoxide radicals or hydrogen peroxide induces in bacterial tester strains the corresponding detoxifying

[1] H. Sies, *Angew. Chem.* **25,** 1058 (1986).

[2] J. A. Imlay and S. Linn, *Science* **240,** 1302 (1988).

[3] W. A. Pryor, *Free Radical Biol. Med.* **4,** 219 (1988).

[4] B. N. Ames, *Science* **204,** 587 (1979).

[5] B. N. Ames, *in* "Mutagens in Our Environment" (M. Sorsa and H. Vainio, eds.), p. 3. Alan R. Liss, New York, 1982.

[6] D. E. Lewin, M. Hollstein, M. F. Christman, E. A. Schwiers, and B. N. Ames, *Proc. Natl. Acad. Sci. U.S.A.* **79,** 7445 (1982).

[7] G. Storz, M. F. Christman, H. Sies, and B. Ames, *Proc. Natl. Acad. Sci. U.S.A.* **84,** 8917 (1987).

[8] J. T. Greenberg and B. Demple, *EMBO J.* **7,** 2611 (1988).

[9] D. E. Levin, M. Hollstein, M. F. Christman, and B. N. Ames, this series, Vol. 105, p. 249.

[10] H. M. Hassan and C. S. Moody, this series, Vol. 105, p. 254.

[11] C. I. Wei, K. Allen, and H. P. Misra, *J. Appl. Toxicol.* **5,** 315 (1985).

enzymes, superoxide dismutase (SOD)[12] or catalase,[13] masking a potential mutagenic effect which could appear in a different environment.

The construction of an *Escherichia coli* K12 double mutant completely lacking both Fe and Mn superoxide dismutase[14] permitted us to assess the role of superoxide radicals in the absence of their scavenger.[15] Results show that the increase in the intracellular flux of superoxide radicals leads to increased mutagenesis, demonstrating the protective role of superoxide dismutase of lowering the rate of spontaneous mutagenesis.[15] Consequently, the superoxide dismutase-deficient *E. coli* mutants constitute a unique tool to specifically assay the role of superoxide radicals in the mutagenic activity of some chemical agents or various forms of radiation.

Material and Methods

Strains. GC4468 is the parental *E. coli* K12 strain, considered as wild type. QC779 is like GC4468, but carries insertional mutations to the *sodA* (for Mn-SOD) and *sodB* (for Fe-SOD) structural genes. Markers encoding resistance factors are carried by the insertions (chloramphenicol resistance for *sodA* and kanamycin resistance for *sodB*), allowing the transfer of *sodA* and *sodB* into a different genetic background, if needed. The transfer can be achieved by two successive P1-mediated transductions selecting for chloramphenicol resistance (associated with *sodA*) and kanamycin resistance (associated with *sodB*).

Cultures are grown in LB medium (10 g Bacto-tryptone, 10 g NaCl, 5 g Bacto yeast extract per liter, pH 7.2) in a New Brunswick Scientific Co. rotary shaker bath (200 rpm) at 37°. The ratio of Erlenmeyer flask volume to culture volume is between 7 : 1 and 10 : 1; 20 μg/ml chloramphenicol and 40 μg/ml kanamycin sulfate are added to QC779 precultures.

Chemicals. Chloramphenicol, kanamycin, trimethoprim, and paraquat (methyl viologen) may be purchased from Sigma (St. Louis, MO).

Mutagenicity Test

Mutagenicity tests using the *sodA sodB* mutant require growth and plating on complete medium, as the strain is unable to grow on minimal medium.[14] The test described below measures forward mutations from Thy$^+$ to Thy$^-$ (thymine-requiring mutants). Thy$^-$ mutants are resistant to

[12] H. M. Hassan and I. Fridovich, *J. Biol. Chem.* **252**, 7667 (1977).

[13] P. C. Loewen, J. Switala, and B. L. Triggs-Raine, *Arch. Biochem. Biophys.* **243**, 144 (1985).

[14] A. Carlioz and D. Touati, *EMBO J.* **5**, 623 (1986).

[15] S. B. Farr, R. D'Ari, and D. Touati, *Proc. Natl. Acad. Sci. U.S.A.* **83**, 8268 (1986).

the drug trimethoprim and can be selected from a Thy^+ population.[16] Other mutagenicity tests performed in rich medium could alternatively be used, following a similar procedure (e.g., the test described by Lucchesi et al.[17]).

Procedure. Cell precultures of GC4468 and QC779 are grown in LB medium at 37° until they reach an optical density at 600 nm (OD_{600}) of 1; the cultures are chilled on ice, and kept in the cold (up to 48 hr) to be used for inoculate. Prewarmed LB medium is inoculated at an OD_{600} of 0.02–0.03 from the preculture. When cells reach an OD_{600} of about 0.1, the culture is subdivided (15 ml/portion) into 100-ml flasks and further incubated until an OD_{600} of 0.2 is reached. Compound to be tested is added at that time, at various concentrations. Cells are further incubated until the corresponding untreated culture reaches an OD_{600} of 1, and the cells are chilled on ice for plating.

Plating. One-tenth milliliter of bacterial dilutions (1/10, 1/40, 1/100) in cold 10 mM $MgSO_4$ is uniformly spread on dried trimethoprim agar plates (1.5% Difco agar, 18 μg/ml trimethoprim, and 50 μg/ml thymine in LB medium; trimethoprim stock solution is made 5 mg/ml in 50% ethanol and kept at −20°). Three plates are made for each dilution. Plates are incubated at 37° for 18 to 24 hr, and Thy^- colonies which appear on a more or less thin lawn of Thy^+ bacteria are counted.

Spreading should be done very carefully to obtain an uniform lawn. A glass spreader can be used. We preferentially use sterile glass beads, 4 mm diameter, usually utilized for separatory columns. Five or six beads are distributed on plates before adding a sample of cells. Plates are gently shaken until the culture sample has been completely adsorbed onto the solid medium. Plates are inverted and the beads removed from the cover before incubation. Up to six plates can be piled up and shaken together.

Alternatively, bacteria can be spread using soft agar. In this method, 0.1 ml bacterial dilution is added to 2.5 ml molten soft agar (0.6% Bacto agar in water) maintained at 48° in a heat block, gently vortexed, and

---

[16] J. Miller, "Experiments in Molecular Genetics." Cold Spring Harbor Laboratory, Cold Spring Harbor, New York 1972.

[17] The test described by Lucchesi et al. [P. Lucchesi, M. Carraway, and G. Marinus, *J. Bacteriol.* **166**, 34 (1986)] measure forward Tet^S to Tet^R mutations. Strains to be tested are transformed with a plasmid, pPY98, a derivative of pBR322 in which the *tet* gene is under the *mnt*-regulated *ant* promotor of P22. Cells containing the wild-type plasmid are Amp^R, Tet^S. Mutations in the *mnt* gene or its operator confer tetracycline resistance to the cell. We have recently developed this test in the laboratory to assay the mutagenic effect of superoxide radicals. The strains in which to use it have been constructed. Protocols and strains are available. This test is easier to handle than the test described here, but its use is somewhat restricted.

poured rapidly onto a plate. The plate is rocked gently to ensure uniform distribution. Plates are left on the bench to harden for 15 min, after which they are inverted and incubated at 37°. Colonies embedded into agar are smaller; therefore, plates should be incubated 36 hr before counting, using this procedure.

Cell Counting. Titration for viable cell counts is obtained by plating on LBT (1.5% Bacto agar and 50 μg/ml thymine in LB), in duplicate, 0.1 ml of 10^{-5} and 2×10^{-6} bacterial dilutions in cold 10 mM MgSO$_4$. Accurate titration is essential, particularly when toxic substances are used. We recommend two independent cell titrations for each experiment. An OD$_{600}$ reading of 1 corresponds to about 4×10^8 cells/ml for untreated cells.

Remarks and Cautions

A major difficulty in using mutants lacking SOD in mutagenesis testing is that such mutants grow more slowly than the corresponding wild-type cells: 45 min generation time for *sodA sodB* versus 28 min for GC4468 in LB at 37°. The mutants are also much more sensitive to compounds that generate reactive oxygen species, and excessive lethality may possibly mask mutagenesis.

For these reasons, the procedure described above does not give quantitative results for the mutation frequency of a compound. However, reliable qualitative results can be obtained, with the following precautions. (1) Thy$^-$ cells should not show any growth advantage or disadvantage over Thy$^+$ cells. This has to be checked after introducing *sodA sodB* in a new genetic background. This was verified for the strains used previously in the assay as follows: Thy$^-$/Thy$^+$ mixed cultures can be grown for more than 12 generations without measurable change in the initial ratio of Thy$^-$ to Thy$^+$. The Thy$^-$/Thy$^+$ ratio is established by replica plating on a minimal medium with or without 50 μg/ml thymine. For *sodA sodB*, minimal medium should be supplemented with 20 amino acids (0.5 mM). (2) Conditions should be chosen to minimize killing by toxic compounds. Concentrations as low as possible should be used for detection. A linear dose–response curve should be obtained. (3) Oxygen increases mutagenesis in *sodA sodB* mutants. Therefore one should be very careful about keeping uniform conditions during culture aeration in experiments, i.e., the same flask/culture volume ratio, same shaking conditions. The initial volume should be large enough so that withdrawing samples during the experiment will not seriously affect the initial flask/culture volume ratio. (4) For each strain used, at least 20 trimethoprim-resistant colonies should be checked for Thy$^-$ phenotype to ensure that Thy$^-$ mutants are counted.

(5) Every experiment should be repeated at least twice with similar results to be taken into consideration.

If enhancement of mutagenicity in the *sodA sodB* versus the wild-type strain is observed under these conditions, further tests to verify the results can be made.

Further Tests for Determining Role of Superoxide Radicals in Mutagenicity

The enhancement of spontaneous mutations in *sodA sodB* mutants is completely oxygen dependent. It is also largely dependent on a functional exonuclease III[18] (encoded by the *xthA* gene), suggesting that this enzyme is specifically involved in converting the premutagenic lesions produced (directly or more likely indirectly) by superoxide radicals to mutations. The introduction of a multicopy plasmid carrying an SOD$^+$ allele into *sodA sodB* reduces the mutation rate to that of the wild type.[15] Therefore, enhancement of mutagenicity of any compound in an *sodA sodB* mutant owing to the intracellular increased flux of superoxide radicals should be suppressed either by performing the assay under anaerobic conditions or in an *sodA sodB* strain deficient in exonuclease III activity (*sodA sodB xthA* mutant), or by introduction of an SOD$^+$ plasmid (*sodA sodB/psod$^+$* strain).

Anaerobic experiments should be performed in an anaerobic chamber. In the absence of an anaerobic chamber, however, microaerobic conditions can be obtained by purging a filled 5-ml flask with nitrogen for 10 min, capping the flask, and allowing the cells to grow at 37° without shaking. One percent glucose should be added to the medium for anaerobic growth. In any case, successive precultures should be grown anaerobically for 48 hr before beginning the experiment to ensure anaerobic conditions.

A set of isogenic strains can be used to test the effect of the *xthA* mutation. They include BW35, the parental strain (wild type); BW295, as BW35 but carrying the *xthA* mutation (the two former strains were provided by B. Weiss); QC910, as BW35 but *sodA sodB;* QC915, as BW295 but *sodA sodB*. Using these strains in the assay, one would expect to find enhanced mutagenicity in QC910 versus wild type but not (or slightly enhanced) in QC915. Both strains have approximately the same generation time.

Strains bearing plasmids expressing Fe-SOD (pHS1-8) or Mn-SOD (pDT1-5) are also available: QC1093 is QC779/pHS1-8 and QC1094 is QC779/pDT1-5. In these strains mutagenicity should be the same as the

---

[18] S. G. Rogers and B. Weiss, this series, Vol. 65, p. 201.

TABLE I

Thy$^+$ → Thy$^-$ Forward Mutation^a

Strain	Relevant genotype	Aerobic	Plus paraquat 50 μM, 200 μM	Plus plumbagin (500 μM)	Oxygenated	Anaerobic
QC4468	Wild type	20	21,19	32	31	15
QC779	sodA sodB	81	105,160	382	1321	13
QC1093	sodA sodB/sodB$^+$	21	NT	NT	NT	NT
QC1094	sodA sodB/sodA$^+$	19	NT	25	NT	NT

^{a} Data are numbers of Thy$^-$ mutants per 10^7 cells (as determined by viable counts) plated on LBT plus trimethoprim. A Forma Scientific Chamber, Model 1024, was used for anaerobic experiments. Hyperoxygenation was achieved by bubbling oxygen constantly through the culture. The efficiency of plating was 90% for oxygen-treated, wild type cells, 50% for sodA sodB; for paraquat-treated cells it was 100% for 50 μM, 70% for 200 μM in wild type, 50% for 50 μM and 35% for 200 μM in sodA sodB, for plumbagin-treated cells it was 70% for wild type, 30% for sodA sodB. NT, Not tested.

wild-type levels. Generation times of sodA sodB/psod$^+$ strains are approximately that of the wild type (see Table I).

Acknowledgments

This work was supported by a grant from the Association pour la Recherche sur le Cancer (No. 6791). We thank A. Eisenstark for careful critical reading of the manuscript.

[69] Measurement of Xanthine Oxidase in Biological Tissues

By Lance S. Terada, Jonathan A. Leff, and John E. Repine

Xanthine oxidase (EC 1.1.3.22) is found in many tissues and a variety of species ranging from bacteria to humans.[1,2] Despite the ubiquity of xanthine oxidase, its exact role in cellular physiology is unclear; however, one of its functions is thought to involve purine metabolism, and this forms the basis for most standard assays which follow the rate of formation of uric acid from xanthine. Additionally, xanthine oxidase has a

[1] D. A. Parks and D. N. Granger, *Acta Physiol. Scand. Suppl.* **548**, 87 (1986).
[2] T. A. Krenitsky, J. V. Tuttle, E. L. Cattau, and P. Wang, *Comp. Biochem. Physiol.* **49B**, 687 (1974).

potential role in disease states, such as ischemia/reperfusion and hyperoxic lung injury, by virtue of its ability to generate toxic O_2 metabolites.[3-6] Several comprehensive reviews are available which discuss the physical properties and structure of xanthine oxidase (e.g., Ref. 7).

Xanthine oxidase appears to exist in nature in two distinct functional forms, an oxidase (XO) which catalyzes reactions (1) and (2) and a dehydrogenase (XD) which catalyzes reaction (3). In addition, xanthine oxidase has a liberal substrate specificity that includes alcohols, aldehydes, and other purines, such as hypoxanthine.[8]

$$\text{Xanthine} + 2\ O_2 + H_2O \rightarrow \text{uric acid} + 2\ O_2^{\overline{\cdot}} + 2\ H^+ \qquad (1)$$
$$\text{Xanthine} + O_2 + H_2O \rightarrow \text{uric acid} + H_2O_2 \qquad (2)$$
$$\text{Xanthine} + NAD^+ + H_2O \rightarrow \text{uric acid} + NADH + H^+ \qquad (3)$$

Although unproved, it is believed that the bulk of XO is derived from posttranslational modification of XD. For instance, oxidation of XD by incubation at 37°, storage at $-20°$, anaerobiosis, or coincubation with oxidized glutathione or sulfhydryl oxidase will cause a functional conversion from XD to "reversible" XO in vitro.[9-12] This conversion can be prevented or reversed by addition of sulfhydryl-reducing agents. Likewise, partial proteolysis of XD will lead to a conversion of XD to "irreversible" XO in vitro.[9] These observations must be considered when preparing and assaying tissue for XO or XD activity.

Tissue Preparation

Most tissues can be assayed for XO and XD activity with very little preparation. However, certain tissue samples, particularly those with low levels of enzyme, must be processed as quickly as possible, as activity can decay considerably within 1 hr. In rabbit myocardium and lung, for

[3] J. M. McCord, N. Engl. J. Med. 312, 159 (1985).

[4] T. C. Rodell, J. C. Cheronis, C. L. Ohnemus, D. J. Piermattei, and J. E. Repine, J. Appl. Physiol. 63, 2159 (1987).

[5] A. Patt, A. H. Harken, L. K. Burton, T. C. Rodell, D. Piermattei, W. J. Schorr, N. B. Parker, E. M. Berger, I. R. Horesh, L. S. Terada, S. L. Linas, J. L. Cheronis, and J. E. Repine, J. Clin. Invest. 81, 1556 (1988).

[6] J. M. Brown, L. S. Terada, M. A. Grosso, G. J. Whitman, S. E. Velasco, A. Patt, A. H. Harken, and J. E. Repine, J. Clin. Invest. 81, 1297 (1988).

[7] R. C. Bray, in "The Enzymes" (P. D. Boyer, ed.), Vol. XII, p. 299. Academic Press, New York, 1975.

[8] T. A. Krenitsky, S. M. Neil, G. B. Elion, and G. H. Hitchings, Arch. Biochem. Biophys. 150, 585 (1972).

[9] F. Stirpe and E. Della Corte, J. Biol. Chem. 244, 3855 (1969).

[10] M. G. Battelli, FEBS Lett. 113, 47 (1980).

[11] E. Della Corte and F. Stirpe, Biochem. J. 126, 739 (1972).

[12] D. A. Clare, B. A. Blakistone, H. E. Swaisgood, and H. R. Horton, Arch. Biochem. Biophys. 211, 44 (1981).

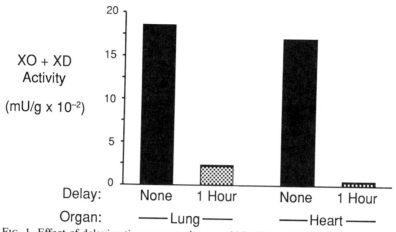

FIG. 1. Effect of delaying tissue processing on rabbit XO and XD activity. Rabbit lung and heart samples had greater activity when processed immediately (i.e., upon sacrifice) than when left at 25° for 1 hr before processing. Tissue preparation and assay were identical in all samples (see text). Values are the average of two samples.

instance, XO and XD activity becomes nearly undetectable if tissues are left for 1 hr at room temperature (Fig. 1). By comparison, in rat tissues which contain larger amounts of XO and XD, rapid decay does not occur during this delay. In addition, XD is converted irreversibly to XO with time in ischemic tissue, the rate of conversion being dependent on the specific organ and ambient temperature.[13] Therefore, tissues should be quickly frozen with liquid nitrogen-cooled clamps and then homogenized in buffer (e.g., phosphate buffer, pH 7.8) containing a protease inhibitor, such as phenylmethylsulfonyl fluoride (PMSF, 1 mM). Tissues with high levels of proteases, such as pancreas, may also require additional protease inhibitors, such as pepstatin A. In addition, EDTA (1 mM) is included to inactivate metalloproteases, inhibit sulfhydryl oxidase (both of which can cause XD to XO conversion), and chelate Cu^{2+} and other metal ions which can both inactivate the enzyme and facilitate conversion of XD to XO.[11,14] Finally, dithioerythritol (DTE, 1 mM) is included to prevent oxidative conversion of XD to XO. A 25,000 g supernatant fraction of this homogenate contains the bulk of enzyme activity.[9,15] Since "low molecular weight inhibitors" have been described with certain tissues,[16] it is best

[13] T. D. Engerson, T. G. McKelvey, D. B. Rhyne, E. B. Boggio, S. J. Snyder, and H. P. Jones, *J. Clin. Invest.* **79,** 1564 (1987).

[14] Z. W. Kaminski and M. M. Jezewska, *Biochem. J.* **181,** 177 (1979).

[15] B. Mousson, P. Desjacques, and P. Baltassat, *Enzyme* **29,** 32 (1983).

[16] B. Schoutsen, J. W. De Jong, E. Harmsen, P. P. DeTombe, and P. W. Achterberg, *Biochim. Biophys. Acta* **762,** 519 (1983).

to desalt the supernatant before assay, for instance, with a 7-ml Sephadex G-25 column.

Assay Method

Principle. The conventional method for determining XO activity involves following the rate of uric acid (UA) formation from xanthine spectrophotometrically at 295 nm in the presence or absence of NAD^+.[9] Although this reaction is physiologically relevant and the substrate (xanthine) is highly specific for XO and XD,[8] this procedure is somewhat limited by its lack of sensitivity. In addition, other compounds which absorb light at 295 nm may reduce the specificity of the assay. An improved procedure involves incubation of the enzyme with the same substrates, assaying for UA levels by HPLC. This allows for a longer incubation, hence increasing sensitivity. HPLC analysis also permits the specific determination of UA formation, as NAD^+, xanthine, EDTA, and perhaps other molecules relevant to the assay system all absorb light at 295 nm.

The reaction pH of 7.8 is higher than physiologic intracellular pH to increase the sensitivity of the assay, as the pH optima of both XO and XD are 8.0 to 9.0.[17]

Because of the prolonged incubation required for the HPLC assay, several modifications are required to prevent inactivation, inhibition, or conversion of the enzyme. First, superoxide dismutase (SOD) and catalase are necessary to prevent oxidative inactivation of the enzyme by its own products, O_2^- and H_2O_2.[18] This possibility of self-inactivation can be seen in Fig. 2. Second, lactate dehydrogenase (LDH) and pyruvate are required to prevent inhibition of XD by NADH.[14] For incubations lasting longer than 2–3 min, each of these products will significantly diminish activity. Third, because significant XD to XO conversion can occur in less than 10 min at 37°,[9] DTE and EDTA must be included to keep critical sulfhydryl groups reduced.

Procedure. Each sample is incubated with xanthine (100 μM) in the presence of ambient oxygen (XO activity) or with xanthine, NAD^+ (100 μM), pyruvate (1.75 mM) and LDH (70 U/ml) (XO + XD activity) in total reaction volumes of 1 ml, with a reaction buffer of 50 mM potassium phosphate (pH 7.8). A baseline tube is prepared (sample but no xanthine) as well as a negative control (sample and all substrates plus allopurinol, 50 μM). All tubes contain SOD (10 μg/ml) and catalase (10 μg/ml). Samples are incubated in a 37° water bath for varying periods of time, depending

[17] T. A. Krenitsky and J. V. Tuttle, *Arch. Biochem. Biophys.* **185,** 370 (1978).
[18] L. S. Terada, C. J. Beehler, A. Banerjee, J. M. Brown, M.-A. Grosso, A. H. Harken, J. M. McCord, and J. E. Repine, *J. Appl. Physiol.* **65,** 2349 (1988).

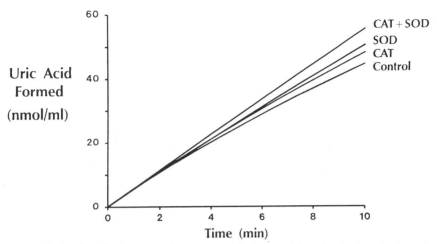

FIG. 2. Autoinactivation of XO by O_2^- and H_2O_2. The activity of purified bovine XO (6 mU/ml, shown here as rate of formation of uric acid) decreased with time in the presence of xanthine (150 μM) (control). Autoinactivation was prevented in additive fashion by catalase (CAT, 10 μg/ml) and superoxide dismutase (SOD, 10 μg/ml).

on the anticipated enzyme concentration (e.g., a sample containing 1 mU/ml should be incubated for about 30 min). Excessive incubation will result in depletion of xanthine, which is easily assessed by observing the size of the xanthine peak on the chromatogram.

Over a broad range of substrate concentrations, formation of UA is linear. The reaction is halted by addition of perchloric acid to a pH of 2.5–3.0. After removal of precipitated protein by centrifugation, UA is quantified at 290 nm by HPLC (Waters C_{18} Resolve reverse-phase column with a mobile phase of 20 mM KH_2PO_4, pH 5.0, plus 3% methanol, at 1 ml/min). The amount of UA produced is determined from the difference between UA in the XO or XO + XD tubes and the baseline tube, as compared to external standards of known UA concentration. A standard containing xanthine alone is also incubated along with the sample tubes, to preclude nonenzymatic oxidation of substrate. XD activity is calculated as the difference between activity in the XO + XD and the XO tubes.

Definition of Specific Activity. One unit of activity is defined as the production of 1 μmol of UA per minute at 37° and pH 7.8. Specific activity is expressed in terms of units per milligram protein.

Inhibitors. The oxidation of xanthine by XO or XD can be inhibited by a number of compounds that are structural purine analogs. Inhibitors include the pyrazolopyrimidines allopurinol and oxypurinol, pterin-6-

aldehyde (a photolytic breakdown contaminant of commercial folic acid preparations), methylxanthines, urea, and guanidinium ion.[1,19] In addition, cyanide, arsenite, formaldehyde, and methanol also interfere with the assay by reacting at or vicinal to the molybdenum active site.[1,20]

Considerations. Although it is possible that HPLC assay of XO could be rendered less sensitive by the presence of uricase in certain tissues, xanthine is a relatively avid competitive inhibitor of uricase (urate oxidase).[21] However, in tissues with substantial uricase activity, it may be necessary to add a uricase inhibitor, such as oxonic acid.[22] Another consideration is that reversible (but not irreversible) XO will be largely converted to XD by DTE, so this assay is relatively insensitive to reversible XO.

The considerable and increasing interest in the possibility that XO-derived O_2 metabolites may contribute to a variety of processes in normal and disease states warrants careful measurement of this enzyme in biological tissues. The techniques described herein offer one approach to this challenge.

[19] I. Fridovich, *J. Biol. Chem.* **239**, 3519 (1964).
[20] V. Massey and D. Edmondson, *J. Biol. Chem.* **245**, 6595 (1970).
[21] J. F. Van Pilsum, *J. Biol. Chem.* **204**, 613 (1953).
[22] J. M. Bagnasco, H. P. Friedl, K. S. Guice, K. T. Oldham, and G. O. Till, *FASEB J.* **3**, A628 (1989).

[70] Oxy Radicals in Disseminated Intravascular Coagulation

By Toshikazu Yoshikawa

Disseminated intravascular coagulation (DIC) is a dynamic, potentially disastrous complication arising from a variety of disease processes. It is a consequence of tissue damage-induced activation of the coagulation cascade. Excess thrombin thus generated overwhelms the normal antithrombin activities of the body and results in consumption of certain clotting elements. Fibrin deposition within the circulation causes ischemic damage, while consumption of clotting factors and activation of the fibrinolytic system lead to a multisite bleeding diathesis.[1]

Increasing experimental and clinical evidence now demonstrates that the generation of oxygen radicals can mediate, and is largely responsible

[1] E. J. W. Bowie and C. A. Owen, *Biol. Haematol.* **49**, 217 (1983).

METHODS IN ENZYMOLOGY, VOL. 186

for, several forms of tissue damage and the resulting organ dysfunction. A generation of oxygen free radicals seems to be implicated in the pathogenesis of DIC caused by endotoxin.[2-5]

Endotoxin-Induced Disseminated Intravascular Coagulation

Endotoxin (*Escherichia coli* 055: B5 lipopolysaccharide B; Difco Lab., Detroit, MI) is dissolved in pyrogen-free physiological saline before every experiment. Experimental DIC is induced by a sustained infusion of 100 mg/kg of endotoxin, diluted in 11.4 ml of saline, into the femoral vein for 4 hr using a syringe. For i.v. infusions the femoral vein of the animal is cannulated in light Nembutal (5 mg/100 g, i.p.) anesthesia.

The severity of DIC is determined with various parameters, such as fibrinogen and fibrin degradation products (FDP), fibrinogen, prothrombin time (PT), partial thromboplastin time (PTT), platelet counts, and percent glomerular fibrin deposits (%GFD). One hundred glomeruli are counted and those having fibrin thrombi are expressed as a percentage.

Four hours after the infusion of endotoxin (100 mg/kg), fibrinogen levels and platelet counts are significantly decreased, PT and PTT are prolonged, and FDP and %GFD are increased.[2,3] Serum thiobarbituric acid (TBA)-reactive substances are significantly increased 3 and 4 hr after the infusion of endotoxin.[3]

TBA-Reactive Substances. TBA reactants are determined in several tissues of rats infused with 100 mg/kg of endotoxin for 4 hr. The levels of TBA-reactive substances in serum, abdominal aortic wall, and ileum mucosa are significantly increased in rats infused for 4 hr with 100 mg/kg of endotoxin[3] (Table I).

Effects of Superoxide Dismutase and Catalase. Superoxide dismutase (SOD) or catalase significantly inhibits the increase in FDP levels and %GFD, the prolongation of PT and PTT, and the reduction of fibrinogen levels and platelet counts[2] (Table II). These findings suggest that active oxygen species, such as superoxide or hydrogen peroxide, can affect the endotoxin-induced DIC.

Effects of Vitamin E. The changes in coagulation parameters are significantly greater in vitamin E-deficient rats when compared to those in

[2] T. Yoshikawa, M. Murakami, N. Yoshida, and M. Kondo, *Thromb. Haemostasis* **50,** 869 (1983).
[3] T. Yoshikawa, M. Murakami, Y. Furukawa, H. Kato, S. Takemura, and M. Kondo, *Thromb. Haemostasis* **49,** 214 (1983).
[4] T. Yoshikawa, M. Murakami, and M. Kondo, *Toxicol. Appl. Pharmacol.* **74,** 173 (1984).
[5] T. Yoshikawa, Y. Furukawa, M. Murakami, K. Watanabe, and M. Kondo, *Thromb. Haemostasis* **48,** 235 (1982).

TABLE I

TBA-Reactive Substances in Serum and Tissues of Rats Induced by Sustained
Infusion of Endotoxin[a]

Source	Control[b]	DIC[c]	p value[d]
Serum[e]	4.0 ± 0.9 (n = 15)[f]	5.8 ± 1.0 (n = 20)	<.001
Aortic wall[g]	28.5 ± 7.9 (n = 12)	58.8 ± 13.2 (n = 12)	<.001
Liver[g]	14.7 ± 2.1 (n = 8)	14.6 ± 1.2 (n = 8)	n.s.
Kidney[g]	17.8 ± 1.9 (n = 8)	17.6 ± 0.8 (n = 8)	n.s.
Gastrointestinal mucosa[g]			
Stomach	6.3 ± 0.7 (n = 8)	6.6 ± 1.2 (n = 8)	n.s.
Jejunum	11.5 ± 2.2 (n = 12)	12.2 ± 2.1 (n = 12)	n.s.
Ileum	14.4 ± 3.1 (n = 10)	19.9 ± 4.3 (n = 10)	<.01
Colon	7.1 ± 2.1 (n = 8)	6.9 ± 1.2 (n = 8)	n.s.

[a] Results are expressed as mean value ± SD. n.s., Not significant.
[b] Control rats were infused continuously with 11.4 ml of saline for 4 hr.
[c] DIC rats were infused for 4 hr with 100 mg/kg of endotoxin diluted in 11.4 ml of saline.
[d] p values were assessed by Student's t test, as compared with control rats.
[e] Serum TBA-reactive substances were measured as nmol of malondialdehyde per ml.
[f] Number of subjects.
[g] Tissue TBA-reactive substances were measured as nmol of malondialdehyde per 100 mg wet weight.

rats supplemented with vitamin E.[4] When compared with rats given HCO-60, a solvent of α-tocopheryl acetate, the significant prevention against DIC was noted in the coagulation parameters in rats treated with 1.0 or 10.0 mg/kg/day of α-tocopheryl acetate for 4 successive days.[5]

Concluding Remarks

Hydrogen peroxide and superoxide radicals have been reported to cause aggregation of blood platelets.[6] Platelet aggregation has been shown to be inhibited by a variety of antioxidants and by scavengers of hydroxyl radicals and singlet oxygen.[7] These findings suggest a role of reactive oxygen species in platelet activation. Endotoxin stimulates the complement pathway to produce large amounts of complement fragment, such as C5a. The complement fragment stimulates polymorphonuclear leukocytes (PMNs) to produce active oxygen species. Furthermore, endotoxin has direct effects on PMNs to release oxygen free radicals that could be relevant to endotoxin-induced endothelial injury. Thromboplastin, one of

[6] O. Higashi and Y. Kikuchi, *Tohoku J. Exp. Med.* **112,** 271 (1974).
[7] M. Steiner and J. Anastasi, *J. Clin. Invest.* **57,** 732 (1976).

TABLE II

EFFECTS OF SOD AND CATALASE ON EXPERIMENTAL DIC IN RATS[a]

Parameter	FDP (μg/ml)	Fibrinogen (g/liter)	PT (sec)	PTT (sec)	Platelet count (×10^9/liter)	GFD[b] (%)	N[c]
Control[d]	56.1 ± 12.1	<0.2[e]	32.4 ± 6.4	152.1 ± 17.5	66 ± 9	87.6 ± 12.4	12
SOD[f]							
0.5 mg/kg	28.4 ± 6.4[h]	<0.2	30.9 ± 8.9	130.6 ± 18.8[i]	79 ± 18[i]	62.7 ± 9.7[h]	8
5.0	20.8 ± 7.7[h]	<0.2	28.9 ± 9.3	93.7 ± 11.5[h]	124 ± 44[h]	58.6 ± 7.2[h]	8
50.0	12.4 ± 6.2[h]	0.58 ± 0.16[h]	19.1 ± 4.2[h]	74.0 ± 12.0[h]	221 ± 71[h]	28.6 ± 5.4[h]	16
Catalase[g]							
0.01 mg/kg	29.6 ± 10.9[h]	<0.2	31.1 ± 13.4	153.8 ± 24.6	135 ± 17[h]	63.8 ± 12.2[h]	8
0.1	17.6 ± 8.1[h]	<0.2	25.0 ± 5.4[i]	81.5 ± 20.2[h]	166 ± 28[h]	60.3 ± 10.8[h]	8
1.0	13.9 ± 6.1[h]	0.48 ± 0.11[h]	19.2 ± 5.7[h]	82.1 ± 19.1[h]	227 ± 35[h]	44.9 ± 8.0[h]	19
Normal rats[g]	3.8 ± 0.8[h]	1.14 ± 0.20[h]	12.7 ± 0.4[h]	59.6 ± 4.8[h]	603 ± 81[h]	0[h]	12

[a] SOD or catalase was injected subcutaneously before the infusion of endotoxin (100 mg/kg for 4 hr).
[b] Glomerular fibrin deposits.
[c] Number of animals.
[d] SOD or catalase was injected once subcutaneously before the infusion of endotoxin (100 mg/kg for 4 hr).
[e] Fibrinogen level was less than 0.2 g/liter.
[f] SOD or catalase was dissolved in 1.0 ml of saline.
[g] Normal rats were infused with 11.4 ml of saline alone over 4 hr.
[h] $p < .001$ versus control rats by Student's t test.
[i] $p < .05$ versus control rats by Student's t test.

the triggers of DIC, is released when the endothelium is damaged. Lipid peroxidation mediated by active oxygen species is believed to be one of the important causes of cell membrane destruction and cell damage. The injury to the endothelial cells, platelets, or any kind of tissues may activate the coagulation system. These findings suggest that active oxygen species and lipid peroxidation may play important roles in the thrombus formation induced by endotoxin.

[71] Oxy Radicals in Endotoxin Shock

By TOSHIKAZU YOSHIKAWA

There is considerable indirect evidence supporting the role of oxygen radicals in circulatory shock.[1,2] Endotoxin is a toxin released from dead gram-negative bacteria, and this toxin produces circulatory depression in septicemic patients concomitantly with a variety of other cellular dysfunctions. Exposure to endotoxin can prime neutrophils to increase their release of superoxide,[3] and endotoxin challenge results in superoxide generation in mice at levels that may ultimately result in death.[4]

Endotoxin-Induced Shock

Experimental shock can be induced by a single injection of 100 mg/kg of endotoxin (*Escherichia coli* 055: B5 lipopolysaccharide B; Difco Lab., Detroit MI), diluted in 1.0 ml of physiological saline, into the femoral vein using a syringe. Immediately after the injection of endotoxin, systolic blood pressure is reduced and heart rate is increased. Serum acid phosphatase and β-glucuronidase activities are increased. These changes are greatest at 45 min after the injection of endotoxin. Serum thiobarbituric acid (TBA)-reactive substances are increased, and the serum α-to-

[1] T. Yoshikawa, M. Murakami, O. Seto, Y. Kakimi, T. Takemura, T. Tanigawa, S. Sugino, and M. Kondo, *J. Clin. Biochem. Nutr.* **1,** 165 (1986).
[2] T. Yoshikawa, O. Seto, K. Itani, Y. Kakimi, S. Sugino, and M. Kondo, *in* "Oxygen Free Radicals in Shock" (G. P. Novelli and F. Ursini, eds.), p. 125. Karger, Basel, 1986.
[3] R. B. Johnson, Jr., L. A. Guthrie, and L. McPhail, *in* "Oxy Radicals and Their Scavenger Systems" (R. A. Greenwald and G. Cohen, eds.), Vol. 2, p. 69. Elsevier, New York, 1983.
[4] B. Gray, *Toxicol. Appl. Pharmacol.* **60,** 479 (1981).

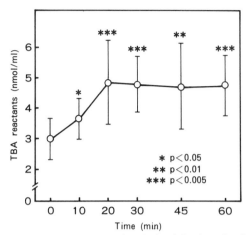

FIG. 1. Changes in serum TBA reactants after the i.v. injection of endotoxin (100 mg/kg). Results are expressed as the mean value ± S.D. for 8 rats.

copherol/cholesterol ratio decreases after the endotoxin injection[5,6] (Fig. 1).

Effect of Superoxide Dismutase and Catalase

The reduction of blood pressure by the injection of endotoxin is significantly inhibited by the administration of superoxide dismutase (SOD) or catalase (Fig. 2). The increases in serum TBA reactants and lysosomal enzymes, such as acid phosphatase, β-glucuronidase, and cathepsin B, are inhibited by the treatment with SOD.

Effect of Polymorphonuclear Leukocytes

Polymorphonuclear leukocytes (PMN)-depleted rats are used to examine the effect of PMNs on experimental shock. Anti-rat PMN antibody is obtained by immunizing rabbits with rat PMNs ($\sim 2 \times 10^7$ cells/rabbit) in Freund's complete adjuvant (Difco), using the methods of Ward and Cochrane.[7] The PMN preparation is obtained by instilling 0.12% oyster

---

[5] T. Yoshikawa, N. Yoshida, H. Miyagawa, T. Takemura, T. Tanigawa, M. Murakami, and M. Kondo, *in* "Lipid Mediators in the Immunology of Shock" (M. Paubert-Braquet, ed.), p. 87. Plenum, New York, 1987.

[6] T. Yoshikawa, T. Takemura, T. Tanigawa, H. Miyagawa, N. Yoshida, S. Sugino, and M. Kondo, *Bioelectrochem. Bioenerg.* **18**, 295 (1987).

[7] P. A. Ward and C. G. Cochrane, *J. Exp. Med.* **121**, 215 (1964).

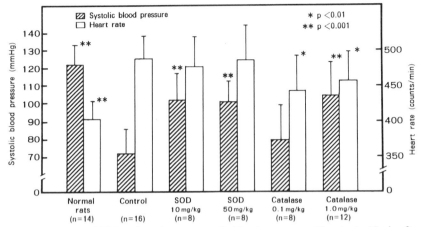

FIG. 2. Effects of SOD and catalase on systolic blood pressure and heart rate 45 min after the injection of endotoxin (100 mg/kg). Results are expressed as the mean value ± S.D. Normal rats were injected with 1.0 ml of physiological saline alone. Control rats were injected s.c. with 1.0 ml of physiological saline 12 and 1 hr before the injection of endotoxin (100 mg/kg). SOD or catalase dissolved in physiological saline was injected s.c. 12 and 1 hr before the administration of endotoxin (100 mg/kg).

glycogen type II (Sigma Chemical Co., St. Louis, MO) intraperitoneally into rats, followed by peritoneal lavage with sterile saline 6 hr later. Anti-rat PMN antibody is injected intraperitoneally into the rats to produce PMN-depleted rats. The number of PMNs is significantly reduced at 18 hr after injection of antibody. The reduction of systolic blood pressure and the increase in serum lysosomal enzymes induced by the injection with endotoxin are significantly inhibited by the pretreatment with anti-rat PMN antibody (Fig. 3).

Effect of Complement

To examine the influence of complement on endotoxin shock, the complement-depleted rats and the anticomplement agent are used. Rats are intraperitoneally injected with 200 U/kg body weight (in 0.5 ml sterile saline) of purified cobra venom factor 6 hr before the endotoxin treatment to induce complement depletion. The anticomplement agent K76-COOH (provided by Otsuka Pharmaceutical Co., Tokushima, Japan), an oxidized derivative of a sesquiterpene compound which inhibits complement activity, especially the C5 step, is injected intraperitoneally at a dose of 3.0 mg/kg, 30 min before the endotoxin treatment. When rats are depleted of complement by the injection of cobra venom factor or are pretreated

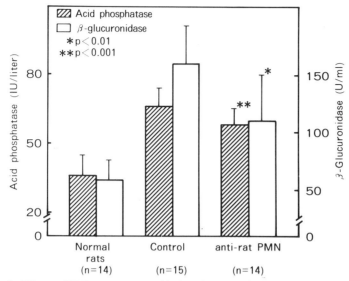

FIG. 3. Effects of PMNs on serum acid phosphatase and β-glucuronidase activities 45 min after the injection of endotoxin (100 mg/kg). Normal rats were injected with 1.0 ml of physiological saline alone. In control rats, normal rabbit serum (10 ml/kg) was injected intraperitoneally 18 hr before the injection of endotoxin (100 mg/kg). In PMN-depleted rats, anti-rat PMN antibody (10 ml/kg) was injected intraperitoneally 18 hr before the injection of endotoxin (100 mg/kg).

with K76-COOH, the reduction in systolic blood pressure and the increase in serum lysosomal enzymes are significantly prevented.

Discussion

Exposure of lysosomes to decomposing hydrogen peroxide induces enzyme release. Lysosomes possess a variety of hydrolytic enzymes, and the release of lysosomal enzymes initiates further damage to the structural and functional regions of the cell. During shock, lysosomal enzymes and serum TBA-reactive substances, which are important and damaging products of free radical lipid peroxidation, are significantly increased.

The serum α-tocopherol/cholesterol ratio is significantly decreased after endotoxin treatment. α-Tocopherol (vitamin E) is a fat-soluble molecule which is concentrated in the interior of membranes and in blood lipoproteins. The functions of vitamin E in living systems are primarily as a lipid antioxidant and free radical scavenger. The serum α-tocopherol concentration is decreased after endotoxin treatment, which suggests that it may be consumed to inhibit the free radical lipid peroxidation. The important factor influencing the serum α-tocopherol concentration appears to be the content of total lipid. To avoid the influence of serum lipid, the α-tocopherol/cholesterol ratio should be examined. Thus, the α-tocopherol/cholesterol ratio may be a good indicator of lipid peroxidation under these pathological conditions.

It is demonstrated that SOD and catalase effectively protected the aggravation of shock induced by endotoxin. This is a very important observation which suggests that superoxide and hydrogen peroxide exert certain effects on the experimental shock state.

The generation of oxygen-derived free radicals is increased in some critical states directly related to shock such as tissue hypoxia, incomplete ischemia and tissue reperfusion, activation of the arachidonic acid cascade, and complement-induced granulocyte aggregation. Human peripheral PMNs generate superoxide anion when exposed to appropriate phagocytosable and nonphagocytosable stimuli. Exposure to endotoxin can prime neutrophils for increases in superoxide and increased release of hydrolytic enzymes on subsequent contact with a variety of stimuli. Depletion of PMNs by pretreatment with anti-rat PMN significantly protected against aggravation of the state of shock, i.e., decreases in systolic blood pressure and increases in serum lysosomal enzymes level are inhibited.

Endotoxin activates the complement system. Complement components, such as C5a, stimulate PMNs to produce a large amount of oxygen

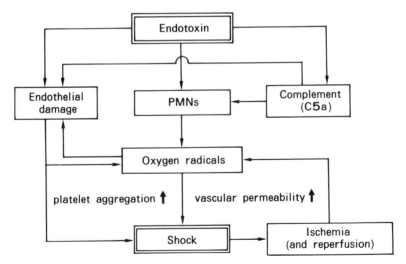

FIG. 4. Role of oxygen radicals in endotoxin-induced shock.

free radicals. Anticomplement agents inhibit the aggravation of endotoxin shock. Depletion experiments also suggest the essential role played by complement-activated PMNs in the pathogenesis of shock. These observations provide the evidence and hypothesis that complement activation products, PMNs, and oxygen radicals are important factors in the pathogenesis of endotoxin-induced shock (Fig. 4).

[72] Preparation of Human Lung Tissue from Cigarette Smokers for Analysis by Electron Spin Resonance Spectroscopy

By DANIEL F. CHURCH, THEODORE J. BURKEY, and WILLIAM A. PRYOR

Introduction

At autopsy, the lungs of cigarette smokers are observed to have varying amounts of a black pigment; this dark pigmentation has been attributed to the accumulation of particle-laden alveolar macrophages in alveoli and small bronchioles.[1] A recent report by Ohta *et al.*[2] describes an electron spin resonance (ESR) study of freeze-dried tissue from smokers and concludes that there is a paramagnetic species associated with the black pigment that is similar to the stable organic radical that we had previously reported to be present in cigarette tar itself.[3] This identification could have important implications for the toxicity of the tar deposits in smokers' lungs in view of the ability of aqueous tar extracts to produce active oxygen species, including hydrogen peroxide, superoxide, and hydroxyl radical.[3-6]

The ESR experiments of Ohta *et al.* must be interpreted with some caution, however, in view of the artifactual ESR signals that can occur in

---

[1] D. E. Niewoehner, J. Kleinerman, and D. B. Rice, *N. Engl. J. Med.* **291**, 755 (1974).

[2] Y. Ohta, H. Shiraishi, and Y. Tabata, *Arch. Environ. Health* **40**, 279 (1985).

[3] W. A. Pryor, B. J. Hales, P. I. Premovic, and D. F. Church, *Science* **220**, 425 (1983).

[4] D. F. Church and W. A. Pryor, *Environ. Health Perspect.* **64**, 111 (1985).

[5] J. P. Cosgrove, E. T. Borish, D. F. Church, and W. A. Pryor, *Biochem. Biophys. Res. Commun.* **132**, 390 (1985).

[6] T. Nakayama, D. F. Church, and W. A. Pryor, *Free Radical Biol. Med.* **7**, 9 (1989).

METHODS IN ENZYMOLOGY, VOL. 186

tissue samples prepared by drying.[7–9] We here report a methodology for the preparation of lung tissue for ESR analysis that takes into account the potential for generation of artifactual signals during sample preparation. We also report preliminary results using this methodology which suggest that the black pigmentation in the lungs of smokers is due to the accumulation of iron in a form that is spectroscopically similar to that in ferritin or hemosiderin.

Materials and Methods

Frozen whole lung specimens are obtained from a light (~60 packs/year) and from a heavy (precise rate of smoking unknown) cigarette smoker. The lungs from the light smoker show only a light mottling, while tissues from the smoker show extensive areas of dark rust-brown pigmentation surrounded by tissue that appears similar in color to that of normal lung tissue.

The lung tissue is kept frozen at $-20°$ until prepared for ESR analysis. The frozen tissue is sliced into thin sections that are then thawed. Pigmented and nonpigmented areas of the tissue are dissected into separate samples with a scalpel. Each of the samples is frozen in liquid nitrogen and powdered while frozen using a mortar and pestle. The frozen powder is then transferred in liquid nitrogen to a standard (4 mm o.d.) ESR tube and immediately placed under vacuum to remove the nitrogen and to dry the tissue. After drying, the samples are left sealed under nitrogen.

The ESR analyses are carried out using a Brucker ESR spectrometer model ER/100. Samples are run at room temperature in a TM_{110} wide-bore cavity using the following instrument settings: microwave power, 2 mW; modulation amplitude, 0.1 mT; time constant, 0.5 sec; sweep width, 500 mT; scan time, 200 sec; temperature, 130 K. Spectra are calibrated for determination of the g values against the stable free radical N,N-diphenyl-N'-picrylhydrazyl (DPPH), using $g = 2.0037$ for DPPH.[10]

Results

Spectra A–D obtained from samples of human lung tissue, prepared as described in Materials and Methods, are shown in Fig. 1. Spectrum A was

[7] R. J. Heckly, *in* "Free Radicals in Biology" (W. A. Pryor, ed.), Vol. 4, p. 136. Academic Press, New York, 1976.

[8] N. F. J. Dodd and H. M. Swartz, *Br. J. Cancer* **42**, 349 (1980).

[9] N. F. J. Dodd and H. M. Swartz, *Br. J. Cancer* **49**, 65 (1984).

[10] J. E. Wertz and J. R. Bolton, "Electron Spin Resonance," p. 465. Chapman & Hall, New York, 1986.

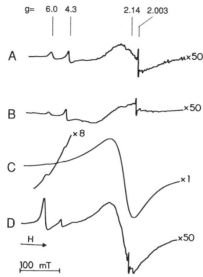

FIG. 1. ESR spectra observed in samples of lung tissue obtained from light (A, B) and a heavy (C, D) smoker. For each pair of spectra, the first corresponds to samples prepared from pigmented regions, while the second spectrum corresponds to a sample prepared from nonpigmented tissue. See text for further details.

obtained from the pigmented regions of lung tissue from a light smoker and shows relatively narrow lines at g values of 2.003, 4.3, and 6.0; the predominant signal is very broad and has a g value of 2.14. Spectrum B was obtained from the nonpigmented lung tissue of the same light smoker; this spectrum shows the same four lines as does spectrum A. The intensity of the lines is similar to the values observed in the pigmented region, although the intensity of the lines with g values of 2.14 and 2.003 is slightly less in the nonpigmented sample.

The ESR spectrum of the sample obtained from the pigmented region of the lung of the heavy smoker is shown in spectrum C. Note that the instrument gain has been reduced 50-fold for this spectrum. Under these conditions, only the species with $g = 2.14$ is observed; the increase in the signal intensity of this species is of the order of 100-fold relative to the intensity of this line in the specimen prepared from the lungs of the light smoker. Superimposed on spectrum C is the low-field portion of that spectrum obtained at a gain setting 8 times more sensitive. This superimposed spectrum shows that the signals at g values of 6.0 and 4.3 are still present, with intensities approximately equal to those observed in spectra A and B. Finally, spectrum D in Fig. 1 was obtained from the nonpigmented tissue of the same lungs that produced the sample which gave spectrum C. This spectrum shows all four of the lines seen in spectra A and B; the intensities of all the signals except the line at $g = 2.003$ appear to be about twice as great as in spectra A or B.

Discussion

The advantage of the procedure we have outlined here is that tissue from the same lung serves in effect as an internal standard. It is therefore more likely that observed changes reflect actual variation in the concentrations of a paramagnetic species.

The preliminary results reported here indicate the presence of at least four paramagnetic species in the lung tissue of cigarette smokers. Three of the ESR signals do not vary significantly, despite the marked variations of the visible degree of black pigmentation in these samples. Two of the ESR signals occur at g values of 6.0 and 4.3; the former signal has been attributed to high-spin heme iron, while the signal at $g = 4.3$ is associated with nonheme iron.[11] The third species that does not appear to vary with pigmentation occurs at $g = 2.003$. One candidate for this species is a soot- or tar-derived radical, as proposed by Ohta et al.[2] However, this identification seems less likely in view of the lack of correlation between its signal intensity and the degree of pigmentation. It is perhaps more likely that the species with a g value of 2.003 is the same organic radical species reported by a number of workers in a variety of types of dried tissues.[7]

The only paramagnetic species that varies in intensity with the degree of pigmentation of the tissue has a g value of 2.1. The most likely identity of the species giving rise to this absorption is an iron hydroxide cluster like that which has previously been reported as giving the ESR signals observed in ferritin and hemosiderin.[11-14]

Thus, our results are fully consistent with the several reports in the literature that the lungs, and especially the alveolar macrophages, of cigarette smokers contain high concentrations of iron.[15-18] Our ESR results

[11] F. X. R. Van Leeuwen, F. J. M. Zuyderhoudt, B. F. Van Gelder, and J. Van Gool, Biochim. Bioophys. Acta 500, 304 (1977).
[12] M. P. Weir, J. F. Gibson, and T. J. Peters, Biochem. Soc. Trans. 12, 316 (1984).
[13] J. F. Boas and J. Troup, Biochim. Biophys. Acta 229, 68 (1971).
[14] M. P. Weir, T. J. Peters, and J. F. Gibson, Biochim. Biophys. Acta 828, 298 (1985).
[15] S. E. McGowan, J. J. Murray, and M. G. Parrish, J. Lab. Clin. Med. 108, 587 (1986).
[16] S. G. Quan and D. W. Golde, Proc. Soc. Exp. Biol. Med. 167, 175 (1981).
[17] W. S. Lynn, J. A. Kylstra, S. C. Sahu, J. Trainer, J. Shelburne, P. C. Pratt, W. F. Gutknecht, R. Shaw, and P. Ingram, Chest 72, 483 (1977).
[18] S. E. McGowan and S. A. Henley, J. Lab. Clin. Med. 111, 611 (1988).

further suggest that this iron either is in ferritin or hemosiderin or is very similar to the form of iron occurring in these storage proteins.

The accumulation of iron, especially if it is ferritin- or hemosiderin-like, in the lungs of smokers may have an important impact on the level of oxidative stress to which the lungs of smokers are exposed.[19] The iron in ferritin is bound in the +3 oxidation state and is unreactive; however, this iron can be easily released from the ferritin core as ferrous iron by the action of reducing agents and, thereby, be made available to initiate oxidative damage such as lipid peroxidation.[20-22] We speculate that this may contribute to an "oxidative cascade" of damage, since damage to lipid membranes is likely to lead to microhemorrhaging, which has been suggested as a source for the iron that accumulates in alveolar macrophages.[16]

Thus, our data and that in the literature lead us to suggest the following hypothesis. Cigarette smoke produces high concentrations of oxidizing radicals in a microdomain in the lung, leading to microhemorrhaging (which has been suggested to be a source for the iron that accumulates in alveolar macrophages[16]). This hemorrhaging leads to increased local concentrations of iron, which catalyzes further oxidative damage, leading to further accumulation of iron. Thus, the observed pigment consists of oxidative debris that contains iron in a +3 redox state in an environment similar to that of the iron in ferritin and hemosiderin.

Acknowledgments

This work was supported by a grant from the National Institutes of Health and by a contract from the National Foundation of Cancer Research, both to W.A.P.

[19] E. R. Pacht, H. Kaseki, J. R. Mohammed, D. G. Cornwell, and W. B. Davis, *J. Clin. Invest.* **77**, 789 (1986).
[20] M. J. O'Connell, R. Ward, H. Baum, and T. J. Peters, *Life Chem. Rep.* **3**, 131 (1985).
[21] C. E. Thomas, L. A. Morehouse, and S. D. Aust, *J. Biol. Chem.* **260**, 3275 (1985).
[22] S. D. Aust, *in* "Oxygen Radicals and Tissue Injury, Proceedings of the Brook Lodge Symposium" (B. Halliwell, ed.), p. 27. Fed. Am. Soc. Exp. Biol., Bethesda, Maryland, 1988.

[73] Assay for Free Radical Reductase Activity in Biological Tissue by Electron Spin Resonance Spectroscopy

By Jürgen Fuchs, Rolf J. Mehlhorn, and Lester Packer

Introduction

A variety of antioxidant defense systems have evolved in aerobic organisms to protect against oxidative stress.[1] Reactive oxygen species include free radicals and electronically excited molecules like singlet oxygen. Both enzymatic and nonenzymatic mechanisms serve to intercept oxidants and repair damage. Free radicals can arbitrarily be divided into three classes: highly reactive radicals, e.g., hydroxyl, moderately reactive radicals, e.g., superoxide, and persistent radicals, e.g., ascorbyl. Kinetic considerations suggest that highly reactive radicals cannot be intercepted selectively by antioxidative systems. Intermediate and persistent free radicals, however, could be detoxified by specific protective mechanisms. Despite their low reactivity, weakly reactive radicals may pose a considerable threat to biomolecules, since they can react more selectively with targets at critical cell locations. If these radicals can be produced at high rates, e.g., like superoxide, then specific detoxifying systems like superoxide dismutase can confer effective protection. It is also plausible that persistent radicals, like those that arise from ascorbic acid, α-tocopherol,[2] and, perhaps, β-carotene during reactions with reactive free radicals, may be reduced by specific enzymes that regenerate the original antioxidant.

We have developed a simple assay for the free radical-reducing activity of tissue that is based on the one-electron reduction of nitroxide radicals. The nitroxides are relatively persistent radicals[3] whose concentration can easily be quantitated by electron spin resonance (ESR) spectroscopy. This assay had been applied to epidermis and skin homogenates, since the skin is a readily accessible tissue and a target organ of oxidative stress[4] which is subject to a variety of pathologic conditions that have been associated with free radical processes.

[1] H. Sies (ed.), "Oxidative Stress." Academic Press, New York, 1985.
[2] P. McCay, *Fed. Proc., Fed. Am. Soc. Exp. Biol.* **45,** 451 (1986).
[3] E. G. Rozentsev (ed.), "Free Nitroxyl Radicals." Plenum, New York, 1970.
[4] Y. Niwa, T. Kanoh, T. Sakane, H. Soh, S. Kawai, and Y. Miyachi, *Life Sci.* **40,** 912 (1987).

METHODS IN ENZYMOLOGY, VOL. 186

Methods

Skin Preparation from Hairless Mice

Female hairless mice from Jackson Laboratory (Bar Harbor, ME), 10–12 weeks old, are sacrificed by CO_2 inhalation followed by decapitation. This procedure avoids anesthetics which interfere with free radical metabolism. After a longitudinal cut through the back skin, the whole skin is manually separated from the body and washed blood-free with an isotonic buffer (130 mM sodium chloride, 5 mM glucose, 1 mM disodium EDTA, 10 mM sodium phosphate, pH 7.0). Adherent subcutis and fascia are removed by gently scraping the dermal side with tissue paper. The epidermal/dermal skin sheet is kept on glass wool pads (Type A/E, 47 mm, Millipore Corp., Bedford, MA). The pads are well saturated with isotonic buffer and ket in petri dishes on ice.

Skin Homogenate. Skin homogenates are prepared by cutting skin from hairless mice (2 g) into small pieces that are homogenized with an Ultra-Turrax tissue blender (Tekmar Company, Cincinnati, OH) in an isotonic buffer (5–10 ml), as described for skin preparation, and subsequently with a Teflon pestle in a tight-fitting glass vessel. During homogenization the sample is continuously flushed with argon gas to prevent oxidation of reactive tissue components. The homogenate can be frozen and stored on liquid nitrogen. Soluble enzymes, like catalase, superoxide dismutase, glutathione peroxidase, and glutathione reductase can be assayed from the supernatant of the homogenate, after separating membranes and other tissue fragments by centrifugation in an Eppendorf centrifuge for 25 min.

Electron Paramagnetic Resonance Spectroscopy

In the examples shown, first-derivative EPR spectra (100 kHz modulation frequency) are recorded at ambient temperature on a Bruker ER 200 D-SCR spectrometer (X-band). Modulation amplitude is 1.25 G, scan range 100 G, and microwave power 10 mW. Nitroxide concentrations are determined from the peak-to-peak height of the low-field line of the first derivative spectrum and calibrated according to a nitroxide standard. Nitroxide reduction in skin homogenate is measured in 75-μl quartz glass capillaries. Skin biopsies are cut with a punch biopsy needle, care being taken to cut biopsies of similar sizes (surface area). Depending on the special characteristics of the EPR instrument and cavity used, skin biopsies 4 mm in diameter are the maximum size that allow for adequate cavity tuning. However, with some dehydration, experiments are feasible with 6 mm biopsies. Reduction of nitroxides in epidermis/dermis biopsies is measured by placing a 4-mm biopsy in the lower end of a quartz glass tube (2.5 mm i.d.) and centering the tissue sample in the middle of the EPR cavity. Repetitive measurements of different samples result in a tuning error of less than 5% of the total signal observed.

Chemicals. The uncharged nitroxide Tempol (2,2,6,6-tetramethyl-1-piperidinoxy-4-ol) and the cationic nitroxides CAT-1 (2,2,6,6-tetramethyl-1-piperidinoxy-4-trimethylammonium bromide) and CAT-4 (2,2,6,6-tetramethyl-1-piperidinoxy-4-*N*,*N*-dimethyl-*N*-butylammonium bromide) are obtained from Molecular Probes (Junction City, OR). *N*-Ethylmaleimide (NEM), ascorbate oxidase (AO), and oxidized and reduced nicotinamide adenine dinucleotide diphosphate (NADP, NADPH) are purchased from Sigma (St. Louis, MO). 1,3-Bis[2-chloroethyl]-1-nitrosourea (BCNU) was a gift from Dr. M. Smith, Berkeley, CA.

Analysis of Nitroxide Reduction in Skin Homogenates

Reduction of piperidinoxy and dihydropyrrolinoxy spin probes can be used to discriminate between major reducing agents in biological tissue.[5] Ascorbic acid, which is usually the most important reductant for nitroxides in biological tissues, is a potent reductant for piperidinoxyl radicals (six-membered rings), but has little effect on dihydropyrrolidinoxyl radicals (five-membered rings). In tissue homogenates, the influence of ascorbic acid in free radical reduction can also be studied by using AO and measuring the residual nitroxide-reducing activity. The involvement of thiol groups in free radical reduction is assessed by treatment of the homogenates with NEM or zinc. The residual nitroxide-reducing activity in skin homogenates after treatment with AO is NEM sensitive (Fig. 1). NEM-sensitive nitroxide reduction can be stimulated by NADPH and NADP, the latter being abolished by BCNU, a glutathione reductase inhibitor (Fig. 2). The involvement of thiol groups in nonascorbic acid-mediated nitroxide scavenging is also demonstrated by a stimulation of reducing activity (data not shown) by mammalian thioredoxin,[6] a protein that works as a disulfide reductant for other enzymes and low molecular weight compounds. Discrimination between enzymatic and nonenzymatic reduction can be achieved by heat treatment, exercising due care to consider the possible activation of transition metal-catalyzed reactions, which could occur when enzyme denaturation releases iron or copper.

---

[5] S. Belkin, R. J. Mehlhorn, K. Hideg, O. Hankovsky, and L. Packer, *Arch. Biochem. Biophys.* **256,** 232 (1987).

[6] A. Holmgren, *Annu. Rev. Biochem.* **54,** 237 (1985).

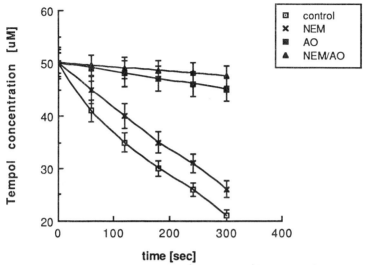

FIG. 1. Reduction of the membrane-permeable spin probe Tempol in skin homogenates and its inhibition by N-ethylmaleimide (NEM) and ascorbate oxidase (AO). Experimental conditions: reaction volume, 100 μl; 22°; 200 mg skin wet weight/ml homogenization medium; Tempol concentration, 100 μM; NEM final concentration, 25 mM; AO, 25 U/ml. Experiments were done in triplicate; mean values and standard deviation are shown.

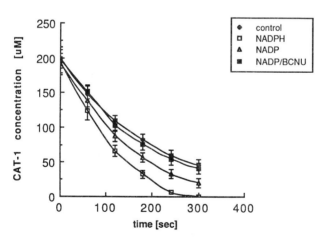

FIG. 2. CAT-1 reduction in skin homogenates and its stimulation by NADP(H). Experimental conditions: reaction volume, 100 μl; 400 mg skin wet weight/ml homogenization medium; CAT-1 concentration, 200 μM (control) (CAT-1 is a cationic membrane-permeable spin probe which partitions in the aqueous phase); NADPH final concentration, 10 mM; NADP, 10 mM; BCNU, 100 μM plus 10 mM NADP. Experiments were repeated at least four times; mean values and standard deviation are shown.

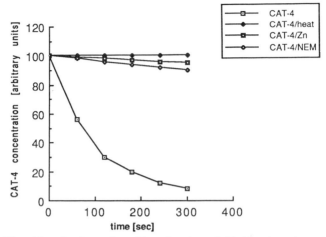

FIG. 3. Nitroxide reduction at the epidermal surface of skin biopsies. The cationic membrane-binding nitroxide CAT-4, which partitions between the membrane surface and aqueous compartment, was applied on the epidermal side of a 4-mm skin biopsy (CAT-4, control). Biopsies were immersed in 70° isotonic saline for 5 min and then exposed to CAT-4 (CAT-4/heat), 10 mM zinc sulfate (CAT-4/Zn), and 50 mM NEM (CAT-4/NEM) prior to nitroxide treatment. Representative experiments are shown.

Free Radical Reduction at Epidermal Surface of Skin Biopsies

The epidermal surface of 4-mm skin biopsies is treated with 2 μl of a 10 mM solution of nitroxide, incubated for 5 min, and washed for 15 sec with isotonic saline. Subsequently, the sample is immediately transferred to the EPR cavity and the nitroxide spectrum recorded. Nitroxide uptake into the epidermal membranes varies considerably and is roughly a function of the octanol–water partition coefficient. The epidermis can be pretreated with skin-permeable thiol group inhibitors, like NEM, treated with zinc sulfate, or heated briefly (70°, 5 min). A heat-, zinc-, and NEM-sensitive free radical reducing component was identified in the epidermis with the membrane-binding cationic nitroxide CAT-4 (Fig. 3). The results confirm the finding of a "free radical reductase" at the surface of the epidermis[7]; the nature of the reductant, however, remains to be characterized.[8–10]

[7] K. U. Schallreuter, M. R. Pittelkow, and J. M. Wood, *J. Invest. Dermatol.* **87,** 728 (1986).
[8] J. Fuchs, *J. Invest. Dermatol.* **91,** 92 (1988).
[9] K. U. Schallreuter, M. P. Pittelkow, and J. M. Wood, *J. Invest. Dermatol.* **91,** 92 (1988).
[10] J. Fuchs, R. J. Mehlhorn, and L. Packer, *J. Invest. Dermatol.* submitted (1988).

[74] Biochemical Pharmacology of Inflammatory Liver Injury in Mice

By ALBRECHT WENDEL

Introduction

The study of liver disease in suitable animal models seems a desirable means of elucidating pathophysiological mechanisms with the aim of developing appropriate drugs. Previous experimental approaches include studies on the metabolism of hepatotoxic compounds in subcellular preparations, in isolated hepatocyte suspensions, and in cultured hepatocytes. Very recently, the improvement of cell separation techniques has permitted the study of the complex morphology of the liver *in vitro* by coculturing parenchymal and nonparenchymal cells. Although these sophisticated techniques allow some mechanistic insight into special cases of xenobiotic-induced hepatotoxicity, none of the models reflects the clinically relevant situation in humans, i.e., the initiation of primary endogenous lesions followed by a progressive development of severe organ impairment. It is the purpose of this chapter to introduce an inflammatory liver disease model which includes a functioning circulation and a counterregulation of the peripheral system and which is intimately connected to leukocyte function. In spite of the complexity of the whole animal system, available basic biochemical knowledge will allow the deduction of detailed conclusions about the sequence of pathophysiological events and the site of action of pharmacological intervention.

Selection of Animals and Pilot Experiments

Reproducible reactions of animals to inflammatory stimuli such as bacterial endotoxins require that any previous exposure of the animals to bacterial contact is minimized, that bacterial infection is strictly excluded, and that any solvent injected is pyrogen-free. The first two points require animal house facilities that meet good laboratory practice rules and provide for the control of the health state of the animal by bacteriological and virological supervision. The third is most conveniently solved if commercially available solutions for clinical application in humans are used. Even if the animals are specified pathogen free, the sensitivity of mice of either sex to endotoxins varies over a wide range and depends not only on the strain, the diet (fish-derived components are to be avoided), and the age of

METHODS IN ENZYMOLOGY, VOL. 186

the animals, but also on the time of day and the season of the year. Because these variables and their mutual interference are essentially unpredictable, the sensitivity of the dose dependence of the animals against endotoxin has to be investigated from time to time in pilot experiments. The course of such experiments is described in detail in the following section. In our experience, the following strains of mice are suitable for this liver model: NMRI, BALB/c, C3H/HeN (not C3H/HeJ), C57BL/10, and DBA/2NCr.

Experimental Protocol

Preparation of Solutions

Commercially available endotoxin from *Salmonella abortus equi* (or other endotoxin-producing species) is dissolved as a 1 mg/ml solution in pyrogen-free phosphate-buffered saline (PBS), pH 7.3. In order to achieve a homogeneous solution, it is absolutely essential to add 20 μl of a 1% (w/v) purest grade solution of hydroxylamine in PBS to 1 ml of the endotoxin solution and to sonicate this preparation for 1 min at room temperature in a immersion bath sonifier. The stock solution can be kept frozen for several weeks and has to be sonified each time after thawing before repeated use. Immediately before the experiment, the stock solution is diluted into a galactosamine solution containing 105 mg/ml D-galactosamine hydrochloride in PBS.

Leukotrienes of commercial grade are checked for purity on a HPLC C_{18} reversed-phase column run isocratically with acetonitrile–methanol–water–acetic acid (33.6:5.4:61:1) at a flow rate of 1 ml/min and detected at 280 nm. Ten micrograms of leukotriene in 10 μl PBS–ethanol (80:20, v/v) are transferred into 2 ml PBS and immediately used for intravenous injection. Recombinant murine tumor necrosis factor α (TNF-α) available in sterile PBS containing 0.1% bovine serum albumin as a carrier protein is appropriately diluted with PBS to about 1.5 μg/ml and used as such.

Administration of Agents

In the routine experiments, the combination of 700 mg/kg galactosamine and 5 to 50 μg/kg endotoxin are intraperitoneally injected into mice in a total volume of 200 μl per animal. Mice 10 to 12 weeks old are used. Alternatively, the animals receive 200 μl galactosamine solution intraperitoneally and 1 hr later either leukotrienes or TNF-α intravenously. A maximum volume of 400 μl is slowly injected via a 0.45 × 12 mm sterile needle within 1 min. Before the injection, the animal is allowed to crawl into a dark tube and is kept immobilized there during the injection.

Pretreatment of Animals with Pharmacologically Active Compounds

Routinely, compounds are given 30 to 60 min before galactosamine/ endotoxin by intraperitoneal administration in PBS or soybean oil or in suspensions in either solvent. If administered more than 60 min prior to the galactosamine/endotoxin challenge, care must be taken to the induction of endotoxin resistance[1] by omnipresent endotoxin in chemicals and solvents. Agents with a short half-life *in vivo* are repeatedly administered every 30 min (e.g., diethylcarbamazine, FPL 55712). Unsuitable solvents in this model are dimethyl sulfoxide or ethanol. For intravenous administration care needs to be taken that isosmolar conditions are given. The pharmacologically most interesting route, i.e., oral administration, is experimentally carried out by intragastric intubation of up to 0.5 ml solution in PBS. Water-insoluble compounds are extensively ground, suspended in 1% methylcellulose (Tylose), and instilled as suspension.

Quantification of Degree of Liver Injury

The animals are sacrificed by cervical dislocation and the abdominal cavity is immediately opened. Thirty microliters of heparin ($\sim$130 units) in PBS is drawn into a 1-ml syringe equipped with a 0.7×30 mm needle. At least 50 μl of blood is withdrawn by heart puncture or by puncturing the inferior caval vein, transferred to an Eppendorf cup, and immediately centrifuged for 4 min at 5000 g. Serum can be frozen for several days or used immediately for determination of serum alanine aminotransferase, serum aspartate aminotransferase, and sorbitol dehydrogenase (L-iditol dehydrogenase) activities as an index for hepatic injury. The enzyme determinations are carried out as described in Ref. 2, preferably on an autoanalyzer. The experimental design of the three variants, i.e., models for endotoxin-, leukotriene D_4-, or TNF-α-induced liver injury, is schematically summarized in Fig. 1.

Experimental Findings

Mice treated according to the protocol in Fig. 1 develop a fulminant hepatitis within 8 hr which is usually lethal after 15 hr. The time course of the release of serum aminotransferases as well as the liver-specific enzyme sorbitol dehydrogenase is illustrated in Fig. 2 for galactosamine/ endotoxin. Essentially similar time courses of hepatic injury are obtained

[1] M. A. Freudenberg and C. Galanos, *Infect. Immun.* **56,** 1352 (1988).
[2] H. U. Bergmeyer, "Methods of Enzymatic Analysis," Vol. 3, 3rd Ed. Verlag Chemie, Weinheim, 1983.

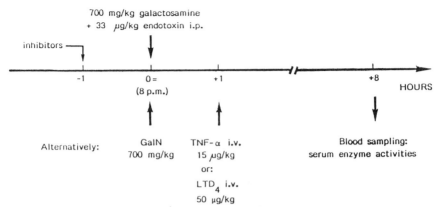

FIG. 1. Experimental design of endotoxin (lipopolysaccharide)-, leukotriene D₄-, or tumor necrosis factor α-induced hepatitis in mice.

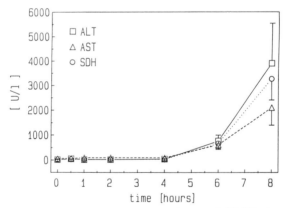

FIG. 2. Time course of serum enzyme activities of 35-g male NMRI mice treated with 700 mg/kg D-galactosamine hydrochloride plus 33 μg/kg *Salmonella abortus equi* endotoxin (lipopolysaccharide). ALT, Alanine aminotransferase (EC 2.6.1.2); AST, aspartate aminotransferase (EC 2.6.1.1); SDH, sorbitol dehydrogenase (EC 1.1.1.14, L-iditol dehydrogenase).

when the galactosamine-sensitized animals are alternatively treated with 50 μg/kg leukotriene D₄ or with more than 15 μg/kg human or murine recombinant TNF-α instead of endotoxin. The severe inflammatory hepatic injury induced by any of these agents is morphologically indistinguishable. Leukocyte infiltration is observed after 1 hr.[3] The final stage is

[3] H. J. Schlayer, H. Laaf, T. Peters, M. Woort-Mencker, H. C. Estler, U. Karck, H. E. Schaefer, and K. Decker, *J. Hepatol.* **7**, 239 (1988).

TABLE I

PHARMACOLOGICAL INTERVENTION IN ENDOTOXIN, LEUKOTRIENE D_4, OR TUMOR NECROSIS FACTOR α-INDUCED HEPATITIS IN GALACTOSAMINE-SENSITIZED MICE[a]

Compound	Type of action	Dose (mg/kg)	Route	GalN/LPS	GalN/LTD$_4$	GalN/TNF-α
Dexamethasone	Antiphospholipase, many other effects	0.2	i.p.	FP	PP	PP
Aspirin	Cyclooxygenase inhibitor	220	i.p.	0	0	0
Ebselen	Lipoxygenase inhibitor	50	p.o.	PF	0	0
FPL 55712	Leukotriene D$_4$ antagonist	13 × 10	i.p.	FP	FP	0
Acivicine	Inhibitor of LTD$_4$ synthesis	50	i.v.	FP	0	0
Iloprost	Vasodilator	0.01	i.p.	FP	FP	0
Pentoxifylline	Improves blood rheology	50	p.o.	FP	FP	0
Allopurinol	Xanthine oxidase inhibitor	2 × 100	i.p.	FP	FP	0
Superoxide dismutase	O$_2^-$ scavenger	10	i.v.	FP	FP	0
Uridine	Counteracts GalN	800	i.p.	FP	FP	FP
Calmidazolium	Ca^{2+}/calmodulin antagonist	6	i.p.	FP	FP	FP

[a] GalN, Galactosamine; LPS, lipopolysaccharide (endotoxin); FP, protection; PP, partial protection; 0, no significant protection.

characterized by complete destruction of the centrolobular and midzonal hepatocytes.

Results obtained in previous studies[4-6] on the influence of pharmacological intervention in the three liver disease models are given in Table I. Combined with the available enzymological evidence, these data allow the deduction of the sequence of pathogenic mediators in this model of inflammatory liver injury. In this particular instance, it is concluded that endotoxin administration leads to the synthesis and/or release of leukotriene D_4.[5] This leukocyte-derived mediator is known to be a strong vasoconstrictor and eventually leads to ischemic microcirculatory episodes.[4] After termination of the action of this short-lived agonist, reoxygenation of the affected areas may lead to superoxide-mediated triggering of TNF production, which is considered a major terminal event.[6]

Comments

Because the pathophysiology of the animal model described here affects a central organ and its microcirculation, the interindividual variability of the results is affected by many endogenous factors. Any kind of stress will result in a hormonal response which affects the reaction of an individual; anesthetics, high molecular weight solubilizers, plasma expanders, or detergents profoundly interfere with the response of the animal. Also, the quality of commercially available galactosamine differs so that each new lot needs to be checked for suitability. It should be also noted that the data in Table I are not "all-or-none" results but rather are based on dose–response curves for the specific agent used. The serum enzyme activities released on liver injury in a group of animals do not follow a normal distribution. Therefore, statistical evaluation of the data using the paired rank order test according to Dunett–Wilcoxon is preferable. Finally, neither in hepatocyte cultures nor in the isolated perfused liver is endotoxin-, LTD$_4$-, or TNF-α-induced cytotoxicity observed beyond the basal cell damage induced by galactosamine. This negative finding stresses the importance of extrahepatic cells, e.g., leukocytes, in the pathogenesis of liver damage.

Acknowledgments

This work was supported by the Deutsche Forschungsgemeinschaft (Grant We 686/11-1).

[4] A. Wendel, G. Tiegs, and C. Werner, *Biochem. Pharmacol.* **36,** 2637 (1987).
[5] G. Tiegs and A. Wendel, *Biochem. Pharmacol.* **37,** 2569 (1988).
[6] G. Tiegs, M. Wolter, and A. Wendel, *Biochem. Pharmacol.* **38,** 627 (1989).

[75] Measurement of Oxidant Stress *in Vivo*

By HELEN HUGHES, HARTMUT JAESCHKE, and JERRY R. MITCHELL

Oxidized glutathione (GSSG) is a physiological indicator of the activity of the intracellular defense system against reactive oxygen, and it can be used to monitor oxidant stress *in vivo*. GSSG can be measured in plasma, bile, lymph, and tissue. The general characteristics of GSSG formation and cellular release are discussed in accompanying chapters.[1-3] As in the isolated perfused liver, secretion of biliary GSSG is the most sensitive index of hepatocellular GSSG formation *in vivo*. Increased levels of GSSG in tissue can also be used qualitatively to detect enhanced detoxification of reactive oxygen in different organs, although concentrations are usually lower than in bile, lymph, or plasma because of the active excretion of GSSG from cells.

Measurement of GSSG in plasma is convenient but requires caution to avoid assay artifacts or data misinterpretation. When released into the blood, glutathione (GSH) and GSSG have a half-life of less than 2 min.[4,5] Both compounds are degraded mainly in the kidney.[6] Thus, actual plasma GSSG concentrations are the result of cellular release and degradation and depend in part on the location of the blood vessel from which the blood is sampled. Since various tissues such as the heart,[7,8] lung,[9] and liver[10,11] and erythrocytes[12] can detoxify reactive oxygen and release

---

[1] H. Jaeschke and J. R. Mitchell, this volume [83].

[2] R. F. Burk and K. E. Hill, this volume [84].

[3] A. Wendel, this volume [74].

[4] A. Wendel and P. Cykrit, *FEBS Lett.* **120**, 209 (1980).

[5] A. Wendel and H. Jaeschke, *Biochem. Pharmacol.* **31**, 3607 (1982).

[6] N. P. Curthoys, *Miner. Electrolyte Metab.* **9**, 236 (1983).

[7] T. Ishikawa and H. Sies, *J. Biol. Chem.* **259**, 3838 (1984).

[8] Y. Xia, K. E. Hill, and R. F. Burk, *J. Nutr.* **115**, 733 (1985).

[9] S. G. Jenkinson, T. H. Spence, R. A. Lawrence, K. E. Hill, C. P. Duncan, and K. H. Johnson, *J. Appl. Physiol.* **62**, 55 (1987).

[10] B. H. Lauterburg, C. V. Smith, H. Hughes, and J. R. Mitchell, *J. Clin. Invest.* **73**, 124 (1984).

[11] J. D. Adams, B. H. Lauterburg, and J. R. Mitchell, *J. Pharmacol. Exp. Ther.* **227**, 749 (1983).

[12] E. Beutler, *in* "Functions of Glutathione: Biochemical, Physiological, Toxicological, and Clinical Aspects" (A. Larsson, S. Orrenius, J. A. Holmgren, and B. Mannervik, eds.), p. 65. Raven, New York, 1983.

GSSG into the plasma, the source of GSSG is not always clear. Determination of GSSG in the arterial versus venous blood supply for an organ and the organ GSSG concentration can be helpful in identifying individual tissue contributions to plasma GSSG.[11]

Analysis of Plasma GSH and GSSG

GSSG can be measured in plasma as an index of drug-induced oxidant stress[11,13,14] by a modification of the enzymatic cycling method of Tietze.[15] A technical problem in determination of plasma GSSG is the 100-fold greater concentration of GSH in red blood cells than in plasma. Additionally, GSH concentration in plasma samples decrease rapidly[16] because of autoxidation to GSSG and formation of mixed disulfides with plasma proteins. A useful method that avoids this artifact is measurement of total GSH, i.e., GSH + GSSG, by addition of 200 μl of fresh blood to 200 μl of 5,5'-dithiobis(2-nitrobenzoic acid) (DTNB, 10 mM) in phosphate buffer (100 mM), pH 7.5, containing disodium ethylenediaminetetraacetic acid (EDTA, 17.5 mM) as an anticoagulant. Plasma is separated by centrifugation, 2000 g for 6 min. An aliquot (50 μl) of DTNB plasma solution is added to a cuvette containing GSSG reductase (0.5 U, Sigma type III) and 5 mM EDTA in potassium phosphate buffer (100 mM), pH 7.5. The solution is allowed to equilibrate for 1 min and the reaction initiated by the addition of NADPH (220 nmol). The change in absorbance at 412 nm is monitored continuously in a dual-beam spectrophotometer. The reference cuvette contains equal concentrations of DTNB, NADPH, and enzyme.

For the assay of GSSG, 200 μl of fresh whole blood is added to 200 μl of N-ethylmaleimide (NEM, 10 mM) in phosphate buffer (100 mM), pH 6.5, containing EDTA (17.5 mM). Plasma is again separated by centrifugation. An aliquot (250 μl) is passed through a prewashed C_{18} Sep-Pak (Waters Associates, Milford, MA). The cartridge is washed with 1 ml buffer, the eluates combined, and an aliquot (750 μl) added to a cuvette containing DTNB (250 nmol) and assayed as described above for total GSH.

The added NEM reacts in whole blood with unoxidized GSH to form a stable adduct and thus minimizes artifactual autoxidation of plasma GSH to GSSG. NEM and the GSH–NEM adduct are retained on the C_{18} Sep-Pak allowing GSSG to be assayed directly in plasma. Since this method relies on the enzyme GSSG reductase, compounds that inhibit the en-

[13] J. D. Adams, B. H. Lauterburg, and J. R. Mitchell, *Res. Commun. Chem. Pathol. Pharmacol.* **46,** 401 (1984).

[14] B. H. Lauterburg and J. R. Mitchell, *J. Hepatol.* **4,** 206 (1987).

[15] F. Tietze, *Anal. Biochem.* **27,** 502 (1969).

[16] M. A. Anderson and A. Meister, *J. Biol. Chem.* **255,** 9530 (1980).

zyme will interfere with the assay. To guard against this source of error, known amounts of GSH or GSSG are added to the cuvette after the initial reading and used for internal standardization. Since erythrocytes contain very high levels of GSH it is essential that the extent of hemolysis also be assessed; this can be done using the benzidine assay described by Tietz.[17]

Analysis of Lipid Peroxidation Products Formed in Vivo during Oxidant Stress

The analysis of lipid peroxidation products is made difficult by the multitude of products formed during autoxidation of polyunsaturated fatty acids and esters. The determination of some of these products is described in earlier chapters. Here we describe the methods we have developed to analyze the hydroxy fatty acids formed in lipids during the peroxidative process.

We have used two methods for analyzing lipid peroxidation products in animal tissue. The first[18] involves extraction of total lipids and isolation of the phosphatidylcholine fraction by thin-layer chromatography. Hydroperoxy derivatives are reduced and the phospholipids hydrolyzed with phospholipase A_2. The released hydroxy fatty acids are methylated and analyzed by normal-phase high-performance liquid chromatography (HPLC), monitoring absorbance at 235 nm. This method proved successful for the analysis of lipid peroxidation products generated in rat and mouse liver following carbon tetrachloride administration. Although carbon tetrachloride treatment initiates massive lipid peroxidation, no oxidant stress (GSSG formation) is observed. In contrast, we have found that the massive oxidant stress generated by drugs such as diquat produces only a low level of lipid peroxidation *in vivo*.[19,20] We therefore developed a more specific and sensitive method for the analysis of lipid hydroxy acids based on gas chromatography–mass spectrometry (GC–MS).[21]

Quantitation of Lipid Hydroxy Acids by GC–MS

Following sacrifice of the animals, tissue is removed, rinsed in ice-cold saline, weighed, and lipids extracted with 20 volumes of chloroform–

[17] N. W. Tietz, *in* "Fundamentals of Clinical Chemistry," pp. 437, 876. Saunders, Philadelphia, Pennsylvania, 1976.

[18] H. Hughes, C. V. Smith, E. C. Horning, and J. R. Mitchell, *Anal. Biochem.* **130**, 431 (1983).

[19] C. V. Smith, H. Hughes, B. H. Lauterburg, and J. R. Mitchell, *J. Pharmacol. Exp. Ther.* **235**, 172 (1985).

[20] C. V. Smith, *Mol. Pharmacol.* **32**, 417 (1987).

[21] H. Hughes, C. V. Smith, J. O. Tsokos-Kuhn, and J. R. Mitchell, *Anal. Biochem.* **152**, 107 (1986).

methanol (2 : 1).[22] Alternatively, subcellular fractions such as microsomes may be prepared and their lipids subsequently extracted in the same manner. Methyl 15-hydroxyarachidate (0.3 nmol) is added to the chloroform–methanol extract, which is filtered and washed with saline (0.9%, 0.2 volume). After standing overnight at 4°, the lower organic phase is evaporated to dryness under nitrogen. Methanol (0.2 ml) is added to the lipid residue followed by triphenylphosphine (0.2 ml, 1 mg/ml in ether), and the samples are kept on ice for 1 hr. The solvent is then evaporated, benzene (0.2 ml) and sodium methoxide (0.5 ml, 0.5 M in methanol) are added, and the resulting solution is left at room temperature. After 1 hr the pH is adjusted to 3 by the addition of HCl (0.27 M) containing sodium chloride (5%), and the transmethylated lipid sample is extracted with ether (10 ml). The ether extract is dried over magnesium sulfate, filtered, and evaporated to dryness. Since free fatty acids are not methylated with sodium methoxide, the sample is then treated with diazomethane. The resulting fatty acid methyl esters are redissolved in hexane, and purified by HPLC. A Porasil column (30 cm × 3.9 mm, Waters Associates) with mobile phase hexane–2-propanol–acetic acid (995 : 4 : 1) at 2.4 ml/min is used. Arachidonic acid and other nonoxidized fatty acid methyl esters elute in the first fraction (1–6 min). A second fraction, eluting between 6 and 13 min and containing methyl 15-hydroxyarachidate and 11-, 12-, and 15-hydroxyeicosatetraenoic acid (HETE) methyl esters, is collected for GC–MS analysis.

Prior to GC–MS analysis the HPLC eluates are evaporated to dryness and derivatized with bistrimethylsilyl(trifluoro)acetamide (BSTFA)–pyridine (5 : 1, 15 μl) for 1 hr at 70°. GC–MS analysis is performed either with a packed column (3 ft × 2 mm i.d., 3% SP 2100, Supelco, Inc., Bellefonte, PA) or a capillary column (10 m × 0.25 mm, DB-1, J&W Scientific, Rancho Cordova, CA) operated between 205 and 250°. Selected ions are monitored at m/z 225, 295, and 335 for the TMS ethers of 11-, 12-, and 15-HETE methyl esters, respectively, and m/z 343 for methyl 15-OTMS-arachidate, the internal standard. The small contribution to the m/z 225 ion current from the 15-hydroxy isomer is computed from the m/z 335 ion peak area after analysis of the standard 15-HETE methyl ester TMS ether and subtracted from the m/z 225 peak area to give the value for the 11-hydroxy isomer. Quantitation from biological samples is accomplished using standard curves prepared from mixtures of 11-, 12-, and 15-HETE methyl esters (0, 15, 30, 60, and 150 pmol) added to chloroform–methanol with internal standard and carried through extraction.

Our previous studies with carbon tetrachloride-treated animals indi-

[22] J. Folch, M. Lees, and S. G. Sloane, *J. Biol. Chem.* **226,** 497 (1967).

TABLE I

EFFECT OF DIQUAT ON LIPID HYDROXY ACID CONTENT OF RAT LIVER[a]

Diquat (mmol/kg)	Time (hr)	n	11-OH 20:4	12-OH 20:4 (nmol/g of tissue)	15-OH 20:4
0	3	5	0.28 ± 0.02	0.22 ± 0.02	0.41 ± 0.05
0.1	3	3	0.37 ± 0.08	0.34 ± 0.05	0.65 ± 0.06[b]
0.1	6	3	0.43 ± 0.04	0.43 ± 0.04[b]	0.75 ± 0.08[b]

[a] Adapted from Ref. 19.
[b] Different from control by Mann–Whitney rank sum test $P < .05$; mean ± S.E.

cated that hydroxyeicosatetraenoates in phospholipids were major products of lipid peroxidation. The GC–MS assay was developed to analyze three of these hydroxylated fatty acid esters, 11-, 12-, and 15-hydroxyeicosatetraenoate, as representatives of this class of products. Table I illustrates the results obtained with this method for reactive oxygen-induced lipid peroxidation following diquat administration to Fischer 344 rats. Although GSSG excretion into bile was increased severalfold by the diquat administration, indicative of massive oxidant stress, only small increases in lipid peroxidation products were observed.[19] To emphasize this point the liver content of these hydroxy acids 30 min following carbon tetrachloride administration to the mouse was 7.9 ± 2.16, 7.93 ± 2.53, and 11.80 ± 5.27 nmol/g for 11-, 12-, and 15-HETEs respectively, i.e., 15 to 20 times the values observed following diquat administration.[18,21]

These findings indicate obvious differences in the extent of lipid peroxidation occurring *in vivo* as a result of organic radical generation, e.g., the trichloromethyl radical formed during the metabolism of carbon tetrachloride, and that occurring as a result of massive oxygen radical generation by hepatotoxic doses of diquat. Similar results[20] (and conclusions) are obtained when the extent of lipid peroxidation is determined *in vivo* by quantitation of expired alkanes derived from β scission of the lipid hydroperoxides rather than by the methods described here for measurement of lipid hydroperoxides as their reduced hydroxy fatty acids.

[76] Radioprotection by Ascorbic Acid, Desferal, and Mercaptoethylamine

By AJIT SINGH,* HARWANT SINGH,* and JAMES S. HENDERSON*

Introduction

It is well known that exposure to high-energy radiation can cause damage to biological systems.[1] This effect has been exploited in treating cancer, since the cancer cells can be killed by exposure to high-energy radiation.[2,3] However, in the process, the adjacent healthy tissue is also damaged. The protection of healthy tissue during radiotherapy for cancer has been one of the strong motivations for continuing research on exogenous radioprotectors.[1,4–7] The main aim of this chapter is to focus on the methodology of determining radioprotection by chemical agents using the mouse as a model animal system. From this point of view, (1) some of our recent results on radioprotection of mice[8] are shown, (2) the mechanisms of radiation damage and their multiplicity are briefly discussed, and (3) the rationale for choosing three chemical agents [ascorbic acid (AA), Desferal (DF), and mercaptoethylamine (MEA)] for radioprotection studies is presented.

Understanding of radiobiological effects in humans requires studies involving model chemical systems, microorganisms, tissue culture, and animals. Experimentation with animals ultimately provides assurance that conclusions from experimental results obtained in simpler systems have a biomedical reality. Laboratory mice, available in many well-de-

* Atomic Energy of Canada Limited.

[1] A. P. Casarett, "Radiation Biology." Prentice-Hall, Engelwood Cliffs, New Jersey, 1968.

[2] K. N. Prasad, "Human Radiation Biology." Harper and Row, Hagerstown, Maryland, 1974.

[3] F. A. Mettler, Jr., and R. D. Moseley, Jr., "Medical Effects of Ionizing Radiation." Grune & Stratton, Orlando, Florida, 1985.

[4] A. Singh and H. Singh, *Prog. Biophys. Mol. Biol.* **39**, 69 (1982).

[5] G. E. Adams, *in* "Advances in Radiation Chemistry" (M. Burton and J. L. Magee, eds.), Vol. 3, p. 125. Wiley, New York, 1972.

[6] D. L. Klayman and E. S. Copeland, *in* "Drug Design" (E. J. Ariens, ed.), Vol. 6, p. 82. Academic Press, New York, 1975.

[7] A. M. Michelson and K. Puget, *in* "Oxygen Radicals in Chemistry and Biology" (W. Bors, M. Saran, and D. Tait, eds.), p. 831. de Gruyter, Berlin, 1984.

[8] A. Singh, H. Singh, J. S. Henderson, R. D. Migliore, J. Rousseau, and J. E. Van Lier, *in* "Oxygen Radicals in Biology and Medicine" (M. G. Simic, K. A. Taylor, J. F. Ward, and C. von Sonntag, eds.), p. 587. Plenum, New York, 1989.

METHODS IN ENZYMOLOGY, VOL. 186

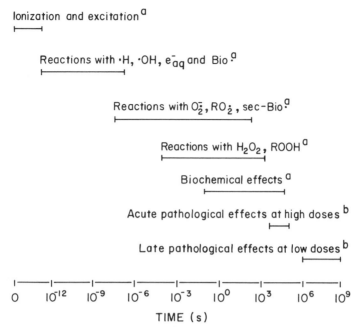

FIG. 1. Time scale of damaging events from the time of an energy deposition event, leading to biological effects of irradiation; a, for details, see A. Singh and H. Singh, *Prog. Biophys. Mol. Biol.* **39,** 69 (1982); b, for details, see F. A. Mettler, Jr., and R. D. Moseley, Jr., "Medical Effects of Ionizing Radiation." Grune & Stratton, Orlando, Florida, 1985.

fined inbred strains, are large enough for easy handling but not too large for economic husbandry and whole-body irradiations in large numbers. However, the experiments should be carefully designed to keep the wastage of animals to a minimum and relevant animal care legislation[9] should be followed.

Mechanisms of Damage and Protection

Radiation damage in biological systems is initiated by the primary ionic, excited, and free radical species formed during the energy deposition events.[4,5] Several mechanisms contribute to the subsequent radiation damage over a wide time scale, as shown in Fig. 1. For example, the primary free radicals [hydrogen atom ($\cdot$ H); hydroxyl radical ($\cdot$ OH); hydrated electron (e_{aq}^{-}), and bioradicals (Bio $\cdot$)] induce damaging reactions

[9] Foundation of Biomedical Research (U.S.A), *Clin. Res. Pract. Drug Regul. Aff.* **5,** 265 (1987).

within a few microseconds of their formation. These reactions are, in turn, followed by reactions induced by the secondary free radicals [superoxide anion ($O_2^{\cdot-}$), peroxy radicals ($RO_2 \cdot$), and secondary bioradicals (sec-Bio $\cdot$)] and the metastable species (hydrogen peroxide and organic hydroperoxides), which may continue over minutes and days, respectively. Biochemically altered macromolecules (DNA, RNA, enzymes) may ultimately engender gross pathological lesions (e.g., neoplasia) over a much longer period.[2–5,10]

Most of the chemical and biochemical components of biological systems are sensitive to and reactive with the primary species and secondary free radicals formed on exposure to high-energy radiation.[11–14] Because of the biochemical complexity of biological systems, a very large number of secondary free radical species form. Many of the reactions between the free radical species and the biochemicals present may protect biological integrity. For example, the reactions of free radicals with sulfhydryl compounds like cysteine[4,5] and the reaction of superoxide anions with superoxide dismutase[7] are protective. However, damage also results from the reactions of the primary and secondary species, through a variety of mechanisms.

Results of many radioprotection studies, two of which are mentioned here, also point to the conclusion that radiobiological damage occurs through more than one mechanism. Storer[15] investigated radioprotection of A/J and C57BL/6J mice, by S-2-aminoethylisothiouronium bromide hydrobromide, mercaptoethylamine (MEA), p-aminopropiophenone (PAPP), and 5-hydroxytryptamine creatinine sulfate. The results showed that (1) although the protection of the bone marrow by each of these chemical agents was similar, the protection of the life span was dependent on the sex and strain of the mice; (2) the protection against lethality offered by the different agents was different for mice of the same strain and sex; and (3) the protection by each agent varied in a different manner with increasing radiation dose. We have also reported similar dose-dependent effects.[8] In the case of BALB/c mice, DF acts as a radiosensitizer at

[10] L. Packer, (ed.), this series, Vol. 105.

[11] B. H. J. Bielski, D. E. Cabelli, R. L. Arudi, and A. B. Ross, J. Phys. Chem. Ref. Data **14**, 1041 (1985).

[12] G. V. Buxton, C. L. Greenstock, W. P. Helman, and A. B. Ross, J. Phys. Chem. Ref. Data **17**, 513 (1988).

[13] P. Neta and A. B. Ross, in "Chemical Kinetics of Small Organic Radicals, Vol. 4: Reactions in Special Systems" (Z. B. Alfassi, ed.), p. 187. CRC Press, Boca Raton, Florida, 1988.

[14] P. Neta, R. E. Huie, and A. B. Ross, J. Phys. Chem. Ref. Data **17**, 1007 (1988).

[15] J. B. Storer, Radiat. Res. **47**, 537 (1971).

700 rads (7 Gy), but becomes a radioprotector at 1000 and 1300 rads (10 and 13 Gy).[8]

Biological systems have intrinsic protective components, for example, the sulfhydryl compounds like cysteine, and the DNA repair enzymes.[4] Exogenous administration of the same, or similar, compounds that provide inherent protection would be expected to provide additional radioprotection. Many studies have confirmed this expectation. Information is now available on the radioprotection of microorganisms and animal systems by many chemicals and biochemicals.[4-7]

Choice of Radioprotectors

Since a variety of mechanisms are responsible for radiobiological damage, mixtures of protectors, each aimed against a different damaging mechanism, should be more effective than the same protectors used individually. Results with mixtures of radioprotectors have been reported[16] which justify this expectation. The concept of the additive effect of a mixture of protectors guided our further work on radioprotection of BALB/c mice[8] using the following three different types of radioprotectors.

Ascorbic Acid

The oxygen effect in radiation biology is well known.[5] Since oxygen enhances radiation-induced biological damage, antioxidants should be radioprotectors. Indeed, many of the radioprotectors are antioxidants.[6] Ascorbic acid is an antioxidant,[17] an essential vitamin for humans,[18] and a prime factor in biological defense.[19] Though the mechanisms of action of AA are not fully understood, it is known to reduce ferric ions to ferrous ions and to react with free radicals.[17,20,21] It is important to understand its role, if any, in modifying radiobiological damage.

Ascorbic acid has been used in radiation protection studies in cells, mice, and rats, with varying results. Radioprotection of *E. coli,* Chinese

[16] J. R. Maisin, G. Mattelin, and M. Lambiet-Collier, *Radiat. Res.* **71,** 119 (1977).

[17] T. L. Dormandy, *Lancet* **1,** 647 (1978).

[18] J. Drummond, *Biochem. J.* **13,** 77 (1919).

[19] A. Szent-Gyorgyi, "The Living State." Academic Press, New York, 1972.

[20] B. H. J. Bielski, *in* "Ascorbic Acid: Chemistry, Metabolism and Uses" (P. A. Seib and B. M. Tolbert, eds.), p. 81. Advances in Chemistry Series 200, American Chemical Society, Washington, D.C., 1982.

[21] R. L. Willson, *in* "Radioprotectors and Anticarcinogens" (O. F. Nygaard and M. G. Simic, eds.), p. 1. Academic Press, New York, 1983.

hamster ovary cells, and rats by AA has been reported.[22–24] In the case of mice (NMRI males, Swiss, and CF_1), however, it was found to be a radiosensitizer rather than a radioprotector.[25,26]

Desferal

Part of the radiobiological damage by radiation-induced hydrogen peroxide and organic hydroperoxides could be due to the Fenton reaction,

$$Fe_2^+ + H_2O_2 \rightarrow \cdot OH + OH^- + Fe_3^+ \tag{1}$$

and its equivalent in the case of the hydroperoxides,

$$Fe_2^+ + ROOH \rightarrow RO \cdot + OH^- + Fe_3^+ \tag{2}$$

brought about by free transition metal ions, particularly iron, in the biological systems.[4,27,28] Desferal forms a very stable water-soluble complex with free ferric ions. The chelated ferric ions are then removed from the body by renal excretion.[29,30] Based on work on inhibition of lipid peroxidation by DF, Gutteridge et al.[28] suggested that DF may act as a radioprotector by chelating iron.

Mercaptoethylamine

Thiols are very reactive toward free radicals[4,5,12,13]:

$$RSH + R' \cdot \rightarrow RS \cdot + R'H \tag{3}$$

The thiyl radicals formed are relatively nonreactive and they mainly produce nontoxic disulfides:

$$RS \cdot + RS \cdot \rightarrow RSSR \tag{4}$$

[22] M. Naslund, L. Ehrenberg, and G. Djalali-Behzad, Int. J. Radiat. Biol. Relat. Stud. Phys. Chem. Med. **30,** 95 (1976).

[23] M. K. O'Conner, J. F. Malone, M. Moriarty, and S. Mulgrew, Br. J. Radiol. **50,** 587 (1977).

[24] L. Ala-Ketola, R. Varis, and K. Kiviniitty, Strahlentherapie **148,** 643 (1974).

[25] J. Forsberg, M. Harms-Ringdahl, and L. Ehrenberg, Int. J. Radiat. Biol. Relat. Stud. Phys. Chem. Med. **34,** 245 (1978).

[26] H. H. Tewfik, F. A. Tewfik, and E. F. Riley, in "Vitamin C: New Clinical Applications in Immunology, Lipid Metabolism and Cancer" (A. Hanck, ed.) p. 265. Huber, Bern, 1982.

[27] A. Singh, in "CRC Handbook of Free Radicals and Antioxidants in Biomedicine" (J. Miquel, A. T. Quintanilha, and H. Weber, eds.), Vol. 1, p. 17. CRC Press, Boca Raton, Florida, 1989.

[28] J. M. C. Gutteridge, R. Richmond, and B. Halliwell, Biochem. J. **184,** 469 (1979).

[29] Desferal, Desferrioxamine, Product Information, Ciba-Giegy Limited, Basel, Switzerland, 1988.

[30] M. Aksoy and G. F. B. Birdwood (eds.), "Hypertransfusion and Iron Chelation in Thalassaemia." Huber, Bern, 1985.

These reactions probably constitute the main mode of the radioprotective mechanism of thiols, though other mechanisms have also been suggested.[4] Mercaptoethylamine (cysteamine, 2-aminoethylthiol), which is a thiol, is a widely used and very effective radioprotector,[6,16,25] in mice, against radiation-induced lethality.

Experimental

Radioprotection Studies with Mice

Mice used in various radiation protection studies[8,15,25,26,31] include NMRI, Swiss, CF_1, A/J, BALB/c, BALB/c$^+$, C57B1, HA(ICR)f, and C57BL/6J strains. Results in experiments testing such complexities as the effects of irradiation on mice vary with sex, age, strain, and husbandry. For example, the LD_{50} (radiation dose which causes death of 50% of the irradiated mice[1] in 30 days) values for CF_1, Swiss, NMRI, and HA(ICR)f mice have been reported as 429, 525, 610, and 679 rads, respectively. Thus, the CF_1 mice appear to be the most radiosensitive of the four strains, and the HA(ICR)f the most radioresistant. The differences in radioprotection of different sexes are exemplified in the study of A/J and C57BL/6J mice by Storer,[15] in which it was found that while MEA protected females better than males in both strains, PAPP protected A/J females more than the males, but C57BL/6J males better than the females. Colonies of the same strain may also differ in their radioresponse significantly, on account of environmental factors, such as the bacterial flora resident in the gut. It is also important to use experimental mice within as narrow a range of age as possible, since the radiation effects can be age-dependent.[1] The details about the BALB/c mice used by us[8] have been published.[32]

Radioprotection studies entail all the complexities of radiation effects plus those from the putative protectors, which are not necessarily simple. Careful planning and rigorously disciplined execution of the experiments are therefore needed to obtain information when dealing with complex biological effects and their modification by exogenous radioprotectors. In the case of inbred mice, it is usual to allot, from animals of the same age and sex, a subset of 10 for each treatment. To obtain reliable results on the effect of each protective protocol, all factors except the controlled variation in the concentrations, and possibly modes of administration, of the radioprotector(s) should be kept constant.

[31] W. F. Ward, A. Shih-Hoellwarth, and P. M. Johnson, *Radiat. Res.* **81,** 131 (1980).
[32] J. S. Henderson and J. L. Weeks, *Ind. Med.* **42,** 10 (1973).

It is possible to obtain mice for such experiments from established suppliers. It is advisable to quarantine them upon receipt to allow adaptation to their new environment. Indeed for the month required for an $LD_{50/30}$ assay it would be best to keep them quarantined to minimize deaths from unfamiliar exposures. In such cases, the recommendations of the suppliers or the procedures given in standard handbooks[33] should be followed.

The irradiated mice should be returned to their own cages and held there in equable conditions with their accustomed diet and water *ad libitum*. Dead mice should be removed and counted during twice daily inspections of the cages. Where possible, necropsies should be done to monitor any deaths that may not be radiation-induced.

Administration of Radioprotectors

Ascorbic acid taken orally is satisfactorily absorbed from the gut but can be injected as necessary. In mice about 0.27 mg/day of AA is endogenously produced.[26] The pharmacokinetics of AA has been discussed by Basu and Schorah.[34] We injected 6 mg of AA per mouse (in 1 ml saline solution alone or in combination with the other radioprotectors) approximately 15 min before irradiation.[8] In comparison, Tewfik *et al.*[26] administered AA in drinking water before and after irradiation, and Forsberg *et al.*[25] injected AA 1 hr before irradiation. We preferred injections since administration of radioprotectors in drinking water or food can change the drinking and eating patterns of treated mice, compared to the controls. Our choice of the injection time of approximately 15 min before irradiation allowed us to administer all three radioprotectors, alone or in mixture, in single injections.

Desferal is poorly absorbed when taken orally. When injected, however, it is quickly distributed through all body fluids and then gradually lost (over several hours[29]) to renal glomerular filtration and tubular secretion. Though DF has not been used as a radioprotector before, Ward *et al.*[31] administered another chelator, penicillamine, intraperitoneally, 15 min before irradiation of mice. The dose used in our work[8] was 1.5 mg/ mouse.

Mercaptoethylamine is distributed quickly, when injected, and is found throughout all organs (except the testes) within 15 min. It is rapidly

[33] C. S. F. Williams, "Practical Guide to Laboratory Animals." Mosby, St. Louis, Missouri, 1976.
[34] T. K. Basu and C. J. Schorah, "Vitamin C in Health and Disease." AVI, Westport, Connecticut, 1982.

degraded in the tissues and rapidly excreted in the urine; hence, most of it is lost within 2 hr. It is toxic to humans in the doses required for radioprotection, but its beneficial effects in animals and in cultures of animal cells are well documented.[6,16,22,25] Our injection of 6 mg/mouse is consistent with the previous usage of this radioprotector.[15,16]

In our work,[8] each mouse receives the desired dose in 1 ml of fluid, injected into the peritoneal sac through the belly wall, which had been shaved and cleansed with alcohol. The control group of mice is injected in the same way and at the same time as the treated groups but with the saline carrier only.

The time of administration of the radioprotectors can be optimized by changing the time interval between the administration of the protector and irradiation, keeping all other factors unchanged. Timing must allow a balance between the attainment of uniform distribution and elimination through normal metabolic processes. Where information is not available from previous studies in the literature, simple tests on the time profile of the level of the radioprotector in the body fluids and organs of mice, subsequent to its administration, may allow the optimum time for the administration of the protector to be determined with economy of time and mice.

The concentration of a radioprotector can be optimized by changing its concentration and monitoring the effect, keeping all other factors unchanged. Again, it may be useful to establish first how the *in vivo* concentration increases with increased amounts of the radioprotector being administered. Many radioprotectors may be toxic to mice at very high concentrations, so their levels should be kept low enough to avoid toxicity but high enough to get the maximum possible benefit. Once the concentrations of various radioprotectors have been individually optimized, their optimization in mixtures can be accomplished by variations in the levels of each one.

Irradiation

Mice can be irradiated with X-rays, γ-rays (^{60}Co), electron beams from an electron accelerator, or heavier energetic particles such as protons, neutrons, helium ions, and other heavier ions.[35,36] The use of the first three is most prevalent. It is important that there be a zone for uniform irradiation of 10 mice at a time, housed in a suitable, ventilated holder.

[35] I. G. Draganic and Z. D. Draganic, "The Radiation Chemistry of Water." Academic Press, New York, 1971.
[36] W. M. Saunders, *Radiat. Phys. Chem.* **24**, 365 (1984).

They should not be unduly warm; however, the holder should not be large enough to permit them too much movement. The mouse holder should preferably be of organic material, e.g., polycarbamate, polycarbonate, or polystyrene, so that the absorption of radiation by it is similar to that of the mice. A metal holder may introduce complex dose contours and shadows owing to its higher density. The uniformity of irradiation and the dose rate in the mouse holder, at the precise site for irradiation, can be determined with many dosimeter systems,[37] the Fricke dosimeter being the most common. We used Theraton F (Manitoba Cancer Foundation) for the γ-irradiation of mice. Its output is frequently checked and calibrated for therapeutic purposes. The ^{60}Co source of the Theratron emits γ-rays as a collimated beam for precise periods by the carefully timed opening of its port. The rate at which the power of the γ-emission decays is well known (~1% per month). From the dosimetry data, the known decay rate for ^{60}Co, and the exposure time, the dose delivered to the mice is calculated.[8]

For irradiation the mice are loaded in groups of 10 into a 10-celled polycarbamate holder, at room temperature. Each cell of the holder is 56 × 28 × 43 mm and is ventilated with three 7-mm holes in the lid and two in the floor. The cells are arranged in two parallel banks of five so that all are encompassed within a cross section of 17.5 × 16.0 cm. The mice receive the required dose, uniformly, within 5 min. Since all the irradiations are done in one evening, the dose rate is constant, for all practical purposes. The entire holder is collared by a 31-mm thickness of solid polycarbamate. This keeps scatter from the incident radiation uniform throughout the holder. Four identical holders are used, one at a time, to enable a large number of mice to be irradiated in a continuous session.

Generation of Survival Curves

The curves obtained by plotting the animal survival data at a given dose against time are called survival curves.[1] Our work is done in the range of 700 to 1300 rads (7 to 13 Gy).[8] For the purpose of discussion here, our detailed data at 1000 rads is given in Fig. 2. These data show that (1) both DF and MEA protect but AA sensitizes mice under the conditions used; (2) the protection offered by a mixture of MEA and AA is less than that offered by MEA alone; and (3) the protection offered by MEA and DF is much greater than that offered by MEA alone. In the case of results such as shown for saline and AA in Fig. 2, it is advisable to obtain a

[37] N. W. Holm and R. G. Berry, "Manual on Radiation Dosimetry." Dekker, New York, 1970.

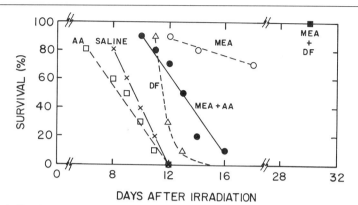

FIG. 2. Survival of BALB/c mice treated with exogenous radioprotectors, on irradiation to 1000 rads (10 Gy): AA, ascorbic acid; DF, Desferal; MEA, mercaptoethylamine. ×, ○, ●, ■, Data from A. Singh, H. Singh, J. S. Henderson, R. D. Migliore, J. Rousseau, and J. E. Van Lier, *in* "Oxygen Radicals in Biology and Medicine" (M. G. Simic, K. A. Taylor, J. F. Ward, and C. von Sonntag, eds.), p. 587. Plenum, New York, 1989. □, △, Unpublished data (A. Singh, H. Singh, J. S. Henderson, R. D. Migliore, J. Rousseau, and J. E. Van Lier, 1985).

greater number of points for statistical analysis,[38] to increase the degree of certainty of the conclusions being drawn.

In Fig. 2, the radioprotection seen with MEA is similar to that reported by others.[15,25] The radiosensitization seen with AA is consistent with that reported by Tewfik *et al.*[26] and Forsberg *et al.*[25] However, the data in Fig. 2 suggest radiosensitization at a dose of 1000 rads, whereas Tewfik *et al.*[26] reported radiosensitization at doses below 700 rads. This difference could be due to the different strains of mice used in the two studies and due to the different modes of administration of AA (injection[8] versus orally in drinking water[26]). The drastic reduction of the protection offered by MEA by the simultaneous administration of AA is also consistent with the previous work by Forsberg *et al.*[25]; however, the effect of AA seems to be greater according to the data in Fig. 2, again perhaps owing to the different strains of mice in the two studies. The radioprotection offered by DF (Fig. 2) supports the view[4,28] that reactions (1) and (2) play important roles in radiobiological damage. The much larger protection offered by DF and MEA administered together as compared to MEA alone is very promising and warrants extension of this work.

The survival curves shown in Fig. 2 can be reproduced, or similar curves obtained with other mice and other radioprotectors, by following

---

[38] P. Armitage, "Statistical Methods in Medical Research." Blackwell, Oxford, 1971.

the details given above. Our investigation of three radioprotectors and their two mixtures were conducted simultaneously on a nearly homogeneous population of mice irradiated during one session. Therefore one control group satisfactorily served all the protective protocols tested. From the data on mortality of mice in 30 days as a function of dose, LD_{50} values for the mice chosen can be obtained.[1] Published LD_{50} values should not be used as a benchmark reference for such work, since many factors can influence them. They should only be used as a guide in the choice of the doses to be used.

Concluding Remarks

The mouse provides a good model animal system to investigate radioprotection by chemical and biochemical agents. A promising approach is to investigate radioprotection by mixtures of protectors, each agent aimed against a different mechanism of radiobiological damage. A good example of this approach is the result obtained with a mixture of MEA and DF, where chelation of ferric ions by DF seems to enhance the protection offered by MEA at a dose of 1000 rads.

[77] Influence of Thiols on Thermosensitivity of Mammalian Cells *in Vitro*

By ROLF D. ISSELS and ARNO NAGELE

Introduction

The cytotoxic effects of a number of chemotherapeutic agents have been reported to be enhanced by heat.[1,2] The exact mechanisms underlying thermosensitization are not understood in detail and might differ with the various compounds. We recently demonstrated that the enhanced toxicity at 37 and 44° of aminothiols like cysteamine is due to the generation of activated oxygen species during the autoxidation of these compounds and is significantly reduced in the presence of catalase[3] (see Fig. 1).

When studying the thermosensitizing capability of thiols, the following observations have to be taken into account. As demonstrated for Chinese

[1] G. M. Hahn, *Cancer Res.* **39**, 2264 (1979).
[2] G. M. Hahn and G. C. Li, *Natl. Cancer Inst. Monogr.* **61**, 467 (1982).
[3] R. D. Issels, J. E. Biaglow, L. Epstein, and L. Gerweck, *Cancer Res.* **44**, 3911 (1984).

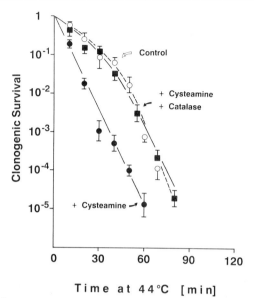

FIG. 1. Survival curves at 44° of exponentially growing CHO cells at low densities (10^2–10^5 cells/cm²), exposed to heat alone (control), to 0.4 mM cysteamine plus heat, and to cysteamine plus heat in the presence of 50 μg/ml catalase. Bars indicate the S.D.

hamster ovary (CHO) cells under normothermic conditions, the cytotoxicity of thiols like cysteamine, cysteine, N-acetylcysteine, and dithiothreitol shows a "paradoxical" dependency on the concentration of the thiols. Maximal toxicity occurs at 0.4–1.0 mM which levels off at higher concentrations.[4] Also, as shown for cysteamine in Fig. 2, the thiol toxicity depends markedly on the population density of the cells during treatment, i.e., cells treated at densities above about 10^6 cells/25 cm² are rather insensitive to cysteamine.

One of the most pronounced metabolic effects of thiol treatment is the intracellular increase of glutathione (GSH) synthesis. Thus, exposure at 37° to 0.8 mM cysteamine, cysteine, N-acetylcysteine, or dithiothreitol increases GSH levels by a factor of 3 within 2 hr, which is still further enhanced at 44°.[4] For cysteamine, N-acetylcysteine,[5] and, more recently, the free thiol form of WR-2721 [S-2-(3-aminopropylamino)ethyl phosphorothioate], we could demonstrate an immediate increase of cyst(e)ine

[4] R. D. Issels, S. Bourier, J. E. Biaglow, L. E. Gerweck, and W. Wilmanns, *Cancer Res.* **45**, 6219 (1985).

[5] R. D. Issels, A. Nagele, K.-G. Eckert, and W. Wilmanns, *Biochem. Pharmacol.* **37**, 881 (1988).

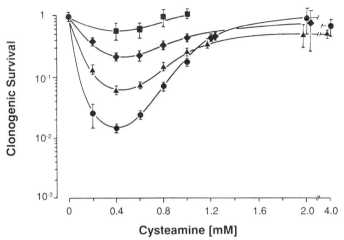

FIG. 2. Effect of cell density on the toxicity of cysteamine for 120 min at 37°. Clonogenic CHO test cells were plated at densities of 10^2–10^4 cells/T_{25} flask 24 hr prior to treatment. At 4.5 hr before treatment, increasing numbers of nonclonogenic, (lethally) irradiated (20–25 Gray), but metabolically active CHO feeder cells were added. Curves show addition of (●) no feeders, (▲) 10^5 feeders, (◆) 3×10^5 feeders, and (■) 10^6 feeders. Values are means ± S.D.

uptake from the medium into the cells and the rapid incorporation of cysteine into newly synthesized GSH. In contrast to the other thiols studied, the free thiol form of WR-2721 leads after prolonged incubation to a significant depletion of intracellular GSH and a concomitant decrease of the glutathione redox status in CHO cells but not in Chinese hamster ovarian carcinoma (OvCa) cells[6] (see Table I). More recently, we found that CHO cells but not OvCa cells are thermosensitized by the free thiol of WR-2721 when treated at high cell densities.[7] This effect might be due to the effort of cells to detoxify breakdown products of this compound by conjugation with GSH.

GSH is the most abundant low molecular weight thiol in mammalian cells. An overview on glutathione has been given by Meister.[8] A collection of fundamental papers on the role of GSH in protection against oxidative stress, in detoxification of xenobiotics, and in the control of cellular redox status can be found in Ref. 9. The important protective role of GSH

[6] R. D. Issels and A. Nagele, *Cancer Res.* **49,** 2082 (1989).

[7] R. D. Issels, S. Kappes, M. Schimak, and W. Wilmanns, *Annu. Meet. N.A. Hyperthermia Group, Philadelphia,* Abstr. Ci-3 (1988).

[8] A. Meister, *Annu. Rev. Biochem.* **52,** 711 (1983).

[9] A. Larsson, S. Orrhenius, A. Holmgren, and B. Mannervik (eds.), "Functions of Glutathione." Raven, New York, 1983.

TABLE I

HPLC MEASUREMENT OF REDUCED AND OXIDIZED INTRACELLULAR GLUTATHIONE[a]

Step	CHO cells			OvCa cells		
	GSH (nmol/mg)[b]	GSSG (nmol/mg)	GSH/GSSG	GSH (nmol/mg)	GSSG (nmol/mg)	GSH/GSSG
Control	23.8 ± 4.4[c]	0.31 ± 0.15	77	18.2 ± 3.1	0.19 ± 0.06	96
0.4 mM Cysteamine	70.6 ± 24.0	0.74 ± 0.41	95	49.7 ± 4.1	0.58 ± 0.15	86
0.4 mM WR-2721	32.3 ± 3.7	0.44 ± 0.12	73	24.6 ± 5.1	0.24 ± 0.05	103
0.4 mM WR-2721 + AP[d]	2.2 ± 1.4	0.12 ± 0.06	18[e]	22.8 ± 5.5	0.28 ± 0.12	81

[a] After 200 min exposure to thiols at 37°. Modified from R. D. Issels and A. Nagele, *Cancer Res.* **49**, 2082 (1989), with permission.
[b] Concentrations are expressed as nmol/mg cell protein.
[c] Means ± S.D. of three or more experiments.
[d] Alkaline phosphatase (AP, 20 U/ml medium) was added immediately in order to convert WR-2721 to its dephosphorylated free thiol form.
[e] This ratio differs significantly from the ratio of the control (a two-tailed t test of the logarithms of individual ratios gave $p < .05$).

against thermosensitization and its role for the development of thermotolerance have been demonstrated in a number of recent papers[10-12] and are not discussed here in detail.

Heat Survival and Thermotolerance

Extensive studies have been carried out to investigate the effect of elevated temperatures on cells *in vitro*. Most mammalian cells respond to a continuous heat treatment at constant temperature (range 40–45°) in a similar and predictable way.[13] During heat treatment (e.g., 44°) the cell survival curve typically passes an initial shoulder region until a constant maximal rate of inactivation is reached (see Fig. 1). This type of survival curve can be parameterized in terms of the shoulder (extrapolation number n) and the slope (D_0).[14] Here D_0 is the time required to reduce cell survival 37% ($1/e$) in the exponential region. A thermodynamical analysis by mapping $\ln(1/D_0)$ versus $1/T(°K)$ (Arrhenius analysis) shows an inflection point at 42.5–43° (so-called breakpoint), indicating that treatments below the breakpoint temperature are not equivalent to those above in terms of activation energies (for review see Ref. 15). Above this breakpoint an increase of 1° in temperature halves the D_0 value, although the absolute values of the inactivation rates are different for each cell line.

The phenomenon of thermotolerance, the acquired and transient heat resistance, is complex (reviewed in Ref. 16). At temperatures above 43°, thermotolerance does not develop during the initial heat treatment but requires time at 37° to be expressed. In "split dose" experiments in which after the first heat treatment cells are allowed to rest at 37°, the degree of the acquired thermotolerance becomes apparent in a second heat treatment and is usually expressed as the thermotolerance ratio [TTR, i.e., the ratio of the D_0 values of preheated to nonpreheated (control) cells]. In continuous treatments below the breakpoint temperature thermotolerance develops during the exposure after about 3–4 hr and the inactivation rate becomes much lower than that seen at the initiation of treatment. The magnitude of thermotolerance and its decay depend on a variety of fac-

[10] J. B. Mitchell and A. Russo, *Radiat. Res.* **95**, 471 (1983).
[11] M. L. Freeman, A. W. Malcolm, and M. J. Meredith, *Cell. Biol. Toxicol.* **3**, 213 (1985).
[12] D. C. Shrieve, G. C. Li, A. Astromoff, and J. W. Harris, *Cancer Res.* **46**, 1684 (1986).
[13] K. J. Henle and L. A. Dethlefsen, *Ann. N.Y. Acad. Sci.* **335**, 234 (1980).
[14] A. Westra and W. C. Dewey, *Int. J. Radiat. Biol. Relat. Stud. Phys. Chem. Med.* **19**, 467 (1971).
[15] K. J. Henle, *in* "Hyperthermia in Cancer Therapy" (K. Storm, ed.), p. 47. G. K. Hall, Medical Publishers, Boston, Massachusetts, 1983.
[16] E. W. Gerner, *in* "Hyperthermia in Cancer Therapy" (K. Storm, ed.), p. 141. G. K. Hall, Medical Publishers, Boston, Massachusetts, 1983.

tors, e.g., triggering temperature,[17] time interval between priming treatment and second dose,[13] pH,[18] and protein synthesis.[19,20] There is a temporal correlation between the acquisition of thermotolerance and the induction of an increased synthesis of heat-shock proteins, a family of proteins which is induced after a variety of environmental stresses.[21] The role of heat-shock proteins for the status of thermotolerance has been reviewed recently by Lindquist.[22]

Clonogenic Assay for Survival of Cells Treated by Heat and Thiols

The clonogenic assay of adherent cells is an established experimental procedure for the quantification of end points in thermobiological studies. It permits the determination of the probability of reproductive survival over three to four orders of magnetude with an error of 10% or less throughout the measurable range.

Precautions. The clonogenic assay has important prerequisites. The homogeneous cell population must steadily divide over an extended range of time and form defined colonies of 50 cells or more. In cell lines, the routine passage easily allows the determination of the population doubling time (dt), which gives sensitive and immediate information on the conditions of the cell culture.

To assure that the assay reflects the properties of the whole population, the cell culture system must allow a high plating efficiency (PE), preferably over 70% (i.e., 70 colonies or more should be formed when 100 cells are plated with no treatment). The PE of the passage is an essential control for the environmental conditions and should be routinely examined. The PE of nontreated cells is an important parameter for the evaluation of a specific treatment effect on cell survival and should not be substituted by the PE of the passage. When the PE depends on the cellular density, an adequate treatment of the survival data becomes questionable. In this case, constant cell densities must be assured by the addition of nonclonogenic (i.e., irradiated) but metabolically active feeder cells.

When the potential colony-forming units (cfu) do not consist of single cells but of groups, the survival data have to be corrected for the average cellular multiplicity (N) at the time of treatment. The information needed here is the *viable cell* multiplicity, which is not easily obtained when the

[17] G. C. Li, G. A. Fisher, and G. M. Hahn, *Radiat. Res.* **89**, 361 (1982).
[18] L. E. Gerweck, W. K. Dahlberg, and B. Greco, *Cancer Res.* **43**, 1163 (1983).
[19] K. J. Henle and D. B. Leeper, *Radiat. Res.* **90**, 339 (1982).
[20] A. Lazlo, *Int. J. Hyperthermia* **4**, 513 (1988).
[21] G. C. Li and Z. Werb, *Proc. Natl. Acad. Sci. U.S.A.* **79**, 3129 (1982).
[22] S. Lindquist, *Annu. Rev. Biochem.* **55**, 1151 (1986).

PE is low. Also, the correction for multiplicity can only be applied when the survival of individual cells is independent of the presence of other cells. This correction is not necessary when cells are treated at high densities and are then replated as single cells for colony formation.

It is self-evident that an unbiased estimate of cell survival requires the simultaneous harvest of both the treated specimens and the untreated controls. Only those colonies that contain more than the generally accepted limit of 50 cells should be counted, and only those flasks in which colonies do not overlap should be evaluated. It is advisable to use at least four replicate flasks to obtain a measure of the experimental error, but it should be kept in mind that the lower limit of the standard error is imposed by the sampling error which is equal to the square root of the combined colony counts.

There are mainly two objections raised against the use of colony formation as an assay for the survival of heated cells. First it has been shown that heat-treated cells may lose their ability to attach and may therefore escape the assay procedure.[23] Second, since the cell membrane is very likely one of the targets of heat damage, there may be an interaction with damage from trypsinization.[24] It seems, however, that both criticisms can be overcome by proper controls and adequate experimental design.[25]

When these conditions and precautions are met, the surviving fraction (*SF*) of the cells is calculated according to the following equation:

$$SF = 1 - (1 - F)^{1/N} \tag{1}$$

where N is the average cellular multiplicity and F is the group survival probability, i.e.,

$$F = \text{No. of colonies}/(\text{No. of cfu plated} \times PE) \tag{2}$$

For different sized groups Eq. (1) underestimates the surviving fraction at cell survivals greater than 0.1. In high-density experiments in which, after the treatment, single cells are replated for colony formation, Eq. (1) simplifies to

$$SF = \text{No. of colonies}/(\text{No. of cells plated} \times PE) \tag{3}$$

A thorough discussion of the theory and the caveats of the clonogenic assay can be found in Elkind and Whitmore.[26] Figure 3 shows the general

[23] P. A. Lin, C. E. Butterfield, and D. F. M. Wallach, *Cell Biol. Int. Rep.* **1,** 51 (1977).
[24] H. Bass, J. L. Moore, and W. T. Cloakely, *Int. J. Radiat. Biol. Relat. Stud. Phys. Chem. Med.* **33,** 57 (1978).
[25] G. M. Hahn, "Hyperthermia and Cancer." Plenum, New York, 1982.
[26] M. M. Elkind and G. F. Whitmore, "The Radiobiology of Cultured Mammalian Cells." Gordon & Breach, New York, 1967.

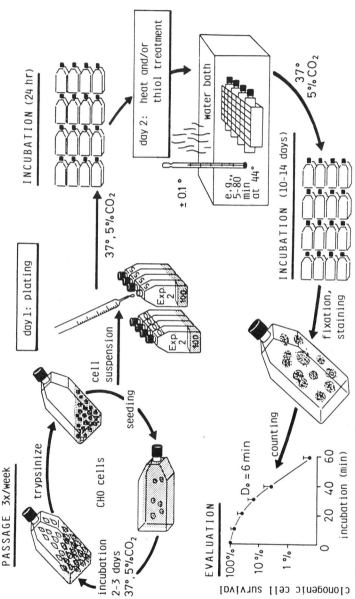

Fig. 3. Assay of CHO cell survival after treatment with heat and/or thiols at low cell densities. Cells of the subculture are trypsinized, plated at known numbers into replicate flasks, and allowed to attach and multiply for 24 hr. Thiols, with or without antioxidant enzymes (catalase, superoxide dismutase), can be added to the medium before the flasks are submersed into a temperature-controlled water bath. After the treatment, cells are exposed to fresh medium and are allowed to form colonies for 10–14 days at 37° in a CO_2 incubator. The stained colonies are counted, and the single cell survival is calculated with corrections for plating efficiency and cellular multiplicity. Finally, the effect of the treatment can be evaluated from a semilogarithmic plot of the survival curve or by computer-fitting of an appropriate mathematical model.

outline of a typical experiment for the determination of clonogenic survival after heat and/or thiol treatment of cells at low density.

Materials for Cell Culture and Reagents. Media, sera, antibiotics, and
trypsin are purchased from Gibco BRL (Eggenstein, FRG). Tissue culture
flasks are from Greiner (Nürtingen, FRG) or from Nunc (Wiesbaden,
FRG). Concentrated stock solutions (e.g., 20 mM) of thiols [usually the
hydrochlorides, Sigma (Deisenhofen, FRG)] are prepared in ice-cold
phosphate-buffered saline (PBS), which was previously gassed with nitrogen to prevent oxidation. The pH is then carefully adjusted to 7.2–7.4
with NaOH and filtered through a 0.2-μm cellulose acetate membrane
filter (Minisart SM 16534, Sartorius, Göttingen, FRG). The sterile thiol
solution is put on ice and used without delay. Catalase from bovine liver
(17,600 U/mg) and superoxide dismutase from bovine blood (2750 U/mg)
are also purchased from Sigma.

Cell Culture. Chinese hamster ovary (CHO) cells are grown in 4.5 ml
McCoy's 5A medium, which is supplemented with 10% (v/v) newborn
calf and 5% (v/v) fetal calf serum, 50 mg/liter penicillin, 50 mg/liter streptomycin, and 0.1 g/liter neomycin sulfate. The culture flasks are tissue
culture grade polystyrene flasks with an area of 25 cm^2 (T$_{25}$ flasks). Cells
are maintained in exponential growth at 37° in a 5% CO$_2$ atmosphere, and
3×10^5 cells are routinely subcultured every 2 or 3 days. Under these
conditions the population doubling time, which is calculated from $dt =$
$\log(2) \times$ hr between passages/log(cells obtained/cell plated), is 13 to
15 hr, and the *PE* is 80 to 90%. It is also important to check the culture
system for contamination with mycoplasms.

Trypsinization. The cell layer is washed with warm PBS (Dulbeccos's
formulation). Five milliliters of trypsin (0.25%, without EDTA) is added,
poured off after a few seconds, and the flask placed into the incubator.
Trypsinization should be as short as possible to assure minimal cell damage and prevent reaggregation of cells. After about 2–3 min a microscopic
control should reveal that all cells have rounded and the process is completed. Then cells are detached and the trypsin inactivated by flushing the
cell layer with 5 ml of serum-containing medium. The cell suspension
should be stored on ice and processed within 30 min.

Heat and Drug Exposure at Low Cell Densities. Cells are trypsinized
24 hr prior to drug exposure at different temperatures and counted in a
Coulter Counter. For each dose, dilutions of known cell numbers (10^2 to 5
$\times$ 10^4 cells) are inoculated in four to six replicate T$_{25}$ flasks (total volume
4.5 ml). For the determination of *PE* four additional flasks inoculated with
100 cells/flask each and two flasks with 2000 cells/flask are prepared for
the control of cellular multiplicity (Fig. 4).

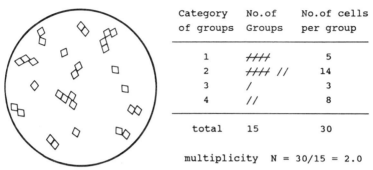

Category of groups	No. of Groups	No. of cells per group
1	////	5
2	//// //	14
3	/	3
4	//	8
total	15	30

multiplicity N = 30/15 = 2.0

FIG. 4. Determination of the average cellular multiplicity. Approximately 2000 cells are plated into T_{25} flasks 24 hr prior to the experiment. Immediately at the beginning of heat and/ or thiol exposure the flasks are fixed and stained. In each flask, the different cell groups are examined in an inverted microscope and classified according to the cell number. The average cellular multiplicity is the total cell number divided by the total number of cell groups examined.

For combined drug and heat treatments, the thiol is added in small volumes (50–100 μl) directly to the warm medium prior to the heat exposure to give the final concentration. To study the protective effect of antioxidant enzymes, catalase (50 μg/ml) or superoxide dismutase (10 μg/ml) is added before thiol exposure.

The flasks are then tightly closed, sealed with Parafilm (American Can Co.), and placed horizontally into a circulating water bath which allows the temperature to be kept constant within $\pm 0.1°$ (Umwaelz-Thermostat W45/EB, Haake AG, Berlin, FRG). The temperature equilibrium in the flasks is reached within approximately 2.5 min. At the end of the treatment, the flasks are dried with a towel, the Parafilm is removed, and the necks of the flasks are wiped with 70% (v/v) 2-propanol. The cell layer is carefully washed twice with fresh medium (37°), and the cells are returned to incubation at 37° in 5% CO_2. After colony development (10–14 days), the flasks are rinsed with 0.9% NaCl solution, fixed, and stained with 20% (v/v) ethanol containing 0.8% ammonium oxalate and 2% crystal violet.

Heat and Drug Exposure at High Cell Densities. Experiments under conditions of high cell densities require an additional trypsinization step for replating of cells after the treatment. Here, a constant number of cells (e.g., 10^6 cells) is inoculated into a T_{25} flask and allowed to attach and grow for 24 hr prior to the treatment. After treatments, which are performed essentially as described above, the cell layer is washed with warm PBS and trypsinized. A possible effect of trypsinization must be controlled in parallel control flasks. Determination of cellular multiplicity is not necessary. Cells are then inoculated in known numbers into T_{25} flasks and immediately returned to 37° and 5% CO_2 for the development of colonies.

Determination of Intracellular Glutathione

A number of sensitive photometric assays exist that permit the determination of total (GSH + GSSG), reduced (GSH), or oxidized (GSSG) glutathione. These are described and critically reviewed in another volume of this series.[27]

A procedure which allows the simultaneous determination of cystine, cysteine, GSH, and GSSG and measurement of their specific radioactivities is the HPLC method of Reed *et al.*[28] It is used in our laboratory with good success. In untreated samples of CHO cells we reproducibly obtain GSH/GSSG ratios larger than 50. In our experience oxidation of glutathione is not, as sometimes suspected, a consequence of the prolonged derivatization procedure, but is due to incautious preparation of the sample. The procedure described here is a slight modification of the original method, allowing the analysis of smaller sample volumes.

Reagents for HPLC. Glutathione, reduced and oxidized form, 1,10-phenanthroline, γ-L-glutamyl-L-glutamic acid, iodoacetic acid, and 2,4-dinitrofluorobenzene are purchased from Sigma. Perchloric acid, potassium hydrogen carbonate, sodium acetate trihydrate, acetic acid, and methanol are obtained from Merck (Darmstadt, FRG) in the highest purity available.

Preparation and Derivatization of Samples. Cells (3×10^6) are trypsinized, put on ice, and centrifuged ($500 g$, 4°) for 5 min. After the addition of 5 μl of 1,10-phenanthroline (50 mg/ml ethanol) the cell pellet is vortexed, 120 μl of ice-cold 1 N perchloric acid containing 20 nmol of the internal standard γ-L-glutamyl-L-glutamic acid added, and the sample again vortexed and shortly centrifuged at high speed. To the clear acid extract 20 μl of iodoacetic acid (95 mg/ml water, freshly prepared) is pipetted, and the extract is adjusted to pH 8.3–8.5 by repetitive addition of small amounts of solid potassium bicarbonate. After 1 hr at room temperature in the dark, when carboxymethylation of reduced thiol groups is complete, 50 μl of a 7.5% solution (v/v) of 2,4-dinitrofluorobenzene in ethanol is added and dinitrophenylation of the amino groups al-

[27] T. P. M. Akerboom and H. Sies, this series, Vol. 77, p. 373.
[28] D. J. Reed, J. R. Babson, P. W. Beatty, A. E. Brodie, W. W. Ellis, and D. W. Potter, *Anal. Biochem.* **106**, 55 (1980).

lowed to proceed for 24 hr at room temperature in the dark. The potassium perchlorate precipitate is removed by high-speed centrifugation, and the reaction mixture is stored for a maximum of 1 week in a refrigerator before HPLC analysis.

Separation of N-Dinitrophenyl-S-carboxymethylaminothiols by HPLC. The sample (100 μl) is injected onto a μBondapak amine column (4 × 250 mm, Waters, Eschborn, FRG) and eluted with a sodium acetate gradient in a water–methanol–acetic acid solvent at pH 4.5. Solvent A and solvent B of the original method are used to establish the gradient. Solvent A is 4 : 1 methanol–water and solvent B is prepared by mixing 200 ml of a solution of 272 g sodium acetate trihydrate, 122 ml water, and 378 ml glacial acetic acid with 800 ml of solvent A. Before use the solvents are filtered through a Schott G4 sintered glass filter. Specifically, with new columns the gradient starts isocratically at 10% B for 10 min and increases linearly to 95% B within 30 min with a flow rate of 1.5 ml/min. Depending on the reduced capacity by hydrolysis of older columns, the elution times are shortened and the gradient and/or flow rates adjusted accordingly. The dinitrophenyl derivatives are detected at 365 nm and quantified by relating peak areas to the internal standard. For example, typical retention times on a new column are as follows: 9.7 min, cystine; 15.5 min, cysteine; 19.3 min, γ-glutamylglutamate; 22.6 min, glutathione; and 27.1 min, glutathione disulfide.

In Vitro Assay of Thiol-Promoted Cystine Uptake

Cells. For cystine uptake studies, CHO cells are trypsinized and plated in 35-mm polystyrene tissue culture petri dishes (5 × 10⁵ cells) containing 2 ml of complete McCoy's 5A medium. At the time of the experiment, 24 hr after plating, the cell number is 1.5–2 × 10⁶ per dish and the confluency is about 50%.

Uptake Medium. The uptake medium is the same as the McCoy's 5A growth medium, except that glutathione and methionine are omitted, the calf sera are dialyzed, and cyst(e)ine is replaced by 0.1 mM L-[³⁵S]cystine (50–120 mCi/mmol, Amersham-Buchler, Braunschweig, FRG).

Assay. Cells are washed twice with warm PBS, and 0.5 ml of the radioactive uptake medium is added together with small volumes (5–20 μl) of the thiol stock solution. The dishes are immediately placed on a temperature-controlled water bath, the bottom of the dishes being submersed in the water. Following incubation for 4 min, the dishes are placed on ice, the radioactive medium is removed, and the cell layer is extensively washed with ice-cold PBS. The cells are then trypsinized (0.25% trypsin) for 2 min at room temperature, the cell number determined in an aliquot, and the rest of the suspension transferred to an Eppendorf reaction vial. After centrifugation (500 g, 5 min at 4°) the cell pellet is precipitated with 0.5 ml of 1 N perchloric acid, put on ice for 10 min, and again centrifuged at high speed. The acid-insoluble pellet, which is solubilized in 0.5 ml of 1 N KOH, and the acid-soluble supernatant are transferred into a xylene-based scintillation cocktail and separately counted in a β-counter which allows for quench correction. If desired, an aliquot of the KOH solution can be used for protein determination.

Conclusions

Low molecular weight thiols have turned out to be valuable model substances for the study of thermosensitization by oxidative stress *in vitro*. Their use in thermobiological studies may lead to better understand the fundamental processes that underlie the action of heat in biological systems.

Section VI

Oxygen Free Radicals in Ischemia and
Reperfusion-Associated Tissue Damage

[78] Oxygen-Derived Free Radicals in Reperfusion Injury

By MARC S. SUSSMAN and GREGORY B. BULKLEY

Introduction

Reperfusion injury can be defined as the damage that occurs to an organ during the resumption of blood flow following an episode of ischemia. This can be distinguished from the injury caused by the ischemia per se (i.e., sustained during the period of ischemia), although the conditions needed to cause a reperfusion injury are generated by the ischemic episode. One hallmark of a reperfusion injury is that it may be ameliorated by interventions which are not initiated until after the period of ischemia but which are begun just prior to the reestablishment of blood flow.

Fundamental Trigger Mechanism

The mechanisms by which toxic oxygen metabolites participate in organ injury caused by reperfusion have been evaluated in a number of individual organs. Granger *et al.*[1] found that a 1-hr period of partial ischemia increased the capillary permeability of the cat small intestine 4-fold. Pretreatment with indomethacin, methylprednisolone, or H_1 and H_2 antagonists failed to prevent this capillary injury, suggesting that prostaglandins, lysosomal enzymes, and histamine are not major mediators of this injury. However, it was found that superoxide dismutase (SOD), the highly specific scavenger of the superoxide radical, almost completely prevented this injury. The fact that SOD was effective when given near the end of the ischemic period, but just before reperfusion, indicated that the injury which had been prevented must have been incurred at the time of reperfusion. This was the first demonstration of free radical-mediated reperfusion injury in any organ. A beneficial effect similar to that seen with SOD was also provided by scavenging either hydrogen peroxide with catalase[2] or the hydroxyl radical (or activated iron oxygen complexes that act like hydroxy radicals[3]) with dimethyl sulfoxide.[4] Moreover, prevent-

---

[1] D. N. Granger, G. Rutili, and J. M. McCord, *Gastroenterology* **81**, 22 (1981).
[2] D. N. Granger, M. E. Hollwarth, and D. A. Parks, *Acta Physiol. Scand.* **548**, 47 (1986).
[3] G. Minotti and S. O. Aust, *Chem. Phys. Lipids* **44**, 191 (1987).
[4] D. A. Parks and D. N. Granger, *Am. J. Physiol.* **245**, G285 (1983).

ing the secondary generation of hydroxyl radicals from superoxide via the Haber–Weiss reaction[5] by chelating iron with either desferrioxamine or transferrin also ameliorated this injury.[6] In these experiments, the beneficial effects of transferrin and desferrioxamine were apparently due to the iron-chelating properties of these compounds, because protection was not seen when the chelators were saturated with iron.

In seeking the source of superoxide generation at reperfusion, the enzyme xanthine oxidase was a leading candidate, particularly in the intestinal mucosa, where it facilitates iron absorption by the epithelium.[7] In the same cat model of intestinal ischemia, inhibition of xanthine oxidase activity with allopurinol,[4,8] pterin aldehyde,[9] or tungsten feeding[10] also provided protection equivalent to free radical ablation. Xanthine oxidase can exist in two forms, both of which catalyze oxidation of the purine metabolites hypoxanthine and xanthine to uric acid (Fig. 1). The dehydrogenase, which uses NAD^+ as an electron receptor, predominates under normal conditions.[11] Following periods of ischemia, however, this form may be converted, reversibly or irreversibly, to the oxidase form, which transfers the electron to molecular oxygen and thereby generates a superoxide anion as a by-product of the oxidation.[12,13] This activation of the oxidase from the dehydrogenase can be mediated by proteolysis and inhibited by serine protease inhibitors *in vitro*.[14] When one of these compounds, soybean trypsin inhibitor, was given *in vivo*, it also protected the cat small intestine from the capillary injury seen following ischemia/reperfusion.[15] This is consistent with the concept that the proteolytic activation of xanthine oxidase during ischemia facilitates the generation of toxic oxygen metabolites when oxygen is reintroduced at reperfusion.

This body of work focused attention on the xanthine oxidase system as central to reperfusion injury in the intestine and supports the hypothe-

[5] F. Haber and J. Weiss, *Proc. R. Soc.* **147**, 332 (1934).

[6] L. A. Hernandez, M. B. Grisham, and D. N. Granger, *Am. J. Physiol.* **253**, G49 (1987).

[7] R. W. Topham, M. C. Walker, M. P. Calisch, and R. W. Williams, *Biochemistry* **21**, 4529 (1982).

[8] D. A. Parks, G. B. Bulkley, D. N. Granger, S. R. Hamilton, and J. M. McCord, *Gastroenterology* **82**, 9 (1982).

[9] D. N. Granger, J. M. McCord, D. A. Parks, and M. E. Hollwarth, *Gastroenterology* **90**, 80 (1986).

[10] D. N. Granger, *Am. J. Physiol.* **255**, H1269 (1988).

[11] D. A. Parks and D. N. Granger, *Acta Physiol. Scand.* **548**, 97 (1986).

[12] J. M. McCord and R. S. Roy, *Can. J. Physiol. Pharmacol.* **60**, 1346 (1982).

[13] D. A. Parks, T. K. Williams, and J. S. Beckman, *Am. J. Physiol.* **254**, G768 (1988).

[14] E. Della Corte and F. Stirpe, *FEBS Lett.* **2**, 83 (1968).

[15] D. A. Parks, D. N. Granger, G. B. Bulkley, and A. K. Shah, *Gastroenterology* **89**, 6 (1985).

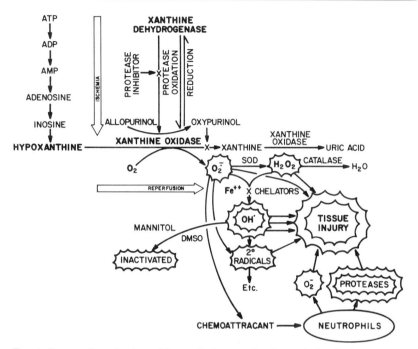

FIG. 1. Proposed mechanism of free radical generation in postischemic tissues. During ischemia, the breakdown of high-energy phosphate compounds results in the accumulation of the purine metabolites hypoxanthine and xanthine. At the same time the predominate form of the enzyme xanthine oxidoreductase, xanthine dehydrogenase, is converted to xanthine oxidase, possibly by proteolysis. At reperfusion, oxygen is reintroduced, permitting the xanthine oxidase-catalyzed oxidation of xanthine and hypoxanthine with the simultaneous reduction of oxygen to the superoxide free radical. The superoxide can then generate secondarily the highly toxic hydroxyl radical via the iron-catalyzed Haber–Weiss reaction. [Modified from D. N. Granger, G. Rutili, and J. M. McCord, *Gastroenterology* **81,** 22 (1981).]

sis of Granger *et al.*,[1] which is summarized in Fig. 1. During ischemia there is a net catabolism of ATP, leading to increased concentrations of purine metabolites, such as hypoxanthine, which can serve as substrates for both xanthine dehydrogenase and xanthine oxidase. At the same time, ischemia appears to mediate the conversion of xanthine dehydrogenase to the oxidase, the only form which can generate superoxide in the presence of molecular oxygen. (Although proteolysis may well be involved, the precise mechanism by which ischemia mediates this step is unclear.[16])

[16] T. G. McKelvey, M. E. Hollwarth, D. N. Granger, T. D. Engerson, U. Landler, and H. P. Jones, *Am. J. Physiol.* **254,** G753 (1988).

Just prior to reperfusion the xanthine oxidase system is lacking only one substrate, oxygen, in order to generate superoxide. This missing oxygen is reintroduced, suddenly and in excess, at the time of reperfusion. While the cascade of events leading to injury is triggered by the generation of superoxide by xanthine oxidase, the actual agent of injury is probably some form of the hydroxyl radical, because scavenging OH · or blocking its secondary generation had the same effect as blockade of the cascade more proximally (Fig. 1).[2,4]

Neutrophil Amplification

Further studies on the cat small intestine suggest that the neutrophil also has an important role in the mediation of injury. Hernandez et al.[17] found that treating cats either with antineutrophil serum or with a monoclonal antibody (MAb 60.1) that prevents neutrophil adherence to endothelial cells also prevented the reperfusion-induced capillary leak. This finding might suggest that free radical-mediated reperfusion injury is entirely explained by the generation of superoxide by neutrophils. This analysis, however, is simplistic. On the one hand, it fails to account for the wide variety of other toxic mediators, such as elastase, collagenase, other proteases, platelet-activating factor, and leukotrienes, which are also released by neutrophils and could be important contributors to the end organ injury.[18] Moreover, it fails to explain how the inhibition of xanthine oxidase can also prevent injury: Neutrophils generate superoxide not with xanthine oxidase (indeed, neutrophils do not contain significant quantities of xanthine oxidase), but with a membrane-bound NADPH oxidase. The latter enzyme is not inhibited by allopurinol. Moreover, the superoxide-generated respiratory burst of neutrophils is not inhibited by allopurinol in vitro.[19] While Moorhouse et al.[20] have found that high concentrations of allopurinol and its active metabolite oxypurinol can act directly as hydroxyl scavengers, Zimmerman et al.[21] subsequently found that the levels achieved in the plasma and intestinal

[17] L. A. Hernandez, M. B. Grisham, B. Twohig, K. E. Arfors, J. M. Harlan, and D. N. Granger, Am. J. Physiol. 253, H699 (1987).

[18] S. J. Weiss, N. Engl. J. Med. 320, 365 (1989).

[19] H. P. Jones, M. B. Grisham, S. K. Bose, V. A. Shannon, A. Schott, and J. M. McCord, Biochem. Pharmacol. 34, 3673 (1985).

[20] P. C. Moorhouse, M. Grootveld, B. Halliwell, J. G. Quinlan, and J. M. C. Gutteridge, FEBS Lett. 213, 23 (1987).

[21] B. J. Zimmerman, D. A. Parks, M. B. Grisham, and D. N. Granger, Am. J. Physiol. 255, H202 (1988).

lymph in the studies of reperfusion injury in the cat small intestine[4,8] were insufficiently high to produce a hydroxyl-scavenging effect but high enough to inhibit xanthine oxidase.

The role of neutrophils in reperfusion injury is further explained by the finding that partial ischemia of the cat small intestine caused a marked accumulation of neutrophils (assayed as myeloperoxidase) within the wall of the intestine during the ischemic period and at reperfusion.[22] This accumulation was prevented by xanthine oxidase inhibition with allopurinol. Hernandez and colleagues[17] have suggested that the accumulation of neutrophils in the postischemic cat small intestine is mediated by a xanthine oxidase-dependent initial event, but that the capillary permeability injury itself is mediated by toxins, including superoxide, that are then generated by the accumulated neutrophils.

Reperfusion Injury in Other Organs

There is evidence for an important role for toxic oxygen metabolites in the pathogenesis of reperfusion injury in splanchnic organs other than the small bowel. In the stomach, superficial mucosal injury secondary to hypotension can be prevented by SOD or allopurinol.[23,24] This is a model of the acute hemorrhagic mucosal necrosis ("stress ulcer") that is seen in the stomach of patients following shock or other forms of severe physiologic insult. Moreover, in an isolated, perfused canine pancreas model of the similar clinical syndrome of ischemic pancreatitis, either allopurinol or SOD at reperfusion provided protection against the injury.[25,26] In the liver, which is rich in xanthine oxidase, SOD and catalase, or allopurinol, have been found to ameliorate postischemic reperfusion injury.[27,28] These animal models mimic the situation both in ischemic hepatitis, a form of liver failure that frequently develops following shock and resuscitation, and in livers preserved (ischemic) for transplantation.

Normothermic ischemia of the kidney, secondary to hypotension, frequently results in renal vascular and tubular injury causing acute renal

[22] M. B. Grisham, L. A. Hernandez, and D. N. Granger, *Am. J. Physiol.* **251,** G567 (1986).

[23] M. Itoh and P. H. Guth, *Gastroenterology* **88,** 1162 (1985).

[24] M. A. Perry, S. Wadhwa, D. A. Parks, W. Pickard, and D. N. Granger, *Gastroenterology* **90,** 362 (1985).

[25] H. Sanfey, G. B. Bulkley, and J. L. Cameron, *Ann. Surg.* **201,** 633 (1985).

[26] H. Sanfey, M. G. Sarr, G. B. Bulkley, and J. L. Cameron, *Acta Physiol. Scand.* **126,** 109 (1986).

[27] D. Adkinson, M. E. Hollwarth, J. N. Benoit, D. A. Parks, J. M. McCord, and D. N. Granger, *Acta Physiol. Scand.* **126,** 101 (1986).

[28] G. Nordstrom, T. Seeman, and P. O. Hasselgren, *Surgery (St. Louis)* **97,** 679 (1985).

failure. Vasko *et al.*[29] found that administering allopurinol to dogs prior to such normothermic renal ischemia markedly decreased subsequent mortality from renal failure. SOD has been found to ameliorate short-term changes in function and morphology following 60 min of warm ischemia in dogs[30] and rats.[31] Baker and colleagues[32] found that SOD administered just prior to reperfusion significantly improved the 7-day survival of rats after 45 min of warm renal ischemia. Kidneys used for transplantation undergo a period of hypothermic ischemia between harvesting and implantation. Either SOD or allopurinol significantly ameliorated the decrease in function caused by 24 hr of cold ischemia.[33] Similarly, the injury to lungs undergoing cold (4°) preservation (ischemia) for transplantation was diminished by the presence of free radical scavengers.[34]

Clinically, ischemia/reperfusion injury in the heart may be manifest as either a global or a regional phenomenon. Global ischemia and reperfusion occur following cardiac arrest and resuscitation, or more often during elective cardiopulmonary bypass for open-heart surgery. While a number of studies have shown protection in animal models of prolonged global cardiac ischemia with either free radical scavengers or (in some species) allopurinol,[35–37] most cases of modern cardiac surgery do not require the prolonged periods of bypass used in these studies. On the other hand, when global ischemia (bypass) must be superimposed on a heart immediately following a regional infarction, a global "stunning" results in decreased myocardial contractility outside of the infarcted area; this may prove to be amenable to free radical ablation in a more clinically useful way. The actual size of a myocardial infarction, itself the direct manifestation of regional ischemia, can be altered by free radical ablation as well.[38] Although initial studies in this area were most promising, it now appears that the optimal benefit to be achieved is small when studies are conducted with appropriate corrections for collateral flow[39] and with appro-

[29] K. A. Vasko, R. A. DeWall, and A. M. Riley, *Surgery (St. Louis)* **71,** 787 (1972).

[30] M. S. Paller, J. R. Hoidal, and T. F. Ferris, *J. Clin. Invest.* **74,** 1156 (1984).

[31] K. Ouriel, N. G. Smedira, and J. J. Ricotta, *J. Vasc. Surg.* **2,** 49 (1985).

[32] G. L. Baker, R. J. Corry, and A. T. Autor, *Ann. Surg.* **202,** 628 (1985).

[33] I. Koyama, G. B. Bulkley, G. M. Williams, and M. J. Im, *Transplantation* **40,** 590 (1985).

[34] R. S. Stuart, W. A. Baumgartner, A. M. Borkon, G. B. Bulkley, J. D. Brawn, S. M. DeLaMonte, G. M. Hutchins, and B. A. Reitz, *Transplant. Proc.* **17,** 1454 (1985).

[35] M. Shlafer, P. F. Kane, and M. M. Kirsh, *J. Thorac. Cardiovasc. Surg.* **83,** 830 (1982).

[36] A. S. Casale, G. B. Bulkley, B. H. Bulkley, J. T. Flaherty, V. L. Gott, and T. J. Gardner, *Surg. Forum* **34,** 313 (1983).

[37] J. R. Stewart, S. L. Crute, V. Loughlin, M. L. Hess, and L. J. Greenfield, *J. Thorac. Cardiovasc. Surg.* **90,** 68 (1985).

[38] D. E. Chambers, D. A. Parks, G. Patterson, S. Yoshida, K. Burton, L. F. Parmley, J. M. McCord, and J. M. Downey, *Fed. Proc., Fed. Am. Soc. Exp. Biol.* **47,** 1093 (1983).

[39] K. A. Reimer and R. B. Jennings, *Circulation* **71,** 1069 (1985).

priate length of follow-up to determine ultimate infarct size.[40] This whole issue is further complicated by a study that was unable to demonstrate measurable levels of xanthine oxidase in homogenized human heart.[41] It remains unclear whether the technique employed in this study could have missed small quantities of xanthine oxidase localized at strategic anatomic sites such as the endothelial cells which constitute less than 1% of the myocardial mass (see below).

A series of experiments in models of island and free skin flaps in rats has been found that both SOD and allopurinol protected against the ischemia/reperfusion injury which can result from inadvertent vascular occlusion.[42,43] In skeletal muscle, swelling that follows an ischemia/reperfusion insult can result in a rise in pressure within the noncompliant fascial compartments of the lower leg. If this hydrostatic pressure becomes greater than the capillary perfusion pressure, secondary ischemia and anoxic injury, known as compartment syndrome, will result. Korthius *et al.*[44] found that the increase in capillary permeability caused by a reperfusion injury to skeletal muscle was blocked with free radical scavengers.

Of all of the organs that have been studied perhaps the greatest potential benefit is in the central nervous system. The low tolerance of the brain for ischemia has limited the value of any attempt at reperfusion, including cardiopulmonary resuscitation, in the treatment of cardiac arrest.[45] However, in a model of regional ischemia in the cat brain, it was found that treatment with SOD improved the acute recovery of electrical function after 30 min of ischemia.[46] This suggests that at least part of the damage done to the brain by such an insult is due to a potentially treatable reperfusion injury.

Endothelial Cell Trigger

The fact that reperfusion injury can be blocked by allopurinol in a wide variety of organs, some of which do not appear to contain measurable

[40] T. Miura, D. M. Yellon, J. Kingma, and J. M. Downey, *Free Radical Biol. Med.* **4,** 25 (1988).

[41] L. J. Eddy, J. R. Stewart, H. P. Jones, T. D. Engerson, J. M. McCord, and J. M. Downey, *Am. J. Physiol.* **253,** H709 (1987).

[42] M. J. Im, P. N. Manson, G. B. Bulkley, and J. E. Hoopes, *Ann. Surg.* **201,** 357 (1985).

[43] P. N. Manson, K. K. Narayan, M. J. Im, G. B. Bulkley, and J. E. Hoopes, *Surgery (St. Louis)* **99,** 211 (1986).

[44] R. J. Korthius, D. N. Granger, M. I. Townsley, and A. E. Taylor, *Circ. Res.* **57,** 599 (1985).

[45] J. S. Beckman, G. A. Campbell, C. J. Hannan, C. S. Karfias, and B. A. Freeman, *in* "Superoxide and Superoxide Dismutase in Chemistry, Biology, and Medicine" (G. Rotilio, ed.), p. 602. Elsevier Science Publishers, New York, 1986.

[46] R. J. Davis, G. B. Bulkley, and R. J. Traystman, *Fed. Proc., Fed. Am. Soc. Exp. Med.* **46,** 799 (1987).

quantities of xanthine oxidase in their parenchymal cells, and that allopurinol can prevent neutrophil accumulation in postischemic intestine focused attention on the endothelial cell. This was suggested by the immunohistochemical studies of Jarasch and others[47] which appeared to localize xanthine oxidase activity in the microvascular endothelium of a number of organs, including the human heart. Subsequently, Ratych and colleagues[48] found that xanthine dehydrogenase in cultured rat pulmonary artery endothelial cells was converted to high levels of xanthine oxidase during periods of anoxia. Moreover, 45 min of anoxia followed by 30 min of reoxygenation was found to produce "cell lysis," which was prevented by either allopurinol or SOD and catalase at the time of reoxygenation.

These results indicate that the entire xanthine oxidase free radical-generating system is present and operative within the endothelial cell itself, in the absence of either neutrophils or parenchymal cells. We have proposed that this endothelial cell trigger mechanism is the ubiquitous initiator of free radical-mediated reperfusion injury (Fig. 2). In some organs this injury to the endothelial cell itself may account for the ultimate loss of organ function via microvascular thrombosis.[49] In other organs the endothelial cell may act as an initial trigger to attract and activate neutrophils, which themselves then cause the major part of the ultimate injury.[17,22] This neutrophil amplifying system appears to be of variable quantitative importance from organ to organ, perhaps corresponding to the variable accumulation of neutrophils in different organs subject to ischemia.[50]

Quantitative Importance of Reperfusion Injury

From the above review, it appears that oxygen-derived free radicals participate in reperfusion injury in a number of models of postischemic organ injury. In other models, the impact of this mechanism seems mimimal.[39,49,51] What elements of these models are critical to produce a measurable reperfusion injury? In the cat small intestine, this mechanism of reperfusion injury appears to mediate the increase in capillary permeability after 1 hr and the necrosis of the villus epithelium after 3 hr of partial

[47] G. Bruder, H. W. Heid, E. D. Jarasch, and I. H. Mather, *Differentiation* **23**, 218 (1983).

[48] R. E. Ratych, R. S. Chuknyiska, and G. B. Bulkley, *Surgery (St. Louis)* **102**, 122 (1987).

[49] L. Marzella, R. R. Jesudass, P. N. Manson, R. A. M. Myers, and G. B. Bulkley, *Plast. Reconstr. Surg.* **81**, 742 (1988).

[50] M. B. Grisham and D. N. Granger, *in* "Splanchnic Ischemia and Multiple Organ Failure" (A. Marston, G. B. Bulkley, R. G. Fiddian-Green, and U. H. Haglund, eds.), p. 140. Arnold, London, 1989.

[51] U. Haglund, G. B. Bulkley, and D. N. Granger, *Acta Chir. Scand.* **153**, 321 (1987).

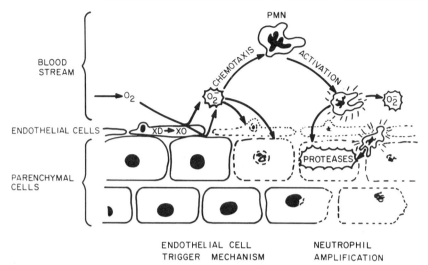

FIG. 2. Endothelial cell trigger mechanism. During ischemia, xanthine dehydrogenase (XD) is converted to xanthine oxidase (XO) within the endothelial cell. Molecular oxygen (O_2) is reintroduced at the time of reperfusion, permitting the xanthine oxidase-catalyzed production of superoxide (O_2^-). Cellular injury may occur as a direct consequence of this endothelial cell radical production or as a result of secondary neutrophil activation that acts as an amplifier of the initial endothelial cell injury. [From R. E. Ratych, R. S. Chuknyiska, and G. B. Bulkley, *Surgery* (*St. Louis*) **102**, 122 (1987).]

ischemia/reperfusion. Longer or more severe degrees of ischemia produce a more massive injury, but this latter injury is unaffected by free radical ablation.[8,51,52]

The quantitative impact of the reperfusion mechanism was evaluated formally in a porcine model of human cadaveric renal transplantation.[53] Hoshino *et al.*[53] studied the effect of varying periods of ischemic preservation on the efficacy of free radical ablation for the preservation of posttransplant renal function. Treatment with SOD significantly improved the function of kidneys that were subjected to 1 or 2 hr but not to 3 hr of warm ischemia (alone) prior to transplantation. Similarly, allopurinol significantly ameliorated the injury that occurred after either 24 or 48 hr of cold ischemia (Fig. 3A). However, allopurinol had no effect on function in kidneys which had undergone either shorter (18 hr) or longer (72 hr)

[52] D. A. Parks, D. N. Granger, and G. B. Bulkley, *Fed. Proc., Fed. Am. Soc. Exp. Biol.* **41,** 1742 (1982).
[53] T. Hoshino, W. R. Maley, G. B. Bulkley, and G. M. Williams, *Transplantation* **45,** 284 (1988).

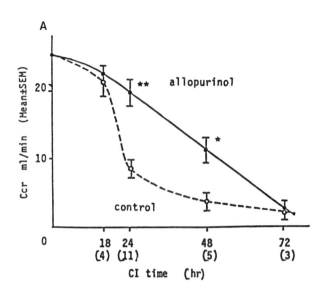

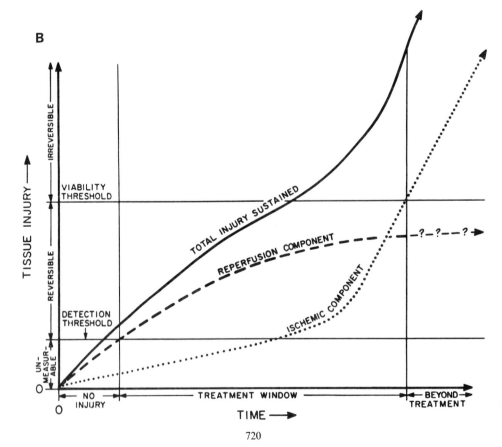

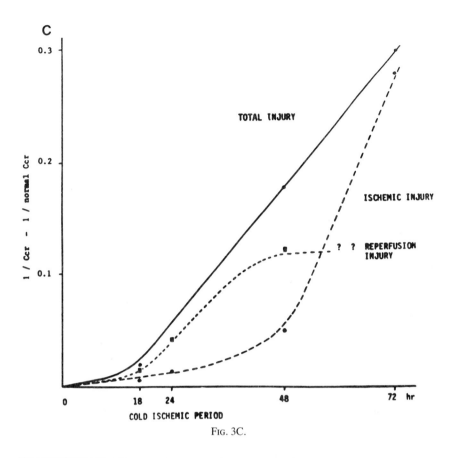

FIG. 3C.

FIG. 3. Therapeutic window. (A) Postischemic renal function. Posttransplant renal function of paired allopurinol-treated and control pig kidneys is shown following various periods of cold ischemia (CI, cold ischemia time; C_{cr}, creatinine clearance; number of experiments is given in parentheses; *, $p < .05$ versus control; **, $p < .01$ versus control). [From T. Hoshino, W. R. Maley, G. B. Bulkley, and G. M. Williams, *Transplantation* **45**, 284 (1988).] (B) Theoretical curve of the components of ischemia/reperfusion injury. Both components of this injury, that due to ischemia per se and the biochemical processes which result in the reperfusion injury (see Fig. 1), begin at the start of ischemia. The component due to ischemia itself appears later but eventually overwhelms any contribution due to reperfusion. Very early in the ischemic period, neither component is significant. In the intermediate period (in most models), about two-thirds of the injury is due to the reperfusion mechanism. Free radical-ablative therapies will only ameliorate the injury following ischemic periods from within this treatment window. [Modified from G. B. Bulkley, *Br. J. Cancer* **55**, 66 (1987).] (C) Ischemic and reperfusion components of injury. Tissue injury in transplanted kidneys is plotted as a function of cold ischemia time. The data from (A) have been recalculated and presented here to allow demonstration of the magnitude of the reperfusion injury. Tissue injury has been defined as $1/C_{cr} - 1/C_{cr}$ (normal). The values for the ischemic injury curve are the results from the allopurinol-treated kidneys. The data for the total injury curve come from the values for the control kidneys. The values for reperfusion injury were calculated by subtracting the value of ischemic from total injury. [From T. Hoshino, W. R. Maley, G. B. Bulkley, and G. M. Williams, *Transplantation* **45**, 284 (1988).]

periods of cold ischemia prior to transplantation. The kidneys preserved only 18 hr functioned well whether or not they were treated, while the kidneys preserved 72 hrs had poor function, regardless of treatment. Thus, there was a defined time period (between 24 and 48 hr) during which ablation of free radical generation at reperfusion provided a measurable, statistically significant, and clinically beneficial effect. As illustrated in Fig. 3B, this results from the fact that at short durations of ischemia the degree of injury due to reperfusion is so small that diminishing it has no noticeable effect. On the other hand, after long periods of ischemia the injury due to ischemia itself is so great that preventing the reperfusion injury has no measurable nor clinically meaningful effect on organ function.

These findings may well explain why different investigators studying the same organ have found vastly different results for free radical scavenging on postischemic injury. Since there is a wide variation in the tolerance to ischemia itself in different organs, the size and timing of the window appear to be quite variable. This variation appears to be more related to the organ itself, and to the ischemic conditions, than to the particular pharmacologic agent that is used for free radical ablation.

Clinical Significance

The clinical importance of reperfusion injury depends fundamentally on the relative width of the therapeutic window. For example, we have found that SOD significantly ameliorates the frostbite injury to a rabbit ear caused by cold ($-21°$) immersion for 60 or 90 sec (P. Manson and G. Bulkley, unpublished data).[49] However, SOD has no beneficial effect on ears subjected to 30 or to 120 sec of freezing. Since it is unlikely that a victim of frostbite will be subjected to less than 2 min of cold, this finding, although interesting, is probably not very relevant clinically.

On the other hand, a recent study of human renal transplant patients did find a striking therapeutic benefit of SOD when it was given after a precise but clinically significant cold–ischemia time window (Fig. 4).[54] In this study, the first clinical trial of free radical ablation for the prevention of reperfusion injury, there was an overall trend toward improved function in kidneys given SOD at implantation (reperfusion). This difference was only statistically significant, however, in kidneys preserved for a time that fell within the above-defined therapeutic window. As the mechanism of free radical-mediated reperfusion injury is investigated in a wider range

[54] H. Schneeberger, W. D. Illner, D. Abendroth, G. Bulkley, F. Rutili, M. Williams, M. Thiel, and W. Land, *Transplant. Proc.* **21**, 1245 (1989).

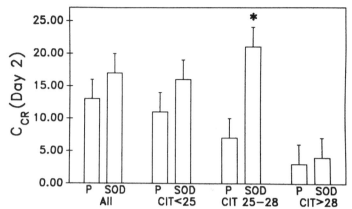

FIG. 4. Quantitative importance of free radical ablation in human renal transplantation. SOD infusion at implantation produced a significant improvement in renal function only in kidneys subjected to 25–28 hr of cold ischemia [H. Schneeberger, W. D. Illner, D. Abendroth, G. Bulkley, F. Rutili, G. M. Williams, M. Thiel, and W. Land, *Transplant. Proc.* **21,** 1245 (1989)]. This effect was less significant in kidneys subjected either to longer or to shorter periods of ischemia prior to transplantation. C_{CR}, Creatinine clearance; P, placebo treatment; CIT, cold ischemia time; *, $p < .05$ versus placebo.

of experimental models, and tested in further clinical trials, it seems likely that this therapeutic window, defined by the conditions of ischemia and by the organ itself, will be the major determinant of both scientific and clinical significance.

[79] Noninvasive Fluorometric Measurement of Mitochondrial Membrane Potential in Isolated Working Rat Hearts during Ischemia and Reperfusion

By JÜRGEN FUCHS, G. ZIMMER, T. THÜRICH, J. BEREITER-HAHN, and LESTER PACKER

Introduction

Direct evidence for the validity of the free radical hypothesis of ischemia/reperfusion injury in perfused hearts has recently been obtained by electron paramagnetic resonance spectroscopy.[1] Mitochondria are target organelles of the ischemia/reperfusion syndrome,[2] and preservation of

[1] J. L. Zweier, J. T. Flaherty, and M. L. Weisfeldt, *Proc. Natl. Acad. Sci. U.S.A.* **84,** 140 (1987).

[2] F. Beyersdorf, C. Gauhl, O. Elert, and P. Satter, *Basic Res. Cardiol.* **76,** 106 (1981).

mitochondrial structure and function is one of the main objectives in cardioprotection.[3] In addition to ^{31}P nuclear magnetic resonance spectroscopy, surface fluorimetry is a suitable method to investigate energy metabolism noninvasively in isolated organs. The fluorochrome DASPMI, 2-[4-(dimethylamino)styryl]-1-methylpyridinium iodine, constitutes a specific in situ probe to monitor mitochondrial electrochemical membrane potential.[4,5] We have used DASPMI to stain mitochondria in the isolated perfused and working rat heart and to measure continuously mitochondrial membrane potentials during ischemia and reperfusion and under the influence of uncouplers of oxidative phosphorylation. This method provides a new, noninvasive tool in studying mitochondrial pathology in an isolated organ under conditions of oxidative stress.

Perfusion Model

Isolated rat hearts are perfused via the left atrium according to the anterograde method described by Taegtmeyer et al.[6] and Neely and Rovetto.[7] Ischemia is induced by gassing the perfusion medium with a nitrogen–carbon dioxide mixture. To avoid temperature-dependent changes in fluorescence intensity, the cardiac preload is not decreased. Within minutes aortic flow ceases and coronary flow decreases to about 50% of its original value. After a period of 15 min the perfusion medium is reoxygenated to provide normoxic reperfusion, and aortic flow recovers to about 50% of its original value. A prolonged ischemia period (60 min) results in an irreversible decline of aortic flow. For detailed description of the perfusion model, see also Ref. 8.

Epifluorescence Technique

The experimental device resembles that described by Chance et al.[9] A 50 W mercury arc lamp (Model HBO 50, Leitz GmbH, Wetzlar, FRG) equipped with a magnetic lock ventile (Model 0, Prontor Werke GmbH, Wildbad, FRG) is used as a light source. Excitation light is passed through a band pass filter (450–490 nm, Schott, Mainz, FRG) and transmitted to the heart via glass fibers. Emitted fluorescence is collected by fibers inter-

[3] G. Zimmer, F. Beyersdorf, and J. Fuchs, Mol. Physiol. 8, 495 (1985).
[4] J. Bereiter-Hahn, Biochim. Biophys. Acta 423, 1 (1976).
[5] J. Bereiter-Hahn, K. H. Seipel, and M. Vöth, Cell Biochem. Funct. 1, 147 (1983).
[6] H. Taegtmeyer, R. Hems, and H. A. Krebs, Biochem. J. 186, 701 (1980).
[7] J. R. Neely and M. J. Rovetto, this series, Vol. 39, p. 43.
[8] J. Fuchs, P. Veit, and G. Zimmer, Basic Res. Cardiol. 80, 231 (1985).
[9] B. Chance, A. Mayevsky, C. Goodwin, and L. Mela, Microvasc. Res. 8, 276 (1974).

mingled with the excitation fibers in the same strand, passed through an OG 530 long pass filter (>520 nm, Schott, Mainz, FRG) and a photomultiplier (Model EM I 6256 A, SKV/S, BN 316 V, S/N, Knott Elektronic, München, FRG). Fluorescence is then recorded by a photon counter (SSR photon counter, Model 1105 data converter, Api Instr. Co.) and an amplifier system (SSR amplifier discriminator, Model 1120, Api Instr. Co.). For recording a single-channel recorder (BD 40, Kipp and Zonen, Kronberg, FRG) is used. The glass fiber strand (Volpi AG, Schlieren, Switzerland) is attached to the left ventricular wall and fixed with Histoacryl tissue glue (Braun Melsungen, Bad Homburg, FRG). A magnetic shutter (Model O, Prontor Werke GmbH) is opened by an impulse generator (function generator 3300A, Hewlett-Packard) every 30 sec for 1 sec using a steering device (Model 1017019, Prontor Werke GmbH) to allow fluorimetric measurements. For schematic description of the apparatus, see Fig. 1. The fluorochrome DASPMI,[10] commercially available through Sigma Chemical Company (St. Louis, MO), is added directly to the perfusion medium at a final concentration of 0.33 μM.

Measurement of Mitochondrial Membrane Potential *in Situ*

A steady-state level in fluorescence intensity at the surface of the isolated working rat heart is reached about 45 min after addition of 0.33 μM DASPMI to the perfusion buffer (Fig. 2). Flavin fluorescence does not contribute significantly to DASPMI fluorescence, and heart performance, monitored by aortic flow, is not affected by the fluorochrome at this concentration. Fluorescence micrographs show a selective accumulation of the dye in mitochondria. Uncouplers of oxidative phosphorylation like CCCP (*m*-chlorocarbonyl cyanide phenylhydrazone) decrease fluorescence intensity within 20 min to baseline levels (Fig. 2). In the initial phase there is a parallel decrease in aortic flow and fluorescence intensity. A short ischemic insult (15 min) induces a rapid decrease of fluorescence to a level of about 50% of the original intensity. This is accompanied by a total decline in aortic flow to zero. If the ischemia period is short (15 min) the partial decrease of fluorescence intensity is fully reversible after reperfusion and is paralleled by a partially recovering aortic flow (Fig. 3). After a longer ischemia time (60 min) neither aortic flow nor fluorescence recover (data not shown). Only mechanically recovering hearts regain preischemic fluorescence levels. Inhibition of anaerobic glycolysis during short-term ischemia (15 min) by oxamate results in irreversible decrease of fluorescence. In these hearts the fluorescence level in ischemia is sig-

---

[10] A. P. Phillips, *J. Org. Chem.* **12**, 333 (1947).

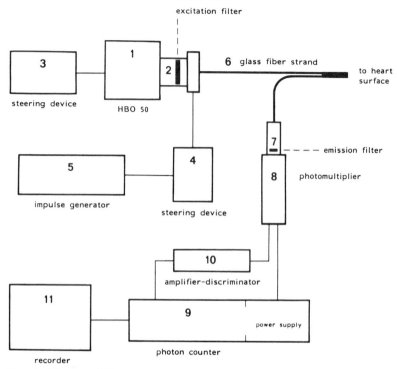

FIG. 1. Diagram of fluorimetry apparatus. Components: 1, 50 W mercury arc lamp (Model HBO 50, Leitz GmbH) equipped with a magnetic lock ventile (Model 0, Prontor Werke GmbH); 2, excitation filter 450–490 nm (Schott); 3, steering device for HBO 50 mercury lamp; 4, steering device (Model 1017019, Prontor Werke GmbH); 5, impulse generator (function generator 3300A, Hewlett-Packard); 6, glass fiber strand (Volpi AG); 7, emission filter OG 530 (>520 nm) (Schott); 8, photomultiplier (Model EM I, 6256 A, SKV/S, BN 316 V, S/N, Knott Electronic); 9, photon counter (SSR photon counter, Model 1105 data converter, Api Instr. Co.); 10, amplifier system (SSR amplifier discriminator, Model 1120, Api Instr. Co.); 11, single-channel recorder BD 40 (Kipp and Zonen).

nificantly lower than in ischemic hearts with noninhibited anaerobic glycolysis (Fig. 4).

Evaluation of Results

After addition of uncouplers, fluorescence levels indicate complete loss of membrane electrochemical potential with breakdown of all proton-pumping activities. During ischemia only partial mitochondrial deenergization is noted, which is aggravated by inhibitors of anaerobic glycolysis

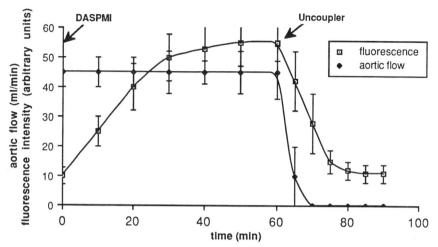

FIG. 2. Aortic flow and DASPMI fluorescence at the surface of the normoxic, isolated working rat heart. DASPMI buffer concentration is 0.33 μM, and the uncoupler CCCP was added at a final concentration of 1.5 μM. Mean values of five experiments $\pm$ standard deviation are shown.

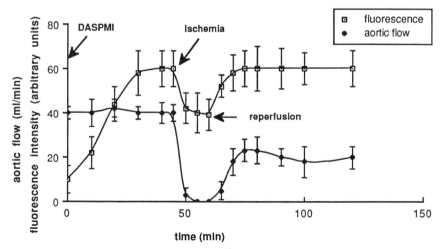

FIG. 3. Aortic flow and DASPMI fluorescence during normoxia (0–50 min), ischemia (50–65 min), and reperfusion (65–120 min) of the isolated working rat heart. DASPMI buffer concentration is 0.33 μM. Mean values of five experiments $\pm$ standard deviation are shown.

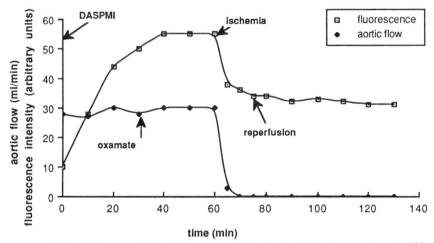

FIG. 4. Aortic flow and DASPMI fluorescence in hearts treated with oxamate, an inhibitor of anaerobic glycosis (addition of neutralized oxamate to the perfusion buffer at a final concentration of 6 mM after 40 min of normoxia), during normoxia (0–60 min), ischemia (60–75 min), and reperfusion (75–140 min). Representative data are shown.

like oxamate. Myocardial reperfusion after a short time of ischemia (15 min) results in restoration of the mitochondrial membrane potential, but heart function (aortic flow) is reduced and high-energy phosphate levels are impaired.[11] Mitochondrial ATP production is not normalized although the inner mitochondrial membrane is reenergized. Ultrastructural studies of the inner membrane in the reperfusion phase reveal a disarrangement of F_1F_0-ATPase particles,[3] an increase in ATPase activity, and a decrease in oligomycin sensitivity.[11] These structural and functional postischemic alterations of the mitochondrial ATPase complex are probably related to inefficient energy conversion in reenergized postischemic heart mitochondria. Inhibition of anaerobic glycolysis during short-term ischemia (15 min) results in an irreversible deterioration of mitochondrial membrane potential and heart function. Reperfusion after a long ischemic insult (60 min) is accompanied neither by mitochondrial reenergization nor by regain of mechanical function. Reestablishing the mitochondrial membrane potential is therefore considered as a prerequisite for production of high-energy phosphates.

In situ measurement of mitochondrial membrane potential provides a new tool to study mitochondrial pathology in isolated organs under oxida-

[11] J. Fuchs and G. Zimmer, *Biochem. Pharmacol.* **35**, 4381 (1986).

tive stress. It provides information about the energetic status of the inner mitochondrial membrane that cannot be obtained by ^{31}P nuclear magnetic resonance spectroscopy. This method may be suitable in studying the influence of drugs on the mitochondrial membrane potential during the ischemia/reperfusion syndrome of isolated working hearts.[12]

[12] J. Fuchs, G. Zimmer, and J. Bereiter-Hahn, *Cell Biochem. Funct.* **5**, 245 (1987).

[80] Assessment of Leukocyte Involvement during Ischemia and Reperfusion of Intestine

By MATTHEW B. GRISHAM, JOSEPH N. BENOIT, and D. NEIL GRANGER

Introduction

Neutrophilic polymorphonuclear leukocytes (neutrophils) have been implicated as possible mediators of mucosal injury in a variety of gastrointestinal disorders. Recent experimental evidence indicates that reperfusion of ischemic intestine initiates a series of events that culminate in an acute inflammatory response.[1] The objective of this chapter is to describe the methods and summarize the experimental findings that have led investigators to reach the conclusion that neutrophils play a role in mediating the mucosal injury observed during ischemia and reperfusion of the small bowel. In order to facilitate communication, the chapter is divided into two major sections. The first describes methods used to quantify leukocyte infiltration into the postischemic intestinal mucosa. In the second section we describe the methods used to determine whether leukocytes mediate the microvascular injury associated with reperfusion of the ischemic small bowel.

Methods for Measurement of Leukocyte Infiltration

Tissue-Associated Myeloperoxidase Activity

The biochemical methods available to study neutrophil infiltration *in vivo* have largely been confined to measuring the appearance of radiolabeled cells or granulocyte-specific enzymes such as myeloperoxidase

[1] M. B. Grisham, L. A. Hernandez, and D. N. Granger, *Am. J. Physiol.* **251**, G567 (1986).

METHODS IN ENZYMOLOGY, VOL. 186

(MPO) in tissue. Although radiolabeled cells are easy to measure, the primary disadvantage of using this technique is the requirement for the *ex vivo* manipulation (isolation and purification) of neutrophils. Such manipulation has been shown to produce subtle alterations in cell structure and function that consequently result in extensive pulmonary and hepatic margination.[2]

Other laboratories have measured tissue-associated MPO activity to quantify neutrophil infiltration.[1,3–9] The major advantages of the MPO technique are that (1) it does not require *ex vivo* manipulation of leukocytes, (2) MPO is relatively easy to measure, and (3) it quantifies the normally inhomogeneous process of inflammation. The major disadvantage is that there is no peroxidase assay available which is specific for neutrophilic MPO. This assay will measure total hemoprotein peroxidase, which could include both monocyte MPO as well as eosinophil peroxidase. This is not a severe limitation when one considers that fact that the acute inflammatory response is characterized by the infiltration of predominantly neutrophils. Because the enzymatic determination of tissue-associated MPO is based on the H_2O_2-dependent oxidation of an artificial electron donor, the assay may be susceptible to interference from certain hemoproteins such as catalase, hemoglobin, or myoglobin. In addition, MPO activity may be inhibited by the presence of naturally occurring electron donors present in tissue such as ascorbate, reduced glutathione (GSH), and possibly α-tocopherol. We have developed a procedure for the preparation of tissue that largely eliminates many of these interfering cytosolic proteins and substrates, yet provides a very sensitive assay for MPO activity that is designed to process large numbers of samples. Although this procedure was originally developed for gastrointestinal tissue, it may easily be adapted to other tissues.

Tissue Preparation. Intestinal mucosa from large animals or whole intestine from rodents is rapidly excised, rinsed with ice-cold saline, blotted dry, and frozen at $-70°$. We have found that freezing intact intestinal

[2] S. H. Saverymuttru, A. M. Peters, H. J. Danpure, H. J. Reavy, S. Osmon, and J. P. Lavender, *Scand. J. Haematol.* **30,** 151 (1983).

[3] A. L. Smith, I. Rosenberg, D. R. Averill, E. R. Moxon, T. Stossel, and D. H. Smith, *Infect. Immun.* **10,** 356 (1974).

[4] J. W. Smith and G. A. Castro, *Am. J. Physiol.* **234,** R72 (1978).

[5] P. P. Bradley, D. A. Priebat, R. D. Christensen, and G. Rothestein, *J. Invest. Dermatol.* **78,** 206 (1982).

[6] J. E. Krawisz, P. Sharon, and W. F. Stenson, *Gastroenterology* **87,** 1344 (1984).

[7] G. Allan, P. Bhattacherjee, C. D. Brook, N. G. Read, and A. J. Parke, *J. Cardiovasc. Pharm.* **7,** 1154 (1985).

[8] K. M. Mullane, R. Kraemer, and B. Smith, *J. Pharmacol. Methods* **14,** 157 (1985).

[9] S. E. Goldblum, K. M. Wu, and M. Jay, *J. Appl. Physiol.* **59,** 1978 (1985).

tissue for periods up to 2 weeks at $-70°$ does not significantly affect enzymatic activity. The tissue is then thawed, weighed, and homogenized in 10 volumes of ice-cold potassium phosphate buffer (pH 7.4) using a 20-sec burst on setting 6 of a Brinkman tissue homogenizer (Model PT10/35). The homogenate is centrifuged at 20,000 g for 20 min at 4°. The supernatant, which contains most of the cytosolic proteins and substrates (hemoglobin, myoglobin, catalase, ascorbate, GSH) but less than 10% of the MPO, is discarded. The pellet, which contains greater than 90% of the total MPO activity, is homogenized in 10 volumes of ice-cold 50 mM potassium phosphate buffer (pH 6.0) containing 0.5% hexadecyltrimethylammonium bromide (HETAB) and 10 mM EDTA. The addition of HETAB to the second homogenization buffer serves two important functions: (1) it solubilizes MPO bound to the granular membranes, and (2) it completely inhibits the pseudoperoxidase activity of hemoglobin and myoglobin that may remain associated with the membranous pellet (Fig. 1). Apparently, HETAB selectively destroys the heme nucleus of hemoglobin and myoglobin without effecting MPO activity (M. B. Grisham, unpublished observations). In the final step, we subject the HETAB-containing homogenate to one cycle of freezing and thawing and a brief period (15 sec) of sonication. We have found that approximately 60–70% of the MPO is completely solubilized using this procedure; therefore, the homogenate is assayed without an additional centrifugation step.

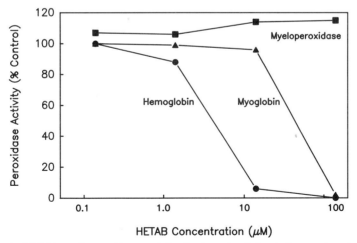

FIG. 1. Inhibition of peroxidase activity by hexadecyltrimethylammonium bromide (HETAB). The concentrations of hemoglobin (human), myoglobin (horse heart), and myeloperoxidase (rat) were adjusted such that each hemoprotein exhibited approximately 0.1 U/ml of peroxidase activity in the absence of detergent.

A recent report[10] suggests that the MPO assay should be used with caution in certain tissues such as liver, kidney, and spleen because homogenates of these tissues inhibit MPO activity. It should be pointed out that these three tissues contain tremendous amounts of catalase and GSH peroxidase, both of which would compete with MPO for H_2O_2. Unfortunately, the authors of this study did not remove these cytosolic proteins as we have proposed, using an initial homogenization and centrifugation step in the absence of detergent. Using our protocol for preparation of intestinal tissue, we can demonstrate virtually complete recovery of exogenously added MPO.

Myeloperoxidase (MPO) Assay. MPO activity is determined using a modification of the method of Suzuki *et al.*[11] in which the enzyme catalyzes the oxidation of 3,3',5,5'-tetramethylbenzidine (TMB) by H_2O_2 to yield a blue chromogen that possesses a wavelength maximum at 655 nm. Briefly, an aliquot (50 μl) of homogenate is added to a 0.5-ml reaction volume containing 80 mM potassium phosphate buffer (pH 5.4), 0.5% (w/v) HETAB, and 1.6 mM TMB added as a 10 mM stock solution dissolved in N,N'-dimethylformamide. The mixture is then warmed to 37°, and the reaction is started by the addition of 0.3 mM H_2O_2. Each tube containing the complete reaction mixture is incubated exactly 3 min at 37°. The reaction is terminated by the sequential addition of catalase (20 μg/ml) and 2 ml of 0.2 M sodium acetate (pH 3.0) spaced 3 min apart and then placed on ice. Any membranous material that may interfere with the spectrophotometric analysis can be removed by centrifugation at this point. The absorbance of each tube is determined at 655 nm and then corrected by subtracting the blank value. We define 1 unit of activity as the amount of enzyme present that produces a change in absorbance per minute of 1.0 at 37° in the final reaction volume containing the sodium acetate. We have found that the final absorbance at 655 nm must be maintained below 1.2 because fully oxidized TMB has a tendency to polymerize and precipitate, thereby producing an artifactual underestimation of MPO activity.

Using this method, we have demonstrated that ischemia and reperfusion of the feline small intestine results in dramatic increases in MPO activity, suggesting infiltration of large numbers of granulocytes into the mucosal interstitium (Fig. 2). Inasmuch as cat eosinophils contain no peroxidase activity,[12] the measured increase in peroxidase activity follow-

[10] D. J. Ormrod, G. L. Harrison, and T. E. Miller, *J. Pharmacol. Methods* **18**, 137 (1987).
[11] K. Suzuki, H. Ota, S. Sasagawa, T. Sakatoni, and T. Fujikura, *Anal. Biochem.* **132**, 345 (1983).
[12] B. Presentey, Z. Jerushalmy, M. Ben-Bassat, and K. Peck, *Anat. Rec.* **196**, 119 (1980).

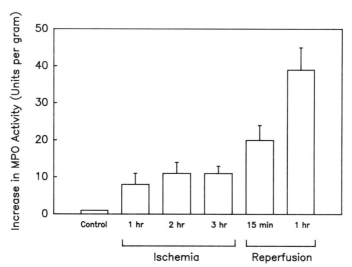

FIG. 2. Effect of ischemia and reperfusion on myeloperoxidase (MPO) activity in the feline intestinal mucosa. Data represent the increase in MPO activity compared to control values. From Zimmerman *et al. Am. J. Physiol.* **258,** G185 (1990).

ing ischemia and reperfusion presumably reflects neutrophil infiltration. Furthermore, the reperfusion-induced neutrophil accumulation is largely prevented by pretreatment with either SOD or allopurinol.[1] These data suggest that XO-derived oxidants play a role in the recruitment of neutrophils into the intestinal mucosa.

Intravital Microscopic Assessment of Leukocyte Adherence and Extravasation

An alternative approach to studying adherence of leukocytes to vascular endothelium and their subsequent infiltration in the tissue is to directly visualize these processes by employing intravital videomicroscopic techniques. With this approach, the tissue to be studied is gently positioned over an optically clear window and the microcirculation viewed directly. The following section details the methodology that we use to study leukocyte adherence in the microvasculature of the cat mesentery. It should be pointed out that these approaches can be applied to nearly all organs including muscle, brain, kidney, heart, stomach, small intestine, large intestine, lungs, and liver. However, thin tissues that can be transilluminated by conventional microscope systems are particularly well suited for study. For detailed description of the application of intravital microscopic techniques to various organs as well as information on video microscopy, the reader is referred to several excellent monographs.[13–16]

Surgical Preparation. Cats of either sex weighing 2–4 kg are anesthetized with ketamine hydrochloride (Ketalar, 50 mg/kg, i.m.) and placed in a supine position on a heated Lucite platform. A tracheostomy is performed to facilitate respiration, and the left jugular vein is cannulated for administration of supplemental sodium pentobarbital anesthetic (W. A. Butler, Columbus, OH). The right femoral artery is cannulated, and arterial pressure is continuously monitored by a pressure transducer (Cobe Electronics, Denver, CO) attached to a physiological recorder (Grass Instruments, Quincy, MA). A midline abdominal incision is made, and a segment of small intestine, jejunum, and/or ileum is surgically isolated from the upstream and downstream intestinal segments. Care is taken to ensure that the blood and lymph vessels supplying the isolated intestinal segment remain intact. The duodenum, cecum, and colon are then surgically resected. The superior mesenteric artery is subsequently dissected from the surrounding tissue and the animal heparinized by intravenous injection of heparin (1000 U/kg). An arterial circuit is then established between the left femoral artery and the superior mesenteric artery. Blood flow through the circuit is measured by an electromagnetic flow probe attached to a flow meter (Carolina Medical Electronics, King, NC). Superior mesenteric arterial pressure is measured via the side port of the flow probe.

Once the arterial circuit is established, a segment of midjejunum with its adjacent mesentery is exteriorized through the abdominal incision and the mesentery draped over an optically clear semicircular viewing pedestal (1.5 × 3 inches). The pedestal is constructed of plexiglass and is water jacketed for temperature regulation. The semicircular viewing window is constructed of clear silicone elastomer (Sylgard 184, Dow Corning, Midland, MI). The exposed small intestine is covered with warm (37°) saline-soaked gauze and Saran wrap (Dow Corning) to prevent drying and evaporative water loss. The exposed mesentery is continuously suffused by warm (37°) bicarbonate-buffered salt solution (131.9 mM NaCl, 4.7 mM

[13] C. H. Baker and W. L. Nastuk, "Microcirculatory Technology," p. 534. Academic Press, Orlando, Florida, 1986.
[14] M. P. Wiedeman, R. F. Tuma, and H. N. Mayrovitz, "An Introduction to Microcirculation," p. 226. Academic Press, New York, 1981.
[15] E. M. Renkin and C. C. Michel, *in* "Handbook of Physiology Section 2: The Cardiovascular System Volume IV (Parts 1 and 2), Microcirculation." American Physiological Society, Bethesda, Maryland, 1984.
[16] S. Inoue, "Video Microscopy," p. 584. Plenum, New York, 1986.

KCl, 2.0 mM CaCl$_2$, 1.2 mM MgSO$_4$, and 20 mM NaHCO$_3$) at a rate of 2.8 ml/min. The buffer is equilibrated with a mixture of 5% CO$_2$ and 95% N$_2$ to maintain the pH at approximately 7.4.

The preparation is then transferred to a videomicroscope (Leitz Ortholux II, Wetzlar, FRG), equipped with a 12-V, 100-W DC-stabilized light source, for further study. Images from the microscope are projected onto a high-resolution monochrome monitor (Cohu Electronics, San Diego, CA) by a Neuvicon tube video camera (Dage, MTI-65). Images from the video camera are recorded with a video cassette recorder (Panasonic NV-8950). A video time–date generator (Panasonic WJ-810) with stopwatch function interposed in the video circuit imprints the elapsed time on the video tape. The microcirculation is visualized with a 40× objective lens (Zeiss UD40, N.A. 0.60) to yield an eyepiece magnification of 400×. A 2.5× projection eyepiece (Zeiss, West Germany) is placed in front of the video camera to yield a magnification of 1400× on the video screen.

Single unbranched arterioles and venules ranging between 25 and 35 μm are selected for study in order to minimize changes in blood flow related to turbulence that occurs at or near points of bifurcation. Microvascular diameter (D) is measured with a video image shearing monitor (I.P.M., La Mesa, CA) or video caliper (Microcirculation Research Institute, Texas A&M University). Red blood cell centerline velocity (V_c) is measured with an optical Doppler velocimeter (Microcirculation Research Institute, Texas A&M University) interposed in the optical path of the microscope. Microvascular blood flow (nl/sec) is calculated from microvessel cross-sectional area (derived from diameter measurements with the assumption of cylindrical vessel geometry) and mean red blood cell velocity (i.e., $V_m = V_c/1.6$).[13,15] Wall shear rate (γ), a force tending to oppose neutrophil adherence to the blood vessel surface, is calculated from the Newtonian definition [$\gamma = 8(V_m/D)$].

The number of leukocytes adherent to the vascular endothelium, the number of leukocytes that traverse the vascular wall (i.e., extravasated leukocytes), and the rolling velocity of the marginated pool of leukocytes are determined during playback of the videotaped image. All data are obtained from venules since adherence of leukocytes to arteriolar endothelium is rarely observed. We define a leukocyte as adherent if it remains stationary for at least 30 sec. A leukocyte is defined as extravasated if the entire cell is located outside of the venular lumen. Rolling leukocytes are the marginated leukocytes that appear to roll along the endothelial surface at a much slower velocity than the red blood cells flowing in the center of the microvessel. Leukocyte rolling velocity is measured during frame-by-frame playback of the videotaped image by determining the time required for a leukocyte to travel a given distance (50 μm) along the length of the venule. In order to minimize variation in adherence values, all numbers were normalized to a vessel length of 100 μm. We have found that normalization in this fashion is suitable when vessel diameter is relatively constant (e.g., 25–35 μm). If a wide range of vessel diameters are studied, then perhaps it would be more appropriate to express adherence as a function of endothelial surface (πD × length).

This method of studying leukocyte adherence and extravasation has recently been employed by us to assess the effects of ischemia and reperfusion on leukocyte kinetics in the cat mesentery. Figure 3 shows the effects of 60 min of intestinal ischemia (produced by lowering superior mesenteric blood flow to 20% of control) and reperfusion on the number of adherent and extravasated leukocytes observed in mesenteric venules. The results indicate that the number of adherent and extravasated leukocytes are increased during both the ischemic and reperfusion periods.

Leukocytes as Mediators of Microvascular Injury

Capillary Permeability Measurements as Index of Microvascular Injury

A sensitive measure of microvascular permeability which is not influenced by capillary surface area is the osmotic reflection coefficient (σ). The osmotic reflection coefficient describes the fraction of the total protein osmotic pressure generated across a membrane (impermeant proteins generate 100% of their maximum osmotic pressure, $\sigma = 1$, whereas freely permeable proteins generate no osmotic pressure, $\sigma = 0$). We have developed a technique for estimating the osmotic reflection coefficient of intestinal capillaries to plasma proteins using lymphatic protein flux data.[17] The reflection coefficient is estimated from the steady-state relationship between the lymph-to-plasma protein concentration ratio (C_L/C_P) and lymph flow. As lymph flow is increased by elevating capillary hydrostatic pressure, C_L/C_P rapidly decreases (filtration rate dependent) and then becomes relatively constant at a minimal value (filtration rate independent) when lymph flow is high. At low capillary pressures, the exchange of macromolecules across the intestinal capillary wall occurs by both diffusion and convection. Elevation of capillary pressure increases the convective movement of macromolecules across the capillary wall while at the same time the diffusive contribution to total exchange is reduced to a negligible level. Theoretical and experimental evidence suggests that $\sigma_d = 1 - C_L/C_P$ when C_L/C_P is filtration rate independent, i.e., when

---

[17] D. N. Granger and A. E. Taylor, *Am. J. Physiol.* **238**, H457 (1980).

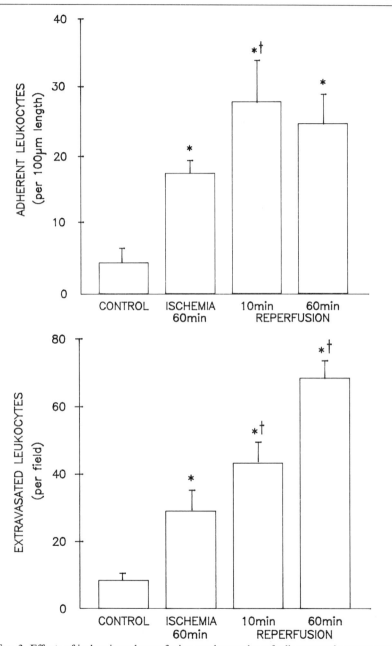

FIG. 3. Effects of ischemia and reperfusion on the number of adherent and extravasated leukocytes in the cat mesentery. Adherent leukocytes are expressed as number per 100-μm length of venule. Extravasated leukocytes are expressed as number per video field (0.0425 mm^2).

TABLE I

MODULATION OF ISCHEMIA/REPERFUSION-INDUCED INCREASE
IN MICROVASCULAR PERMEABILITY[a]

Condition	Microvascular permeability $(1 - \sigma)$
Control	0.08 ± 0.005
Ischemia (1 hr)	0.15 ± 0.03
Ischemia (1 hr) + reperfusion	0.41 ± 0.02
I/R + treatment with	
Allopurinol	0.18 ± 0.01
Folic acid	0.16 ± 0.04
Pterin aldehyde	0.15 ± 0.02
Tungsten-supplemented diet	0.20 ± 0.02
Superoxide dismutase	0.14 ± 0.01
Cu-DIPS	0.19 ± 0.03
Catalase	0.19 ± 0.01
Dimethyl sulfoxide	0.19 ± 0.02
Desferrioxamine	0.15 ± 0.01
Iron-loaded desferrioxamine	0.44 ± 0.03
Apotransferrin	0.17 ± 0.01
Antineutrophil serum	0.13 ± 0.01
Monoclonal antibody 60.3	0.12 ± 0.01

[a] Modified from Ref. 18.

diffusive exchange is negligible. In our studies microvascular permeability was expressed as $1 - \sigma$.

We have used the osmotic reflection coefficient measurements to define the influence of ischemia and reperfusion on the integrity of the intestinal microvasculature.[18] The experimental model employed in these studies is identical to that described above for intravital microscopy measurements of leukocyte adherence with the exception that the vein and lymphatic draining the small bowel are cannulated. Cannulation of the superior mesenteric vein is necessary for manipulation of intestinal capillary pressure. Lymphatic cannulation allows for measurement of lymph flow and lymph protein concentration. The experimental protocol simply involves measurements of lymph flow and lymph and plasma protein concentrations at intestinal venous pressures of 0, 10, 20, 30, and 40 mm Hg.

If the cat small bowel is subjected to 1 hr of ischemia (blood flow reduced to 20% of control) without reperfusion, a doubling of microvascular permeability is noted (Table I). However, the same period of ischemia

[18] D. N. Granger, *Am. J. Physiol.* **255,** H1269 (1988).

followed by reperfusion causes a 5-fold increase in microvascular permeability. In a number of studies we have demonstrated that xanthine oxidase inhibitors (allopurinol, pterin aldehyde) and a variety of antioxidants (e.g., superoxide dismutase, catalase, dimethyl sulfoxide) largely prevent the increased microvascular permeability induced by ischemia/reperfusion (Table I). Based on these observations we proposed that xanthine oxidase-derived oxidants formed during reoxygenation of the ischemic (hypoxic) bowel mediate the microvascular injury induced by ischemia/reperfusion. The assumption that reoxygenation accounts for the greater rise in permeability after reperfusion is supported by the consistent observation that antioxidants and inhibitors of oxy radical formation (e.g., allopurinol) attenuate only that component of the increased permeability that is manifested after reperfusion.

Manipulation of Leukocyte Function in Vivo

Two approaches have been used to define the role of circulating neutrophils in mediating reperfusion-induced increases in microvascular permeability in the gastrointestinal tract: (1) rendering animals neutropenic with polyclonal antibodies and (2) prevention of neutrophil adherence to microvascular endothelium with monoclonal antibodies directed against specific leukocyte adhesion molecules. We have developed antisera to cat and rat neutrophils by immunization of New Zealand White rabbits. After absorption, the antineutrophil sera (ANS) have agglutination titers against neutrophils of 1 : 1024. *In vitro* and *in vivo* characterization of the ANS reveals minimal cross-reactivity with red blood cells, lymphocytes, monocytes, and platelets. Figure 4 illustrates the time course of neutropenia induced by rat ANS after intraperitoneal administration. Approximately 6 hr is required for reduction of circulating neutrophils to less than 10% of control after intraperitoneal administration. However, we have rendered cats neutropenic (circulating level less than 5% of control) within 1 hr if ANS is administered intraarterially.[19]

We have demonstrated that neutropenia induced by ANS affords significant protection against reperfusion-induced increases in microvascular permeability (Table I). The degree of protection afforded by neutropenia is comparable to that observed with xanthine oxidase inhibitors and oxy radical-scavenging enzymes. In our experience circulating neutrophil levels must be reduced below 20% of control in order for ANS to offer significant protection. We have also used agents that interfere with leukocyte production (hydroxyurea) to render cats neutropenic. While hy-

[19] L. A. Hernandez, M. B. Grisham, B. Twohig, K. E. Arfors, J. M. Harlan, and D. N. Granger, *Am. J. Physiol.* **253**, H699 (1987).

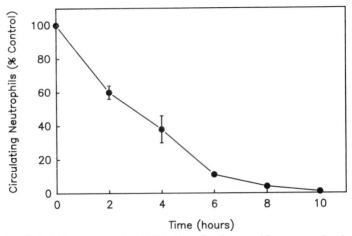

FIG. 4. Effect of intraperitoneal administration of antineutrophil serum on the circulating number of neutrophils.

droxyurea administration for 4 days reduces circulating neutrophils to less than 5% of control, there is significant mucosal injury, presumably owing to the influence of these agents on the rapidly proliferating intestinal epithelium.

The neutrophil membrane glycoprotein CD18 has been shown to play an important role in mediating neutrophil adhesiveness. Monoclonal antibodies directed to a function-related epitope on CD18 inhibit the spreading on plastic, chemotaxis, aggregation, and adherence to endothelial monolayers of human and rabbit neutrophils.[20] We have determined that MAb 60.3 and IB$_4$ (which is directed to the same epitope as 60.3) are very effective in preventing the adherence of cat neutrophils to a glass surface (Fig. 5). Furthermore, both monoclonal antibodies reduce the number of leukocytes adherent to mesenteric venules after 1 hr of reperfusion by 65–80%. In order to assess the importance of leukocyte adherence in reperfusion-induced injury to the intestinal microvasculature, we administered MAb 60.3 to cats subjected to our standard bowel ischemia/reperfusion regimen. The results obtained from these experiments (Table I) demonstrate that prevention of neutrophil adherence to microvascular endothelium is as effective as neutropenia in attenuating reperfusion-induced microvascular injury. These findings indicate that adherence to microvascular endothelium is an essential step in neutrophil-mediated reperfusion injury.

[20] J. M. Harlan, B. R. Schwartz, W. J. Wallis, and T. H. Pohlman, in "Leukocyte Emigration and Its Sequelae" (H. Z. Movat, ed.), p. 94. Karger, Basel, 1987.

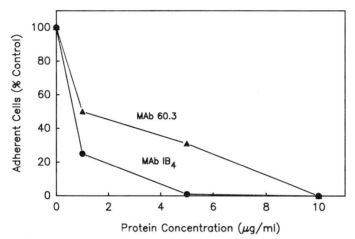

FIG. 5. Effect of monoclonal antibody (MAb) IB_4 or 60.3 on adherence of cat neutrophils *in vitro*.

Working Hypothesis

Based on the results of our studies on microvascular permeability and leukocyte kinetics in the postischemic bowel, we have proposed a hypothesis to explain the relative roles of xanthine oxidase-derived oxidants

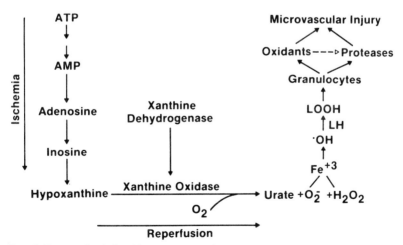

FIG. 6. Proposed relationship among xanthine oxidase-generated oxidants, neutrophil infiltration, and microvascular injury. Reperfusion and reoxygenation of ischemic intestine initiate the production of xanthine oxidase-derived oxidants, which promote the formation of proinflammatory mediators that attract and activate circulating neutrophils. Adherent and/or extravasated granulocytes may mediate microvascular injury using oxidative and/or nonoxidative mechanisms.

and neutrophils in reperfusion-induced injury to the intestinal microvasculature (Fig. 6). We propose that during the ischemic period, ATP is catabolized to yield hypoxanthine. The ischemic stress also triggers the conversion of NAD-reducing xanthine dehydrogenase to the oxygen radical-producing xanthine oxidase. During reperfusion molecular oxygen is reintroduced into the tissue where it reacts with hypoxanthine and xanthine oxidase to produce a burst of superoxide anion and hydrogen peroxide. In the presence of iron, superoxide anion and hydrogen peroxide react via the Haber–Weiss reaction to form hydroxyl radicals. This highly reactive and cytotoxic radical then initiates lipid peroxidation of cell membrane components and the subsequent release of substances that attract, activate, and promote the adherence of granulocytes to microvascular endothelium. The adherent granulocytes then cause endothelial cell injury via the release of superoxide and various proteases.

Acknowledgments

Supported by a grant from the National Institutes of Health (DK 33594).

[81] Oxygen Scavengers in Myocardial Preservation during Transplantation

By JAMES R. STEWART, WILLIAM H. FRIST, and WALTER H. MERRILL

Introduction

Heart transplantation has emerged as an important therapy in the treatment of end-stage heart disease. Advances in surgical techniques, myocardial preservation, and the advent of cyclosporin immunosuppression have been responsible for the widespread application of cardiac transplantation. As more experience has been gained with preoperative evaluation, operative techniques, and postoperative care of patients, indications have been expanded and restrictions have been relaxed, resulting in increased numbers of patients who would potentially benefit from this form of therapy.

Current estimates are that approximately 10,000 people in the United States would benefit from heart transplantation each year. Despite the increasing number of centers offering this therapeutic modality, in 1987 the number of heart transplants reached a plateau of 1,500 owing to the relatively constant and limited supply of acceptable donors. Since the introduction of distant organ procurement in 1977, the generally accepted limit of hypothermic preservation of the human heart for transplantation has been approximately 4 hr. Prolongation of acceptable donor ischemic time holds the potential of increasing the allowable travel distance between the recipient and the donor, thereby enlarging the potential pool of donors available to any one recipient.

Oxy Radicals in Heart Transplantation

A number of investigators have evaluated the potential benefits of oxygen radical scavengers in the setting of experimental heart transplantation.[1-8] For example, we previously reported our experience using superoxide dismutase in an experimental model of ovine orthotopic cardiac transplantation.[1] Hearts received standard cardioplegic solution, were excised, then stored for 6 hr at 4° in a balanced electrolyte solution. After the ischemic interval, the recipient animal was placed on total cardiopulmonary bypass. A recipient cardiectomy was performed, and orthotopic cardiac transplantation was carried out. With the restoration of coronary blood flow, hearts were rewarmed, defibrillated, and maintained in the beating, vented, and nonworking state for 45 min. Cardiopulmonary bypass was terminated, and functional measurements were repeated. Platelet deposition was determined using ^{111}In-labeled autologous platelets, and coronary blood flow measurements were repeated. At the termination of the study, a sample of left ventricular (LV) free wall was obtained for determination of lipid peroxidation products using the thiobarbituric acid assay (TBA). There were two experimental groups: the first group served as an experimental control, receiving only standard treatment, and the

[1] J. R. Stewart, E. B. Gerhardt, C. J. Wehr, T. Shuman, W. H. Merrill, J. W. Hammon, and H. W. Bender, Ann. Thorac. Surg. **42**, 390 (1986).

[2] F. Gharagozloo, F. J. Melendez, R. A. Hein, R. G. Laurence, M. S. Rosenzweig, R. J. Shemin, V. J. DiSesa, and I. H. Cohn, Surg. Forum **37**, 243 (1986).

[3] F. Gharagozloo, F. J. Melendez, R. A. Hein, R. J. Shemin, V. J. DiSesa, and L. H. Cohn, J. Thorac. Cardiovasc. Surg. **95**, 2008 (1988).

[4] M. J. Jurmann, H.-J. Schaefers, L. Dummenhayn, and A. Haverich, J. Thorac. Cardiovasc. Surg. **95**, 368 (1988).

[5] K. Bando, M. Tago, and S. Teramoto, J. Thorac. Cardiovasc. Surg. **95**, 465 (1988).

[6] H. Teraoka, M. Tago, K. Bando, S. Seno, Y. Senoo, and S. Teramoto, J. Heart Transplant. **7**, 53 (1988).

[7] J. Bergsland, L. LoBalsamo, P. Lajos, M. J. Feldman, and B. Mookerjee, J. Heart Transplant. **6**, 137 (1987).

[8] H. G. Davtyan, A. F. Corno, H. Lahs, S. Bhata, W. M. Flynn, C. Laidig, P. Chang, and D. Drinkwater, J. Thorac. Cardiovasc. Surg. **96**, 44 (1988).

second group received identical treatment plus the addition of 10 mg superoxide dismutase (SOD) to the cardioplegic solution, and 20 mg SOD was given into the ascending aorta immediately before reperfusion.

Our results documented improved postischemic LV function in the group receiving SOD. Left ventricular elastance, as determined by the slope of the end-systolic pressure–volume relationship (E_{max}), was significantly better in SOD-treated hearts (Fig. 1). Similarly, the load-independent relationship of stroke-work versus left ventricular end-diastolic volume showed improved postischemic recovery in hearts receiving SOD compared to controls (0.42 ± 0.17 in Group 2, and 0.18 ± 0.13 in Group 1). There was an increase in coronary blood flow during reperfusion in both groups due to the hemodilution of cardiopulmonary bypass, but the redistribution of blood flow away from the endocardium (reflected in the endocardial versus epicardial ratios), was significantly less in SOD-treated hearts. The ratio of tissue-to-blood scintigraphic platelet counts (Fig. 2) was significantly greater in the Group 1 controls compared to hearts receiving SOD. In addition, TBA-reactive lipid peroxidation products were significantly greater in the untreated group compared to the group receiving SOD.

The cardiac surgical group at Brigham and Women's Hospital has reproduced these results using an *ex vivo* asanguinous perfusion system for extended preservation of the ovine heart.[2,3] After 8 hr of preservation,

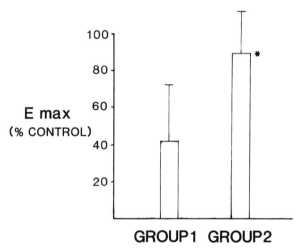

FIG. 1. Left ventricular function in control (Group 1) and SOD-treated hearts (Group 2). Left ventricular elastance, E_{max}, as determined by the slope of the end-systolic pressure–volume relationship, is presented for both groups as the percentage of baseline.

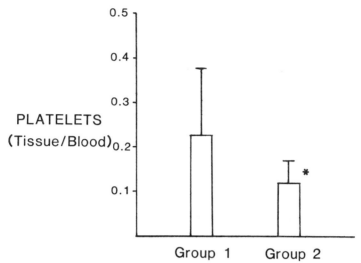

FIG. 2. Platelet deposition, measured as the ratio of tissue-to-blood ^{111}In-labeled platelet counts for control (Group 1) and SOD-treated hearts (Group 2).

measurements of cardiac function and metabolism were repeated. Hearts receiving SOD and catalase (CAT) had improved biventricular systolic function and compliance after the 8-hr ischemic interval compared to a control group. In addition, hearts receiving free radical scavengers had higher myocardial oxygen consumption with less lactate production than controls.

More recently, several other investigators have evaluated the potential application of xanthine oxidase inhibitors SOD and/or CAT as potential adjuvants for prolonged myocardial preservation during experimental heart transplantation. Jurmann et al.[4] evaluated the potential benefits of oxy radical scavengers in the setting of 3 hr of hypothermic ischemia and subsequent reperfusion using an isolated, perfused porcine heart model. Improved postischemic left ventricular developed pressure, dP/dt_{max}, and coronary blood flow were noted in the group receiving SOD and CAT compared to controls. No difference in lactate/pyruvate ratios in coronary sinus effluent was detected. In contrast to *in vivo* orthotopic transplantation models, this model is flawed as postischemic function shows a continuous decline after a relatively brief ischemic interval. The model appears inherently unstable, which makes any conclusions regarding ventricular function of doubtful significance. Very low dose SOD and CAT were used, which may account for the relatively modest and temporary functional difference between the experimental group receiving oxygen free radical scavengers and controls. Serum SOD and CAT levels were not measured.

Bando and co-workers[5] used a canine model of orthotopic heart transplantation to study animals pretreated with allopurinol for 72 hr. One group received transplants immediately after excision, the other after 18 hr of hypothermic perfusion using modified Collin's solution. There were appropriate untreated control groups. Allopurinol pretreatment slowed the appearance of TBA-reactive products and the release of creatine kinase MB (CK-MB) isoenzymes in the coronary sinus effluent of both treated groups. In addition, LV systolic function was improved in groups receiving allopurinol as compared to the respective control groups.

These investigators have recently published further studies evaluating leukocyte depletion in the setting of canine orthotopic heart transplantation.[6] Use of in-line arterial filters in the cardiopulmonary bypass circuit for leukocyte depletion results in improved functional recovery and delays release of CK-MB and TBA-reactive species in the coronary effluent. A similar investigation using an isolated Langendorff perfused rat heart has demonstrated improved functional recovery and coronary blood flow in hearts receiving SOD, CAT, and allopurinol.[7]

These studies have several major flaws which make extrapolation from the experimental to the clinical setting very difficult. Measurement of TBA-reactive species, even from the coronary sinus, may be unreliable when general anesthesia is achieved using intravenous barbiturates. In their choice of experimental animals, the investigators have chosen animals with high levels of myocardial xanthine oxidase. We have shown previously that there is no detectable xanthine oxidase activity in human myocardium,[9] and this cannot be a significant source of oxygen free radicals in the human heart. However, Jarasch et al.[10] have demonstrated the presence of xanthine oxidase in human microvascular coronary endothelium using an ultrasensitive radioimmunoassay. The presence of xanthine oxidase in human microvascular coronary endothelium may function as a triggering mechanism for leukocyte adherence and further free radical formation. In this regard, the potential application of leukocyte depletion is important, but more appropriate in other experimental animals. All future investigators of free radicals in myocardial ischemia must be carried out in xanthine oxidase-deficient species such as the rabbit and the pig.

Davtyan et al.[8] using neonatal piglet hearts, have shown that the addition of SOD and CAT to a modified blood perfusate results in full func-

---

[9] L. Eddy, J. Stewart, H. Jones, D. Yellon, J. McCord, and J. Downey, Am. J. Physiol. **253,** H709 (1987).

[10] E.-D. Jarasch, G. Bruder, and H. W. Heid, Acta Physiol. Scand. Suppl. **548,** 39 (1986).

tional recovery after 12 hr of hypothermic global ischemia. They used an *ex vivo* perfusion circuit and an adult support pig to mimic transplantation. Altered storage solutions and modifications of blood reperfusate were used, then functional recovery and ultrastructural changes were compared. Long-term storage in Sacks solution with supplemental glucose, followed by perfusion with blood modified with the addition of amino acids, glucose, superoxide dismutase, and catalase, resulted in significantly improved recovery.

Discussion and Future Directions

In all published reports to date, the use of oxygen-derived free radical-scavenging agents in experimental transplantation has resulted in improved postischemic ventricular function, suggesting indirectly that oxygen free radicals may be responsible for some component of ischemic/reperfusion injury. While most evidence is inferential, measurement of lipid peroxidation products in the coronary effluent is more direct evidence of ongoing lipid peroxidation. Unfortunately, measurement of TBA-reactive species *in vivo* as a marker of oxygen free radical-mediated phospholipid oxidation is subject to numerous chemical interferences and is not specific to oxygen-dependent free radical lipid peroxidation.[11–13]

Given that xanthine oxidase is either absent or present in undetectable levels in human myocardium, questions remain about the potential role of oxygen-derived free radicals in human myocardial ischemia. If oxygen radicals are important mediators of postischemic ventricular dysfunction in humans, the proposed mechanisms of generation, be they leukocyte derived, oxidation of catecholamines, etc., must be further clarified. Unequivocal demonstration of oxygen radicals in pathologic concentrations during the ischemic/perfusion process, and their elimination with scavenging systems, must follow. Currently available methods using measurement of hydroxy-conjugated dienes, a chemical by-product and signature of oxygen free radical-mediated injury,[14] or use of electron spin-trapping agents with electron paramagnetic resonance spectroscopy may have some application.

As more information becomes available with ongoing randomized controlled studies of recombinant human SOD and human renal transplantation, investigation must eventually extend to human heart transplanta-

[11] R. P. Bird and H. H. Draper, this series, Vol. 105, p. 299.
[12] B. Samuelson, *Science* **220**, 568 (1983).
[13] F. A. Kuehl, J. Humes, and M. L. Torchiana, *Adv. Inflammation Res.* **1**, 419 (1979).
[14] A. D. Ramachin, I. Rebeyka, G. J. Wilson, and D. A. G. Mickle, *J. Mol. Cell. Cardiol.* **19**, 289 (1987).

tion. Such an effort will require a pathologic confirmation of injury, detection of free radicals or their products (hydroxy-conjugated dienes), as well as reliable, universally applicable, and precise measurements of global ventricular function to make meaningful conclusions.

We have recently concluded pharmacokinetic and dose–response studies using 90 min of hypothermic global ischemia in dogs.[15,16] Because of potential interspecies differences related to xanthine oxidase, there may be a difference in the magnitude of oxygen-derived free radicals produced during myocardial ischemia and reperfusion. Therefore, such dose–response studies will have to be repeated in xanthine oxidase-deficient species, using a prolonged cold ischemic interval followed by orthotopic transplantation.

The possibility of directly measuring oxygen free radicals or their immediate by-products in human hearts is feasible in the clinical setting of heart transplantation. Myocardial biopsies have been made in many patients, and the safety of the procedure has been confirmed. The possibility of making such direct measurements in patients receiving enzymes or drugs thought to reduce production or ameliorate damage of oxygen free radicals may provide additional evidence for efficacy in the setting of human heart transplantation.

[15] R. B. Lee, J. R. Stewart, S. A. Morley, W. H. Merrill, W. H. Frist, J. W. Hammon, and H. W. Bender, *Surg. Forum* **39**, 216 (1988).
[16] R. B. Lee, J. R. Stewart, S. A. Morley, W. H. Merrill, W. H. Frist, J. W. Hammon, and H. W. Bender, *Circulation* **80**, Suppl. III, 25 (1989).

[82] Quantitation of Free Radical-Mediated Reperfusion Injury in Renal Transplantation

By G. MELVILLE WILLIAMS

Reperfusion injury connotes a preceding ischemic event. It suggests that this ischemic event alone may produce relatively little injury and that major damage may occur at reperfusion in the reactions mediated by singlet oxygen reduction. To challenge or confirm this hypothesis, conditions must be established to define ischemia, which is difficult to do in complex biological systems. However, study of the kidney offers a unique opportunity to control for ill-defined variables active prior to, during, and following ischemia because kidneys are paired. It is possible to remove both kidneys from an animal or a human cadaver donor and treat one but

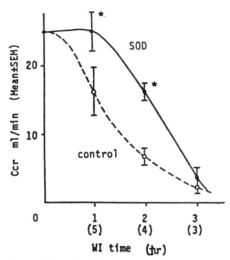

FIG. 1. Postischemic renal function following various periods of warm ischemia (WI). C_{cr} is creatinine clearance, and the number of pairs is given in parentheses. *, $p < .05$ versus corresponding control.

not the other with the pharmacological agent designed to block the formation of or detoxify oxygen free radicals. Moreover, in the experimental animal it is possible to render the pair ischemic for exact periods of time, reimplant them into a common nephrectomized recipient, and measure the function of each organ separately at a standard time following transplantation by collecting urine from cutaneous ureterostomies.

Quantitation of Reperfusion Injury in the Pig Kidney

Hoshino and co-workers studied reperfusion injury after defined periods of occlusion of the renal peticle in pigs.[1] In these experiments, the renal blood supply was isolated and occluded on the control kidney 30 min before it was occluded on the test kidney. After set periods of time, clamps were removed and flow restored to the control kidney. Thirty minutes later, superoxide dismutase (SOD) was perfused into the renal artery coincident with the release of the clamp to the test kidney. Forty-eight hours later, urine was collected from cutaneous ureterostomies to enable the calculation of creatinine clearance. SOD normalized creatinine clearance after 1 hr of warm ischemia (Fig. 1). After 2 hr, renal function could not be normalized but was still significantly better than in control

[1] T. Hoshino, W. R. Maley, G. B. Bulkley, and G. M. Williams, *Transplantation* **45**, 284 (1988).

animals. After 3 hr, the injury capable of being blocked by SOD was negligible compared to the very severe injury produced by ischemia itself. The window of reperfusion injury may be defined as the difference between these two curves and is clearly significant.

A large window was also found employing either allopurinol added to the standard human renal preservation fluid or SOD given intraarterially in kidneys subjected to long periods of cold ischemia. Here it is important to note that control kidneys sustained excellent function after 18 hr of cold ischemia, but considerable metabolic change must have occurred in the interval of 18 to 24 hr, rendering kidneys preserved for 24 hr sensitive to reperfusion injury. Control kidneys lost two-thirds of their functional ability to clear creatinine, whereas kidneys treated with either allopurinol

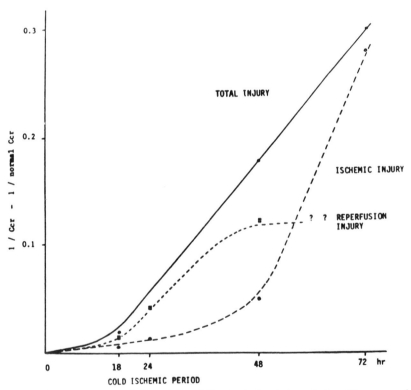

FIG. 2. Two components of tissue injury following ischemia/reperfusion. The curve of total injury was constructed from actual measurement of C_{cr} values of control kidneys, that of ischemic injury from allopurinol-treated kidneys. The curve of reperfusion injury was made by subtraction of the other above two curves.

or SOD had nearly normal function. A significant but less substantial difference occurred after 48 hr of cold ischemia. Neither SOD nor allopurinol was able to prevent the very severe injury occurring to kidneys after 72 hr of cold ischemia. Reperfusion injury can be plotted in this experimental system as shown in Fig. 2.

Reperfusion Renal Damage in Humans

Schneeberger et al. reported on 100 cadaver donor kidneys in which one, but not the other, of the pair of kidneys was treated with SOD at reperfusion.[2] The study was randomized and double-blinded. Considering the data overall, there were no significant differences in the function of SOD- and placebo-treated kidneys. However, more detailed analysis of the pairs revealed that SOD-treated kidneys had double the creatinine clearance at days 2 and 7, and the time for serum creatinine to fall without dialysis was reduced from 8 to 4 days. Because of small numbers, these differences did not prove to be statistically significant. Considering the function of the second of the two kidneys of a pair transplanted, by selecting kidneys that were ischemic in the cold for longer periods of time, more dramatic improvement was noted in all parameters in test kidneys, but the differences still fell short of statistical significance ($p = .08$). Thus, there is no conclusive data in humans documenting a window of reperfusion damage. However the existing data are encouraging enough to promote two larger studies. One is analyzing the effect of allopurinol added to Euro-Collin's solution for renal preservation while the second is using larger doses of SOD employed intravenously 15 min prior to reperfusion. When these are completed in 1 year, a window of effectiveness is likely to be defined.

Conclusions

Injury at reperfusion defined by the ability of allopurinol and SOD to improve renal function is significant in controlled studies in swine. Intraarterial administration of SOD at the time of reperfusion does not prevent acute tubular necrosis in the wide spectrum of ischemia times (both warm and cold) present in human renal transplantation. Acute renal failure is caused by many factors which obscure the influence of any one. Large studies enabling the assay of the effects individual variables on the incidence of nonfunction are essential to evaluate the role of reperfusion injury in this clinical setting.

---

[2] H. Schneeberger, W. D. Illner, D. Abendroth, G. Bulkley, F. Rutili, G. M. Williams, M. Thiel, and W. Land, *Transplant Proc.* **21**, 1245 (1989).

[83] Use of Isolated Perfused Organs in Hypoxia and Ischemia/Reperfusion Oxidant Stress

By HARTMUT JAESCHKE and JERRY R. MITCHELL

Introduction

Studies by many laboratories have shown a protective effect of antioxidants against ischemia/reperfusion injury in heart, liver, intestine, and kidney, suggesting an important role of reactive oxygen species in the pathogenesis.[1] However, to define the exact contribution of reactive oxygen to ischemia/reperfusion injury, it is necessary to detect and quantify the formation of reactive oxygen from potential intracellular sources, such as mitochondria and xanthine oxidase, or extracellular sources, such as neutrophils and tissue macrophages. Isolated, blood-free perfused organs provide a useful tool to study selectively reactive oxygen formation from intracellular sources. Additionally, isolated organs can be examined in different pathophysiological situations such as no-flow ischemia versus various forms of low-flow ischemia and hypoxia with retention of organ architecture and more precise control of experimental conditions than is possible *in vivo*.

Several methods are available to detect reactive oxygen species in isolated organs. The physicochemical methods electron spin resonance (ESR) and chemiluminescence are described elsewhere in this volume.[2,3] A sensitive method without the necessity of additional chemicals interfering with the biological system or the use of expensive equipment is to monitor the formation and cellular release of glutathione disulfide (GSSG) as an index for the activity of the endogenous defense system against reactive oxygen species. Understanding the specific problems involved with the use of this method is essential for interpretation of the data.

General Properties of the Glutathione Defense System

The method is based on the very rapid enzymatic dismutation of intracellularly generated superoxide to molecular oxygen and hydrogen perox-

---

[1] J. R. Mitchell, C. V. Smith, H. Hughes, M. Lenz, H. Jaeschke, S. Shappell, L. Michael, and M. L. Entman, *Trans. Assoc. Am. Physicians* C, 54 (1987).

[2] G. R. Buettner and R. P. Mason, this volume, [9].

[3] M. E. Murphy and H. Sies, this volume, [63].

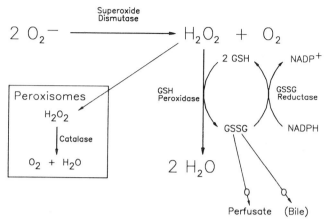

FIG. 1. Enzymatic defense mechanisms against reactive oxygen (details are discussed in the text).

ide, which is then reduced to water through glutathione peroxidase (Fig. 1). Glutathione (GSH) provides the reducing equivalents for this reaction and is oxidized to its disulfide (GSSG). Most of the GSSG formed (between 95 and 99%) is immediately reduced back to GSH through glutathione reductase with the cofactor NADPH. In the liver GSSG is secreted mainly into bile against a steep concentration gradient.[4] The biliary portion of the total GSSG export is a relatively constant value of about 80 to 85% as long as no major interference occurs with bile formation.[5] GSSG is released mainly into the perfusate in other organs such as heart[6,7] and lung[8,9] in contrast to the 15 to 20% released into the perfusate in the liver.[5,10]

Although only a small percentage of the GSSG generated is released from the cells, any change in the overall GSSG formation is reflected by similar changes in the cellular efflux. Thus, a significant increase in the efflux of GSSG into perfusate or bile (liver) indicates an enhanced activity of the defense system against reactive oxygen species (oxidant stress). Determination of the total GSSG efflux from cells allows additionally a quantitative estimation of the amount of superoxide (O_2^-) and hydrogen

[4] T. Akerboom, M. Bilzer, and H. Sies, *J. Biol. Chem.* **257,** 4248 (1982).

[5] H. Jaeschke, *Am. J. Physiol.* **258,** 6499 (1990).

[6] T. Ishikawa and H. Sies, *J. Biol. Chem.* **259,** 3838 (1984).

[7] Y. Xia, K. E. Hill, and R. F. Burk, *J. Nutr.* **115,** 733 (1985).

[8] J. R. Dunbar, A. J. DeLucia, and L. R. Bryant, *Biochem. Pharmacol.* **33,** 1343 (1983).

[9] S. G. Jenkinson, T. H. Spence, R. A. Lawrence, K. E. Hill, C. A. Duncan, and K. H. Johnson, *J. Appl. Physiol.* **62,** 55 (1987).

[10] H. Jaeschke, C. V. Smith, and J. R. Mitchell, *J. Clin. Invest.* **81,** 1240 (1988).

peroxide (H_2O_2) generated intracellularly. In the isolated perfused liver infusion of glutathione peroxidase-specific substrates demonstrated that a constant value of about 3 to 5% of the GSSG formed is actually exported from the cell when the infusion rate is higher than 500 nmol *tert*-butyl hydroperoxide/min/g liver.[10,11] Below these infusion rates, the fraction of GSSG that is released declines.

Another factor to be considered in quantitative calculations is the amount of H_2O_2 detoxified by catalase. Experiments with infusion of the common substrate H_2O_2 and the peroxidase-specific substrate *tert*-butyl hydroperoxide indicated that catalase increasingly contributes to the metabolism of H_2O_2 with infusion of increasing H_2O_2 concentrations (up to 50% at 2000 nmol H_2O_2/min/g liver).[12] The intracellular generation of O_2^- in livers with inhibition of catalase by aminotriazole demonstrated that the contribution of catalase may be underestimated when tested with peroxide infusion experiments, especially with low infusion rates.[13]

Nevertheless, the glutathione peroxidase system is the major defense system against intracellular formation of reactive oxygen, and quantitative considerations permit an estimation of hepatic capacity to detoxify reactive oxygen species. Livers of male Fischer rats detoxified superoxide continuously generated with the redox cycling compound diquat at a rate of 0.8 to 0.9 μmol O_2^-/min/g liver for 1 hr without impairment of the detoxification rate or production of functional or structural cell damage.[13] When superoxide formation was increased to about 2 μmol O_2^-/min/g, the cellular defense system initially responded with a similar increase in metabolism but was not able to maintain these high rates, eventually leading to cell damage after about 40 to 50 min. Thus, the detoxification capacity for reactive oxygen of a liver isolated from a fed male Fischer rat is between 50 and 120 μmol O_2^-/g tissue, indicating a considerable protective capacity even in this susceptible strain.[14] Comparable data for other organs are not available, but similar experimental approaches should be applicable.

Biochemistry of No-Flow Ischemia versus Low-Flow Ischemia/Hypoxia

The different models used to study the effect of oxygen deprivation and reoxygenation in isolated organs cause specific metabolic and physio-

[11] H. Sies and K. H. Summer, *Eur. J. Biochem.* **57**, 503 (1975).
[12] N. Oshino and B. Chance, *Biochem. J.* **162**, 509 (1977).
[13] H. Jaeschke, C. V. Smith, and J. R. Mitchell, unpublished.
[14] C. V. Smith, H. Hughes, B. H. Lauterburg, and J. R. Mitchell, *J. Pharmacol. Exp. Ther.* **235**, 172 (1985).

logical changes in cells that affect the detoxification capacity of the organ against reactive oxygen and hence GSSG efflux as an index for the activity of the defense system. During no-flow ischemia the hepatocyte tries to maintain cellular ATP levels by massive glucose mobilization (glycogenolysis) with enhanced anaerobic glycolysis, thereby generating large amounts of lactate and moderate amounts of ATP. Since lactate is not removed (no-flow), the accumulating lactate shifts the redox potential of the lactate/pyruvate redox couple to more negative values until even high concentrations of NADH are no longer able to reduce pyruvate. As a consequence, no NAD^+ is available for the glycolytic enzyme glyceraldehyde-phosphate dehydrogenase, resulting in a block of glycolysis with accumulation of glucose. Hepatocytes seem to tolerate this metabolic block for hours with only little damage and have the potential for good recovery. The initial reperfusion phase, however, causes major problems since the accumulated metabolites (lactate and especially glucose) lead to a temporary influx of water with massive cell swelling and enzyme release.[10] In contrast to total ischemia, if some flow is maintained even at a rate inadequate to maintain total organ oxygenation (low-flow ischemia), periportal areas are well oxygenated but the centrilobular part is anoxic.[15] Under these conditions, lactate is constantly removed, the glycolysis pathway can generate moderate amounts of ATP, and massive cell swelling does not contribute to the injury when normal reperfusion is restored.

Thus, low-flow ischemia in this model resembles normal-flow hypoxia–reoxygenation injury (perfusate gassed with 95% N_2–5% CO_2). The critical factor for maintaining adequate ATP levels during hypoxia is the availability of glycolysis substrates.[16,17] If the hepatocyte mobilizes most of its glycogen (which happens after about 2 hr of hypoxia in a fed rat but after only 20 min in a fasted rat), the glycolysis rate slows down, the hepatic ATP levels decline to critical values, and the rate of cell damage increases exponentially and is accompanied by an irreversible cholestasis and irreversible mitochondrial damage.[17,18] This massive injury can be greatly delayed in onset by providing glycolysis substrates such as fructose in a normal-flow hypoxia model.[16,18]

[15] M. E. Marotto, R. G. Thurman, and J. J. Lemasters, *Hepatology* **8**, 585 (1988).

[16] I. Anundi, J. King, D. A. Owens, H. Schneider, J. J. Lemasters, and R. G. Thurman, *Res. Commun. Chem. Pathol. Pharmacol.* **55**, 111 (1987).

[17] H. Jaeschke, C. V. Smith, and J. R. Mitchell, *Biochem. Biophys. Res. Commun.* **150**, 568 (1988).

[18] H. Jaeschke and J. R. Mitchell, *Biochem. Biophys. Res. Commun.* **160**, 140 (1989).

Special Aspects of Glutathione Defense System
during Hypoxia or Ischemia

The effect of metabolic and physiological alterations caused by ischemia or hypoxia and reoxygenation on the function of the glutathione defense system was investigated in the liver. Many problems discussed here apply also to other organs.

1. The availability of NADPH to maintain the reducing capacity of the cells is not limited during ischemia since the metabolic block occurs in the glycolysis pathway, leaving sufficient substrates for NADPH regeneration in the pentose phosphate pathway. Thus, no change of the hepatic GSSG content was found during no-flow ischemia of rat liver at various times up to 4 hr.[10,19] Hypoxia enhanced the generation of reducing equivalents in the cytosol and caused a significant decline of the tissue GSSG content accompanied by a drastically diminished release of GSSG into bile and perfusate.[5,17,20] A detailed analysis revealed that the decreased GSSG efflux rates under such conditions reflect reduced formation of GSSG and thus reactive oxygen, not enhanced re-reduction to GSH.[5]

2. Hypoxia and ischemia inhibit bile formation to various degrees, that is, reversible cholestasis occurs during short-term hypoxia,[20] irreversible cholestasis during long-term hypoxia,[17,18] and impaired recovery during reoxygenation after no-flow ischemia.[10] More than 80% of total exported GSSG is released into bile under normal conditions, but any impairment of bile formation can decrease the biliary export of GSSG. In such situations the enhanced sinusoidal efflux of GSSG appears to almost completely compensate for the loss of biliary excretion capacity.[5]

3. Cellular ATP levels decline gradually during hypoxia[17] and drop exponentially during no-flow ischemia.[10] Marked depletion of ATP content stops bile flow and biliary excretion of GSSG completely. However, a 50% reduction of ATP levels with respective reduction in bile flow as seen during short-term hypoxia had only marginal effects on the biliary secretion of glutathione conjugates of sulfobromophthalein and 1-chloro-2,4-dinitrobenzene, compounds used to test selectively biliary and sinusoidal export mechanisms for GSSG.[5]

4. In some cases enhanced GSSG release from the liver may not reflect enhanced intracellular formation of reactive oxygen. During the

[19] W. Siems, B. Mielke, M. Mueller, C. Heumann, L. Raeder, and G. Gerber, *Biomed. Biochim. Acta* **42,** 1079 (1983).
[20] S. W. Cummings, K. E. Hill, R. F. Burk, and D. M. Ziegler, *Biochem. Pharmacol.* **37,** 967 (1988).

development of severe hypoxic injury, the sinusoidal release of GSSG, GSH, and lactate dehydrogenase increased severalfold, indicating a leakage of cell contents into the perfusate owing to hepatocellular damage.[17] Since no change of the perfusate GSH/GSSG ratio was seen and no change of tissue GSSG levels was observed, the higher GSSG concentration in the perfusate does not indicate more reactive oxygen formation. A similar situation occurs during the first seconds of reoxygenation after hepatic no-flow ischemia when a brief increase in GSSG release is observed. This increase vanishes rapidly with return to control rates of efflux by 1 min.[10] This GSSG efflux from the liver appears to represent GSH released into the vascular space and autoxidized in the absence of glutathione reductase. During the initial reperfusion phase these metabolites were washed out from the vascular compartment of the organ, giving rise to a brief increase in apparent efflux rates.[10]

In summary, the excretion of GSSG from cells is a useful and sensitive index of intracellular formation of reactive oxygen in isolated perfused organs. Different models of hypoxia or ischemia with subsequent reoxygenation cause metabolic and physiologic changes that need to be considered for qualitative and especially for quantitative interpretation of experimental data.[5] This approach revealed a significant resistance of the liver against reactive oxygen toxicity even after ischemia,[10] a finding in agreement with the high resistance of the liver *in vivo* to oxidant stress generated by administration of diquat and other redox-cycling toxins.[14]

Assay of Glutathione and Glutathione Disulfide

A detailed discussion of methods available for assay of GSH and GSSG and the specific problems involved was published in this series.[21] Here we describe a procedure that is adapted to the very low levels of GSSG in perfusate.

Preparation of Samples

An aliquot of the effluent perfusate is immediately pipetted into a 10 mM solution of N-ethylmaleimide (NEM) in 100 mM potassium phosphate buffer (pH 6.5) to trap GSH. Since excess NEM interferes with the enzymatic assay for GSSG, NEM is removed by chromatographic separation on a C_{18} SepPak column (Waters Associates, Division of Millipore

---

[21] T. P. M. Akerboom and H. Sies, this series, Vol. 77, p. 373.

Corporation, Milford, MA).[22] A portion of the NEM sample is passed through the column followed by 1 ml of phosphate buffer without NEM. GSSG is determined in the combined eluates. Since the levels of GSSG in perfusate are already very low (basal values 20 to 50 nM GSSG for the liver), dilution of the sample must be kept to a minimum. For the determination of GSSG in tissue or bile a similar procedure can be followed. An aliquot of the acidic homogenate of a freeze-clamped tissue or an aliquot of acidified bile is mixed with the NEM–phosphate buffer and processed as described.

Enzymatic Assay for GSH and GSSG

GSSG and the sum of GSH and GSSG are determined with a modified kinetic assay[23,24] originally described by Tietze.[25] GSSG is reduced enzymatically with NADPH to GSH, which then reduces 5,5'-dithiobis(2-nitrobenzoic acid) (DTNB) spontaneously to 5-thio-2-nitrobenzoate (TNB). The formation of TNB within a given time is determined at 412 nm. The cycling rate of the reaction is proportional to the concentration of GSSG in the test. Since the reaction rate is also influenced by the glutathione reductase activity and other factors such as ionic strength of the buffer and chemicals present in the perfusate, it is essential to use a calibration curve prepared in the same solution mixture as the sample will be in the test. The use of internal standards also reduces potential interference problems of other factors.

Reagents

Solution A: 600 mM potassium phosphate buffer (pH 7.2), 1.2 mM DTNB, 0.08% serum albumin (BSA), 30 mM Na-EDTA (this solution is stable for weeks at 4°)

Solution B: 100 mM imidazole buffer (pH 7.2), 2 mM Na-EDTA, 0.04% BSA (this solution is stable for weeks at 4°)

Solution C: Add 5 mg NADPH and glutathione reductase (final activity 0.16 U/ml) to 10 ml of solution B (prepare this solution daily)

Procedure. Pipette 500 μl of buffer, appropriately diluted sample, or calibration solution into a microcentrifuge tube and add 150 μl of solution A. The timed reaction is started with the addition of solution C (150 μl). Read the absorbance at 412 nm against a blank after exactly 10 min (cali-

---

[22] J. D. Adams, B. H. Lauterburg, and J. R. Mitchell, *J. Pharmacol. Exp. Ther.* **227,** 749 (1983).

[23] J. E. Brehe and H. B. Burch, *Anal. Biochem.* **74,** 189 (1976).

[24] A. Wendel, S. Feuerstein, and K. H. Konz, *Biochem. Pharmacol.* **28,** 2051 (1979).

[25] F. Tietze, *Anal. Biochem.* **27,** 502 (1969).

bration curve 50 to 500 nM GSSG) or 30 min (calibration curve 5 to 50 nM GSSG) and calculate concentration from calibration curve.

Acknowledgments

Excellent technical assistance by Michael Fisher and Bradley Black is gratefully acknowledged. Supported by National Institutes of Health Grants GM-34120, GM-42957, and RR-05425.

[84] Use of Perfused Organs in Measurement of Drug-Induced Oxidant Stress

By RAYMOND F. BURK and KRISTINA E. HILL

Many drugs and chemicals raise the concentration of oxidizing molecules in tissue and thereby cause an oxidant stress. Such effects can often be studied to advantage in a perfused organ preparation. This approach permits the imposition of graded oxidant stresses and allows measurements to be made free from the influence of other tissues.

The liver has been the most frequently used organ for studies of drug-induced oxidant stress. This can be traced to its role as the principal drug-metabolizing site in the body and to the ease with which liver perfusion can be carried out. We have studied oxidant stress in the perfused liver and describe our approach.

Choice of Perfusion System

Most perfused organ systems used to study oxidant stress employ cell-free perfusate and avoid its recirculation. Single-pass design allows control of perfusate constituents, including the drug and its metabolites. Avoidance of cells in the perfusate eliminates another site of metabolism and a potentially confounding source of marker molecules. A disadvantage of the single-pass perfusion is the large volume of perfusate required, which can increase the cost of the experiment. In addition, hemoglobin-free systems generally require a higher flow rate for adequate oxygenation than do red cell-containing systems.[1]

---

[1] H. Sies, *Hoppe-Seyler's Z. Physiol. Chem.* **358,** 1021 (1977).

There are other differences between the liver *in vivo* and the hemoglobin-free, nonrecirculating perfused liver which are potentially important in studies of oxygen-dependent processes. Oxidant defenses are present in the blood, and their absence from the perfused liver system could affect experimental results. Oxygen concentration has a profound effect on enzymatic and nonenzymatic oxidative processes, and *in vivo* oxygen concentrations are difficult to reproduce in perfused organs. Moreover, sole reliance on dissolved oxygen in hemoglobin-free systems is likely to produce sinusoidal oxygen gradients which are different from those in systems containing hemoglobin. Enzymes involved in oxidative metabolism are not distributed uniformly along the hepatic sinusoid, so changes in oxygen gradients are likely to affect their function.

Technique of Hemoglobin-Free, Nonrecirculating Liver Perfusion

Variations of the hemoglobin-free perfusion technique have been employed by many investigators, and ours differs only in detail from most others. We use a Krebs–Henseleit bicarbonate buffer (118 mM NaCl, 5.9 mM KCl, 4.3 mM MgSO$_4$, 1.2 mM Na$_2$SO$_4$, 1.3 mM NaH$_2$PO$_4$, 3.3 mM CaCl$_2$, 25 mM NaHCO$_3$) at pH 7.4 as perfusate. The apparatus needed is a pump, an oxygenator, a perfusate-warming device, a bubble trap, and an infusion block. In our system a peristaltic pump is located between the perfusate reservoir and the oxygenator. The oxygenator we use is a Plexiglas cylinder (16.5 × 4 cm diameter) with gas inflow and outflow connections. Approximately 30 feet of Dow Corning Silastic tubing (0.058 inches i.d. by 0.077 inches o.d.) is wrapped around a grooved aluminum core through which water from a heated water bath is circulated. Thus, we use a single device to oxygenate and to warm the buffer which we perfuse at a flow rate of 4 ml/g liver/min. The gas mixture passing through the oxygenator is kept under 80 mmHg pressure, and the temperature of the water bath is adjusted so the perfusate temperature is 37° when it enters the liver. Upon leaving the oxygenator, the perfusate passes sequentially through a bubble trap and an infusion block with three connectors for infusing drugs. Syringe pumps are used for drug infusion.

Drugs are dissolved at high concentrations in water or perfusion buffer. This minimizes infusion effects on perfusate composition, temperature, and oxygen concentration. Drugs which are not readily water soluble may require carriers such as albumin[2] or liposomes.[3]

---

[2] L. B. LaCagnin, H. D. Connor, R. P. Mason, and R. G. Thurman, *Mol. Pharmacol.* **33,** 351 (1988).

[3] A. R. Nicholas and M. N. Jones, *Biochim. Biophys. Acta* **860,** 600 (1986).

Occasionally experiments are carried out which require two different gas mixtures. We then employ two oxygenators in parallel with a three-way stopcock between the oxygenators and the bubble trap to allow immediate switching from one oxygenator to the other. If low oxygen tensions are needed, glass tubing is used between the oxygenator and the liver to avoid oxygen diffusion from the air into the perfusate.

The abdomen is opened after induction of pentobarbital anesthesia. The bile duct is cannulated with a 10.5-cm-length of PE-10 tubing. Care is taken not to advance the tip of the tube past the bifurcation of the bile duct. A tapered glass cannula is then inserted into the portal vein with perfusate flowing to avoid interruption of the oxygen supply to the liver. The cannula is secured with ligatures and the aorta is cut to allow efflux of blood and perfusate. Then the chest is opened and the inferior vena cava is cannulated through the right atrium. Ligatures are placed around the cannula and around the vena cava between the liver and kidneys to isolate the liver circulation. Most of our experiments are performed with the liver *in situ*. The liver must be removed from the animal for organ spectrophotometry.[4]

Effluent perfusate is sampled at intervals for analysis. Bile is collected with the tip of the bile cannula immersed in 0.25 ml of 3% metaphosphoric acid in a preweighed snap-top plastic tube. This prevents oxidation of GSH to GSSG. Timed collections are made, and the tubes are reweighed to determine bile flow.

GSSG in bile is determined using dual-wavelength spectrophotometry. Acidified bile (0.1 ml) is neutralized with 0.4 ml of 0.3 M Na$_2$HPO$_4$ immediately prior to GSSG measurement. An aliquot of neutralized bile (containing 0.25 to 5 nmol GSSG) is added to a cuvette containing 0.13 μmol NADPH, 0.1 M potassium phosphate, 1 mM EDTA (pH 7.0) to give a final volume of 0.7 ml. The absorbance ($A_{340} - A_{380}$) of the solution is determined before and after addition of 0.2 units (10 μl) of glutathione reductase (yeast type IV, Sigma Chemical Co., St. Louis, MO). The change in absorbance takes place over 1–3 min and is a measure of the amount of NADPH oxidized and reflects the quantity of GSSG reduced. A standard curve (0.25 to 5 nmol GSSG) is determined daily.

Although it is not an indicator of oxidant stress, total glutathione in the bile and in the effluent perfusate can be measured by a modification[5] of the recirculating method of Tietze.[6] Depending on the experiment, other assays may be performed on these samples.

[4] H. Sies, this series, Vol. 52, p. 48.
[5] O. W. Griffith, *Anal. Biochem.* **106**, 207 (1980).
[6] F. Tietze, *Anal. Biochem.* **27**, 502 (1969).

Measurement of Oxidant Stress

Assessment of oxidant stress is complicated by the lack of a universal oxidant stress product or marker. The nature of the stress and the conditions of the experiment dictate the measurements which can be used as indices of the oxidant stress. Figure 1 lists four categories of oxidant stress indices.

Oxidizing Molecules. The most specific indication that a drug or chemical causes an oxidant stress is the detection of oxidizing molecules which arise from it or from its metabolism. Direct detection of excited species is sometimes possible with measurement of organ chemiluminescence. This is a sensitive technique but it provides minimal characterization of the oxidant species. Dual-wavelength organ spectrophotometry allows the determination of heme occupancy of catalase from which hydrogen peroxide concentration can be estimated.[7] Its application has been restricted to a few laboratories because of the apparatus required and the availability of simpler methods such as GSSG release (see below).

Because oxidant molecules are usually very reactive, metabolites of them are often measured to indicate their presence. An example of this is the use of $CHCl_3$ to indicate the presence of $CCl_3 \cdot$.[8] Another technique used recently to identify and measure primary oxidant molecules is infusion of spin traps which allows detection of free radicals by electron spin resonance.[2]

Detoxification Products. Evidence that cellular oxidant defenses are functioning can indicate that an oxidant stress is occurring. Measurement of products of these defenses is commonly employed for this purpose. Many investigators use GSSG release by organs as an index of oxidant stress.[9] The rationale for this is shown in Fig. 2. GSH inside the cell is oxidized to the disulfide by oxidant molecules acting through glutathione peroxidases. Most GSSG is reduced by glutathione reductase, but a small fraction, usually less than 5% of that formed, is transported into the extracellular fluid. The liver transports GSSG into the bile and the heart transports it into the perfusate. This allows intracellular GSSG to be monitored by measurement of GSSG release from the organ. Linear correlation of intracellular GSSG concentration with biliary GSSG release has been demonstrated under a variety of conditions.[10] A similar relationship of tissue GSSG concentration with GSSG release has been shown in the heart.[11] However, saturation of the transport mechanism in the heart

[7] H. Sies and B. Chance, *FEBS Lett.* **11,** 172 (1970).

[8] T. C. Butler, *J. Pharmacol. Exp. Ther.* **134,** 311 (1961).

[9] H. Sies and T. P. M. Akerboom, this series, Vol. 105, p. 445.

[10] T. P. M. Akerboom, M. Bilzer, and H. Sies, *J. Biol. Chem.* **257,** 4248 (1982).

[11] Y. Xia, K. E. Hill, and R. F. Burk, *J. Nutr.* **115,** 733 (1985).

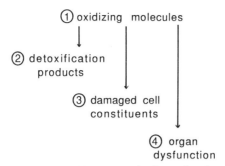

FIG. 1. Measurable indices of oxidant stress in the perfused organ. The indices are numbered from most specific and sensitive to least specific and sensitive.

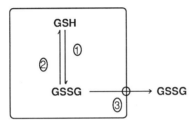

FIG. 2. General scheme of glutathione redox cycle of the cell. (1) The formation of GSSG from GSH is mediated by glutathione peroxidases, selenium-dependent and non-selenium-dependent. Other processes besides oxidant stress can cause this reaction and complicate the use of GSSG as a marker of oxidant stress when they occur. (2) Approximately 95% of the GSSG formed in the perfused liver is reduced to GSH by glutathione reductase. (3) A small fraction of GSSG formed in the cell is transported across the cell membrane by the same mechanism that transports glutathione conjugates. GSSG is released in the bile by the perfused liver and in the perfusate by the perfused heart.

has been observed. This saturation appears to be related to a limited capacity for GSSG reduction which leads to very high cardiac concentrations of GSSG.

The fraction of GSSG formed which is released by the organ can be influenced by a number of factors. Inhibition of GSSG reduction by lowering NADPH availability or by inhibiting glutathione reductase causes increased GSSG release.[12] Formation of glutathione conjugates reduces GSSG release because conjugates compete for the carrier which transports GSSG.[13] A hypothetical factor which could affect release is GSSG gradients within the cell, similar to those described for oxygen.[14] GSSG

[12] C. V. Smith and J. R. Mitchell, *Biochem. Biophys. Res. Commun.* **133,** 329 (1985).

[13] T. P. M. Akerboom, M. Bilzer, and H. Sies, *FEBS Lett.* **140,** 73 (1982).

[14] D. P. Jones, *Am. J. Physiol.* **250,** C663 (1986).

formed at different sites within the cell might diffuse to the bile canaliculus with different efficiencies. These examples point to the need for considering efficiency of release in the interpretation of results. In some experiments this can be estimated if the rate of metabolism of the oxidant species is known.[15]

Quantitation of GSSG released is complicated by extracellular oxidation of GSH, which is also released in the bile,[16] and by glutathione degradation mediated by biliary γ-glutamyltransferase.[17] Oxidation is prevented by collecting bile in an acid medium.[16] Oxidation does not appear to be a problem in heart perfusions because GSH release by that organ is very low.[11] Degradation of glutathione in the biliary system occurs so rapidly in some species that none is detectable in bile.[18] About 50% appears to be degraded in rat bile.[17] Inhibitors of γ-glutamyltransferase increase measurable biliary GSH and GSSG[17] but have effects on glutathione metabolism which preclude their routine use. They may be required, however, for experiments in species with very high biliary γ-glutamyltransferase that prevents detection of glutathione in the bile.

Information on the nature of an oxidant stress can be obtained by studying selenium-deficient organs as illustrated in Fig. 3. There are two glutathione peroxidases in rat liver.[19] One is selenium-dependent and can metabolize hydrogen peroxide. Non-selenium-dependent glutathione peroxidase cannot use hydrogen peroxide as a substrate. Infusion of hydrogen peroxide into selenium-deficient liver does not cause release of GSSG as it does in control liver.[20] Thus, if an oxidant stress does not cause GSSG release from selenium-deficient liver, but does from control liver, it can be concluded that hydrogen peroxide is the cause of the GSSG release. Heart contains little or no non-selenium-dependent glutathione peroxidase,[11] so this strategy cannot be used with heart.

Release of GSSG in the bile is not a specific indicator of oxidant stress. Several drug-related mechanisms for GSSG formation in liver have been described which are unrelated to glutathione peroxidase. One of these is the GSH-dependent metabolism of organic nitrates,[21] and another is metabolism of thiocarbamide substrates by the flavin-containing monooxygenase.[22] Therefore, some chemicals can cause GSSG release without

[15] H. Sies and K.-H. Summer, *Eur. J. Biochem.* **57,** 503 (1975).
[16] D. Eberle, R. Clarke, and N. Kaplowitz, *J. Biol. Chem.* **256,** 2115 (1981).
[17] N. Ballatori, R. Jacob, and J. L. Boyer, *J. Biol. Chem.* **261,** 7860 (1986).
[18] N. Ballatori, R. Jacob, C. Barrett, and J. L. Boyer, *Am. J. Physiol.* **254,** G1 (1988).
[19] R. A. Lawrence and R. F. Burk, *Biochem. Biophys. Res. Commun.* **71,** 952 (1976).
[20] R. F. Burk, K. Nishiki, R. A. Lawrence, and B. Chance, *J. Biol. Chem.* **253,** 43 (1978).
[21] J. H. Keen, W. H. Habig, and W. B. Jakoby, *J. Biol. Chem.* **251,** 6183 (1976).
[22] P. A. Krieter, D. M. Ziegler, K. E. Hill, and R. F. Burk, *Mol. Pharmacol.* **26,** 122 (1984).

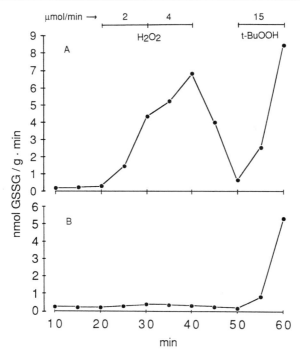

FIG. 3. Use of selenium-deficient liver to show that GSSG release is caused by hydrogen peroxide. Both experiments used the same infusion protocol, which is shown at top. (A) Biliary GSSG release (per g liver) by a control liver; (B) biliary GSSG release by a selenium-deficient liver. Both released GSSG when *tert*-butyl hydroperoxide, a substrate of non-selenium-dependent glutathione peroxidase, was infused. Only the control liver released GSSG in response to hydrogen peroxide infusion.

imposing an oxidant stress. Other chemicals can impose an oxidant stress without causing GSSG release. CCl_4 causes lipid peroxidation but no GSSG release.[23]

The source of biliary GSSG in the unstressed liver is not known, but it may not be hydrogen peroxide detoxification because selenium-deficient livers release as much or more GSSG as do control livers.[24] However, basal GSSG release is oxygen dependent because it falls during hypoxic perfusion.[25]

[23] B. H. Lauterburg, C. V. Smith, H. Hughes, and J. R. Mitchell, *J. Clin. Invest.* **73,** 124 (1984).

[24] P. A. Krieter, D. M. Ziegler, K. E. Hill, and R. F. Burk, *Biochem. Pharmacol.* **34,** 955 (1985).

[25] S. W. Cummings, K. E. Hill, R. F. Burk, and D. M. Ziegler, *Biochem. Pharmacol.* **37,** 967 (1988).

Damaged Cell Constituents. Oxidant stress which escapes suppression by defense mechanisms causes injury to the cell. The injury can take several forms, with the most common ones being adduct formation between the oxidant species and a cell constituent and fragmentation of cell constituents by processes such as lipid peroxidation.

Several markers of lipid peroxidation have been used in studies with perfused organs. Thiobarbituric acid-reactive substances are sought in the effluent or in freeze-clamped tissue. Ethane and pentane have been detected in closed chambers containing perfused livers.[26] Quantitation of lipid peroxidation and thus oxidant stress using these markers is difficult because their formation is highly oxygen dependent and they can be metabolized by the organ.[27] For these reasons, measurement of lipid peroxidation has been of limited value in assessing oxidant stress in perfused organs.

Organ Dysfunction and Injury. When oxidant stress in an organ is severe, organ injury and dysfunction may supervene. Measurements of bile flow, myocardial contractility, trypan blue uptake, and enzyme release have been used to gauge the severity of oxidant stress in perfused organs. This type of measurement has low specificity but may be useful in assessing *in vivo* effects of oxidant stress.

Acknowledgments

Research by the authors is supported by National Institutes of Health Grants ES 02497 and HL 36371.

[26] A. Müller and H. Sies, *Eur. J. Biochem.* **134,** 599 (1983).
[27] R. F. Burk and T. M. Ludden, *Biochem. Pharmacol.* **38,** 1029 (1989).

Author Index

Numbers in parentheses are footnote reference numbers and indicate that an author's work is referred to although the name is not cited in the text.

L

Subject Index

generation of radicals to stimulate lipid
    peroxidation, 52–53
as radical initiator, 101
Azo compounds, 199, 303–304
    hydrophilic, 101
    lipophilic, 101
    as radical initiators, 100–108
    as radical sources, for *in vivo* system,
        106
    safety of, 108
    thermal decomposition, 106, 303
Azo dyes
    carcinogenicity of, 580
    genotoxicity of, 580
    reaction with DT-diaphorase, 294

B

Bacillus subtilis, exocellular superoxide
    dismutase, purification, 256–257
Bacteria
    colonies of, superoxide dismutase assay
        in, 240–242
    extracts of, preparation, 500–501
    photosynthetic blue-green, 301
    protein degradation in, studies of, 496
    superoxide dismutase assay in, 237–242
    suspensions of, DMSO-pretreated,
        extraction of sulfinic acid from, 141
    toluenized cells, superoxide dismutase
        assay in, 238–240
Bathophenanthroline
    complexes with iron salts, 42
    as Fe(II) chelator, 459
Bathophenanthroline assay, 123
Benzene, myelotoxicity induced by, 580
Benzo[*a*]pyrene quinones, reaction with
    DT-diaphorase, 293–294, 296
Benzoic acid, in detection of hydroxyl
    radicals, 137
p-Benzoquinone
    amperometric detection of, 182–187
    effects on intracellular thiol homeostasis,
        628–635
    half-wave potentials of, effects of sub-
        stituents on, 187–189
    hydroxyl- and glutathionyl-disubstituted,
        half-wave potentials of, 188, 192–
        194

hydroxyl-substituted
    half-wave potentials of, 188–191
    origins of, 190
    methyl-substituted, half-wave potentials
        of, 188–189
    reaction with DT-diaphorase, 295
p-Benzoquinone epoxides
    half-wave potentials of, 188
    reaction with DT-diaphorase, 295
BHA, reaction of radicals with, rate con-
    stants for, 99
BHT, reaction of radicals with, rate con-
    stants for, 99
Biflavanoids, 343–344
Bile pigments
    analysis of, 309
    antioxidant activities of, 301–302
    occurrence, 301
    peroxyl radical scavenging activities of,
        measurement of, 305–308
    purification, 305
    structure, 301
Bilins, 301
Bilirubin
    albumin-bound, antioxidant activity of,
        307
    antioxidant activities of, 301–302
    conjugated
        antioxidant activity of, 307–308
        purification, 305
    ditaurate, purification, 305
    high-performance liquid chromatography
        analysis of, 309
    membrane-bound, antioxidant activity
        of, 306
    peroxyl radical scavenging activity of,
        303
        measurement, 306
    physicochemical properties, 302
    properties of, 302
    purification, 305
(4*Z*,15*Z*)-Bilirubin IXα, configurational
    photoisomers of, separation and quan-
    titation, 309
Bilirubin : NAD(P)$^+$ oxidoreductase. *See*
    Biliverdin reductase
Biliverdin
    antioxidant activities of, 301–302
    conjugated, antioxidant activity of, 307–
        308

E

Ebselen, effect on inflammatory liver
injury in mice, 679
EDDA. *See* Ethylenediaminediacetic acid
EDRF. *See* Endothelium-derived relaxing
factor
EDTA
  as chelating agent, 41–42
  effects of, on lipid peroxidation, 48
  iron complexes of, oxidation of, 152
  and solution metal concentrations, 122
  in spin-trapping experiments, with super-
    oxide-generating systems, 132
EDTA–MnCl$_2$, solutions
  kinetics of coenzyme oxidation, 217
  and NADH oxidation rates, 219
  preparation, 212
  stability of, 212–213
Eicosanoids
  conversion of arachidonic acid to, role
    of prostaglandin H synthase in,
    283–284
  formation of
    and cyclooxygenase- and lipoxy-
      genase-catalyzed reactions in
      endothelium, 64
    and oxidized LDL, 64
Eicosatetraenoic acid hydroperoxide,
  positional isomers of, initiation of
  cyclooxygenase by, 435–438
Electrochemical detection
  multielectrode, 186
  oxidative, 182–183
  reductive, 182–183, 186
Electron spin resonance spectroscopy, 89,
  93, 127, 320
  of free radical reductase in biological
    tissue, 670–674
  of human lung tissue from cigarette
    smokers, 665–669
  spectrometer
    low-frequency, 614–616
    sensitivity, 617
  spin trapping, 17–18
  of tocopheroxyl free radicals, 197–205
Electron transfer, 196
Emphysema, oxidatively modified proteins
  in, 576

Endothelium-derived relaxing factor, 79
Endotoxin
  DIC caused by, 657–660
  effects of, 74
  liver injury in mice caused by, 677–680
  sensitivity of mice to, 675–676
  and superoxide generation, 660
Endotoxin shock
  experimental induction of, 660–661
  oxidants and eicosanoids in, 74
  oxy radicals in, 660–665
  pathogenesis of, 664–665
Enolase, rabbit muscle
  DTT/Fe^{3+}/O$_2$MFO inactivation reaction,
    483
  mercaptan-dependent MFO inactivation,
    478–479
Enterocytes
  metabolism of menadione, 611–612
  production of CO$_2$ by, effect of mena-
    dione on, 612
Enzymes, binding of copper ions to, 40
Eosin Y, as enhancer of chemiluminescence, 602–603
Epinephrine, autoxidation of, superoxide
  dismutase assay based on, 221
2,3-Epoxy-1,4-*p*-benzoquinone, half-wave
  potentials of, 188
2,3-Epoxy-*p*-benzoquinone, 194
  reaction with DT-diaphorase, 295
5,6α-Epoxy-5α-cholestan-3β-ol. *See* Cholesterol α-oxide
5,6β-Epoxy-5β-cholestan-3β-ol. *See* Cholesterol β-oxide
2,3-Epoxy-2,6-dimethyl-*p*-benzoquinone,
  half-wave potentials of, 188
2,3-Epoxy-2,3-dimethyl-1,4-naphtho-
  quinone, reaction with DT-diaphorase,
  295
2,3-Epoxy-2-methyl-*p*-benzoquinone, half-
  wave potentials of, 188
2,3-Epoxy-2-methyl-1,4-naphthoquinone,
  reaction with DT-diaphorase, 295
2,3-Epoxy-1,4-naphthoquinone, 194
  half-wave potentials of, 188
  reaction with DT-diaphorase, 295
Ergothioneine
  as antioxidant, 310–318
  assays for, 318

Superoxide radical, 5–8
  in aqueous solution, 6
  chemical reactivity of, 6
  cytotoxicity, 24
  damage done by, 10
  decay, 95
  determination of, based on chemi-
      luminescence of luciferin analogs,
      585–591
  dismutation reaction, 6
  formation, 5–6
  generation, 9, 95
    at reperfusion, 712–714
    from xanthine/xanthine oxidase sys-
        tem, spin trapping of, 619, 621
  in hydroxyl radical production, 20
  kinetics and reaction mechanisms of,
      95–96
  lifetime, 95
  nomenclature, 5
  in organic solvents, 6
  production, by neutrophils and other
      phagocytic cells, kinetic microplate
      assay for, 567–575
  pulse radiolysis studies of, 90
  reactions with hemoglobin, 266
  scavengers, 221
  spin-trapping of, 127–128
    in cell organelles, intact cells, and
        organs, 131
    in vivo and *in vitro*, 131
    intracellular, effect of oxygen concen-
        tration on, 612–613
    during metabolism of xenobiotics by
        rat enterocytes, 611–621
  as tocopherol oxidant, 200
  toxicity, site-specific nature of, 40
Superoxide radical anion, rate constants
    for, in aqueous solution, 114
Sweat
  bleomycin-detectable iron in, 37
  phenanthroline-detectable copper in,
      41
Synovial fluid, bleomycin-detectable iron
    in, 37
Synthetic hydroperoxides, peroxidation
    stimulated by, 52–53
Systemic lupus erythematosus, clastogenic
    activity in plasma of patients with,
    556

T

Taurine chloramine, and xenobiotic activa-
    tion by PMNs, 583
Tetraborate, assay, 251
2,2,6,6-Tetramethylpiperidine, as singlet
    oxygen trap, 604
Tetroxides
  decay, 96
  formation, 96
Thermotolerance, 700–701
α-Thioalkyl radicals
  chemistry of, 178
  optical detection of, 173–174
Thiobarbituric acid
  malondialdehyde determination with,
      407–413
  reactivity toward, by DNA sugar frag-
      ments, 518
Thiobarbituric acid–malondialdehyde
    complex
  generation, as simple assay for hydroxyl
      radical generation, 26–27
  high-performance liquid chromatogra-
      phy, prior to absorptiometry, 426
  MDA determined as, using high-perfor-
      mance liquid chromatography, 427
Thiobarbituric acid-reactive substances,
    410–414, 426–427, 766
  as marker of oxygen free radical-medi-
      ated phospholipid oxidation, 747
Thiobarbituric acid test, 43, 67–70
  lack of specificity, 69
  for measuring lipid peroxidation, 371
  peroxidation during, 68
Thiol compounds
  and O_2–dependent hydroxyl radical
      formation, 21
  one-electron oxidation of, to thiyl radi-
      cal, 320
  radioprotective effect afforded by, 319
  reaction of, with nitro compounds, 325
  thiyl radical metabolites of, 318–329
Thiol/disulfide systems
  radical-mediated processes in, chemistry
      of, 175
  radical species in, 168–169
Thiol-promoted cystine uptake, *in vitro*
    assay of, 707–708
Thiol pumping, 325–328